Fundamentals of Building Construction

Cover image: © Amana Images/Alamy
Cover design: Michael Rutkowski

This book is printed on acid-free paper. ∞

Copyright © 2014 by John Wiley & Sons, Inc. All rights reserved.

Published by John Wiley & Sons, Inc., Hoboken, New Jersey.

Published simultaneously in Canada.

No part of this publication may be reproduced, stored in a retrieval system, or transmitted in any form or by any means, electronic, mechanical, photocopying, recording, scanning, or otherwise, except as permitted under Section 107 or 108 of the 1976 United States Copyright Act, without either the prior written permission of the Publisher, or authorization through payment of the appropriate per-copy fee to the Copyright Clearance Center, 222 Rosewood Drive, Danvers, MA 01923, (978) 750-8400, fax (978) 646-8600, or on the web at www.copyright.com. Requests to the Publisher for permission should be addressed to the Permissions Department, John Wiley & Sons, Inc., 111 River Street, Hoboken, NJ 07030, (201) 748-6011, fax (201) 748-6008, or online at www.wiley.com/go/permissions.

Limit of Liability/Disclaimer of Warranty: While the publisher and author have used their best efforts in preparing this book, they make no representations or warranties with the respect to the accuracy or completeness of the contents of this book and specifically disclaim any implied warranties of merchantability or fitness for a particular purpose. No warranty may be created or extended by sales representatives or written sales materials. The advice and strategies contained herein may not be suitable for your situation. You should consult with a professional where appropriate. Neither the publisher nor the author shall be liable for damages arising herefrom.

For general information about our other products and services, please contact **our** Customer Care Department within the United States at (800) 762-2974, outside the United States at (317) 572-3993 or fax (317) 572-4002.

Wiley publishes in a variety of print and electronic formats and by print-on-demand. Some material included with standard print versions of this book may not be included in e-books or in print-on-demand. If this book refers to media such as a CD or DVD that is not included in the version you purchased, you may download this material at http://booksupport.wiley.com. For more information about Wiley products, visit www.wiley.com.

Library of Congress Cataloging-in-Publication Data:
Allen, Edward, 1938-
   Fundamentals of building construction : materials and methods /
Edward Allen and Joseph Iano. – Sixth Edition.
      pages cm
   Includes index.
   ISBN 978-1-118-13891-5 (hardback); ISBN 978-1-118-41919-9 (ebk); ISBN 978-1-118-42086-7 (ebk)
1. Building.   2. Building materials.   I. Iano, Joseph.   II. Title.
   TH145.A417 2013
690–dc23
                                                                                          2013002846

Printed in the United States of America

10 9 8 7 6 5 4 3 2

FUNDAMENTALS OF BUILDING CONSTRUCTION

MATERIALS AND METHODS

SIXTH EDITION

Edward Allen and **Joseph Iano**

WILEY

CONTENTS

Preface to the Sixth Edition   xi

1   Making Buildings   3

Learning to Build   4
Buildings and the Environment   5
The Work of the Design Professional   11
The Work of the Construction Professional   19
Trends in the Delivery of Design and
    Construction Services   24

2   Foundations and Sitework   31

Foundation Requirements   32
Earth Materials   33
**SUSTAINABILITY OF FOUNDATIONS
AND THE BUILDING SITE   39**

Earthwork and Excavation   40
Foundations   51
Protecting Foundations from Water, Heat Flow,
    and Radon Gas   66
Sitework   73
GEOTEXTILES   76
Designing Foundations   77
Foundation Design and the Building Code   78

3   Wood   83

Trees   84
SUSTAINABILITY OF WOOD   90
Lumber   90
Wood Products   103
**A NATURALLY GROWN STRUCTURAL
    MATERIAL   108**
Plastic Lumber   110
Wood Panel Products   110
Wood Chemical Treatments   118
Wood Fasteners   120
**WOOD PRODUCT ADHESIVES AND
    FORMALDEHYDE   127**
Prefabricated Wood Components   128
Types of Wood Construction   130

FROM CONCEPT TO REALITY   135
French American School

4 Heavy Timber Frame Construction 139

Fire-Resistive Heavy Timber
Construction 142
■ SUSTAINABILITY IN HEAVY TIMBER
CONSTRUCTION 143
Heavy Timber in Other Construction
Types 150
Lateral Bracing 152
Cross-Laminated Timber
Construction 152
Accommodating Building Services 156
Wood-Concrete Composite
Construction 156
Longer Spans in Heavy Timber 157
■ FOR PRELIMINARY DESIGN OF HEAVY
TIMBER STRUCTURES 164
Heavy Timber and the Building Codes 164
Uniqueness of Heavy Timber Framing 165

5 Wood Light Frame Construction 171

History 173
Platform Frame 174
■ SUSTAINABILITY IN WOOD LIGHT FRAME
CONSTRUCTION 176
Foundations for Light Frame
Structures 176
Building the Frame 184
Variations on Wood Light Frame
Construction 219
■ FOR PRELIMINARY DESIGN OF A WOOD
LIGHT FRAME STRUCTURE 222
Wood Light Frame Construction and the
Building Codes 222
Uniqueness of Wood Light Frame
Construction 224

6 Exterior Finishes for Wood Light Frame Construction 231

Protection from the Weather 232
Roofing 233
Windows and Doors 240
■ PAINTS AND COATINGS 244
Siding 248
Corner Boards and Exterior Trim 257
Sealing Exterior Joints 258
■ SUSTAINABILITY OF PAINTS AND OTHER
ARCHITECTURAL COATINGS 260
Exterior Painting, Finish Grading, and
Landscaping 260
Exterior Construction 260

7 Interior Finishes for Wood Light Frame Construction 265

Completing the Building Enclosure 273
■ SUSTAINABILITY OF
INSULATION MATERIALS FOR
WOOD LIGHT FRAME CONSTRUCTION 282
Wall and Ceiling Finish 285
Millwork and Finish Carpentry 285
■ PROPORTIONING FIREPLACES 286
■ PROPORTIONING STAIRS 300
Flooring and Ceramic Tile Work 302
Finishing Touches 304

8 Brick Masonry 309

History 310
Mortar 313
■ SUSTAINABILITY OF BRICK MASONRY 316
Brick Masonry 316
Masonry Wall Construction 339

9 **Stone and Concrete Masonry 349**

Stone Masonry 350
SUSTAINABILITY IN STONE AND CONCRETE MASONRY 362
Concrete Masonry 370
Other Types of Masonry Units 381
Masonry Wall Construction 382

10 **Masonry Wall Construction 387**

Types of Masonry Walls 388
FOR PRELIMINARY DESIGN OF A LOADBEARING MASONRY STRUCTURE 396
Spanning Systems for Masonry Bearing Wall Construction 396
Detailing Masonry Walls 400
Some Special Problems of Masonry Construction 406

MOVEMENT JOINTS IN BUILDINGS 408
Masonry Paving 414
Masonry and the Building Codes 415
Uniqueness of Masonry 415

11 **Steel Frame Construction 421**

History 422
The Material Steel 424
FOR PRELIMINARY DESIGN OF A STEEL STRUCTURE 427
Joining Steel Members 435
Details of Steel Framing 441
The Construction Process 451
Fire Protection of Steel Framing 468
Longer Spans and High-Capacity Columns in Steel 473
FABRIC STRUCTURES 482
Industrialized Systems in Steel 486
SUSTAINABILITY IN STEEL FRAME CONSTRUCTION 487
Steel and the Building Codes 488
Uniqueness of Steel 488

12 **Light Gauge Steel Frame Construction 499**

The Concept of Light Gauge Steel Construction 500
SUSTAINABILITY IN LIGHT GAUGE STEEL FRAMING 501
Light Gauge Steel Framing 502
Other Uses of Light Gauge Steel Framing 511
FOR PRELIMINARY DESIGN OF A LIGHT GAUGE STEEL FRAME STRUCTURE 513
Insulating Light Gauge Steel Frame Structures 513
Advantages and Disadvantages of Steel Framing 514

Light Gauge Steel Framing and the Building
Codes   514

Finishes for Light Gauge Steel Framing   514

■ **METALS IN ARCHITECTURE**   516

FROM CONCEPT TO REALITY   522
Camera Obscura at Mitchell Park,
Greenport, New York

**13 Concrete
Construction   527**

History   528

Cement and Concrete   529

■ **SUSTAINABILITY IN CONCRETE
CONSTRUCTION   532**

Making and Placing Concrete   535

Formwork   540

Reinforcing   541

Concrete Creep   555

Prestressing   555

ACI 301   560

Innovations in Concrete Construction   560

**14 Sitecast Concrete
Framing Systems   565**

Casting a Concrete Slab on Grade   567

Casting a Concrete Wall   571

Casting a Concrete Column   577

One-Way Floor and Roof Framing
Systems   578

Two-Way Floor and Roof Framing Systems   587

Other Uses of Sitecast Concrete   592

Sitecast Posttensioned Framing Systems   592

Selecting a Sitecast Concrete Framing
System   594

Innovations in Sitecast Concrete
Construction   594

■ **FOR PRELIMINARY DESIGN OF A SITECAST
CONCRETE STRUCTURE   597**

Architectural Concrete   600

■ **CUTTING CONCRETE, STONE, AND
MASONRY   604**

Longer Spans in Sitecast Concrete   608

Designing Economical Sitecast Concrete
Buildings   611

Sitecast Concrete and the Building Codes   611

Uniqueness of Sitecast Concrete   612

**15 Precast Concrete
Framing Systems   621**

Precast, Prestressed Concrete Structural
Elements   624

■ **FOR PRELIMINARY DESIGN OF A PRECAST
CONCRETE STRUCTURE   625**

Assembly Concepts for Precast Concrete
Buildings   626

Manufacture of Precast Concrete Structural
Elements   627

Joining Precast Concrete Members   633

■ **FASTENING TO CONCRETE   634**

Composite Precast/Sitecast Concrete
Construction   647

The Construction Process   647

■ **SUSTAINABILITY IN PRECAST CONCRETE
CONSTRUCTION   648**

Precast Concrete and the Building Codes   649

Uniqueness of Precast Concrete   649

16 Roofing 661

Low-Slope Roofs   663
BUILDING ENCLOSURE ESSENTIALS:
THERMAL INSULATION AND VAPOR
RETARDER   668
Steep Roofs   688
SUSTAINABILITY IN ROOFING   702
Cool Roofs   702
Green Roofs   705
Photovoltaic Systems   706
Roofing and the Building Codes   707
BUILDING ENCLOSURE ESSENTIALS:
DISSIMILAR METALS AND THE GALVANIC
SERIES   708

17 Glass and Glazing 717

History   718
The Material Glass   720
SUSTAINABILITY OF GLASS   722
Glazing   734
Glass and Energy   744
Glass and the Building Codes   747

FROM CONCEPT TO REALITY   750
Skating Rink at Yerba Buena Gardens

18 Windows and Doors 755

Windows   756
PLASTICS IN BUILDING CONSTRUCTION   766
SUSTAINABILITY OF WINDOWS AND
DOORS   777
Doors   777
Other Window and Door Requirements   783

19 Designing Exterior Wall Systems 791

Design Requirements for the Exterior
   Wall   792
SUSTAINABILITY OF EXTERIOR WALL
SYSTEMS   797
Conceptual Approaches to Watertightness in
   the Exterior Wall   798
Sealing Joints in the Exterior Wall   803
Loadbearing Walls and Curtain Walls   807
BUILDING ENCLOSURE ESSENTIALS: AIR
BARRIER   808
The Exterior Wall and the Building
   Codes   812

20 Cladding with Masonry and Concrete 817

Masonry Veneer Curtain Walls   818
Stone Curtain Walls   825
Precast Concrete Curtain Walls   830
Exterior Insulation and Finish System   836
Keeping Water Out with Masonry and Concrete
   Cladding   840

FROM CONCEPT TO REALITY   842
Seattle University School of Law

21 Cladding with Metal and Glass 847

Aluminum Extrusions 848
■ SUSTAINABILITY OF ALUMINUM CLADDING COMPONENTS 852
Aluminum and Glass Framing Systems 854
Double-Skin Facades 868
Sloped Glazing 869
The Curtain Wall Design and Construction Process 869
Metal Panel Cladding 871

22 Selecting Interior Finishes 877

Installation of Mechanical and Electrical Services 878
The Sequence of Interior Finishing Operations 880
■ SUSTAINABILITY OF INTERIOR FINISHES 882
Selecting Interior Finishes 883
Trends in Interior Finish Systems 888

23 Interior Walls and Partitions 891

Types of Interior Walls 892
Framed Partition Systems 893
■ SUSTAINABILITY OF GYPSUM PRODUCTS 898
■ PLASTER ORNAMENT 910
Masonry Partition Systems 924
Wall and Partition Facings 924

24 Finish Ceilings and Floors 931

Finish Ceilings 932
Types of Ceilings 932
■ SUSTAINABILITY OF FINISH CEILINGS AND FLOORS 943
Finish Flooring 944
Types of Finish Flooring Materials 948
Flooring Thickness 961

Appendix 966

Glossary 968

Index 999

PREFACE TO THE SIXTH EDITION

First published a quarter-century ago, *Fundamentals of Building Construction: Materials and Methods*, now in its sixth edition, has wrought a revolution in construction education. It has been instrumental in making a previously unpopular area of study not merely palatable but vibrant and well liked. It has taken a body of knowledge once characterized as antithetical to design excellence and made it widely recognized as being centrally relevant to good building design. It has replaced dry, unattractive books with a well-designed, readable volume that students value and keep as a reference work. It was the first book in its field to be even-handed in its coverage and profusely and effectively illustrated throughout. It was the first to release the teacher from the burden of explaining everything in the subject, thereby freeing class time for discussions, case studies, field trips, and other enrichments.

Gaining a useful knowledge of the materials and methods of building construction is crucial and a necessity for the student of architecture, engineering, or construction, but it can be a daunting task. The field is huge, diverse, and complex, and it changes at such a rate that it seems impossible ever to master it. This book has gained its preeminent status as an academic text in this field because of its logical organization, outstanding illustrations, clear writing, pleasing page layouts, and distinctive philosophy.

It is integrative, presenting a single narrative that interweaves issues of building science, materials science, legal constraints, and building craft so that the reader does not have to refer to separate parts of the book to make the connections among these issues. Building techniques are presented as whole working systems rather than component parts.

It is selective rather than comprehensive. This makes it easy and pleasant for the reader to gain a basic working knowledge that can later be expanded, without piling on so many facts and figures that the reader becomes confused or frightened away from wanting to learn about construction. Reading other texts was once like trying to drink from a fire hose; reading this one is like enjoying a carefully prepared meal.

It is empowering because it is structured around the process of designing and constructing buildings. The student of architecture will find that it features the design possibilities of the various materials and systems. Students interested in building or managing the construction process will find its organization around construction sequences to be invaluable.

The book is necessarily complex without being complicated. It avoids the dilemma of having to expand ad infinitum over time by presenting the basic construction systems, each in sufficient detail that the student is brought to an operational level of knowledge. It deals, as its subtitle indicates, with fundamentals.

In this latest edition, we have updated and revised many portions of the text. To name a few, in Chapter 1, an expanded discussion of buildings and the environment reflects the continuing evolution and maturation of sustainable building practice. Attention to the role of the constructor and to considerations of construction management continues to receive more in-depth consideration. Many facets of foundations and sitework have been updated in Chapter 2. Chapters 3 through 7, covering wood materials and construction systems, have been extensively updated: An entirely new construction system, cross-laminated timber, is introduced; a greater emphasis is given to manufactured wood products, reflecting this trend within the industry; and the influence of the most recent energy codes and sustainability standards on the enclosure of these building systems is reflected in revised text and updated illustrations. Throughout the remaining portions of the text, new developments in materials and methods, sustainable practices, and building regulations have been incorporated. Finally, a special effort was made to update illustrations and photographs, both to ensure currency of information and to provide the greatest possible visual interest for the reader.

We continue to take maximum advantage of the World Wide Web. The text's encyclopedic details, along with an array of additional resources for both students and instructors, are readily available via its dedicated web site (www.wiley.com/go/constructioneducation6e). A Respondus test bank, PowerPoint lecture slides, *Instructor's Manual*, and more can be found there for instructors. For students, there are flashcards and interactive self-test questions, as well as SketchUp exercises and animations, which are indicated by icons found throughout the text. Coauthor Joseph Iano's construction blog (www.ianosbackfill.com) provides an outlet for additional content and coverage of new developments in the field. The selected list of Web site addresses included in the reference section at the end of each chapter provides links to the other most relevant resources that are available on the Web, which provide starting points for students' further explorations.

The updated companion *Exercises in Building Construction* and its answer key continue to provide a unique and invaluable tool for helping students to understand the real-world application of building construction knowledge to the design and construction of buildings.

In this edition, a special thank-you goes to illustrators Heather McArthur and Terrel Broiles for their patience with the authors, perseverance in their work, and success in maintaining the high standards established by this book's previous illustrators. Alexander Schreyer also deserves mention for his contribution of interactive exercises and other supporting materials on the companion Web site.

The authors are, as always, grateful to the publisher, John Wiley & Sons, without whom the continued improvement of this text and its supporting materials would not be possible. Amanda L. Miller, Vice President and Publisher, has for many years been a source of wisdom and support. Paul Drougas, Editor, has been invaluable for his industry knowledge, patience, and sense of humor. He remains a true friend. Lauren Olesky, Developmental Editor, and Mike New, Editorial Assistant, have been hard working and helpful through all stages of this revision. Donna Conte, Senior Production Editor, continues, as in previous revisions, to oversee the most difficult task of managing production and schedules with grace and perseverance. Diana Cisek, and too many others to name here, also deserve our thanks for their parts in helping bring this effort to fruition.

We especially offer our thanks to the many teachers, students, and professionals who have purchased and used this work. Your satisfaction is our greatest reward, your loyalty is greatly appreciated, and your comments are always welcome!

—E.A., South Natick, Massachusetts
—J.I., Seattle, Washington

A registration code to access the resources included on the
Interactive Resource Center, shown on the previous page,
is included with each new print copy of
Fundamentals of Building Construction, Sixth Edition.
If you've purchased another version and wish to purchase
access to the Interactive Resource Center, you can go to
www.wiley.com/go/constructioneducation6e,
click on "Student Companion Site" and then "Register,"
which will allow you to enter a code or to purchase access
if you do not have a code.

How to get the most out of this text for students and instructors

The following pages give students tools and tips on maximizing their study time using the integrated text and media within *Fundamentals of Building Construction*. Both students and instructors will appreciate the outstanding resources available to seamlessly link in class activity with lecture presentation and study time.

Sidebars addressing sustainability considerations

70 new photorealistic renderings, also available in full color as part of the Interactive Resource Center and in e-book formats

Icons indicate availability of Interactive Exercises and Animations on the online Interactive Resource Center to expand learning on difficult concepts

"From Concept to Reality" case studies demonstrate how technical decisions are made in real-world building projects

Sidebars covering the building envelope address thermal insulation, vapor retarders, air barriers, dissimilar metals, and related topics

Entirely updated back-of-chapter sections, including selected references, websites, key terminology, review questions, and exercises

Interactive Resource Center
for Students www.wiley.com/go/constructioneducation6e

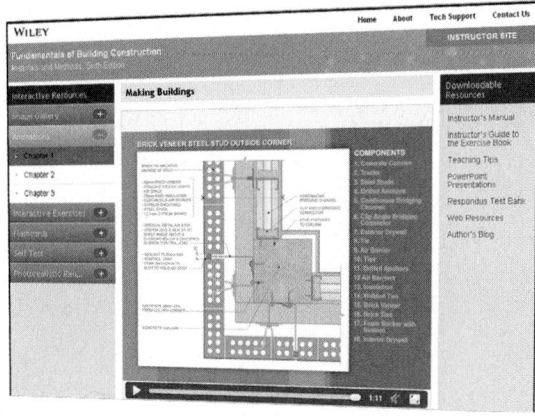

45 Interactive Exercises tied to key concepts, including all required objects, as well as a tutorial on the basics of SketchUp and how to complete exercises—also available in standard 3D CAD format for use in other software, such as Rhino or 3ds max

20 Animations revealing complex construction processes, including wood light frame construction, brick veneer, and masonry wall construction, all designed to demonstrate sequencing and other concepts

Interactive Flashcards for learning key terms and concepts

Interactive self-test for learning assessment

Supplementary image gallery demonstrating construction operations

---

Instructor Resources

- Instructor's Manual
- Test Bank available in Respondus
- Instructor's Guide to *Exercises in Building Construction*
- Supplementary image bank
- PowerPoint presentations by chapter

Additional Format Options

- Available in eBook formats for iOS, Android, Mac and PC, Kindle, and all other eReaders and tablets, including full-color images
- Complete Access Pack including Wiley E-Text version of full text, printed Exercise Book, and access to online Interactive Resource Center
- Available in Wiley CustomSelect www.wiley.com/college/custom

Exercises in Building Construction

- Companion resource featuring 46 real-world construction problems, challenges students with the kind of everyday issues faced by professional architects, available in both print and e-Book formats

Fundamentals of Building Construction

MAKING BUILDINGS

- **Learning to Build**

- **Buildings and the Environment**
  Sustainable Building Materials
  Assessing Sustainable Buildings

- **The Work of the Design Professional**
  Zoning Ordinances
  Building Codes
  Other Constraints
  Construction Standards and
      Information Resources

- **The Work of the Construction Professional**
  Providing Construction Services
  Construction Scheduling
  Managing Construction

- **Trends in the Delivery of Design and Construction Services**
  Increasing Collaboration Among
      Team Members
  Improving Efficiency in Production
  Improving Information Management

An ironworker connects a steel wide-flange beam to a column.
(Courtesy of Bethlehem Steel Company)

We build because not all human activity can take place outdoors. We need shelter from sun, wind, rain, and snow. We need dry, level surfaces for our activities. Often we need to stack these surfaces to multiply available space. On these surfaces, and within our shelter, we need air that is warmer or cooler, more or less humid, than outdoors. We need less light by day, and more by night, than is offered by the natural world. We need services that provide energy, communications, and water and dispose of wastes. So, we gather materials and assemble them into the constructions we call buildings in an attempt to satisfy these needs.

LEARNING TO BUILD

This book is about the materials and methods of building construction. Throughout it, alternative ways of building are described: different structural systems, different methods of building enclosure, and different interior finishes. Each has characteristics that distinguish it from the alternatives. Sometimes a material is selected chiefly for its visual qualities, as in choosing one type of granite over another, selecting a particular color of paint, or specifying a special pattern of tile. Visual distinctions can extend beyond surface qualities. A designer may prefer the massive appearance of a masonry bearing wall building to that of a more slender exposed steel frame on one project, yet would choose the steel for another. Choices may be made for functional reasons, as in selecting a highly durable and water-resistant polished concrete instead of carpet or wood for a restaurant kitchen floor. Or, choices can be made on purely technical grounds, as, for example, in selecting a construction system that is noncombustible, so as to achieve a suitable level of building fire safety.

A building designer's choices are frequently constrained by regulations intended to protect public safety and welfare. Choices may be influenced by considerations of environmental

sustainability. And frequently, selections are made on economic grounds. Sometimes one system is chosen over another because its first cost is less. Other times the full life-cycle costs—including first cost, maintenance, energy consumption, useful lifetime, and replacement—of competing systems are compared.

In describing the major systems of building construction, this textbook presents concerns that fall into two broad categories: building performance and building construction. Performance concerns relate to the inescapable problems that must be confronted in every building: fire; the flow of heat, air, and water vapor through the building enclosure; the small, but nonetheless important, movements of the building and its parts; water leakage; acoustical performance; aging and deterioration of materials; cleanliness; building maintenance; and so on. Construction concerns relate to the practical problems of getting a building built safely, on time, within budget, and to the required standards of quality: sequencing of construction operations for maximum productivity; optimum use of building trades; division of work between the shop and the building site; convenient and safe worker access to construction operations; effects of weather; making building components fit together; quality testing of materials and components during construction; and much more. To the novice, these matters may seem of minor consequence

when compared to the larger and often more interesting themes of building form and function. To the experienced building professional, who has seen buildings fail both aesthetically and functionally for want of attention to one or more of these concerns, these are issues that must be resolved as a matter of course to ensure a successful project outcome.

To gain a thorough knowledge of building construction, it is incumbent upon the student to go beyond what can be presented here—to other books, product literature, trade publications, professional periodicals, and especially the design office, workshop, and building site. One must learn how materials feel in the hand; how they look in a building; how they are manufactured, worked, and put in place; how they perform in service; how they deteriorate with time. One must become familiar with the people and organizations that produce buildings—the architects, engineers, materials suppliers, contractors, subcontractors, workers, inspectors, managers, and building owners—and learn to understand their respective methods, problems, and points of view. There is no other way to gain the breadth of information and experience necessary than to get involved in the art and practice of building.

In the meantime, this long and hopefully enjoyable process of education in the materials and methods of building construction can begin with the information presented in this textbook.

Go into the field where you can see the machines and methods at work that make the modern buildings, or stay in construction direct and simple until you can work naturally into building-design from the nature of construction.

—Frank Lloyd Wright, "To the Young Man in Architecture," 1931

BUILDINGS AND THE ENVIRONMENT

In constructing and occupying buildings, we expend vast quantities of the earth's resources and generate a significant portion of its environmental pollution. Buildings account for 30 to 40 percent of the world's energy consumption and carbon dioxide gas (CO_2) emissions. In the United States, buildings consume approximately 35 percent of this country's energy, 65 percent of its electricity, 12 percent of its potable water, and 30 percent of its raw materials. Building construction and operation together are responsible for roughly 40 percent of U.S. total greenhouse gas emissions and a third of its solid waste stream. Buildings are also significant emitters of particulates and other air pollutants. In short, building construction and operation cause many forms of environmental degradation and place a heavy burden on the earth's resources.

One simple definition of *sustainability* is building to meet the needs of the present generation without compromising the ability of future generations to meet their needs. By consuming irreplaceable fossil fuels and other nonrenewable resources, by building in sprawling urban patterns that cover extensive areas of prime agricultural land, by using destructive forestry practices that degrade natural ecosystems, by allowing topsoil to be eroded by wind and water, by generating substances that pollute water, soil, and air, and by generating copious amounts of waste materials that are eventually incinerated or buried in the earth, we have been building in a manner that will make it increasingly difficult for our children and grandchildren to meet their needs for communities, buildings, and healthy lives.

Sustainable building construction demands a more symbiotic relationship between people, buildings, communities, and the natural environment. Sustainable buildings—in both their operation and construction—must use less energy, consume fewer resources, cause less pollution of the air, water, and soil, reduce waste, discourage wasteful land development practices, contribute to the protection of natural environments and ecosystems, provide healthier interiors for building occupants, and minimize adverse social impacts.

The practice of sustainable design and construction, also called *green building*, continues to mature. The understanding of the interplay between buildings and the environment has deepened and standards for sustainability continue to evolve. Interest in and adoption of green building has broadened among public agencies, private owners, and building occupants. The design and construction industry has become more skillful at applying green practices, and sustainable building is becoming more integrated with mainstream practice. As a result, sustainable building performance is improving while the premium in cost and effort to design and construct such buildings in comparison to conventional buildings is declining or disappearing completely.

Sustainable Building Materials

Building sustainably requires access to information about the environmental impacts of the materials used in construction. For example, when selecting a material, the designer must ask: Does its manufacture depend on the extraction of nonrenewable resources, or is it made from recycled or rapidly renewable materials? Is additional energy required to ship the material from a distant location, or is it produced locally? Does the material contain toxic ingredients or generate unhealthful emissions, or is it free of such concerns? To enable meaningful decision making, reliable product information must be readily available to all the parties involved in the selection of materials for sustainable building design.

A series of international standards, designated as ISO 14020, distinguish three types of *environmental labels* that define expectations for comprehensiveness and reliability of sustainable materials and product information:

- *Type I Ecolabels* are independent, third-party certifications of environmental performance. Their accuracy and comprehensiveness are intended to ensure that the information provided is unbiased, relevant, and reliable.

- *Type II Self-Declared Environmental Claims* are provided directly by product manufacturers, without independent verification. They may also be more limited in scope than Type I labels. Type II labels may provide useful information, but users must employ their own judgment in evaluating the appropriateness of these products for a particular application.

- *Type III Environmental Impact Labels* provide the most comprehensive assessments of products and their environmental impacts on a comprehensive life-cycle basis. However, they do not in themselves provide environmental ratings or judgments—it is up to the user to interpret the data for this purpose. The information in Type III labels is independently verified, but the label itself may be prepared by the product manufacturer.

An example of a Type I Ecolabel is Green Seal Standard GS-11 for Paints and Coatings. Green Seal is a not-for-profit, independent organization that develops sustainability standards and certifications. For a paint or coating product to be certified to the GS-11 standard, it must:

- Meet minimum performance requirements, such as adhesion, ease of application, hiding power, washability, and fade resistance

- Be free of highly hazardous ingredients (for example, carcinogens)

- Not exceed permitted amounts for other less hazardous ingredients such as volatile organic compounds (a class of chemicals that contribute to air pollution and can act as irritants to building occupants)

- Be sold with instructions to the end user for safe application and responsible methods of disposal

By relying on this Green Seal certification, the paint specifier can easily and confidently identify environmentally responsible products from which to choose, without having to perform in-depth investigations of individual products.

Recycled materials content is an example of product information that is often provided in the form of a Type II Self-Declared Environmental Claim. That is, this information is usually reported directly by the product manufacturer, without third-party verification. This places more burden on the user of that information to determine its applicability. For example, the LEED® rating system (discussed later in this section) calculates recycled materials content as the sum of postconsumer content (materials recycled after use) plus one-half of preconsumer content (materials recycled during manufacturing). When relying on a Type II claim to determine recycled content for possible LEED credit, the designer must verify that the manufacturer's content claims accurately correspond to this standard's calculation method. At present, in North America, Type I and II environmental labels are the types in most common use.

An example of Type III labeling is the Western Red Cedar Association's Typical Red Cedar Decking Product Declaration. This 10-page document describes this product's material characteristics and quantifies—in detail—the environmental impacts of the product throughout its life. For example, for every 100 square feet of decking harvested, milled, trucked to the construction site, installed, maintained through its useful life, and then disposed of at the end of its life, this declaration reports the following impacts:

- 2,500,000 BTU (2600 MJ) of energy consumed

- 0.1 gallons (0.3 liters) of fresh water consumed
- 180 lb (80 kg) of nonhazardous waste generated

Additional information in the report quantifies materials consumption, *global warming potential* (total contribution to global warming), smog production, ozone depletion, acidification and eutrophication potential, and more. Information about the standards to which this information is prepared and independent verification of the reported results are also included. This document does not, in itself, provide an environmental rating of the product. But it can be used, for example, in comparing Western red cedar to some other decking material, such as recycled plastic decking, to assess the relative environmental consequences of choosing one of these materials over the other.

Not all sustainable product information necessarily fits neatly into one of these three label types. But considerations of comprehensiveness, independence, and relevance of information are appropriate to the review of building materials data from any source.

The Material Life Cycle

To most fully account for the environmental effects of a building material, its effects on the environment must be considered throughout its life cycle. This begins with raw materials extraction, continues with production and use, and finishes at end of life when a material is disposed of or put to an entirely new use. Such a *life-cycle analysis (LCA)* or *cradle-to-grave analysis* is considered the most comprehensive method for describing and quantifying environmental impacts associated with building materials. Through each life-cycle stage, environmental impacts are tallied: How much fossil fuel, electricity, water, and other materials are consumed? How much solid waste, global warming gasses, and other air and water pollutants

are generated? The total of all these impacts describes the environmental footprint of the material. As noted earlier in this section, this type of comprehensive life-cycle analysis is an essential part of a Type III Environmental Impact Label (Figure 1.1).

The concept of embodied energy also derives from life-cycle analysis. *Embodied energy* is the sum total of energy consumed during the material's life cycle. Because energy consumption tends to correlate with the consumption of nonrenewable resources and the generation of greenhouse gasses, it is easy to assume that materials with lower embodied energy are better for the environment than others with greater embodied energy. However, in making such comparisons, it is important to be sure that functionally equivalent quantities of materials are considered. For example, a material with an embodied energy of 10,000 BTU per pound is not necessarily environmentally preferable to another with an embodied energy of 15,000 BTU per pound, if 2 pounds of the first material are required to accomplish the same purpose as 1 pound of the second. The types of energy consumed for each material, such as fossil, nuclear, or renewable, should be considered, as impacts differ from one energy source to another. Differences in the life span of materials must also be accounted for.

Embodied energy and other life-cycle effects may sometimes be calculated for only a part of the material life cycle. A *cradle-to-gate analysis* begins with materials extraction but extends only as far as when the material leaves the factory, excluding the effects of transportation to the building site, installation, use, maintenance, and disposal or recycling. In other cases, data may be reported from cradle to the construction site. Though less comprehensive, such analyses can in many cases still provide a useful basis for comparison between products. For example, for most materials, the difference in

*Western Red
Cedar Decking
Life Cycle*

Extraction
Construction of
  logging roads
Operation of
  logging
  equipment
Felling of trees
Delimbing
Log transport to
  mill

*Manufacture or
Processing*
Log storage
Sorting
Debarking
Sawing
Seasoning
Planing
Packaging

Transportation
Shipping to
  construction site

Installation
Sawing
Nailing
Finishing
Disposal of cutoffs
  and waste

*Use and
Maintenance*
Cleaning
Refinishing
Repair

*Disposal or
Recycling*
Removal
Transport
Disposal in landfill

*Environmental
Impacts*

Fossil fuel
Nuclear energy
Renewable
  energy
Biomass energy
Fresh water
Ancillary
  materials

Nonhazardous
  waste
  materials
Hazardous
  waste
  materials
Global warming
Acidification
Eutrophication
Smog
Ozone

FIGURE 1.1

Life-cycle analysis of Western red cedar decking. The underlined life-cycle stages (Extraction, Manufacture or Processing, etc.) are applicable to any building construction material LCA. The activities listed under each stage here are specific to the example of Western red cedar decking. For other materials, other activities would be listed. The right-hand column lists the types of environmental impacts associated with this material, both resources consumed (such as energy and water) and pollutants and wastes emitted (such as global warming gasses and nonhazardous waste). Though not included here, the LCA also quantifies these impacts so that one material can be readily compared with another.

Embodied carbon refers to the total carbon-related greenhouse gas emissions associated with a building material.

While life-cycle analysis represents the most comprehensive materials assessment method currently available, it does not necessarily address all environmental impacts. LCA of wood products, for example, does not capture the loss of biodiversity, decreased water quality, or soil erosion caused by poor forestry practices. These concerns are better addressed by sustainable forestry certification programs. As another example, although global warming potential is quantified in materials LCA, its ultimate consequences for ecosystems and wildlife populations are not described.

Unhealthy and Toxic Materials
Life-cycle analysis does not fully address human health impacts of materials used in the construction of buildings. For example, although LCA may describe a material's contribution to various forms of air pollution, it will not account for the increase in incidence of asthma or shortening of life expectancy that may result therefrom. To address such concerns, green building

programs explicitly discourage the use of materials known to contain harmful ingredients or that generate such ingredients as byproducts of their manufacturing, use, or disposal.

As an example, historically, formaldehyde commonly has been used as an ingredient in binders and adhesives for many kinds of manufactured wood products. However, this chemical is now a recognized carcinogen and associated with a variety of additional adverse human health impacts. As a consequence, the use of materials with added formaldehyde in buildings is discouraged, and where such materials must be used, strict limits are set on acceptable formaldehyde emission levels. Lead, cadmium, and asbestos are other examples of once-common ingredients that are now discouraged or banned from use in building materials due to their toxicity. These and other examples are discussed in more detail throughout this book.

Assessing Sustainable Buildings

In the United States, the most widely applied system for evaluating building sustainability is the U.S. Green Building Council's Leadership in Energy and Environmental Design, or *LEED*®, rating system. LEED for New Construction and Major Renovation groups sustainability goals into eight broad categories addressing areas such as site selection and development, energy efficiency, conservation of materials and resources, and others (Figure 1.2). Within each category are mandatory *prerequisites* and optional *credits* that contribute points toward a building's overall sustainability rating. Depending on the total number of points achieved, four levels of sustainable design are recognized, including, in order of increasing performance, Certified, Silver, Gold, and Platinum. The LEED rating system is voluntary. It is used when adopted by a private building owner or mandated by a public building agency.

embodied energy between a cradle-to-cradle and cradle-to-construction site analysis is small, as most of the energy expenditure occurs prior to the material's installation, use, and disposal.

The concept of embodied effects can also be applied to any other measured input or output from a material life-cycle analysis. For example, *embodied water* refers to the fresh water consumed as a consequence of building with a particular material.

LEED for New Construction and Major Renovation
Project Checklist

Project Name

Date

Y	?	N			
			Credit 1	Integrative Process	1

			Location and Transportation	**Possible Points:**	**16**
			Credit 1	LEED for Neighborhood Development Location	16
			Credit 2	Sensitive Land Protection	1
			Credit 3	High Priority Site	2
			Credit 4	Surrounding Density and Diverse Uses	5
			Credit 5	Access to Quality Transit	5
			Credit 6	Bicycle Facilities	1
			Credit 7	Reduced Parking Footprint	1
			Credit 8	Green Vehicles	1

			Sustainable Sites	**Possible Points:**	**10**
Y			Prereq 1	Construction Activity Pollution Prevention	Required
			Credit 1	Site Assessment	1
			Credit 2	Site Development--Protect or Restore Habitat	2
			Credit 3	Open Space	1
			Credit 4	Rainwater Management	3
			Credit 5	Heat Island Reduction	2
			Credit 6	Light Pollution Reduction	1

			Water Efficiency	**Possible Points:**	**11**
Y			Prereq 1	Outdoor Water Use Reduction	Required
Y			Prereq 2	Indoor Water Use Reduction	Required
Y			Prereq 3	Building-Level Water Metering	Required
			Credit 1	Outdoor Water Use Reduction	2
			Credit 2	Indoor Water Use Reduction	6
			Credit 3	Cooling Tower Water Use	2
			Credit 4	Water Metering	1

			Energy and Atmosphere	**Possible Points:**	**33**
Y			Prereq 1	Fundamental Commissioning and Verification	Required
Y			Prereq 2	Minimum Energy Performance	Required
Y			Prereq 3	Building-Level Energy Metering	Required
Y			Prereq 4	Fundamental Refrigerant Management	Required
			Credit 1	Enhanced Commissioning	6
			Credit 2	Optimize Energy Performance	18
			Credit 3	Advanced Energy Metering	1
			Credit 4	Demand Response	2
			Credit 5	Renewable Energy Production	3
			Credit 6	Enhanced Refrigerant Management	1
			Credit 7	Green Power and Carbon Offsets	2

			Materials and Resources	Possible Points: 13
Y			Prereq 1 Storage and Collection of Recyclables	Required
Y			Prereq 2 Construction and Demolition Waste Management Planning	Required
			Credit 1 Building Life-Cycle Impact Reduction	5
			Credit 2 Building Product Disclosure and Optimization - Environmental Product Declarations	2
			Credit 3 Building Product Disclosure and Optimization - Sourcing of Raw Materials	2
			Credit 4 Building Product Disclosure and Optimization - Material Ingredients	2
			Credit 5 Construction and Demolition Waste Management	2

			Indoor Environmental Quality	Possible Points: 16
Y			Prereq 1 Minimum Indoor Air Quality Performance	Required
Y			Prereq 2 Environmental Tobacco Smoke Control	Required
			Credit 1 Enhanced Indoor Air Quality Strategies	2
			Credit 2 Low-Emitting Interiors	3
			Credit 3 Construction Indoor Air Quality Management Plan	1
			Credit 4 Indoor Air Quality Assessment	2
			Credit 5 Thermal Comfort	1
			Credit 6 Interior Lighting	2
			Credit 7 Daylight	3
			Credit 8 Quality Views	1
			Credit 9 Acoustic Performance	1

			Innovation	Possible Points: 6
			Credit 1 Innovation	5
			Credit 2 LEED Accredited Professional	1

			Regional Priority	Possible Points: 4
			Credit 1 Regional Priority: Specific Credit	1
			Credit 2 Regional Priority: Specific Credit	1
			Credit 3 Regional Priority: Specific Credit	1
			Credit 4 Regional Priority: Specific Credit	1

			Total	Possible Points: 110

Certified 40 to 49 points   Silver 50 to 59 points   Gold 60 to 79 points   Platinum 80 to 110

FIGURE 1.2

The LEED-NC v4 Project Checklist. *(Courtesy of U.S. Green Building Council.)*

The process of achieving LEED certification for a building begins at the earliest stages of project conception, continues throughout the design and construction of the project, and involves the combined efforts of the owner, design team, builder, subtrades, and materials suppliers. Its successful implementation requires a high level of cooperation among all of these parties. During this process, the achievement of individual credits is documented and submitted to the Green Building Council, which then makes the final certification of the project's LEED compliance after construction is completed.

The Green Building Council has also developed related rating systems for other types of construction, including existing buildings, commercial interiors, building core and shell construction, schools, retail building, healthcare facilities, homes, and neighborhood development. Through affiliated organizations, LEED is also implemented in Canada and other countries.

A second sustainability standard, the International Living Building Institute's *Living Building Challenge™*, sets a more ambitious standard for sustainable building. The Living Building Challenge aspires to move society beyond making buildings that do less environmental harm, to constructing buildings that do no harm at all, or even provide benefit, to the natural environment. For example, a building constructed and operated to this standard will (when considered on an annualized basis) generate all its own energy from on-site renewable resources, consume no fresh water, and have no net carbon greenhouse gas emissions.

The Living Building Challenge consists of 20 *imperatives* (e.g., net zero energy or appropriate sourcing of materials), grouped into 7 broad categories, called Petals, such as Site, Water, Materials, Social Equity, and so on. For a new building to meet this standard, it must meet all 20 imperatives. There are two certification levels: "Living," for buildings that fully meet the standard; and "Petal Recognition," for buildings that meet minimum partial requirements. Certification occurs after a building has been operational for at least one year, when its real-life performance can be assessed.

With regard to construction materials and methods, the most relevant imperatives in this standard are found in the Materials category, including:

• Red List: Materials considered toxic or highly detrimental to the environment are entirely excluded from buildings.
• Embodied Carbon Footprint: Considered on a life-cycle basis, the building must not be a net emitter of carbon greenhouse gasses.
• Responsible Industry: Materials must meet responsible third-party sustainability standards.
• Appropriate Sourcing: Depending on material density, materials must be manufactured within certain distances of the building construction site.
• Conservation and Reuse: Materials waste must be minimized throughout all project life phases.

The Living Building Challenge can also be applied to neighborhoods, landscape and infrastructure projects, and building renovations. It is a testament to the progress of the sustainable building industry that buildings meeting this ambitious standard and achieving Living certification are now a reality.

Other green building programs and standards offer a variety of pathways to sustainable building construction, suitable to various building types, building owner objectives, and markets. The U.S. National Green Building Standard addresses residential building types, from single-family homes to multistory apartment buildings and hotels. The International Green Construction Code is a model code that puts green building standards into a legally enforceable format, useful for municipalities that wish to make sustainable construction mandatory. CALGreen is the sustainable construction code for the state of California. A variety of professional organizations and government agencies offer programs to support sustainable building, such as the 2030 Challenge, ASHRAE's high-performance building standards, and the U.S. EPA's green programs. Green Globes certifies new and existing commercial buildings in the United States, Canada, and other countries. The Building Research Establishment Environmental Assessment Method, or BREEAM, is an environmental assessment system for buildings constructed in the United Kingdom and other European countries.

Sustainable building practice is producing measurable, positive results in building performance. An evaluation of sustainable facilities completed by the U.S. General Services Administration in the first decade of this century showed reductions in energy consumption and greenhouse gas emissions in the range of 25 to 35 percent in comparison to conventional building stock. Meaningful improvements in other sustainability metrics were achieved as well. Sustainable buildings being designed to meet today's best practices are capable of even higher performance levels.

Along with its tangible success, sustainable building also presents new challenges and risks to the design and construction industry. Reformulated or entirely new materials may prove to be less durable than those they replace. Products from unique sources may be susceptible to supply shortages or price instability. Buildings constructed with more airtight enclosures may become vulnerable to problems of indoor air quality or moisture accumulation within exterior walls and roofs. Green roofs may be more vulnerable to leakage, and if they do leak, more expensive to repair. Inexperience with green building technologies may lead to design or construction errors. Completed buildings may not meet the performance goals to which they were designed or the heightened expectations of their owners. Uncertainties regarding the costs of green building design, construction, or operation may create financial uncertainty for the parties involved.

It is incumbent upon those involved with sustainable building to recognize and minimize these risks. All parties must collaborate effectively, to ensure that the broad goals of sustainability are understood and effectively implemented. Designers and builders must adequately educate themselves so that they properly apply new technologies. The selection of green materials and systems must be done with care, and without losing sight of traditional concerns such as durability, practicality, and safety. Further, the contractual agreements between designers, builders, and owners must appropriately set expectations and fairly balance risks.

Considerations of sustainability are included throughout this book. In addition, a sidebar in nearly every chapter describes the major issues of sustainability related to the materials and methods discussed in that chapter. For more information on sustainable design and construction resources, see the references listed at the end of this chapter.

THE WORK OF THE DESIGN PROFESSIONAL

A building begins as an idea in someone's mind, a desire for new and ample accommodations for a family, many families, an organization, or an enterprise. For any but the smallest buildings, the next step for the owner of the prospective building is to engage, either directly or through a hired construction manager, the services of building design professionals. An architect helps to organize the owner's ideas about the new building while various engineering specialists work out concepts and details of foundations, structural support, and mechanical, electrical, and communications services.

> **[T]he architect should have construction at least as much at his fingers' ends as a thinker his grammar.**
>
> —Le Corbusier, "Towards a New Architecture," 1927

This team of designers, working with the owner, then develops the scheme for the building in progressively finer degrees of detail. *Drawings*, primarily graphic in content, and *specifications*, mostly written, are produced by the architect–engineer team to describe how the building is to be made and of what. These drawings and specifications, collectively known as the *construction documents*, are submitted to the local government building authorities, where they are checked for conformance with zoning ordinances and building codes before a permit is issued to build. A general contractor is selected, who then plans the construction work in detail. Once construction begins, the general contractor oversees the construction process and hires the subcontractors who carry out many portions of the work, while the building inspector, architect, and engineering consultants observe the work at frequent intervals to be sure that it is completed according to plan. Finally, construction is finished, the building is made ready for occupancy, and that original idea—which may have been initiated years earlier—is realized.

Zoning Ordinances

The legal restrictions on buildings begin with local *zoning ordinances*, which govern the types of activities that may take place on a given piece of land, how much of the land may be covered by buildings, how far buildings must be set back from adjacent property lines, how many parking spaces must be provided, how large a total floor area may be constructed, and how tall the buildings may be. In larger cities, zoning ordinances may include fire zones with special fire-protection requirements, neighborhood enterprise districts with economic incentives for new construction or revitalization of existing buildings, or other special conditions.

Building Codes

In addition to their zoning ordinances, local governments regulate building activity by means of *building codes*. Building codes protect public health and safety by setting minimum standards for construction quality, structural integrity, durability, livability, accessibility, and especially fire safety.

Most building codes in North America are based on one of several *model building codes*, standardized codes that local jurisdictions may adopt for their own use as a simpler alternative to writing their own. In Canada, the *National Building Code of Canada* is published by the Canadian Commission on Building and Fire Codes. It is the basis for most of that country's provincial and municipal building codes. In the United States, the *International Building Code® (IBC)* is the predominant model code. This code is published by the International Code Council, a private, nonprofit organization whose membership consists of local code officials from throughout the country. It is the basis for most U.S. building codes enacted at the state, county, and municipal levels. The International Building Code is the first unified model building code in U.S. history. First published in the year 2000, it was a welcome consolidation of a number of previous competing regional model codes.

Building-code-related information in this book is based on the IBC. The IBC begins by defining *occupancies* for buildings as follows:

- A-1 through A-5 Assembly: public theaters, auditoriums, lecture halls, nightclubs, restaurants, houses of worship, libraries, museums, sports arenas, and so on

- B Business: banks, administrative offices, college and university buildings, post offices, banks, professional offices, and the like

- E Educational: schools for grades K through 12 and some types of child day-care facilities

- F-1 and F-2 Factory Industrial: industrial processes using moderate-flammability or noncombustible materials, respectively

- H-1 through H-5 High Hazard: occupancies in which toxic, corrosive, highly flammable, or explosive materials are present

- I-1 through I-4 Institutional: occupancies in which occupants under the care of others may require assistance during a fire or other building emergency, such as 24-hour residential care facilities, hospitals, nursing homes, prisons, and some day-care facilities

• M Mercantile: stores, markets, service stations, salesrooms, and other retail and wholesale establishments

• R-1 through R-4 Residential: apartment buildings, dormitories, fraternity and sorority houses, hotels, one- and two-family dwellings, and assisted-living facilities

• S-1 and S-2 Storage: facilities for the storage of moderate- and low-hazard materials, respectively.

• U Utility and Miscellaneous: agricultural buildings, carports, greenhouses, sheds, stables, fences, tanks, towers, and other secondary buildings.

The IBC's purpose in describing occupancies is to establish different levels of life-safety hazard in buildings. For example, a hospital, in which patients are bedridden and cannot escape a fire without assistance from others, must be designed to a higher standard of safety than a hotel or motel occupied by able-bodied residents. A large retail mall building, containing large quantities of combustible materials and occupied by many users varying in age and physical capacity, must be designed to a higher standard than a warehouse storing noncombustible masonry materials and occupied by relatively few people who are all familiar with their surroundings. An elementary school requires more protection for its occupants than a university building. For a theater, with patrons densely packed in dark spaces, more attention must be paid to emergency egress provisions than for an ordinary office building.

These occupancy classifications are followed by a set of definitions for *construction types*. At the head of this list is Type I construction, made with highly fire-resistant, noncombustible materials. At the foot of it is Type V construction, which is built from combustible light wood framing—the

least fire-resistant of all construction types. In between are Types II, III, and IV, with levels of resistance to fire falling between these two extremes.

With occupancies and construction types defined, the IBC proceeds to match the two, stating which occupancies may be housed in which types of construction, and under what limitations of building height and area. Figure 1.3 is reproduced from the IBC. This table gives starting values for the maximum building height, in both feet and number of stories above grade, and the maximum area per floor for every possible combination of occupancy and construction type. Once the values in this table are adjusted according to other provisions of the code, the maximum permitted size for a building of any particular use and type of construction can be determined.

Consider, for example, an office building. Under the IBC, this building is classified as Occupancy B, Business. Reading across the table from left to right, we find immediately that this building may be built to any desired size, without limit, using Type I-A construction.

Type I-A construction is defined in the IBC as consisting of only non-combustible materials—masonry, concrete, or steel, for example, but not wood—and meeting requirements for resistance to the heat of fire. Looking at the upper table in Figure 1.4, also reproduced from the IBC, we find under Type I-A construction a listing of the required *fire resistance ratings*, measured in hours, for various parts of our proposed office building. For example, the first line states that the structural frame, including such elements as columns, beams, and trusses, must be rated at 3 hours. The second line also mandates a 3-hour resistance for *bearing walls*, which serve to carry floors or roofs above. *Nonbearing walls or partitions*, which carry no load from above, are listed in the third line, referring to Table 602, which gives fire resistance rating requirements for exterior walls

FIGURE 1.3

Height and area limitations of buildings of various types of construction, as defined in the 2012 IBC. These base values are modified according to various code provisions to arrive at the final allowable height and area for any particular building. For the purposes of this book, many of these modifications are simplified or ignored. (*Table 503 excerpted from the 2012 International Building Code, Copyright 2011. Washington, D.C.: International Code Council. Reproduced with permission. All rights reserved. www.ICCSAFE.org*)

of a building based on their proximity to adjacent buildings. (Table 602 is included in the lower portion of Figure 1.4.) Requirements for floor and roof construction are defined in the last two lines of Table 601.

Looking across Table 601 in Figure 1.4, we can see that fire resistance rating requirements are highest for Type I-A construction, decrease to 1 hour for various intermediate types, and fall to zero for Type V-B construction. In general, the lower the construction type numeral, the more fire-resistant the construction system is. (Type IV construction is somewhat of an anomaly, referring to *Heavy Timber construction* consisting of large wooden members that are relatively slow to catch fire and burn.)

Once fire resistance rating requirements for the major parts of a building have been determined, the design of these parts can proceed, using building assemblies meeting these requirements. Tabulated fire resistance ratings for common building materials and assemblies come from a variety of sources, including the IBC itself, as well as from catalogs and handbooks issued by building material manufacturers, construction trade associations, and organizations concerned with fire protection of buildings. In each case, the ratings are derived from large-scale laboratory tests of building components carried out in accordance with an

TABLE 503
ALLOWABLE BUILDING HEIGHTS AND AREAS[a,b]
Building height limitations shown in feet above grade plane. Story limitations shown as stories above grade plane.
Building area limitations shown in square feet, as determined by the definition of "Area, building." per story

GROUP		TYPE I A	TYPE I B	TYPE II A	TYPE II B	TYPE III A	TYPE III B	TYPE IV HT	TYPE V A	TYPE V B
HEIGHT (feet)		UL	160	65	55	65	55	65	50	40
		STORIES(S) AREA (A)								
A-1	S	UL	5	3	2	3	2	3	2	1
	A	UL	UL	15,500	8,500	14,000	8,500	15,000	11,500	5,500
A-2	S	UL	II	3	2	3	2	3	2.	1
	A	UL	UL	15,500	9,500	14,000	9,500	15,000	11,500	6,000
A-3	S	UL	11	3	2	3	2	3	2	1
	A	UL	UL	15,500	9,500	14,000	9,500	15,000	11,500	6,000
A-4	S	UL	11	3	2	3	2	3	2	1
	A	UL	UL	15,500	9,500	14,000	9,500	15,000	11,500	6,000
A-5	S	UL	UL	UL	UL	UL	UL	UL	UL	UL
	A	UL	UL	UL	UL	UL	UL	UL	UL	UL
B	S	UL	11	5	3	5	3	5	3	2
	A	UL	UL	37,500	23,000	28,500	19,000	36,000	18,000	9,000
E	S	UL	5	3	2	3	2	3	1	1
	A	UL	UL	26,500	14,500	23,500	14,500	25,500	18,500	9,500
F-1	S	UL	11	4	2	3	2	4	2	1
	A	UL	UL	25,000	15,500	19,000	12,000	33,500	14,000	8,500
F-2	S	UL	11	5	3	4	3	5	3	2
	A	UL	UL	37,500	23,000	28,500	18,000	50,500	21,000	13,000
H-1	S	1	1	1	1	1	1	1	1	NP
	A	21,000	16,500	11,000	7,000	9,500	7,000	10,500	7,500	NP
H-2	S	UL	3	2	1	2	1	2	1	1
	A	21,000	16,500	11,000	7,000	9,500	7,000	10,500	7,500	3,000
H-3	S	UL	6	4	2	4	2	4	2	1
	A	UL	60,000	26,500	14,000	17,500	13,000	25,500	10,000	5,000
H-4	S	UL	7	5	3	5	3	5	3	2
	A	UL	UL	37,500	17,500	28,500	17,500	36,000	18,000	6,500
H-5	S	4	4	3	3	3	3	3	3	2
	A	UL	UL	37,500	23,000	28,500	19,000	36,000	18,000	9,000
I-1	S	UL	9	4	3	4	3	4	3	1
	A	UL	55,000	19,000	10,000	16,500	10,000	18,000	10,500	4,500
I-2	S	UL	4	2	1	1	NP	1	1	NP
	A	UL	UL	15,000	11,000	12,000	NP	12,000	9,500	NP
I-3	S	UL	4	2	1	2	1	2	2	1
	A	UL	UL	15,000	10,000	10,500	7,500	12,000	7,500	5,000
I-4	S	UL	5	3	2	3	2	3	1	1
	A	UL	60,500	26,500	13,000	23,500	13,000	25,500	18,500	9,000
M	S	UL	11	4	2	4	2	4	3	1
	A	UL	UL	21,500	12,500	18,500	12,500	20,500	14,000	9,000
R-1	S	UL	11	4	4	4	4	4	3	2
	A	UL	UL	24,000	16,000	24,000	16,000	20,500	12,000	7,000
R-2	S	UL	11	4	4	4	4	4	3	2
	A	UL	UL	24,000	16,000	24,000	16,000	20,500	12,000	7,000
R-3	S	UL	11	4	4	4	4	4	3	3
	A	UL	UL	UL	UL	UL	UL	UL	UL	UL
R-4	S	UL	11	4	4	4	4	4	3	2
	A	UL	UL	24,000	16,000	24,000	16,000	20,500	12,000	7,000
S-1	S	UL	11	4	2	3	2	4	3	1
	A	UL	48,000	26,000	17,500	26,000	17,500	25,500	14,000	9,000
S-2	S	UL	11	5	3	4	3	5	4	2
	A	UL	79,000	39,000	26,000	39,000	26,000	38,500	21,000	13,500
U	S	UL	5	4	2	3	2	4	2	1
	A	UL	35,500	19,000	8,500	14,000	8,500	18,000	9,000	5,500

For SI: 1 foot = 304.8 mm, 1 square foot = 0.0929 m².
A = building area per story, S = stories above grade plane, UL = Unlimited, NP = Not permitted.
a. See the following sections for general exceptions to Table 503:
   1. Section 504.2, Allowable building height and story increase due to automatic sprinkler system installation.
   2. Section 506.2, Allowable building area increase due to street frontage.
   3. Section 506.3, Allowable building area increase due to automatic sprinkler system installation.
   4. Section 507, Unlimited area buildings.
b. See Chapter 4 for specific exceptions to the allowable height and areas in Chapter 5.

TABLE 601
FIRE-RESISTANCE RATING REQUIREMENTS FOR BUILDING ELEMENTS (HOURS)

BUILDING ELEMENT	TYPE I		TYPE II		TYPE III		TYPE IV	TYPE V	
	A	B	A[d]	B	A[d]	B	HT	A[d]	B
Primary structural frame[g] (see Section 202)	3[a]	2[a]	1	0	1	0	HT	1	0

Bearing walls
  Exterior[f,g]
  Interior
	3
	3[a]
		2
		2[a]
			1
			1
				0
				0
					2
					1
						2
						0
							2
							1/HT
								1
								1
									0
									0

Nonbearing walls and partitions Exterior	See Table 602								
Nonbearing walls and partitions Interior[e]	0	0	0	0	0	0	See Section 602.4.6	0	0
Floor construction and associated secondary members (see Section 202)	2	2	1	0	1	0	HT	1	0
Roof construction and associated secondary members (see Section 202)	1½[b]	1[b,c]	1[b,c]	0[c]	1[b,c]	0	HT	1[b,c]	0

For SI: 1 foot = 304.8 mm.

a. Roof supports: Fire-resistance ratings of primary structural frame and bearing walls are permitted to be reduced by 1 hour where supporting a roof only.

b. Except in Group F-1, H, M and S-1 occupancies, fire protection of structural members shall not be required, including protection of roof framing and decking where every part of the roof construction is 20 feet or more above any floor immediately below. Fire-retardant-treated wood members shall be allowed to be used for such unprotected members.

c. In all occupancies, heavy timber shall be allowed where a 1-hour or less fire-resistance rating is required.

d. An approved automatic sprinkler system in accordance with Section 903.3.1.1 shall be allowed to be substituted for 1-hour fire-resistance-rated construction, provided such system is not otherwise required by other provisions of the code or used for an allowable area increase in accordance with Section 506.3 or an allowable height increase in accordance with Section 504.2. The 1-hour substitution for the fire resistance of exterior walls shall not be permitted.

e. Not less than the fire-resistance rating required by other sections of this code.

f. Not less than the fire-resistance rating based on fire separation distance (see Table 602).

g. Not less than the fire-resistance rating as referenced in Section 704.10

TABLE 602
FIRE-RESISTANCE RATING REQUIREMENTS FOR EXTERIOR WALLS BASED ON FIRE SEPARATION DISTANCE[a, e, h]

FIRE SEPARATION DISTANCE = X (feet)	TYPE OF CONSTRUCTION	OCCUPANCY GROUP H[f]	OCCUPANCY GROUP F-1,M, S-1[g]	OCCUPANCY GROUP A, B, E, F-2, I, R, S-2[g], U[b]
X < 5[c]	All	3	2	1
5 ≤ X < 10	IA Others	3 2	2 1	1 1
10 ≤ X < 30	IA, IB IIB, VB Others	2 1 1	1 0 1	1[d] 0 1[d]
X ≥ 30	All	0	0	0

For SI: 1 foot = 304.8 mm.

a. Load-bearing exterior walls shall also comply with the fire-resistance rating requirements of Table 601.

b. For special requirements for Group U occupancies, see Section 406.3.

c. See Section 706.1.1 for party walls.

d. Open parking garages complying with Section 406 shall not be required to have a fire-resistance rating.

e. The fire-resistance rating of an exterior wall is determined based upon the fire separation distance of the exterior wall and the story in which the wall is located.

f. For special requirements for Group H occupancies, see Section 415.5.

g. For special requirements for Group S aircraft hangars, see Section 412.4.1.

h. Where Table 705.8 permits nonbearing exterior walls with unlimited area of unprotected openings, the required fire-resistance rating for the exterior walls is 0 hours.

FIGURE 1.4

Fire resistance of building elements as required by the IBC. Types I and II construction allow the building structure to be made only of noncombustible materials, that is, steel, concrete, and masonry. Type V construction allows any material, including wood. Types III and IV allow combinations of internal wood structure surrounded by noncombustible exterior walls. (*Tables 601 and 602 excerpted from the 2012 International Building Code, Copyright 2011. Washington, D.C.: International Code Council. Reproduced with permission. All rights reserved. www.ICCSAFE.org*)

(Reprinted with permission of Underwriters Laboratories Inc.)

FIGURE 1.5
Fire resistance ratings for a steel floor structure (*above*) and column (*below*), taken from the Underwriters Laboratories *Fire Resistance Directory*. In the floor assembly, the terms "restrained" and "unrestrained" refer to whether or not the floor is connected to its supporting structure in such a way that it is, or is not, prevented from expanding longitudinally when subjected to the heat of a fire. *(Reprinted with permission of Underwriters Laboratories Inc.)*

Design No. A814
Restrained Assembly Rating—3 Hr.
Unrestrained Assembly Rating—3 Hr.
Unrestrained Beam Rating—3 Hr.

Section A-A

Beam—W 12 × 27, min size.
1. **Sand-Gravel Concrete**—150 pcf unit weight 4000 pcf compressive strength.
2. **Steel Floor and Form Units***—Non-composite 3 in. deep galv units. All 24 in. wide, 18/18 MSG min cellular units. Welded to supports 12 in. O.C. Adjacent units button-punched or welded 36 in. O.C. at joints.
3. **Cover Plate**—No. 16 MSG galv steel.
4. **Welds**—12 in. O.C.
5. **Fiber Sprayed***—Applied to wetted steel surfaces which are free of dirt, oil or loose scale by spraying with water to the final thickness shown above. The use of adhesive and sealer and the tamping of fiber are optional. The min and density of the finished fiber should be 11 pcf and the specified fiber thicknesses require a min fiber density of 11 pcf. For areas where the fiber density is between 8 and 11 pcf, the fiber thickness shall be increased in accordance with the following formula:

$$\text{Thickness, in.} = \frac{(11)\,(\text{Design Thickness, in.})}{\text{Actual Fiber Density, pcf.}},$$

Fiber density shall not be less than 8 pcf. For method of density determination refer to General Information Section.

*Bearing the UL Classification Marking.

Design No. X511
Rating—3 Hr.

1. **Steel Studs**—1⅝ in. wide with leg dimensions of 1-5/16 and 1-7/16 in. with a ¼-in. folded flange in legs, fabricated from 25 MSG galv steel, ¾- by 1¾-in. rectangular cutouts punched 8 and 16 in. from the ends. Steel stud cut ½ in. less in length than assembly height.
2. **Wallboard, Gypsum***—½ in. thick, three layers.
3. **Screws**—1 in. long, self-drilling, self-tapping steel screws, spaced vertically 24 in. O.C.
4. **Screws**—1⅝ in. long, self-drilling, self-tapping steel screws, spaced vertically 24 in. O.C., except on the outer layer of wallboard on the flange, which are spaced 12 in. O.C.
5. **Screws**—2¼ in. long, self-drilling, self-tapping steel screws, spaced vertically 12 in. O.C.
6. **Tie Wire**—One strand of 18 SWG soft steel wire placed at the upper one-third point, used to secure the second layers of wallboard only.
7. **Corner Beads**—No. 28 MSG galv steel, 1¼ in. legs or 27 MSG uncoated steel, 1⅜ in. legs,

accepted standard protocol to ensure uniformity of results. (This test, ASTM E119, is described more fully in Chapter 22 of this book.) Figures 1.5 and 1.6 show examples of how such ratings are commonly presented.

In general, when determining the level of fire resistance required for a building, the greater the degree of fire resistance, the higher the cost. Most frequently, therefore, buildings are designed to the lowest level of fire resistance permitted by the building code. Our hypothetical office building could be built using Type IA construction, but does it really have to be constructed to this high standard?

Let us suppose that the owner desires a three-story building with 30,000 square feet per floor. Reading across the table in Figure 1.2, we can see that in addition to Type I-A construction, the building can be of Type I-B construction, which permits a building of 11 stories and unlimited floor area; or of Type II-A construction, which permits a building of 5 stories and 37,500 square feet per floor. But it cannot be of Type II-B construction, which allows a building of only 3 stories and 23,000 square feet per floor. It can also be built of Type IV construction but not of Type III or Type V.

Other factors also come into play in these determinations. If a building is protected throughout by a fully automatic sprinkler system for suppression of fire, the tabulated

Fire Rating	Sound Rating STC	GA File No.	DETAILED DESCRIPTION	SKETCH AND DESIGN DATA	
				Fire	Sound
1 HR	30 to 34	WP 3620	**Construction Type: Gypsum-Veneer Base, Veneer Plaster, Wood Studs** One layer ½″ type X gypsum veneer base applied at right angles to each side of 2 × 4 wood studs 16″ o.c. with 5d etched nails, 1¾″ long, 0.099″ shank, ¼″ heads, 8″ o.c. Minimum ¹⁄₁₆″ gypsum-veneer plaster over each face. Stagger vertical joints 16″ and horizontal joints each side 12″. Sound tested without veneer plaster. (LB)	Thickness: 4⅞″ Approx. Weight: 7 psf Fire Test: UC, 1-12-66 Sound Test: G & H IBI-35FT. 5-26-64	

FIGURE 1.6

A sample of fire resistance ratings published by the Gypsum Association, in this case for an interior partition consisting of wood studs and fire-resistant gypsum wallboard. *(Courtesy of the Gypsum Association.)*

area per floor may be tripled for a multistory building or quadrupled for a single-story building. The rationale for this permitted increase is the added safety to life and property provided by such a system. A one-story increase in allowable height is also granted under most circumstances if such a sprinkler system is installed. If the 3-story, 30,000-square-foot office building that we have been considering is provided with such a sprinkler system, a bit of arithmetic will show that it can be built of any construction type shown in Figure 1.2 except Type V.

If more than a quarter of the building's perimeter walls face public ways or open spaces accessible to firefighting equipment, an additional increase of up to 75 percent in allowable area is granted in accordance with another formula. Furthermore,

if a building is divided by fire walls having the fire resistance ratings specified in another table (Figure 1.7), each divided portion may be considered a separate building for purposes of computing its allowable area, which effectively permits the creation of a building many times larger than Figure 1.2 would, at first glance, indicate. (For the sake of simplicity, additional considerations in determining allowable building height and area in the IBC have been omitted from these examples.)

The IBC also establishes standards for natural light; ventilation; means of egress (exiting); structural design; construction of floors, walls, and ceilings; chimney construction; fire protection systems; accessibility for disabled persons; and many other important aspects of building design. In addition to the IBC, the Interna-

tional Code Council also publishes the *International Residential Code (IRC)*, a simplified model code addressing the construction of detached one- and two-family homes and townhouses of limited size. Within any particular building agency, these codes may be adopted directly in their model form. Or, as is more common, they may be adopted with amendments, adjusting the code to better suit the needs of that jurisdiction while still retaining its overall structure and intent.

The building code is not the only code with which a new building must comply. Energy codes establish standards of energy efficiency for buildings, affecting a designer's choices of windows, heating and cooling systems, and many aspects of the construction of a building's enclosing walls and roofs. Because of the significant environmental impacts associated with

FIGURE 1.7

Fire resistance requirements for fire walls, according to the IBC. *(Table 706.4 excerpted from the 2012 International Building Code, Copyright 2011. Washington, D.C.: International Code Council. Reproduced with permission. All rights reserved. www.ICCSAFE.org)*

TABLE 706.4
FIRE WALL FIRE-RESISTANCE RATINGS

GROUP	FIRE-RESISTANCE RATING (hours)
A, B, E, H-4, I, R-1, R-2, U	3[a]
F-1, H-3[b], H-5, M, S-1	3
H-1, H-2	4[b]
F-2, S-2, R-3, R-4	2

a. In Type II or V construction, walls shall be permitted to have a 2-hour fire-resistance rating.
b. For Group H-1, H-2 or H-3 buildings, also see Sections 415.6 and 415.7.

building energy consumption, the development of higher-performance energy codes that require buildings to consume less energy is one of the most important contributors to improving building sustainability.

Health codes regulate aspects of design and operation related to sanitation in public facilities such as swimming pools, food-service operations, schools, or healthcare facilities. Fire codes regulate the operation and maintenance of buildings to ensure that egress pathways, fire protection systems, emergency power, and other life-safety systems are properly maintained. Electrical and mechanical codes regulate the design and installation of building electrical, plumbing, and heating and cooling systems. Some of these codes may be locally written, but, like the building codes discussed earlier, most are based on national models. In fact, an important task in the early design of any major building is determining what agencies have jurisdiction over the project and what codes and regulations apply.

Other Constraints

Other types of legal restrictions must also be observed in the design and construction of buildings. Along with the accessibility provisions of the IBC, the *Americans with Disabilities Act (ADA)* makes accessibility to public buildings a civil right of all Americans, and the *Fair Housing Act* does the same for much multifamily housing. Together, these *access standards* regulate the design of entrances, stairs, doorways, elevators, toilet facilities, public areas, living spaces, and other parts of many buildings to ensure that they are usable by physically handicapped members of the population. The U.S. *Occupational Safety and Health Administration (OSHA)* controls the design of workplaces to minimize hazards to the health and safety of workers. OSHA sets safety standards under which a building must be constructed and also has an important role in the design of industrial and commercial buildings.

Conservation laws protect wetlands and other environmentally sensitive areas from encroachment by buildings. Fire insurance companies exert a major influence on construction standards. Through their testing and certification organizations (Underwriters Laboratories and Factory Mutual, for example) and the rates they charge for building insurance coverage, these companies offer financial incentives to building owners to build hazard-resistant construction. Federal labor agencies, building contractor associations, and construction labor unions have standards, both formal and informal, that affect the ways in which buildings are built. Contractors have particular types of equipment, certain kinds of skills, and customary ways of going about things. All of these affect a building design in myriad ways and must be appropriately considered by building designers.

Construction Standards and Information Resources

The tasks of the architect and the engineer would be much more difficult to carry out without the support of dozens of standards-setting agencies, trade associations, professional organizations, and other groups that produce and disseminate information on materials and methods of construction, some of the most important of which are discussed in the following sections.

Standards-Setting Agencies

ASTM International (formerly the American Society for Testing and Materials) is a private organization that establishes specifications for materials and methods of construction accepted as standards throughout the United States. Numerical references to ASTM standards—for example, ASTM C150 for portland cement, used in making concrete—are found throughout building codes and construction specifications, where they are used as a precise shorthand for describing the quality

of materials or the requirements of their installation. Throughout this book, references to ASTM standards are provided for the major building materials presented. Should you wish to examine the contents of the standards themselves, they can be found in the ASTM references listed at the end of this chapter. In Canada, corresponding standards are set by the *Canadian Standards Association (CSA)*. The *International Organization for Standardization (ISO)*, an organization with more than 160 member countries, performs a similar role internationally.

The *American National Standards Institute (ANSI)* is another private organization that develops and certifies North American standards for a broad range of products, such as exterior windows, mechanical components of buildings, and even the accessibility requirements referenced within the IBC itself. Government agencies, most notably the U.S. Department of Commerce's *National Institute of Science and Technology (NIST)* and the National Research Council Canada's *Institute for Research in Construction (NRC-IRC)*, also sponsor research and establish standards for building products and systems.

Construction Trade and Professional Associations

Design professionals, building materials manufacturers, and construction trade groups have formed a large number of organizations that work to develop technical standards and disseminate information related to their respective fields of interest. The Construction Specifications Institute, whose MasterFormat™ standard is described in the following section, is one example. This organization is composed both of independent building professionals, such as architects and engineers, and of industry members. The Western Wood Products Association, to choose an example from among hundreds of *trade associations*, is made up of producers of lumber and wood products.

It carries out research programs on wood products, establishes uniform standards of product quality, certifies mills and products that conform to its standards, and publishes authoritative technical literature concerning the use of lumber and related products. Associations with a similar range of activities exist for virtually every material and product used in building. All of them publish technical data relating to their fields of interest, and many of these publications are indispensable references for the architect or engineer. In some cases, the standards published by these organizations are even incorporated by reference into the building codes, making them, in effect, legal requirements. Selected publications from professional and trade associations are identified in the references listed at the end of each chapter in this book. The reader is encouraged to obtain and explore these publications and others available from these various organizations.

MasterFormat and Other Systems of Organizing Building Information

The *Construction Specifications Institute (CSI)* of the United States, and its Canadian counterpart, *Construction Specifications Canada (CSC)*, have evolved over a period of many years a comprehensive outline called *MasterFormat* for organizing information about construction materials and systems. MasterFormat is used as the outline for construction specifications for the vast majority of large construction projects in these two countries. It is frequently used to organize construction cost data, and it forms the basis on which most trade associations' and manufacturers' technical literature is cataloged. In some cases, MasterFormat is used to cross-reference materials information on construction drawings as well.

MasterFormat is organized into 50 primary *divisions* intended to cover the broadest possible range of construction materials and buildings systems. The portions of

MasterFormat relevant to the types of construction discussed in this book are as follows:

Procurement and Contracting Requirements Group
  Division 00—Procurement and Contracting Requirements

Specifications Group
  General Requirements Subgroup
    Division 01—General Requirements
  Facility Construction Subgroup
    Division 02—Existing Conditions
    Division 03—Concrete
    Division 04—Masonry
    Division 05—Metals
    Division 06—Wood, Plastics, and Composites
    Division 07—Thermal and Moisture Protection
    Division 08—Openings
    Division 09—Finishes
    Division 10—Specialties
    Division 11—Equipment
    Division 12—Furnishings
    Division 13—Special Construction
    Division 14—Conveying Equipment
  Facilities Services Subgroup
    Division 21—Fire Suppression
    Division 22—Plumbing
    Division 23—Heating, Ventilating, and Air Conditioning (HVAC)
    Division 25—Integrated Automation
    Division 26—Electrical
    Division 27—Communications
    Division 28—Electronic Safety and Security
  Site and Infrastructure Subgroup
    Division 31—Earthwork
    Division 32—Exterior Improvements
    Division 33—Utilities

These broadly defined divisions are further subdivided into *sections*, each describing a discrete scope of work usually provided by a single construction trade or subcontractor. Individual sections are identified by six-digit codes, in which the first two digits correspond to the division number and the remaining four digits identify subcategories and individual units within the division. Within Division 05—Metals, for example, some commonly referenced sections are:

Section 05 12 00—Structural Steel Framing
Section 05 21 00—Steel Joist Framing
Section 05 31 00—Steel Decking
Section 05 40 00—Cold-Formed Metal Framing
Section 05 50 00—Metal Fabrications

Every chapter in this book gives MasterFormat designations for the information it presents to help the reader know where to look in construction specifications and other technical resources for further information.

MasterFormat organizes building systems information primarily according to work product, that is, the work of discrete building trades. This makes it especially well suited for use during the construction phase of building. For example, Section 06 10 00—Rough Carpentry specifies the materials and work of rough carpenters who erect a wood light frame building structure. However, finish carpentry, such as the installation of interior doors and trim, occurs later during construction, requires different materials, and is performed by different workers with different skills and tools. So it is specified separately in Section 06 20 00—Finish Carpentry. Defining each of these aspects of the work separately allows the contractor to most efficiently and accurately manage the work's costing and execution.

In contrast to MasterFormat, the *UniFormat*™ standard organizes building systems information into functional groupings. For example, Uniformat defines eight *Level 1 categories*:

- A Substructure
- B Shell
- C Interiors
- D Services
- E Equipment and Furnishings
- F Special Construction and Demolition
- G Building Sitework
- Z General

Where greater definition is required, these categories are subdivided into so-called *Level 2 classes, Level 3 and 4 subclasses*, and even Level 5 or higher-numbered sub-subclasses, each describing more finely divided aspects of a system or assembly. For example, wood floor joist framing can fall under any of the following UniFormat descriptions:

- Level 1: B Shell
- Level 2: B10 Superstructure
- Level 3: B1010 Floor Construction
- Level 4: B1010.10 Floor Structural Frame
- Level 5: B1010.10.WF Wood Floor Framing
- Etc.

UniFormat provides a more systems-based view of construction in comparison to MasterFormat and is most useful where a broader, more flexible description of building information is needed. This includes, for example, description of building systems and assemblies during project definition and early design, or the performance specification of building systems, such as discussed later in this chapter for design/build project delivery. UniFormat is also well suited to organizing construction data in computer-aided design and building information modeling systems, which naturally tend to aggregate information into functional groupings. (Building information modeling is also discussed at greater length later in this chapter.)

The *OmniClass*™ *Construction Classification System* is an overarching scheme that attempts to incorporate multiple existing building information organizational systems, including MasterFormat, UniFormat, and others, into one system. OmniClass consists of 15 *Tables*, some of which include:

- Table 13: Spaces by Function
- Table 21: Elements
- Table 22: Work Results
- Table 23: Products
- Table 31: Phases
- Table 32: Services
- Table 35: Tools
- Table 41: Materials
- Table 49: Properties

For example, Table 13—Spaces by Function merges a number of existing systems for the management of information about rooms and spaces within buildings, useful to building owners and facilities managers. Table 21—Elements is based on UniFormat, and Table 22—Work Results is based on MasterFormat. OmniClass is an open standard that is described broadly by its authors as "a strategy for classifying the built environment." It is based on an international standard for organizing construction information, called "ISO 120006-2," and it continues to undergo active development.

The increasing attention given to organizational systems like Uni-Format and OmniClass reflects the building industry's need to manage increasingly complex sets of data and efficiently share that data between disciplines, across diverse information technology platforms, and throughout the full building life cycle, from conception to extended occupancy.

For more information on Master-Format, UniFormat, and OmniClass, see the references at the end of this chapter.

THE WORK OF THE CONSTRUCTION PROFESSIONAL

Providing Construction Services

An owner wishing to construct a building hopes to achieve a finished project that functions as required, meets expectations for design and quality, costs as little as possible, and will be completed on a predictable schedule. A contractor offering its construction services hopes to produce quality building, earn a profit, and complete the project in a timely fashion. Yet, the process of building itself is fraught with uncertainty: It is subject to the vagaries of the labor market, commodity prices, and the weather; despite the best planning efforts, unanticipated conditions arise, delays occur, and mistakes are made; not infrequently, requirements change over the course of the project; and the pressures of schedule and cost inevitably minimize the margin for miscalculation. In this high-stakes environment, the relationship between the owner and contractor must be structured to share reasonably between them the potential rewards and risks.

Construction Project Delivery Methods

In conventional *design/bid/build* project delivery (Figure 1.8), the owner first hires a team of architects and engineers to perform design services, leading to the creation of construction documents that comprehensively describe the facility to be built. Next, construction firms are invited to bid on the project. Each bidding firm reviews the construction documents and proposes a cost to construct the facility. The owner evaluates the

FIGURE 1.8

FIGURE 1.8
In design/bid/build project delivery, the owner contracts separately with the architect/engineer (A/E) design team and the construction general contractor (GC). In a design/build project, the owner contracts with a single organizational entity that provides both design and construction services.

submitted proposals and awards the construction contract to the bidder deemed most suitable. This selection may be based on bid price alone, or other factors related to bidders' qualifications may also be considered. The construction documents then become part of the construction contract, and the selected firm proceeds with the work. On all but small projects, this firm acts as the *general contractor*, coordinating and overseeing the overall construction process but frequently relying on smaller, more specialized *subcontractors* to perform significant portions or even all of the construction work. During construction, the design team continues to provide services to the owner, helping to ensure that the facility is built according to the requirements of the documents as well as answering questions related to the design, changes to the work, verification of payments to the contractor, and similar matters.

Among the advantages of design/bid/build project delivery are its easy-to-understand organizational scheme, well-established legal precedents, and relative simplicity of management. The direct relationship between the owner and the design team ensures that the owner retains control over the design and provides a healthy set of checks and balances during the construction process. With design work completed before the project is bid, the owner starts construction with a well-defined scope of work, a fixed price bid, and a high degree of confidence regarding the project schedule and final costs.

In design/bid/build project delivery, the owner contracts with two entities, and design and construction responsibilities remain divided between these two throughout the project. In *design/build* project delivery, one entity ultimately assumes responsibility for both design and construction (Figure 1.8). A design/build project begins with the owner developing a conceptual design or program that describes the functional or performance requirements of the proposed facility but does not detail its form or how it is to be constructed. Next, using this conceptual information, a design/build organization is selected to complete the remainder of the design and construction of the project. Selection of the designer/builder may be based on a competitive bid process similar to that for design/bid/build projects, on negotiation and evaluation of an organization's qualifications for the proposed work, or on some combination of these. Design/build organizations themselves can take a variety of forms: a single firm encompassing both design and construction expertise; a construction management firm that subcontracts with a separate design firm to provide those services; or a joint venture between two firms, one specializing in construction and the other in design. Regardless of the internal structure of the design/build organization, the owner contracts with this single entity throughout the remainder of the project, and this entity assumes responsibility for all remaining design and construction services.

Design/build project delivery gives the owner a single source of accountability for all aspects of the project. It also places the designers and constructors in a collaborative relationship, introducing construction expertise into the design phases of a project and allowing the earliest possible consideration of constructability, cost control, construction scheduling, and similar matters. This delivery method also readily accommodates fast track construction, a scheduling technique for reducing construction time that is described later in this chapter.

Other delivery methods are possible: An owner may contract separately with a design team and a *construction manager*. As in design/build construction, the construction manager participates in the project prior to the onset of construction, introducing construction expertise during the design stage. Construction management project delivery can take a variety of forms and is frequently associated with especially large or complex projects (Figure 1.9). In *turnkey* construction, an owner contracts with a single entity that provides not only design and construction services, but financing for the project as well. Or design and construction can be undertaken by a *single-purpose entity*, of which the owner, architect, and contractor are all joint members. Aspects of these and other project delivery methods can also be intermixed, allowing many possible organizational schemes for the delivery of design

Construction Management at Fee

Construction Management at Risk

FIGURE 1.9
In its traditional role, a construction manager (CM) *at fee* provides project management services to the owner and assists the owner in contracting directly for construction services with one or more construction entities. A CM at fee is not directly responsible for the construction work itself. A CM *at risk* acts more like a general contractor and takes on greater responsibility for construction quality, schedule, and costs. In either case, the A/E design team also contracts separately with the owner.

and construction services that are suitable to a variety of owner requirements and project circumstances.

Paying for Construction Services

With *fixed-fee* or *lump-sum compensation,* the general contractor or other construction entity is paid a fixed dollar amount to complete the construction of a project regardless of that entity's actual costs to perform the work. With this compensation method, the owner begins construction with a known, fixed cost and assumes minimal risk for unanticipated cost increases. In contrast, the construction contractor assumes most of the risk of unforeseen costs, but also stands to gain from potential savings. Fixed-fee compensation is most suitable to projects where the scope of the construction work is well defined when the construction fee is set, as is the case, for example, with design/bid/build construction.

With *cost plus a fee* compensation, the owner agrees to pay the construction entity for the actual costs of construction—whatever they may turn out to be—plus an additional amount to account for overhead and profit. In this case, the construction contractor is shielded from most cost uncertainty, and it is the owner who assumes most of the risk of added costs and stands to gain the most from potential savings. Cost plus a fee compensation is most often used with

projects for which the scope of construction work is not fully known at the time compensation is established, a circumstance most frequently associated with construction management or design/build contracts.

Cost plus a fee compensation may also include a *guaranteed maximum price (GMAX or GMP).* In this case, there is a maximum fee that the owner is required to pay. While the contractor's compensation remains under the guaranteed amount, compensation is made in the same manner as with a standard cost plus a fee contract. However, once the compensation reaches the guaranteed maximum, the owner is no longer required to make additional payments and the contractor assumes responsibility for all additional costs. This compensation method retains some of the scope and price flexibility of cost plus a fee compensation while also establishing a limit on the owner's cost risks.

Incentive provisions in owner/contractor agreements can be used to more closely align owner and contractor interests. For example, in simple cost plus a fee construction, there may be an incentive for a contractor to add costs to a project, as these added costs will generate added fees. To eliminate such a counterproductive incentive, a bonus fee or profit-sharing provision can provide for some portion of construction cost

savings to be returned to the contractor. In this way, the contractor and owner jointly share in the benefits of reduced construction cost. Bonuses and penalties for savings or overruns in costs and schedules can be part of any type of construction contract.

Surety bonds are another form of legal instrument used to manage financial risks of construction, most frequently with publicly financed or very large projects. The purpose of a surety bond is to protect an owner from the risks of default, such as bankruptcy, by the construction contractor. For a fixed fee, a third party (surety) promises to complete the contractual obligations of the contractor if that contractor should for any reason fail to do so. Mostly commonly, two separate bonds are issued, one for each of the general contractor's principal obligations: a *performance bond* to assure completion of the construction and a *payment bond* to assure full payment to suppliers and subcontractors.

With competitive bidding and fixed-fee compensation, the owner is assured of competitive pricing for construction services and the contractor assumes most of the risk for unanticipated costs. With a negotiated contract and simple cost plus a fee compensation, the risks of noncompetitive pricing and unanticipated costs are shifted mostly to the owner. By adjusting projecting delivery and compensation methods, these and other construction-related risks can be allocated in varying degrees between the two parties to best suit the requirements of any particular project.

Sequential versus Fast Track Construction

In *sequential construction* (Figure 1.10), each major phase in the design and construction of a building is completed before the next phase begins, and construction does not start until all design work has been completed. Sequential construction can take place under any of the project delivery methods described previously. It is frequently associated with design/bid/build construction, where the separation of design and construction phases fits naturally with the contractual separation between design and construction service providers.

Phased construction, also called *fast track construction*, aims to reduce the time required to complete a project by overlapping the design and construction of various project parts (Figure 1.10). By allowing construction to start sooner and by overlapping the work of design and construction, phased construction can reduce the total time required to complete a project. However, phased construction also introduces its own risks. Because construction on some parts of the project begins before all design is complete, an overall cost for the project cannot be established until a significant portion of construction is underway. Phased construction also introduces more complexity into the design process and increases the potential for design errors (for example, if foundation design does not adequately anticipate the requirements of the not yet fully engineered structure above). Phased construction can be applied to any construction delivery method discussed earlier. It is frequently associated with design/build and construction management project delivery methods, where the early participation of the construction entity provides resources that are helpful in managing the complex

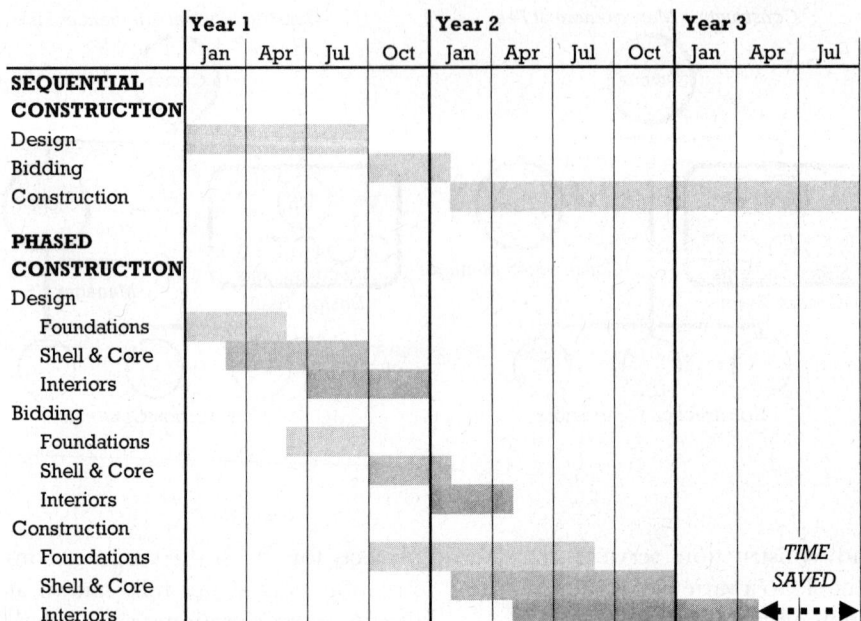

FIGURE 1.10

In sequential construction, construction does not begin until design is complete. In phased construction, design and construction activities overlap, with the goal of reducing the overall time required to complete a project.

coordination of overlapping design and construction activities.

Construction Scheduling

Constructing a building of any significant size is a complex endeavor, requiring the combined efforts of countless participants and the coordination of myriad tasks. Managing this process requires an in-depth understanding of the work required, of the ways in which different aspects of the work depend upon each other, and of the constraints on the sequence in which the work must be performed.

Figure 1.11 captures one moment in the erection of a tall building. The process is led by the construction of the building's central, stabilizing core structures (in the photograph, the pair of concrete towerlike structures extending above the highest floor levels). This work is followed by construction of the surrounding floors, which rely, in part, on the previously completed cores for support. Attachment of the exterior skin can follow

only after the floor plates are in place and structurally secure. As the building skin is installed and floor areas become enclosed and protected from the weather, further operations, such as the roughing in of mechanical and electrical systems, and eventually, the installation of finishes and other elements, can proceed in turn. This simple example illustrates considerations that apply to virtually every aspect of building construction and at every scale from a building's largest systems to its smallest details: Successful construction requires a detailed understanding of the tasks required and their interdependencies in time and space.

The construction project schedule is used to analyze and represent construction tasks, their relationships, and the sequence in which they must be performed. Development of the schedule is a fundamental part of construction project planning, and regular updating of the schedule throughout the life of the project is essential to its successful

FIGURE 1.11

In this photo, the construction sequence of a tall building is readily apparent: A pair of concrete core structures leads the construction, followed by concrete columns and floor plates and, finally, the enclosing curtain wall. *(Photo by Joseph Iano.)*

management. In a *Gantt chart* or *bar chart*, a series of horizontal bars represents the duration of various tasks or groups of tasks that make up the project. Gantt charts provide an easy-to-understand representation of construction tasks and their relationships in time. They can be used to provide an overall picture of a project schedule, with only a project's major phases represented (Figure 1.10), or they can be expanded to represent a larger number of more narrowly defined tasks at greater levels of project detail (Figure 1.12).

FIGURE 1.12

In a Gantt chart, varying levels of detail can be represented. In this example, roughly the top three-quarters of the chart is devoted to a breakdown of preconstruction and procurement activities such as bidding portions of the work to subtrades, preparing cost estimates, and making submittals to the architect (*a*). Construction activities, represented more broadly, appear in the bottom portion (*b*).

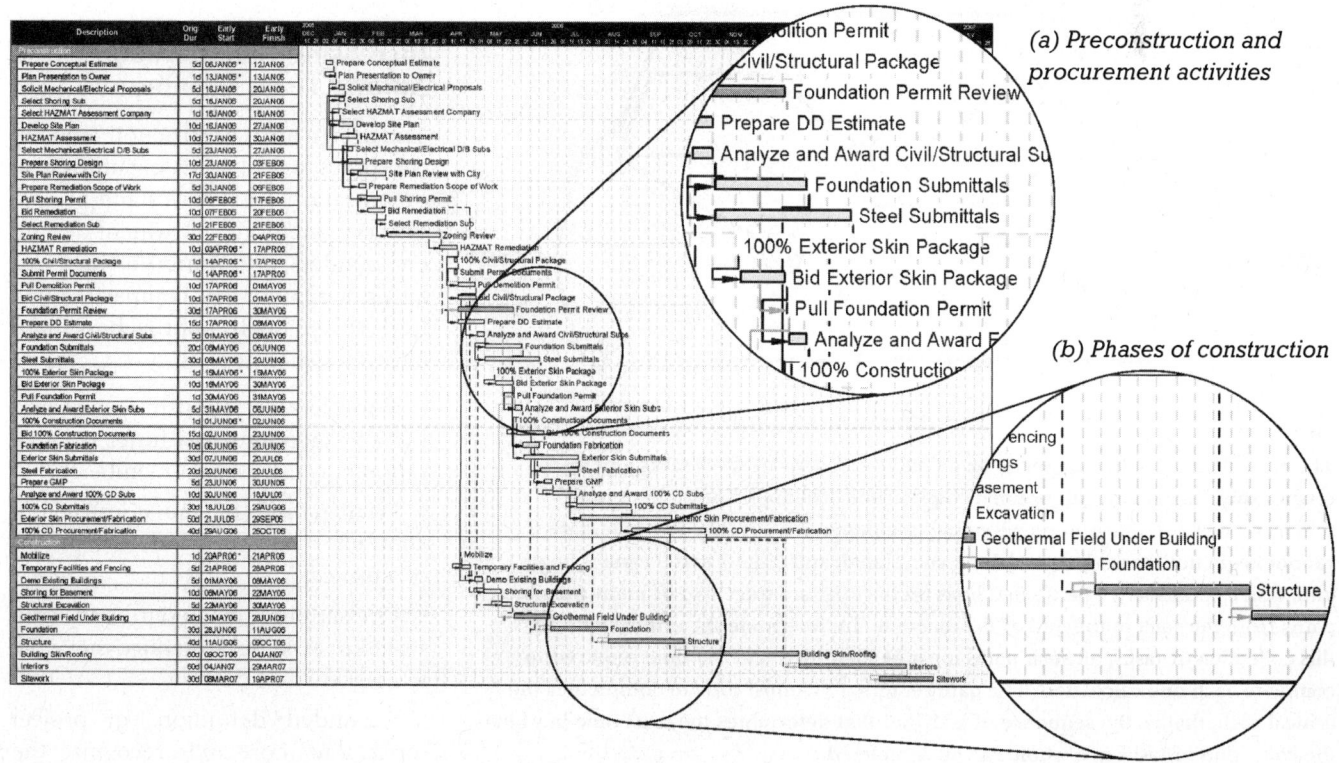

(a) Preconstruction and procurement activities

(b) Phases of construction

The *critical path* of a project is the sequence of activities that determines the least amount of time in which a project can be completed. For example, the construction of a building's primary structural system is commonly on the critical path of a project schedule. If any of the activities on which the completion of this system depends—such as design, shop drawing production and review, component fabrication, materials delivery, or erection on site—are delayed, then the final completion date of the project will be extended. In contrast, other systems not on the critical path have more flexibility in their scheduling, called *float*, and delays (within limits) in their execution will not necessarily affect the overall project schedule.

The *critical path method* is a technique for analyzing collections of activities and optimizing the project schedule to minimize the duration and cost of a project. This requires a detailed breakdown of the work involved in a project and the identification of dependencies among the parts (Figure 1.13). This information is combined with considerations of cost and resources available to perform the work, and then analyzed, usually with the assistance of computer software, to identify optimal scenarios for scheduling and worker and resource allocation. Once the critical path of a project has been established, the elements on this path are likely to receive a high degree of scrutiny during the life of the project, as delays in any of these steps will have a direct impact on the overall project schedule.

Managing Construction

Once a construction project is underway, the general contractor assumes responsibility for day-to-day oversight of the construction site, management of trades and suppliers, and communications between the construction team and other major parties, such as the owner and the designer. On projects of any significant size, this may include responsibility for filing construction permits, securing the project site, providing temporary power and water, setting up office trailers and other support facilities, providing insurance coverage for the work in progress, managing personnel on site, maintaining a safe work environment, stockpiling materials, performing testing and quality control, providing site surveying and engineering, arranging for cranes and other construction machinery, providing temporary structures and weather protection, disposing or recycling of construction waste, soliciting the work of subtrades and coordinating their efforts, submitting product samples and technical information to the design team for review, maintaining accurate records of the construction as it proceeds, monitoring costs and schedules, managing changes to the work, protecting completed work, and more.

TRENDS IN THE DELIVERY OF DESIGN AND CONSTRUCTION SERVICES

Increasing Collaboration Among Team Members

The design and construction industry continues to evolve, testing innovative organizational structures and project delivery methods in which designers, builders, and owners assume less adversarial and less compartmentalized roles. Such approaches share characteristics such as:

• Contractual relationships and working arrangements that foster collaboration between project members

• Participation of the construction contractor during the design phases of a project

• Overlapping of design and construction activities to reduce the "time to market"

• Expanded definitions of project services to more fully recognize the

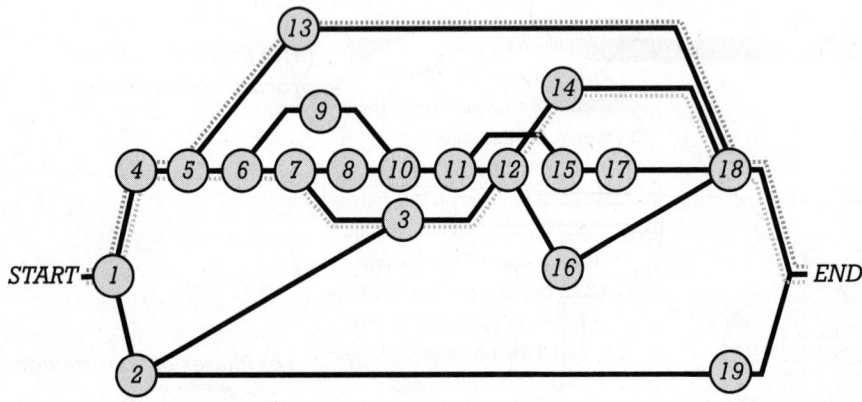

FIGURE 1.13

The critical path method depends on a detailed analysis of work tasks and their relationships to generate an optimal construction schedule. Shown here is a schematic network diagram representing task dependencies. For example, task 6 cannot begin until tasks 1, 4, and 5 are completed, and tasks 7 and 9 cannot begin until task 6 is finished. The dashed lines on the diagram trace two of many possible paths from the start to the end of the diagram. To determine the critical path for this collection of tasks, all such paths must be identified and the time required to complete each one calculated. The path requiring the most time to complete is the critical path, that is, the sequence of activities that determines the least time in which the collection of tasks as a whole can be completed.

full life cycle of a project—from its original conception, through planning, design and construction, to postconstruction occupancy—to best serve the needs of the building owner

The growth of design/build in the construction marketplace is one example of this trend: Between 1980 and 2010, the share of private, non-residential construction work performed as design/build construction increased from roughly 5 percent of the total market to 40 percent.

The current state of the art in collaborative project delivery is *integrated project delivery (IPD)*. In IPD, the key parties—including at least the design team, construction team, and owner agency—share mutually the responsibilities, decision making, and financial risks and rewards of the project. In its purest form, all major parties to the project share one agreement, binding them all to the same goals and outcomes. In other cases, a shared *joining agreement* may be used to mutually bind parties contracted under separate agreements. The goal of IPD is to increase project efficiency, improve project outcomes, and reduce conflict and litigation. In every case, IPD project delivery depends on a high level of collaborative and open communication, respect, and trust between the parties involved.

Improving Efficiency in Production

Other efforts within the construction industry focus on improvements in the efficiency of construction methods themselves. Unlike factory production, most building construction takes place outdoors, is performed within constrained and often physically challenging work areas, and is executed by a highly fragmented workforce. Despite the differences in these production environments, the construction industry is looking to lessons learned in factory production for approaches to improving its own

processes. Such so-called *lean construction* methods attempt to:

- Reduce complexity
- Eliminate wasteful activities
- Structure the supply of materials and methods of production to achieve the quickest and most reliable workflow
- Decentralize information and decision making, to put control of processes into the hands of those most familiar with the work and most capable of improving it

A typical home may be made of more than 100,000 separate pieces, assembled by as many as 1000 workers. Estimates of labor inefficiency in building construction in general run as high as 35 to 40 percent, and of materials wastage, 20 percent or more. The challenge of lean construction is to restructure all aspects of construction—its design, materials manufacturing, product transport, component assembly, and workforce integration—to reduce these inefficiencies and deliver a finished product of the highest possible quality.

Improving Information Management

Developments in information technology also are influencing the way buildings are designed and constructed. Most notable is *building information modeling (BIM)*, the computerized, three-dimensional modeling of building systems. Unlike the two-dimensional representation of building systems characteristic of conventional *computer-aided design (CAD)*, BIM involves an intelligent model. Components are not only represented geometrically and spatially but are linked to data describing their intrinsic properties and their relationships to other components as well. Originally developed for use in highly capital-intensive industries such as aerospace and automobile manufacturing, this technology has now established a firm presence in

the design and building construction sector.

BIM has the potential to affect all phases of the building life cycle. It can aid the design team in the effective communication of design concepts or the description of complex geometries. It can improve coordination between design disciplines, for example, performing *clash detection* to find *collisions* between mechanical system ductwork, structural framing, and other systems designed by separate teams. It can facilitate the modeling of building energy use and the performance of other building systems. For the builder, BIM can be used to improve coordination of trades, drive the automated fabrication or pre-assembly of building components, and integrate cost and schedule data more closely with building design and construction activities. For the building owner, information accumulated in the model during design and construction can be carried forward for use with postconstruction building operations and facilities planning.

Implementation of BIM also profoundly impacts design and construction work processes and the communication and sharing of information between disciplines. A key component to successful implementation is the *BIM execution plan*. This plan defines the role of the building model and its level of development at various project stages, identifies the sources of data that will contribute to the model, assigns responsibilities for authoring and managing the model, establishes protocols for information exchange among parties, and defines the technical and project infrastructure required to support these activities. As BIM becomes increasingly integrated into design and construction activities and the integrated building model is shared across traditional discipline boundaries, the boundaries of responsibility between the designers, constructors, and owners will also evolve, with new, more integrated relationships likely to evolve (Figure 1.14).

(a) Uses for the model

(b) Data exchange among disciplines

FIGURE 1.14

A sample high-level diagram of the BIM execution process through design and construction phases. Note how the model is used for numerous purposes at each phase by a variety of disciplines, while data must be exchanged regularly between disciplines in order to support the continued development of the model. *(Excerpted from: The Computer Integrated Construction Research Group.* Building Information Modeling Execution Planning Guide (Version 2.1). *The Pennsylvania State University, State College, PA, 2011. This work is licensed under the Creative Commons Attribution-Share Alike 3.0 United States License. To view a copy of this license, visit http:// creativecommons.org/licenses/by-sa/3.0/ us/ or send a letter to Creative Commons, 171 Second Street, Suite 300, San Francisco, California, 94105, USA.)*

CSI/CSC MasterFormat Sections for Procurement of Construction and General Requirements

00 10 00	**SOLICITATION**
00 11 00	Advertisements and Invitations
00 20 00	**INSTRUCTIONS FOR PROCUREMENT**
00 21 13	Instructions to Bidders
00 40 00	**PROCUREMENT FORMS AND SUPPLEMENTS**
00 41 00	Bid Forms
00 50 00	**CONTRACTING FORMS**
00 52 00	Agreement Forms
00 60 00	**PROJECT FORMS**
00 61 13	Performance and Payment Bond Form
00 70 00	**CONDITIONS OF THE CONTRACT**
01 10 00	**SUMMARY**
01 11 00	Summary of the Work
01 30 00	**ADMINISTRATIVE REQUIREMENTS**
01 31 13	Project Management and Coordination
01 32 13	Scheduling of Work
01 40 00	**QUALITY REQUIREMENTS**
01 41 00	Regulatory Requirements
01 45 00	Quality Control
01 50 00	**TEMPORARY FACILITIES AND CONTROLS**
01 70 00	**EXECUTION AND CLOSEOUT REQUIREMENTS**
01 80 00	**PERFORMANCE REQUIREMENTS**
01 81 13	Sustainable Design Requirements

The end of each chapter includes a list of MasterFormat sections relevant to the topics discussed in that chapter. In this list, Division 0 includes sections relevant to the procurement of construction services. Division 1 addresses general project requirements that apply broadly to all aspects of the work.

KEY TERMS

sustainability
green building
environmental labels
Type I Ecolabel
Type II Self-Declared Environmental Claim
Type III Environmental Impact Label
global warming potential
life cycle analysis (LCA), cradle-to-grave analysis
embodied energy
cradle-to-gate analysis
embodied water
embodied carbon
LEED
LEED prerequisite
LEED credit
Living Building Challenge
Living Building Challenge imperative
drawings
specifications
construction documents
zoning ordinance
building code
model building code
National Building Code of Canada
International Building Code (IBC)
occupancy
construction type
fire resistance rating
bearing wall

nonbearing wall, partition
Heavy Timber construction
International Residential Code (IRC)
Americans with Disabilities Act (ADA)
Fair Housing Act
access standard
Occupational Health and Safety Administration (OSHA)
ASTM International
Canadian Standards Association (CSA)
International Organization for Standardization (ISO)
American National Standards Institute (ANSI)
National Institute of Science and Technology (NIST)
Institute for Research in Construction (NRC-IRC)
trade association
Construction Specifications Institute (CSI)
Construction Specifications Canada (CSC)
MasterFormat
specification division
specification section
UniFormat
(UniFormat) Level 1 category
(UniFormat) Level 2 class
(UniFormat) Level 3 and 4 subclasses
OmniClass Construction Classification System

(OmniClass) Tables
design/bid/build
general contractor
subcontractor
design/build
construction manager
turnkey
single-purpose entity
fixed-fee compensation, lump-sum compensation
cost plus a fee compensation
guaranteed maximum price (GMAX, GMP)
incentive provisions
surety bond
performance bond
payment bond
sequential construction
phased construction, fast track construction
Gantt chart, bar chart
critical path
float
critical path method
integrated project delivery (IPD)
joining agreement
lean construction
building information modeling (BIM)
computer-aided design (CAD)
system collision

REVIEW QUESTIONS

1. What is sustainable building? Why is it important?

2. What are the three types of ISO 14020 environmental labels? Describe each.

3. What is a life-cycle analysis? What are the major life-cycle stages in such an analysis?

4. What is the embodied energy of a material?

5. Who are the three principal team members involved in the creation of a new building? What are their respective roles?

6. What are construction documents? What two items are they comprised of?

7. What types of subjects are covered by zoning ordinances? By building codes?

8. What is a building code occupancy? What is a construction type? How are they related in a building code?

9. In what units is fire resistance measured? How is the fire resistance of a building assembly determined?

10. What is the MasterFormat? What is it used for?

11. Compare and contrast design/bid/build and design/build construction.

12. What is the difference between lump-sum and cost plus a fee compensation?

13. What are the two common types of surety bonds? What are they used for?

14. What is fast track construction, and what types of contracts and fee compensation is it mostly commonly associated with?

15. What is the critical path? Why is it important to construction scheduling?

16. You are designing a three-story office building (Occupancy B) with 20,000 square feet per floor. What types of construction will you be permitted to use under the IBC if you do not install sprinklers? How does the situation change if you install sprinklers? In the second, sprinklered case, what is the least fire resistant construction type permitted. With this construction type, what level of fire resistance is required for the structural frame of the building?

EXERCISES

1. Choose any one building material. Obtain available product and environmental data for that material. Determine what LEED credits this material can contribute to.

2. Have each class member choose two or three trade associations at the beginning of the term. Obtain lists of publications from each and then obtain several of these publications. Display and discuss the publications.

3. Obtain copies of your local zoning ordinances and building code (they may be in a local library). Look up the applicable provisions of these documents for a specific site and building. What setbacks are required? How large a building is permitted? What construction types may be employed? What are the minimum fire-resistance ratings for the structural and nonstructural parts of the building?

4. Arrange permission to shadow an architect or CM during visits to a construction site or during project meetings related to a construction project. Take notes. Interview the architect or CM about their role and the challenges they have encountered. Report back to the class what you have learned.

SELECTED REFERENCES

Allen, Edward. *How Buildings Work* (3rd ed.). New York, Oxford University Press, 2005.

Describes what buildings do and how they do it. This book sets forth, in easily understandable terms, the physical principles by which buildings stand up, enclose a piece of the world, and modify it for human use.

Allen, Edward, and Joseph Iano. *The Architect's Studio Companion* (5th ed.). Hoboken, NJ, John Wiley & Sons, Inc., 2012.

This design reference simplifies the determination of construction type and building size for any building according to the IBC or the Canadian Building Code. It also gives extensive rules of thumb for structural systems, mechanical systems, egress planning, and sustainable design.

American Institute of Architects. *The Architect's Handbook of Professional Practice* (14th ed.). Hoboken, NJ, John Wiley & Sons, Inc., 2008.

ASTM International. *ASTM Standards in Building Codes*. Philadelphia, updated regularly.

Contains most of the ASTM standards referenced in standard building construction practice.

Canadian Commission on Building and Fires Codes. *National Building Code of Canada*. Ottawa, National Research Council of Canada, updated regularly.

The model building code used as the basis for most Canadian provincial and municipal building codes.

Clough, Richard H., et al. *Construction Contracting* (7th ed.). Hoboken, NJ, John Wiley & Sons, Inc., 2005.

A thorough coverage of the essentials of construction contracting and management.

Construction Specification Institute. *The Project Resource Manual, CSI Manual of Practice* (5th ed.). Alexandria, VA, 2004.

Provides industry-standard guidelines for organization, management, and execution of design and construction projects.

Construction Specifications Institute and Construction Specifications Canada. *MasterFormat*™. Alexandria, VA, and Toronto, updated regularly.

Includes the full list of MasterFormat numbers and titles under which construction information is most commonly organized.

International Code Council, Inc. *International Building Code®*. Falls Church, VA, updated regularly.

The model building code used as the basis for the majority of U.S. state, county, and municipal building codes.

U.S. Green Building Council. *LEED Reference Guide for Green Building Design and Construction*. Washington, DC, updated regularly.

Provides essential information for the design and construction of buildings meeting the requirements of the U.S. Green Building Council's LEED for New Construction and Major Renovations, Schools New Construction and Renovations, and Core & Shell Development rating systems.

WEB SITES

Making Buildings

Author's supplementary web site: **www.ianosbackfill.com/01_making_buildings**

Whole Building Design Guide: **www.wbdg.org**

Buildings and the Environment

Architecture 2030: **architecture2030.org**

Athena Sustainable Materials Institute: **www.athenasmi.org**

BEES (Building for Environmental and Economic Sustainability): **www.nist.gov/el/economics/BEESSoftware.cfm**

Declare: **www.declareproducts.com**

The Green Building Information Gateway: **www.gbig.org**

International Living Building Institute: **www.living-future.org**

Living Building Challenge: **www.livingbuildingchallenge.org**

Pharos: **www.pharosproject.net**

U.S. Environmental Protection Agency, Green Building: **www.epa.gov/greenbuilding/**

U.S. Green Building Council: **www.usgbc.org**

Worldwatch Institute: **www.worldwatch.org**

The Work of the Design Professional

American Institute of Architects: **www.aia.org**

American National Standards Institute (ANSI): **www.ansi.org**

ASTM International: **www.astm.org**

Canadian Standards Association (CSA): **www.csa.ca**

Construction Specifications Canada (CSC): **www.csc-dcc.ca**

Construction Specifications Institute (CSI): **www.csinet.org**

International Code Council: **www.iccsafe.org**

National Institute of Building Sciences (NIBS): **www.nibs.org**

NRC Institute for Research in Construction: **www.nrc-cnrc.gc.ca/eng/ibp/irc.html**

OmniClass: **www.omniclass.org**

Uniformat: **www.csinet.org/uniformat**

The Work of the Construction Professional

Associated General Contractors of America (AGC): **www.agc.org**

Building Owners and Managers Association (BOMA): **www.boma.org**

Construction Management Association of America (CMAA): **cmaanet.org**

Design-Build Institute of America (DBIA): **www.dbia.org**

Engineers Joint Contract Documents Committee: **www.ejcdc.org**

Lean Construction Institute: **www.leanconstruction.org**

FOUNDATIONS AND SITEWORK

- **Foundation Requirements**

- **Earth Materials**
  Classifying Earth Materials
  Properties of Soils
  Soils for Building Foundations
  Subsurface Exploration and Soils
    Testing

  SUSTAINABILITY OF FOUNDATIONS AND
  THE BUILDING SITE

- **Earthwork and Excavation**
  Excavation
  Excavation Support
  Dewatering

- **Foundations**
  Shallow Foundations
  Deep Foundations
  Seismic Base Isolation
  Underpinning
  Up–Down Construction

- **Protecting Foundations from
  Water, Heat Flow, and Radon
  Gas**
  Waterproofing and Drainage
  Basement Insulation
  Radon Gas Control

- **Sitework**
  Retaining Walls

  GEOTEXTILES

  Filling and Finish Grading

- **Designing Foundations**

- **Foundation Design and the
  Building Code**

Foundation work in progress for a midrise hotel and apartment building. The earth surrounding the excavation is retained with steel sheet piling supported by steel walers and tiebacks. Equipment enters and leaves the site via the earth ramp at the bottom of the picture. Although a large backhoe at the right continues to dig around old piles from a previous building on the site, the installation of pressure-injected concrete pile footings is also well underway, with several piledrivers at work in the near and far corners and clusters of completed piles visible in the center of the picture. Concrete pile caps and column reinforcing are under construction in the center of the excavation. *(Courtesy of Franki Foundation Company.)*

The function of a *foundation* is to transfer structural loads reliably from a building into the ground. Every building must have a foundation of some kind: A backyard toolshed will not be damaged by slight shifting of its foundation, and thus may need only wooden skids to spread its load across an area of the ground surface sufficient to support its weight. A wood-framed house needs greater stability than a toolshed, so its foundation reaches through the unstable surface to underlying soil that is free of organic matter and unreachable by winter frost. A larger building of masonry, steel, or concrete weighs many times more than a house, and its foundations must penetrate the earth until they reach soil or rock that is capable of carrying its massive loads; on some sites, this means going 100 feet (30 m) or more below the surface. Foundation design is a specialized field that must account for the interaction of building loads with the various soil, rock, and water conditions encountered below the surface of the ground. The choice of foundation type can have a significant impact on building costs, construction schedule, and choice of structural systems for the remainder of the building.

FOUNDATION REQUIREMENTS

The most important role of the foundation is to prevent building collapse. The foundation must receive the various loads acting on the building and transfer these loads into the underlying earth in a manner such that the building remains upright and stable (Figure 2.1). These loads may include:

• *Dead load,* the combined weight of all the permanent components of the building, including its own structural frame, floors, roofs, and walls, major permanent electrical and mechanical equipment, and the foundation itself

• *Live loads,* nonpermanent loads caused by the weights of the building's occupants, furnishings, and movable equipment

• *Rain and snow loads,* which act primarily downward on building roofs

• *Wind loads,* which can act laterally (sideways), downward, or upward on a building

• *Seismic loads,* highly dynamic horizontal and vertical forces caused by the motion of the ground relative to the building during an earthquake

• Loads caused by soil and hydrostatic pressure, including *lateral soil pressure loads* consisting of horizontal pressures of earth and groundwater against basement walls; in some instances, *buoyant uplift* forces from underground water, identical to the forces that cause a boat to float; in other instances, lateral force *flood loads* that can occur in areas prone to flooding

• In some buildings, *horizontal thrusts* from long-span structural components such as arches, rigid frames, domes, vaults, or tensile structures

Foundations must limit *settlement.* All foundations settle to some extent as the surrounding earth compresses and adjusts to the loads imposed by the building above. Over the life of the building, settlement must not exceed amounts that would cause structural distress, damage nonstructural components, or interfere with building functions.

Foundations on bedrock settle a negligible amount. Foundations in other types of soil may settle more, but are normally designed to limit settling to amounts measured in millimeters or fractions of an inch. In rare cases, buildings may settle by significantly greater amounts. Mexico City's Palace of Fine Arts, for example, has sunk more than 15 feet (4.5 m) into the clay soil on which it is founded since it was constructed in the early 1930s.

DEAD LOAD

RAIN, SNOW, AND LIVE LOADS

WIND LOAD

FIGURE 2.1

Some of the loads that act on buildings. Under any combination of possible load scenarios, the foundation must transmit the forces acting upon the building into the ground, and the ground must react with equal and opposite forces so that the building remains stable.

> **We must never trust too hastily to any ground. . . . I have seen a tower at Mestre, a place belonging to the Venetians, which, in a few years after it was built, made its way through the ground it stood upon . . . and buried itself in earth up to the very battlements.**
>
> —Leon Battista Alberti, *Ten Books on Architecture*, 1452

Where settling occurs at roughly the same rate throughout all parts of a building, it is termed *uniform settlement*. When different parts of a building settle at differing rates, it is called *differential settlement*. Differential settlement is most likely to occur where soil types, building loads, or structural systems vary among different parts of a single building. This can lead to distortion of the building frame, sloped floors, cracked walls and glass, or inoperable doors and windows (Figure 2.2). Most foundation failures are attributable to excess differential settlement. Gross failure of a foundation, in which the soil fails completely to support the building, is rare.

Where foundations enclose basements or other usable space, they must keep those spaces dry and at a comfortable temperature. Where foundations are constructed close to other existing buildings, they must not impose new loads or alter groundwater conditions in ways that could adversely affect those nearby buildings. Furthermore, foundations must be feasible to construct, both technically and economically. These and other aspects of foundation design are discussed at more length in this chapter.

EARTH MATERIALS

Classifying Earth Materials

For the purposes of foundation design, *earth materials* are classified according to particle size, the presence of organic content, and, in the case of finer-grained soils, sensitivity to moisture content.

Consolidated rock or *bedrock* is a dense, continuous mass of mineral materials that can be removed only by drilling, fracturing, or blasting. Rock is rarely completely monolithic and may vary in composition or structure, or be crossed by systems of joints (cracks). Despite such variations, bedrock is generally the strongest and most stable material on which a building can be founded.

Soil is a general term referring to any earth material that is particulate. Particulate soils are further defined according to ASTM D2487, Unified Soil Classification System (Figure 2.3), as follows:

- *Boulders* are greater than 12 inches (300 mm) in diameter.
- *Cobbles* are smaller than boulders but greater than 3 inches (75 mm) in diameter.
- *Gravel* is from 3 inches to 0.187 inches (75 mm to 4.75 mm) in diameter.
- *Sand* is from 0.187 inches to 0.003 inches (4.75 mm to 0.07 mm) in diameter.
- Gravel and sand are also collectively referred to as *coarse-grained soils*.
- *Silt* particles are smaller than 0.0029 inches (0.075 mm). Like sand and gravel, silt particles are roughly spherical in shape.
- *Clay* particles are also defined as smaller than 0.0029 inches (0.075 mm), though typically they are an order of magnitude (10 times) or more smaller. Also, unlike larger-grained particles, they are flat or plate-shaped rather than spherical.
- Both silts and clays are also referred to as *fine-grained soils*.

In the field, major soil types can be roughly distinguished with simple hand tests. It takes two hands to lift a boulder, and one to lift a cobble. If you can easily lift just one particle between two fingers, it is gravel. If individual soil particles are large enough to be seen, but too small to be picked up singly, they are sand. If particles are too small to see with the unaided eye, they are silt or clay. When wet, clay soils are putty-like; when dry they are hard. Silts are not sticky when wet and have little or no cohesiveness when dry.

Peat, topsoil, and other *organic soils* are not suitable for the support of building foundations. Their high organic matter content makes them spongy and sensitive to changes in water content or biological activity within the soil.

Properties of Soils

Particle Size

Coarse-grained soils—sands and gravels—consist of relatively large mineral particles with little or no attractive or repulsive forces acting between

(a) *Building before settlement occurs*   (b) *Uniform settlement*   (c) *Differential settlement*

FIGURE 2.2

Uniform settlement (*b*) is usually easily controlled and of little consequence in a building, but differential settlement (*c*) can cause severe structural damage.

			Group Symbol	Descriptive names of soil within this group
Coarse-Grained Soils	Gravels	Clean Gravels	GW	Well-graded gravel or well-graded gravel with sand, little or no fines
			GP	Poorly graded gravel or poorly graded gravel with sand, little or no fines
		Gravels with Fines	GM	Silty gravel, silty gravel with sand
			GC	Clayey gravel, clayey gravel with sand
	Sands	Clean Sands	SW	Well-graded sand or well-graded sand with gravel, little or no fines
			SP	Poorly graded sand or poorly graded sand with gravel, little or no fines
		Sands with Fines	SM	Silty sand, silty sand with gravel
			SC	Clayey sand, clayey sand with gravel
Fine-Grained Soils	Silts and Clays	Liquid Limit < 50	ML	Silt or silt-sand-gravel mixtures, low plasticity
			CL	Lean clay or clay-sand-gravel mixtures, low plasticity
			OL	Organic clay or silt (clay or silt with significant organic content), or organic clay- or silt-sand-gravel mixtures, low plasticity
		Liquid Limit ≥ 50	MH	Elastic silt, silt-sand-gravel mixtures
			CH	Fat clay or clay-sand-gravel mixtures, high plasticity
			OH	Organic clay or silt (clay or silt with significant organic content), or organic clay- or silt-sand-gravel mixtures, high plasticity
	Highly Organic Soils		PT	Peat, muck, and other highly organic soils

FIGURE 2.3
The Unified Soil Classification System, from ASTM D2487. The Group Symbols are a universal set of abbreviations for soil types, as seen, for example, in Figure 2.6.

them. The ability of these soils to support building loads without shifting depends primarily on friction between the particles to keep the particles from sliding past one another. This resistance to internal sliding, called *shear strength*, varies with the degree of interlocking between particles and the confining force of the surrounding soil. Where coarse-grained soils are densely packed with little space between particles and securely confined by surrounding soils, it is relatively difficult for particles to move past one another. Soils such as these exhibit relatively high strength and can support greater loads. Where coarse-grained soils are loosely packed or poorly confined, particles can more easily slide past one another, and less load can be safely supported. Soils that rely primarily on internal friction for strength are termed *frictional* or *cohesionless*.

Smaller-grained soils may be subject to a wider array of interparticle forces. As particle size decreases, surface area increases in relation to weight and size, and the spaces between the particles, called *soil pores*, get smaller. In essence, the particles become lighter and more easily pushed and pulled by electrostatic forces, chemical interactions, and forces related to the presence of water in the soil.

For example, whereas gravels are generally little affected by moisture in the soil, the properties of sand can vary noticeably with moisture content. As any beachgoer knows, wet sand makes a stronger sand castle than dry sand, as capillary forces acting between particles help to hold the particles in place. And wet sand responds more firmly to the pressure of our feet as we walk on the beach than does dry sand, as the hydrostatic pressure of the water helps to distribute the load exerted on the soil. A dramatic example of the effects of moisture on smaller-grained soils is a phenomenon called *soil liquefaction*. Water-saturated sands or silts may lose virtually all of their strength and behave as a liquid when subjected to sudden, large changes in load, such as may occur during an earthquake.

Clay particles are extremely small in size and flatter in shape, making the ratio of their surface area to volume hundreds or thousands of times greater than even silts. They also differ in mineral composition from larger-grained soils and tend to arrange themselves into more complex internal structures, called *fabric*, as particles aggregate into sheetlike or other geometric arrangements.

As a result, clay soils tend to stick together, and are characterized as *cohesive* rather than frictional. For example, it may be possible to dig a vertical-walled excavation in clay soil without temporary soil support (Figure 2.4). There is sufficient shear strength in the unconfined soil to prevent the excavation walls from collapsing. In contrast, a cohesionless soil such as sand must be excavated at a shallower angle to avoid the collapse of unsupported walls. Cohesive soils tend to be hard when dry and moldable, or *plastic*, when moist. They also have what is called a higher *liquid limit* than more coarsely grained soils. That is, they can sustain a higher moisture content before arriving at a flowable consistency. Silts may also exhibit cohesive properties, but to a lesser extent than clays.

Clay soils may also be *expansive*, that is, prone to expand or contract

EXCAVATION IN FRICTIONAL SOIL

EXCAVATION IN HIGHLY COHESIVE SOIL

FIGURE 2.4

Excavations in frictional and highly cohesive soils.

with changes in moisture content. Clays with very small particle size and high liquid limits are most susceptible to this behavior. Highly expansive soils can increase in volume by 10 percent or more with increased moisture content, and with sufficient force to cause damage to the building

structure if not properly accounted for in the foundation design.

The unique properties of clay soils tend to cause water to pass through them very slowly, or in some cases not at all. In fact, some clay soils are fabricated into sheets and used as waterproofing for basements and other underground structures.

Gradation

Within any soil sample, the range of particle sizes present, or *gradation*, may vary. A *well graded* soil includes a broad, well-distributed range of particle sizes. A *poorly graded* soil consists of particles more limited in range of sizes. Well graded soils contain less empty space between particles than poorly graded soils, as smaller particles fill in gaps between larger ones. Broadly speaking, well graded soils tend compact more effectively than poorly graded ones, but also tend to drain water less efficiently.

When deliberately prepared for use in earthwork, soil materials may be purposefully graded in specific ways: A *uniformly graded* material is composed of particles within a limited, narrow size range (Figure 2.5). This produces the maximum possible volume of empty space within the material. A *gap graded* soil contains a broader range

FIGURE 2.5

Two gravel samples, illustrating differences in gradation. The left-hand sample, with a broad range of particle sizes, comes from a well graded sandy gravel. On the right is a uniformly graded gravel sample in which there is little variation in size among particles.

(Photos by Joseph Iano.)

TABLE 1804.2
ALLOWABLE FOUNDATION AND LATERAL PRESSURE

CLASS OF MATERIALS	ALLOWABLE FOUNDATION PRESSURE (psf)[d]	LATERAL BEARING (psf/f below natural grade)[d]	LATERAL SLIDING	
			Coefficient of friction[a]	Resistance (psf)[b]
1. Crystalline bedrock	12,000	1,200	0.70	—
2. Sedimentary and foliated rock	4,000	400	0.35	—
3. Sandy gravel and/or gravel (GW and GP)	3,000	200	0.35	—
4. Sand, silty sand, clayey sand, silty gravel and clayey gravel (SW, SP, SM, SC, GM and GC)	2,000	150	0.25	—
5. Clay, sandy clay, silty clay, clayey silt, silt and sandy silt (CL, ML, MH and CH)	1,500[c]	100	—	130

For SI: 1 pound per square foot = 0.0479kPa, 1 pound per square foot per foot = 0.157kPa/m.
a. Coefficient to be multiplied by the dead load.
b. Lateral sliding resistance value to be multiplied by the contact area, as limited by Section 1804.3.
c. Where the building official determines that in-place soils with an allowable bearing capacity of less than 1,500 psf are likely to be present at the site, the allowable bearing capacity shall be determined by a soils investigation.
d. An increase of one-third is permitted when using the alternate load combinations in Section 1605.3.2 that include wind or earthquake loads.

FIGURE 2.6

Allowable bearing values for various soil types, from the IBC. The soil class abbreviations in rows 3, 4, and 5 make reference to the group symbols in Figure 2.3. *(Excerpted from the 2012 International Building Code, Copyright 2011. Washington, D.C.: International Code Council. Reproduced with permission. All rights reserved. www.ICCSAFE.com)*

of particle sizes, but with certain sizes omitted. For example, aggregates for pervious concrete, as described in Chapter 13, may be gap graded. This ensures a sufficient particle size distribution to produce a strong concrete that also includes sufficient void space that stormwater can effectively drain through the finished pavement.

The term *sorting* can also be used to describe particle size distribution within a soil, but with the opposite sense of grading. That is, a well graded soil is *poorly sorted* and poorly graded soil is *well sorted*.

Soils for Building Foundations

Generally, soil groups listed toward the top of Figure 2.3 are better suited for supporting building foundations than those listed further down. Those closer to the top of the list exhibit greater loadbearing capacity and stability, and are less sensitive to moisture content. Consolidated rock is usually the strongest material on which to found a building. Usually, however, such rock is too deep to be reached economi-

cally, and the foundation is designed to bear on some other particulate stratum closer to the ground surface. An often-cited example of the suitability of continuous rock for large building foundations is the clustering of tall buildings in New York City toward the central portion of Manhattan Island. This is the portion of the island where the bedrock is closest to the surface and the massive foundations needed for these buildings can be constructed most easily and inexpensively.

Figure 2.6 gives values for *allowable foundation pressures* (or *allowable soil pressures*) for common classes of soil materials. These figures may be used for the design of small building foundations where analysis of soil samples from the site is deemed unnecessary. The allowable pressures are significantly less than the maximum loadbearing capacities of the soils, accounting for uncertainties in soil composition and properties, design factors of safety, and acceptable settlement limits. For larger buildings or on sites with more questionable soil conditions, soil

properties are determined through site investigation and laboratory tests.

Rock and coarse-grained soils are also generally the most stable materials for supporting foundations, behaving more consistently under varying moisture content than fine-grained soils. Clay soils, in particular, can present unique challenges. When clay with high moisture content is put under continuous pressure, water can be slowly pressed out of it, with a corresponding gradual reduction in soil volume, a behavior called *consolidation*. Where such a soil stratum underlies a foundation, the possibility of long-term settlement must be considered. Or, as noted earlier, where highly expansive clays are present, provisions may be required to allow the clay to expand without causing damage to the building substructure.

In regions of significant earthquake risk, soil stability during seismic events must be considered. For example, soil liquefaction can cause excessive pressures on foundation walls, or in extreme cases, complete loss of support for a building foundation.

In addition to foundation support, *imported soil* materials (those brought from offsite) may be used for backfilling excavations, construction of base layers under pavements and concrete slabs at grade, finish grading for landscaping, or replacement of unsuitable *native soils* (those already present at the building site). In each case, materials with appropriate strength and stability, drainage capacity, compaction characteristics, and organic content are selected.

For example, for *general purpose fill*, good compaction and stability are important. Here, a well graded, coarse-grained soil is likely a good choice. However, directly around the foundation, a more porous *drainage fill* may be required, and a more uniformly graded material is usually preferable. In this way, water passing through the soil toward the building is readily intercepted by this layer, where it drains downward and can be drawn off in perforated pipes (see Figure 2.56).

Subsurface Exploration and Soils Testing

For all but the smallest buildings, foundation design is preceded by investigation of soil conditions at the site. *Test pits* can be dug when the foundation will not extend deeper than roughly 16 feet (3 m), the maximum practical reach of small excavating machines. Soil strata can be observed in the pit and samples taken for evaluation.

The *water table* is the elevation at which the soil is normally fully saturated. If this level lies within the depth of a test pit dug in coarse-grained soils, it will be readily apparent, as the pit will quickly fill with water up to that level. Test pits excavated in less permeable fine-grained soils may not easily reveal the water table level, as groundwater may seep only slowly through the soil. In such cases, the water table level can be determined with a separately drilled observation well or special devices that measure water pressure within the soil.

Where open test pits are not practical or information at greater

FIGURE 2.7

A truck-mounted drilling rig performing test borings at a site planned for future construction. Equipment of this type is capable of boring into the ground 1000 ft (300 m) or more. The drilling auger can be seen just off the back of the truck bed. Hydraulic jacks, just slightly forward and to either side of the auger, level and stabilize the drilling platform. A pair of drill operators is preparing to connect a new length of drilling pipe while, to the left, a technician oversees the drilling and collects soil samples. *(Photo by Joseph Iano.)*

depths is required, *test borings* are performed with portable drilling rigs (Figure 2.7). As boring progresses, soil conditions are evaluated at regular intervals and especially as differing soil strata are encountered. Soil density and potential bearing capacity are evaluated by counting hammer blows on an open-ended hollow tube called a *penetration sampler* as it is advanced a standard distance into the bore hole. Samples extracted from the boring are examined in the field and frequently also taken for laboratory testing. Test borings can return information on each soil stratum, such as soil type, depth, and thicknesses, as well as groundwater conditions (Figure 2.8). Usually, a number of borings are taken around a site. The information from each boring is mapped and the results interpolated to create a fuller picture of subsurface conditions for use by the foundation engineer.

Load tests also may be performed on the construction site to further evaluate the bearing capacity and stability of the soil. Where bearing soils are exposed by test pits, a temporary framework supporting large concrete blocks may be constructed to apply a static (constant) load to the exposed soils and observe their response over a period of days or weeks. Where deeper foundations—for example, friction piles, as discussed later in this chapter—are used, static or dynamic (controlled impact) loading of these elements may be used to verify that the intended bearing strata have been reached and that the strata and foundation elements are behaving as predicted.

In the laboratory, soil samples are dried and then particle gradation is determined using a collection of *sieves* with wire mesh screens of varying spacing. As a soil sample is passed through consecutively finer sieves, particles of differing size ranges are separated and the relative proportions of each quantified. For very fine

EXPLORATORY BORING LOG EB-6
SURFACE ELEVATION: 118 FT

DESCRIPTION	CONSISTENCY	CLASSIFICATION	DEPTH (FT)
Dark brown, Sandy Lean Clay, moist, moderate plasticity, fine to medium sand, reddish-brown mottling, the upper portion appears to be fill or reworked native soil.	Hard	CL	0
Gray, Sandy Lean Clay, moist, moderate plasticity, fine to coarse sand, tan and light orange mottling. LL=38, PL=20	Very Stiff	CL	
Gray to brown, Clayey Sand, moist, fine to coarse sand, fine rounded to angular gravel, light orange mottling, moderately plastic fines. 40% passing No. 200 Sieve. ▼ Ground water measured at 6 ft after drilling. ▼ Ground water encountered during driling at 8.5 ft.	Medium Dense to Dense	SC	5 ▼ ▼ 10
Gray, Sandy Lean Clay, moist, moderate plasticity, fine sand.	Stiff	CL	15
Brown, Poorly Graded Sand, moist, fine to coarse sand. Bottom of boring at 20.5 ft.	Dense	SP	20

FIGURE 2.8
An example log from a test boring indicating the type of soil in each stratum and the depth at which it was encountered. Depths at which groundwater were found are also recorded. The Classification abbreviations correspond to the Unified Soil Classification System as explained in Figure 2.3.

soils, particle size is determined by mixing the soil with water and observing how rapidly particles precipitate out of liquid suspension (larger particles will fall to the bottom of the mixture more rapidly than smaller ones).

For cohesive soils, properties such as the liquid limit and *plastic limit* (the water content at which the soil transitions from solid to plastic) are determined. Additional tests can determine soil water content, permeability, liquefaction potential, chemical constituents, expansion potential, strength in shear and compression, and consolidation potential (Figure 2.9).

The information gained through subsurface exploration and laboratory testing is summarized in a written *geotechnical report*. This report may include recommendations for allowable bearing loads for the various soil strata, appropriate foundation types, estimated rates of foundation settlement, soil drainage and foundation waterproofing, and other relevant information. This report is used by engineers in the design of excavations, excavation support systems, dewatering, building foundations, and substructure. It is also used by contractors in the planning and execution of their sitework during construction.

(a)

(b)

FIGURE 2.9
Two laboratory procedures used to determine various strength-related soil properties. (a) In a triaxial load test, a soil sample is loaded axially by the piston and circumferentially by water pressure in the transparent cylinder. (b) In a direct shear test, a rectangular prism of soil is placed in a split box and sheared by applying pressure in opposite directions to the two halves of the box. (*Courtesy of Ardaman and Associates, Inc., Orlando, Florida.*)

SUSTAINABILITY OF FOUNDATIONS AND THE BUILDING SITE

Building sites should be selected and developed to protect and conserve natural habitats and resources, enhance sustainable community development patterns, promote biodiversity, preserve quality open space, and minimize pollution and unnecessary energy consumption.

Site Selection

Renovating an existing building rather than building a new one saves energy and building materials. If it saves the existing building from demolition, renovation also diverts enormous quantities of material from landfill.

Building within densely populated areas with existing infrastructure and amenities rather than in areas unconnected to community resources promotes neighborhood vitality, reduces private motor vehicle travel, and protects undeveloped open space.

Building on a previously developed, damaged, or polluted *brownfield site*, and designing the building so that it restores that site, helps to mitigate previous environmental degradation.

Selecting a building site that is well connected to existing networks of public transportation, bicycle routes, and pedestrian paths reduces fuel consumption, air pollution, and other adverse impacts of private motor vehicle use.

Avoiding construction on prime agricultural land preserves productive land for the future.

Avoiding construction on undeveloped, environmentally sensitive land protects the wildlife and natural habitats that such land supports. This includes floodplains, land that provides habitat for endangered or threatened species, wetlands, mature forest lands and prairies, and land adjacent to natural bodies of water.

Avoiding construction on public parkland or land adjacent to recreational bodies of water preserves public open space and recreational resources.

Foundations

Properly waterproofed and drained foundations keep building interiors dry, and prevent the growth of mold and mildew.

Adequately insulated basements reduce building energy consumption and improve occupant comfort.

In areas of high radon soil-gas risk, gas venting systems protect building occupants from high concentrations of this unhealthful gas.

On building sites polluted by previous development, soil mixing techniques can remediate in-ground contaminants. Where it is safe to encapsulate soil pollutants on site, foundation systems that minimize soil displacement, such as piles or rammed aggregate piers, minimize the need to handle and dispose of hazardous soils.

Site Design

Preservation of mature trees, distinctive topographic formations, recreational pathways, and other unique site features prevents the loss of irreplaceable site assets.

Minimizing the building footprint preserves open space. Protecting and enhancing unbuilt portions of the site with natural vegetation preserves green space and animal habitat, and helps maintain biodiversity.

Appropriate landscape design and plant selection minimize or eliminate the need for landscape irrigation and unneeded water consumption.

Minimizing impervious ground surface area and providing drainage systems that allow the reabsorption of stormwater into the ground work to maintain the natural hydrology of the building site, reduce demand on municipal storm sewer systems, and protect natural waterways from overloading and pollution.

Providing vegetated or reflective roof surfaces, and shaded or reflective paving, reduces heat island effects and creates an improved microclimate for both humans and wildlife.

Minimizing nighttime light pollution is a benefit to humans and nocturnal wildlife alike.

Siting a building for optimal exposure to sun and wind maximizes opportunities for passive heating and cooling, and natural daylighting, all of which are strategies that reduce building energy consumption and enhance occupant comfort and productivity.

Avoiding unnecessary shading of adjacent buildings protects those buildings' sources of natural illumination and useful solar heat.

Construction Process

Complying with local and regional conservation laws contributes to the protection of natural resources and animal habitat.

Maintaining proper landscape protection prevents the loss of difficult- or impossible-to-replace trees and vegetation.

Erosion control measures minimize the loss of soil by wind or water action, prevent sedimentation of streams and sewers, and minimize pollution of the air with dust or particulates.

Stockpiling and protection of topsoil and subsoils makes these materials available for reuse at the end of construction or at other sites nearby.

Well-marked and properly prepared construction vehicle access ways prevent overcompaction of soils, and minimize noise, dust, air pollution, and inconvenience to neighboring buildings and sites.

Recycling of construction wastes diverts solid waste from permanent landfills.

EARTHWORK AND EXCAVATION

Virtually all building construction is accompanied by at least some form of *earthwork* during construction. On undeveloped sites, construction may begin with *grubbing and clearing*, in which trees and plants, stumps, large roots, and other surface materials are removed with heavy machinery. Next, organically rich topsoil may be scraped away and stockpiled to one side, to await reuse at the end of construction.

Excavation

Excavation is necessary for basement construction, to reach undisturbed, adequately firm soil for shallow footings, for trenches for buried utilities, and to remove native soils that are contaminated or too weak or unstable to build over.

In particulate soils, any of a wide variety of machines, such as bulldozers, backhoes, bucket loaders, scrapers, trenching machines, and others, may be used to loosen and lift the soil from the ground. If the soil must be moved more than a short distance, dump trucks come into use.

In rock, excavation is slower and many times more costly. Weak or highly fractured rock may be broken up with power shovels, tractor-mounted pneumatic hammers, or other specialized tools. Blasting, in which explosives are placed and detonated in lines of closely spaced holes drilled into the rock, may also be used. Where blasting is impractical, hydraulic splitters, devices inserted into similarly drilled holes but which rely on driven wedges to split the rock, may be used.

Excavation Support

If the construction site is sufficiently larger than the area to be covered by the building, the edges of the excavation can be sloped back or *benched* at a low enough angle that the soil will not slide back into the hole. This angle, called the *maximum allowable slope* or *angle of repose*, can be steep for cohesive soils such as stiff clays, or shallower for frictional soils such as sand and gravel. On constricted sites, the soil surrounding an excavation must be held back by some kind of *excavation support* capable of resisting the pressures of earth and groundwater (Figure 2.10). Excavation support can take many forms, depending on the soil type, depth of excavation, type of construction to follow, equipment and preferences of the contractor, proximity of surrounding roads or buildings, and presence of groundwater.

Shoring

Shoring is construction used to support the sides of an excavation and prevent its collapse. For large excavations, the most common types of shoring are soldier beams and lagging, and sheet piling. With soldier beams and lagging, steel columns called *H-piles* or *soldier beams* are driven vertically into the earth at close intervals around an excavation site before digging begins. As earth is removed, the *lagging*, usually consisting of heavy wood planks, is placed against the flanges of the columns to retain the soil outside the excavation (Figures 2.11 and 2.12). *Sheet piling* or *sheeting* consists of vertical sheets of various materials that are aligned

SECTION THROUGH BENCHED EXCAVATION

Bracing is required to resist soil pressure

SECTION THROUGH SHEETED EXCAVATION

FIGURE 2.10

On a spacious site, an excavation can be benched. When excavating close to property lines or nearby buildings, some form of excavation support, such as sheeting, is used to retain the soil around the excavation.

Soil

Steel H-pile

Wooden planks (lagging) are inserted between the piles to retain the soil as excavation progresses

FIGURE 2.11

Soldier beams and lagging, shown in horizontal section.

Earthwork and Excavation / **41**

FIGURE 2.12
Soldier beams and lagging. Lagging planks are added at the bottom as excavation proceeds. The drill rig is boring a hole for a tieback (explained later in this chapter) to brace a soldier beam. *(Courtesy of Franki Foundation Company.)*

TIMBER SHEET PILING

STEEL SHEET PILING

PRECAST CONCRETE SHEET PILING — Grout key

FIGURE 2.13
Horizontal sections through three types of sheet piling. The shading represents the retained earth.

tightly against one another edge-to-edge and driven into the earth to form a solid wall, also before excavation begins (Figures 2.13 and 2.14). The most common material for sheet piling is steel, but wood, aluminum, PVC plastic, composite polymers, or precast concrete may also be used. For trench work, a variety of easy-to-deploy, reusable support systems are also used.

Most often, shoring is temporary, and is removed as soil is replaced in the excavation. However, it may also be left in place to become a permanent part of the building's substructure. This may be necessary, for example, where shoring is located close to a property line and there is no practical way to remove it after completion of construction without disturbing the adjacent property.

Where soil is sufficiently cohesive to hold an adequate slope at least temporarily, excavations can be stabilized with *pneumatically applied concrete*, also called *shotcrete*. This stiff concrete mixture is sprayed directly from a hose onto the soil shortly after the soil is excavated. The hardened concrete reinforces the slope and protects against erosion (Figure 2.15).

FIGURE 2.14
Steel sheet piling being installed with a vibratory driver. This type of driver imparts a rapid up-and-down motion to the sheets. This causes temporary liquefaction of the soil directly under the leading edge of the sheet, allowing the sheet to descend easily into the soil. Depending on soil conditions, impact hammers or hydraulic presses may also be used. The circular hole in the top of each of sheet is used when lifting the sheets by crane. *(Photo by Joseph Iano.)*

FIGURE 2.15
Where excavation support turns the corner and the soil can be sloped at a lesser angle, less expensive shotcrete stabilizes the soil. *(Photo by Joseph Iano.)*

Soil Mixing

With *soil mixing*, the sides of an excavation are strengthened by blending portland cement and water with the existing soil. Mixing occurs in place, using rotating augers or paddles lowered into the ground. The soil-cement mix hardens into a series of underground, abutting vertical cylinders of low-strength concrete against which excavation can proceed (Figure 2.16). Where bracing is required (as discussed later in this chapter), soldier piles can be inserted into the soil-cement mix before the mix hardens, to become part of the bracing structure (Figure 2.17). Soil-mixed excavation support always remains in place, becoming a permanent part of the subgrade construction.

Soil mixing is one method of so-called *ground improvement* or *earth improvement;* these terms refer to any of a variety of methods for altering the properties of soil in place. Soil mixing, for example, can also be used to create cutoff walls to prevent water seepage into excavations, to stabilize or strengthen areas of weak soil around or under buildings, or to remediate biologically or chemically contaminated soil by adding chemicals that neutralize the contamination. Another form of ground improvement, rammed aggregate piers, is discussed later in this chapter.

Slurry Walls

A *slurry wall* is a more complex method of constructing a complete, steel-reinforced, concrete wall in the ground, even many stories below the surface, before excavation takes place. It is a relatively expensive form of excavation support that is usually economical only if it becomes part of the permanent foundation of the building. The excavation for a slurry wall is performed with a narrow *clamshell bucket,* operated by crane, and guided by temporary guides at the ground surface defining the edges of the wall (Figures 2.18 and 2.19). As the trench deepens, the tendency of the earth walls to collapse is counteracted by maintaining the trench full at all times with a viscous mixture of water and bentonite clay, called a *slurry.* This slurry exerts pressure against the earth walls, holding them in place.

When the trench has been excavated to its full depth, steel tubes with diameters equal to the width of the trench are inserted vertically at intervals to divide the trench into sections of a size that can be reinforced and concreted conveniently. Into each section, a steel cage of reinforcing steel is lowered, and concrete is poured from the bottom up, using a funnel-and-tube arrangement called a *tremie.* As the concrete rises in the trench, the slurry is displaced and pumped out into

FIGURE 2.16
Soil mixing.

FIGURE 2.17
Soil-mixed excavation support. *Above:* Excavation proceeds after the soil mixture has hardened. Bracing consists of soldier piles, walers, and tiebacks. The soldier piles are inserted during mixing, before the soil-cement mixture hardens. The walers and tiebacks are installed later, as excavation progresses. *Below:* Fully excavated soil-mixed excavation support, strong enough to resist the pressures from adjacent buildings. *(Photographs courtesy of Schnable Foundation Company.)*

holding tanks, where it is stored for reuse. After the concrete in one section reaches the top of the trench and has hardened, the vertical pipes on either side are withdrawn, and adjoining sections can be poured. This process is repeated until concreting of all sections of the wall is completed. When the concrete wall has cured to its intended strength, earth removal begins inside the wall, which serves as shoring for the excavation. In most cases, this wall will also become a permanent part of the

future building foundation or substructure.

Slurry walls can also be constructed from precast, rather than sitecast, concrete. Prestressed wall sections are produced in a precasting plant (see Chapter 15) and then trucked to the construction site. The slurry for precast walls may include portland cement in addition to the water and bentonite clay. Before a section is lowered by crane into the slurry, one face is coated with a release compound that will prevent

the clay-cement slurry from adhering. The wall sections are installed side by side in the trench, joined by tongue-and-groove edges or synthetic rubber gaskets. After the portland cement has caused the slurry to harden, excavation begins, with the hardened slurry on the exposed face of the wall dropping away from the coated surface as soil is removed.

Compared to sitecast concrete, a precast concrete slurry wall can have a smoother, flatter, more attractive surface; may be thinner due to the

FIGURE 2.18
Steps in constructing a slurry wall.
(*A*) Temporary concrete guide walls are installed at the surface, and the clamshell bucket begins excavating the trench through a bentonite clay slurry. (*B*) The trench is dug to the desired depth, with the slurry serving to prevent collapse of the walls of the trench. (*C*) A welded cage of steel reinforcing bars is lowered into the slurry. (*D*) The trench is concreted from the bottom up with the aid of a tremie. The displaced slurry is pumped from the trench, filtered, and stored for reuse. (*E*) The reinforced concrete wall is tied back as excavation progresses.

structural efficiency of prestressing; and may have greater watertightness because of the continuous layer of solidified clay on the unexcavated side.

Bracing

As an excavation deepens, its support system must be braced against earth and water pressures (Figure 2.20). *Crosslot bracing* uses temporary steel wide-flange columns that are driven into the earth at points where braces will cross. As the earth is excavated down around the shoring and the columns, tiers of horizontal bracing, usually of steel, are added to support *walers*, which are beams that span across the face of the sheeting. Where the excavation is too wide for crosslot bracing, sloping *rakers* are used instead, bearing against temporary footings.

Both rakers and crosslot bracing, especially the latter, are a hindrance to the excavation process, as their presence within the excavation places limitations on earth removal methods and equipment. Where soil conditions permit, *tiebacks* can be used instead of bracing to support the shoring while maintaining a fully open excavation. At each level of walers, holes are drilled at intervals through the shoring and the surrounding soil into rock or a stratum of stable soil. Steel cables or rods are inserted into the holes, grouted to anchor them in place, and stretched tight with hydraulic jacks (posttensioned) before they are fastened to the walers (Figures 2.21, 2.22, and 2.23).

Excavations in fractured rock can often do without any sheeting, either by injecting grout into the joints of the rock to stabilize it or by drilling into the rock and inserting *rock anchors* that fasten the blocks together (Figure 2.24).

In some cases, vertical walls of particulate soils can be stabilized by *soil nailing*. A soil nail is similar to a rock anchor. A length of steel bar is inserted into a nearly horizontal hole drilled deep into the soil. Grout is injected into the hole to bind the nail to the surrounding soil. Large numbers of closely spaced nails are used to knit a large block of soil together so that it behaves more like weak rock than particulate soil.

Bracing and tiebacks in excavations are usually temporary. Their function is eventually taken over by the floor structure of the basement levels of the building, which is designed to resist the lateral loads from the surrounding earth acting on the substructure walls.

Dewatering

During construction, excavations must be kept free of standing water from precipitation or groundwater seepage. Some shallow excavations in relatively dry soils may remain free of water without any intervention. But most excavations require some form of *dewatering*, or removal of water from the excavation or surrounding soil. The most common method of dewatering is by pumping as the

FIGURE 2.19
Constructing a slurry wall. (*a*) Concrete guide walls are formed and poured in a shallow trench. (*b*) The narrow clamshell bucket discharges a load of soil into a waiting dump truck. Most of the trench is covered with wood pallets for safety. (*c*) A detail of the clamshell bucket. (*d*) Hoisting a reinforcing cage from the area where it was assembled and getting ready to lower it into the trench. The depth of the trench and height of the slurry wall are equal to the height of the cage, which is about four stories for this project. (Concreting of the wall after placement of reinforcing is not shown.) (*Photo a courtesy of Soletanche. Photos b, c, and d courtesy of Franki Foundation Company.*)

CROSSLOT BRACING

RAKERS

TIEBACKS

FIGURE 2.20
Three methods of bracing excavation support, drawn in cross section. The connection between the waler and the brace, raker, or tieback requires careful structural design. The broken line between rakers indicates the mode of excavation: The center of the hole is excavated first with sloping sides, as indicated by the broken line. The heel blocks and uppermost tier of rakers are installed. As the sloping sides are excavated deeper, more tiers of rakers are installed. Notice how the tiebacks leave the excavation totally free of obstructions.

FIGURE 2.21
Three steps in the installation of a tieback to a soil anchor. (*A*) A rotary drill bores a hole through the sheeting and into stable soil or rock. A steel pipe casing keeps the hole from caving in where it passes through noncohesive soils. (*B*) Steel prestressing tendons are inserted into the hole and grouted under pressure to anchor them to the soil. (*C*) After the grout has hardened, the tendons are tensioned with a hydraulic jack and anchored to a waler.

(*a*)

FIGURE 2.22
Installing tiebacks. (*a*) Drilling through
a slurry wall for a tieback. The ends
of hundreds of completed tiebacks
protrude from the wall.

(*b*)

(*c*)

(*b*) Inserting prestressing tendons into the steel casing for a
tieback. The apparatus in the center of the picture is for pressure-
injecting grout around the tendons. (*c*) After the tendons have
been tensioned, they are anchored to a steel chuck that holds
them under stress, and the cylindrical hydraulic jack is moved to
the next tieback. (*Photos courtesy of Franki Foundation Company.*)

FIGURE 2.24
Rock anchors are similar to tiebacks but are used to hold jointed rock formations in place around an excavation.

FIGURE 2.23
Slurry wall and tieback construction used to support historic buildings around a deep excavation for a station of the Paris Metro. *(Photo courtesy of Soletanche.)*

WELL POINTS

WATERTIGHT BARRIER

FIGURE 2.25
Two methods of keeping an excavation dry, viewed in cross section. The water sucked from well points depresses the water table in the immediate vicinity to a level below the bottom of the excavation. Watertight barrier walls work only if their bottom edges are inserted into an impermeable stratum that prevents water from working its way under the walls.

water accumulates in pits, called *sumps*, located at low points in the excavation.

Where the volume of groundwater flowing into the excavation is great, or with soils, such as sand or silt, that may be softened by constant seepage, it may be necessary to keep groundwater from entering the excavation at all. This can be done either by pumping water from the surrounding soil to depress the water table below the bottom of the excavation or by erecting a watertight barrier around the excavation (Figure 2.25).

Well points are commonly used to depress the water table. These are vertical pipes inserted into the ground with screened openings at the bottom that keep out soil particles while allowing water to enter. Closely spaced well points are driven into the soil around the perimeter of the excavation and connected to pumps that continually draw water from the ground and discharge it away from the building site. Once pumping has drawn down the water table in the area of the excavation, work can proceed "in the dry" (Figures 2.26

and 2.27). Suction pumps stationed at ground level can only draw water from a depth of about 20 feet (6 m). For deeper excavations, two or more rings of well points may be required, the inner rings working at deeper levels than the outer ones. Or, a single ring of deep wells with submersible pumps at their bottoms may be installed.

Well points are not always practical: They may have insufficient capacity to keep the excavation dry, restrictions on the discharge of groundwater may preclude their use, or lack of reliability

FIGURE 2.26
Well point dewatering. Closely spaced vertical well points connect to the larger-diameter header pipe. *(Courtesy of Griffin Dewatering Corporation.)*

FIGURE 2.27
An excavation is kept dry despite the close proximity of a large body of water. Two dewatering pumps are visible in the foreground. A pair of header pipes and numerous well points can be seen surrounding the excavation. *(Courtesy of Griffin Dewatering Corporation.)*

due to power outages may be a concern. In some locations, lowering of the water table could adversely affect neighboring buildings, causing soil consolidation under their foundations or exposing wood foundation piles, previously protected by immersion in water, to decay once they contact air. In these cases, a *watertight barrier wall* made from a slurry wall or soil-mixed wall may be used (Figure 2.25). (Sheet piling can also work, but it tends to leak at the joints.) *Soil freezing* is also possible; with this strategy, an array of vertical pipes similar to well points is used to circulate coolant at temperatures low enough to freeze the soil around an excavation area, resulting in a temporary but reliable barrier to groundwater. Watertight barriers must resist the hydrostatic pressure of the surrounding water, which increases with depth, so for deeper excavations, a system of bracing or tiebacks is required. A watertight barrier also works only if it reaches into a stratum of impermeable soil at its bottom, such as bedrock or water-impermeable clay. Otherwise, water can flow underneath the barrier and rise up from the bottom of the excavation.

FOUNDATIONS

It is convenient to think of a building as consisting of three major parts: the *superstructure*, the above-ground portion of the building; the *substructure*, the habitable portion below ground; and the foundations, the below-ground components of the building devoted solely to the transfer of loads into the soil (Figure 2.28).

There are two basic types of foundations: shallow and deep. *Shallow foundations* transfer building loads to the earth close to the base of the substructure. *Deep foundations*, either piles or caissons, extend downward through layers of weak or unstable strata to reach more competent soil or rock deeper within the earth. Shallow foundations are less expensive than deep ones and are used wherever possible.

FIGURE 2.28

Superstructure, substructure, and foundation. The substructure in this example contains two levels of basements, and the foundation consists of bell caissons. (In some buildings, the substructure and foundation may be partly or wholly the same.)

The best choice of foundation type for any particular building is sometimes obvious, especially where shallow foundations will work. In other cases, in-depth investigation and evaluation may be required to determine the optimal design. Subsurface soil types, groundwater conditions, and the structural requirements of the superstructure are primary considerations. Additionally, local construction practices; environmental considerations of noise, traffic, and disposal of earth materials and water; regulatory restrictions; potential impacts on adjacent properties; construction schedules; and other considerations may come into play.

Shallow Foundations

Spread Footings
Most shallow foundations are simple concrete *spread footings*. Spread footings take concentrated loads from above and spread them out across an area of soil large enough that the allowable soil pressure is not exceeded. A *column footing* is a square block of concrete, with or without steel reinforcing, that distributes a column load to the soil below. A *wall footing* or *strip footing* is a continuous strip of concrete that serves the same function for a loadbearing wall (Figures 2.29 and 2.30).

To minimize settlement, spread footings must be placed on undisturbed soil. Alternatively, where there are areas of unsuitable soil at the bearing level, native soil may be removed and replaced with *engineered fill*, properly formulated higher-strength, more stable soil material brought from offsite. This material is placed in layers and compacted to a specified density, usually under the supervision of a soils engineer, to ensure that the required loadbearing capacity and stability are achieved.

In cold climates, footings must also be placed below the *frost line*, the level to which the ground freezes in winter. Foundations exposed to freezing temperatures can be lifted and damaged by soil that expands as it freezes or by *ice lenses*, thick layers of ice that form as water vapor migrates upward from the soil and is trapped under the footing.

In climates with little or no ground freezing, the thickened edges of a concrete *slab on grade* (see Chapter 14 for further information on slabs on grade) can function as simple, inexpensive spread footings for one- and two-story buildings. Where footings must be deeper, or floors are raised over a *crawlspace* or *basement*, concrete or masonry walls resting on strip footings provide support for the structure above (Figure 2.31). When building on slopes, strip footings are stepped, to maintain the required depth of footing at all points around the building (Figure 2.32). If a sloping site or earthquake precautions require it, column footings may be linked together with reinforced concrete *tie beams* to maintain stability of the footings when subjected to lateral forces.

COLUMN FOOTING   WALL FOOTING

FIGURE 2.29
A column footing and a wall footing of concrete. The steel reinforcing bars have been omitted from this illustration for clarity. The role of steel reinforcing in the structural performance of concrete elements is explained in Chapter 13.

FIGURE 2.30
These concrete foundation walls for an apartment building, with their steel formwork not yet stripped, rest on wall footings. For more extensive illustrations of wall and column footings, see Figures 5.11, 14.5, 14.7, and 14.14. (*Reprinted with permission of the Portland Cement Association from* Design and Control of Concrete Mixtures; *Photos: Portland Cement Association, Skokie, IL.*)

Footings cannot legally extend beyond a property line, even for a building built tightly to that line. If the outer toe of the footing were simply cut off at the property line, the footing would be unbalanced by the off-center column or wall above and tend to rotate and fail. *Combined footings* and *cantilever footings* solve this problem by tying the footings for the outside row of columns to those of the next row in such a way that any imbalance is neutralized (Figure 2.33).

Shallow Frost-Protected Foundations
Where the frost line is deep, excavation costs can be saved by constructing *shallow frost-protected foundations.* These are footings placed closer to the ground surface, but insulated in such a way that the ground underneath them cannot freeze. Continuous layers of insulation board are placed around the perimeter of the building in such a way that heat

SLAB ON GRADE

CRAWLSPACE

BASEMENT

FIGURE 2.31
Three types of substructures with shallow foundations. The slab on grade, where adequate, is the most economical under many circumstances. A crawlspace is used under a raised floor structure and gives easier access to underfloor piping and wiring than a slab on grade. Basements provide usable space.

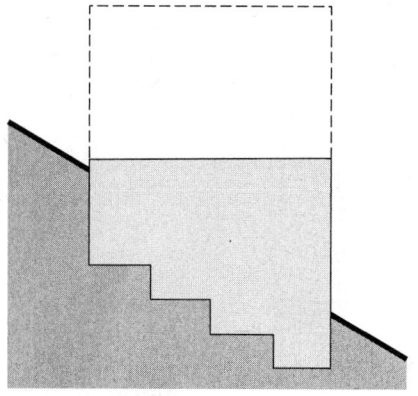

STEPPED FOOTING

FIGURE 2.32

Foundations on sloping sites, viewed in a cross section through the building. The broken line indicates the outline of the superstructure. Wall footings are stepped to maintain the necessary distance between the bottom of the footing and the surface of the ground. Separate column foundations, whether caissons (as shown here) or column footings, are often connected with reinforced concrete tie beams to reduce differential movement between the columns. A grade beam differs from a tie beam by being reinforced to distribute the continuous load from a bearing wall to separate foundations.

Tie beams between columns

TIE BEAMS

Grade beam

GRADE BEAM

FIGURE 2.33

Either a combined footing (*top*) or a cantilevered footing (*bottom*) is used when columns must abut a property line. By combining the foundation for the column against the property line, at the left, with the foundation for the next interior column to the right in a single structural unit, a balanced footing design can be achieved. The concrete reinforcing steel has been omitted from these drawings for the sake of clarity.

flowing into the soil in winter from the interior of the building maintains the soil beneath the footings at a temperature above freezing (Figure 2.34). Even beneath unheated buildings, properly installed thermal insulation can trap enough geothermal heat around shallow foundations to prevent freezing. The insulation boards for shallow frost-protected footings are made from foam plastic or other material that can withstand the effects of ground moisture and earth pressures.

Mat Foundations

In situations where the bearing capacity of the soil is low in relation to building loads, column footings may become so closely spaced that it is more effective to merge them into a single *mat* or *raft foundation* that supports the entire building. Mat foundations for very tall buildings are heavily reinforced and may be 6 feet (1.8 m) or more in thickness (Figure 2.35).

Wall structure

Protective coating

Plastic foam insulation

Shallow footing in inorganic soil

FIGURE 2.34
A typical detail for a shallow frost-protected footing.

FIGURE 2.35
Pouring a large foundation mat. Six truck-mounted pumps receive concrete from a continuous procession of transit-mix concrete trucks and deliver this concrete to the heavily reinforced mat. Concrete placement continues nonstop around the clock until the mat is finished, to avoid "cold joints," weakened planes between hardened concrete and fresh concrete. The soil around this excavation is supported with a sitecast concrete slurry wall. Most of the slurry wall is tied back, but a set of rakers is visible at the lower right.
(Courtesy of Schwing America, Inc.)

Especially where settlement must be carefully controlled, a *floating* or *compensated foundation*—a variation on a mat footing—may be used. A compensated foundation is a mat foundation placed at a depth such that the weight of the soil removed from the excavation is close to the weight of the building constructed above. In this way, the load on the underlying soil changes very little, and settlement is minimized. As a rule of thumb, one story of excavated soil weighs about the same as five to eight stories of superstructure, depending on the density of the soil and the construction of the building (Figure 2.36). A compensated foundation for a 30-story building, therefore, would require an excavation 4 to 6 stories deep to achieve the necessary balance between removed soil and imposed building load.

Deep Foundations

Caissons

A *caisson*, or *drilled pier* (Figure 2.37), is similar to a column footing in that it spreads the load from a column over a large enough area of soil that the allowable pressure in the soil is not exceeded. It differs from a column footing in that it extends through strata of unsatisfactory soil beneath the substructure of a building until it reaches a more suitable stratum.

A caisson is constructed by drilling a hole, *belling* (flaring) the hole out at the bottom as necessary to achieve the required bearing area, and filling the hole with concrete. Large *auger drills* (Figures 2.38 and 2.39) are used for drilling caissons. Occasionally hand excavation is used where the soil is too full of boulders

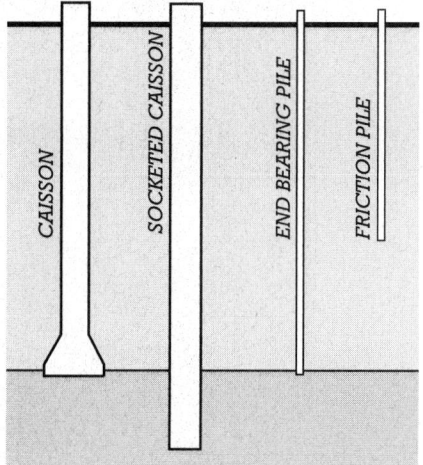

FIGURE 2.37

Deep foundations. Caissons are concrete cylinders poured into drilled holes. They reach through weaker soil (light shading) to bear on competent soil beneath. The end bearing caisson at the left is belled as shown when additional bearing capacity is required. The socketed caisson is drilled into a hard stratum and transfers its load primarily by friction between the soil or rock and the sides of the caisson. Piles are driven into the earth. End bearing piles act in the same way as caissons. The friction pile derives its load-carrying capacity from friction between the soil and the sides of the pile.

FIGURE 2.36

A cross section through a building with a floating foundation. The building weighs approximately the same as the soil excavated for the substructure, so the stress in the soil beneath the building is the same after construction as it was before.

FIGURE 2.38

A 6-foot- (1828-mm-) diameter auger on a telescoping 70-foot (21-m) bar brings up a load of soil from a caisson hole. The auger will be rotated rapidly to spin off the soil before being reinserted in the hole. (*Courtesy of Calweld Inc.*)

FIGURE 2.39
For cutting through hard material, the caisson drill is equipped with a carbide-toothed coring bucket. *(Courtesy of Calweld Inc.)*

FIGURE 2.40
The bell is formed at the bottom of the caisson shaft by a belling bucket with retractable cutters. The example shown here is for an 8-foot- (2.44-m-) diameter shaft and makes a bell 21 feet (6.40 m) in diameter. *(Courtesy of Calweld Inc.)*

to drill. Depending on soil conditions, the soil around the drilled hole may be temporarily supported with a cylindrical steel casing lowered around the drill as it progresses, or by temporarily filling the hole with water or a slurry, similar to slurry wall construction. When a firm bearing stratum is reached, the bell, if required, is created at the bottom of the shaft either by hand excavation or by a special *belling bucket* on the drill (Figure 2.40). The bearing surface of the soil at the bottom of the hole is then inspected to be sure it is of the anticipated quality. Finally, the hole is filled with concrete, with any temporary casing withdrawn or water or slurry pumped out as the concrete rises. Reinforcing is seldom used in the concrete except near the top of the caisson, where it joins the building structure above.

Caissons are large, heavy-duty foundation components. Their shaft diameters range from 18 inches (460 mm) up to 12 feet (3.6 m) or more. *Belled caissons* are practical only where the bell can be excavated in a cohesive soil that will retain its shape at least until the concrete is poured. Where groundwater is present, the temporary steel casing can prevent flooding of the caisson hole during its construction. But where the bearing stratum is permeable, water may fill the hole from below and caisson construction may not be practical.

A *socketed caisson* (Figure 2.37) is drilled into rock at the bottom rather than belled. Its bearing capacity comes not only from its end bearing, but also from friction between the sides of the caisson and the rock. Figure 2.41 shows the installation of a rock caisson or drilled-in caisson, a special type of socketed caisson with a steel H-section core.

Piles

Piles (Figure 2.37) are more slender than caissons, and forcibly driven into the earth rather than drilled and poured. They are used where noncohesive soils, subsurface water, or excessive depth of bearing strata make caissons impractical. If a pile is driven until its tip encounters firm resistance from a suitable bearing stratum such as rock, dense sands, or

(a)

(b)

FIGURE 2.41

Installing a rock caisson. (a) The shaft of the caisson has been drilled through softer soil to the rock beneath and cased with a steel pipe. A churn drill is being lowered into the casing to begin advancing the hole into the rock. (b) When the hole has penetrated the required distance into the rock stratum, a heavy steel H-section is lowered into the hole and suspended on steel channels across the mouth of the casing. The space between the casing and the H-section is then filled with concrete, producing a caisson with a very high load-carrying capacity because of the composite structural action of the steel and the concrete.
(Courtesy of Franki Foundation Company.)

gravels, it is an *end bearing pile.* If no firm bearing layer can be reached, a pile may still develop a considerable load-carrying capacity through frictional resistance between the sides of the pile and the soil through which it is driven; in this case, it is known as a *friction pile.* (Often piles rely to some degree on a combination of end bearing and friction for their strength.) Piles are usually driven closely together in clusters that contain 2 to 25 piles each. The piles in each cluster are later joined at the top by a reinforced concrete *pile cap,* which distributes the load of the column or wall above among the piles (Figures 2.42 and 2.43).

If . . . solid ground cannot be found, but the place proves to be nothing but a heap of loose earth to the very bottom, or a marsh, then it must be dug up and cleared out and set with piles made of charred alder or olive wood or oak, and these must be driven down by machinery, very closely together.

—Marcus Vitruvius Pollio (Roman architect), *The Ten Books on Architecture,* 1st century BC

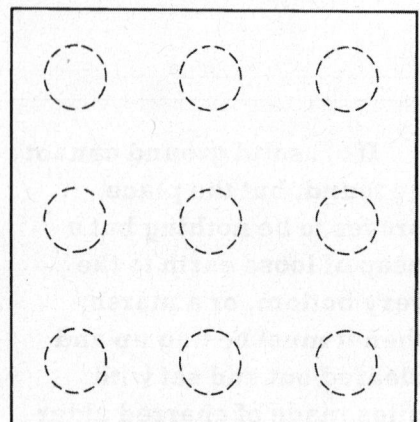

FIGURE 2.42
Clusters of two, three, four, and nine piles with their concrete caps, viewed from above. The caps are reinforced to transmit column loads equally into all the piles in the cluster, but the reinforcing steel has been omitted here for the sake of clarity.

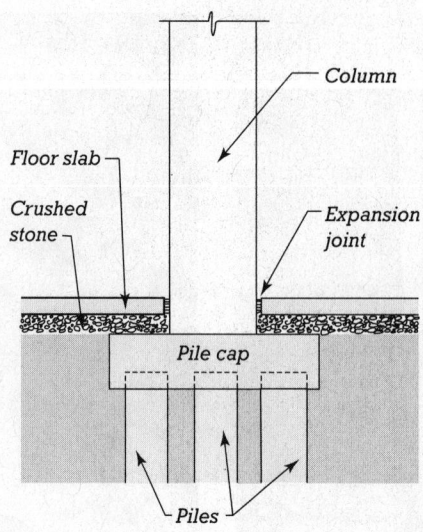

FIGURE 2.43
An elevation view of a pile cap, column, and floor slab.

End bearing piles work essentially the same as caissons and are used on sites where a firm bearing stratum can be reached, even sometimes at depths of 150 feet (45 m) or more. Each pile is *driven to refusal*, the point at which little additional penetration is made with continuing blows of the hammer, indicating that the pile is firmly embedded in the bearing layer. Friction piles are driven either to a predetermined depth or until a certain level of resistance to hammer blows is encountered, rather than to refusal as with end bearing piles. Clusters of friction piles have the effect of distributing a concentrated load from the structure above into a large volume of soil around and below the cluster, at stresses that lie safely within the capacity of the soil (Figure 2.44). The loadbearing capacities of piles are calculated in advance based on soil test results and the properties of the piles. To verify these calculations, as well as to determine the rate and ease with which piles can be installed, test piles are often driven and loaded on the building site before foundation work begins.

Where piles are used to support loadbearing walls, reinforced concrete *grade beams* are constructed between the pile caps to transmit the wall loads to the piles (Figure 2.45). Grade beams are also used with caisson foundations for the same purpose.

Pile Driving

Conventional piles are driven with *pile hammers*, heavy weights lifted by the energy of steam, compressed air, compressed hydraulic fluid, or diesel fuel combustion, then dropped on the top of the pile. The hammer travels on vertical rails called *leads* (pronounced "leeds") at the front of a piledriver (Figure 2.46). It is first hoisted up the leads to the top of each pile as driving commences, then follows the pile down as it penetrates the earth. The *piledriver* includes additional hoisting machinery for raising the piles themselves into position before driving.

In certain types of soil, piles can be driven more efficiently by vibration than by hammer blows alone, using a vibratory hammer mechanism. Also, as discussed later in this chapter, some lightweight pile systems are installed by rotary drilling or hydraulic pressing.

Pile Materials

Piles may be made of timber, steel, concrete, or various combinations of these materials (Figure 2.47). The simplest are *timber piles*, made from tree trunks with their branches and bark removed and driven into the ground small end first. Timber piles have been used since Roman times, when they were driven by large hammers hoisted by muscle power. Their main advantage is that they are economical for lightly loaded foundations. A primary disadvantage is that they cannot be spliced during driving and are, therefore, limited to the length of available tree trunks, approximately 65 feet (20 m). Unless pressure treated with a chemical preservative or completely submerged below the water table, they will decay (the lack of free oxygen in the water prohibits organic growth). Relatively small hammers must be used when driving timber piles, to avoid splitting

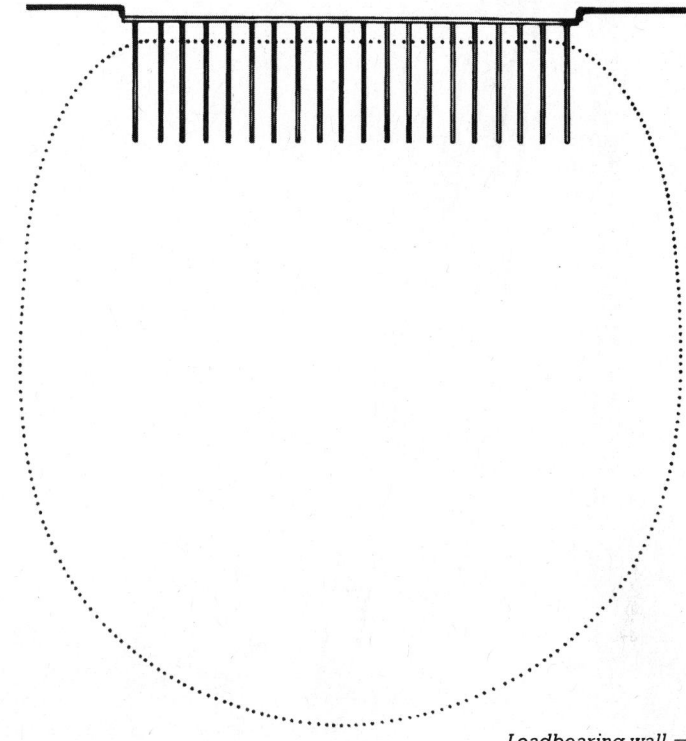

FIGURE 2.44

A single friction pile (*left*) transmits its load into the earth as an equal shear pressure along the bulb profile indicated by the dotted line. As the size of the pile cluster increases, the piles act together to create a single larger bulb of higher pressure that reaches deeper into the ground. A building with many closely spaced clusters of piles (*right*) creates a very large, deep bulb. Care must be taken to ensure that large-pressure bulbs do not overstress the soil or cause excessive settlement of the foundation. The settlement of a large group of friction piles in clay, for example, will be considerably greater than that of a single isolated pile.

Loadbearing wall

Reinforced concrete grade beam with integral pile caps

Piles

FIGURE 2.45

To support a loadbearing wall, pile caps are joined by a grade beam. The reinforcing in the grade beam is similar to that in any ordinary continuous concrete beam and has been omitted for clarity. In some cases, a concrete loadbearing wall can be reinforced to act as its own grade beam.

FIGURE 2.46
A piledriver hammers a precast concrete pile into the ground. The pile is supported by the lead and driven by a heavy diesel piston hammer. The hammer follows the pile down the leads as the pile progresses deeper into the soil. *(Photo by Joseph Iano.)*

STEEL H-PILE

STEEL PIPE PILE

PRECAST
CONCRETE PILE

WOOD PILE

FIGURE 2.47
Cross sections of common types of piles. Precast concrete piles may be square or round instead of the octagonal section shown here and may be hollow in the larger sizes.

the piles. Capacities of individual timber piles lie in the range of 10 to 55 tons (90 to 490 kN).

Steel piles made be made from H-sections or pipes. H-piles are special wide-flange sections (see Chapter 11), 8 to 14 inches (200 to 355 mm) deep, which are approximately square in cross section. They are used in end bearing applications. H-piles displace relatively little soil during driving. This minimizes the upward displacement of adjacent soil, called *heaving*, that may occur with other pile types. Soil displacement can be a problem on urban sites, where it can disturb adjacent buildings.

H-piles can be brought to the site in any convenient lengths, welded together as driving progresses to form any necessary length of pile, and cut off with an oxyacetylene torch when the required depth is reached. The cutoff ends can then be welded onto other piles to avoid waste. Corrosion can be a problem in some soils, however, and unlike closed pipe piles and hollow precast concrete piles, H-piles cannot be inspected after driving to be sure they are straight and undamaged.

Steel *pipe piles* have diameters of 8 to 24 inches (200 to 600 mm) or more. The lower, driven end of the pipe may be either open or closed. An open pile is easier to drive than a closed pile and, depending on soil conditions, may displace less soil. After driving, the pipe is cleared of soil, inspected to ensure that it is straight and undamaged, and then filled with concrete. A closed pile can be inspected and concreted immediately after driving. Pipe piles generally displace larger amounts of soil during driving than H-piles, increasing the possibility of upward heaving of nearby soil and disturbance of nearby buildings. The larger sizes of pipe piles require a heavier hammer for driving as well. H-piles and steel pipe piles can carry loads up to approximately 200 to 300 tons (1800 to 2700 kN).

Conventional steel piles normally have no added corrosion protection.

The low levels of free oxygen in undisturbed soils generally limit corrosion to rates that can be accounted for in the thickness of the steel sections themselves. Or, where soil conditions are unusually corrosive, a concrete pile may chosen instead.

Minipiles, also called *pin piles* or *micropiles*, are made from steel bar or pipe 2 to 12 inches (50 to 300 mm) in diameter. They are pressed or rammed into holes drilled in the soil and then grouted in place. Where vertical space is limited, such as when working in the basement of an existing building, minipiles can be installed in individual sections as short as 3 feet (1 m) that are coupled end-to-end as driving progresses. *Helical piles* or *screw piles* are similar to minipiles, but with one or more helical boring blades, up to 24 inches (600 mm) in diameter, attached (Figure 2.48). A helical pile is installed by rotation, causing it to auger into the ground without predrilling.

Minipiles and helical piles are installed without hammering, thereby avoiding much of the vibration and noise associated with conventional pile installation. This makes these pile types good choices for work close to existing buildings or for the improvement of existing foundations where excessive vibration could damage existing structures or noise could disrupt ongoing activities. Because of their slenderness, their installation also entails little or no soil displacement, minimizing the risk of disturbance to nearby foundations. Bearing capacity for minipiles and helical piles ranges from as little as 2 tons (18 kN) to as much as 200 tons (1800 kN). See Figure 2.54 for an illustration of minipiles used to reinforce or underpin an existing foundation. Larger-capacity drilled minipiles, using pipes with diameters up to 24 inches (610 mm), blur the distinction between traditional piles and caissons, while retaining the advantages of low-vibration installation and ability to be installed in confined spaces.

FIGURE 2.48
A helical pile. Rows of piles transfer loads from the shallow footing, located in a relatively weak soil (lighter in tone in the illustration), into the stronger underlying stratum (darker tone in the illustration).

Minipiles and helical piles, being thinner in section than conventional steel piles, sometimes require added corrosion protection. For minipiles, the surrounding grouting is usually adequate. Ungrouted helical piles and minipiles may be galvanized (zinc-coated) or coated with plastic or epoxy.

Precast concrete piles are square, octagonal, or round in section, and in large sizes may have open cores to allow inspection after driving (Figure 2.49). Most are prestressed, but some for smaller buildings may be only conventionally reinforced (for an explanation of concrete reinforcing and prestressing, see Chapter 13). Typical cross-sectional dimensions range from 10 to 30 inches (250 to 800 mm) or more. Advantages of precast concrete piles include high load capacity and freedom from corrosion or decay. Precast piles must

FIGURE 2.49
Precast, prestressed concrete piles. Lifting loops are cast into the sides of the piles as crane attachments for hoisting them into a vertical position. In the background can be seen driven piles, ready for cutting off and capping. (*Courtesy of Lone Star/San-Vel Concrete.*)

be handled carefully to avoid bending and cracking before installation. Splices between lengths of precast piling can be made with mechanical fastening devices that are cast into the ends of the sections. Bearing capacities range up to approximately 500 tons (4400 kN).

Sitecast concrete piles are usually installed by driving a hollow steel shell into the ground and then filling the shell with concrete. The shell may remain in place as part of the finished pile, or it may be removed and reused for the installation of subsequent piles (Figure 2.50). When left in place, the shell may be corrugated to increase its stiffness. Alternatively, an auger with

| Steel Point | Concrete Plug | Open Ended | Fluted Tapered | Compressed Base |
| CASED PILES |

| Compressed Concrete | Pedestal Pile | Steel Point |
| UNCASED PILES |

FIGURE 2.50
Various types of sitecast concrete piles. Most are cast into steel casings that have been driven into the ground. The uncased piles are made by withdrawing the casing as the concrete is poured and saving it for subsequent reuse. (Compared to steel pipe piles, the metal shells in cased concrete piles are thinner and contribute less to the bearing capacity of the pile.)

a hollow stem may be used to drill the hole for the pile; then concrete is deposited through the stem as the auger is withdrawn. Load capacities for sitecast concrete piles range up to approximately 200 tons (1800 kN).

In a concrete *pressure-injected footing* or *compaction grouted footing,* a dry, stiff mixture of concrete or grout is compacted into a predrilled hole. The compaction of the concrete or grout in place also densifies and strengthens the surrounding soil. *Rammed aggregate piers* or *stone columns* are similar, but are constructed solely with crushed rock (Figure 2.51). Like soil mixing, these types of foundation elements are a form of ground improvement. When used for foundation work, arrays of such piers are installed to improve the structural properties of soil relatively close to the ground surface. This permits shallow, less expensive foundations to be constructed above the improved soil, in place of the deeper and more expensive foundation types that would otherwise be required. These pier types can also be used to reduce the risk of liquefaction in soils prone to this behavior, for slope stabilization, and for other types of soil strengthening.

Seismic Base Isolation

In areas where very strong earthquakes are common, large buildings may be placed on *base isolators.* When significant ground movement occurs, the base isolators flex or yield to absorb a significant portion of this movement. As a result, movement of the building structure is lessened, and the magnitude of the forces acting on the structure and the potential for damage are reduced. One type of base isolator consists of a multilayered sandwich of rubber and steel plates (Figure 2.52). When subjected to lateral forces, yielding of the rubber layers allows the isolator to deform. A lead core provides damping action and keeps the layers of the sandwich aligned.

STEPS IN AGGREGATE PIER INSTALLATION

SECTION

Figure 2.51

Above, four steps in the construction of rammed aggregate piers. From left to right: (1) The pier hole is drilled with a rotating auger or high-energy vibrating probe. (2) The first lift of aggregate is compacted at the base of the hole. (3) Succeeding lifts are compacted one after the other. As lifts are compacted in place, the surrounding soil is also densified. (4) The pier is completed. Below, a diagrammatic section illustrating shallow spread footings and a slab on grade bearing on soil strengthened with aggregate piers. Finished piers may be up to 36 in. (900 mm) in diameter and 30 ft (9 m) deep.

Displacement of building

Base isolators

Motion of ground

Column base plate

Lead core

Multiple layers of steel plates and rubber

Isolator base plate

FIGURE 2.52
Base isolation.

Underpinning

Underpinning is the strengthening and stabilizing of an existing foundation. It may be required where a foundation was improperly designed and has proved inadequate, when a change in use or increase in building size increases the loads on an existing foundation, or when new nearby construction disturbs the soil around an existing foundation and requires that the foundation be carried deeper. Underpinning methods vary, but generally rely on one of three approaches: The existing foundation elements may be enlarged to distribute loads over a greater soil area; new, deeper foundations can be inserted under existing ones to carry loads to a deeper, stronger stratum of soil; or the soil itself can be strengthened by grouting or chemical treatment. Figures 2.53 and 2.54 illustrate

in diagrammatic form some selected concepts of underpinning.

Up–Down Construction

Normally, the substructure of a building is completed before work begins on its superstructure. If the building has several levels of basements, however, substructure work can take many months, or on very complex projects a year or more. In such a case, *up–down construction* may be an economical option, even if its first cost is somewhat more than that of the normal procedure, because it can save considerable construction time.

As diagrammed in Figure 2.55, up–down construction begins with installation of a perimeter slurry wall. Internal steel columns for the substructure are lowered into drilled, slurry-filled holes, and concrete foot-

ings are tremied beneath them. After the ground-floor slab is in place and connected to the substructure columns, erection of the superstructure may begin. Construction continues simultaneously on the substructure, largely by means of mining techniques: A story of soil is excavated from beneath the ground-floor slab and a level mud slab of CLSM (see section on "Filling and Finish Grading" in this chapter) is poured. Working on the mud slab, workers reinforce and pour a concrete structural slab for the floor of the topmost basement level and connect this floor to the columns. When the slab is sufficiently strong, another story of soil is removed from beneath it, along with the mud slab. The process is repeated until the substructure is complete, by which time the superstructure has simultaneously been built many stories into the air.

A. ELEVATION   *SECTION*   *B. ELEVATION*   *SECTION*

FIGURE 2.53

Two methods of supporting a building while carrying out underpinning work beneath its foundation, each shown in both elevation and section. (*A*) Trenches are dug beneath the existing foundation at intervals, leaving the majority of the foundation supported by the soil. When portions of the new foundations have been completed in the trenches, using one of the types of underpinning shown in Figure 2.54, another set of trenches is dug between them and the remainder of the foundations is completed. (*B*) The foundations of an entire wall can be exposed at once by needling, in which the wall is supported temporarily on needle beams threaded through holes cut in the wall. After underpinning has been accomplished, the jacks and needle beams are removed and the trench is backfilled.

A. ELEVATION   *SECTION*   *B. ELEVATION*   *SECTION*   *C. SECTION*

FIGURE 2.54

Three types of underpinning. (*A*) A new foundation wall and footing are constructed on either side of the existing foundation. (*B*) New piles or caissons are constructed on either side of the existing foundation. (*C*) Minipiles are inserted through the existing foundation. This may be accomplished without temporary support of the building and only minimal excavation.

Steel columns placed
in slurry-filled shafts

Slurry wall around
perimeter

(a)

FIGURE 2.55
Up–down construction. (*a*) Preliminary slurry wall and
column construction prior to excavation. (*b*) Construction
proceeds both upward and downward simultaneously.

PROTECTING FOUNDATIONS FROM WATER, HEAT FLOW, AND RADON GAS

Waterproofing and Drainage

Where building substructures enclose basements, parking garages, or other usable space, groundwater must be kept out. Concrete alone is rarely adequate for this purpose. Moisture can migrate through its microscopic pores, or through pathways created by shrinkage cracks, form tie holes (see Chapter 14), utility penetrations, or the joints that occur between separate pours. Two basic approaches are used to resist water entry: drainage and waterproofing. Drainage draws groundwater away from a foundation, reducing the volume and pressure of water acting on the foundation's walls and slabs. Waterproofing acts as a barrier to the passage of water through the foundation, preventing it from reaching the interior.

Superstructure
erection up

Mining
excavation
down

(*b*)

Drainage typically consists of some combination of porous back-fill material (such as a well sorted gravel), drainage mat, and perforated drain piping (Figure 2.56). *Drainage mat* is a manufactured sheet product, usually about ½ inch (12 mm) thick, made of a plastic egg-crate-like structure or some other very open, porous material. It is faced on the outside with *filter fabric* that allows water to pass easily but prevents fine soil particles from entering and clogging its drainage passages. Subterranean water that approaches the wall falls through the mat to *perforated drain piping* near the bottom of the foundation wall. The drain piping is laid around the outside perimeter of the building foundation. The pipes are 4 or 6 inches (100 or 150 mm) in diameter, with several parallel rows of perforations that allow the inflow of water that arrives at that level. Water in the pipes then flows by gravity either "to daylight" at a lower surface elevation

FIGURE 2.56
Two methods of relieving water pressure around a building substructure by drainage. A drainage system that relies solely on crushed stone or gravel placed against the foundation (*left*) is hard to do well, because of the difficulty of depositing the drainage aggregate and backfill soil in neatly separated, alternating layers. The drainage mat (*right*) is easier and often more economical to install.

on a sloping site, a municipal storm sewer system, or a sump pit that can be automatically pumped dry whenever it fills. The perforations in the pipes face downward so that as the water level in the soil rises, it enters the pipes at the lowest possible level. Where groundwater conditions are severe, rows of perforated pipe may be installed under the basement slab as well (Figure 2.57).

On many substructures, a barrier of some kind is added to increase the level of protection against water entry. *Dampproofing* is a moisture-resistant cement plaster or asphalt compound applied to basement walls where groundwater conditions are mild or waterproofing requirements are not critical. *Cement plaster damp-proofing*, or *parge coating*, is light gray in color and is troweled on. *Asphalt* or *bituminous dampproofing* is almost black in color and is applied as a liquid by spray, roller, or trowel.

Waterproofing, unlike dampproofing, resists the passage of water even under the more demanding conditions of hydrostatic pressure. It is more costly than dampproofing, and is used where groundwater conditions are severe or the need to protect subgrade space from moisture is critical (Figure 2.58).

Waterproofing membranes are formulated from plastics, asphalt compounds, synthetic rubbers, and other materials. They come in a variety of forms and their selection is often made by consultants with specialized knowledge of these materials and their use. Some considerations in their selection include:

• *Liquid-applied membrane waterproofing* materials are applied by spray or roller as viscous liquids and then allowed to cure in place. They are seamless and easy to form around intricate shapes. However, they may be subject to inconsistent application thickness or poor curing depending on ambient conditions.

BELOW-SLAB DRAINAGE

ABOVE-SLAB DRAINAGE

FIGURE 2.57
For a high degree of security against substructure water entry, drainage both around and under the basement is required, as seen here in a section view. Above-slab drainage is used in buildings with mat foundations.

Protection board
Membrane
Drainage mat
Grade

Basement wall

Basement floor slab
Protection board
Membrane
Mud slab

Footing

FIGURE 2.58
A diagrammatic representation of
the placement of sheet membrane
waterproofing around a basement. A
mud slab of low-strength concrete was
poured to serve as a base for placement
of the horizontal membrane. Notice that
the vertical and horizontal membranes
join to wrap the basement completely in
a waterproof enclosure. However, the
section of membrane running between
the bottom of the basement wall and
the top of the footing may weaken the
connection between these two elements,
so this detail should be reviewed by
the structural engineer to verify its
compatibility with foundation design
requirements.

• *Sheet membrane waterproofing* materials are manufactured in the factory. They are more consistent in material quality and thickness than liquid-applied membranes. However, they are more difficult to apply around complex shapes and careful attention must be given to sealing the seams between sheets, which is often the most vulnerable point in these systems.

Some sheet membrane materials are fully adhered to the foundation wall and others are allowed to lie loosely against it. Loosely laid sheet membranes are less vulnerable to tearing in the case of movement or cracking in the underlying wall. But leaks in loosely laid membranes can permit moisture to travel under the membrane, to appear inside a sub-structure far from where the defect in the outer membrane occurs, making repairs more difficult. Fully adhered sheet membranes (as well as liquid-applied membranes) confine leaks close to where the defect in the membrane occurs. But they are also potentially more vulnerable to stress and tearing when cracking or movement occurs in the wall.

FIGURE 2.59
Waterproofing in progress on a concrete foundation. *Leftmost:* The bare foundation wall remains exposed. *Middle:* Bentonite panels are fastened in place. These panels are lined on the outer face with a black-colored, high-density plastic that adds to the waterproofing qualities of the panel. *Right:* Drainage mat has been installed over the waterproofing. The mat's outer face of filter fabric is lightly dimpled, telegraphing the egg-crate structure of the underlying molded plastic panel. The top edge of the mat is secured in place with an aluminum termination bar that holds the mat in place and keeps dirt and debris from falling behind it. *Lower right:* White, plastic perforated drain piping can be seen, temporarily supported on wood blocking and running alongside the footing. *(Photos by Joseph Iano.)*

• *Bentonite waterproofing* is made from sodium bentonite, a naturally occurring, highly expansive clay. It is most often applied as preformed sheets consisting of dry clay sandwiched within corrugated cardboard, geotextile fabric, or plastic sheets (Figure 2.59). When the bentonite is contacted by moisture, it swells to several times its dry volume and forms a barrier that is impervious to the further passage of water.

• *Integral waterproofing* describes a variety of ingredients added directly into concrete when it is mixed. These materials stop up the pores in the cured concrete and render the wall more watertight.

• *Electro-osmotic waterproofing* relies on electrical current flow induced into the concrete wall to drive water molecules toward the outer face of the wall. It is most commonly used to remediate leakage in existing foundations where more conventional waterproofing methods are impractical or overly expensive.

• *Blind-side waterproofing* is installed prior to the pouring of concrete walls.

This occurs most commonly when a substructure wall is built close to a property's edge, and excavation cannot be enlarged beyond the property line to permit workers access to the outer face of the wall after its completion. Drainage matting is first applied directly to the excavation sheeting, and then any of a number of possible waterproofing membrane types are applied over the drainage mat. Later, the concrete wall is poured against the membrane. The sheeting remains permanently in place (Figure 2.60).

Waterproof membrane

Drainage blanket

Mat slab

Protection slab

Mud slab

Waterproof membrane

Crushed stone

Filter fabric

Plastic drain line (perforated)

FIGURE 2.60
Blind-side waterproofing is used where there is no working space between a sheeted excavation and the outside of the foundation wall. The drainage mat and waterproof membrane are applied to the sheeting. Then the basement wall is poured against them.

Joints in foundation construction, such as those occurring between separate concrete pours, require attention to ensure watertightness as well. Preformed *waterstops* made of plastic, synthetic rubber, metal, or materials that swell when they come in contact with water can be cast into the mating concrete edges to block the passage of water through these vulnerable locations (Figures 2.61, 2.62, and 2.63).

Most waterproofing systems are inaccessible once building construction is complete, and they are expected to perform for the life of the building. However, even small defects can allow the passage of significant amounts of water. For these reasons, waterproofing membranes are inspected carefully during installation. Horizontal membranes may also be *flood tested* (submerged for an extended period of time while leak-checking is performed) to find any defects while repairs can still be easily made. Once inspection and testing are complete, membranes are covered with a *protection board*, insulation board, or drainage matting to shield the membrane from prolonged exposure to sunlight and to prevent physical damage during soil backfilling or subsequent construction operations.

Basement Insulation

Occupant comfort and energy efficiency require that occupied basements be thermally insulated to limit their loss of heat to the surrounding soil. On the inside of the basement

SECOND POUR   FIRST POUR

FIGURE 2.61
A synthetic rubber waterstop is used to seal against water penetration at joints in concrete construction. The type shown here is split on one side so that its halves can be placed flat against the formwork where another wall will join the one being poured. After the concrete has been poured and cured and the formwork has been removed from the first wall, the split halves are folded back together before the next wall is poured. (This type of waterstop can be used for both movement and nonmovement joints in the wall.)

FIGURE 2.62
A synthetic rubber waterstop ready for the next pour of a concrete wall, as diagrammed in Figure 2.60. *(Courtesy of Vulcan Metal Products, Inc., Birmingham, Alabama.)*

FIGURE 2.63
A swelling bentonite waterstop is adhered to a concrete footing prior to casting of the concrete wall above. Later, if groundwater seeps into this area, the bentonite will swell to fully seal the joint. The waterstop is positioned to the side of the steel reinforcing bars closer to the wall's exterior, also protecting the reinforcing from moisture and corrosion. However, because of bentonite's expansive force, the waterstop must not be positioned too close to the surface of the wall, or it could cause portions of concrete to split away when it swells.
(Photo by Joseph Iano.)

wall, insulation may be attached directly to the wall, installed between furring strips as shown in Figure 23.3, or installed in a separately framed wall built close to the inside of the wall. On the outside of the wall, water-resistant materials with good compressive strength, such as extruded polystyrene foam or glass or mineral fiber insulation boards, typically 2 to 4 inches (50 to 100 mm) thick, may be placed against the wall, held by adhesive, fasteners, or the pressure of the soil (Figure 5.8). Insulation may also be integrated into the foundation wall itself, such as with insulated concrete forms (Figure 14.13) or within the hollow cores of a concrete masonry unit wall (Figures 10.16, 10.17).

Radon Gas Control

Radon is a cancer-causing gas that occurs naturally within soils in concentrations that vary by region and locality. Where soil emissions are high, gas seeping through cracks and unsealed penetrations in the foundation can reach unhealthful concentration levels within the building, endangering the well-being of occupants.

Where the radon gas risk is high, buildings are constructed with *passive radon control* methods designed to minimize gas infiltration into the building. For a typical basement with concrete slab on grade, these include:

• A layer of gravel or other gas-permeable material beneath the concrete slab

• Over the gravel layer, beneath the concrete slab, a gas-impermeable plastic sheet or membrane

• After the pouring of the concrete foundation walls and slab on grade, sealing of all joints, penetrations, and cracks in these components

• Coating of the outside of basement walls with dampproofing or waterproofing

• A vent pipe extending vertically from the gas-permeable gravel layer through the roof of the building.

These passive methods work to minimize pathways for radon gas to enter the building through its foundation and provide a ready path to vent such gas directly to the exterior. They are most frequently applied to residential and school buildings in areas of known high radon gas risk. Because many of the necessary components for passive radon control are frequently included for the purposes of waterproofing and drainage, the added cost to provide passive radon control at the time of construction is usually small.

If, at some time after completion of the building, passive protection proves inadequate to control gas entry into the structure, *active subslab depressurization* can be added by installing a small electric fan in the vertical vent pipe(s). This fan continuously draws air from underneath the gas-impermeable membrane. By reducing the air pressure in the gas-permeable layer, radon gas is more effectively extracted from the soil beneath the structure and the risk of gas entry into the building is further reduced.

SITEWORK

Retaining Walls

A *retaining wall* holds back soil where an abrupt change in ground elevation occurs. The wall must resist the pressure of the earth and groundwater that presses against it from the uphill side. Retaining walls may be made of masonry, preservative-treated wood, coated steel, precast concrete, or, most commonly, sitecast concrete.

The design of a retaining wall must take into account the height of the wall, the pressures acting on the wall, and the character of the soil on which the wall will rest. If inadequately structured, failure can occur by overturning, sliding, or undermining (Figure 2.64).

For small retaining walls, unreinforced construction methods are used that rely on the mass of the wall, interlocking units, or other simple techniques to develop adequate resistance to soil pressures (Figures 2.65 and 2.66). For taller walls or ones subject to greater loads, more complex, reinforced solutions are required (Figure 2.67).

An alternative to a conventional retaining wall is *earth reinforcing* (Figure 2.68). Soil is compacted in layers, each sandwiched between strips or meshes of galvanized steel, polymer fibers, or glass fibers, which stabilize the soil in much the same manner as the roots of plants. *Gabions* are another form of earth retention in which corrosion-resistant wire baskets are filled with cobble- or boulder-sized rocks and then stacked to form retaining walls and slope protection. Soil mixing, as discussed earlier in this chapter, and a variety of grout injection techniques can also be used to strengthen and stabilize underground soils without requiring excavation.

FIGURE 2.64

Three failure mechanisms in retaining walls. The high water table shown in these illustrations creates pressure against the walls that contributes to their failure. The undermining failure is directly attributable to groundwater running beneath the base of the wall, carrying soil with it.

STONE GRAVITY WALL

*VERTICAL TIMBER
CANTILEVERED WALL*

*HORIZONTAL TIMBER
WALL WITH DEADMEN*

FIGURE 2.65
Three types of simple retaining walls, usually used for heights not exceeding 3 feet (900 mm). The deadmen in the horizontal timber wall are timbers embedded in the soil behind the wall and connected to it with timbers inserted into the wall at right angles. The timbers, which should be pressure treated with a wood preservative, are held together with very large spikes or with steel reinforcing bars driven into drilled holes. The crushed-stone drainage trench behind each wall is important as a means of relieving water pressure against the wall to prevent wall failure. With proper engineering design, any of these types of construction can also be used for taller retaining walls.

FIGURE 2.66
A segmental retaining wall consisting of specially made concrete blocks designed to interlock and prevent sliding. The wall leans back against the soil it retains; this reduces the amount of soil the wall must retain and makes it more stable against the lateral push of the soil. *(Photo courtesy of VERSA-LOK Retaining Wall Systems.)*

Crushed stone drainage layer

Weep holes

Key to prevent sliding

REINFORCED CONCRETE

Perforated drainage pipe

REINFORCED CONCRETE MASONRY

FIGURE 2.67
Cantilevered retaining walls of concrete and concrete masonry. The footing is shaped to resist sliding and overturning, and drainage behind the wall reduces the likelihood of undermining. The pattern of steel reinforcing (broken lines) is designed to resist the tensile forces in the wall.

FIGURE 2.68
Two examples of earth reinforcing. The embankment in the top section was placed by alternating thin layers of earth with layers of synthetic mesh fabric. The retaining wall in the lower section is made of precast concrete panels fastened to long galvanized steel straps that run back into the soil.

GEOTEXTILES

Geotextiles are fabrics made of chemically inert plastics that are resistant to deterioration in the soil. They are used for a variety of purposes relating to site development and foundation construction. As described in the accompanying text, plastic mesh fabrics or grids are used as *earth reinforcement* or *soil reinforcement* to support earth embankments and retaining walls. The same type of mesh may be utilized in layers to stabilize engineered fill beneath a shallow footing (see the accompanying figure), or to stabilize marginal soils under driveways, roads, or airport runways, acting very much as the roots of plants do to prevent the movement of soil particles.

Another geotextile discussed elsewhere in this chapter is drainage matting, an open matrix of plastic filaments with a feltlike filter fabric laminated onto one side to keep soil particles from entering the matrix. In addition to providing free drainage around foundation walls, drainage matting is often used beneath the soil in the bottoms of planter boxes and under heavy paving tiles on rooftop terraces, where it maintains free passage for the drainage of water above a waterproof membrane.

Synthetic filter fabrics are wrapped over and around subterranean crushed stone drainage layers such as the one frequently used around a foundation drain. In this position, they keep the stone and pipe from gradually becoming clogged with fine soil particles carried by groundwater. Similar fabrics are used at grade during construction, acting as temporary barriers to filter soil out of water that runs off a construction site, thereby preventing contamination of lakes, streams, or stormwater systems.

Special geotextiles are staked down on freshly cut slopes to prevent soil erosion and encourage revegetation; some of these are designed to decay and disappear into the soil as plants take over the function of slope stabilization. Another type of geotextile is used for weed control in landscaped beds, where it allows rainwater to penetrate the soil but blocks sunlight, preventing weeds from sprouting.

Column

Footing

Soil reinforcing grids

Filling and Finish Grading

Filling refers broadly to any placing of earth material (for example, to raise an existing grade), whereas *backfilling* refers more specifically to the replacement of soil materials in an excavation to restore it close to its finished level. Backfilling occurs around foundations and substructures, in utility trenches, and behind retaining walls.

For any type of filling, an appropriate type of replacement soil is added in layers, or *lifts*, which may range from 4 inches to roughly a foot (100 to 300 mm) in depth. Each lift is compacted before the next is added. Compaction may be performed by heavy rolling, vibrating, or ramming, depending on soil type and depth of lift. Small, walk-behind machines are used in confined areas, and bigger, driven machinery is used over larger areas.

To achieve optimal compaction and minimize future settlement, fill material must be moist enough to readily compress into a dense particle arrangement, but not so wet as to become unstable or soupy. Where backfill material will support slabs, pavements, or foundations, its strength and settlement properties are critical, and the soil materials, moisture content, and compaction work are all monitored by on-site engineers to ensure that the required performance will be achieved.

Around foundations, waterproofing, insulation, and drainage components must be in place before backfilling can begin. In addition, sufficient internal structure must be constructed to provide bracing for the foundation wall, so that the wall can resist the lateral loading from the backfilled soil. For example, in residential construction, the first-floor platform must completed before foundation backfilling occurs, as this floor structure acts to laterally brace the top of the foundation wall (Figure 5.7).

Controlled low-strength material (CLSM) is a manufactured fill material made from portland cement and/or fly ash (a byproduct of coal-burning power plants), sand, and water. CLSM, sometimes called "flowable fill," is delivered in concrete mixer trucks and poured into the excavation, where it compacts and levels itself, then hardens. CLSM may be used to pour *mud slabs*, which are weak concrete slabs used to create temporary, level, dry bases over irregular, often wet excavations. Such a temporary slab can serve as a working surface for the installation of reinforcing and pouring of a foundation mat or basement floor slab. It can also serve as a surface to which a waterproofing membrane is applied. CLSM may be used to replace pockets of unstable soil encountered beneath a substructure or to backfill around basement walls. The strength of CLSM is matched to the situation. For example, for a utility trench, CLSM is formulated so that it is weak enough to be excavated easily by ordinary digging equipment when the pipe requires servicing, yet is as strong as a normal, good-quality compacted backfill.

Finish grading refers to the final leveling and smoothing of soil surfaces to their required contours and elevations. Where plant materials are planned, finish grading includes the application of nutrient-rich, organic topsoil to depths suitable for supporting the growth of the types of grass, plants, shrubs, or trees that have been specified by the landscape architect.

DESIGNING FOUNDATIONS

In foundation design, there are thresholds that, when crossed, cause significant, sudden increases in construction costs:

• *Building below the water table.* If the substructure and foundations of a building are above the water table, minimal effort will be required to keep the excavation dry during construction. Once the water table is reached, more expensive efforts will be needed to dewater the site, strengthen excavation support systems, waterproof the foundation, and protect the finished foundation against permanent hydrostatic pressures.

• *Building close to an existing structure.* If the excavation can be kept well away from adjacent structures, the foundations of these structures will remain undisturbed and no effort or special expense will be required to protect them. When digging close to an existing structure, and especially when digging deeper than that structure's foundations, temporary bracing or permanent underpinning may be required to prevent disturbance.

• *Increasing the column or wall load from a building beyond what can be supported by a shallow foundation.* Shallow foundations are far less expensive than piles or caissons under most conditions. If the building grows too high, or is structured so that individual column loads are too high, a shallow foundation may no longer be able to carry the load and a more expensive, deep foundation system may be required.

For one- and two-family dwellings, foundation design is usually much simpler than for larger buildings, because foundation loadings are low. The uncertainties in foundation design can be reduced with reasonable economy by adopting a large factor of safety in calculating the bearing capacity of the soil. Unless the designer has reason to suspect poor soil conditions, the footings are usually designed using rule-of-thumb allowable soil stresses and standardized footing dimensions. The designer then examines the actual soil when the excavations have been made. If it is not of the quality that was expected, the footings can be hastily redesigned using a revised estimate of soil-bearing capacity before construction continues. If unexpected groundwater is encountered, better drainage may have to be provided around the foundation or the depth of the basement decreased.

In contrast, as discussed throughout this chapter, foundation design for larger buildings may require extensive investigation, analysis, design, and site supervision during construction.

FOUNDATION DESIGN AND THE BUILDING CODE

Because of the public safety considerations involved, building codes contain numerous provisions relating to the design and construction of excavations and foundations. The IBC defines which soil types are considered satisfactory for bearing the weight of buildings and establishes a set of requirements for subsurface exploration, soil testing, and submission of soil reports to the local building inspector. It goes on to specify the methods of engineering design that may be used for the foundations. It sets forth maximum loadbearing values for soils that may be assumed in the absence of detailed test procedures (refer back to Figure 2.6). It establishes minimum dimensions for footings, caissons, piles, and foundation walls and contains lengthy discussions relating to the installation of piles and caissons and the drainage and waterproofing of substructures. The IBC also requires engineering design of retaining walls. In all, the building code attempts to ensure that every building will rest upon secure foundations and a dry substructure.

CSI/CSC MasterFormat Sections for Earthwork, Foundations, Waterproofing, and Site Improvements

02 20 00	**SUBSURFACE INVESTIGATION**
02 32 13	Subsurface Drilling and Sampling
07 10 00	**DAMPPROOFING AND WATERPROOFING**
07 11 00	Dampproofing
07 13 00	Sheet Waterproofing
07 14 00	Fluid-Applied Waterproofing
07 16 00	Cementitious and Reactive Waterproofing
07 17 00	Bentonite Waterproofing
31 10 00	**SITE CLEARING**
31 20 00	**EARTH MOVING**
31 22 00	Grading
31 23 00	Excavation and Fill
31 23 19	Dewatering
31 32 00	**EARTHWORK METHODS**
31 33 00	Rock Stabilization
31 34 00	Soil Reinforcement
31 40 00	**SHORING AND UNDERPINNING**
31 50 00	**EXCAVATION SUPPORT AND PROTECTION**
31 60 00	**SPECIAL FOUNDATIONS AND LOAD-BEARING ELEMENTS**
31 62 00	Driven Piles
31 63 00	Bored Piles
31 64 00	Caissons
31 66 13.13	Rammed Aggregate Piers
31 66 15	Helical Foundation Piles
32 30 00	**SITE IMPROVEMENTS**
32 32 00	Retaining walls

KEY TERMS

foundation
dead load
live load
rain load
snow load
wind load
seismic load
lateral soil pressure load
buoyant uplift
flood load
horizontal thrust

settlement
uniform settlement
differential settlement
earth material
consolidated rock
bedrock
soil
boulder
cobble
gravel
sand

coarse-grained soil
silt
clay
fine-grained soil
organic soil
shear strength
frictional soil, cohesionless soil
soil pore
soil liquefaction
soil fabric
cohesive soil

plastic soil
liquid limit
expansive soil
gradation
well graded soil
poorly graded soil
uniformly graded soil
gap graded soil
sorting
poorly sorted soil
well sorted soil
allowable foundation pressure
allowable soil pressure
consolidation
imported soil
native soil
general purpose fill
drainage fill
test pit
water table
test boring
penetration sampler
load test
sieve
plastic limit
geotechnical report
brownfield site
earthwork
grubbing and clearing
excavation
benched excavation
maximum allowable slope, angle of
    repose
excavation support
shoring
H-pile
soldier beam
lagging
sheet piling, sheeting
pneumatically applied concrete,
    shotcrete
soil mixing
ground improvement, earth
    improvement
slurry wall
clamshell bucket
slurry

tremie
crosslot bracing
waler
raker
tieback
rock anchor
soil nailing
dewatering
sump
well point
watertight barrier wall
soil freezing
superstructure
substructure
shallow foundation
deep foundation
spread footing
column footing
wall footing, strip footing
engineered fill
frost line
ice lens
slab on grade
crawlspace
basement
tie beam
combined footing
cantilever footing
shallow frost-protected foundation
mat foundation, raft foundation
floating foundation, compensated
    foundation
caisson, drilled pier
belling
auger drill
belling bucket
belled caisson
socketed caisson
pile
end bearing pile
friction pile
pile cap
driven to refusal
grade beam
pile hammer
piledriver lead
piledriver

timber pile
heaving
pipe pile
minipile, pin pile, micropile
helical pile, screw pile
precast concrete pile
sitecast concrete pile
pressure-injected footing
compaction grouted footing
rammed aggregate pier, stone column
base isolator
underpinning
up–down construction
drainage
drainage mat
filter fabric
perforated drain piping
dampproofing
cement plaster dampproofing, parge
    coating
asphalt dampproofing, bituminous
    dampproofing
waterproofing
liquid-applied membrane
    waterproofing
sheet membrane waterproofing
bentonite waterproofing
integral waterproofing
electro-osmotic waterproofing
blind-side waterproofing
waterstop
flood test
protection board
radon (gas)
passive radon control
active subslab depressurization
retaining wall
earth reinforcing
gabion
geotextile
earth reinforcement, soil reinforcement
filling
backfilling
soil lift
controlled low-strength material (CLSM)
mud slab
finish grading

REVIEW QUESTIONS

1. What is the nature of the most common type of foundation failure? What are its causes?

2. Explain the differences among sand, silt, and clay, both in their physical characteristics and their behavior in relation to building foundations.

3. Explain the difference between a cohesionless soil and a cohesive soil. Give one example of each soil type.

4. Explain the difference between well graded and poorly graded soil. How does their behavior differ?

5. What is a test soil boring and why is it performed?

6. What is excavation sheeting used for? List three different types of excavation sheeting.

7. Under what conditions would you use a watertight barrier instead of well points when digging below the water table?

8. In cold climates, how does the frost line affect the placement of shallow footings? What footing type is an exception to this general principle?

9. What conditions might lead to the choice of a mat foundation for a building?

10. If shallow foundations are substantially less costly than deep foundations, why do we use deep foundations?

11. What soil conditions favor piles over caissons? What type of pile is especially well suited to repair or improvement of existing foundations and why?

12. List and explain some cost thresholds frequently encountered in foundation design.

13. Explain the difference between waterproofing and dampproofing. When is one or the other an appropriate choice for protecting a foundation from moisture?

14. List two types of waterproofing and describe one possible advantage of each.

15. List the components of a typical foundation drainage system and their functions.

16. What is filling? Why is fill placed in lifts?

EXERCISES

1. Obtain the foundation drawings and soils report for a nearby building. Look first at the log of test borings. What sorts of soils were found beneath the site? How deep is the water table? What types of foundations do you think might be suitable for use in this situation? Now look at the foundation drawings. What type was actually used? Can you explain why?

2. What types of foundation and substructure are normally used for houses in your area? Why?

3. Look at several excavations for major buildings under construction. Note carefully the arrangements made for excavation support and dewatering. How is the soil being loosened and carried away? What is being done with the excavated soil? What type of foundation is being installed? What provisions are being made to keep the substructure permanently dry?

SELECTED REFERENCES

Ambrose, James E. *Simplified Design of Building Foundations* (2nd ed.). Hoboken, NJ, John Wiley & Sons, Inc., 1998.

Discusses soil properties and provides simplified computation procedures for both shallow and deep foundations.

Cheng, Liu, and Jack Evett. *Soils and Foundations* (7th ed.). Upper Saddle River, NJ, Prentice Hall, Inc., 2008.

Provides detailed discussion of the engineering properties of soils, subsurface exploration techniques, soil mechanics, and shallow and deep foundations.

Kubal, Michael. *Construction Waterproofing Handbook* (2nd ed). New York, McGraw-Hill, 2008.

Covers all aspects of building waterproofing, both above and below grade.

National Roofing Contractors Association. *The NRCA Waterproofing Manual.* Rosemont, IL, updated regularly.

Provides industry-standard guidelines and details for the application of waterproofing and dampproofing to building substructures.

WEB SITES

Foundations

Author's supplementary web site: **www.ianosbackfill.com/02_foundations/**

Whole Building Design Guide, Foundation Walls: **www.wbdg.org/design/env_bg_wall.php**

Earth Retention and Deep Foundations

Case Foundation Company: **www.casefoundation.com**

Geopier Foundation: **www.geopier.com**

Hayward Baker Geotechnical Construction: **www.haywardbaker.com**

Nicholson Construction Company: **www.nicholsonconstruction.com**

Schnabel Foundation Company: **www.schnabel.com**

Waterproofing

CETCO Building Materials Group: **buildingmaterials.cetco.com**

Grace Construction Products: **www.na.graceconstruction.com**

WOOD

- **Trees**
  Tree Growth
  Softwoods and Hardwoods
  Environmentally Certified Wood

  SUSTAINABILITY OF WOOD

- **Lumber**
  Sawing
  Seasoning
  Surfacing
  Lumber Defects
  Lumber Grading
  Structural Properties of Lumber
  Lumber Dimensions
  Veneer

- **Wood Products**
  Glue-Laminated Wood
  Cross-Laminated Timber
  Structural Composite Lumber
  Wood I-Joists
  Wood-Plastic Composites
  Finger-Jointed Wood Trim

  A NATURALLY GROWN STRUCTURAL
  MATERIAL

- **Plastic Lumber**

- **Wood Panel Products**
  Structural Panel Types
  Plywood Production

- Specifying Structural Panels
  Other Wood Panel Products
  Non-Wood Fiber Panels

- **Wood Chemical Treatments**
  Preservative Chemicals
  Naturally Decay-Resistant Woods
  Durable Construction Techniques

- **Wood Fasteners**
  Nails
  Screws
  Bolts
  Timber Connectors
  Toothed Plates
  Sheet Metal and Metal Plate Framing
     Devices
  Wood Adhesives

  WOOD PRODUCT ADHESIVES AND
  FORMALDEHYDE

- **Prefabricated Wood
  Components**
  Trusses
  Prefabricated Panels
  Factory-Built Housing

- **Types of Wood Construction**

FROM CONCEPT TO REALITY
French American School

A logger fells a large Douglas fir. *(Weyerhaeuser Company photo.)*

Wood is perhaps the best loved of all the materials that we use for building. It delights the eye with its endlessly varied colors and grain patterns. It invites the hand to feel its subtle warmth and varied textures. When it is fresh from the saw, its fragrance enchants. We treasure its natural, organic qualities and take pleasure in its genuineness. Even as it ages, bleached by the sun, eroded by rain, worn by the passage of feet and the rubbing of hands, we find beauty in its transformations of color and texture.

Wood earns our respect as well as our love. It is strong and stiff, yet by far the least dense of the materials used for the beams and columns of buildings. It is worked and fastened easily with small, simple, relatively inexpensive tools. It is readily recycled from demolished buildings for use in new ones, and when finally discarded, it biodegrades rapidly to become natural soil. It is our only renewable building material—one that will be available to us for as long as we manage our forests with an eye to the perpetual production of wood.

But wood, like a valued friend, has its idiosyncrasies. A piece of lumber is never perfectly straight or true, and its size and shape can change significantly with changes in the weather. Wood is peppered with defects that are relics of its growth and processing. Wood can split; wood can warp; wood can give splinters. If ignited, wood burns. If left in a damp location, it decays and harbors destructive insects. The skillful designer and the seasoned carpenter, however, know all these things and understand how to build with wood to bring out its best qualities, while neutralizing or minimizing its problems.

TREES

Wood comes from trees and is produced through natural growth processes. Understanding tree physiology is essential to knowing how to build with wood.

Tree Growth

The trunk of a tree is covered with a protective layer of dead *bark* (Figure 3.1). Inside the dead bark is a layer of living bark composed of hollow longitudinal cells that conduct nutrients downward from the leaves to the roots and other living parts of the tree. Inside this layer of living bark lies a very thin layer, the *cambium*, which creates new bark cells toward the outside of the trunk and new wood cells toward the inside. The thick layer of living wood cells inside the cambium is the *sapwood*. In this zone of the tree, nutrients are stored and sap is pumped upward from the roots to the leaves and distributed laterally in the trunk. At the inner edge of this zone, sapwood dies progressively and becomes *heartwood*. In many species of trees, heartwood is easily distinguished from sapwood by its darker color. Heartwood no longer participates in the life processes of the tree but continues to contribute to its structural strength. At the very center of the trunk, surrounded by heartwood, is the *pith* of the tree, a small zone of weak wood cells that were the first year's growth.

An examination of a small section of wood under a low-powered microscope shows that it consists primarily of tubular cells whose long axes are parallel to the long axis of the trunk. The cells are structured of tough *cellulose* and are bound together by a softer cementing substance called *lignin*. The direction of the long axes of the cells is referred to as the *grain* of the wood. Grain direction is important to the designer of wooden buildings because the appearance and physical properties of wood parallel to grain and perpendicular to grain are very different.

In temperate climates, the cambium begins to manufacture new sapwood cells in the spring, when the air is cool and groundwater is plentiful, conditions that favor rapid growth. Growth is slower during the heat of the summer, when water is scarce. *Springwood* or *earlywood* cells are therefore larger and less dense in substance than *summerwood* or *latewood* cells. Concentric bands of springwood and summerwood make up the annual growth rings in a trunk that can be counted to determine the age of a tree. The relative proportions of springwood and summerwood have a direct bearing on the structural properties of the wood a given tree will yield, because summerwood is stronger and stiffer than springwood. A tree grown under continuously moist, cool conditions grows faster than another tree of the same species grown under warmer, drier conditions, but its wood is not as dense or as strong.

Softwoods and Hardwoods

Softwoods come from coniferous trees and *hardwoods* from broad-leafed trees. Most softwoods are cone-bearing, with needle-like leaves

FIGURE 3.1

Summerwood rings are prominent and a few rays are faintly visible in this cross section of an evergreen tree. But the cambium, which lies just beneath the thick layer of bark, is too thin to be seen, and heartwood cannot be distinguished visually from sapwood in this species. *(Courtesy of Forest Products Laboratory, Forest Service, USDA.)*

that remain on the trees during the colder winter months. Most hardwoods drop their leaves seasonally. Also, as the names imply, softwoods are usually less dense and softer than hardwoods. But there are exceptions. The softwoods Douglas fir and longleaf pine, for example, are at least as dense as some hardwoods.

Nevertheless, the distinction between these two types of woods remains useful. Softwood trees have a relatively simple microstructure, consisting mainly of large longitudinal cells (*tracheids*) together with a small percentage of radial cells (*rays*). The tracheids provide long-distance transport of sap within the tree and account for most of the tree's structural strength. Rays provide for the storage and radial transfer of nutrients (Figure 3.2). Hardwood trees are more complex in structure, with a larger percentage of rays and two different types of longitudinal cells: small-diameter *fibers* and large-diameter *vessels* or *pores* (Figure 3.3).

When cut into lumber, softwoods generally have a relatively coarse and plain grain structure or *grain figure*, whereas many hardwoods show finer, more attractive patterns (Figure 3.4). Most of the lumber used today for the building structural frame comes from softwoods, which are comparatively plentiful and inexpensive. For furniture, cabinetry, interior paneling, flooring, and other fine woodwork, hardwoods (and some of the denser softwoods) are often chosen for their better stability, attractiveness, finishability, and resistance to wear. Some of the softwood and hardwood species most widely used in North America are listed in Figure 3.5, along with their principal uses. However, it should be borne in mind that thousands of species of wood are used in construction around the world and that available species vary considerably with geographic location. In

Cell structure of a softwood

FIGURE 3.2
Vertical cells (tracheids, labeled TR) dominate the structure of a softwood, seen here greatly enlarged. But rays (WR), which are cells that run radially from the center of the tree to the outside, are clearly in evidence. An annual ring (labeled AR) consists of a layer of smaller summerwood cells (SM) and layer of larger springwood cells (S). Simple pits (SP) allow sap to pass from ray cells to longitudinal cells and vice versa. Resin is stored in vertical and horizontal resin ducts (VRD and HRD), with the horizontal ducts centered in fusiform wood rays (FWR). Border pits (BP) allow the transfer of sap between longitudinal cells. The face of the sample labeled RR represents a radial cut through the tree, and TG represents a tangential cut. *(Courtesy of Forest Products Laboratory, Forest Service, USDA.)*

FIGURE 3.3
Rays (WR) constitute a large percentage of the mass of a hardwood, as seen in this sample, and are largely responsible for the beautiful grain figures associated with many species. The vertical cell structure is more complex than that of a softwood, with large pores (P) to transport the sap and smaller wood fibers (F) to give the tree structural strength. Pore cells in some hardwood species end with crossbars (SC), while those of other species are entirely open. Pits (K) pass sap from one cavity to another. *(Courtesy of Forest Products Laboratory, Forest Service, USDA.)*

Cell structure of a hardwood

(a) (b)

(c) (d)

FIGURE 3.4
**The grain figures of two softwood species (*top*) and two hardwoods (*bottom*)
demonstrate the difference in cellular structure between the classes of woods: (*a*) The
cells in sugar pine are so uniform that the grain structure is almost invisible except
for scattered resin ducts. (*b*) Vertical-grain Douglas fir shows very pronounced dark
bands of summerwood. (*c*) Red oak exhibits large open pores amid its fibers. (*d*) This
quartersliced mahogany veneer has a pronounced "ribbon" figure caused by varying
light reflections off its fibers.** (*Photos by Edward Allen.*)

North America, the major lumber-producing forests are in the western and eastern mountains of both the United States and Canada. Other regions, chiefly in the southeastern United States, also produce significant quantities.

Environmentally Certified Wood

Environmentally certified wood comes from forests that are managed according to the guidelines of one of several organizations that set standards for long-term ecological sustainability, resource conservation, and other considerations. The *Forest Stewardship Council (FSC)* certification program encompasses ecological, economic, and social principles. It ensures responsible forest management practices that

SOFTWOODS	HARDWOODS
Used for Framing, Structural Panels	*Used for Trim, Paneling, Cabinetwork, Furniture*
Alpine fir	American beech
Balsam fir	American chestnut
Black spruce	Aspen
Douglas fir	Basswood
Eastern hemlock	Benge*
Eastern spruce	Birch
Eastern white pine	Black ash
Englemann spruce	Black cherry
Idaho white pine	Black walnut
Jack pine	Butternut
Larch	Elm
Loblolly pine	Hard maple
Lodgepole pine	Hickory
Longleaf pine	Lauan*
Mountain hemlock	Magnolia
Northern pine	Mahogany*
Pitch pine	Pecan
Ponderosa pine	Red alder
Red pine	Red oak
Red spruce	Rosewood*
Shortleaf pine	Sassafras
Sitka spruce	Soft maple
Slash pine	Sweetgum
Sugar pine	Teak*
Sycamore	Tupelo
Tamarack	White oak
True fir	White poplar
Virginia pine	Yellow poplar
Western hemlock	
Western larch	
Western white pine	
White spruce	

FIGURE 3.5
Selected species or species groups of woods commonly used in building construction in North America, listed alphabetically according to end use.

prevent overharvesting, maintain biological diversity, conserve natural resources, and protect the environment. It encourages efficient, economically sound use of forest products and services. Importantly, it also addresses respect for governing laws and international treaties, long-term tenure and land ownership rights and responsibilities, rights of indigenous peoples, community relations, and workers' rights.

The FSC establishes certification criteria but does not itself certify forests or wood products. Rather, it accredits other independent certifying bodies that implement the certification process. FSC *Forest Management Certification* is applied to the forests where trees are harvested; and *Chain of Custody Certification* is applied to the manufacturers and distributors that process the wood after it leaves the forest, before it reaches

SOFTWOODS	HARDWOODS
Used for Trim, Paneling, Window and Door Frames	
	Used for Finish Flooring
Bald cypress	Hard maple
Douglas fir	American beech
Eastern red cedar	Pecan
Ponderosa pine	Red oak
Port-Orford cedar	Tanoak
Redwood	White oak
Sitka spruce	
Sugar pine	
True fir	
White pine	
Yellow cedar	
	Decay-Resistant, Used for Outdoor Structures
Used for Finish Flooring	
Douglas fir	Ipe*
Longleaf pine	
Decay-Resistant, Used for Shingles, Siding, Outdoor Structures	
Atlantic white cedar	
Bald cypress	
Incense cedar	
Northern white cedar	
Port-Orford cedar	
Redwood	
Western red cedar	
Yellow cedar	

*All species are domestic except benge (Western Africa), ipe (Central and South America), lauan (Asia), mahogany (Central and South America, Africa), rosewood (South America, Africa), and teak (Asia). Many more imported species, not listed, are also available.

the retailer or consumer. Specifying wood with Chain of Custody Certification ensures that the wood arriving on the construction site originated from a certified forest and was harvested and processed according to FSC criteria.

FSC Controlled Wood is a less stringent certification that assures that wood products were not illegally harvested or associated with other highly objectionable practices. Wood with the *FSC Mixed Sources* label may include Chain of Custody Certified wood, Controlled Wood, and recycled wood. Forests in more than 80 countries around the globe are FSC certified. FSC Chain of Custody Certified wood is eligible for points in the LEED rating system discussed in Chapter 1.

Other certification systems with a significant presence in the North American lumber market include the Sustainable Forestry Initiative, American Tree Farm System, Canadian Standards Association Sustainable Forests Management System, and Programme for the Endorsement of Forest Certification schemes.

SUSTAINABILITY OF WOOD

Wood and Agrifiber: Renewable Resources

Wood is the only major structural material that is renewable. In North America, tree growth each year greatly exceeds the volume of harvested trees.

In other countries, some forests are being depleted by poor management practices and slash-and-burn agriculture.

The increased use of recycled timbers, recovered wood fiber, agricultural fibers, and bamboo in the manufacture of structural and finish products helps to reduce the demand for newly harvested wood.

Agrifiber products are manufactured from agricultural waste materials that would otherwise be disposed of in landfills or by burning.

Energy Content

Much of the energy consumed in the production of solid lumber comes from the sun, while the trees grow in the forest, and from the combustion of wood and paper manufacturing waste byproducts in the mill. Wood production consumes significantly less energy from nonrenewable sources than steel or concrete.

Solid lumber has a cradle-to-gate embodied energy of roughly 1000 to 3000 BTU per pound (2.3 to 7.0 MJ/kg). An 8-foot-long 2×4 (2.4-m-long 38×89 mm) has an embodied energy of about 17,000 BTU (19 MJ).

Manufactured wood products have higher embodied energy content than solid lumber, due to the additional ingredients and energy required in their manufacture. The embodied energy of such products ranges from about 3000 to 7500 BTU per pound (7.0 to 17 MJ/kg).

Wood construction involves large numbers of steel fasteners and connectors of various kinds. Because steel is produced by relatively energy-intensive processes, fasteners add considerably to the total energy embodied in a wood-frame building.

Wood does not have the lowest embodied energy of the major structural materials when measured on a pound-for-pound basis. However, when buildings of comparable size, structured with either wood, steel, or concrete, are compared, those with wood have the lowest embodied energy.

Wood and Carbon

Wood is unique among structural materials in its participation in the Earth's natural carbon cycles. Around the globe, tens of billions of tons of carbon are stored in the trees, ground litter, and soils of forests. Hundreds of millions of tons of carbon are sequestered annually by forests. At the same time, deforestation, especially in tropical areas, is responsible for roughly 20 percent of human-caused carbon dioxide emissions.

On a pound-for-pound basis, carbon emissions from the manufacture of solid lumber are roughly 6 to 8 times lower than those from the manufacture of recycled steel or concrete.

While wood products remain in use, they are frequently carbon-negative. That is, more carbon is stored within their wood fiber than is emitted during their manufacture. For example, the U.S. housing stock is estimated to account for the sequestration of roughly 750 million (680 million metric) tons of carbon. Only when wood products are burned or fully decayed is this sequestered carbon released back into the atmosphere.

Forestry and Agricultural Practices

In a clearcut forest, all of the harvestable trees in an area are removed at one time, leaving the stumps, tops, and limbs to decay and become compost. New trees are planted and tended until they are ready for harvest, and then the cycle is repeated.

Environmental problems often associated with logging of forests include loss of wildlife habitat, soil erosion, pollution of waterways, and air pollution from machinery exhaust and burning of tree wastes.

In sustainable forestry, trees are harvested selectively from the forest in such a way as to minimize damage to the forest ecosystem.

The mountain pine beetle blight afflicting much of North America's pine forests in the first decade of the 21st century has left large stands of trees that, though dead, remain suitable as an economical source of sound, structural building material. Harvesting and building with this wood removes it from the forests where it otherwise

LUMBER

Sawing

The production of *lumber*—lengths of squared wood for use in construction—begins with the felling of trees and the transportation of the logs, or *roundwood*, to a sawmill (Figure 3.6). Sawmills range in size from tiny family operations to giant semiautomated factories. But the process of lumber production is much the same regardless of scale. Each log is stripped of its bark, then passed repeatedly through a large *headsaw*, which may be either a circular saw or a bandsaw, to reduce the log to untrimmed slabs of lumber. The *sawyer* judges (with the aid of a computer in the larger mills) how to obtain the maximum marketable wood from presents a fire hazard and prevents its decay and the return of its sequestered carbon to the atmosphere.

Mill Practices

Skilled sawyers working with modern computerized systems ensure the highest possible utilization of each log. In North America, over the past four decades, recovery rates of marketable wood products from logs have improved by as much as 25 percent.

In comparison to solid lumber, manufactured wood products efficiently utilize more of the wood fiber in a tree and can be produced from younger-growth forests.

Kiln drying uses large amounts of fuel but produces more stable, uniform lumber than air drying. Air drying requires only sunlight and wind, but requires more time.

Mill wastes are voluminous: Bark may be shredded and sold as a landscape mulch, composted, burned, or buried in a landfill. Sawdust, chips, and wood scraps may be burned to generate steam to power the mill, used as livestock bedding, composted, burned, or buried in a landfill.

Transportation

Because the major commercial forests are located in concentrated regions of the United States and Canada, much lumber is shipped considerable distances. Fuel consumption is lessened by planing and drying the lumber before it is shipped, which reduces both weight and volume.

Some wood products can be harvested or manufactured locally or regionally.

Construction Process

A significant fraction of the lumber delivered to a construction site is wasted: It is cut off when each piece is sawed to size and shape and ends up as scrap, which may be burned, taken to a landfill, or recycled. On-site cutting of lumber also generates considerable quantities of sawdust.

Construction site waste can be reduced by designing buildings that utilize full standard lengths of lumber and full sheets of wood panel materials and by planning lumber orders carefully to correspond to the requirements of the project.

Wood construction lends itself to various types of prefabrication that can reduce waste and improve the efficiency of material usage in comparison to on-site building methods.

Indoor Air Quality (IAQ)

Wood itself seldom causes air quality problems. Very few people are sensitive to the odor of wood.

Some adhesives and binders used in the manufacture of wood products and composites can cause serious IAQ problems by their emission of volatile organic compounds. The most notable of these is formaldehyde gas emissions by the urea-formaldehyde used in some binders and adhesives. Selecting products manufactured with alternative low-emitting or formaldehyde-free resins protects building occupants from this hazard.

Some paints, varnishes, stains, and lacquers for wood also emit fumes that are unpleasant and/or unhealthful.

In damp locations, molds and fungi may grow on wood members, creating unpleasant odors and releasing spores to which many people are sensitive or allergic.

Building Life Cycle

If the wood frame of a building is kept dry, it can last indefinitely. However, if the building is poorly maintained and wood elements frequently get wet, wood components may decay and require replacement.

Wood is combustible and gives off toxic gases when it burns. It is important to keep sources of ignition away from wood and to provide protection, such as fire-resistant coverings, chemical treatments, or fire-sprinkler systems appropriate to a building's size and use.

When a building is demolished, wood framing members can be recycled directly into the frame of another building, sawn into new boards or timbers, or shredded as raw material for oriented-strand materials. There is a growing industry whose business is purchasing and demolishing old barns, mills, and factories and selling their timbers as *reclaimed lumber*.

---

each log and uses hydraulic machinery to rotate and advance the log in order to achieve the required succession of cuts. As the slabs fall from the log at each pass, a conveyor belt carries them to smaller saws where they are reduced to square-edged pieces of the desired widths (Figure 3.7). The sawn pieces at this stage of production have rough-textured surfaces and may vary slightly in dimension from one end to the other.

As a log is sawn, the positioning of the saw relative to the log and the sequencing of cuts determines the grain orientation within the finished lumber pieces. For softwoods, lumber is described as either plainsawn or quartersawn. With *plainsawn* lumber, significant portions of the growth rings are oriented roughly

FIGURE 3.6
Loading logs onto a truck for their trip to the sawmill. *(Photo by Donald K. O'Brien.)*

FIGURE 3.7
Sawn lumber is sorted into stacks according to its cross-sectional dimensions and length. *(Courtesy: Western Wood Products Assn.)*

flat relative to the board's broader face. Plainsawing requires the least repositioning of the log during sawing and produces the maximum yield of useful pieces, making it the most economical sawing method. For this reason, most structural lumber is plainsawn. With *quartersawn* softwood lumber, the growth rings are consistently aligned at an angle of approximately 45 degrees or steeper relative to the board's broader face. Quartersawing requires more handling of the log during sawing, produces smaller pieces from a log of the same size, and generates greater waste. But it also results in boards that are more dimensionally stable and have more pleasing grain figure. Quartersawn flooring and exterior siding wear better, as there are no broad areas of soft springwood exposed on the face as there are in plainsawn lumber. In some wood species, quartersawn lumber also holds paint better than plainsawn. Quartersawn softwood lumber is used for flooring, trim, siding, and other types of finish woodwork where its advantages justify the added expense. Plainsawn and quartersawn lumber may also be referred to as *flat-grain* or *vertical-grain*, respectively, reflecting the patterning of the grain on the board's broader face produced by each of these sawing methods (Figures 3.8 and 3.9).

Hardwoods may be quartersawn, riftsawn, or plainsawn. A quartersawn hardwood board has growth rings aligned at an angle of approximately 60 degrees or greater to the board's broader face. In woods with strong ray structures, such as oak, this cut produces an often-sought-after appearance in which the rays appear prominently as contrasting elements, called *flecks* or *flakes*, within the grain figure. With *riftsawn* lumber, growth rings align at an angle of roughly 30 to 60 degrees to the board's broader face. This cut produces a tightly spaced and uniform grain figure. As with softwood lumber, these cuts are reserved for uses where the better appearance, durability, and stability of the finished boards justify the greater cost in comparison to plainsawn boards.

Seasoning

The amount of water present in wood is called its *moisture content (MC)* and is described as the weight of the water in the wood as a percentage of the weight of the dry wood. To determine moisture content, wood is first weighed in its wet state (W_{wet}). It is then placed in

PLAINSAWING   QUARTERSAWING   *TYPICAL SAWING OF A LARGE LOG*

FIGURE 3.8

Left: Plainsawing produces boards with a broad, irregular grain figure.
Middle: Quartersawing produces a more tightly spaced, regular grain structure.
Right: A large softwood log may be sawn to produce a variety of sizes and cuts of lumber, some large timbers, some plainsawn dimension lumber, and, in the horizontal row of small pieces seen just below the heavy timbers, some pieces of vertical-grain decking.

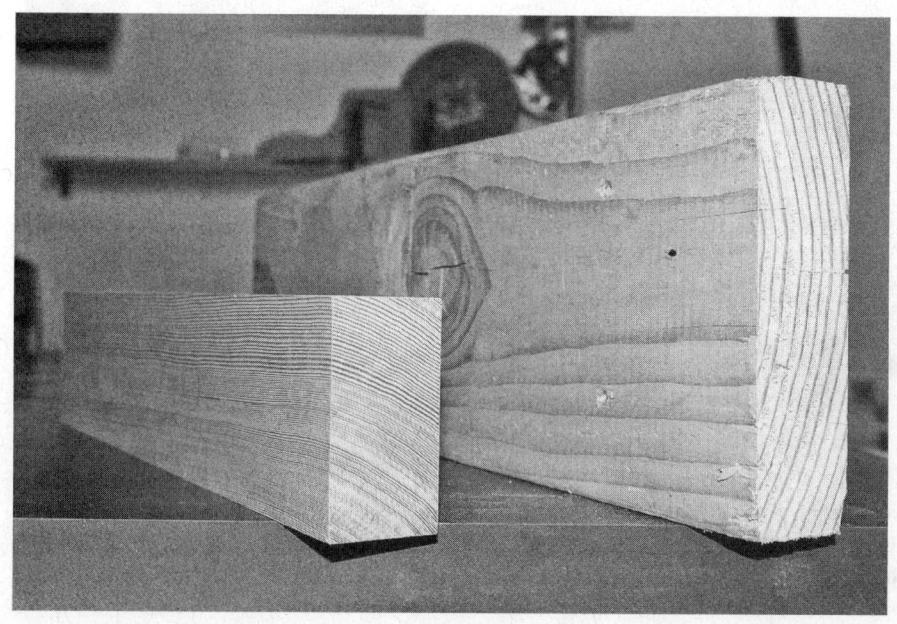

FIGURE 3.9

Left: Vertical-grain Douglas fir. Tightly spaced growth rings and quartersawing produce a distinctive, consistent grain pattern on the face of this piece of lumber. *Right:* Flat-sawn pine. More widely spaced growth rings and plainsawing produce a broader, more irregular grain figure. *(Photo by Joe Iano.)*

an oven, dried until all moisture in the wood has evaporated, and weighed again (W_{dry}). Moisture content can then be calculated as follows:

$$MC = \frac{W_{wet} - W_{dry}}{W_{dry}} \times 100$$

The moisture content of growing wood can vary from about 30 percent to 200 percent or more. After a tree is cut, water starts to evaporate from the wood. First to leave is the *free water*, water stored within the cell cavities. The evaporation of free water has little effect on the physical properties of the wood, other than to lighten it. When the free water is gone, only *bound water*, water held more tightly within the cellulose of the cell walls, remains. This moisture condition is termed the wood's *fiber saturation point*, and averages around 30 percent. Reductions in moisture content below the fiber saturation point have significant effects on the mechanical properties of the wood, causing it to shrink, and gain strength and stiffness. Eventually, as water continues

to evaporate from the wood, the wood arrives at a moisture content that brings it into equilibrium with its ambient surroundings—dryer, for wood used indoors or in dry outdoor climates, and more moist, for wood used outdoors and in higher humidity or wetter climates. This final moisture condition is called the *equilibrium moisture content (EMC)* of the wood.

In North America, the equilibrium moisture content for exterior wood averages around 12 percent, and for interior wood, 8 percent. These values are lower for dry climate areas, and higher for more humid areas. Ideally, wood would always be close to its anticipated final equilibrium moisture content when installed in the building. In this way, distortions or movement caused by subsequent changes in moisture would be minimized. However, drying wood takes time and energy. So, in practice, the extent of drying, or *seasoning*, varies with the application. Standard framing lumber is seasoned to a moisture content of 19 percent at the mill. For structural applications that require

better control of wood shrinkage, lumber seasoned to a moisture content of 15 percent, labeled "MC 15," is produced. Woods used for finish carpentry and fine woodwork are seasoned to within a few percentage points of their expected final equilibrium moisture contents prior to installation.

Most lumber is seasoned at the sawmill, as the lighter weight of dried lumber makes it less expensive to ship. Seasoning may be done by air-drying in loose stacks for a period of months or by drying within kilns under controlled conditions of temperature and humidity for a period of days (Figures 3.10 and 3.11). Though it requires more energy, *kiln drying* is generally preferred to *air drying* because it can be done faster, and it produces lumber with fewer distortions and more uniform quality. In areas where trees are harvested and milled close to their final point of use, lumber may delivered to the construction site *green* and allowed to season in place. Larger-dimension timbers, which are very slow to dry, are

FIGURE 3.10
For proper air drying, lumber is supported well off the ground. The *stickers*, which keep the boards separated for ventilation, are carefully placed above one another to avoid bending the lumber, and a watertight roof protects each stack from rain and snow. *(Courtesy of Forest Products Laboratory, Forest Service, USDA.)*

FIGURE 3.11
Measuring moisture content in boards in a drying kiln. *(Courtesy: Western Wood Products Assn.)*

also frequently allowed to complete their seasoning in place.

Wood does not shrink and swell uniformly with changes in moisture content. Moisture shrinkage along the length of the log (*longitudinal shrinkage*) is negligible for practical purposes. Shrinkage in the radial direction (*radial shrinkage*) is many times larger by comparison, and shrinkage around the circumference of the log (*tangential shrinkage*) is larger again by half or more (Figure 3.12). If an entire log is seasoned before sawing, it will shrink very little along its length, but it will grow noticeably smaller in diameter, and the difference between the tangential and radial shrinkage will cause it to check, that is, to split open along its length (Figure 3.13).

Nor do I ever come to a lumber yard with its citylike, graduated masses of fresh shingles, boards and timbers, without taking a deep breath of its fragrance, seeing the forest laid low in it by processes that cut and shaped it to the architect's scale of feet and inches. . . .

—**Frank Lloyd Wright, in *Architecture Record*, December, 1928**

When lumber is sawn from the log, its position in the log determines in large part how it will distort as it dries. Figure 3.14 shows how the differences between tangential and radial shrinkage cause this to happen. These effects are pronounced, and are readily predicted and observed in everyday practice.

The shrinkage behavior of woods cannot be ignored in building design. For example, in constructing building frames of plainsawn lumber, a simple distinction is made between parallel-to-grain shrinkage, which is negligible, and

FIGURE 3.12

Shrinkage of a typical softwood with decreasing moisture content. Longitudinal shrinkage, not shown on this graph, is so small by comparison to tangential and radial shrinkage that it is of no practical consequence in wood buildings. *(Courtesy of Forest Products Laboratory, Forest Service, USDA.)*

FIGURE 3.13

Because tangential shrinkage is so much greater than radial shrinkage, high internal stresses are created in a log as it dries, inevitably resulting in the formation of radial cracks called *checks*.

FIGURE 3.14
The difference between tangential and radial shrinkage also produces seasoning distortions in lumber. The nature of the distortion depends on the position the piece of lumber occupied in the tree. The distortions are most pronounced in plainsawn lumber (*upper right, extreme right, lower right*). *(Courtesy of Forest Products Laboratory, Forest Service, USDA.)*

perpendicular-to-grain shrinkage, which is considerable. The difference between radial and tangential shrinkage is not considered because the orientation of the annual rings in plainsawn lumber is random and unpredictable. As we will see in Chapter 5, wood building frames should be designed to equalize the amount of wood loaded perpendicular to grain from one side of the structure to the other, to avoid the tilting of floors or tearing of wall finish materials that can otherwise occur.

Surfacing

Lumber used in building construction is normally *surfaced* to make it smooth, more dimensionally precise, and safer to handle. Surfacing is done by high-speed automatic machines, called *planes*, whose rotating blades smooth the surfaces of the piece and round the edges slightly. Most lumber is surfaced on all four sides, *surfaced four sides (S4S)*, but hardwoods are often *surfaced two sides (S2S)*, leaving the two rougher edges to be finished by

the finish woodworker. Rough (*unsurfaced*) lumber is also available commercially. For example, planking for temporary scaffolding benefits from the rougher texture of unsurfaced lumber (making it less slippery when wet) and from the slightly greater overall dimensions of the plank (making it slightly stronger and stiffer).

Lumber is usually seasoned before it is surfaced, which allows the planing process to remove some of the distortions that occur during seasoning. For some framing lumber, though, this order of operations is reversed. The designation *S-DRY* in a lumber grade stamp indicates that the piece was surfaced when in a seasoned (dry) condition, and *S-GRN* indicates that it was surfaced when green.

Lumber Defects

Almost every piece of lumber contains discontinuities in its structure caused by *growth characteristics* of the tree from which it came or *manufacturing characteristics* that were created

FIGURE 3.15 (*opposite page*)
Surface features often observed in lumber include, in the left-hand column from top to bottom, a knot cut crosswise, a knot cut longitudinally, and a bark pocket. To the right are a gradestamp, wane on two edges of the same piece, and a small check. The gradestamp indicates that the piece was graded according to the rules of the American Forest Products Association, that it is #2 grade Spruce–Pine–Fir, and that it was surfaced after drying. The 27 is a code number for the mill that produced the lumber. (*Photos by Edward Allen.*)

at the mill (Figures 3.15 and 3.16). Among the most common growth characteristics are *knots*, which are places where branches joined the trunk of the tree; *knotholes*, which are holes left by loose knots dropping out of the wood; *decay*; and *insect damage*. Knots and knotholes reduce the strength of a piece of lumber, make it more difficult to cut and shape, and are often considered detrimental to its appearance. Decay and insect damage that occurred during the life of the tree may or may not affect the useful properties of the piece of lumber, depending on whether the organisms are still alive in the wood and the extent of the damage.

Manufacturing characteristics arise largely from changes that take place during the seasoning process because of the differences in rates of shrinkage with varying orientations to the grain. *Splits* and *checks* are usually caused by shrinkage stresses. *Crooking, bowing, twisting,* and *cupping* all occur because of nonuniform shrinkage. *Wane* is an irregular rounding of edges or faces that is caused by sawing pieces too close to the perimeter of the log. Experienced carpenters judge the extent of these defects and distortions in each piece of lumber and decide accordingly where and how to use the piece in the building. Checks are of little consequence in

framing lumber, but a joist or rafter with a crook in it is usually placed with the convex edge (the "crown") facing up to allow the floor or roof loads to straighten the piece. Badly bowed wall studs, floor joists, or roof rafters may be straightened by sawing or planing away the crown before being covered by wallboard, subflooring, or sheathing. Badly twisted pieces are put aside to be cut up for blocking. The effects of cupping in flooring and interior baseboards and trim are usually minimized by using well-seasoned, quartersawn stock and by shaping the pieces so as to reduce the likelihood of distortion (Figure 3.17).

FIGURE 3.17
The effects of seasoning distortions can often be minimized through knowledgeable detailing practices. As an example, this wood baseboard, seen in cross section, has been formed with a relieved back, a broad, shallow groove that allows the piece to lie flat against the wall even if it cups (broken lines). The sloping bottom on the baseboard ensures that it can be installed tightly against the floor despite the cup. The grain orientation in this piece is the worst possible with respect to cupping. If quartersawn lumber were not available, the next best choice would have been to mill the baseboard with the center of the tree toward the room rather than toward the wall.

FIGURE 3.16
Four types of seasoning distortions in dimension lumber.

CROOK

BOW

CUP

TWIST

Relieved back

Lumber Grading

Before each piece of lumber leaves the mill, it is graded either for appearance or structural properties, depending on its intended use. Grading offers the architect and the engineer the opportunity to build as economically as possible by using only as high a grade as is required for a particular use. For example, the main beams or columns of a building may require a stronger, higher structural grade of lumber, while the remainder of the framing members may perform adequately with an intermediate, less expensive grade. For blocking, the least expensive lowest grade is perfectly adequate. For interior trim that will be coated with a clear finish, a high appearance grade is desirable; for painted trim, a less expensive lower grade will suffice.

Structural grading is used to rate the strength and stiffness properties of a piece of lumber. It is done either visually or by machine. In *visual grading,* trained inspectors examine each piece for growth-ring density and for growth and manufacturing characteristics, judge it according to standard grading rules, and then stamp it accordingly (Figure 3.18). In *machine grading,* an automatic device assesses the structural properties of the wood and stamps a grade on the piece. This assessment is made either by flexing each piece between rollers and measuring its resistance to bending (*machine stress-rated*) or by scanning the wood electronically to determine its density.

Much structurally graded lumber is rated by *species group* or *species combination,* that is, by collections of individual wood species that are sufficiently similar in their properties that they may be used interchangeably. By allowing the intermingling of similar species within one group, harvesting, production, and distribution costs are reduced. For example, Hem-Fir, one of the highest-volume species groups marketed in North America, may include any combination

FIGURE 3.18

A grader marks the grade on a piece with a lumber crayon preparatory to applying a grade stamp. *(Courtesy: Western Wood Products Assn.)*

of California red fir, Grand fir, Noble fir, Pacific silver fir, Western hemlock, and white fir. Other commonly encountered structural grade species groups include Douglas Fir-Larch, Spruce-Pine-Fir, and Southern pine.

Appearance grading is used to rank the visual qualities of lumber intended for flooring, trim, cabinetry, and other finish, nonstructural uses. Naturally, it is done visually, with boards having the fewest knots, checks, splits, staining, and other defects receiving the highest grades.

Figure 3.19 outlines a typical grading scheme for framing lumber, and Figure 3.20 outlines example appearance grades for nonstructural softwood lumber. Light framing lumber for houses and other small buildings is usually ordered as "#2 and better" (a mixture of #1 and #2 grades) for floor joists and roof rafters and as "Stud" grade for wall framing. "Economy" grade, not shown in the table, is reserved for lumber not intended

FIGURE 3.19
Standard structural grades for western softwood lumber. For each species of wood, the allowable structural stresses for each of these grades are tabulated in the structural engineering literature, a few examples of which are shown in Figure 3.21. *(Courtesy: Western Wood Products Assn.)*

for structural purposes. For fine hardwoods used in architectural woodwork, even more stringent appearance criteria than those indicated in Figure 3.20 may be applied.

Structural Properties of Lumber

The strength of a piece of wood varies with the direction in which the load acts in relation to the grain of the wood, the duration of the stress, and the species and grade of the wood. For example, the compressive strength of a wood member with the load applied parallel to its grain may be as much as 3 or 4 times as great as its strength with the load applied across the grain. In any given orientation, wood is also normally stronger in compression than in tension, and stronger under short-term loading than under longer-lasting or permanent loads. Comparative strengths of selected wood species and grades are listed in Figure 3.21.

Figure 3.22 compares the average structural properties of framing lumber to those of some other common structural materials: brick masonry, steel, and concrete. Of the four materials, only wood and steel have useful tensile strength. Defect-free wood is comparable to steel on a strength-per-unit-weight basis. But with the ordinary run of defects, an average piece of lumber is somewhat inferior to steel by this yardstick.

When designing a wooden structure, the architect or engineer determines the maximum stresses expected to occur in the various structural members and selects the appropriate species groups and

DIMENSION LUMBER GRADES   Table 2.1

Product	Grades	WWPA Western Lumber Grading Rules Section Reference	Uses
Structural Light Framing (SLF) 2" to 4" thick 2" to 4" wide	SELECT STRUCTURAL NO.1 NO.2 NO.3	(42.10) (42.11) (42.12) (42.13)	Structural applications where highest design values are needed in light framing sizes.
Light Framing (LF) 2" to 4" thick 2" to 4" wide	CONSTRUCTION STANDARD UTILITY	(40.11) (40.12) (40.13)	Where high-strength values are not required, such as wall framing, plates, sills, cripples, blocking, etc.
Stud 2" to 4" thick 2" and wider	STUD	(41.13)	An optional all-purpose grade designed primarily for stud uses, including bearing walls.
Structural Joists and Planks (SJ&P)	SELECT STRUCTURAL NO.1 NO.2 NO.3	(62.10) (62.11) (62.12) (62.13)	Intended to fit engineering applications for lumber 5" and wider, such as joists, rafters, headers, beams, trusses, and general framing.

STRUCTURAL DECKING GRADES   Table 2.2

Product	Grades	WWPA Western Lumber Grading Rules Section Reference	Uses
Structural Decking 2" to 4" thick 4" to 12" wide	SELECTED DECKING	(55.11)	Used where the appearance of the best face is of primary importance.
	COMMERCIAL DECKING	(55.12)	Customarily used when appearance is not of primary importance.

TIMBER GRADES   Table 2.3

Product	Grades	WWPA Western Lumber Grading Rules Section Reference	End Uses
Beams and Stringers 5" and thicker, width more than 2" greater than thickness	DENSE SELECT STRUCTURAL* DENSE NO. 1* DENSE NO. 2* SELECT STRUCTURAL NO.1 NO.2	(53.00 & 170.00) (53.00 & 170.00) (53.00 & 170.00) (70.10) (70.11) (70.12)	Grades are designed for beam and stringer type uses when sizes larger than 4" nominal thickness are required.
Post and Timbers 5" x 5" and larger, width not more than 2" greater than thickness	DENSE SELECT STRUCTURAL* DENSE NO. 1* DENSE NO. 2* SELECT STRUCTURAL NO.1 NO.2	(53.00 & 170.00) (53.00 & 170.00) (53.00 & 170.00) (80.10) (80.11) (80.12)	Grades are designed for vertically loaded applications where sizes larger than 4" nominal thickness are required.

*Douglas Fir or Douglas Fir-Larch only.

	APPEARANCE LUMBER GRADES			Table 2.5
	Product	**Grades[1]**	**Equivalent Grades in Idaho White Pine**	**WWPA Grading Rules Section Number**
Highest Quality Appearance Grades	Selects *(all species)*	B & BTR SELECT C SELECT D SELECT	SUPREME CHOICE QUALITY	10.11 10.12 10.13
	Finish *(usually available only in Doug Fir and Hem-Fir)*	SUPERIOR PRIME E		10.51 10.52 10.53
	Special Western Red Cedar Pattern Grades	CLEAR HEART A GRADE B GRADE		20.11 20.12 20.13
General Purpose Grades	Common Boards (WWPA Rules) *(primarily in pines, spruces, and cedars)*	1 COMMON 2 COMMON 3 COMMON 4 COMMON 5 COMMON	COLONIAL STERLING STANDARD UTILITY INDUSTRIAL	30.11 30.12 30.13 30.14 30.15
	Alternate Boards (WCLIB Rules) *(primarily in Doug Fir and Hem-Fir)*	SELECT MERCHANTABLE CONSTRUCTION STANDARD UTILITY ECONOMY		**WCLIB[3]** 118-a 118-b 118-c 118-d 118-e
	Special Western Red Cedar Pattern[2] Grades	SELECT KNOTTY QUALITY KNOTTY		**WCLIB[3]** 111-e 111-f

[1]Refer to WWPA's *Vol. 2, Western Wood Species* book for full-color photography and to WWPA's *Natural Wood Siding* for complete information on siding grades, specification, and installation.

FIGURE 3.20

Nonstructural boards are graded according to appearance. *(Courtesy: Western Wood Products Assn.)*

Species Group	**Stud Grade**	**No. 2**	**Select Structural Grade**
Spruce-Pine-Fir (South)	625 psi (4.3 MPa)	1000 psi (6.9 MPa)	1200 psi (8.3 MPa)
Hem-Fir	800 psi (5.5 MPa)	1300 psi (9.0 MPa)	1500 psi (10.3 MPa)
Douglas Fir-Larch	850 psi (5.9 MPa)	1350 psi (9.3 MPa)	1700 psi (11.7 MPa)
Southern Pine	625 psi (4.3 MPa)	1100 psi (7.6 MPa)	2100 psi (14.5 MPa)

FIGURE 3.21

Comparative strengths of four common species groups and three structural grades. These figures are for allowable compressive stress (maximum design stress that includes factors of safety) parallel to the grain of the member.

grades from which to build. In a given locale, a limited number of species and grades are usually available in retail lumberyards, and it is from these that the selection is made. It is common practice to use a stronger but more expensive species group (Douglas Fir-Larch or a high grade of Southern Pine, for example) for highly stressed major members and to use a weaker, less expensive group (Hem-Fir, Spruce-Pine-Fir) for other parts of the structure. Within each species group, the designer selects grades based on published tables of allowable stresses. The higher the structural grade, the higher the allowable stress. But the lower the grade, the less costly the lumber.

Other factors that influence the useful strength of wood include, the

Material	Working Strength in Tension[1]	Working Strength in Compression[1]	Density	Modulus of Elasticity
Wood (framing lumber)	300–1000 psi (2.1–6.9 MPa)	600–2000 psi (4.1–14 MPa)	30 pcf (480 kg/m³)	1,000,000–1,900,000 psi (6900–13,000 MPa)
Brick masonry (including mortar, unreinforced)	0	250–1300 psi (1.7–9.0 MPa)	120 pcf (1900 kg/m³)	700,000–3,700,000 psi (4800–25,000 MPa)
Structural steel	24,000–43,000 psi (170–300 MPa)	24,000–43,000 psi (170–300 MPa)	490 pcf (7800 kg/m³)	29,000,000 psi (200,000 MPa)
Concrete (unreinforced)	0	1000–12,000 psi (6.9–83 MPa)	145 pcf (2300 kg/m³)	3,000,000–4,500,000 psi (21,000–31,000 MPa)

[1]Allowable stress or approximate maximum stress under normal working conditions.

FIGURE 3.22

Comparative physical properties of four common structural materials: wood (shaded row), brick masonry, steel, and concrete. Wood has significant strength in both tension and compression. The ranges of values of wood strength and stiffness reflect differences among species and grades of lumber.

temperature and moisture conditions under which it serves, and the size and shape of the piece. Chemical treatments applied to wood to increase its resistance to decay or fire may also reduce its strength slightly. All these factors are taken into account when engineering a building structure of wood.

Lumber Dimensions

Lumber sizes in the United States are given as *nominal dimensions* in inches, such as 1 × 2 ("one by two"), 2 × 10, and so on. At one time, sawn lumber may have approached these actual dimensions. Today, however, subsequent to sawing, seasoning, and surfacing, true sizes are less. By the time a kiln-dried 2 × 10 reaches the lumberyard, its *actual dimensions* are close to 1½ by 9¼ inches (38 by 235 mm). The relationship between nominal lumber dimensions (which are always written without inch marks) and actual dimensions (which are written with inch marks) is given in simplified form in Figure 3.23 and in more complete form in Figure 3.24. Anyone who designs or constructs wooden buildings soon commits the simpler of these relationships to memory. Because of changing moisture content and manufacturing tolerances, however, it is never wise to assume

that a piece of lumber will conform precisely to its expected dimensions. Wood members vary in size seasonally with changes in humidity and temperature. In hot, dry locations such as attics, wood framing may shrink to dimensions substantially below its original measurements. Members in older buildings may have been manufactured to full nominal dimensions or to earlier standards of actual dimensions such as 1⅝ inches or 1¾ inches (41 or 44 mm) for a nominal 2-inch member.

Pieces of lumber that are less than 2 inches in nominal thickness (38 mm actual thickness) are called *boards*. Pieces ranging from 2 to 4 inches in nominal thickness (38 to 89 mm actual thickness) are referred to as *dimension lumber*. Pieces nominally 5 inches (actual size 114 mm) and more in thickness are termed *timbers*.

Dimension lumber is usually supplied in 2-foot (610-mm) increments of length. The most commonly used lengths are 8, 10, 12, 14, and 16 feet (2.44, 3.05, 3.66, 4.27, and 4.88 m), but retailers frequently stock rafter material in lengths to 24 feet (7.32 m). Actual lengths are usually slightly longer than nominal lengths.

Lumber in the United States is priced by the *board foot*. Board foot measurement is based on nominal dimensions, not actual dimensions.

A board foot of lumber is defined as a solid volume 12 square inches in nominal cross-sectional area and 1 foot long. For example, a 1 × 12 or 2 × 6 10 feet long contains 10 board feet. A 2 × 4 10 feet long contains 6.67 board feet [(2 × 4)/12] × 10 = 6.77], and so on. Prices of dimension lumber and timbers in the United States are usually quoted on the basis of

Nominal Dimension	Actual Dimension
1 inch	¾" (19 mm)
⁵⁄₄ inch	1" (25 mm)
2 inch	1½" (38 mm)
3 inch	2½" (64 mm)
4 inch	3½" (89 mm)
5 inch	4½" (114 mm)
6 inch	5½" (140 mm)
8 inch	7¼" (184 mm)
10 inch	9¼" (235 mm)
12 inch	11¼" (286 mm)
Over 12 inch	¾" (19 mm) less

FIGURE 3.23

The relationship between nominal and actual dimensions for the most common sizes of kiln-dried lumber is given in this simplified chart, which is extracted from the complete chart in Figure 3.24.

STANDARD SIZES—FRAMING LUMBER
Nominal & Dressed (Based on *Western Lumber Grading Rules*)
Table 2.4

Product	Description	Nominal Size		Dressed Dimensions		
				Thicknesses & Widths (inches)		
		Thickness (inches)	Width (inches)	Surfaced Dry	Surfaced Unseasoned	Length (feet)
DIMENSION	S4S	2 3 4	2 3 4 5 6 8 10 12 over 12	1½ 2½ 3½ 4½ 5½ 7¼ 9¼ 11¼ off ¾	1⁹⁄₁₆ 2⁹⁄₁₆ 3⁵⁄₈ 4⁵⁄₈ 5½ 7¾ 9½ 11½ off ½	6' and longer, generally shipped in multiples of 2'
TIMBERS	Rough or S4S (shipped unseasoned)	5 and larger		**Thickness (unseasoned)**	**Width (unseasoned)**	6' and longer, generally shipped in multiples of 2'
				1/2 off nominal (S4S). See 3.20 of WWPA Grading Rules for Rough.		
DECKING	2" (Single T&G)	Thickness	Width	Thickness (dry)	Width (dry)	6' and longer, generally shipped in multiples of 2'
		2	5 6 8 10 12	1½	4 5 6¾ 8¾ 10¾	
	3" and 4" (Double T&G)	3 4	6	2½ 3½	5¼	

Abbreviations: FOHC—Free of Heart Center T&G—Tongued and grooved Rough Full Sawn—Unsurfaced lumber cut to full specified size
S4S—Surfaced four sides

FIGURE 3.24
A complete chart of nominal and actual dimensions for both framing lumber and finish lumber. *(Courtesy: Western Wood Products Assn.)*

dollars per thousand board feet. In other parts of the world, lumber is sold by the cubic meter. Because the board foot is based on lumber nominal dimensions and metric sizes use actual dimensions, there is no direct conversion between board feet and cubic meters.

Veneer

Wood produced in very thin sheets (about ⅛ inch or 3 mm in thickness or less) is called *veneer*. *Rotary-cut* or *rotary-sliced veneer* is produced by inserting a log into a large lathe and spinning the log against a long knife edge (Figures 3.25 and 3.34). As the knife moves steadily into the log, a continuous strip of veneer is peeled away, much as paper is unwound from a roll. Rotary-cut veneer is analogous to plainsawn lumber: It is the least expensive cut to produce, and finds its greatest use in structural applications where appearance is not a primary concern. The most common application for rotary-cut veneer is in the manufacture of structural plywood panels,

as explained in more detail later in this chapter.

Sliced veneer is produced by pressing a log against a knife, without rotating the log. Individual sheets, each only as wide as the log, are sliced off in succession. Or, veneer may be *sawn* in a similar manner. Veneer slicing and sawing are more expensive processes than rotary slicing, but they also produce veneers with grain patterns considered more attractive for fine woodwork. For this reason, hardwood veneers used in fine woodworking applications are

ROTARY SLICING

PLAIN SLICING

QUARTER SLICING

FIGURE 3.25
Veneers for structural panels are rotary-sliced, which is the most economical method. For better control of grain figure in veneers destined for finish woodwork, logs may be *plainsliced* or *quartersliced*. Though more expensive to produce, these veneers have more attractive grain figures than rotary-cut veneers.

commonly produced by one of these methods.

The log from which veneer is taken is called a *flitch*. In fine paneling, cabinetry, or other veneered applications, all the veneers may be required to come from a single flitch, to maintain uniformity of appearance among them. Veneers from a flitch can also be *sequenced*, that is, arranged in the finished work in the same order in which they came from the log, so that grain patterns on adjacent pieces match as closely as possible.

WOOD PRODUCTS

Much of the wood used in modern building construction is processed into manufactured products, rather than used directly as sawn lumber. Manufactured products can exceed the structural efficiency and quality of sawn lumber. They can overcome the size limitations of lumber extracted from the younger trees typical of modern forests. And they make economic use of raw wood materials that previously may have been treated as waste. As discussed in the following sections, sawn lumber may be laminated into structural elements with higher strength and greater spanning ability. Wood veneers are laminated to make structural panels that sheath walls, floors, and roofs more efficiently than traditional wood boards. Wood strands, particles, and cellulose fibers are used to make a variety of structural and nonstructural products.

Glue-Laminated Wood

Large wood structural members can be produced by joining many smaller strips of wood together with glue to form *glue-laminated wood* (called *glulam* for short). There are three major reasons to laminate: size, shape, and quality. Any desired size of structural member can be laminated, up to the capacities of the hoisting and transportation machinery needed to deliver and erect it, without having to search for a tree of sufficient girth and height. Wood can be laminated into shapes that cannot be obtained in nature: curves, angles, and varying cross sections (Figure 3.26). Quality can be specified and closely controlled in laminated members because defects can be cut out of the wood before laminating. Seasoning is carried out before the wood is laminated, largely eliminating the checks and distortions that characterize solid timbers, and the strongest, highest-quality wood can be placed in the parts of the member that will be subjected to the highest structural stresses. The fabrication of laminated members adds to the cost per board foot, but provides structural members that are smaller in size than solid timbers of equal load-carrying capacity. In many cases, solid timbers are simply not available in the required size, shape, or quality—at any price.

The manufacturing and structural performance of glue-laminated wood members are defined by the ANSI A190.1 and ASTM D3737 standards. Individual laminations are most commonly 1½ inches (38 mm) thick (2-inch nominal) except in curved members with small bending radii, where ¾-inch (19-mm) stock may be used. The laminations are seasoned to a lower-than-usual moisture content of 12 percent, to minimize shrinkage or checking in the finished glulam. End joints between individual pieces are either *finger jointed* or *scarf jointed*. These types of joints allow the glue to transmit tensile and compressive forces longitudinally from one piece to the next within a lamination (Figure 3.27). Adhesives are chosen according to the moisture conditions under which the member will serve. In residential construction, standard sizes range from 3⅛ to 6¾ inches (79 to 171 mm) in width and from 9 to 36 inches (229 to 914 mm) in depth. For larger buildings, standard sizes

FIGURE 3.26

Glue-laminating a U-shaped timber for a ship. Several smaller members have also been glued and clamped and are drying alongside the larger timber. *(Courtesy of Forest Products Laboratory, Forest Service, USDA.)*

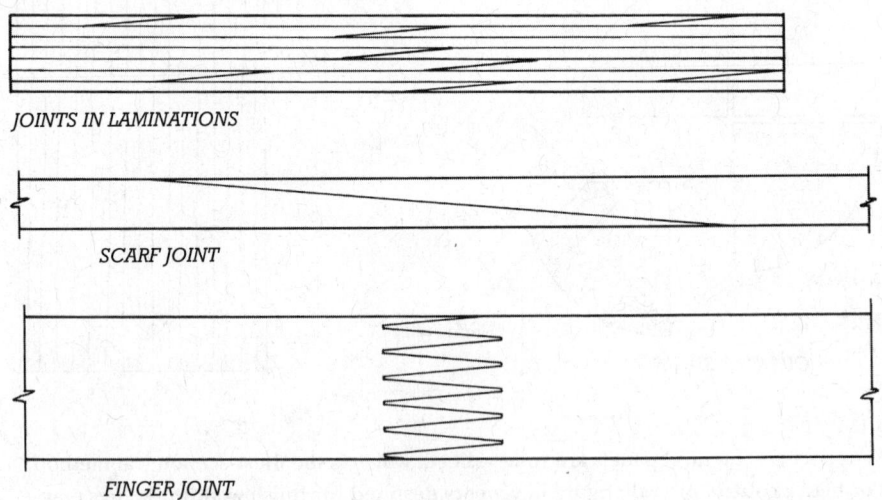

JOINTS IN LAMINATIONS

SCARF JOINT

FINGER JOINT

FIGURE 3.27

Joints within a lamination of a glue-laminated beam, seen in the upper drawing in a small-scale elevation view, must be scarf jointed or finger jointed to transmit the tensile and compressive forces from one piece of wood to the next. The individual pieces of wood are prepared for jointing by high-speed machines that mill the scarf or fingers with rotating cutters of the appropriate shape.

up to 14 inches (356 mm) wide and 75 inches (1905 mm) deep are not uncommon. Where glue-laminated beams will be exposed to the weather or to high levels of moisture in the completed construction, the laminations may be preservative-treated, as explained later in this chapter, to protect against decay.

"Your chessboard, sire, is inlaid with two woods: ebony and maple. The square on which your enlightened gaze is fixed was cut from the ring of a trunk that grew in a year of drought: you see how its fibers are arranged? Here a barely hinted knot can be made out: a bud tried to burgeon on a premature spring day, but the night's frost forced it to desist . . . . Here is a thicker pore: perhaps it was a larvum's nest." . . . **The quantity of things that could be read in a little piece of smooth and empty wood overwhelmed Kublai; Polo was already talking about ebony forests, about rafts laden with logs that come down the rivers, of docks, of women at the windows. . . .**
—Italo Calvino, *Invisible Cities*, 1978

Hybrid glulam beams substitute composite laminated veneer lumber (described later in this chapter) for the usually solid wood top and bottom laminations in the beam. Because the highest stresses in the beam occur at these edges, substituting a stronger material in just these locations results in a beam that is stronger and stiffer overall. In *FRP reinforced glulam* beams, structural capacity is increased by gluing a thin strip of high-strength fiber-reinforced plastic (FRP) between the first and second laminations nearest the edge of the beam that acts in tension (usually the lower edge). The fibers used—aramid, glass, carbon, or high-performance polyethylene—are much stiffer and stronger than wood. They are oriented longitudinally and embedded in a plastic matrix before being fabricated into the beam. The result is a savings of 25 to 40 percent in the volume of wood in comparison to a conventional glulam.

Cross-Laminated Timber

Cross-laminated timbers (CLTs) are structural panels laminated from solid lumber, with the orientation of members in each layer alternating from those above and below. The resulting panels are suitable for use as structural wall, floor, or roof components.

CLTs are assembled from dimension lumber $\frac{5}{8}$ to 2 inches (16 to 50 mm) in actual thickness and up to $9\frac{1}{2}$ inches (240 mm) in width. As with glulams, the lumber is seasoned to a 12 percent moisture content to minimize dimensional changes and checking in the finished product. Finger-jointed stock is used where longer continuous lengths are needed. Prior to assembly, the gluing faces of the lumber are planed to ensure good wood-to-adhesive bonding.

Panel assembly consists of stacking layers of lumber, with each layer oriented perpendicular to the previous one and adhesive applied in between. The full stack is compressed in a hydraulic or vacuum press and the adhesive allowed to cure. After assembly, panel faces are planed or sanded to produce a flatter surface. Finally, cutting of openings, preparing panel edges for joining, and other finish operations are completed, mostly using precision, computer-driven machinery. Fabrication tolerances range from $\frac{1}{16}$ in. (1.6 mm) in thickness to $\frac{1}{4}$ in. (6.4 mm) in length.

CLTs are usually made up of an odd number of layers, to create balanced structural behavior in the finished panel (Figure 3.28). Available sizes vary with the manufacturer. Common thicknesses range from $2\frac{1}{4}$ in. to 16 in. (57 to 400 mm). Widths range from 4 ft to 15 ft (1220 to 4500 mm), and maximum lengths from 40 ft to 60 ft (12 to 18 m).

As with glue-laminated timbers, the structural grades of lumber used in CLTs can be varied, with stronger grades used at the panel faces, where stresses are highest, and less expensive material used closer to the panel center. Structural composite lumber products (discussed later in this chapter) can be incorporated. Research is also underway to reduce panel costs by eliminating adhesives from the manufacturing process. Instead, lumber members may be mechanically interconnected by doweling or interlocking of dovetail and tongue-and-groove joints between members. In North America, CLT manufacturing and performance requirements are defined by the ANSI/APA PRG 320 standard.

CLT panels share many of the advantages of glue-laminated wood. Structural components can be produced in sizes and shapes not otherwise available, and the quality of panels can be carefully controlled. CLTs are lightweight and structurally efficient. The CLT fabrication process allows a high degree of prefabrication and rapid on-site assembly. Compared to concrete construction of equal capacity, CLT wall and floor panels can be as much as 4 to 6 times lighter in weight. Unlike glulams, the alternating orientation of lumber in the layers of the CLT panel creates two-way structural action. This permits relatively large openings to be made in a panel while maintaining the structural integrity of the panel as a whole. The high volume of solid wood used in a CLT structure brings with it the environmental benefits

FIGURE 3.28
CLT panels, both 5-layer and 3-layer, stacked on a truck bed. The alternating direction of cross-laminations is clearly visible. To keep the panels dry, they may be coated in wax, wrapped in plastic, or covered with tarps while being transported. Transportation restrictions should be considered when planning maximum panel widths. *(Photo courtesy Structurlam Inc.)*

LVL is similar in appearance to plywood except without crossbands (Figure 3.29). *Parallel strand lumber (PSL)* is made from long, thin strips of wood veneer glued and pressed in a process similar to that for LSL and OSL, but with the veneer strips arranged more uniformly parallel than the strands in those other products. PSL is the heaviest, strongest, and most expensive of the composite lumber products (Figure 3.30). LVL and PSL are most commonly used for longer-span headers and floor beams.

Studs and posts may also be made of any of these composite lumber types, as well as from *structural finger-jointed lumber* in which short lengths of solid lumber scrap are finger-jointed and glued end-to-end into longer lengths. Such manufactured components are well suited for use in the framing of walls that are especially tall or that have very large openings, or wherever else long, straight,

discussed earlier in this chapter, that is, utilization of a renewable resource, low embodied energy, and favorable carbon footprint.

Structural Composite Lumber

Structural composite lumber, also called *engineered lumber*, products are substitutes for solid lumber and are made from wood veneers or wood fiber strands and glue. *Laminated strand lumber (LSL)* and *oriented strand lumber (OSL)* are made from shredded wood strands, coated with adhesive, pressed into a rectangular cross section, and cured under heat and pressure (the wood strands used in the manufacture of LSL are longer than those used in OSL). LSL and OSL are the least strong and least expensive of the composite lumber products. They are used mainly for rim boards and short-span headers. *Laminated veneer lumber (LVL)* is made from thin wood veneer sheets, as wide as the member is deep, that are glued and laminated into thicker members.

FIGURE 3.29
An LVL beam, made of veneers similar to those used in the manufacture of plywood, resting on its side in preparation for installation on a concrete foundation wall. When placed in its final position, the beam will be rotated so that the laminations are oriented vertically in the beam. Also visible in the photograph are a preservative-treated wood sill (*left*) and an LSL rim board (*right*). *(Photo by Joseph Iano.)*

FIGURE 3.30
A view of corner framing for a residential garage. The lower beam is a PSL. A high-strength member such as this is needed to span the roughly 9-foot (2.7-m)-wide garage opening while supporting the floor load above. There are two solid wood top plates above the PSL. Above the top plates is an LSL rim board that encloses the floor framing behind it. This lower-strength member is adequate for this nonstructural function. Examples of wood products discussed later in this chapter are also in view: Above the rim board, the laminated edge of the plywood subfloor is visible, and to the left of the rim board, an I-joist is exposed where the floor framing projects beyond the plane of the wall below. *(Photo by Joseph Iano.)*

FIGURE 3.31
A bundle of I-joists delivered to the construction site. The webs are made from OSB and the flanges from solid lumber. *(Photo by Joseph Iano.)*

binders used in their manufacture and the potential for offgassing of volatile organic compounds such as formaldehyde gas.

Wood I-Joists

Manufactured wood I-shaped members, called *I-joists*, are used for framing of both roofs and floors (Figure 3.31). The top and bottom flanges of the members may be made from solid lumber, laminated veneer lumber, or laminated strand lumber. The thinner webs, which connect the top and bottom flanges, may be plywood or OSB (discussed later in this chapter). OSB is the most common choice because of its greater shear strength, an advantage in this application. These components use wood more efficiently than conventional

and especially strong, stiff members are needed.

Structural composite lumber products make productive use of wood materials that are fast-growing or that might otherwise be treated as waste, and they offer many of the same benefits as glue-laminated wood: dimensional stability, structural strength up to three times that of conventional solid material, availability of large sizes and long lengths, and consistent quality. Where these products are used in interior applications, however, attention should be paid to the types of adhesives and dimension lumber, and they can span farther between supports. They are also lighter in weight than corresponding solid members, lack crooks and bows, are more dimensionally stable, and are available in lengths up to 40 feet (12.2 m). Because I-joists span farther than conventional framing lumber, floors framed with these components are sometimes more prone to uncomfortable vibration when subjected to normal occupant loads. Extra stiffening elements or oversized joists may be used to counteract this tendency. Chapter 5 describes the use of I-joists in wood framing.

Wood-Plastic Composites

Wood-plastic composite (WPC) products are made from blends of plastic with wood or agricultural fibers, usually in roughly equal proportions. Smaller quantities of ultraviolet stabilizers, pigments, lubricants, and biocides may be added. The mixture is heated and pressed, extruded, or injection-molded into final form. In comparison to their solid wood counterparts, WPC materials offer more consistent material quality, freedom from defects and distortion, and (depending on their formulation) possibly superior resistance to

moisture. They are used most often for exterior decking, exterior railing systems, and finish trim, both interior and exterior. Like structural composites, WPCs make productive use of rapidly renewable or waste materials. Some WPCs also have high recycled-materials content.

WPC decking, most commonly made from blends of polyethylene or polypropylene and wood fiber, is available in sizes matching conventional solid wood decking in lengths up to 20 feet (6.1 m). It may be fastened with corrosion-resistant nails or screws, or with concealed hardware that engages the edges of the boards.

A NATURALLY GROWN STRUCTURAL MATERIAL

Wood is the only major structural material that is organic in origin. Trees grow naturally in the forest, and most of the work of manufacturing wood is done for us by the solar-powered biological processes within the tree itself and the forest ecosystem. This makes wood a renewable resource and relatively inexpensive to produce: Trees need only be harvested, cut to size, and dried before they are ready for use in construction. However, we have only limited control over the quality and properties of wood. Unlike steel, concrete, or masonry, we can do little to refine solid wood to suit our needs. Rather, we must accept its natural strengths and limitations.

The tree itself is a wooden structure, a tower, erected for the purpose of displaying leaves to the sun.[1] The tree is subject to much the same forces as the buildings we erect: It supports itself against the pull of gravity; it withstands the pressures of wind and the accumulation of snow and ice on its limbs; and it resists the natural stresses of the environment, such as changes in temperature and moisture, and attack by other organisms. Wood has evolved to perform these jobs well in the tree. In many cases, it performs them well in our buildings as well.

In nature, the greatest forces acting on a tree, such as the pressure of the wind, tend to act in bursts of relatively short duration. In response, wood has evolved to be strongest when resisting forces that act briefly rather than

those that act over longer periods of time. When we build with wood, we account for this unique behavior. In the engineering design of a wooden structure, the maximum loads that the structure may carry are increased as the expected duration of these loads decreases. This strength increase, called the *load duration factor*, ranges from 15 to 100 percent over the basic design strength of the material and is applied to wooden structures when considering the effects of wind, snow, earthquake, and impact—the same types of loads the tree has evolved to withstand most effectively.

The structural form of the tree is also well suited to its environment. Tree branches are supported at one end only—an arrangement called a *cantilever*. The joint supporting a cantilevered branch must be strong and stiff, efficiently utilizing the material's structural capabilities to the utmost. Yet despite the stiffness of the joint, the branch it supports can deflect relatively easily, as it is fastened at one end only. Under heavy loads, such as wet snow or strong winds, the branch can droop or deflect and shed much of the load. In this way, the tree's capacity to bend and sway helps it to survive forces that might otherwise break it.

When we build with wood, we cannot directly exploit the tree's efficient structure. The large deflections characteristic of this form are unsuitable for buildings, where they would cause discomfort to occupants and place undue stress on other building components. Nor can we easily exploit the strong joints that connect tree limbs to each other and the trunk, despite their benefits in stiffness and economy of material. In preparing wood for our

[1] Brayton F. Wilson and Robert R. Archer, "Tree Design: Some Biological Solutions to Mechanical Problems," *BioScience*, vol. 29, no. 5, p. 293, 1979.

A variety of maintenance-free colors and textured finishes are available, some remarkably similar in appearance to genuine woods.

Ingredients and processes used in the manufacture of *composite wood trim* vary widely, as do the workability, surface qualities, and durability of the finished products. Blends of plastic and wood similar to those used for the manufacture of composite decking may be used. Other formulations with a higher proportion of wood fiber—more similar in composition to traditional wood panel products, such as plywood, OSB, and fiberboard, discussed later in this chapter—may be used. Products may be prefinished, factory primed, encapsulated within a dense plastic outer shell, or coated with a resin-impregnated paper that improves the quality of field-applied finishes.

In comparison to solid lumber, nonstructural composite lumber expands and contracts more with changes in temperature, so greater allowance for thermal movement must be made during installation. In the case of spanning members such as decking, the lesser stiffness of composites necessitates closer spacing of the supporting members.

Finger-Jointed Wood Trim

Finger-jointed wood trim, made from short lengths of solid wood finger-jointed and glued into longer lengths, competes in the same market with WPC trim as an alternative to conventional nonstructural finish lumber. Finger-jointed stock is more stable and more consistently free of defects than solid lumber. It also makes productive use of short-length scraps that might otherwise be treated as waste. This material is typically finished with paint, rather than a clear or translucent coating, to hide the joints and variation in appearance among the short lengths from which it is made.

use, we cut the tree into pieces of convenient size and shape, removing the continuity between its parts. When we reassemble these pieces, we must devise new ways to join them. Despite the variety of methods available for connecting wood members, joints with strength and stiffness comparable to the tree's are mostly too difficult or expensive to be practical. Rather, we rely on simpler connection types that suffice when a beam is supported at both ends. (The structural continuity in long, glue-laminated wood members is one notable exception to this generalization.)

Wood in a tree performs many functions that wood in a building does not, because wood is involved with all aspects of the tree's life processes and growth. For both biological and structural reasons, wood has evolved a complex internal structure that is highly directional. The fact that wood is anisotropic—that is, it has different qualities in different directions—affects virtually all of its properties. The direction of the grain in a piece of wood influences its production, shaping, fastening, strength, durability, aging, finishing, and beauty. No other major structural material, perhaps with the exception of some specialized reinforced concrete systems, has such directional characteristics.

Consider two wood building systems, the log cabin and the heavy timber frame. Log construction is one of the simplest forms of wood construction. Logs found on or near the site are prepared for use with a minimum of labor and simple tools. Once stacked one upon another, the logs provide structural support, interior finish, exterior cladding, and insulation. Though a practical solution to providing shelter, log structures use wood inefficiently and result in a structure that is dimensionally unstable due to the large stacks of cross-grain material. Where more sophisticated tools and methods are available, heavy timber takes the place of log construction. In this system, wood members are assembled in ways that more closely resemble the tree form and that result in a more efficient building structure. A simple change in the orientation of the wood members produces a completely different building system: one that can be built higher, uses wood more efficiently, and is better suited to producing a durable, comfortable finished building.

Modern developments in the wood construction materials industry reflect efforts to overcome the limitations of wood in its natural state and to make the most efficient use of the material. Plywood panels are larger and more stable than those that can be produced from solid boards while minimizing the adverse effects of wood grain and natural defects. OSB, particleboard, and fiberboard make use of wood waste that might otherwise be discarded. Glue-laminated timbers, composite lumber, and cross-laminated timbers can be produced in sizes, shapes, and quality exceeding those available in solid timbers. Chemical treatments protect wood from fire and decay. Thus, the trend is toward using wood less in its natural state and more as a raw material in manufacturing processes to produce more refined and higher-quality construction components. Yet despite such advances, the tree itself remains an inspiration in its grace and strength, as well as in the lessons it offers for our understanding of building materials and methods.

PLASTIC LUMBER

Lumberlike products made entirely or mostly from plastic resins are called *plastic lumber*. When manufactured from recycled plastics, they may be called *recycled plastic lumber (RPL)*. The use of plastic lumber reduces the demand for harvested wood and, where recycled ingredients are used, diverts solid waste from landfills.

In the manufacturing of plastic lumber, one or more resins may be used alone, or mixed with rubber, wood waste, glass fiber, or other materials, and then molded or extruded into solid or hollow shapes mimicking those of conventional solid lumber. In its finished state, plastic lumber is resistant to sun, water, and insects; does not require protective coatings or finishes; and is nontoxic, durable, and maintenance free, making it an especially attractive alternative to preservative-treated lumber for exterior applications. However, like WPCs, plastic lumber is more flexible than solid wood and must be supported at more closely spaced intervals. In addition, plastic lumber expands and contracts more with changes in temperature and so must be installed with greater allowances for thermal movement. The most common material used in the manufacture of plastic lumber, high-density polyethylene (HDPE), is obtained from the recycling of postconsumer waste such as milk jugs and detergent bottles. Polystyrene and polyvinyl chloride are also used. Exterior decking, trim, fencing, and site furniture claim the largest share of the plastic lumber market in building construction.

Structural-grade plastic lumber (SGPL), most commonly made from HDPE reinforced with glass fibers, can be formulated to be at least as strong as conventional solid wood, though less stiff and more prone to long-term creep under permanent loads. SGPL planks, joists, beams, posts, piles, and other manufactured elements are used in the construction of decks, docks, piers, other types of exterior and marine structures, and even vehicular-capacity bridges.

WOOD PANEL PRODUCTS

Wood in panel form is useful for many building applications (Figure 3.32). The panel dimensions are usually 4 by 8 feet (1220 by 2440 mm). Panels require less labor for installation than individual boards because fewer pieces must be handled, and wood panel products are fabricated in such a way that they minimize many of the limitations of boards and dimension lumber: Panels are more nearly equal in strength in their two principal directions than solid wood. Shrinking, swelling, checking, and splitting are greatly reduced. Additionally, panel products make more efficient use of forest resources than solid wood products through less wasteful ways of reducing logs to building products and through utilization (in some types of panels) of material that would otherwise be thrown away—branches, undersized trees, and mill wastes. Many wood panel products are made largely from recovered or recycled wood waste, and panels made from rapidly renewable vegetable fibers are also available.

Structural Panel Types

Structural wood panels includes plywood, composite panels, and nonveneered panels (Figure 3.33). *Plywood* panels are made up of thin layers of wood veneer glued together. The grain on the front and back face veneers runs in the long direction of the sheet, whereas the grain in one or more interior crossbands runs perpendicular, in the shorter direction. There is always an odd number of layers in plywood, which equalizes the effects of moisture-related movement, but an interior layer may be made up of a single veneer or of two veneers with their grains running in the same direction. *Composite panels* have two parallel face veneers bonded to a core of reconstituted wood fibers. *Nonveneered panels* are made from various formulations of reconstituted wood fiber materials as follows:

- *Oriented strand board (OSB)* is made of long shreds (strands) of wood compressed and glued into three to five layers. The strands are oriented in the same manner in each layer as the grains of the veneer layers in plywood. Because of the length and controlled orientation of the strands, OSB is the strongest and stiffest type of nonveneered panel. Because it can be produced from small trees and even branches, OSB is generally more economical than plywood. It is the material most commonly used for sheathing and subflooring of light frame wood buildings.

- *Particleboard* is manufactured in different density ranges, and is made up of smaller wood particles than OSB that are compressed and bonded into panels. It is used mainly as a base material for wood veneer and plastic laminate. It is also used as an *underlayment panel* to create a very smooth base for the application of resilient flooring.

- *Fiberboard* is a very fine-grained board made of wood fibers and synthetic resin binders. It is generally limited to interior uses. The processing of the raw wood products in fiberboard manufacturing is more intensive than in the manufacture of particleboard, resulting in a panel that is dimensionally more stable, stiffer, better able to hold fasteners, and superior in its working and finishing characteristics. The most commonly used form of fiberboard is *medium-density fiberboard (MDF)*. MDF is used in the production of cabinets, furniture, moldings, paneling, and many other manufactured products.

Standard structural panels are 4 by 8 feet (1220 by 2440 mm) in width and length and range in thickness

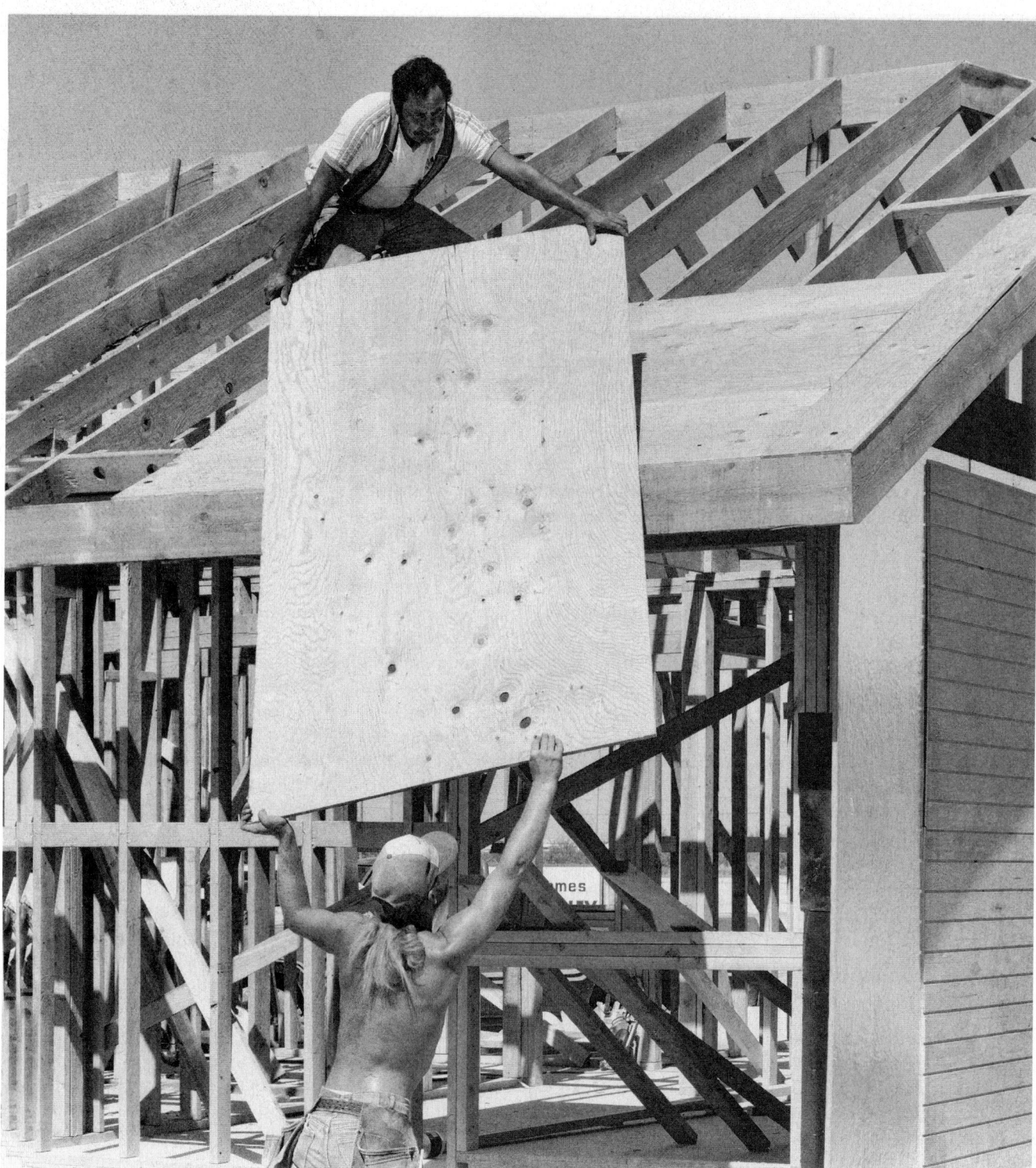

FIGURE 3.32
Plywood is made of veneers selected to give the optimum combination of economy
and performance for each application. This sheet of roof sheathing plywood is faced
with a D veneer on the underside and a C veneer on the top side. These veneers,
though unattractive, perform well structurally and are much less costly than the higher
grades incorporated into plywood made for uses where appearance is important.
(Courtesy of APA—The Engineered Wood Association.)

FIGURE 3.33
Three common wood panel products. From bottom to top: ¾-inch (19-mm) plywood, ½-inch (12-mm) OSB, and ¾-inch (19-mm) MDF. Particleboard, not shown, is visually distinguishable from fiberboard by its coarser, more varied particle composition.
(Photo by Joe Iano.)

from ¼ to 1⅛ inch (6.5 to 28.5 mm). Longer panels are manufactured for siding and industrial uses. The width and length of a structural panel is normally ⅛ inch (3 mm) less than nominal, permitting panels to be installed with gaps between them to allow for moisture expansion. Panels intended for use as subflooring can be manufactured with tongue-and-groove edges that eliminate unevenness in the subfloor that could telegraph through the finish flooring.

Plywood Production

Veneers for structural plywood and composite panels are rotary sliced (Figure 3.34). The strip of veneer is clipped into sheets that pass through a drying kiln where, in a few minutes, their moisture content is reduced to roughly 5 percent. The sheets are then assembled into larger pieces, repaired as necessary with patches glued into the sheet to fill open defects, and graded and sorted according to quality (Figure 3.35). A machine spreads glue on the veneers as they are laid atop one another in the required sequence and grain orientations. The glued veneers are transformed into plywood in presses that apply elevated temperatures and pressures to create dense, flat panels. The panels are trimmed to size, sanded as required, and graded and gradestamped before shipping. Veneers of Grade B and higher are always sanded smooth, but panels intended for sheathing are left unsanded because sanding slightly reduces their thickness, which diminishes the structural strength of the panel. Panels intended for subfloors and floor underlayment are lightly *touch sanded* to produce a flatter, smoother surface without significantly reducing their structural performance.

Where an especially smooth and durable surface is required, plywood may be finished with a resin-treated overlay on one or both sides to make a *medium-density overlay (MDO)* or *high-density overlay (HDO)* panel. Overlaid panels are used in the construction of concrete forms, cabinetwork, furniture, exterior siding, signage, and other applications where the greatest durability and high surface finish quality are required. (Though the names are similar, care should be taken not to confuse MDO plywood with the MDF nonveneered panels discussed previously.)

Composite panels and nonveneered panels are manufactured by analogous processes to the same set of standard sizes as plywood and to some larger sizes as well. Panels of all kinds used for sheathing and subflooring can also be manufactured with water-resistant coatings and edge sealers, and in the case of OSB, special resins, to provide better resistance to extended periods of wetting during construction.

Specifying Structural Panels

For structural uses such as subflooring and sheathing, wood panels may be specified either by thickness or by *span rating*. The span rating is determined by laboratory load testing and is given on the gradestamp on the back of the panel, as shown in Figures 3.36 and 3.37. The purpose of the span rating system is to permit the use of many different species of woods and types of panels while achieving the same structural objectives. Under normal loading conditions, any panel with a span rating of 32/16 may be used as roof sheathing over rafters spaced 32 inches (813 mm) apart or as subflooring over joists spaced 16 inches (406 mm) apart. The long dimension of the sheet must be placed perpendicular to the length of the supporting members. A 32/16 panel may be plywood, composite, or OSB; may be composed of any accepted wood species; and may be any of several thicknesses, so long as it passes the structural requirements for a 32/16 rating.

When specifying panels by thickness, industry standards refer to nominal thickness, called *performance category*. Like the nominal dimensions used to describe solid lumber size, performance category is a conveniently expressed value, around which actual panel thickness may vary. Depending on the overall thickness of a structural panel, its actual thickness may deviate by $\frac{1}{32}$ in. to $\frac{2}{32}$ in. (0.8 mm to 1.6 mm) from the performance category nominal thickness.

The designer must also select from two *bond classifications* for structural wood panels, either Exterior or Exposure 1. Panels marked "Exterior" are made with waterproof glues and the highest-quality veneers. They are the most durable panel type, suitable for use as exterior siding or in other applications permanently exposed to the weather. Exposure 1 panels are also made with waterproof glue but a broader range of quality of veneers. They are commonly used for structural sheathing and subflooring, which will be protected from the weather once the building is finished, but must often endure long periods of wetting during construction. Exposure 1 panels are also used for interior applications. About 95 percent of structural panel products are classified as Exposure 1. (A past classification, Exposure 2, intended for panels fully protected from weather, is no longer used.) The complete performance specifications for structural wood panels are contained in the APA Voluntary Product Standard PS 2.

Other Wood Panel Products

Hardboard is a thin, dense panel made of highly compressed wood fibers. It is available in various thicknesses and surface finishes, and in some formulations is durable against weather exposure. Hardboard is produced in configurations for residential siding as well as in general-purpose panels of standard dimension.

Insulating fiberboard sheathing is a low-density panel, usually ½ or ¾ inch (13 or 19 mm) thick, made of wood (or vegetable) fibers and binders, and coated with asphalt for water resistance. It has some thermal insulating value and is used in wood construction chiefly as nonstructural wall sheathing, although some panels of this type are strong enough to also contribute modest lateral force resistance to a building frame. Other *cellulosic fiberboard* panels are made from finely processed recycled paper waste and are used for wall sheathing, acoustical isolation, carpet underlayment, and even structural roof decking. Panels of these types are low in cost, make productive use of waste or recycled materials, and help to conserve forest resources.

Hardwood plywood panels, made from birch, maple, poplar, or alder veneers, are popular for use in cabinetry and other finish carpentry. For higher-quality architectural woodwork, these panels may be faced with fine, hardwood face veneers.

Non-Wood Fiber Panels

Agrifiber or *biocomposite panels* are made from agricultural waste products, such as the stalks or chaff of wheat, rice, hemp, sorghum, and other crop residues that are otherwise usually disposed of as solid waste or by open-air burning.

Structural *strawboard panels* are made from wheat or rice straw. The straw is subjected to high pressure and temperature, causing the release of natural resins that bind the material into a solid mass roughly half as dense as wood. The compressed core is faced both sides with OSB or other material to create a finished panel 2½ to 8 inches (64 to 200 mm) thick. These panels have significant structural capacity and can be used in place of conventional wood stud framing.

I shall always remember how as a child I played on the wooden floor. The wide boards were warm and friendly, and in their texture I discovered a rich and enchanting world of veins and knots. I also remember the comfort and security experienced when falling asleep next to the round logs of an old timber wall; a wall which was not just a plain surface but had a plastic presence like everything alive. Thus sight, touch, and even smell were satisfied, which is as it should be when a child meets the world.

—Christian Norberg-Schulz, *Wooden Houses*, 1979

Nonstructural panels made from a variety of compressed agrifiber and binder mixtures are manufactured in thicknesses ranging from about ¼ in. to 1⅛ inches (6.5 to 30 mm). They can be used in place of MDF, particleboard, and interior plywood, for cabinetry and other interior woodworking applications.

Bamboo, a rapidly growing woody grass, is used in the manufacture

(a)

(b)

(c)

(d)

FIGURE 3.34
Plywood manufacture. (*a, b*) A 250-horsepower lathe spins a softwood log as a knife peels off a continuous sheet of veneer for plywood manufacture. (*c*) An automatic clipper removes unusable areas of veneer and trims the rest into sheets of the proper size for plywood panels. (*d*) The clipped sheets are fed into a continuous forced-air dryer, along whose 150-foot (45-m) path they will lose about half their weight in moisture. (*e*) Leaving the dryer, the sheets have a moisture content of about 5 percent. They are graded and sorted at this point in the process. (*f*) The higher grades of veneer are patched on this machine, which punches out defects and replaces them with tightly fitted wood plugs. (*g*) In the layup line, automatic machinery applies glue to one side of each sheet of veneer and alternates the grain direction of the sheets to produce loose plywood panels. (*h*) After layup, the loose panels are prepressed with a force of 300 tons per panel to consolidate them for easier handling. (*i*) Following prepressing, panels are squeezed individually between platens heated to 300°F (150°C)

to cure the glue. (*j*) After trimming, sanding, or grooving as specified for each batch, the finished plywood panels are sorted into bins, ready for shipment. (*Photos b and i courtesy of Georgia-Pacific; others courtesy of APA–The Engineered Wood Association.*)

(*e*)

(*f*)

(*g*)

(*h*)

(*i*)

(*j*)

TABLE 1

VENEER GRADES

A Smooth, paintable. Not more than 18 neatly made repairs, boat, sled, or router type, and parallel to grain, permitted. Wood or synthetic repairs permitted. May be used for natural finish in less demanding applications.

B Solid surface. Shims, sled or router repairs, and tight knots to 1 inch across grain permitted. Wood or synthetic repairs permitted. Some minor splits permitted.

C Plugged — Improved C veneer with splits limited to 1/8-inch width and knotholes or other open defects limited to 1/4 × 1/2 inch. Wood or synthetic repairs permitted. Admits some broken grain.

C Tight knots to 1-1/2 inch. Knothole to 1 inch across grain and some to 1-1/2 inch if total width of knots and knotholes is within specified limits. Synthetic or wood repairs. Discoloration and sanding defects that do not impair strength permitted. Limited splits allowed. Stitching permitted.

D Knots and knotholes to 2-1/2-inch width across grain and 1/2 inch larger within specified limits. Limited splits are permitted. Stitching permitted. Limited to Exposure 1 or Interior panels.

FIGURE 3.35
Veneer grades for softwood plywood. Plywood with A grade face veneer is costly and limited in availability. It is used in applications demanding the highest-quality appearance (for example, cabinet construction). Plywood with B face veneer is used for concrete formwork construction and for less demanding appearance applications. Plywood with C-plugged face veneer is used for underlayment and combination subflooring/underlayment, where a uniform, smooth substrate is required to avoid telegraphing irregularities through thin finish flooring materials. Plywood for sheathing and subflooring is most commonly specified as "CDX," consisting of a C grade face, a D grade face, and constructed with exterior glue (the panel is installed with the C face oriented toward the weather). C grade veneer is also the lowest grade permitted in any ply of an Exterior rated panel, that is, a siding panel intended for permanent exposure to the weather. *(Courtesy of APA–The Engineered Wood Association.)*

1 Panel grade
2 Span Rating
3 Tongue-and-groove
4 Exposure durability classification
5 Product Standard
6 Thickness
7 Mill number
8 APA's Performance Rated Panel Standard
9 Siding face grade
10 Species group number
11 HUD/FHA recognition
12 Panel grade, Canadian standard
13 Panel mark – Rating and end-use designation, Canadian standard.
14 Canadian performance-rated panel standard
15 Panel face orientation indicator

FIGURE 3.36
Typical gradestamps for structural wood panels. Gradestamps are found on the back of each panel. *(Courtesy of APA–The Engineered Wood Association.)*

TABLE 2

GUIDE TO APA PERFORMANCE RATED PANELS[a][b]
FOR APPLICATION RECOMMENDATIONS, SEE FOLLOWING PAGES.

APA RATED SHEATHING Typical Trademark	*(APA trademark stamps:)* **APA** THE ENGINEERED WOOD ASSOCIATION RATED SHEATHING 40/20 19/32 INCH SIZED FOR SPACING EXPOSURE 1 000 PS 2-92 SHEATHING PRP-108 HUD-UM-40C **APA** THE ENGINEERED WOOD ASSOCIATION RATED SHEATHING 24/16 7/16 INCH SIZED FOR SPACING EXPOSURE 1 000 PRP-108 HUD-UM-40C	Specially designed for subflooring and wall and roof sheathing. Also good for a broad range of other construction and industrial applications. Can be manufactured as plywood, as a composite, or as OSB. EXPOSURE DURABILITY CLASSIFICATIONS: Exterior, Exposure 1, Exposure 2. COMMON THICKNESSES: 5/16, 3/8, 7/16, 15/32, 1/2, 19/32, 5/8, 23/32, 3/4.
APA STRUCTURAL 1 RATED SHEATHING[c] Typical Trademark	*(APA trademark stamps:)* **APA** THE ENGINEERED WOOD ASSOCIATION RATED SHEATHING STRUCTURAL I 32/16 15/32 INCH SIZED FOR SPACING EXPOSURE 1 000 PS 1-95 C-D PRP-108 **APA** THE ENGINEERED WOOD ASSOCIATION RATED SHEATHING 32/16 15/32 INCH SIZED FOR SPACING EXPOSURE 1 000 STRUCTURAL I RATED DIAPHRAGMS-SHEAR WALLS PANELIZED ROOFS PRP-108 HUD-UM-40C	Unsanded grade for use where shear and cross-panel strength properties are of maximum importance, such as panelized roofs and diaphragms. Can be manufactured as plywood, as a composite, or as OSB. EXPOSURE DURABILITY CLASSIFICATIONS: Exterior, Exposure 1. COMMON THICKNESSES: 5/16, 3/8, 7/16, 15/32, 1/2, 19/32, 5/8, 23/32, 3/4.
APA RATED STURD-I-FLOOR Typical Trademark	*(APA trademark stamps:)* **APA** THE ENGINEERED WOOD ASSOCIATION RATED STURD-I-FLOOR 24 OC 23/32 INCH SIZED FOR SPACING T&G NET WIDTH 47-1/2 EXPOSURE 1 000 PS 2-92 SINGLE FLOOR PRP-108 HUD-UM-40C **APA** THE ENGINEERED WOOD ASSOCIATION RATED STURD-I-FLOOR 20 OC 19/32 INCH SIZED FOR SPACING T&G NET WIDTH 47-1/2 EXPOSURE 1 000 PRP-108 HUD-UM-40C	Specially designed as combination subfloor-underlayment. Provides smooth surface for application of carpet and pad and possesses high concentrated and impact load resistance. Can be manufactured as plywood, as a composite, or as OSB. Available square edge or tongue-and-groove. EXPOSURE DURABILITY CLASSIFICATIONS: Exterior, Exposure 1, Exposure 2. COMMON THICKNESSES: 19/32, 5/8, 23/32, 3/4, 1, 1-1/8.
APA RATED SIDING Typical Trademark	*(APA trademark stamps:)* **APA** THE ENGINEERED WOOD ASSOCIATION RATED SIDING 24 OC 19/32 INCH SIZED FOR SPACING EXPOSURE 1 000 PRP-108 HUD-UM-40C **APA** THE ENGINEERED WOOD ASSOCIATION RATED SIDING 303-18-S/W 16 OC 11/32 INCH GROUP 1 SIZED FOR SPACING EXTERIOR 000 PS 1-95 PRP-108 FHA-UM-64	For exterior siding, fencing, etc. Can be manufactured as plywood, as a composite or as an overlaid OSB. Both panel and lap siding available. Special surface treatment such as V-groove, channel groove, deep groove (such as APA Texture 1-11), brushed, rough sawn and overlaid (MDO) with smooth- or texture-embossed face. Span Rating (stud spacing for siding qualified for APA Sturd-I-Wall applications) and face grade classification (for veneer-faced siding) indicated in trademark. EXPOSURE DURABILITY CLASSIFICATION: Exterior. COMMON THICKNESSES: 11/32, 3/8, 7/16, 15/32, 1/2, 19/32, 5/8.

(a) Specific grades, thicknesses and exposure durability classifications may be in limited supply in some areas. Check with your supplier before specifying.

(b) Specify Performance Rated Panels by thickness and Span Rating. Span Ratings are based on panel strength and stiffness. Since these properties are a function of panel composition and configuration as well as thickness, the same Span Rating may appear on panels of different thickness. Conversely, panels of the same thickness may be marked with different Span Ratings.

(c) All plies in Structural 1 plywood panels are special improved grades and panels marked PS 1 are limited to Group 1 species. Other panels marked Structural 1 Rated qualify through special performance testing. Structural II plywood panels are also provided for, but rarely manufactured. Application recommendations for Structural II plywood are identical to those for APA RATED SHEATHING plywood.

FIGURE 3.37

A guide to specifying structural panels. Plywood panels for use in wall paneling, furniture, and other applications where appearance is important are graded by the visual quality of their face veneers rather than their structural properties. *(Courtesy of APA–The Engineered Wood Association.)*

of panels and planks that can serve as alternatives to interior wood panel products and hardwood flooring.

Agrifiber and other bio-based alternatives to wood products are eligible for points under the LEED rating system when such products meet the Sustainable Agriculture Network's *Sustainable Agriculture Standard.* Similar in intent to Forest Stewardship Council certification for wood products, this standard promotes environmentally and socially sustainable agricultural practices.

WOOD CHEMICAL TREATMENTS

Chemical treatments are used to counteract two major weaknesses of wood: its combustibility and its susceptibility to attack by decay and insects. *Fire-retardant treatment (FRT)* is accomplished by placing lumber in a vessel and impregnating it under pressure with certain chemical salts that greatly reduce its combustibility. Fire-retardant-treated wood is expensive. Its major uses are roof sheathing in attached houses and framing for nonstructural partitions and other interior components in buildings of fire-resistant construction.

Preservative-treated wood is used where decay or insect resistance is required, such as with wood that is used in or near the ground; that is exposed to moisture in outdoor structures such as fences, decks, and marine docks; or that is used in areas of high termite risk. Preservative-treated wood is frequently referred to as *pressure-treated wood,* though this latter term can more accurately refer to either fire-retardant or preservative treatments, as both are typically applied using pressure impregnation processes.

Preservative Chemicals

Creosote is an oily derivative of coal that is widely used to treat wood in engineering structures. But the odor, toxicity, and unpaintability of creosote-treated wood make it unsuitable for most purposes in building construction. *Pentachlorophenol* is also impregnated as an oil solution, and as with other oily preservatives, wood treated with it cannot be painted.

The most widely used wood preservatives in building construction are waterborne salts, which permit subsequent painting or staining. For many decades, the most common of these was *chromated copper arsenate (CCA)*. However, due to concerns over its toxicity, CCA-treated lumber has been phased out of most residential and much commercial building construction in favor of lumber treated with salts free from arsenic.

Many of the preservatives on the market today rely primarily on high concentrations of copper solutions for their preservative properties. Examples include *alkaline copper quat (ACQ)* and *copper boron azole (CBA and CA)*. Though arsenic-free, these treatments are still considered potentially hazardous and must be handled with appropriate precautions. The heavy concentrations of copper used in these preservatives also make the wood treated with them corrosive to many metals. Plain steel or aluminum should not be used in contact with such wood. Rather, nails, screws, bolts, joist hangers, hold-downs, column bases, and other metal components must be stainless steel, heavily galvanized (zinc-coated) steel, or copper alloy that are not adversely affected by the copper content.

Micronized copper preservatives also contain high concentrations of copper, but in the form of very small solid particles, rather than in solution. These preservatives offer comparable levels of decay resistance but are somewhat less corrosive to metals than ACQ, CBA, and CA. More recently introduced *carbon-based PTI* (propiconazole, tebuconazole, imidacloprid) preservatives are completely free of copper and other metals, and are expected to be even less corrosive to metal fasteners, flashings, and hardware.

Borate compounds, such as *sodium borate (SBX)* and *disodium octaborate tetrahydrate (DOT)*, are considered nontoxic to humans. Their use is limited to above-ground applications that are protected from the weather. For example, in high-termite-risk areas, all the structural lumber in a building may be borate treated to protect against termite infestation.

For long-lasting protection, most preservative processes rely on *pressure impregnation,* which drives the preservative chemicals deeply into the fibers of the wood. To improve absorption, some wood species are punctured with an array of cuts in the wood's surface, called *incising,* before the preservative treatment is applied. Incising improves retention of the chemical preservative but also somewhat lowers the structural capacity of the wood member (Figure 3.38).

Depending on the treatment chemical and wood species, drying of wood before treatment may be necessary to obtain adequate penetration of the preservative chemicals. However, water-based treatments significantly increase the moisture content of the treated wood. Most such wood is *kiln-dried after treatment* to restore it to a lower moisture content, making the wood lighter, more stable, and better suited for use in building.

The concentration of preservative necessary to protect any given lumber product varies depending on the treatment chemical, the species of wood, and the severity of the environment in which the product will be used. To simplify the specification and selection of treated lumber, the American Wood Protection Association (AWPA) *Use Category System* provides designations corresponding to varying degrees of exposure (Figure 3.39). For example, wood deck posts, with ends in direct contact with the earth, should be treated to a higher Use Category (UC4A) than wood decking intended for use solely above ground (UC3B). It is then left to the treated wood producer to ensure that, for any particular use category, a suitable combination of wood species, chemical treatment, and treatment concentration is provided.

Chemical preservative treatments continue to evolve. Nonmetallic treatments and heat-treating processes are appearing on the market that can match the levels of protection

FIGURE 3.38
Pressure preservative-treated lumber. Incising marks are visible on the board surfaces. Each piece has a label stapled to one end indicating the degree of preservative treatment, in this case sufficient for above-ground use. This means that these boards are suitable, for example, for exterior decking or railings, or for use as foundation sill plates, but they are not appropriate for use in direct contact with soil. As part of the treatment process, this lumber has also been stained a light brown color to provide a more attractive appearance. (*Photo by Joseph Iano.*)

of currently established products but that are also safer to handle, environmentally less hazardous, less energy intensive to produce, and less corrosive.

Naturally Decay-Resistant Woods

The heartwood of some species of wood is naturally resistant to decay and insects and can often be used where preservative-treated wood would otherwise be required. The IBC recognizes redwood, cedar, black locust, and black walnut as decay-resistant species and redwood and Eastern red cedar as termite-resistant species. Because the sapwood of these species is no more resistant to attack than the wood of any other tree, "All-Heart" grade should be specified. A more comprehensive listing of wood species and their relative decay resistance can be found in the USDA Forest Products Laboratory's *Wood Handbook*, listed in the references at the end of this chapter. WPCs and plastic lumber, which are immune

Use Category	Service Conditions	Typical Building Uses
UC1	Interior construction, above ground, dry	Interior framing, woodwork, and furnishings where resistance to insect attack is required
UC2	Interior construction, above ground, damp	Interior construction where resistance to insect attack and/or moisture is required
UC3	Above ground exterior	
UC3A	Not exposed to prolonged wetting, finish coated, readily sheds water	Exterior painted or stained siding, millwork, and trim
UC3B	Exposed to prolonged wetting, unfinished or poorly drained	Exterior decking, deck framing, railings, and uncoated trim
UC4	Ground contact	
UC4A	Normal exposure conditions, noncritical components	Fence, deck, and guardrail posts in areas with relatively low decay risk
UC4B	High decay potential, critical or difficult-to-replace components	Permanent wood foundations, ground-contact wood columns
UC4C	Extreme decay potential, critical components	Pilings, ground-contact wood in severe decay areas

FIGURE 3.39
Part of the AWPA Use Category System, listing the categories most commonly specified for pressure preservative-treated wood used in building construction. The lumber in Figure 3.38, indicated as suitable for above-ground use, has been treated to meet the requirements of Use Category UC3B. (*Copyrighted material reproduced with permission from the American Wood Preservers Association, www.awpa.com*)

to decay or insect attack, are also becoming popular alternatives to preservative-treated lumber.

Durable Construction Techniques

Most wood-attacking organisms need both air and moisture to live. Accordingly, most can be kept out of wood by constructing and maintaining a building so that its wood components are kept dry. This includes keeping wood well clear of the soil, ventilating attics and crawlspaces to remove moisture, using good construction detailing to shed water and keep it out of building assemblies, using air and vapor retarders properly in conjunction with insulation to prevent the accumulation of condensation within exterior walls and roofs, and repairing roof and plumbing leaks as soon as they occur. Somewhat counterintuitively, wood that is fully and continuously submerged in fresh water is immune to decay because the water does not provide sufficient oxygen for decay-causing organisms to survive. However, where wood is only periodically or partially submerged and both moisture and oxygen are available, it is highly vulnerable to decay. Submersion in salt water does not prevent deterioration because marine organisms can attack wood under these conditions.

WOOD FASTENERS

From a structural standpoint, fasteners are often the weak link in wood construction. The interlocking timber connections of the past, laboriously mortised and pegged, were weaker than the members they connected because of the volume of wood removed to make the connection. Even with modern nails, screws, and bolts, it is usually impossible to insert enough of these fasteners in a connection to develop the full strength of the members being joined. And while adhesive joints and toothed-plate connections are capable of achieving the full strength of the connected members, they are largely limited to factory installation. As a result, most connections in wood structures depend primarily on direct bearing of one member on another for strength, while a variety of relatively simple fasteners serve to fix members in position and keep the connection tight.

Nails

Nails are sharp-pointed metal pins that are driven into wood with a hammer or a mechanical gun. They are the favored means of fastening wood because they are inexpensive, require no predrilling of holes under most conditions, and can be installed extremely rapidly. *Common nails* and the slightly slenderer *box* and *sinker nails* have flat heads and are used for structural fastening in light frame construction. Nails that are smaller in diameter and with smaller heads, such as *finish* and *casing nails* and *brads*, are used to fasten finish woodwork, where they are less obtrusive than common nails (Figure 3.40).

In North America, the size of a nail is measured in *pennies*, abbreviated "d." One explanation for this strange unit is that it originated long ago as the price of 100 nails of a given size. It persists today despite the loss of its original meaning. Figure 3.41 shows the dimensions of the various sizes of common nails. Box nails and finish nails are thinner in diameter but have the same length as common nails of equal penny size.

Sinkers, designed for ease of driving, are both slightly shorter and thinner than common nails of the same penny size. The nail head is usually stamped with a cross-hatch pattern, and is slightly funnel-shaped to make it easier to drive flush with the surface of the wood. A coating on the shaft acts as a lubricant to ease installation and as an adhesive, that bonds with the wood after the nail is set, to increase withdrawal resistance. Because of the sinker's slightly lesser size in comparison to a common nail, its structural capacity is less. To achieve the same strength as a joint fastened with common nails, either more sinkers of the same penny size are required, or nominally larger sinkers must be used. *Coolers* are another nail type that is slightly shorter and slenderer than the common nail. With the more recent profusion of variations on the traditional common nail, it has become important to prevent confusion between fasteners of the same nominal size but differing in structural capacity. To avoid this problem, nails for use in structural connections may be specified by actual length and shaft diameter rather than conventional name and penny size.

The three methods of fastening with nails, *face nailing, end nailing,* and *toe nailing,* are shown in Figure 3.42. Each of these methods has its uses in building construction, as illustrated in Chapters 5 and 6.

Joist hanger nails are the same diameter as the equivalent common nail, but much shorter. For example, a nail the same diameter as a 10d common nail but only 1½ inches (38 mm) long (sometimes designated "N10") may be used when nailing a joist hanger directly into the side of a single nominal 2-inch (actual size 38 mm) wood member, where a longer nail would protrude from the member's opposite side (Figure 3.43).

Other types of nails may have oversized heads, deformed shafts that provide increased resistance to shear and withdrawal compared to conventional nails, or other unique features. They should be used only in locations permitted by the building code or as recommended by the fastener manufacturer.

Nails are ordinarily furnished *bright*, meaning that they are made of plain, uncoated steel. Nails that will be exposed to the weather should be corrosion resistant, such as *hot-dip galvanized* (zinc-coated), aluminum, or stainless steel. (*Hot-galvanized nails,*

COMMON NAIL

BOX NAIL

SINKER NAIL

CASING NAIL

FINISH NAIL

BRAD

DEFORMED SHANK NAIL

HARDENED CONCRETE NAIL

CUT NAIL

ROOFING NAIL

FIGURE 3.40

Most nailed framing connections are made with common nails, box nails, sinkers, or their machine-driven equivalents. The more slender shafts of sinkers and box nails make them easier to drive than common nails and also good alternatives where splitting of the fastened pieces may be a concern (such as in a toe nailing). Box nails are also used for fastening wood shingles and other types of siding. Casing nails, finish nails, and brads are used for attaching finish components; their heads are set below the surface of the wood with a steel punch, and the holes are filled before painting. Deformed shank nails, which are more resistant to withdrawal from the wood than smooth shank nails, are used for attaching sheathing, subflooring, and floor underlayment, materials that might otherwise work loose in service. The most common deformation pattern is the ring-shank pattern shown here. Hardened-steel concrete nails can be driven into masonry or concrete for attaching furring strips and sleepers. Cut nails, long ago used for framing connections, are still sometimes used for attaching finish flooring where their blunt ends punch through the wood rather than wedge through, lessening splitting of brittle woods. Roofing nails have large heads to prevent tearing of soft asphalt shingles.

not to be confused with hot-dip galvanized, have a thinner zinc coating applied by a mechanical process rather than by dipping in molten zinc.) Corrosion resistance is particularly important for nails in exterior siding, trim, and decks, which could be stained by rust leaching from bright nails. For more information on the corrosion resistance of metals, see the Chapter 12 sidebar, "Metals in Architecture."

Machine-Driven Nails

Today, much nailing, both for framing and finish work, is done with powered *nail guns* rather than by hand. These guns use *collated nails* that are joined in linear arrays for swift loading into the magazine of the gun (Figure 3.43). Guns may be pneumatic, with air delivered to the gun via hose from a compressor; or be powered by internal combustion, with flammable gas stored in a cartridge within the gun. When

the nose of the nail gun is pressed against a solid surface and the trigger is pulled, a piston drives a nail with a single instantaneous blow. Various sizes of nail guns drive everything from 16d common nails to inch-long finish nails, with significant productivity gains over hand nailing. With a nail gun, large nails can be driven home with a single trigger pull, sheathing fasteners can be set at a rate of one or more per second, and finish nails can be accurately and reliably set below the surface of the wood, ready for filling and painting.

Nail gun fasteners may mimic the size and shape of hand-driven nails, or they may have modified heads, or take the form of T-shaped pins or U-shaped *staples.* Modified-head fasteners fire more reliably with less frequent gun jams or bent nails. Less costly, lightweight staples are used to attach sheathing paper and thermal insulation. Heavier staples may be used for fastening cabinet components, wood flooring, shingles, underlayment panels, and even structural components. Wherever machine-driven nails or staples are used to join structural components, care must be taken to ensure that fasteners of the appropriate type and adequate load capacity are used.

Screws

Screws are spiral-threaded fasteners installed by turning action whereby the threads draw the screw tightly into the material being fastened (Figure 3.44). In comparison to nails, screws cost more and take longer to install, but they can be inserted with greater precision, can exert greater clamping force between joined pieces, have greater holding power, and can be backed out and reinserted if a component has to be adjusted or remounted. Traditional *wood screws* require predrilled *pilot holes* into which the screw is inserted and then tightened with a screwdriver. Common uses include joining of cabinetry parts, installation of wide-plank flooring, mounting of hardware such as hinges, and other finish woodworking applications. Larger *lag screws* are used for heavier structural connections. They have square or hexagonal heads and are driven with a wrench rather than a screwdriver.

Screws may be made of steel, with or without a variety of metallic platings, such as black oxide, statuary bronze, yellow zinc, bright brass, clear zinc, and others. Where screws remain exposed, the finish plating is chosen based on appearance, and

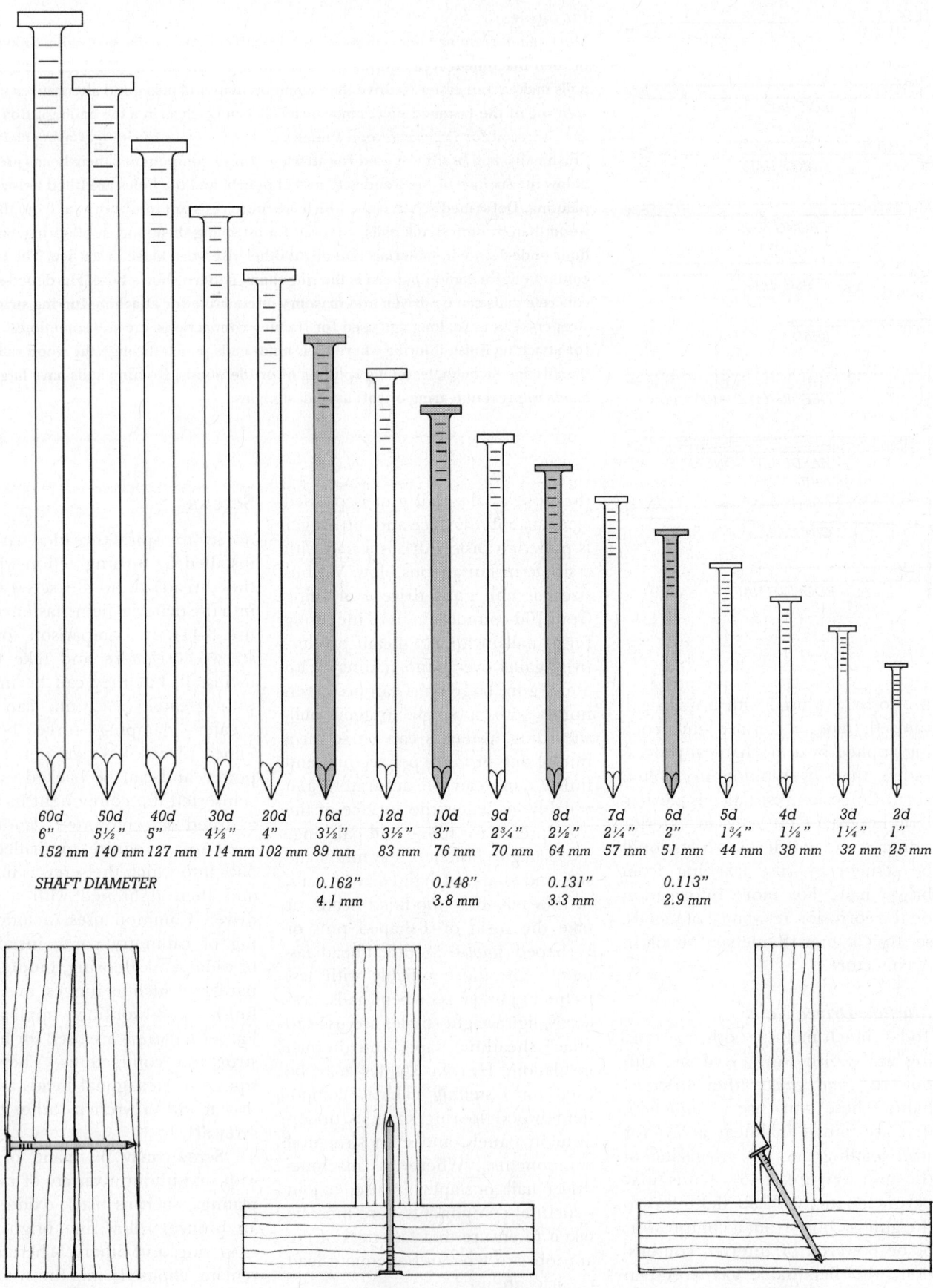

60d	50d	40d	30d	20d	16d	12d	10d	9d	8d	7d	6d	5d	4d	3d	2d
6"	5½"	5"	4½"	4"	3½"	3½"	3"	2¾"	2½"	2¼"	2"	1¾"	1½"	1¼"	1"
152 mm	140 mm	127 mm	114 mm	102 mm	89 mm	83 mm	76 mm	70 mm	64 mm	57 mm	51 mm	44 mm	38 mm	32 mm	25 mm

SHAFT DIAMETER

| 0.162" | 0.148" | 0.131" | 0.113" |
| 4.1 mm | 3.8 mm | 3.3 mm | 2.9 mm |

FACE NAIL

END NAIL

TOE NAIL

the desire for the fastener either to blend in or contrast with the materials being fastened. Some platings also offer modest corrosion resistance. However, where screws are used in exterior applications or will be exposed to a corrosive environment (for example, fasteners for door hardware in an indoor swimming pool), more corrosion-resistant metals (such as stainless steel, aluminum, or silicon bronze) are used.

Self-drilling wood screws do not require pilot holes and can be installed more quickly with power drivers. They are used for attaching subflooring to floor framing (rather than nailing, to reduce floor squeaking), mounting gypsum wallboard to wall studs (to avoid nail popping), attaching exterior decking to deck framing (to resist loosening of deck boards caused by moisture-related expansion and contraction), and wherever else the precision and holding power of screws are benefits. Like

FIGURE 3.43
From left to right, a 16d common nail, 16d box nail (note the more slender shaft but almost equal length), a coated 16d sinker (more slender and shorter than the common nail), an N10 joist hanger nail (hot-dip galvanized for greater corrosion resistance), and a short length of collated 10d nails for use in a nail gun (discussed below). The left sides of the heads on the collated nails are slightly clipped, to aid in firing reliability.

FIGURE 3.41
Standard sizes of common nails, reproduced here full size. The abbreviation "d" stands for "penny." The length of each nail is given below this size designation, in both inches and millimeters. The four sizes most used in light frame construction, 16d, 10d, 8d, and 6d, are shown shaded, with their shaft diameters also listed.

FIGURE 3.42
Face nailing is the strongest of the three methods of nailing. End nailing is relatively weak and is useful primarily for holding framing members in alignment until gravity forces and applied sheathing make a stronger connection. Toe nailing is used in situations where access for end nailing is not available. When properly installed, toe nails are surprisingly strong. Load tests show them to carry about five-sixths as much load as face nails of the same size.

Oval head

Round head

Slotted head

Hex head

Phillips head

Flat head

Washer

WOOD SCREW

SELF-DRILLING SCREW

LAG SCREW

nails, some self-drilling screw types are available in collated strips for use in self-feeding power guns.

Screws seem to come in a never-ending variety of types (Figure 3.45). Special driver shapes, such as square or star-shaped, engage and hold screws more reliably and can transmit greater torque than slotted or Phillips drivers. Wider, steeper thread patterns improve holding strength and allow faster driving. Contoured heads recess neatly below the surface of the wood. Multipitch thread patterns

FIGURE 3.44
Some common screw types. Flat-head screws are used without washers and are driven flush with the surface of the wood. Round-head screws are used with flat washers and oval-head screws with countersunk washers. Slotted head and Phillips head screws, driven by flat-blade and Phillips drivers, respectively, are the most common. Self-drilling screws, such as the drywall screw shown here, do not require predrilled pilot holes.

Washer

Washers

MACHINE BOLT   **CARRIAGE BOLT**

FIGURE 3.45
A manufacturer's chart illustrating variations in styles of power-driven screw heads, shanks, threads, and points. *(Courtesy of Simpson Strong-Tie Company Inc.)*

FIGURE 3.46
Both machine bolts and carriage bolts are used in wood construction. Carriage bolts have a broad button head that needs no washer and a square shoulder under the head that is forced into the drilled hole in the wood to prevent the bolt from turning as the nut on the opposite end is tightened.

improve a screw's ability to draw fastened pieces tightly together. Steel screws with organic coatings can be matched to the finish color of the fastened material and provide corrosion resistance at less cost than fasteners made of stainless steel. Slender, small-headed screws can be installed almost as unobtrusively as finish nails. In structural applications, longer, larger-diameter self-drilling screws can be installed more quickly and easily than conventional lag screws or bolts.

Bolts

Bolts are used mainly for structural connections in heavy timber framing and in wood light framing for fastening ledgers, beams, or other heavy applications. They are installed into predrilled holes. A wrench is used to turn the *nut*, opposite the bolt head, thereby tightening the bolt in place. Flat steel disks called *washers*, inserted under the head and nut, distribute the clamping force from the bolt across a greater area of wood and reduce crushing of the wood fibers. Commonly used sizes range in diameter from ⅜ to 1 inch (9.5 to 25 mm) in almost any desired length (Figure 3.46).

Timber Connectors

Various types of specially designed connectors can provide increased load-carrying capacity over bolted, screwed, or nailed connections, particularly when fastening larger-dimension wood members. The *split-ring connector* (Figure 3.47) is inserted in matching circular grooves to mate pieces of wood clamped together with a central bolt. This connector provides greater capacity by spreading the load across a larger area of wood than can be done with one or a few bolts. The split permits the ring to adjust to wood shrinkage.

Timber rivet connections are formed by fastening steel plates to large wood members with spike-like rivets. Unlike nails, timber rivets are oval in cross-section, and driven so that the wider axis of the rivet is always parallel to the grain of the timber. By driving arrays of rivets in closely spaced but carefully controlled patterns, an area of wood becomes compacted and prestressed, creating a connection between the riveted plate and the wood that is stronger than possible with ordinary nails and

less labor intensive than a bolted or split-ring connection. Additional examples of connectors for heavy timber construction are illustrated in Chapter 4.

Toothed Plates

Sheet metal *toothed plates* (Figure 3.48), similar to timber rivet connections but lighter, are used in factory-produced lightweight roof and floor trusses (Figure 3.51). They are driven into the wood with hydraulic presses, pneumatic presses, or mechanical rollers and act as splice plates, each with a very large number of built-in nails. They are effective

FIGURE 3.47

Split rings are high-capacity connectors used in heavily loaded joints of timber frames and trusses. After the center hole has been drilled through the two pieces, they are separated and the matching grooves are cut with a special rotary cutter driven by a large power drill. The joint is then reassembled with the ring in place.

FIGURE 3.48

Manufacturers of toothed-plate connectors also manufacture the machinery to install them and provide computer programs to aid truss fabricators in designing and detailing trusses for specific buildings. The truss drawing in this photograph was generated by such a program. The small rectangles on the drawing indicate the positions of toothed-plate connectors at the joints of the truss. *(Courtesy of Gang-Nail Systems, Inc.)*

and economical connectors: No drilling or gluing is required, they can be installed rapidly, and their multiple closely spaced points interlock tightly with the fibers of the wood.

Sheet Metal and Metal Plate Framing Devices

Dozens of ingenious sheet metal and metal plate devices are manufactured for joining wood members or

strengthening their joints. The most frequently used is the *joist hanger*, but all of the devices shown in Figures 3.49 and 3.50 find extensive use. There are two parallel series of this type of device, one made of sheet metal for use in light framing and one made of thicker metal plate for heavy timber framing. The devices for light framing are attached with nails and the heavier devices with bolts or lag screws. Where such devices will be

JOIST HANGER

BEAM HANGER

RAFTER ANCHOR

POST CAP

POST CAP

FRAMING ANCHORS

POST BASE

POST BASE

ANGLE

FIGURE 3.49
Joist hangers are used to make strong connections in floor framing wherever wood floor and roof framing members bear on one another at right angles. They are attached with special short nails driven through the holes punched in the hangers. Heavier steel plate hangers are used primarily in heavy timber construction. Post bases serve the twofold function of preventing water from entering the end of the post and anchoring the post to the foundation. The bolts and lag screws used to connect the wood members to the heavier connectors are omitted from this drawing.

FIGURE 3.50
The sheet metal connectors shown in this diagram are less commonly used than those in Figure 3.49, but are invaluable in solving special framing problems and in reinforcing frames against wind uplift and earthquake forces.

exposed to the weather or to lumber treated with corrosive chemical treatments, they should be made from corrosion-resistant metal such as heavily galvanized steel or stainless steel. To prevent galvanic corrosion between dissimilar metals, fasteners for metal framing devices should match the framing device material. That is, galvanized framing devices should be fastened with galvanized fasteners, and stainless steel devices should be fastened with stainless steel fasteners.

Wood Adhesives

Structural wood adhesives form bonds that are at least as strong, stiff, and durable as the members they connect. Adhesives rated for *full exterior exposure* are capable of repeated, long-term soaking and drying while maintaining acceptable levels of performance. Others, rated for *limited exterior* or *interior exposure*, are less water resistant, and their use is restricted to applications correspondingly less exposed to wetting.

Adhesives for Manufactured Wood Products

Since the middle of the 20th century, adhesives used in wood product manufacturing have been derived mostly from petroleum and natural gas. These synthetic polymers are consistent in formulation and performance, high in strength and stiffness, and (where needed) capable of providing excellent water resistance. More recently, efforts to reduce reliance on nonrenewable resources and concern for the human health effects of some synthetic adhesives (see sidebar) have led to renewed interest in traditional *bio-based adhesives* derived from agricultural and animal byproducts. Soy-protein-based structural adhesives, suitable for exterior exposure, have gained a significant foothold in wood product manufacturing. Adhesives derived from tannin and lignin (wood and paper manufacturing byproducts), and various animal proteins, are also gaining renewed attention.

WOOD PRODUCT ADHESIVES AND FORMALDEHYDE

Formaldehyde is an organic compound with many natural sources, including, in small amounts, wood itself. It is an ingredient in some adhesives and binders used in the manufacture of wood products. It is also a recognized carcinogen that, in high concentrations, can cause a range of adverse health effects in humans.

Phenol-formaldehyde (PF) is a structural adhesive, rated for full exterior exposure, used in the production of Exterior rated panels and other wood products. Formaldehyde gas emissions from PF after manufacture are low, below the exposure limits established by prevailing air quality and health standards. Though rated for exterior exposure, panels manufactured with PF are sometimes also specified for interior use, because of their relatively low emissions.

Urea-formaldehyde (UF) is a structural wood adhesive with less moisture resistance, rated for interior (dry) exposure only. Due to lower manufacturing costs and other advantages associated with its use, UF resin has historically been a popular choice in the manufacture of interior plywood, MDF, particleboard, doors, engineered wood flooring, and other wood products. However, UF, unlike PF, can continue to release high concentrations of formaldehyde gas long after manufacturing is complete and the products have been put into service. Because these products are destined almost exclusively for the building interior, their emissions can adversely affect indoor air quality and occupant health. A third resin type, melamine-formaldehyde, or MF, has emissions levels that fall somewhere between UF and PF.

Alternative adhesives are made without formaldehyde and can substitute for UF and PF in many applications. Described as *no added formaldehyde (NAF)*, examples include polyvinyl acetate, isocyanurates such as polymeric MDI, urethanes, and bio-based formulations. CLTs manufactured with urethane adhesive, structural agrifiber panels made with MDI binder, and Interior rated panels manufactured with soy adhesive exemplify the types of products becoming increasingly available as substitutes for products traditionally manufactured with PF and UF.

Over the past decade, formaldehyde gas emissions from manufactured wood products have come under increased scrutiny and government standards for safe emission levels have become more stringent. The LEED rating system awards points for selection of wood products that comply with *ultra low emitting formaldehyde (ULEF)* standards established by the California Air Resources Board or that are made with no added formaldehyde resins. The Living Building Challenge restricts products with added formaldehyde from its certified buildings as well.

Wood Adhesives on the Construction Site

Wood adhesives are used less on the construction site, where it is more difficult to clamp and hold glued joints and to maintain controlled temperature and humidity until the adhesive has cured. In rough carpentry work, subflooring panels are frequently adhered to their supporting wood framing (Figure 5.28). The adhesive is applied using a sealant gun. The nails or screws used to fasten the panels serve as the primary structural connection and also clamp the adhesive joint. Once cured, the adhesive reduces squeaks in the finished floor that could otherwise result from slippage between panels and joists as the floor deflects under live loads.

Finish carpenters use adhesives, usually in combination with nails or screws, to improve the strength and stability of joints in finish trim, wood flooring installations, and other site-built woodwork. In the woodworker's fabrication shop, adhesives may be used in the veneering of wood paneling and the fabrication of cabinets and furniture.

PREFABRICATED WOOD COMPONENTS

Wood products of all types can be used in the manufacture of *prefabricated wood components*. Factory fabrication eliminates weather-related delays and produces components that are structurally efficient, precise in dimension, and consistent in quality. On the construction site, prefabricated components can be erected faster, with smaller work crews and less waste.

Trusses

Trusses for both roof and floor construction are manufactured in small, highly efficient plants in every part of North America. Most are based on 2 × 4s and 2 × 6s joined with toothed-plate connectors. The designer or builder need only specify the span, roof pitch, and desired overhang detail. The truss manufacturer then uses a preengineered design for the specified truss or custom engineers a truss design and develops the necessary cutting patterns for its constituent parts. The manufacture and transportation of trusses are shown in Figure 3.51, and several uses of trusses are depicted in Chapter 5.

Roof trusses use less wood than a comparable frame of conventional rafters and ceiling joists. Like floor trusses, they span the entire width of the building in most applications, allowing the designer greater freedom to locate interior partitions anywhere they are needed. The chief disadvantages of roof trusses as they are most commonly used are that they make the attic space unusable and generally restrict the designer to the spatial monotony of a flat ceiling throughout the building.

Prefabricated Panels

Framed panels are simple sections of conventional dimension lumber framing, sheathed with plywood or OSB, trucked to the construction site, and rapidly nailed together into a complete building frame. Wall, floor, and roof panels with greater structural capacity are made with sheets of plywood or OSB applied to both panel faces. In *structural insulated panels (SIPs)*, the panels are adhered to a stiff plastic foam core. In *stressed-skin panels (SSPs)*, the panels are attached to dimension lumber framing. SIPs (pronounced "sips"), in particular, are a popular choice for the construction of highly energy-efficient homes and small buildings (Figure 3.52).

In *panelized construction*, whole sections of walls or floors are conventionally framed and sheathed in the factory, and then trucked to the construction site and installed in rapid succession. Although prefabrication of more complex assemblies—including, for example, insulation, wiring, windows, doors, and exterior and interior finishes—is technically feasible, it is less frequently done due to the added complexities of joining up all such systems in the field and obtaining the needed building code inspections in the factory setting.

Factory-Built Housing

Houses may be built entirely in a factory, at times even complete with furnishings, and then transported to prepared foundations, where they are set in place and made ready for occupancy in a matter of hours or days. A *manufactured home* (referred to in the past as a *mobile home*) is built on its own permanent, towable chassis and is designed to comply with a federal building code administered by the U.S. Department of Housing and Urban Development. A factory-built *modular home* is designed to comply with the building code in effect where the home will be located and is transported on a conventional flat-bed tractor-trailer. Both manufactured and modular homes are constructed in units not wider than 14 to 16 feet (4.27 to 4.88 m) to permit transport on public roads. Multiple units may be combined side-by-side or stacked vertically to make the complete structure.

At their least expensive, manufactured homes are sold at a fraction of the price of conventionally constructed homes of the same floor area. This is due in part to the economies of factory production and mass marketing and in part to the use of components that are lighter and less costly, and therefore shorter in life expectancy. However, both manufactured and modular homes may also be constructed to levels of cost and quality equal to those conventionally constructed, but with the potential to significantly reduce the duration of construction on site.

(*a*)

(*b*)

(*c*)

FIGURE 3.51
Manufacturing and transportation of wood trusses. (*a*) Factory workers align the wood members of a roof truss and position toothed-plate connectors over the joints, tapping them with a hammer to embed them slightly and keep them in place. The roller marked "Gantry" then passes rapidly over the assembly table and presses the plates firmly into the wood. (*b*) The trusses are transported to the construction site on a special trailer. (*c*) Trusses can be designed and produced in almost any configuration. (*Photos courtesy of the Wood Truss Council of America.*)

FRAMED PANEL   STRESSED-SKIN PANEL   STRUCTURAL INSULATED PANEL

FIGURE 3.52

Three types of prefabricated wood panels. The framed panel is identical to a segment of a conventionally framed wall, floor, or roof. The facings on the stressed-skin panel are bonded by adhesive to wood framing members to form a structural unit in which the facings carry the major stresses. A SIP, also sometimes called a *sandwich panel*, functions structurally in the same way as a stressed-skin panel, but its facings are bonded to a core of insulating foam.

Though most frequently associated with residential construction, modular techniques may be applied to the construction of larger building types as well. See, for example, the feature at the end of this chapter, "From Concept to Reality: French American School," for a case study of modular institutional building construction.

TYPES OF WOOD CONSTRUCTION

Wood construction has evolved into two major systems of on-site construction: The hand-hewn frames of centuries past have become the heavy timber frames of today, used both for single-family houses and for larger buildings. From the heavy timber frame has sprung wood light frame construction, which is the dominant system for houses, small commercial structures, and apartment buildings even as tall as four or five stories. These two systems of framing are detailed in Chapters 4 and 5.

CSI/CSC MasterFormat Sections for Wood, Wood-Plastic, and Plastic Lumber Materials

06 65 00	**COMMON WORK RESULTS FOR WOOD, PLASTICS, AND COMPOSITES**
06 05 23	Wood, Plastic, and Composite Fastenings
06 05 73	Wood Treatment
06 11 00	**WOOD FRAMING**
06 11 13	Engineered Wood Products
06 12 00	**STRUCTURAL PANELS**
06 12 16	Stressed-Skin Panels
06 16 00	**SHEATHING**
06 16 23	Subflooring
06 16 26	Underlayment
06 16 36	Wood Panel Product Sheathing
07 17 00	**SHOP-FABRICATED STRUCTURAL WOOD**
06 17 13	Laminated Veneer Lumber
06 17 23	Parallel Strand Lumber
06 17 33	Wood I-Joists
06 17 53	Shop-Fabricated Wood Trusses
06 18 00	**GLUE-LAMINATED CONSTRUCTION**
06 20 00	**FINISH CARPENTRY**
06 22 00	Millwork
06 40 00	**ARCHITECTURAL WOODWORK**
06 50 00	**STRUCTURAL PLASTICS**
06 80 00	**COMPOSITE FABRICATIONS**

KEY TERMS

bark
cambium
sapwood
heartwood
pith
cellulose
lignin
grain
springwood, earlywood
summerwood, latewood
softwood
hardwood
tracheid
ray
fiber
vessel, pore
grain figure
environmentally certified wood
Forest Stewardship Council (FSC)

Forest Management Certification
Chain of Custody Certification
FSC Controlled Wood
FSC Mixed Sources
reclaimed lumber
lumber
roundwood
headsaw
sawyer
plainsawn, flat-grain
quartersawn, vertical-grain
fleck, flake
riftsawn
moisture content (MC)
free water
bound water
fiber saturation point
equilibrium moisture content (EMC)
seasoning

kiln drying
air drying
green lumber
sticker
longitudinal shrinkage
radial shrinkage
tangentical shrinkage
check
surfaced
plane
surfaced four sides (S4S)
surfaced two sides (S2S)
unsurfaced
S-DRY
S-GRN
growth characteristics
manufacturing characteristics
knot
knothole

decay
insect damage
split
check
crook
bow
twist
cup
wane
structural grading
visual grading
machine grading
machine stress-rated
species group, species combination
appearance grading
nominal dimension
actual dimension
board
dimension lumber
timber
board foot
veneer
rotary-cut veneer, rotary-sliced veneer
sliced veneer
sawn veneer
plainsliced veneer
quartersliced veneer
flitch
sequenced veneer
glue-laminated wood, glulam
finger jointed
scarf jointed
hybrid glulam beam
FRP reinforced glulam
cross-laminated timber (CLT)
structural composite lumber, engineered
    lumber
laminated strand lumber (LSL)
oriented strand lumber (OSL)
laminated veneer lumber (LVL)
parallel strand lumber (PSL)
structural finger-jointed lumber
I-joist
wood-plastic composite (WPC)
composite wood trim
finger-jointed wood trim
load duration factor

cantilever
plastic lumber
recycled plastic lumber (RPL)
structural-grade plastic lumber (SGPL)
structural wood panel
plywood
composite panel
nonveneered panel
oriented strand board (OSB)
particleboard
underlayment panel
fiberboard
medium-density fiberboard (MDF)
touch sanded
medium-density overlay (MDO)
high-density overlay (HDO)
span rating
performance category
exposure bond classification
hardboard
insulating fiberboard sheathing
cellulosic fiberboard
hardwood plywood
agrifiber panel, biocomposite panel
strawboard panel
bamboo
Sustainable Agriculture Standard
fire-retardant treatment (FRT)
preservative-treated wood
pressure-treated wood
creosote
pentachlorophenol
chromated copper arsenate (CCA)
alkaline copper quat (ACQ)
copper boron azoles (CBA and CA)
micronized copper
carbon-based PTI
sodium borate (SBX)
disodium octaborate tetrahydrate (DOT)
pressure impregnation
incising
kiln-dried after treatment
AWPA Use Category System
nail
common nail
box nail
sinker nail

finish nail
casing nail
brad
penny (d)
cooler
face nailing
end nailing
toe nailing
joist hanger nail
bright nail
hot-dip galvanized nail
hot-galvanized nail
nail gun
collated nail
staple
screw
wood screw
pilot hole
lag screw
self-drilling wood screw
bolt
nut
washer
split-ring connector
timber rivet connection
toothed plate
joist hanger
structural wood adhesive
full exterior exposure adhesive
limited exterior exposure adhesive
interior exposure adhesive
bio-based adhesive
formaldehyde
phenol-formaldehyde (PF)
urea-formaldehyde (UF)
no added formaldehyde (NAF)
ultra low emitting formaldehyde
    (ULEF)
prefabricated wood components
truss
framed panel
structural insulated panel (SIP),
    sandwich panel
stressed-skin panel (SSP)
panelized construction
manufactured home, mobile home
modular home

REVIEW QUESTIONS

1. In the tree, what are springwood and summerwood cells? How do they differ?

2. Name three softwood species and three hardwood species. What is one common use for softwood lumber? What is one for hardwood lumber?

3. What are the differences between plainsawn and quartersawn lumber? What applications are appropriate for each?

4. Discuss the changes in moisture content and the effects of these changes on a

piece of dimension lumber from the time the tree is cut, through its processing, and until it has been in service in a building for an entire year.

5. A sample 8-foot-long 2 × 4 weighs 10 pounds. After drying in an oven, it

weighs 8 pounds. What was its moisture content before drying? In its original condition, was it adequately seasoned for use as a structural framing member?

6. Give three examples of growth characteristics and three examples of manufacturing characteristics.

7. Give an example of a structural grade wood species group. Why are species groups used in structural grading?

8. Give the actual cross-sectional dimensions of the following pieces of kiln-dried lumber: $1 \times 4, 2 \times 4, 2 \times 6, 2 \times 8, 4 \times 4, 4 \times 12$.

9. What method of slicing is used for structural plywood panels? What is a common method for hardwood veneers destined for fine architectural woodwork?

10. What are the advantages of using structural composite lumber rather than solid lumber?

11. Give three advantages of wood-plastic composites in comparison to solid lumber. What are two common uses for wood-plastic composites?

12. What are three common structural panel products? What is the most commonly available panel size (length and width).

13. Name the most appropriate plywood veneer grade for each of the following applications: (a) the outer face of a wall or roof sheathing panel, (b) underlayment to receive a thin, resilient vinyl floor covering, (c) plywood to receive a paint finish in an application demanding the highest possible finish quality.

14. What is meant by a structural panel span rating of 32/16? What types of wood products are rated this way?

15. Give one example of an agrifiber panel. What is one reason to select an agrifiber panel rather than one manufactured from wood products?

16. For what reasons might you specify preservative-treated wood?

17. Name three wood species that have naturally decay- or insect-resistant heartwood.

18. Why are nails the most common fastener type used in wood construction?

19. Which is longer, a common nail or a joist hanger nail, both made from the same diameter wire?

20. When are screws preferable to nails for fastening wood?

21. List two possible advantages of a bio-based adhesive over a synthetic polymer adhesive.

EXERCISES

1. Visit a nearby lumberyard or building materials supply center. Examine and list the species, grades, and sizes of lumber carried in stock. For what uses are each of these intended? While at the yard, look also at the available range of fasteners.

2. Pick up a number of scraps of dimension lumber from a shop or construction site. Examine each to see where it was located in the log before sawing. Note any drying distortions in each piece: How well do these correspond to the distortions you would have predicted? Measure the width and thickness of each scrap and compare your measurements to the specified actual dimensions for each.

3. Assemble samples of as many different species of wood as you can find. Learn how to tell the different species apart by color, odor, grain figure, ray structure, relative hardness, and so on. What are the most common uses for each species?

4. Visit a construction site and list the various types of lumber and wood products being used. Look for a gradestamp on each and determine why the given grade is being used for each use. If possible, look at the architect's written specifications for the project and see how the lumber and wood products were specified.

SELECTED REFERENCES

APA–The Engineered Wood Association. *2012 Architect Reference Literature CD.* Tacoma, WA, 2010.

A digital collection of the more than 50 of this association's most useful technical guides to manufactured wood products, including glulam, I-joist, OSB, and plywood. Many of these individual publications are also free to download from the Association's web site, www. apawood.org

Hoadley, R. Bruce. *Understanding Wood: A Craftsman's Guide to Wood Technology* (2nd ed.). Newtown, CT, Taunton Press, 2000.
Vividly explains the properties of wood as a material of construction.
USDA Forest Products Laboratory. *Wood Handbook: Wood as an Engineering Material.* Madison, WI, Author, 2010.
Provides comprehensive coverage of wood properties and the use of wood in construction. This reference has been in print publication since the 1930s. It can also be downloaded free from the Forest Product Laboratory's web site, www.fpl.fs.fed.us

Western Wood Products Association. *WWPA Library.* Portland, OR, updated regularly.

A bound collection containing most of this association's available literature on Western woods.

WEB SITES

Wood

Author's supplementary web site: **www.ianosbackfill.com/03_wood**

Forest Stewardship Council: **www.fsc.org**

Lumber

American Wood Council: **www.awc.org**

Canadian Wood Council: **www.cwc.ca**

Hardwood Manufacturers Association: **www.hardwoodinfo.com**

U.S. Forest Products Laboratory: **www.fpl.fs.fed.us**

Western Wood Products Association: **www.wwpa.org**

Wood Products

American Institute of Timber Construction: **www.aitc-glulam.org**

APA–The Engineered Wood Association: **www.apawood.org**

Composite Panel Association: **www.pbmdf.com**

Hardwood Plywood & Veneer Association: **www.hpva.org**

Agrifiber Products

Sustainable Agriculture Network: **sanstandards.org**

Wood Chemical Treatments

AWPA–The American Wood Protection Association: **www.awpa.com**

Manufactured Wood Components

Manufactured Housing Institute: **www.manufacturedhousing.org**

Modular Building Institute: **www.modular.org**

Structural Insulated Panel Association: **www.sips.org**

PROJECT: French American School

ARCHITECT: Weinstein AIU

For the French American School, the move from temporary facilities to a permanent structure on the same site meant that construction had to take place, in full, between the end of one academic year and the start of the next. Modular construction was the only way to complete the work within the available four-month time span and have any hope of being ready for the school's September move-in date.

Architect Weinstein A|U's design for the building was driven by the nature of the modular construction process, in which the factory-assembled building parts are trucked to the site and lifted into place by crane. The size of the modules was determined by highway regulations. In this case, width and height were strictly limited to 14 feet (4.3 m).

The architectural response was to accept the nature of the modules as simple box forms. The programmatic functions of the school were arranged into four "pods" containing classrooms, offices, and other major spaces, separated by smaller, connecting "gaskets" with bathrooms, stairs, and other service uses. The arrangement created a logical modulation of the building as a whole. Three of the pods would contain classrooms, while the fourth, nearest the entry, would be dedicated to public and school-wide functions: administrative offices, library, and the multipurpose room (Figure A). Exterior, site-built elements, such as horizontal trim pieces to hold vines for a green wall and metal canopies to shade south-facing windows, were used to add scale and interest to the simplicity of the modular boxes themselves.

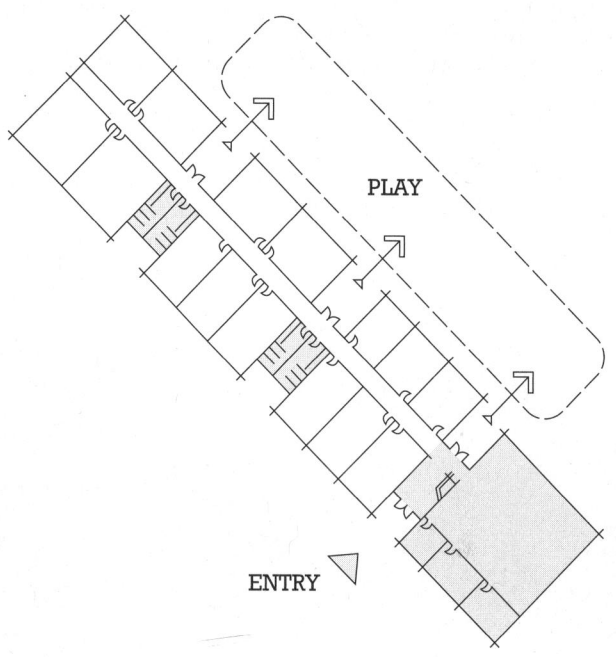

FIGURE A

The plan diagram consists of four pods connected by service links. Three of the pods house classrooms, while the fourth contains administrative and public functions.

FIGURE B

In Phase 1, four one-story pods were erected, with service modules in between. Phase 2 added a second story (lightly shaded modules) to three of the four pods. Exterior horizontal and entry canopies added texture to the main facade. The right-most, single-story pod in this view contains administrative spaces, the school library, and a multipurpose room. The heavily shaded module is the same one seen in Figures C and D.
(Drawing by Adrienne Watkins, Weinstein A|U.)

The modular construction also lent itself to the project's need to be phased. The growing school needed only a portion of the building completed in the first few years, and fundraising also necessitated stretching out development. The first phase would build out the entire planned footprint, but with only a single story. The later, second phase would add more second-floor classrooms to the three classroom pods. Because the second story was designed from the outset, the foundation and ground-floor modules were built to accommodate the stairs, elevator, plumbing, and structural loads that would come later when these additional modules were added (Figure B).

The stresses imposed on modules during transportation and lifting created additional design constraints. The number and size of openings for doors and windows were limited to maintain a sufficiently rigid structure. The design team worked closely with the manufacturer to maximize the size of important corner windows. Additionally, each module required full-depth floor and ceiling joist framing. As a result, when units were stacked, there would be redundant structure and added depth between floors that could have been avoided with site-built construction methods.

The modular units were built off site, beginning in early spring. The day the students left in June, the work began

FIGURE C

A module arriving on site. Each of 25 modules in Phase 1 and 19 modules in Phase 2 was individually trucked to the site from a factory located 30 miles away. Because there was no room on site to store modules, the modules were delivered "just in time" to be lifted into position shortly after their arrival.

Temporary bracing straps and plywood stiffen the module during transportation and lifting. These are removed after the modules have been set in final position. Seen in this module is a corridor wall with doors and interior windows opening into classrooms. (*All photos by Joseph Iano.*)

FIGURE D

The same module seen in Figure B being lifted into position. This module was part of the Phase 2 expansion, so all work was performed without disturbing finished landscaping or site amenities already in place. This module contains half of three classrooms. The remainder of the classroom spaces is contained in an adjoining module not yet in place. The space above the finished ceiling houses mechanical equipment, ductwork, and the roof structural framing. Note the temporary cross bracing still in place during lifting.

on site. Difficulties in gaining building permits from the local building department were an unexpected challenge for the design team. The small, suburban jurisdiction had never before seen an application for construction of this type. Because the units would arrive on site with interior walls finished, state agencies would inspect and approve structural, electrical, and plumbing systems that would be concealed by the time local inspectors first saw them. The design team had to work closely with the local agency to gain their confidence and assure them that all issues of concern would be properly addressed.

For the contractor, limited space available for staging construction on site, combined with overhead telecommunication and high-power electrical lines, presented special logistical difficulties. Module delivery was carefully scheduled so that modules could be lifted into final position as soon as they arrived at the construction site. Positioning and lifting of modules also had to be carefully planned to avoid conflicts with the overhead utilities that could not be relocated or interrupted.

The photograph of the completed project shows the modulation of the pod-and-gasket arrangement. Because the school is approached at an angle, this massing is easily visible. The corner windows were important features, to maximize light on the interior and reduce the "boxy" feel of the modules. If the project had been built using a conventional method, the size and location of the windows would not have been constrained. However, in a modular design, they created a weak point in the module, and thus required special structural attention. This size was the largest possible within the system.

Special thanks to Weinstein A|U and Lesley Bain, Principal, for their contributions to this case study.

FIGURE E

A projecting canopy marks the main entrance to the building. Smaller canopies provide shade for south-facing windows. The horizontal and vertical trim cover joint lines between modules and support growing vines that will, over time, create a green wall on the building facade. Note the corner windows, important for introducing daylight into the classrooms, which created structural difficulties that required special coordination between the design team and the module manufacturer.

## 4

HEAVY TIMBER FRAME CONSTRUCTION

- **Fire-Resistive Heavy Timber Construction**

  SUSTAINABILITY IN HEAVY TIMBER CONSTRUCTION

  Managing Wood Shrinkage
  Anchoring Beams and Walls
  Floor and Roof Decks

- **Heavy Timber in Other Construction Types**

- **Lateral Bracing**

- **Cross-Laminated Timber Construction**

- **Accommodating Building Services**

- **Wood-Concrete Composite Construction**

- **Longer Spans in Heavy Timber**
  Large Beams
  Rigid Frames
  Trusses
  Arches and Domes

  FOR PRELIMINARY DESIGN OF HEAVY TIMBER STRUCTURES

- **Heavy Timber and the Building Codes**

- **Uniqueness of Heavy Timber Framing**

Architect Nils Finne combines heavy timber with robust stonework and finely detailed exterior finish carpentry in this well-crafted residence. *(Architect: FINNE Architects, www.FINNE.com; photograph by Art Grice.)*

Wood beams have been used to span roofs and floors of buildings since the beginning of civilization. The first timber-framed buildings were crude pit houses, lean-tos, teepees, and basketlike assemblages of bent saplings. In earliest historic times, roof and floor timbers were combined with masonry loadbearing walls to build houses and public buildings. In the Middle Ages, braced wall frames of timber were built for the first time (Figures 4.1 and 4.2). The British carpenters who emigrated to North America in the 17th and 18th centuries brought with them a fully developed knowledge of how to build efficiently braced frames, and for two centuries North Americans lived and worked almost exclusively in buildings framed with hand-hewn wooden timbers joined by interlocking wood-to-wood connections (Figure 4.3). Nails were rare and expensive, so they were used only in door and window construction and, sometimes, for fastening siding boards to the frame.

Until two centuries ago, logs could be converted to boards and timbers only by human muscle power. To make timbers, axemen skillfully scored and hewed logs to reduce them to a rectangular profile. Boards were produced slowly and laboriously with a long, two-man pit saw, one man standing in a pit beneath the log, pulling the saw down, and the other standing above, pulling it back up. At the beginning of the 19th century, water-powered sawmills began to take over the work of transforming tree trunks into lumber, squaring timbers and slicing boards in a fraction of the time that it took to do the same work by hand.

Most of the great industrial mills of 19th-century North America, which manufactured textiles, shoes, machinery, and all the goods of civilization, consisted of heavy sawn timber floors and roofs supported by masonry exterior walls (Figures 4.4 and 4.5). The house builders and barn raisers of the early 19th century switched from hand-hewn to sawn timbers as soon as they became available. Many of these early mills, barns, and houses still survive, and with them survives a rich tradition of heavy timber building that continues to the present day.

FIGURE 4.1
Braced wall framing was not developed until the late Middle Ages and early Renaissance, when it was often exposed on the face of the building in the style of construction known as *halftimbering*. The space between the timbers was filled with brickwork or *wattle and daub*, a crude plaster of sticks and mud, as seen here in Wythenshawe Hall, a 16th-century house near Manchester, England. *(Photo by James Austin, Cambridge, England (GB).)*

140

FIGURE 4.2
European timber house forms generally followed a progression of development from crude pit dwellings, made of earth and tree trunks, to cruck frames, to braced frames. The *crucks* (curved timbers), hewn by hand from appropriately shaped trees, were precursors to the laminated wood arches and rigid frames that are widely used today.

Cruck

PIT DWELLING

CRUCK FRAME

BRACED FRAME

This is the way they built the barns............

① *The* FOUNDATION
Place for main Girder ˣ
Girder posts

② *Laying* SILLS *over the* Oak Main Girder

Laying the
③ FLOOR BEAMS *or* JOISTS
and FLOORING (*which is usually laid temporarily till after the Raising*).
oak
Floor Beams (joists)
Floor joists
Sill Mortise

④ RAISING *the first bent with Pikes*
a pike pin
Stay
Pike (about 16 feet long)

a Two brace bent

⑤ *All bents up and ready for the Plate and Rafters.*

FIGURE 4.3
The European tradition of heavy timber framing was brought to North America by the earliest settlers and was used for houses and barns until well into the 19th century. (*Drawing by Eric Sloane; courtesy of the artist.*)

FIGURE 4.4
Most 19th-century industrial buildings in North America were constructed of heavy timber roofs and floors supported at the perimeter on masonry loadbearing walls, a construction method that came to be known as Mill construction. This impressive group of textile mills stretches for a distance of 2 miles (3 km) along the Merrimac River in Manchester, New Hampshire. *(Photo by Randolph Langenbach.)*

> **The old country builder, when he has to get out a cambered beam or a curved brace, goes round his yard and looks out the log that grew in the actual shape, and taking off two outer slabs by handwork in the sawpit, chops it roughly to shape with his side-axe and works it to the finished face with the adze, so that the completed work shall for ever bear the evidence of his skill. . . .**
>
> —*Gertrude Jekyll, English garden designer, writing in 1900*

FIRE-RESISTIVE HEAVY TIMBER CONSTRUCTION

Large timbers, because of their greater capacity to absorb heat, are much slower to catch fire and burn than smaller pieces of wood. When exposed to fire, a heavy timber beam, though deeply charred by gradual burning, will continue to support its load long after an unprotected steel beam exposed to the same conditions has collapsed. If the fire is not prolonged, a fire-damaged heavy timber beam or column can be sandblasted afterward to remove the surface char and continue in service. For these reasons, building codes in many circumstances recognize heavy timber framing as roughly equivalent to steel, concrete, or masonry with a one-hour fire-resistance rating.

FIGURE 4.5
The generously sized windows in the mills provided plenty of daylight to work by. Columns were of wood or cast iron. Most New England mills, like this one, were framed very simply: Their floor decking is carried by beams running at right angles to the exterior walls, supported at the interior on two lines of columns. Notice that the finish flooring runs perpendicular to the structural decking. Overhead sprinklers provide additional fire safety to a construction method that already has inherent fire-resistive qualities. *(Photo by Randolph Langenbach.)*

SUSTAINABILITY IN HEAVY TIMBER CONSTRUCTION

General considerations of wood sustainability are addressed in Chapter 3. With regard to Heavy Timber construction in particular:

It is wasteful to saw large, solid timbers from logs. In most instances, only one or two timbers can be obtained from a log, and it may be difficult to saw smaller boards from the remaining slabs.

Glue-laminated and cross-laminated timber utilize wood fiber more efficiently than solid timbers.

Some wood finishes, and some glues and binders used with glue-laminated timbers and other manufactured wood products, may give off gases such as formaldehyde that can cause indoor air quality (IAQ) problems.

Heavy timber and cross-laminated timber structures have lower embodied energy and a more favorable carbon footprint than steel, concrete, or masonry structures of equal size.

Solid timbers can be recycled from mills, factories, and barns when they are demolished. Most of these originated from old-growth forests in which trees grew slowly, producing fine-grained, dense wood with structural properties superior to those of new-growth timbers. Recycled timbers may be used as is, resurfaced to give them a new appearance, or resawn into smaller members. However, any old metal fasteners retained in the timbers must be meticulously found and removed, because they can cause expensive damage to saw blades and planer knives.

Continuous bending action of heavy timber beams may be created by splicing beams at points of inflection rather than over supports, as shown in Figures 4.12 and 4.22. This reduces maximum bending forces in the beam, allowing timber sizes to be reduced.

Wood structural members do not lose strength with age, although they do sag progressively if they are overloaded. When a heavy timber building is demolished at some time in the future, its timbers can be recycled, even if they were obtained as recycled material for the building that is being demolished.

The long-span post-and-beam structural system typical of heavy timber framing lends itself to easy future adaptability for varying uses, potentially extending the useful lifetime of the building. Many 19th- and early 20th-century heavy timber buildings remain productively useful to this day.

A heavy timber frame enclosed with foam-core sandwich or stressed-skin panels is easily made airtight and well insulated, with few thermal bridges. Heating and cooling of the building will consume relatively little energy.

In the International Building Code, for a building to be classified as *Type IV Heavy Timber (HT) construction*, its wooden structural members must meet minimum size requirements and its exterior walls must be constructed of noncombustible materials. The minimum permitted sizes for solid wood timbers are summarized in Figure 4.6. Exterior walls may be constructed of concrete, masonry, or a steel frame with metal or other noncombustible cladding. Historically, this combination of fire-resistive wood framing and noncombustible exterior was called *Mill construction*, reflecting its origins in 19th-century brick masonry mill structures, or *Slow-burning construction* (Figures 4.7, 4.8, and 4.9).

In contemporary Heavy Timber construction, the wood framing itself may remain remarkably similar to traditional methods, with little more than mild steel connectors replacing those previously made of cast or wrought iron. However, exterior wall and roof construction reflect modern construction practice, such as providing higher insulation values and replacing solid masonry with cavity wall construction (Figure 4.10). Also, as shown in this figure, the exposed edges of timbers may be *chamfered* (beveled at 45 degrees) to eliminate the thin edges of wood that catch fire most easily. While it was once a code requirement for fire-resistant construction, this is now done mainly for aesthetic reasons or to reduce the hazard from splinters.

	Supporting Floor Loads	Supporting Roof and Ceiling Loads Only
Columns	8 × 8 (184 × 184 mm)	6 × 8 (140 × 184 mm)
Beams and Girders	6 × 10 (140 × 235 mm)	4 × 6 (89 × 140 mm)
Trusses	8 × 8 (184 × 184 mm)	4 × 6 (89 × 140 mm)
Decking	3-in. (64-mm) decking plus 1-in. (19-mm) finish flooring	2-in. (38-mm) decking, or 1⅛-in. (29-mm) wood structural panel

FIGURE 4.6
Minimum sizes for solid wood members used in Type IV Heavy Timber construction, as specified in the IBC. Glue-laminated members used in this type of construction must meet similar requirements.

ROOF BEAMS ANCHORED TO GIRDERS, ROOF GIRDERS TO TOP STORY COLUMNS AND TO EXTERIOR WALLS

ROOF PLANKING

ROOF BEAM

Ⓖ

FLUSH TYPE ROOF USED WHERE PARAPET WALL PROTECTION IS NOT REQUIRED.

WALL PLATE ANCHORED TO WALL

FLOOR

ROOF BEAM

PAPER

FINISH ROOF BEAM

Ⓚ

ROOF GIRDER

COLUMN

FLOOR PLANKING

Ⓚ

FLOOR BEAM

WOOD BOLSTER

Ⓒ

ROOF GIRDER

TOP STORY COLUMNS ANCHORED TO FLOOR GIRDERS

FLOOR BEAM

FLO... BEAM

Ⓓ

FLOOR BEAM

FINISH FLOOR

Ⓙ

DOGS

FLOOR GIRDER

FLOOR BEAM

ANCHOR BOLT

Ⓞ

PAPER

SCUPPER

...OR BEAM

FLOOR GIRDER

CAST IRON PINTLE

MALLEABLE IRON POST CAP (SEE DETAILS FOR OTHER TYPES OF CONNECTIONS)

CORNERS CHAMFERED

COLUMN

FLOOR PLANKING

Ⓙ

FLOOR GIRDER

Ⓙ

FLOOR BEAM

FIGURE 4.7

(*opposite page*) Traditional Mill construction. Exterior walls are constructed of noncombustible, solid masonry. Interior framing is heavy timber. Wood girder-column connections are designed so that cross-grain shrinkage in girders does not affect columns, and floor and roof framing remain level. Iron dogs tie beams and girders together at their ends. A long steel strap anchors the roof girder sufficiently low in the outside wall that the weight of the masonry above the anchor point is enough to resist wind uplift on the roof. (*Heavy Timber construction details courtesy of the National Forest Products Association, Washington, D.C.*)

FIGURE 4.8

(*left*) Heavy timber girder-column connections designed so that the large shrinkage typical of beams and girders does not affect the columns. Detail B-1 avoids resting the column on the beam with a cast iron pintle. In detail B-3, the upper column passes through the beam with only a steel bearing plate between it and the column below. Also in B-3, split-ring connectors are used to form a connection between the bearing blocks and the columns strong enough to support the loads from the beams; it would take a much larger number of bolts to do the same job. (*Heavy Timber construction details courtesy of the National Forest Products Association, Washington, D.C.*)

FIGURE 4.9

(*right*) Details for the bearing of a beam on masonry in traditional Mill construction. In each case, the beam end is firecut (see Figure 4.13) and anchored to the wall by means of either lag screws or a lug on the iron bearing plate. (*Heavy Timber construction details courtesy of the National Forest Products Association, Washington, D.C.*)

Stone or precast concrete coping

Drip

Continuous flashing

Metal dowels retain the coping in place

Continuous counterflashing

Roof membrane

Wood beam with firecut end

Concrete block backup

Bituminous dampproofing on face of concrete block

Brick facing

Wire ties and joint reinforcing

Cavity

Plastic foam insulation

Vertical reinforcing in filled cores

A metal strap anchor and bolts tie each beam to the wall

Bond beam with horizontal reinforcing bars under each beam bearing

Continuous flashing and weep holes

Rigid insulation

Metal lath retains the grout in the cavity above

Wood base

Wood finish flooring

Wood structural decking

Rigid insulation

Vapor retarder

Wood structural decking

Chamfering removes the thin, easily ignited edges of beams and columns for greater fire resistance

Column sections rest directly upon one another, to minimize vertical wood shrinkage

Metal beam/column connector and bolts

2'

500 mm

1'

250 mm

0

0

12"

FIGURE 4.10

An example of contemporary heavy timber construction, based on steel plate connectors and an insulated masonry cavity wall. For an explanation of the differences between solid masonry and cavity wall construction, see Chapter 10.

Managing Wood Shrinkage

Where heavy timber floor and roof framing is supported partly by wood, and partly by other structural materials such as concrete, masonry, or steel, care must be taken to avoid uneven shrinkage between different parts of the structure. Compared to other materials, wood expands and contracts more with changes in moisture content, particularly in the direction perpendicular to its grain. In large beams and girders, significant dimensional changes can occur with seasonal changes in ambient conditions or the gradual drying of unseasoned members.

To minimize the impact of these size changes, wood columns at each level do not rest directly on floor beams. Rather, they pass through the beams to bear on columns below. In this way, changes in the depth of the floor framing are not imparted into the columns. Particularly in multilevel structures, this prevents dimensional changes from accumulating floor-to-floor, which could otherwise lead to loss of level in upper floors and roofs. In traditional Mill construction, this type of connection is achieved with cast iron pintels or plates, as illustrated in Figures 4.7 and 4.8. In contemporary practice, columns may be butted directly to one another at the floor level with the aid of a variety of metal connector types (Figure 4.10) or, if longer, they may pass through the column/floor connection as a single piece (Figures 4.11 and 4.12). Longer column lengths are especially easy to achieve with glue-laminated timbers.

Anchoring Beams and Walls

Where heavy timber beams join masonry or concrete walls, three problems must be solved: First, the beam must be protected from decay caused by moisture seeping through the wall. This is done by leaving a ventilating airspace of at least ½ inch (13 mm) between the masonry and all

SHOULDER BEARING   BEARING BLOCKS   SPLIT RINGS   BEARING BLOCKS WITH SPLIT RINGS

FIGURE 4.11
Some beam–column connections for contemporary Heavy Timber construction. The first three examples are for double beams sandwiched on either side of the column. The fourth shows single beams supported in the same plane as the column. In the shoulder bearing connection, the beams are recessed into the column by an amount that allows the load to be transferred safely by wood-to-wood bearing; the bolts serve only to keep the beams in the recesses. Bearing blocks allow more bolts to be inserted in a connection than can fit through the beams, and each bolt in the bearing blocks can hold several times as much load as one through the beam because it acts parallel to the grain of the wood rather than perpendicular. It is generally impossible to place enough bolts in a beam–column joint to transfer the load successfully without bearing blocks unless split rings are used, as shown in the third example. The steel straps and bolts in the fourth example hold the beams on the bearing blocks.

sides of the beam except the bottom, unless the beam is chemically treated to resist decay. The second and third problems have to be solved together: The beam must be securely anchored to the wall so that it cannot pull away under normal service, yet it must be able to rotate freely so that, in the event of a severe building fire, it does not pry the wall apart if it burns through and collapses (Figure 4.13). Traditional and more contemporary methods of meeting these dual needs are shown in Figures 4.9 and 4.10, respectively.

Floor and Roof Decks

Building codes require that Type IV Heavy Timber buildings have floors and roofs of solid wood construction without concealed combustible cavities. Figure 4.14 shows four different types of decking used for this purpose. Minimum permissible thicknesses of decking are given in Figure 4.6. To meet code requirements, floor decking must also be covered with a finish floor consisting of nominal 1-inch (19-mm) tongue-and-groove boards laid diagonally or at right angles to the structural decking. In some circumstances, ½-inch (13-mm) structural wood panels are also permitted as the finish layer. Cross-laminated timber panels, discussed later in this chapter, may also be used as decking in Heavy Timber construction.

FIGURE 4.12
Connections for a glue-laminated wood
frame. The tall column passes by the
beam and remains unaffected by beam
shrinkage. The cantilevered beam and
hinge connector save wood by connecting
the beams at points of zero bending
moment rather than at the columns
to take full advantage of continuous
bending action in the beams.

Lag screws keep
the beam on the
haunch

Cantilevered
beam

A welded steel
hinge connector
joins the beams at
a point of zero
moment

A haunch
laminated as part
of the column
support the
beam

*ELEVATION OF
FRAME*

FIGURE 4.13
A timber floor or roof structure that burns through in a prolonged fire might topple the portion of the wall above it (*left*) unless the beam ends are *firecut* (*right*) to allow the beams to rotate freely out of their wall pockets. In this illustration, the beam is anchored to the wall by a steel strap anchor.

TONGUE AND GROOVE

SPLINE

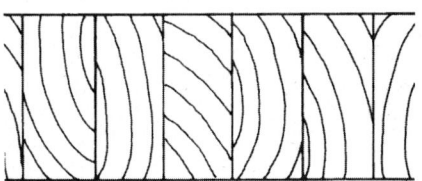

LAMINATED DECK

FIGURE 4.14
Large-scale cross sections of four types of heavy timber decking. *Tongue-and-groove decking* is the most common. *Splined decking* is slightly more economical of lumber because wood is not wasted in the milling of tongues. *Laminated decking* is a traditional type for longer spans and heavier loads; it consists of ordinary dimension lumber laid on edge and spiked together (spikes are not shown in this illustration). *Glue-laminated decking* is a modern type. In the example shown here, five separate boards are glued together to make each piece of decking. Decking of any type is usually installed in random lengths. The end joints do not necessarily line up over beams; rather, they are staggered to avoid creating zones of structural weakness. The tongues, splines, or nails allow individual pieces of decking to share concentrated loads, so that together they act as a continuous plane of solid wood.

GLUE-LAMINATED DECKING

HEAVY TIMBER IN OTHER CONSTRUCTION TYPES

Heavy timber members may also be used in buildings that do not meet the building code definition of Type IV Heavy Timber construction. Any construction type that permits smaller wood members (for example, Type V, wood light frame construction) also permits the use of heavy timbers. In this way, the architectural qualities and structural performance of heavy timber framing can appear in all manner of smaller, residential, commercial, religious, and institutional buildings. In these applications, there are no restrictions (other than structural) on the minimum size of timbers, minimum thickness of decking, or exterior wall materials, and light framing of nominal 2-inch (38-mm) lumber can be incorporated as desired (Figure 4.15). As discussed toward the end of this chapter, limited use of heavy timbers is also permitted in some noncombustible construction types.

(a)

(b)

(c)

(d)

(e)

FIGURE 4.15
A contemporary building constructed using traditional heavy timber framing techniques. (a) Assembling a *bent* (a plane of columns, beams, rafters, and braces). (b) The completed bents are laid out on the floor, ready for raising. (c) Raising the bents, using a truck-mounted crane, and installing floor framing and roof *purlins* (smaller, secondary framing members that span across the primary beams or rafters). (d) The completed frame. (e) Enclosing the timber-framed house with sandwich panels consisting of OSB faces bonded to an insulating foam core. Because the exterior walls of this structure are combustible, the building code classifies this as Type V construction, rather than Type **IV HT.** (*Courtesy of Benson Woodworking, Inc., Alstead, NH.*)

FIGURE 4.16
A glue-laminated wood building frame with diagonal bracing to resist lateral forces. This five-story University of British Columbia classroom building was permitted to be constructed in significant part of heavy timber based on modeling of the wood structure's performance that demonstrated fire resistance comparable to alternative noncombustible structural systems.
Floor plates are composite concrete-CLT construction, discussed later in this chapter.
(Photo courtesy Structurlam, Inc.)

LATERAL BRACING

A heavy timber frame building with an exterior masonry or concrete bearing wall is normally braced against wind and seismic forces by the lateral resistance of its exterior walls and interior elevator or exit stair enclosures. In areas of high seismic risk, walls are heavily reinforced, and floor and roof decks are strengthened with heavily nailed structural panels securely anchored to the walls, creating a rigid, unified structure. In buildings with framed exterior walls, diagonal bracing or interior shear walls must be provided (Figure 4.16). Seismic upgrades to historic heavy timber and masonry buildings often require the insertion of new steel-braced frames or reinforced concrete shear walls and strengthening of floor plates in order to meet contemporary lateral force resistance requirements.

CROSS-LAMINATED TIMBER CONSTRUCTION

Cross-laminated timber panels may serve as an alternative to conventional floor and roof structural systems within any conventional structural system (where heavy timber components are permitted). Or, loadbearing wall and decking CLT panels may be used together to create a complete building structure constructed solely of this panel type.

CLTs are fabricated to the required size and shape at the factory. To the greatest extent possible, window and door openings in walls; stair and shaft openings in floors; panel edge details; channels and cutouts for conduit, piping, and other building services; and other modifications are completed before panels are shipped to the construction site.

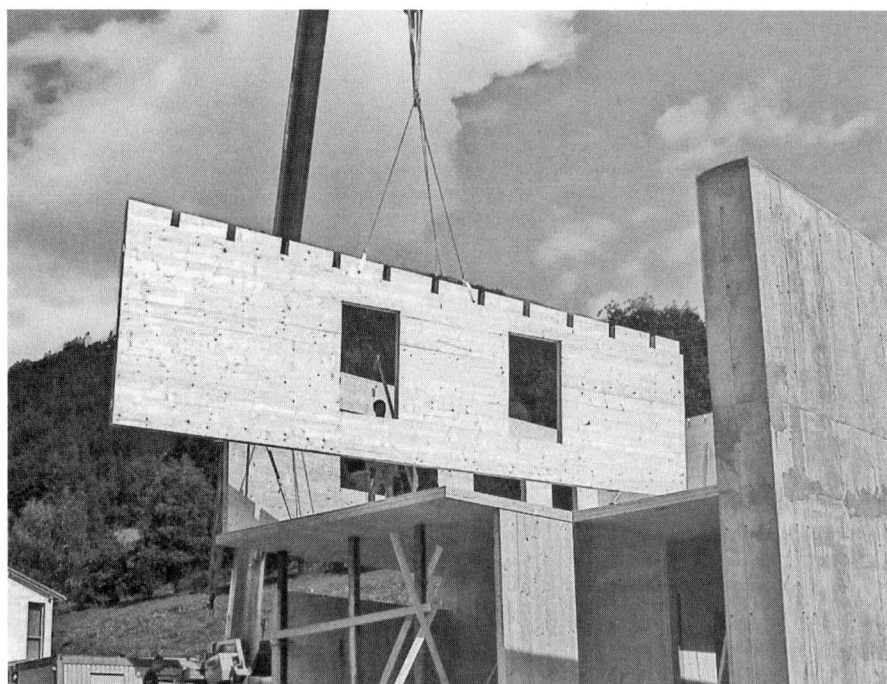

FIGURE 4.17
A CLT wall panel being lifted into position. Note the cutouts already completed in the panel for window rough openings and roof beam pockets. *(Photo courtesy Structurlam, Inc.)*

Once panels arrive on site, they are erected rapidly with small crews, simple tools, and relatively lightweight lifting machinery. In addition to its economic benefits, rapid construction reduces the length of time that panels remain exposed to the elements. This helps to prevent excessive moisture absorption that could cause instability in the panels. Connections rely on simple combinations of long, self-tapping screws, conventional bolts, sheet metal connectors, and other common types of connection devices. Heavy timber connectors such as split rings or shear plates may also be used. Panel-to-panel joints at the exterior are made airtight with the application of foam tape or continuous lines of sealant between panels as they are assembled. Interior panel joints may be similarly sealed to

reduce sound transmission through walls and floors (Figures 4.17, 4.18, and 4.19).

In the completed building, the CLT structure must be protected from the weather and appropriately finished on the interior. Exterior cladding and suitable weather barrier layers must be applied to the exterior side of the panels. Insulation may be applied to the inside or outside of the panels—outside insulation provides the benefit of keeping the wood panels within the heated, controlled environment of the building interior where they are less vulnerable to moisture damage. On the interior, panels can be left exposed, or covered with gypsum wallboard or some other finish material. Furred finishes, which create a narrow hollow space behind the finish, allow easier routing of power and communications wiring.

(For examples of furred wall finish systems, see Chapter 23.) Where necessary, cutouts and channels can also be routed into the panels themselves, either in the factory or on the construction site, to accommodate building services.

CLT construction is well established on the European continent. One of the world's tallest mixed-use wood-framed buildings is the London Stadthaus, a nine-story CLT structure that was erected in just nine weeks by a crew of four carpenters. However, CLT construction is relatively new to North America. Until the necessary standards are developed and integrated into building codes, projects utilizing CLT construction must be submitted to regulatory agencies using alternative materials and methods provisions, as described later in this chapter.

FIGURE 4.18
A partial diagram of CLT construction illustrating some of the possible assembly and connection strategies. The circles labeled *A*, *B*, and *C* refer to details in Figure 4.19.

The upper wall panel bears on the floor panel below and is anchored with angles screwed into both panels.

Gaskets or sealant reduce air leakage at exterior panel joints.

The floor panel bears on the wall panel below and is fastened with long, self-drilling screws.

Gaskets or sealant reduce air leakage at exterior panel joints.

Metal plate connectors anchor wall panels to the foundation.

Non-shrink grout levels the top of the foundation wall.

A structural steel angle bolted to the foundation wall supports floor panel ends.

Floor panel joints may also be lapped, but at the expense of slightly reduced panel-to-panel width.

Floor panels may be joined with plywood spines and self-drilling screws.

A gasket or sealant may be inserted into the joint to reduce sound transmission between floors.

At interior connections, a gasket or sealant may be inserted to reduce sound transmission across the wall.

Self-drilling screws are used at wall panel intersections.

At exterior connections, a gasket or sealant material reduces air leakage.

A. BEARING WALL   B. FLOOR PANEL CONNECTIONS   C. WALL CORNER CONNECTIONS

FIGURE 4.19
Example details for connecting CLT panels. As much as possible, connections rely on ordinary tools and readily available fasteners and hardware.

FIGURE 4.20
When floors and roofs are heavy timber framed, much of the lighting, ductwork, fire sprinklers, and other building services remain exposed to view and must be designed with appropriate care. *(Photo by Joseph Iano.)*

ACCOMMODATING BUILDING SERVICES

Heavy timber and CLT structures pose special problems for the designer because these exposed or solid structural systems do not necessarily provide the concealed cavities that are present in light frame structures and other conventional building systems.

Roof thermal insulation cannot be hidden in spaces between ceiling joists or roof rafters and, instead, must be placed on top of the roof deck. If the roof is nearly flat, it can be insulated and roofed in the manner shown for low-slope roofs in Chapter 16; or, if the roof is steeply pitched, a nailing surface for the shingles or other roofing must be added above the insulation. Unless a suspended ceiling is provided, electrical wiring for overhead lighting fixtures must either run through exposed metal conduits below the deck, which may be visually unsatisfactory, or be channeled within the deck or through the insulation above the deck. Ductwork for heating and cooling must also remain exposed (Figure 4.20).

Where walls are made of masonry, cross-laminated timber, or other solid panel types (such as SIPs or SSPs), similar accommodations must be made. Services may remain exposed or be routed within channels cut or built into these systems, or a hollow space may be added behind the finish wall surface within which these services can be concealed.

WOOD-CONCRETE COMPOSITE CONSTRUCTION

Heavy timber beams and CLT panels can be combined with cast-in-place concrete to create floor systems in which the wood and concrete act together as unified structural components with greater structural efficiency.

In wood-concrete composite systems, rows of *shear connectors* are installed into the upper face of the heavy timber beams or CLT panels. These connectors may be closely spaced, heavy lag screws; continuous angles made of perforated metal; or other components that will bridge between the wood and concrete and mechanically engage with both materials. The shear connectors are installed so that they project 1 to 2 inches (25 to 50 mm) above the wood surface. Concrete is cast on top of the panels or over the beams, fully covering the connectors. Once the concrete has cured, the shear connectors engage with both the wood and the concrete so that under structural loading these

FIGURE 4.21
Installing tongue-and-groove roof decking over laminated beams and girders. *(Photo courtesy of American Institute of Timber Construction.)*

components share stresses and deflect together as one element. This more efficient structural configuration can span further or carry greater loads in comparison to the same structural components acting without composite action. (For a further explanation of composite structural members, see the subsection "Composite Construction" in Chapter 11.)

LONGER SPANS IN HEAVY TIMBER

For buildings that require spans longer than 20 to 30 feet (6 to 9 m), the maximum usually associated with framing of sawn timbers, the designer may select from among several alternative types of timber structural devices.

Large Beams

With large, old-growth trees no longer readily available, very large timber beams are usually built up of laminations or strands rather than sawn. Such beams are stronger and more dimensionally stable than sawn wood beams and can be made in the exact size and shape desired (Figures 4.21 and 4.22).

FIGURE 4.22
Glue-laminated beams support proprietary long-span trusses made of wood with steel tube diagonals. The glulams are hinged in the manner illustrated in Figure 4.12. The trusses, roof joists, and plywood roof deck are prefabricated into panels on the ground and then lifted into final position to reduce installation time. *(Photos provided by Trus Joist MacMillan.)*

Interactive Exercise

Rigid Frames

The cruck (Figure 4.2), cut from a bent tree, was a form of *rigid frame*. Today's rigid frames are glue-laminated to shape and find wide use in longer-span buildings. Standard configurations are readily available (Figures 4.23 and 4.24), or the designer may order a custom shape. Rigid frames exert a horizontal thrust, so they must be tied together at the base with *steel tension rods*, also called *tie rods*. In laminated wood construction, rigid frames are often called *arches*, acknowledging that the two structural forms act in very nearly the same manner.

Trusses

The majority of wood trusses built each year are light roof trusses of nominal 2-inch (38-mm) lumber joined by toothed plates (see Figures 3.48, 3.51, 5.65, and 5.66). For larger buildings, however, *heavy timber trusses* may be used. Their joints are made with steel bolts and welded steel plate connectors, split-ring connectors, or other proprietary connection hardware. Sawn, laminated, and structural composite timbers may be used, sometimes in combination with steel members. Many shapes of truss are possible, and spans of more than 100 feet (30 m) are easily achieved (Figures 4.25, 4.26, 4.27, 4.28, and 4.29).

FIGURE 4.23
Three-hinged arches of glue-laminated wood carry laminated wood roof purlins. The short crosspieces of wood on top of the arches are temporary ladders for workers. (*Photo courtesy of American Institute of Timber Construction.*)

Shear plates and rod

Slide plates and bolts

C, D

B. DETAIL OF CROWN

Welded steel shoe

Anchor bolts

A. DETAIL OF BASE

Tie rod

ELEVATION OF ARCH

A

B

FIGURE 4.24
Typical details for three-hinged arches of laminated wood. The tie rod is later covered by the floor slab. The shear plate in the crown connection is similar to a split-ring connector (see Chapter 3) and serves to spread forces in the connection across a larger surface area of wood.

Bent steel plate

Roof deck spans across purlins

Purlins span across trusses

Steel angle and lag screws at each purlin

Steel pins

Steel bearing plate and anchor bolt

Threaded steel rod with nuts and washers

Bolts

FIGURE 4.25
A conventional heavy timber roof truss with steel rod tension members. This type of truss is easy to construct but cannot be used if wind uplift forces are sufficiently strong to cause the forces in the tension members to reverse (such slender steel rods cannot resist compressive forces). The center splice in the lower chord of the truss is required only if it is impossible to obtain single pieces of lumber long enough to reach from one end of the truss to the other.

FIGURE 4.26
This contemporary heavy timber fastening system uses proprietary self-drilling steel dowels (*top*) in combination with embedded steel *knife plates* spanning the joint between members to create high-strength connections. *(Courtesy of SFS intec, Inc., Wyomissing, PA, www.sfsintecusa.com)*

FIGURE 4.27
A completed heavy timber truss joint using the fastening system illustrated in Figure 4.26. *(Courtesy of SFS intec, Inc., Wyomissing, PA, www.sfsintecusa.com)*

FIGURE 4.28
Laminated wood roof trusses with steel connector plates. The steel rods are for lateral stability, to keep the bottom chords of the trusses from moving sideways. *(Courtesy of Woo & Williams, Architects. Photographer: Richard Bonarrigo.)*

FIGURE 4.29
A contemporary heavy timber truss made from solid wood members. At the truss peak, connections rely on knife plates and through-bolts. At the foot of the truss, a steel shoe anchors the truss to the supporting beam. *(Photo by Joseph Iano.)*

Arches and Domes

Long curved timbers for making *heavy timber vaults* or *arches* and *heavy timber domes* are easily fabricated in laminated wood and are widely used in athletic arenas, auditoriums, suburban retail stores, warehouses, and factories (Figures 4.30, 4.31, and 4.32). Like rigid frames, arched and domed structures exert lateral thrusts that must be countered by tie rods or suitably designed foundations.

FIGURE 4.30
Semicircular laminated wood arches support the timber roof of the Back Bay Station in Boston. Notice the use of horizontal steel tie rods to resist the thrust of the arches at the base. *(Architects: Kallmann, McKinnell & Wood. Photo © Steve Rosenthal.)*

Welded steel plates

Hinge pin

Anchor bolts into concrete foundation

ELEVATION OF ARCH

FIGURE 4.31
A typical hinged foundation connection for an arch or dome, made of welded steel plates. The hinge pin allows for rotation between the arch and the foundations, which avoids additional forces that might otherwise be placed on the structure by wood shrinkage or foundation movement.

FIGURE 4.32
This laminated wood dome spans 530 feet (161.5 m) to cover a 25,000-seat stadium and convention center in Tacoma, Washington.
(Architects: McGranahan, Messenger Associates. Structural engineers: Chalker Engineers, Inc. Photo by Gary Vannest. Courtesy of American Wood Council.)

FOR PRELIMINARY DESIGN OF HEAVY TIMBER STRUCTURES

• Estimate the nominal depth of **wood roof decking** at ¹⁄₄₅ of its span. Estimate the depth of **wood floor decking** at ¹⁄₃₅ of its span. Standard nominal depths of wood decking are 2, 3, 4, 6, and 8 inches (actual size 38, 64, 89, 140, and 184 mm).

• Except in tall or heavily loaded structures, **cross-laminated timber wall panels** are usually made from three laminations, approximately 4 in. (100 mm) in total thickness. The number of laminations and thickness of floor and roof panels vary depending on loads and spans. For preliminary design, estimate the depth of **cross-laminated timber floor and roof decks** at ¹⁄₃₀ of their span.

• Estimate the depth of **solid wood beams** at ¹⁄₁₅ of their span and the depth of **glue-laminated beams** at ¹⁄₂₀ of their span. Add a nominal 6 inches (150 mm) to these depths for girders. The width of a solid wood beam or girder is usually ¼ to ½ of its depth. The width of a glue-laminated beam typically ranges from ¹⁄₇ to ¼ of its depth.

• Estimate the depth of timber **triangular roof trusses** at ¹⁄₅ to ½ of their span and the depth of **bowstring trusses** at ½ to ⅔ of their span.

• To estimate the size of a **wood column**, add up the total roof and floor area supported by the column. A nominal 6-inch (actual size 140 mm) column can support up to about 400 square feet (37 m²) of area, an 8-inch (actual size 184 mm) column 1000 square feet (93 m²), a 10-inch (actual size 235 mm) column 1500 square feet (140 m²), a 12-inch (actual size 286 mm) column 2500 square feet (230 m²), and a 14-inch (actual size 337 mm) column 3500 square feet (325 m²). Wood columns are usually square or nearly square in proportion.

For actual sizes of solid timbers, see Figures 3.23 and 3.24. Standard sizes of glue-laminated timbers are given in subsection "Glue-Laminated Wood" in Chapter 3. For a building that must qualify as Type IV Heavy Timber construction under the IBC, minimum timber sizes are given in Figure 4.6.

These approximations are valid only for purposes of preliminary building layout, and must not be used to select final member sizes. They apply to the normal range of building occupancies such as residential, office, commercial, and institutional buildings. For manufacturing and storage buildings, use somewhat larger members.

For more comprehensive information on preliminary selection and layout of a structural system and sizing of structural members, see Edward Allen and Joseph Iano, *The Architect's Studio Companion* (5th ed.), Hoboken, NJ, John Wiley & Sons, 2012.

HEAVY TIMBER AND THE BUILDING CODES

In the table in Figure 1.3, the column labeled "Type IV HT" shows the International Building Code height and area limits for buildings of different occupancies built of traditional, fire-resistive, heavy timber construction, that is, of heavy timber framing combined with noncombustible exterior walls. Notice that when considering any particular occupancy, the allowable heights and areas for this construction type are roughly comparable to those for buildings made of steel, concrete, or masonry with a fire-resistance rating of one hour (Type IIA). This is an explicit recognition of the inherent fire-resistive qualities of the large-dimension wood members in this construction type.

When heavy timbers are used in other construction types, the height and area limitations of those other types apply. So, for example, an Occupancy B office building constructed as Type IV HT may be five stories, 65 ft (20 m) in height, and as much as 36,000 sf (3300 m²) in area per floor. However, if heavy timbers are mixed with smaller-dimensioned wood framing without noncombustible exterior walls or added fire protection, the structure is considered to be Type VB, and then may be only 2 stories, 40 ft (12 m) in height, and up to 9000 sf (840 m²) per floor.

As indicated in the footnotes to the table in Figure 1.3, heavy timber framing is also an acceptable substitute for noncombustible 1-hour rated roof framing.

Conventional building code requirements for Heavy Timber construction are *prescriptive*. That is, they dictate minimum sizes of wood members considered sufficiently fire-resistant and the maximum building size considered safe for building occupants. Alternative *performance-based* methods for demonstrating structural fire safety rely on the modeled behavior of a particular structure, without presumed limits on the size of the building or the members from which it is made. If the performance of a structure under fire conditions can be demonstrated to be at least as safe as that of an acceptable baseline, then the proposed structure may be accepted as well. As performance-based alternatives to prescriptive code requirements continue to gain acceptance among regulatory bodies, it is expected that heavy timber structures exceeding the current prescriptive size limits will become more prevalent.

UNIQUENESS OF HEAVY TIMBER FRAMING

Heavy timber cannot span as far or with such delicacy as steel, and it cannot match the structural continuity or smooth shell forms of concrete, yet many people respond more positively to the idea of a timber building than they do to one of steel or concrete. To some degree, this response may stem from the color, grain figure, and warmer feel of wood. In part, it may also come from the pleasant associations that people have with the sturdy, satisfying houses our ancestors erected from hand-hewn timbers only a few generations ago. Most people today live in dwellings where none of the framing is exposed. Exposed ceiling beams in one's house or apartment are considered an amenity, and shopping or restaurant dining in a converted mill building of heavy timber and masonry construction is generally thought to be a pleasant experience. There is something in all of us that derives satisfaction from seeing wood beams at work (Figures 4.33, 4.34, 4.35, 4.36, and 4.37).

FIGURE 4.33
Architects Greene and Greene of Pasadena, California, were known for their carefully wrought expressions of timber framing in houses such as this one, built for David B. Gamble in 1909. *(Photo by Wayne Andrews.)*

FIGURE 4.34
Dramatic exposed timber scissor trusses support a steeply sloping roof structure in this vacation cabin, designed by Nils Finne, in the North Cascade mountains of Washington State. *(Architect: FINNE Architects, www.FINNE.com; photograph by Art Grice.)*

FIGURE 4.35
Each of the attached dwellings in Sea Ranch Condominium #1 at Sea Ranch, in northern California, is framed with a simple cage of unplaned timbers sawn from trees taken from another portion of the site. The diagonal members act as wind and seismic braces. *(Architects: Moore, Lyndon, Turnbull, and Whitaker. Photo by Edward Allen.)*

FIGURE 4.36
The exterior of Sea Ranch Condominium #1 is sheathed with vertical 2-inch (38-mm) unplaned tongue-and-groove decking and clad with ¾-inch (19-mm) tongue-and- groove redwood siding. *(Photos © Morley Baer.)*

Even as the quality of forests has diminished and other building materials have become more readily available, heavy timber remains economically competitive in the marketplace. Technological advancements such as CLTs, wood-concrete composite construction, hybrid glulams, and more efficient connection methods have improved the performance of heavy timber and opened up new applications. Alternative approaches to code acceptance are expanding the range of buildings for which heavy timber may be considered. The natural economy of wood construction, the compelling case for wood sustainability, and the attractiveness and feel of large wood members are unique attributes that continue to make Heavy Timber construction a viable choice among the major structural systems.

FIGURE 4.37
Metropol Parasol, Plaza de la Encarnacion, Seville, Spain, by J. Mayer H. Architects and ARUP engineers. This laminated veneer lumber structure covers 490 ft by 245 ft of area and is close to 100 ft tall (150 × 75 × 30 m). It shades the public square and supports a high-level café and panoramic walkway. The LVL members are joined with a steel rod and epoxy connector developed specifically for this structure. The largest single wood member measures 54 ft × 11 ft × 6 in. (16.5 × 3.5 × 0.14 m). *(Photo by Jan-Peter Koppitz.)*

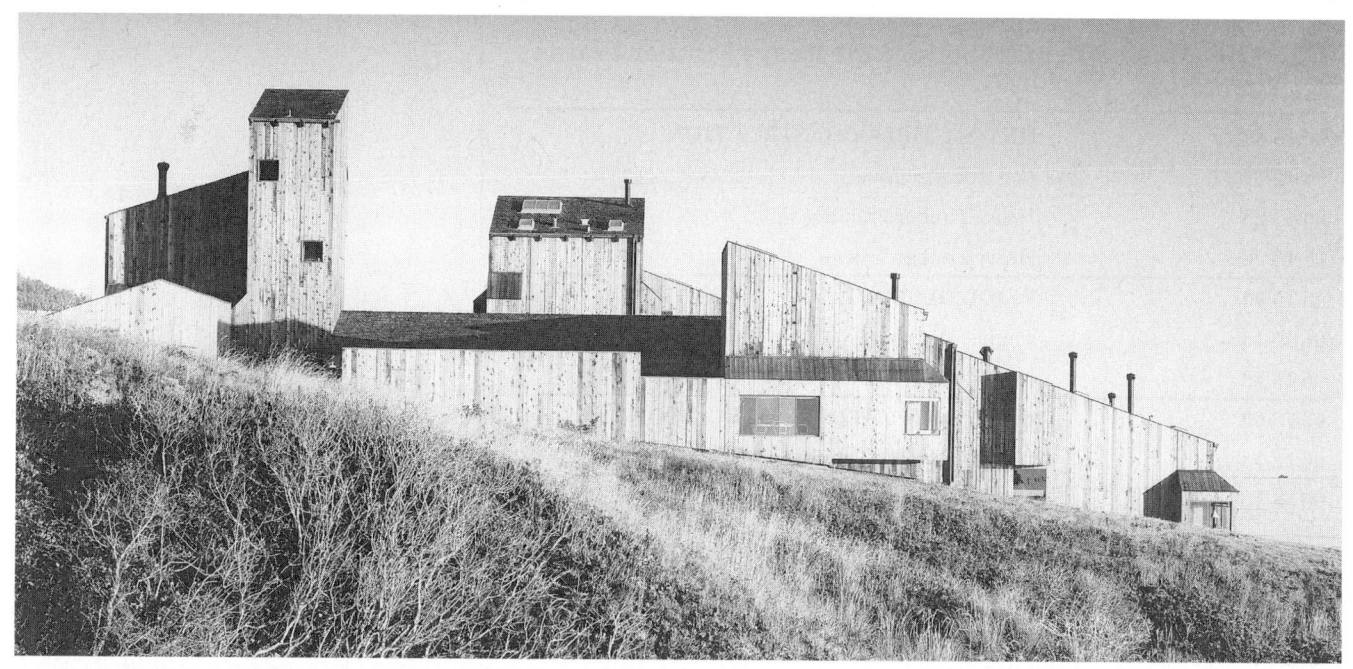

CSI/CSC MasterFormat Sections for Heavy Timber Frame Construction	
06 13 00	**HEAVY TIMBER CONSTRUCTION**
06 13 13	Log Construction
06 13 23	Heavy Timber Framing
06 13 26	Heavy Timber Trusses
06 15 00	**WOOD DECKING**
06 15 19	Timber Decking
06 15 23	Laminated Wood Decking
06 18 00	**GLUE-LAMINATED CONSTRUCTION**
06 18 13	Glue-Laminated Beams
06 18 16	Glue-Laminated Columns

KEY TERMS

halftimbering

wattle and daub

cruck

Type IV Heavy Timber (HT) construction

Mill construction, Slow-burning construction

chamfer

hinge connector

firecut

tongue-and-groove decking

splined decking

laminated decking

glue-laminated decking

bent

purlin

shear connector

rigid frame

steel tension rod, tie rod

arch

heavy timber truss

knife plate

heavy timber vault, heavy timber arch

heavy timber dome

prescriptive code requirement

performance-based code method

REVIEW QUESTIONS

1. Why does Heavy Timber construction receive relatively favorable treatment from building codes and insurance companies?

2. What are the two basic requirements for Type IV Heavy Timber construction in the International Building Code?

3. Assume that heavy timber beams will shrink approximately 2 percent in the cross-grain dimension as they continue to dry after installation in a building. If there are four framed floor levels, supported by 24-inch-deep (600 mm) beams at each level, what would be the accumulated shrinkage at the roof if columns at each floor rested on these beams, rather than being supported as recommended in this chapter?

4. What are the important factors in detailing the junction of a wood beam with a masonry loadbearing wall?

5. Draw from memory one or two typical details for the intersection of a wood column with a floor of a building of Heavy Timber construction.

6. What are three options for routing electrical services along CLT wall panels?

7. What is the difference between a prescriptive code and a performance-based code? What is the potential advantage of a performance-based code?

EXERCISES

1. Determine from the code table in Figure 1.2 whether a building you are currently designing could be built of Type IV Heavy Timber construction. If not, what modifications can you make in your design so that it will conform to these requirements?

2. Find a barn or mill that was constructed in the 18th or 19th century and sketch some typical connection details. How is the structure stabilized against wind and other lateral forces?

3. Obtain a book on traditional Japanese construction from the library and compare Japanese timber joint details with 18th- or 19th-century American practice.

SELECTED REFERENCES

American Institute of Timber Construction. *Timber Construction Manual* (5th ed.). Hoboken, NJ, John Wiley & Sons, 2004.

This is a comprehensive design handbook for timber structures, including detailed engineering procedures as well as general information on wood and its fasteners.

American Institute of Timber Construction. *Typical Construction Details* (AITC 104–2003). Englewood, CA, Author, 2003.

This 32-page reference contains dozens of examples of how to detail connections in heavy timber buildings. Especially instructive are the bad examples that are presented as lessons in what to avoid. This reference can also be viewed for free on the American Institute of Timber Construction's web site, www.aitc-glulam.org

Benson, Tedd. *Timber-Frame Home: Design, Construction and Finishing.* Middletown, CT, Taunton Press, 1997.

The author is one of the most experienced traditional timber framers in the world. This is a comprehensive, richly illustrated, authoritative guide to timber house construction.

Goldstein, Eliot E. *Timber Construction for Architects and Builders.* New York, McGraw-Hill, 1999.

This thoroughly practical book is based on the author's firsthand experience in designing timber structures, as well as recognized timber engineering practice.

WEB SITES

Heavy Timber Frame Construction

Author's supplementary web site: **www.ianosbackfill.com/04_heavy_timber_frame_construction**
American Institute of Timber Construction: **www.aitc-glulam.org**
American Wood Council: **www.awc.org**
Canadian Wood Council: **www.cwc.ca**
Innovative Timber Systems, LLC: **www.smartwoods.com**
Structurlam Innovative Wood Specialists: **www.structurlam.com**
Timber Framers Guild: **www.tfguild.org**

5

WOOD LIGHT FRAME CONSTRUCTION

• **History**

• **Platform Frame**

SUSTAINABILITY IN WOOD LIGHT FRAME CONSTRUCTION

• **Foundations for Light Frame Structures**
Other Foundation Materials

• **Building the Frame**
Planning the Frame
Erecting the Frame
Attaching the Frame to the Foundation
Floor Framing
Wall Framing
Lateral Force Resistance and Shear Walls
Upper-Level Floor and Wall Framing
Roof Framing

• **Variations on Wood Light Frame Construction**
Framing for Increased Thermal Efficiency
Framing for Optimal Lumber Usage
Prefabricated Framing Assemblies

FOR PRELIMINARY DESIGN OF A WOOD LIGHT FRAME STRUCTURE

• **Wood Light Frame Construction and the Building Codes**

• **Uniqueness of Wood Light Frame Construction**

A New England school of arts and crafts is housed in a cluster of small buildings of wood light frame construction that cling to a dramatic mountainside site overlooking the ocean. *(Architect: Edward Larrabee Barnes. Photo by Joseph W. Molitor.)*

171

Wood light frame construction is the most versatile of all building systems. There is scarcely a shape it cannot be used to construct, from a plain rectangular box to cylindrical towers to complex foldings and shapes of every description. During the century and a half since it first came into use, wood light framing has served to construct buildings ranging from reinterpretations of nearly all the historical styles to uncompromising expressions of every contemporary architectural philosophy. During this same period, it has assimilated without difficulty a bewildering and unforeseen succession of technical improvements in building: central heating, air conditioning, gas lighting, electricity, thermal insulation, indoor plumbing, prefabricated components, and electronic communications cabling.

Light frame buildings are easily and swiftly constructed with a minimal investment in tools. Many observers of the building industry have criticized the supposed inefficiency of light frame construction, which is carried out largely by hand methods on the building site, yet it has successfully fought off competition from industrialized building systems of every sort, partly by incorporating their best features, to remain the least expensive form of durable construction. It is the common currency of small residential and commercial buildings in North America today.

Wood light frame construction has its deficiencies: If ignited, it burns rapidly; if exposed to dampness, it decays. It expands and contracts by significant amounts in response to changes in humidity, sometimes causing chronic difficulties with cracking plaster, sticking doors, and buckling floors. The framing itself is so unattractive that it is seldom left exposed in a building. These problems can be controlled, however, by clever design and careful workmanship, and there is no arguing with success: Frames made by the monotonous repetition of wooden joists, studs, and rafters are likely to remain the primary system of building in North America for a long time to come.

FIGURE 5.1
A wood light frame single-family home under construction. At bottom right is a concrete slab-on-grade garage floor. At the left, concrete basement walls have been covered with a dimpled plastic drainage mat. Exterior walls are framed with solid lumber 2 × 6 studs, sheathed with OSB panels. Floors are framed with I-joists (also with OSB subflooring not visible in this photograph). On top of the upper story walls, prefabricated roof trusses have been stacked in preparation for their installation. (*Photo by Joseph Iano.*)

HISTORY

Wood light frame construction was the first uniquely American building system. It was developed in the first half of the 19th century when builders recognized that the closely spaced vertical members used to infill the walls of a heavy timber building frame were themselves sufficiently strong that the heavy posts of the frame could be eliminated. Its development was accelerated by two technological breakthroughs of the period: Boards and small framing members of wood had recently become inexpensive for the first time in history because of the advent of the water-powered sawmill; additionally, machine-made nails had become remarkably cheap compared to the hand-forged nails that preceded them.

The *balloon frame* was the earliest wood framing system to be constructed exclusively of slender, closely spaced wooden members: *joists* for the floors, *studs* for the walls, *rafters* for the sloping roofs. Heavy posts and beams were completely eliminated and, with them, the difficult, expensive mortise-and-tenon joinery they required. There was no structural member in a frame that could not be handled easily by one or two carpenters, and each of the hundreds of joints was made with lightning rapidity with two or three nails. The impact of this new building system was revolutionary: In 1865, G. E. Woodward could write that "[a] man and a boy can now attain the same results, with ease, that twenty men could on an old-fashioned frame . . . . [T]he Balloon Frame can be put up for forty percent less money than the mortise and tenon frame" (Woodward, 1989, p. 164).

The balloon frame (Figure 5.2) used full-length studs that ran continuously for two stories from foundation to roof. In time, it became apparent that these were too long to erect efficiently. Furthermore, the tall, hollow spaces between studs acted as multiple chimneys in a fire,

Attic framing is identical in the two systems

Studs in platform framing extend only from the top of one level of framing to the bottom of the next

Studs in balloon framing extend unbroken from the foundation to the roof

Fireblocking is required in balloon framing to close the cavities in the wall against the passage of fire

Floor joists in balloon framing rest on a wood ribbon (also called a ribband) recessed into the studs

In platform framing the studs and sole plates rest on the top of the floor platforms

Studs in balloon framing rest on the sill

PLATFORM FRAMING

BALLOON FRAMING

FIGURE 5.2
Comparative framing details for platform framing (*left*) and balloon framing (*right*). Platform framing is much easier to erect and is the only light framing system used today. However, a platform frame settles considerably as the wood dries and shrinks. If nominal 12-inch (286-mm) joists are used to frame the floors in these examples, the total amount of loadbearing cross-grain wood between the foundation and the attic joists is 33 inches (838 mm) for the platform frame and only 4½ inches (114 mm) for the balloon frame.

spreading the blaze rapidly to the upper floors, unless they were closed off with wood or brick *fireblocking* at each floor line. Several modified versions of the balloon frame were subsequently developed in an attempt to overcome these difficulties. The final, most successful of these, the *platform frame*, is now the universal standard for wood light frame construction.

PLATFORM FRAME

Although sometimes complex in its details, the platform frame is simple in concept (Figure 5.3). A floor platform is built. Loadbearing walls are erected upon it. A second-floor platform is built upon these walls and a second set of walls upon this platform. The attic and roof are then built upon the second set of walls. There are, of course, many variations: A concrete slab that lies directly on the ground is sometimes substituted for the ground-floor platform; a building may be only one story tall, or three or more stories in height; and roofs of various types are possible. The essentials, however, remain: A floor platform is completed at each level, and the walls bear upon that platform rather than directly upon the walls of the story below.

The advantages of the platform frame over the balloon frame are several: It uses short, easily handled lengths of lumber for the wall framing. Fireblocking is naturally provided by the platform construction at each

floor level. Its platforms are convenient working surfaces for the carpenters who build the frame. The major disadvantage of the platform frame is that each platform constitutes a thick layer of wood whose grain runs horizontally. This leads inevitably to a relatively large amount of vertical shrinkage in the frame as excess moisture dries from the wood, which, if not properly accounted for, can cause distress in the exterior and interior finish surfaces.

A conventional platform frame is made entirely of nominal 2-inch members, which are actually 1½ inches (38 mm) in thickness. These may be ordered and delivered cut to the nearest 2-foot (600-mm) length, then measured and sawn to exact length on the building site. In larger-volume work, wall studs are frequently precut to the precise required length at the lumberyard and delivered to the site ready to install. Connections are made with nails, alone or in combination with metal connectors as required by the characteristics of each joint (Figures 3.42, 3.49, and 3.50). Nails are driven either by hammer or nail gun. In either case, connections are quickly made because the nails are installed without drilling holes or other time-consuming joint preparation.

Each plane of structure in a platform frame is made by aligning pieces of framing lumber parallel to one another at specified intervals, nailing these to crosspieces at either

end to maintain their spacing and flatness, and then covering the plane of framing with *sheathing*, a facing layer of boards or panels that join and stabilize the pieces into a single structural unit, ready for the application of finish materials inside and out (Figure 5.4). In the floor structure, the parallel pieces are the floor joists, and the crosspieces at the ends of the joists are called *rim boards* or *band joists*. The sheathing applied over the floor framing is called the *subfloor*. In a wall structure, the parallel pieces are the studs, the crosspiece at the bottom of the wall is the *sole plate* or *bottom plate*, and the crosspiece at the top (which is frequently doubled for strength) is called the *top plate*. In a sloping roof, the wall top plates play the role of the lower crosspiece and a *ridge board* is provided at the roof peak.

Openings are required in all these planes of structure, such as for windows and doors in the walls, stairs and chimneys in the floors, and chimneys, skylights, and dormers in the roofs. In each case, additional framing is required to support the ends of the shortened framing members. Openings in floors and roofs are framed with *headers* and *trimmers* (see Figure 5.15). For large openings, these members are doubled, to account for the higher loads they carry. In walls, *sills* head off the bottoms of window and door openings, while *trimmer studs* on the sides provide support to loadbearing headers across the tops (see Figure 5.32).

FIGURE 5.3
The concept of platform framing, shown in cross section, reading from left to right: A foundation wall is constructed. A ground-floor platform is framed and sheathed. Ground-floor wall frames are assembled horizontally on the platform, then tilted into their final positions. A second-floor platform is constructed on top of the ground-floor walls, and the process of wall construction is repeated. The attic floor and roof are added.

FIGURE 5.4
The basic components of platform frame construction. (*a*) Walls are framed with repetitive vertical studs, connected top and bottom by horizontal plates. (*b*) Floors are framed with repetitive joists connected at their ends by rim boards or band joists. (*c*) Sloped roofs are framed with rafters, with a ridge board at the peak. Every surface is sheathed, mostly commonly with plywood or OSB panels.

SUSTAINABILITY IN WOOD LIGHT FRAME CONSTRUCTION

In addition to the issues of sustainability of wood production and use discussed in Chapter 3, the following pertain particularly to wood light frame construction:

• A wood light frame building can be designed to minimize waste in several ways. It can be dimensioned to utilize full sheets and lengths of wood products. Most small buildings can be framed with studs 24 inches (610 mm) o.c. (on center) rather than 16 inches (406 mm). A stud can be eliminated at each corner by using small, inexpensive metal clips to support the interior wall finish materials. If joists and rafters are aligned directly over studs, the top plate can be a single member rather than a double one. Floor joists can be spliced at points of inflection rather than over girders; this reduces bending moments and allows use of smaller joists. Roof trusses often use less wood than conventional rafters and ceiling joists.

• Construction waste can be reduced by careful planning and ordering of the minimum quantity of framing materials, or by a greater reliance on panelized or modular construction systems.

• On the construction site, carpenters can waste less lumber by saving cutoffs and reusing them rather than automatically throwing them on the scrap heap. Scrap lumber can be recycled by shredding it for use in OSB production. Construction scrap should not be burned, because of the air pollution it generates.

• Structural composite lumber, I-joists, finger-jointed studs, and OSB sheathing are all materials that utilize trees more efficiently than solid lumber.

• Although the thermal efficiency of wood light frame construction is inherently high, it can be improved substantially by various means, as shown in Chapter 7. Wood framing is much less conductive of heat than light gauge steel framing. Steel framing of exterior walls is not a satisfactory substitute for wood framing unless the heat flow path through the steel framing members can be broken with a substantial thickness of continuous insulation on one side or the other of the framing.

Throughout the platform frame, the floor joists, wall studs, and roof rafters are consistently arranged so these members fall on a 4-foot (1220-mm) module. This ensures that standard 4-foot by 8-foot (1220-mm × 2440-mm) panels used to cover the inner and outer faces of the framing are readily supported at their ends (see, for example, Figure 5.17). Most commonly, framing members are spaced at 16 in. or 24 in. (406 or 610 mm) o.c.; *o.c.* or *on center* meaning that the spacing is measured from center to center of the members.

The sheathing, usually plywood or OSB (Figure 3.33) nailed over the outside face of the framing, is a key component, tying the various parts of the frame into a structural whole. In the wall frame, for example, the end-nailed connections between top or bottom plates and the studs, by themselves, offer little resistance to wind uplift forces acting on the roof. It is the wall sheathing spanning across these parts that provides the structural continuity needed to carry these forces from the roof to the foundation. Similarly, subflooring and roof sheathing provide important structural continuity to these parts of the structure. As discussed later in this chapter, structural sheathing plays a critical role in the platform frame's resistance to lateral forces exerted by wind or earthquake. Sheathing also furnishes a surface to which shingles, boards, and flooring are nailed for finish surfaces.

FOUNDATIONS FOR LIGHT FRAME STRUCTURES

Most residential and small commercial wood light frame buildings are built on shallow foundations consisting of spread footings, basement or crawlspace walls, and slabs on grade. Once these would have been constructed of stone or brick; today stronger, more durable, and more easily waterproofed steel-reinforced sitecast concrete and concrete masonry are the materials of choice. Examples of these foundation types are illustrated and explained in more detail in Figures 5.5, 5.6 through 5.12. For larger buildings (for example, multistory apartments) or where soil conditions close to the surface are problematic, deeper foundation types, as discussed in Chapter 2, may also be used.

For simple basements and crawlspaces, protection from water entry usually takes the form of bituminous or cement plaster dampproofing applied to the outside of the foundation walls, along with drainage materials surrounding the basement walls and perimeter footing drains to draw water away (Figures 5.6 and 5.7). Where groundwater conditions are severe or there is a need to keep the basement interior as dry as possible, more expensive

The construction of a platform frame building begins with the driving of stakes to fix its position on the site, and the placing of batter boards as reference marks for the builder.

2. *Because the corner stakes will be lost during excavation, batter boards are erected just beyond the area to be excavated. A saw kerf or nail is placed on the batter boards in the plane of each wall, so strings may be stretched to define each plane after excavation is complete.*

1. *Corner stakes for the foundation are located and squared accurately with surveying instruments or 3:4:5 right triangles.*

3. *A plumb bob is used initially to align the marks on the batter boards with the stakes. Later a plumb bob is suspended again from the intersecting strings to locate the corners of the walls.*

FIGURE 5.5

(1) Step one in the construction of a simple platform frame building: establishing the location of the building on the site. After the corners of the foundation have been staked, *batter boards* are erected safely beyond the limits of the excavation area. String lines crossing over the corner stakes are stretched from the batter boards, and their positions on the boards are preserved with notches or nails in the boards. Later, after excavation, the strings are stretched again, and the foundation corners can be relocated at the bottom of the excavation. This figure begins a series of isometric drawings that will follow the erection of a wood light frame building step by step throughout the course of this chapter.

A. TOP OF FOUNDATION WALL

L-shaped anchor bolts will be used to attach the foundation sill plate.

Steel reinforcing within the foundation wall varies with structural requirements.

Dampproofing or waterproofing applied to the outside of the wall resists the passage of ground moisture into the basement.

Protection board or drainage mat protects the moisture-resistant coating from damage during backfilling. Drainage board also conducts groundwater toward the bottom of the wall and the perimeter drain.

Backfilling does not occur until after the first-floor platform is in place and can brace the wall against later soil pressures.

Perforated drain pipe surrounded by coarse gravel conducts groundwater away from the foundation. Filter fabric around the gravel prevents gradual silting of the gravel and drain pipe.

A joint filler strip at the slab perimeter allows expansion and contraction of the slab.

B. PERIMETER FOOTING

A water-resistant coating or membrane, sometimes called a capillary break, applied to the top of the footing resists moisture wicking from the soil up into the foundation wall.

Hooked bars and longitudinal reinforcing connect the footing to the foundation wall and stregthen the footing.

Steel pipe columns, to support the floor framing above, are installed before the concrete slab is poured.

The concrete floor slab, 3″ to 4″ (75 to 100 mm) thick, is reinforced with welded wire reinforcing to resist cracking.

A heavy plastic sheet vapor retarder resists the passage of moisture from the ground below.

C. INTERIOR FOOTING

A layer of gravel under the vapor retarder provides a firm base for the slab and reduces moisture wicking up from the soil below.

FIGURE 5.6
Typical details for a sitecast concrete foundation and basement for a platform frame building. Details A, B, and C are keyed to the circled portions of the drawing in Figure 5.7.

After excavation, concrete footings are poured to spread the load of the building across the surface of the soil, and to make a level surface on which the wall forms can be placed.

The wall forms are located using lines stretched across the marks on the batter boards. Marks are made inside the forms to indicate the level to which the concrete will be poured.

A

C

B

Pockets are provided for steel beams

The formwork panels are held together by steel rods that pass through the concrete. After the forms are removed, the protruding tie rods are snapped off and the tie holes are filled with mortar.

Window openings are made with special form inserts.

2

FIGURE 5·7
Step two in the construction of a typical platform frame building: excavation and foundations. The letters *A*, *B*, and *C* indicate portions of the foundation that are detailed in Figure 5.6.

Exterior insulation and moisture barrier

Batt insulation

Vapor retarder

Sill seal

Flashing

Protective finish

Drainage mat, rigid foam insulation, and waterproofing

Rigid foam insulation

Vapor retarder

A. BASEMENT WITH EXTERIOR INSULATION

Spray foam insulation

Batt insulation

Rigid foam insulation

B. BASEMENT WITH INTERIOR INSULATION

Batt insulation

Vapor retarder

C. CRAWLSPACE WITH INTERIOR INSULATION

Batt insulation

Rigid foam insulation

Vapor retarder

D. CRAWLSPACE WITH NATURAL VENTILATION

Protective finish

Vapor retarder

Rigid foam insulation

Vapor retarder

Drainage mat, rigid foam insulation, and dampproofing

Drainage mat and dampproofing

E. SLAB ON GRADE WITH EXTERIOR INSULATION

F. SLAB ON GRADE WITH INTERIOR INSULATION

FIGURE 5.8

Example foundation insulation strategies. The amount of insulation required varies with the severity of the climate. Exterior cladding and interior finishes are not shown. Wall vapor retarders are shown in their most common location, on the interior (warm) side of the insulation. *A.* Foam plastic on the outside of foundation walls and under slabs on grade should be avoided where the risk of termite infestation is high. This material can provide concealed pathways for these insects between the ground and the building structure. *B.* When facing the building interior, foam plastic insulation must be covered with gypsum wallboard or some other fire-resistant material (not shown here) so that, in the event of a building fire, the release of toxic gasses from the foam plastic is delayed. The use of spray foam insulation, as shown at the rim joist here, is discussed in Chapter 7. *C.* This simple application of insulation batts to a heated crawlspace is suitable for a very mild climate. Other approaches, similar to those shown here for full basements, are also possible. *D.* When a crawlspace is ventilated to the exterior, the floor platform above must be insulated. Rigid insulation underneath the floor joists reduces thermal bridging and protects the joists and batt insulation from high humidity in the crawlspace. The foam plastic must be protected from direct exposure to fire (indicated here by the dashed line). Requirements for a vapor retarder over the soil depend on the amount of ventilation provided. *E.* When a slab on grade is connected to the foundation wall, insulation must be applied to the exterior. *F.* When the slab on grade and foundation wall are independent, insulation may be applied directly under and around the edges of the slab.

Interactive Exercise

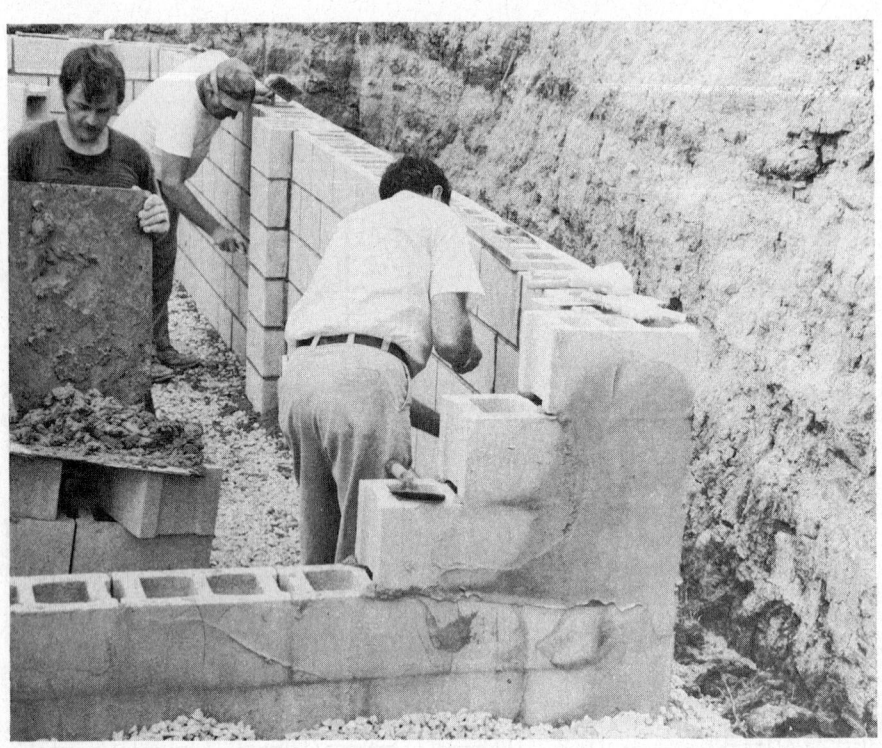

FIGURE 5.9
Masons construct a foundation of concrete masonry. The first coat of parging, which is portland cement plaster applied to help dampproof the foundation, has already been applied to the outside of the wall, and the drainage layer of crushed stone has been backfilled in place. The projecting pilaster in the center of the wall will support a beam under the center of the main floor. After a second coat of parging, the outside of the foundation will be coated with an asphaltic dampproofing compound. *(Reprinted with permission of the Portland Cement Association from* Design and Control of Concrete Mixtures, *12th ed.; Photos: Portland Cement Association, Skokie, IL.)*

FIGURE 5.10
Erecting formwork for a sitecast concrete foundation wall. The footing has already been cast and its formwork removed. It is visible in front of the worker at the left. *(Photo by Joseph Iano.)*

FIGURE 5.11
Concrete foundation work in progress. The foundation wall has not yet been cast on top of the spread footing in the foreground. In the background, crawlspace walls are complete. Openings in the walls will have louvers installed to provide natural ventilation for an unconditioned crawlspace. The rectangular spread footing within the crawlspace is for a masonry chimney. *(Photo by Joseph Iano.)*

FIGURE 5.12
Concrete basement construction in progress. Anchor bolts are omitted from portions of the foundation wall where openings for doors or floor-level windows will be located in the wood framed wall above. Liquid-applied waterproofing and a drainage mat (both dark in color), and polystyrene foam insulation have been applied to the outside of the wall.

and resistant waterproofing treatments may take the place of dampproofing. Slabs on grade should be protected from moisture in the ground below by a heavy plastic sheet vapor retarder. In areas where radon soil gas is a concern, measures for protecting against the entry of this gas into the structure (also as discussed in Chapter 2) may be part of the foundation construction as well.

Energy conservation codes require foundations to be insulated to reduce the exchange of heat with the surrounding soil (Figure 5.8). Some insulation materials (such as boards made of rigid plastic foam or mineral wool) are unaffected by moisture and sufficiently dense that they can be placed on the outside of the foundation wall prior to backfilling or under a slab on grade before

the slab is poured. Many more insulation types can be applied to the inside of the foundation. Crawlspaces may either be ventilated to the exterior or treated as part of the heated or cooled building interior. When ventilated, the floor above the crawlspace is insulated. When the crawlspace is conditioned (heated or cooled), its walls and floor are insulated in manners similar to those for a basement.

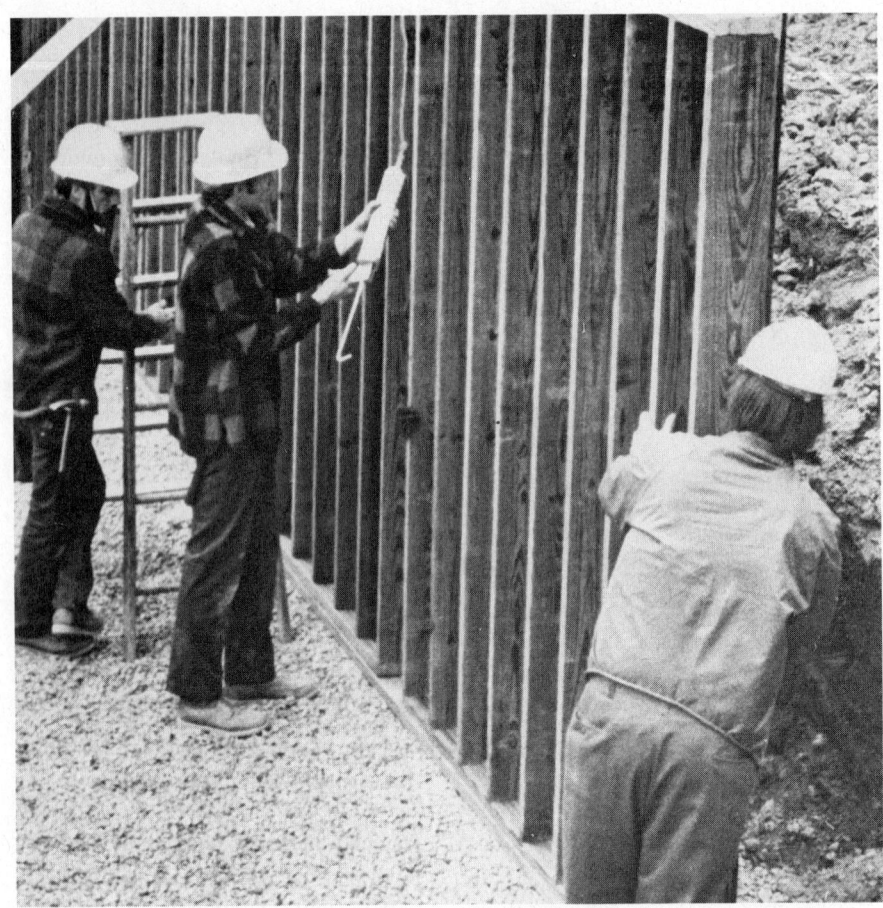

FIGURE 5.13
Erecting a permanent wood foundation.
One worker applies a bead of sealant
to the edge of a panel of preservative-
treated wood components as another
prepares to push the next panel into
position against the sealant. The panels
rest on a horizontal preservative-
treated plank, which, in turn, rests on a
drainage layer of crushed stone. Wood
foundations can be constructed in any
weather and can be insulated in the same
way as the superstructure of the building.
*(Courtesy of APA–The Engineered Wood
Association.)*

Other Foundation Materials

In extremely cold regions, where concrete and masonry construction methods are not practical, foundations may be made entirely of preservative-treated wood (Figure 5.13). Such *permanent wood foundations* can be constructed in any weather by the same crew of carpenters that will frame the remainder of the building. These foundations are readily insulated in the same manner as the frame of the house they support, and they easily accommodate the installation of electrical wiring, plumbing, and interior finish materials. Insulating concrete formwork (ICF) foundation systems using permanent insulating forms are easy to construct, eliminate the time and effort required to remove formwork, and provide integral insulation (see Figure 14.13). Precast concrete foundation systems relying on factory-fabricated reinforced concrete panels can be erected on site more rapidly than traditional sitecast concrete construction, although special attention is also required to seal the joints between adjacent panels. They may be manufactured with insulation integral to the precast panel or designed to accept insulation applied on site after the panels are erected.

BUILDING THE FRAME

Planning the Frame

An experienced carpenter can frame a simple building from the most minimal drawings, but the framing for a larger or custom-designed structure may have to be planned as carefully as for a steel or concrete structure (Figure 5.14). The architect or engineer determines an efficient layout and the appropriate sizes for joists and rafters, and communicates this information to the carpenters by means of *framing plans* (Figures 5.15 and 5.52). For most purposes, member sizes can be determined using standardized structural tables that are part of residential building codes. Or, for more complex framing or special conditions, project-specific engineering may be required. Larger-scale *section details*, similar to those seen throughout this chapter, are prepared for major connections in the building system. The *architectural floor plans* indicate the locations and dimensions of walls, partitions, and openings, and the *exterior elevations* show the outside faces of the building, with vertical dimensions or elevations indicated as required. For most buildings, *building sections* are also drawn that cut completely through the building, showing the dimensional relationships of the various floor levels and roof planes and

SECTION A
1/8′ = 1 foot

FIGURE 5.14

A floor plan and building section are two important components of the construction drawings for a simple house with wood light framing. The ground floor is a concrete slab on grade. Notice that portions of the walls have been designated on the floor plan as shear walls; these are discussed later in this chapter.

FIRST FLOOR PLAN
1/4″ = 1 foot
576 Sq. Ft.

the slopes of the roof surfaces. *Interior elevations* are often prepared for kitchens, bathrooms, and other rooms with elaborate interior features.

Erecting the Frame

The building of the platform frame structure is referred to as *rough carpentry.* The concept of platform frame construction, outlined in Figure 5.3, is illustrated in more detail in this chapter's sequential isometric diagrams, beginning with Figure 5.17, the first-floor platform. In succeeding illustrations, notice how each floor platform provides a ready work surface for the next step in the construction process. As a result, most of the work is accomplished without the use of ladders or scaffolding, and temporary bracing is needed only to support the walls until the next level of framing is installed and sheathed.

The details of a platform frame building—the sizes, spacing, and connections of its members, and even the size and number of nails for each connection (Figure 5.19)—are constructed according to industry standards and closely regulated by building codes. The most common of these details are shown in the figures accompanying the sequential isometrics, beginning with Figure 5.16.

Bridging at midspan is required by some codes

An extra joist is inserted to support the corner of the cantilevered bay

The joists bear on a steel beam in the interior of the house

Fireplace opening

Stair opening

Regular joist spacings of 16″ or 24″ (406 mm or 610 mm) are maintained so as to align with joints in the plywood subfloor

Double header joists support the ends of tail joists at floor openings

Double trimmer joists support header joists

Sheet metal joist hangers are used wherever joists support one another at right angles

FLOOR FRAMING PLAN

FIGURE 5.15
A framing plan for the ground-floor platform of the building shown in Figure 5.17.

FIGURE 5.16
Ground-floor framing details for both solid lumber (*above*) and I-joist framing (*below*), keyed to the lettered circles in Figure 5.17. Continuous pieces of lumber are drawn with an X inside and intermittent blocking with a single diagonal. In both details *B*, framing members that cantilever beyond the foundation must either be preservative-treated or the deck construction must be carefully detailed to protect the framing from moisture and decay. The I-joist version shows cantilevered, preservative-treated solid lumber joists overlapped with the interior I-joist framing; several alternative framing strategies are also shown. (For clarity, webs of I-joists seen in elevation are shaded.)

The sill is bolted down and leveled on top of the foundation wall. The floor joists are toenailed to the sill.

A. EDGE DETAIL AT END WALLS

Solid blocking between joists prevents the joists from overturning.

B. EDGE DETAIL AT CANTILEVERED JOISTS

In areas of high termite risk, a continuous metal shield prevents termites from traveling undetected from the foundation into the wood framing above.

C. EDGE DETAIL AT SIDE WALLS

A wood sill fastened to the steel beam provides nailing for the joists and equalizes shrinkage with the edge detail.

D. DETAIL OF INTERIOR JOIST BEARING

Either an I-joist or a structural composite rim board can be used in this location.

A. EDGE DETAIL AT SIDE WALLS

Section

Cantilevers may be made with various combinations of solid lumber, I-joists, and plywood or OSB, as shown.

Section

Alternate sections

B. EDGE DETAIL AT CANTILEVERED JOISTS

A structural composite rim board, which matches the height and shrinkage rate of the I-joist, is used in place of solid lumber.

C. EDGE DETAIL AT SIDE WALLS

Blocking between joists is made from short sections of I-joists.

D. DETAIL OF INTERIOR JOIST BEARING

Attaching the Frame to the Foundation

The *foundation sill plate* (sometimes also called a *mudsill*), made of preservative-treated wood for resistance to insects and moisture, is attached to the top of the foundation to serve as a base for the wood framing to follow (Figure 5.16, *A, B,* and *C*). A single nominal 2-in. (38-mm) sill, as shown in the details here, may be all that is required. In better-quality work or where seismic or wind design forces are high, the sill may be doubled or made from thicker stock for added stiffness and to create a stronger connection between the foundation and the wood frame above. At a minimum, the sill is fastened in place with ½-in. (13-mm) anchor bolts, embedded in the top of the foundation wall and spaced not more than 6 ft (1829 mm) apart. Alternatively, similarly spaced proprietary sheet metal anchor straps may be used instead of bolts.

Because the top of a foundation wall is likely to be somewhat uneven, the sill is shimmed up at low spots with wood shingle wedges or plastic shims to provide a more level base for subsequent framing. (The term "mudsill" is a carryover from times when the sill was set in a bed of cement mortar, or "mud," to accomplish the same purpose.) A *sill seal* or *sill gasket*, made of any of a number of compressible or resilient materials, is inserted between the sill and

Plywood sheets are considerably stiffer along their length than across their width, so they must be laid with their long dimension perpendicular to the joists. The end joints are staggered to avoid lines of weakness.

When the foundation is complete, basement beams are placed, sills are bolted to the foundation, and the first-floor joists and subfloor are installed.

Joist bridging

FIGURE 5.17

③ Step three in erecting a typical platform frame building: the ground-floor platform. Compare this drawing with the framing plan in Figure 5.15. Notice how the direction of the joists changes to construct the cantilevered bay on the end of the building. A cantilevered bay on a long side of the building could be framed by merely extending the existing floor joists over the foundation. The bridging shown between joists is discussed later in this chapter. The letters *A*, *B*, and *C* represent portions of the framing detailed in Figure 5.16.

A. STEEL BEAM UNDER JOISTS

Steel beams span further
between supports than solid
wood beams, and eliminate
unequal floor subsidence due
to beam shrinkage

B. GLULAM OR STRUCTURAL COMPOSITE BEAMS UNDER JOISTS

Structural composite or glulam
wood beams, which shrink less
than solid wood, are also
suitable for placing underneath
the floor joists.

C. SOLID LUMBER BEAM FLUSH WITH JOISTS

A solid lumber beam should be framed
flush with the floor joists, to minimize
the effects of beam shrinkage. The joist
ends are supported with sheet metal
joist hangers.

D. STEEL BEAM AND I-JOISTS WITH FILLER BLOCKS

Framing a steel beam flush with the
joists increases headroom under the
beam. Filler blocks on either side of
the beam web are required where the
bottom of the joists do not reach the
lower flange of the beam.

E. STEEL BEAM RECESSED WITH JOISTS

Solid lumber joists can be framed into
the steel beam in the same manner as
shown in D or as shown here. A wood
"scab" joins joists and supports the
subfloor.

F. STEEL BEAM AND I-JOISTS WITHOUT FILLER BLOCKS

Where the bottom of the joists reach
the lower flange of the steel beam, no
filler blocks are required.

FIGURE 5.18
Methods of beam support for lumber
joists (*left column*) and I-joists (*right
column*) between foundation walls.
Where joists pass over the top of the
beam (*A, B*), beams made of steel,
glue-laminated wood, or structural
composite lumber should be used.
Solid lumber beams should be avoided
in this application, as their greater
drying shrinkage can cause noticeable
subsidence in the floor. Solid lumber
beams should be framed flush with the
joists (*C*) so that their shrinkage occurs
in parallel with the joists. Ways to frame
joists flush with steel beams are shown
in *D, E,* and *F.* In *D,* the bottoms of the
joists are above the bottom flange of the
beam, and filler blocks are bolted to the
web of the beam to restrain the joists
from sidesway in the case of a seismic
event. Filler blocks also allow the use of
face-mounted joist hangers, rather than
top-mounted as shown here. In *F,* the
joists are deep enough to be restrained
by the bottom flange of the steel beam
and no filler blocks are required. In *E,* a
small space must be left over the top of
the beam to allow for drying shrinkage in
the joists.

Connection	Nail	Minimum Length and Shaft Diameter	Nailing
Girders and beams built up from nominal 2-inch (38-mm) lumber	10d common or box	3 × 0.128″ (76 × 3.3 mm)	Face nail, 32″ (800 mm) o.c. top and bottom edges, nails staggered; plus 2 nails at each end or splice
Floor joists to plate or beam below	8d common or box	2½ × 0.113″ (64 × 2.9 mm)	3 toe nails
Joists to headers, headers to trimmers	Use joist hangers		
Joist and rafter blocking, to plate or beam below	8d common or box	2½ × 0.113″ (64 × 2.9 mm)	3 toe nails
Bottom plate to joist or blocking below	16d common or box	3½ × 0.135″ (89 × 3.4 mm)	Face nail, 16″ (400 mm) o.c.
Stud to bottom plate	8d common or box	2½ × 0.113″ (64 × 2.9 mm)	3 toe nails
Stud to bottom plate or top plate	16d common or box	3½ × 0.135″ (89 × 3.4 mm)	2 end nails
Double studs, built-up corner studs	10d common or box	3 × 0.128″ (76 × 3.3 mm)	Face nail, 24″ (600 mm) o.c.
Built-up header with ½-in. (13-mm) spacer	16d common or box	3½ × 0.135″ (89 × 3.4 mm)	Face nail, 16″ (400 mm) o.c. top and bottom edges
Let-in bracing	8d common or box	2½ × 0.113″ (64 × 2.9 mm)	2 face nails or staples at each stud or plate
Double top plate	10d common or box	3 × 0.128″ (76 × 3.3 mm)	Face nail 24″ (600 mm) o.c.
Double top plate at end laps	16d common or box	3½ × 0.135″ (89 × 3.4 mm)	8 face nails in minimum 24-in. (600-mm) lap
Double top plate corner and intersection laps	10d common or box	3 × 0.128″ (76 × 3.3 mm)	2 face nails
Ceiling joists to top plate	8d common or box	2½ × 0.113″ (64 × 2.9 mm)	3 toe nails
Ceiling joist laps over beam or wall below	10d common or box	3 × 0.128″ (76 × 3.3 mm)	3 face nails
Rim joist to top plate	8d common or box	2½ × 0.113″ (64 × 2.9 mm)	Toe nail 16″ (400 mm) o.c.
Rafter to top plate	16d common or box	3½ × 0.135″ (89 × 3.4 mm)	2 toe nails
Rafter to ridge board, valley, or hip rafter	16d common or box	3½ × 0.135″ (89 × 3.4 mm)	4 toe nails, or 3 face or end nails
Rafter to ceiling joist	10d common or box	3 × 0.128″ (76 × 3.3 mm)	3 face nails
Rafter to collar tie	10d common or box	3 × 0.128″ (76 × 3.3 mm)	3 face nails
Structural sheathing up to ½″ (13 mm), subfloors and interior walls	6d common only	2 × 0.113″ (51 × 2.9 mm)	Face nails 6″ (150 mm) o.c. at edges and 12″ (300 mm) o.c. at intermediate supports
Structural sheathing up to 1″ (25 mm), subfloors, interior walls, exterior walls, and roofs	8d common only	2½ × 0.131″ (64 × 3.3 mm)	Same as above

the foundation to reduce air infiltration through this gap and to restrict moisture wicking up from the foundation into the wood framing (Figure 5.20). Where the risk of termite infestation is high, metal *termite shields* may be inserted between the foundation wall and the sill. These prevent termites from traveling undetected from cracks in the concrete up into the wood framing. As discussed later in this chapter, where wind and earthquake design forces are high, additional measures may be taken to further strengthen the frame-to-foundation connection.

Floor Framing

Solid lumber floor joists typically range in size from 2×6 (38×140 mm) for lightly loaded floors and short joist spans, to 2×12 (38×286 mm) for heavier loads and longer spans. They are most commonly spaced at 16 or 24 in. (406 or 610 mm) o.c. As can be seen in Figure 5.17, this ensures support for the ends of the subflooring at 4-ft (1220-mm) intervals, as occurs when 8-ft (2440-mm)-long panels are laid with their end joints staggered in adjacent rows. For heavy loads or long spans, but where a deeper joist is not practical, a closer joist spacing, such as 12 in. (305 mm) o.c., may also be used.

FIGURE 5.20
Carpenters apply a preservative-treated wood sill to a sitecast concrete foundation. Compressible glass fiber sill seal has been placed on the top of the concrete wall, and the sill has been drilled with snugly fitting holes at anchor bolt locations. Before each section of sill is bolted tightly, it is leveled as necessary with wood shingle or plastic shims between the concrete and the wood. As the anchor bolts are tightened, the sill seal squeezes down to a negligible thickness. A section of bolted-down sill is visible at the upper right. The basement windows were clamped into reusable steel form inserts and placed in the formwork before the concrete was cast. After the concrete was cast and the formwork was stripped, the steel inserts were removed, leaving a neatly formed concrete frame around each window. A pocket in the top of the wall for a steel beam can be seen at the upper left. *(Photo by Edward Allen.)*

FIGURE 5.19
Standard nailing for platform frame construction. These requirements are incorporated into building codes, and experienced framing carpenters know them by heart. For fastening of framing members, either common or box nails are permitted. Structural sheathing must be nailed with common nails only. Where sinkers are used, larger penny sizes than indicated may be needed to meet the minimum length and shaft diameter requirements. Despite the apparent complexity of this list, it is possible to simplify the number of nails types required for rough framing work. For example, all of the above can be accomplished with just a 16d box nail and an 8d common nail. Where this table indicates 10d nails face-nailed into two nominal 2-inch members (3 in. or 76 mm in combined thickness), a 16d nail (3½-in. or 89-mm long) can be used, driven at an angle so that its pointed end does not protrude from the opposite side. A more comprehensive listing of design values and fastening requirements for nails and staples used in structural applications is provided in the industry-standard report ESR-1539.

FIGURE 5.21
Installing solid lumber floor joists. Also visible in the face of the rim joist are the toe nails into the sill and the end nails into the floor joists. *(Photo by Edward Allen.)*

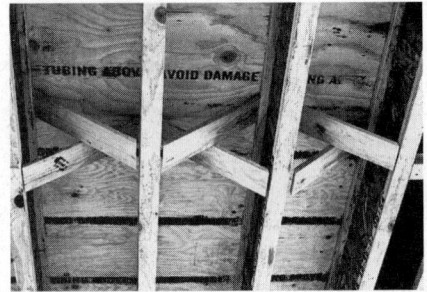

FIGURE 5.22
Wood crossbridging for I-joists. Bridging installation begins from above, before the subflooring is laid. Fastening of the bottom ends must be completed from below, so, as seen in this photograph, it is frequently deferred until later. Note the printing "TUBING ABOVE AVOID DAMAGE" on the underside of the subflooring panels, indicating hydronic heating tubes installed in upper side of the panels. *(Photo by Joseph Iano.)*

Joists, which rest on their narrow edges, must be restrained from overturning. At a minimum, this is accomplished by end nailing through the rim joists into the joist ends (Figure 5.21). In areas subject to high seismic forces, additional overturning resistance is provided by lines of solid blocking inserted between joists over floor beams, bearing walls, and other intermediate lines of joist support. Additionally, where solid joists exceed 12 in. (286 mm) in depth, *bridging* (solid blocking, cross-bracing, or strapping) must be inserted at intervals not exceeding 8 feet (2.4 m) (Figure 5.22). In better-quality construction, blocking or bridging may be installed into all floor framing, regardless of seismic or joist depth requirements, to improve stiffness and reduce vibration in the floor system.

Where floor openings occur, such as for stairs or chimneys, headers and trimmers are used to frame the opening and carry the floor loads around the opening to supporting walls or beams. For larger openings, these members are doubled to account for the greater loads that they must carry. Where the ends of joists meet a supporting header, nailing alone cannot be relied upon to transfer loads between the members, so sheet metal joist hangers are used.

Each hanger provides a secure pocket for the end of the joist and punched holes into which properly sized nails are driven to make the connection. Extra joists may also be inserted into the floor framing anywhere concentrated loads from above are expected, such as underneath bathtubs or load-bearing walls.

Today, manufactured I-joists are as likely to be used for floor framing as solid lumber (Figure 5.24).

FIGURE 5.23
Metal crossbridging for solid wood joists. The steel strip is manufactured in a folded V-shape for increased stiffness. One or both ends of the brace may be toothed, to allow installation directly into the joist without the need for nails. *(Courtesy of APA–The Engineered Wood Association.)*

FIGURE 5.24
**A floor platform being framed with
manufactured I-joists. The joists in this
photograph are made with laminated
veneer lumber (LVL) flanges and
plywood webs. Note the ease with which
one carpenter can lift this relatively long
but lightweight member.** *(Photos provided
by Trus Joist MacMillan.)*

Compared to solid wood joists, these components are lighter in weight, can span greater distances, provide greater straightness and uniformity, and reduce drying shrinkage of the floor platform. I-joists can also more easily accommodate piping, ductwork, and wiring within the floor platform, as openings (sized and located so as not to impair the structural capacity of the joist) can be made in the relatively thin web of the I-joist more easily than in solid lumber. Some I-joists are manufactured with preformed web knockouts, ensuring that openings are both easy to make and properly located and sized.

I-joists range from 9½ to 24 in. (241 to 610 mm) in depth, with flanges from 1¾ to 3½ in. (44 to 90 mm) wide. They are capable of spanning up to approximately 30 ft (9 m). For normal loads and spans, joists are most commonly spaced at 24 in. (610 mm) o.c. However, closer spacings may be used for longer spans, greater floor loads, increased floor stiffness, or where a subfloor panel with a lesser span rating is used.

The relatively thin webs of the I-joist may require special consideration in joist connection details. For example, where a joist hanger is not tall enough to restrain both the top and bottom flanges of the joist, *web blocking* is inserted on either side of the web to restrain the joist from tipping sideways (Figure 5.25).

Where the loads applied to I-joists at their ends or at points of intermediate support are high enough to risk buckling or crushing of the joist web, blocking of various types, as shown in Figure 5.47, may also be used. Bridging and blocking, installed to resist joist overturning or stiffen longer spans, are used with I-joists in the same manner as just described for solid lumber joists. Manufactured floor trusses (Figure 5.26), like I-joists, can span farther than solid wood joists, reduce shrinkage in the floor platform, and provide natural openings to accommodate building services.

In comparison to solid wood joists that provide only 1½ in. (38 mm) of supporting surface for the subfloor above, the top surfaces of I-joists and floor trusses may be as wide as 3½ in. (64 to 89 mm). This allows an opportunity for additional possible economy in the use of these member types. Where solid joists are spaced at 16 in. (406 mm) o.c., it may be possible to space trusses or I-joists at 19.2 in. (488 mm) without having to increase the thickness or strength of the subflooring. This is possible because the wider top surface effectively reduces the span of the subfloor panels between the joists. This small change in joist spacing, which requires one less joist over every 8 ft (2440 mm), results in a 17 percent reduction in joist framing materials and installation labor.

Subflooring is installed after floor framing is complete. The subfloor panels are laid with their longer dimension perpendicular to the framing on which they are supported, as the panels are considerably stiffer in this orientation (Figure 5.27). With plywood subflooring, which typically has different veneer grades on either side, the panel is laid with the better grade facing up. A ⅛-in. (3-mm) gap is maintained around all panel edges, to prevent buckling caused by the expansion of rain-wetted panels. To reduce squeaking in the finished floor and increase floor stiffness, adhesive may be applied to the tops of joists before the panels are laid, and deformed shank nails or self-drilling screws with greater withdrawal resistance than plain common nails may be used for fastening (Figure 5.28).

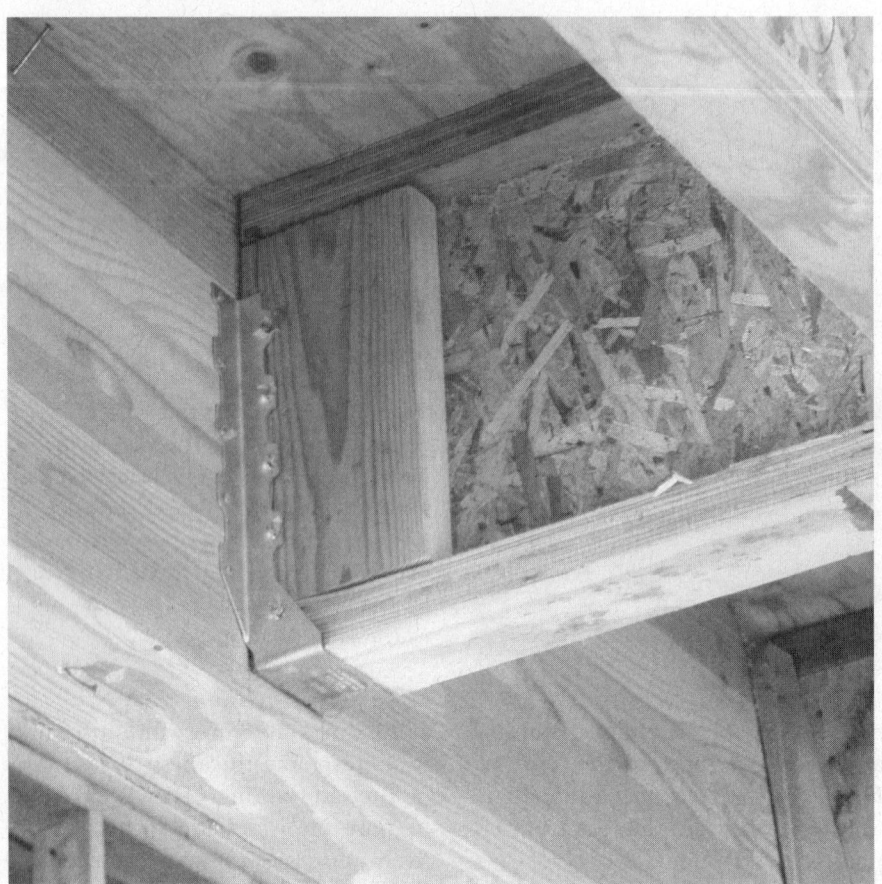

FIGURE 5.25
I-joist web blocking at a sheet metal joist hanger. The blocking is applied to both sides of the web, to restrain the joist from rotating within the hanger. Alternatively, if a taller hanger capable of restraining both the top and bottom flanges of the joist is used, blocking is not required for this purpose. Web blocking may also be used to stiffen the relatively slender web, to prevent it from buckling under heavy loads, such as when a bearing wall rests on the platform directly above. As can be seen in the photograph, web blocking should be cut slightly shorter than the joist web itself, so that if the web compresses slightly under load, the blocking does not pry the top flange apart from the web. *(Photo by Joseph Iano.)*

FIGURE 5.26
These floor trusses (shown here being set up for a demonstration house in a parking lot) are made of sawn lumber members joined by toothed-plate connectors. The OSB web at each end provides a section of truss that can be sawn by workers to easily adjust the length of the truss as necessary. Trusses are deeper than solid wood joists or I-joists but can span farther. Even though they are deeper than other joist products, they may result in a thinner floor/ceiling assembly, as pipes and even relatively large sections of ductwork can run through the passages within the truss rather than below it. *(Photo courtesy of the Wood Truss Council of America.)*

FIGURE 5.27
Applying OSB subflooring. Note that the longer dimension of the panel runs perpendicular to the joist framing. *(Courtesy of APA–The Engineered Wood Association.)*

FIGURE 5.28
For a stiffer, quieter floor, subflooring may be glued to the joists. The adhesive is a thick mastic that is squeezed from a sealant gun on the tops of the joists just before the subflooring is laid in place. *(Courtesy of APA–The Engineered Wood Association.)*

Wall Framing

Walls are usually framed with 2 × 4 (38 × 89 mm) or 2 × 6 (38 × 140 mm) members—the deeper 2 × 6 used to support greater loads or provide more space for insulation in the wall. Solid lumber studs are most common. However, structural composite or finger-jointed lumber may also be used. Like floor joists, wall studs are spaced at 16 or 24 inches (406 or 610 mm) to coordinate with the 4-foot (1.2-m) module of exterior sheathing and interior wallboard panels. Studs are nailed in place between top and bottom plates made from the same size lumber as the studs. With loadbearing walls, the top plate is doubled for greater strength and stiffness, to help with the transfer of the loads from above into the studs below. Walls are constructed in sections lying down, using the previously built floor platform as a work surface. When ready, they are tilted up, plumbed (set vertical), and nailed into position, with temporary bracing applied to hold them in position until adjoining framing is completed (Figures 5.29 through 5.36).

A

B

C

D

E

F

SIDE WALL

INTERIOR BEARING WALL

END WALL

FIGURE 5.29
Typical ground-floor wall framing details, keyed by letter to Figure 5.30.

Openings for windows and doors are formed with headers, trimmers, and sills. Headers, spanning across the top of an opening, take the loads from above the opening and carry them to either side. They are made from two members on edge, with material sandwiched between so that their total width is equal to that of the wall. The depth of the header varies according to the work it must perform: shallow for light loads and short spans, and deeper for larger loads and longer spans. Where they are needed, structural composites with greater strength, such as LVLs or PSLs, may be used. Prefabricated headers that include integrated thermal insulation to reduce heat loss through these difficult-to-insulate areas may also be used. At either end, headers rest on shortened studs called trimmer or *jack studs*, which themselves are nailed to full-height *king studs*. At the bottom of a window opening, the *rough sill* is supported on *cripple studs* (Figure 5.32).

Where walls intersect, nailing surfaces must be provided for supporting the edges of the exterior sheathing and interior wallboard. This requires at least three studs at each intersection, unless special metal clips are

The upper top plate overlaps the lower top plate at corners to join the walls.

The subfloor makes a convenient platform on which to assemble the first-floor wall frames. The assembled frames are tilted up into place, nailed to the floor and to one another, and supported by temporary braces.

④ FIGURE 5.30
Step four in erecting a platform frame building: The ground-floor walls are framed. The letters *A, C, D,* and *F* indicate portions of the framing that are detailed in Figure 5.29.

used to reduce the number to two (Figure 5.32).

Walls exceeding 10 ft (3 m) in height must have solid blocking inserted at midheight, to limit the volume of the wall cavities and reduce the ease with which fire can spread within the wall. Where long, consistently straight pieces of solid lumber are not readily available for the construction of unusually tall walls, structural composite or finger-jointed stock may be used.

Wall sheathing, most frequently plywood or OSB, provides a nailing surface for exterior cladding materials and stiffens the wall against the lateral forces of wind or earthquake (Figure 5.1). Other sheathing panel types, made from wood or paper fiber, plastic foam, or glass or mineral fiber, are intended principally as thermal insulation and to provide a base for building paper, housewrap, or other weather-resistant coverings. Where walls are sheathed with such nonstructural panels, *let-in diagonal* bracing can be used to provide lateral force resistance. Let-in bracing may be made of 1 × 4 (19 × 89 mm) wood boards, or light steel members that are recessed into the outer face of the studs of the wall before it is sheathed

Single top plate

Studs

Joists

2 × 4 (38 mm × 89 mm) blocking 48" (1200 mm) apart supports the top of the partition

A 1 × 8 (19 mm × 184 mm) nailer provides nailing surfaces for the ceiling material edges

Sole plate

Subfloor

Joists

Solid blocking under studs

A nonbearing partition perpendicular to the joists is framed very simply

A nonbearing partition parallel to the joists may be supported by a double joist beneath, or by transverse blocking between joists as shown

FIGURE 5.31
Framing details for nonloadbearing interior partitions.

1. This is the layout of a typical exterior wall. It meets two other exterior walls at the corners, and a partition in the middle. It has two rough openings, one for a window and one for a door.

2. The framer begins by marking all the stud and opening locations on the sole plate and top plate. The "special" studs are cut and assembled first: two corner posts, a partition intersection, and full-length studs and supporting studs for the headers over the openings.

3. The wall is next filled with studs on a regular 16" (400 mm) or 24" (600 mm) spacing, to provide support for edges of sheathing panels.

4. Diagonal bracing, usually 1 × 4 (19 mm × 89 mm), is let into the face of the frame if the building will not have rigid sheathing. The second top plate may be added before the wall is tilted up, or after.

1 × 8 (19 mm × 184 mm)

2 × 4 (38 mm × 89 mm) blocks 24" (600 mm) apart

A. Two alternative ways of making a corner post. Each provides both an exterior and an interior nailing surface for each wall plane

B. Two alternative ways of making a partition intersection. Each provides a nailing surface for each interior wall plane

Nailing surfaces

Nailing surfaces

Nailing surfaces

Cripple studs support the double top plate

Top plate

The header is a sandwich of two 2" (38 mm) members around a ½" (13 mm) plywood spacer, to equal the stud depth of 3½" (89 mm)

Supporting studs support the header

Rough opening width

Rough sill

Sole plate

Header height

Rough opening height

C. SECTION THROUGH A WINDOW OPENING

D. ELEVATION OF A WINDOW OPENING

FIGURE 5.32

Steps in the framing of a typical wall and details at wall intersections and a window opening. In *D*, the short studs above the header and below the rough sill are called "cripple studs."

Watch the Animation

FIGURE 5.33
The wall frame is constructed while it lies flat on the already completed floor platform. Here, studs for a partition intersection are nailed to a plate, using a pneumatic nail gun. *(Photograph courtesy of Senco Products, Inc.)*

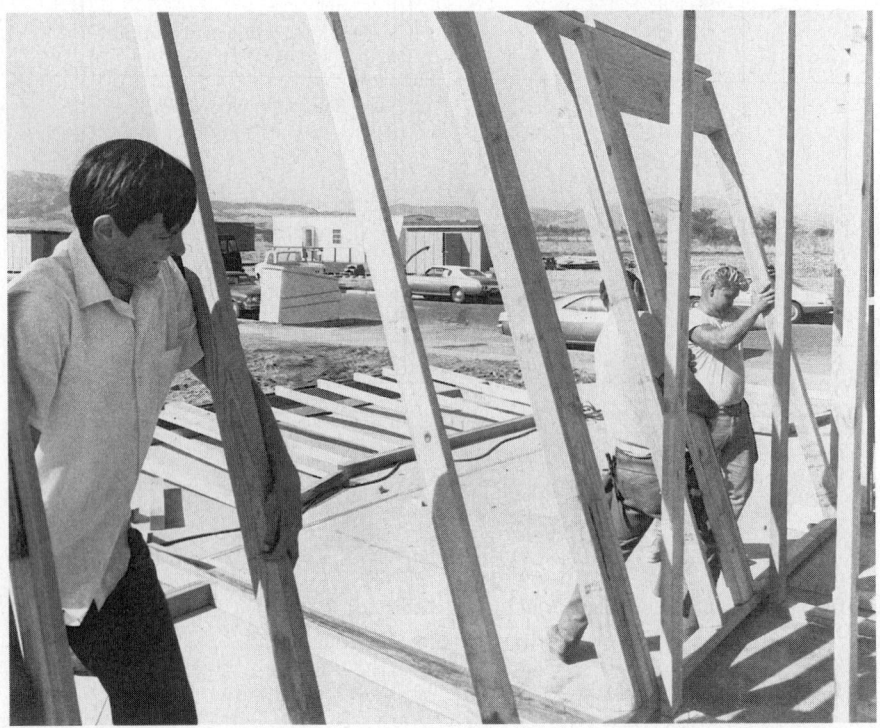

FIGURE 5.34
Tilting an interior partition into position. The gap in the upper top plate will receive the projecting end of the upper top plate from another partition that intersects at this point. *(Courtesy of APA– The Engineered Wood Association.)*

FIGURE 5.35
Fastening a wall to the floor platform. Horizontal blocking between studs, as seen toward the left in this photo, is installed to provide solid fastening for bathroom hardware, exterior panels, or any number of possible other items to come later. *(Photograph courtesy of Senco Products, Inc.)*

FIGURE 5.36
Wall framing is held in place by temporary bracing until the floor framing is in place above and wall sheathing is complete, after which the frame becomes self-bracing. The outer walls of this building are framed with 2 × 6 (38 × 140 mm) studs to allow for a greater thickness of thermal insulation, whereas the interior partitions are framed with 2 × 4 (38 × 89 mm) studs. *(Photo by Joseph Iano.)*

(Figures 5.32 and 5.37). However, let-in bracing is less effective at resisting lateral forces than structural panel sheathing, and it is not suitable for use where these forces are high.

Lateral Force Resistance and Shear Walls

To a greater or lesser degree, all structures must resist the horizontally acting forces of wind and earthquake. In areas prone to very strong winds or severe earthquakes, as well as for taller wood light frame structures, special attention is given to the design of the platform frame and the detailing of its connections to ensure that it can safely withstand the effects of these *lateral forces* (Figure 5.38).

Lateral forces acting on a building can cause the structure to slide off its foundation, overturn, wrack, or suffer failures among its parts. Sliding at the top of the foundation is resisted by the anchor bolts and foundation sill plate, as described earlier in this chapter. Where the forces are high, this connection can be strengthened with thicker plates made of stronger lumber, and anchor bolts that are larger in diameter, more closely spaced, and fastened with larger washers.

FIGURE 5.37
Applying a panel of insulating foam sheathing. Because this type of sheathing is too weak to brace the frame, diagonal bracing is inserted into the outside faces of the studs at the corners of the building. Steel let-in bracing, nailed at each stud, is used in this frame and is visible just to the right of the carpenter's leg. *(Courtesy of The Celotex Corporation.)*

SLIDING

OVERTURNING

FIGURE 5.38
Lateral forces acting on a building can cause sliding (*top*), overturning (*middle*), or wracking (*bottom*). An additional consideration, failure between structural parts, is discussed in the accompanying text.

WRACKING

To resist overturning, *hold-downs* are installed to prevent the structure from lifting off the foundation (as their name suggests). These devices may also be used higher up in the structure to prevent the upper stories or roof from separating from the portions of the structure below (Figures 5.39, 5.40, and 5.41).

A wall without sheathing or bracing has no useful resistance to *wracking*. Because the nailed connections between studs and plates are flexible, the wall can easily deform. In areas of high wind force or earthquake risk, plywood or OSB sheathing panels tightly nailed to the framing create *shear walls* that provide the necessary rigidity. As lateral force design values increase, shear walls are made stronger by lengthening the wall, using thicker structural panels, adding a second panel to the opposite side of the wall, increasing the size of nails used to attach the panels to the framing, and spacing nails more closely, especially around panel edges. Shear walls frequently also require extra studs or thicker posts at their ends, to prevent localized crushing or failure where forces tend to concentrate (Figure 5.41).

Both interior and exterior walls can act as shear walls. Separate sections of wall must be arranged perpendicularly so that the structure as a whole can resist forces acting from different directions. Shear walls must also be distributed in plan so as to ensure a reasonably balanced overall response to such forces.

Openings in shear walls, such as for doors and windows, significantly reduce their lateral force resistance. Where such openings occur, they are reinforced with blocking and strapping. Where openings take up too large a percentage of the wall area and conventional construction methods cannot provide the necessary resistance, stronger and stiffer welded steel frames or factory-fabricated shear

FIGURE 5·39
Strap tie hold-downs are made of galvanized steel straps with hooked or deformed ends cast into the concrete foundation wall (*left*). After the wall is framed, the exposed length of strap may be nailed directly to framing or, as seen here, nailed through the sheathing into the studs or posts behind (*right*). The number, size, and spacing of nails used to fasten the strap depend on the magnitude of the loads that must be resisted and the capacity of the wood member to hold nails without splitting. In the left image, anchor bolts cast into the top of the foundation wall that will be used to secure the sill plate in place are also visible. (*Photos by Joseph Iano.*)

Upper floor platform

Steel angle with welded stiffener plate

Bolts

Corner studs

Threaded anchor rod

Ground-floor platform

Foundation

FIGURE 5.40

Hold-downs made from threaded rod and steel plate anchors can resist much greater forces than strap tie hold-downs. They may be used at each floor level to securely tie the building frame through its full height to the foundation. For the type shown here, the nuts at the ends of the threaded rods may require retightening after the first heating season to compensate for wood shrinkage, which can mean that access holes must be provided through the interior wall surfaces. Other models are made self-adjusting with the use of spring-loaded, tapered shims or other compensation mechanisms.

FIGURE 5.41

A heavy-duty seismic hold-down similar to that illustrated in the previous figure. The anchor rod with one end cast into the concrete foundation wall protrudes through the preservative-treated wood sill where its threaded end is bolted to the anchor. The anchor in turn is bolted to a 4 × 4 (89 × 89 mm) post with five bolts of substantial diameter. A conventional foundation anchor bolt with an oversized square washer is also partially visible to the right of the hold-down. Also note the thicker-than-normal, 3-inch nominal (64-mm) sill plate, a common feature in wood light framing designed for high seismic forces.

(Photo by Joseph Iano.)

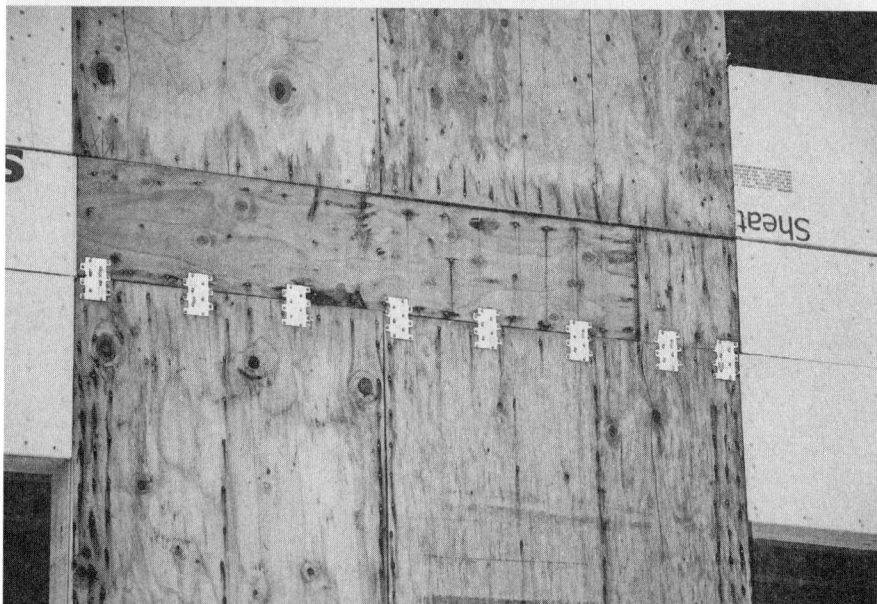

FIGURE 5.42
Above, shear walls for a multistory, wood light frame structure. The darker hue of the plywood sheathing used at the two shear walls distinguishes them from the surrounding framing covered with gypsum panel sheathing. Below, a close-up of the shear wall at the second-floor platform level. The metal clips increase the shear force transfer capacity between the top of the first-floor wall and the second-floor platform. The panel nailing pattern can be observed by the fastener weathering stains. Fasteners are more closely spaced at panel edges. The multiple rows of fasteners at the extreme left and right of the shear wall are an indication of the thicker framing members used at these locations. *(Photos by Joseph Iano.)*

panels made of wood or steel components may be used (Figure 5.43).

In the International Building Code, the term "shear wall" is reserved for a lateral-force-resisting wall designed by a professional engineer, for use in a single building. Where the design conditions are not too severe, the code also provides for more general purpose, preengineered, prescriptive wall designs. These so-called *braced walls* or *braced panels* are designed and built according to the same principles as shear walls, the difference being that

no professional analysis is required. Where design forces are low, sheathing panels made from fiberboard, particle board, or even gypsum wallboard, rather than plywood or OSB, may provide adequate rigidity.

With the components necessary to resist sliding, overturning, and wracking incorporated into the building structure, it still remains necessary to ensure that the various building parts are adequately interconnected. *Collectors* (also called *drag struts* or *drag ties*) are components that transfer

lateral forces from larger areas, such as floors or roofs, to the parts of the structure designed to resist these forces, such as the shear walls. For example, wall top plates typically collect lateral forces that accumulate in the floor platform above and carry these forces to the top of the nearest shear wall. Strategically located and properly fastened joists, rafters, other framing members, and (where necessary) metal strapping or solid wood blocking may also perform in this way (Figure 5.44). Collectively, these components ensure continuous *load paths* so that forces acting within the various parts of the structure will be carried reliably and securely to the building foundation and surrounding soil.

Upper-Level Floor and Wall Framing

After the first-floor platform and wall framing are complete, the process and details of floor framing followed by wall framing may continue for each additional level of the structure, as shown in Figures 5.45 through 5.50.

FIGURE 5.43
Prefabricated shear panels are especially useful where large openings in walls leave little solid area for shear wall construction, such as in garage walls. They may be made of all wood components or, where even greater strength is required, of wood and metal, as shown here. The corrugated galvanized steel in this panel is more than ⅛ inch (3.5 mm) thick, and the panel's capacity to resist lateral forces is several times greater than that of comparable all-wood prefabricated panels. The bottom of the panel will be anchored with bolts embedded 21 inches (533 mm) or more into the concrete foundation, and the sides and top will be screw-fastened to the surrounding wood framing. Holes in the panel can accommodate wiring runs within the wall. *(Courtesy of Simpson Strong-Tie Company Inc.)*

FIGURE 5.44
In addition to supporting gravity loads over the garage opening, this glulam beam acts as a collector of lateral forces in the deck framing that it supports. These forces are carried either to the narrow section of solid wall framing on the right, or via the metal strap and adjacent garage opening beam, to additional shear wall framing beyond the edge of this photograph. *(Photo by Joseph Iano.)*

When the first-floor walls are complete and sheathed, much of the temporary bracing can be removed. The second-floor platform is framed the same as the first, with its joists resting on the double top plates of the first-floor walls.

FIGURE 5.45

(5) Step five in erecting a two-story platform frame building: building the upper-floor platform. The letters *A*, *B*, and *C* indicate portions of the framing that are detailed in Figures 5.46 and 5.47.

A 2 × 4 (38 mm × 89 mm) nailer provides a nailing surface for the ceiling material at the end walls

A. SIDE WALL   *B. INTERIOR BEARING WALL*   *C. END WALL*

FIGURE 5.46

Details of the second-floor platform, keyed to the letters in Figure 5.45. The extra piece of lumber attached to the top plate in *C* provides a nailing surface for the edge of the finish ceiling material, which is usually either gypsum board or veneer plaster base.

A. SIDE WALL WITH
WEB STIFFENER

B-1. INTERIOR BEARING WITH
I-JOIST BLOCKING PANEL

B-2. SQUASH BLOCK FOR
CONCENTRATED LOADS

FIGURE 5.47
I-joist framing details analogous to those for solid lumber shown in Figure 5.46. For clarity, the web of the I-joist is rendered light blue and blocking with a darker shade. The web stiffener shown in *A* is cut slightly shorter than the height of the web of the joist, and is installed so that a small space remains between the top of the blocking and the underside of the top flange. This prevents the blocking from prying the flanges apart if the I-joist itself shrinks or is compressed slightly (see also Figure 5.25). The blocking in *B-1* functions identically to the solid blocking shown in Figure 5.46*B*. A *squash block* is shown in *B-2*. This is a short section of 2-in. (38-mm) framing, installed vertically like a very short stud on either side of the I-joist. Squash blocks are used under points of concentrated load, such as under loadbearing posts or studs on either side of a large opening in a wall above. They are cut slightly longer than the full height of the joist to ensure that the loads are transmitted through the blocks and not the joist.

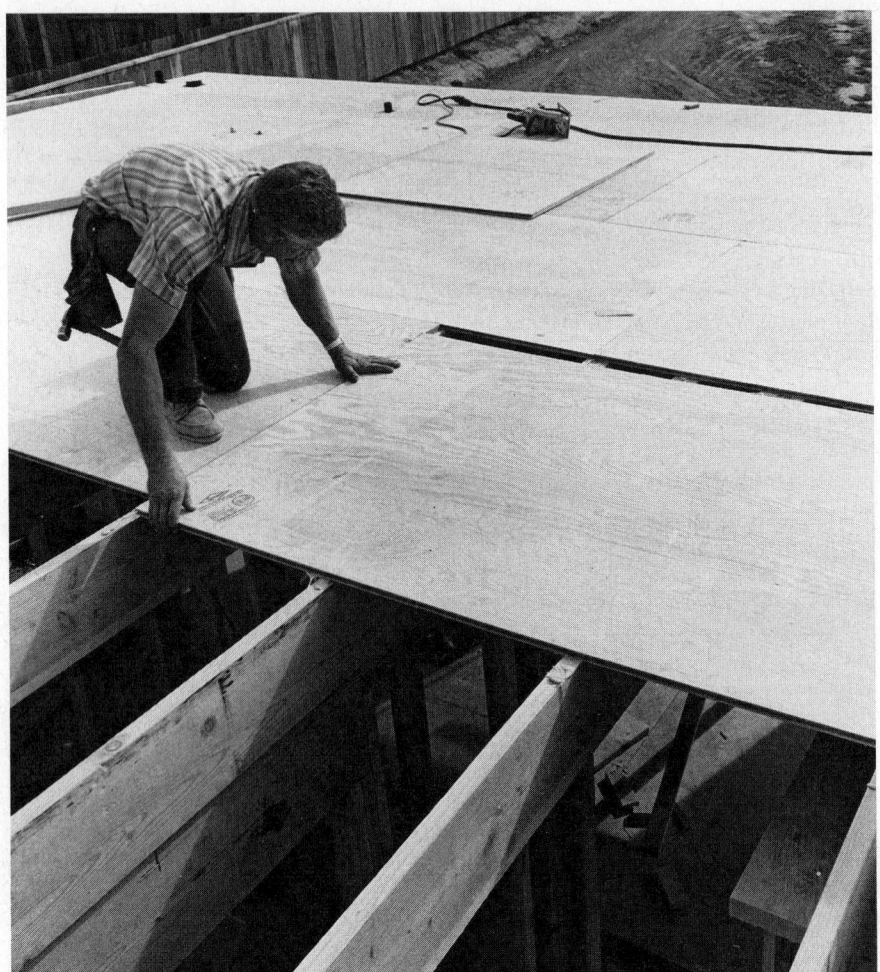

FIGURE 5.48
Installing upper-floor plywood combination subflooring/underlayment. The plywood panel grade is C–C Plugged. Its top (plugged) face has all surface voids filled and is lightly sanded, making it sufficiently smooth to allow carpeting to be installed directly over the subfloor without additional underlayment. The long edges of the panels have interlocking tongue-and-groove joints to prevent deflection of panel edges under heavy loads. (*Courtesy of APA–The Engineered Wood Association.*)

Wall framing procedures for the second floor are identical to those for the first floor.

6

FIGURE 5.49
Step six: The second-story walls are framed. The circle labeled *A* corresponds to the top of stair detail shown in Figure 5.50.

The top of each stringer is often suspended on a 1 × 2 (19 mm × 38 mm) nailed into its edge. An ordinary stair usually has three or four parallel stringers to support the treads and risers

The stringer is cut from a 2 × 12 (38 mm × 286 mm). The cuts are laid out with a framing square, using previously determined tread and riser dimensions

A 2 × 4 (38 mm × 89 mm) thrust block keeps the stringers from sliding

Double joists support the top and bottom of the stair

FIGURE 5.50
Interior stairways are usually framed as soon as the upper-floor platform is completed. This gives the carpenters easy up-and-down access during the remainder of the work. Temporary treads of joist scrap or plywood are nailed to the stringers. These will be replaced by finish treads after the wear and tear of construction are finished.

Roof Framing

The generic roof shapes for wood light frame buildings are shown in Figure 5.51. Simple shapes are combined into more complex arrangements suitable for covering any building plan shape and volume.

For structural stability, rafters in gable and hip roofs must be securely tied together at the top of the supporting walls by well-nailed *ceiling joists* to make what is, in effect, a series of triangular trusses. If the designer wishes to eliminate the ceiling joists and expose the sloping underside of the roof as the finished ceiling surface, a *ridge beam* or bearing wall must be inserted at the ridge or a system of exposed horizontal ties must be used in place of the joists. Sometimes a designer wishes to raise the ceiling joists or exposed *rafter ties* to a higher elevation than the tops of the wall plates. This greatly increases the stresses in the rafters and should be done only within the prescriptive limits of the building code or after consultation with a structural engineer. As with floor framing, ceiling joists and rafters must be protected against overturning. Requirements vary with the span, depth, and support conditions of the members and may include end nailing, blocking, bridging, or strapping, similar to those shown earlier in this chapter for floor joist framing.

Flat roofs

Shed or single-pitch roof

Gable roof

Hip roof

Gambrel roof

Mansard roof

Flat and shed roofs exert no lateral thrust

Ceiling joist

Ridge beam

Gable and hip rafters must be either tied with ceiling joists, or supported by a structural ridge beam

Ceiling joist

Knee wall

Gambrel and mansard roofs require both knee walls and ceiling joists for structural stability

FIGURE 5.51
Basic roof shapes for wood light frame buildings.

Hip rafter

Ridge board

Chimney opening — Dormer

ROOF FRAMING PLAN

FIGURE 5.52
A roof framing plan for the building illustrated in Figure 5.54. The *dormer* and chimney openings are framed with doubled header and trimmer rafters. The dormer is then built as a separate structure that is nailed to the slope of the main roof.

Although a college graduate architect or engineer would find it difficult to use the necessary trigonometry to lay out the cuts for a rafter in a sloping roof, a skilled carpenter, without resorting to mathematics, has little problem making the layout when the *pitch* (slope) is specified as a ratio of *rise* (vertical dimension) to *run* (horizontal dimension). In the United States, pitch is usually given as inches of rise per foot (12 inches) of run. The carpenter uses these two figures on the two edges of a *framing square* to lay out the rafter, as shown in Figure 5.53. The actual length of the rafter is never figured, nor does it have to be, because all the measurements are made as horizontal and vertical distances with the aid of the square. Today, many carpenters prefer to do rafter layout with the aid of tables that give actual rafter lengths for various pitches and horizontal distances; these tables are stamped on the framing square itself or printed in pocket-size booklets. Also available are hand-held calculators that are

specially programmed to find dimensions of rafters.

Hips and *valleys* introduce another level of trigonometric complexity in rafter layout, but the experienced carpenter has little difficulty even here: Again, he or she can use published tables for *hip rafters* and *valley rafters* or do the layout the traditional way, as illustrated in Figure 5.55. The head carpenter lays out only one rafter of each type by these procedures. This then becomes the *pattern rafter* from which other rafters can be traced and cut (Figure 5.58).

In areas subject to hurricanes, special care must be taken to ensure that rafters are securely attached to their supporting walls with sheet metal rafter anchors such as the one shown in Figure 3.50. The type, size, and spacing of the nails that attach the roof sheathing to the rafters are also closely controlled. The intent of both of these measures is to reduce the likelihood that the roof will be blown off in high winds.

The balloon frame is closely connected with the level of industrialization which had been reached in America [in the early 19th century]. Its invention practically converted building in wood from a complicated craft, practiced by skilled labor, into an industry. . . . This simple and efficient construction is thoroughly adapted to the requirements of contemporary architects. . . . [E]legance and lightness [are] innate qualities of the balloon-frame skeleton.

—Sigfried Giedion, *Space, Time and Architecture: The Growth of a New Tradition*, 1967

Roof pitches in the U.S. are usually specified as inches of rise per 12 inches of run. If the rise is 10 in 12, the framer holds the square as shown to mark cuts on the pattern rafter

3-9. The run of the rafter is stepped off in 1-ft horizontal increments

10. The plumb cut for the ridge board is laid out, and the pattern rafter is ready to cut. The rest of the rafters are traced from the pattern rafter

A. RIDGE

The ridge board aligns the tops of the rafters and supports the top edges of the roof sheathing

The birdsmouth cut gives the rafter a level bearing on the top plate of the wall

Collar ties are required near the ridge on steep roofs to prevent uplift of the roof planes in high winds

2. The plumb and level cuts are laid out for the birdsmouth

1. The plumb cut and level cut for the eave are laid out first, and the horizontal distance to the birdsmouth measured along the blade of the square

The ceiling joists are securely face-nailed to the rafters to resist the outward thrust of the roof structure

The details of the rafter cuts outside the birdsmouth are determined solely by the desired overhang and fascia details. Many variations are possible

The interior junction of the ceiling joists must be sufficiently strong to transmit the roof thrust

B. EAVE

C. INTERIOR BEARING OF CEILING JOISTS

FIGURE 5·53

Roof framing details and procedures: The lettered details are keyed to Figure 5.54. The remainder of the page shows how a framing square is used to lay out a pattern rafter, reading from the first step at the lower end of the rafter to the last step at the top. The rafter detailed here is a *common rafter*, as distinguished from hip, valley, and jack rafters as shown in Figure 5.55.

The ceiling joists above the second floor (which also serve as the attic floor joists) are toenailed to the tops of the second floor walls. A few rafters are then erected to support the ridge board, and remainder of the rafters are put up. Double headers and trimmers are used around openings in the roof.

7

FIGURE 5.54
Step seven: framing the attic floor and roof. The outer ends of attic floor joists are not usually headed off, but instead are face nailed to the rafter pairs that overlap them.

Valley rafter

Common rafters

Hip rafter

Valley jack rafters

Hip jack rafters

Fascia

Jack rafters are common rafters cut off at varying lengths to meet a hip or valley rafter. The jacks meet the hip or valley at a compound angle that is easily laid out with a framing square

12"

16.97"

12"

The diagonal of a 12" square is 16.97", or very nearly 17". In laying out a hip or valley rafter, the framer simply aligns the framing square to the rise per foot of a common rafter on the tongue, and 17" rather than 12" on the blade. The marking and stepping-off operations are otherwise identical to those for a common rafter

Ridge

Hip

Valley

FRAMING PLAN

17"

10"

17"

10"

FIGURE 5·55
Framing for a hip roof. The difficult geometric problem of laying out the diagonal hip rafter is solved easily by using the framing square in the manner shown. Toward the bottom of the rafter shown in the lower part of this figure is a *birdsmouth cut*, an angled notch in the rafter that allows the rafter to seat securely on the top plate of the wall.

Interactive Exercise

FIGURE 5.56
Framing for an overhanging *rake*.

Lookouts

Fly rafter

The sloping edge of a gable or shed roof is called the rake. A rake overhang is framed with lookouts and fly rafters. The lookouts are supported by a top plate over the gable end studs

FIGURE 5.57
A framing square being used to mark rafter cuts. The run of the roof, 12 inches, is aligned with the edge of the rafter on the blade (the wider leg) of a square, and the rise, 7 inches in this case, is aligned on the tongue (the narrower leg) of the square. A pencil line along the tongue will be perfectly vertical (a *plumb cut*) when the rafter is installed in the roof, and one along the blade will be horizontal (a *level cut*). True horizontal and vertical distances can be measured on the blade and tongue, respectively. The layout of these types of cuts can also be seen in Figures 5.53 and 5.55. (*Photo by Edward Allen.*)

FIGURE 5.58
Tracing a pattern rafter to mark cuts for the rest of the rafters. The corner of the building behind the carpenters has let-in corner braces on both floors, and most of the rafters are already installed. *(Courtesy of Southern Forest Products Association.)*

FIGURE 5.59
I-joists may be used as rafter material instead of solid lumber. *(Photos provided by Trus Joist MacMillan.)*

FIGURE 5.60
Applying plywood roof sheathing to a half-hipped roof. Blocking between rafters at the wall line (far left in the photo) has been drilled with large holes for attic ventilation. The line of horizontal blocking between studs is to support the edges of plywood siding panels applied in a horizontal orientation. *(Courtesy of APA–The Engineered Wood Association.)*

FIGURE 5.61
Fastening roof sheathing with a pneumatic nail gun. *(Photograph courtesy of Senco Products, Inc.)*

The framing of the building is completed with installation of the roof sheathing, the gable end walls, and the dormer.

(8) **FIGURE 5.62**
Step eight: The sheathed frame is completed.

FIGURE 5.63
A wood light frame structure fully sheathed with plywood structural panels. Installation of wood trim around the edges of the roof has also begun. (*Photo by Joseph Iano.*)

VARIATIONS ON WOOD LIGHT FRAME CONSTRUCTION

Framing for Increased Thermal Efficiency

The 2×4 (38×89 mm) has been the standard-size wall stud since light framing was invented. In recent years, however, pressures for greater conservation of fuel for heating and cooling have led to energy code requirements for more thermal insulation than can be inserted in the cavities of a wall framed with members only 3½ inches (89 mm) deep. One solution is to frame walls with 2×6 (38×140 mm) studs, usually at a spacing of 24 inches (610 mm), creating an insulation cavity 5½ inches (140 mm) deep. Alternatively, 2×4-framed walls may be covered either inside or out with insulating plastic foam sheathing, thus reaching an insulation value about the same as that of a conventionally insulated 2×6-framed wall. For even greater insulation performance, 2×6 studs and insulating sheathing may be used in tandem. (Increasing the stud spacing or adding insulation to one side or the other of the studs also helps to reduce energy losses due to thermal bridging at the studs.) In very cold climates, even more heavily insulated wall assemblies that can achieve greater energy savings may be constructed. Some of these construction methods are illustrated in Figures 7.17 through 7.21.

Framing for Optimal Lumber Usage

Using *advanced framing techniques* (also called *optimum value engineering*), redundant or structurally superfluous wood members are minimized, thereby reducing the amount of lumber needed to construct the frame and, once the frame is insulated, increasing its thermal efficiency (Figure 5.64). A variety of techniques may be used, including:

• Spacing framing members at 24 inches (610 mm) rather than 16 inches (406 mm) o.c.: Wider spacing of framing members reduces the amount of lumber required. In exterior walls, thermal efficiency is improved in comparison to walls framed

A. Conventional wall framing with studs @ 16 o.c.

B. Wall framed with advanced framing techniques

FIGURE 5.64

Comparison of walls framed with conventional and advanced techniques. Wall *A* is framed as explained in Figure 5.32. Studs are spaced at 16 inches (406 mm) o.c., the layout of the wall and its openings is not coordinated with the framing module, and standard details are used for corners, openings, and other features. In wall *B*, studs are spaced at 24 inches (610 mm) o.c., the length of the wall and the location and size of its openings have been coordinated to the greatest extent possible with the 24-inch module, and redundant framing members have been eliminated. Note that with the single top plate, floor joists or roof rafters (shown dashed in these figures) bearing on wall *B* must fall directly over studs below. The total length of framing lumber used in wall *B* is half that required for wall *A*. Though only a foot and a half shorter in length, wall *B* also can be sheathed with five standard-sized sheathing panels, whereas wall *A* requires six. Even if wall *A* is constructed of 2 × 4 studs and wall *B* is constructed of 2 × 6 studs, the overall savings in materials and reduction in waste in wall *B* are substantial.

with more closely spaced members by the reduction in thermal bridging that occurs at the studs.

• Designing to a 24-inch (610-mm) module: When the outside dimensions of a framed structure conform to a 24-inch module, sheathing panel waste is minimized. Planning rough opening sizes and locations in floors, walls, and roofs to conform, where possible, to this module can reduce waste even further. Designing to modular dimensions also reduces wastage of interior wallboard.

• Using single top plates in all walls, both bearing and nonbearing: In the case of bearing walls, this requires floor or roof framing members to

align directly over studs in the walls that support them.

• Minimizing other unneeded framing members: Avoid headers over openings in nonbearing walls, as they are not needed; in bearing walls, use headers only as deep as required for the loads and span. Where corner studs serve only to provide nailing surfaces or support for wallboard, use other, less wasteful blocking techniques or metal clips designed for this purpose instead. Replace jack studs supporting headers at window and door openings with metal hangers; eliminate unneeded cripple studs under rough sills. All of these techniques save lumber and, in exterior

walls, increase energy efficiency by reducing thermal bridging through solid framing.

• Eliminating unneeded plywood and OSB wall sheathing: Where let-in bracing is structurally adequate for lateral force resistance, eliminate structural panel wall sheathing entirely and cover walls with insulating sheathing for better thermal efficiency. Where structural panels are required, use the minimum number of panels necessary.

Advanced framing techniques rely on unconventional framing methods and greatly reduce redundancy in the building frame. For these reasons, they should not be

FIGURE 5.65
Roof trusses are typically lifted to the roof by a boom mounted on the delivery truck. This is one of a series of identical attic trusses that will frame a habitable space under the roof. *(Photo by Rob Thallon.)*

used without guidance from a structural engineer or other qualified design professional, and special review and approval from local building authorities may be required. Nevertheless, where these techniques are used, significant benefits can be realized. According to the National Association of Home Builders Partnership for Advancing Technology in Housing, advanced framing techniques can reduce the amount of lumber used in a wood light frame structure by up to 19 percent and improve the energy efficiency of the insulated structure by as much as 30 percent.

Prefabricated Framing Assemblies

Roof trusses, and to a lesser extent *floor trusses*, are used in platform frame buildings because of their speed of erection, economy of material usage, and long spans. Though some floor trusses are light enough to be lifted and installed by two carpenters, most truss assemblies are erected with the aid of a small crane that often is attached to the truck on which the trusses are delivered (Figure 5.65). Roof trusses are particularly slender in proportion, usually only 1½ inches (38 mm)

thick and capable of spanning 24 to 32 feet (7.5 to 10 m). They must be temporarily braced during construction to prevent buckling or the domino-like collapse of all the trusses until they are adequately secured permanently by the application of internal bracing, roof sheathing panels, and interior finishes (Figure 5.66). Panelized walls—framed and sheathed sections of wall assembled in the factory and delivered to the construction site ready to be erected—are used mostly by larger, high-volume builders who construct hundreds or thousands of houses per year.

FIGURE 5.66
A roof framed with prefabricated trusses. Approximately midway up the upper chords of the trusses, temporary strapping ties the trusses to one another for bracing. Other diagonal bracing, not visible in this photograph, ties the trusses to floor or ceiling framing to prevent the entire row from collectively tipping sideways. *(Photo by Joseph Iano.)*

FOR PRELIMINARY DESIGN OF A WOOD LIGHT FRAME STRUCTURE

Estimate the depth of **wood rafters** on the basis of the horizontal (not slope) distance from the outside wall of the building to the ridge board in a gable or hip roof and the horizontal distance between supports in a shed roof. A 2 × 4 rafter spans approximately 7 feet (2.1 m); a 2 × 6, 10 feet (3.0 m); a 2 × 8, 14 feet (4.3 m); and a 2 × 10, 17 feet (5.2 m).

The depth of **wood light roof trusses** is usually based on the desired roof pitch. A typical depth is one-quarter of the width of the building, which corresponds to a ⁶⁄₁₂ pitch in a gable truss. Trusses are generally spaced 24 inches (600 mm) o.c. and can span up to approximately 65 feet (20 m).

Estimate the depth of **wood floor joists** as follows: 2 × 6 joists span up to 9 feet (2.7 m), 2 × 8 joists 11 feet (3.4 m), 2 × 10 joists 14 feet (4.3 m), and 2 × 12 joists 17 feet (5.2 m).

Estimate the depth of manufactured **wood I-joists** as follows: 9½-inch (240-mm) joists span 16 feet (4.9 m), 11⅞-inch (300-mm) joists span 19 feet (5.8 m), 14-inch (360-mm) joists span 23 feet (7.0 m), and 16-inch (400-mm) joists span 25 feet (7.6 m).

Estimate the depth of **wood floor trusses** as ¹⁄₁₈ of their span. Typical depths of floor trusses range from 12 to 28 inches (305–710 mm) in 2-inch (51-mm) increments.

2 × 4 **wood studs** 24 inches (600 mm) o.c. can support attic and roof loads only. Either 2 × 4 studs 16 inches (400 mm) o.c. or 2 × 6 studs 24 inches o.c. can support one floor plus attic and roof. 2 × 6 studs 16 inches o.c. can support two floors plus attic and roof.

Framing members in light frame buildings are usually spaced either 16 or 24 inches (400 or 600 mm) o.c. For actual sizes of dimension lumber in both conventional and metric units, see Figures 3.23 and 3.24.

These approximations are valid only for purposes of preliminary building layout and must not be used to select final member sizes. They apply to the normal range of building occupancies, such as residential, office, commercial, and institutional buildings. For manufacturing and storage buildings, use somewhat larger members.

For more comprehensive information on preliminary selection and layout of a structural system and sizing of structural members, see Edward Allen and Joseph Iano, *The Architect's Studio Companion* (5th ed.), Hoboken, NJ, John Wiley & Sons, Inc., 2012.

WOOD LIGHT FRAME CONSTRUCTION AND THE BUILDING CODES

As shown in the table in Figure 1.3, the International Building Code allows buildings of almost every occupancy group to be constructed with wood platform framing, classified as Type V construction, but with relatively strict restrictions on height and floor area. For example, an unsprinklered, commercial office building, Occupancy Group B, built of Type V-B (unprotected) construction may be two stories in height and 9000 square feet (835 m²) in area per floor. In comparison, if built of Type IV heavy timber construction, the same building may be five stories tall and up to 36,000 square feet (3345 m²) per floor and, if built of more fire-resistive noncombustible construction types, even larger. (Recall also that allowable floor areas in this table can in many circumstances be increased by installing an automatic fire suppression sprinkler system in the building, as outlined in Chapter 1 of this book.)

Building size may also be increased by subdividing a structure with fire walls, an approach commonly used in the construction of attached dwellings and row townhouses. A fire wall separates a single structure into separate portions, each of which is treated by the building code as a independent structure and may have a floor area as large as that normally permitted for an entire building. The required fire resistance of the fire wall is given in a separate table in the code, reproduced in this book as Figure 1.7. Additionally, the fire wall must extend from the foundation through the roof and must be constructed so that it remains standing even if the construction on either side burns completely away. The traditional fire wall is made of brick or concrete masonry, but lighter, less expensive systems using metal framing and gypsum board may also be used (Figures 5.67 and 5.68).

Though platform frame construction is the least fire-resistive of all construction types, building code limits placed on it are sufficiently flexible to allow its use for a diverse

FIGURE 5.67

This proprietary fire wall consists of light gauge metal framing, noncombustible insulation, and gypsum board. The metal framing is attached to the wood structure on either side with special clips (not shown here) that break off if the wood structure burns through and collapses, leaving the wall supported by the undamaged clips on the adjacent building. *(Photographs of Area Separation Wall supplied by Gold Bond Building Products, Charlotte, NC.)*

FIGURE 5.68

This townhouse has collapsed completely as the result of an intense fire, but the houses on either side, protected by fire walls of the type illustrated in Figure 5.67, are essentially undamaged. *(Photographs of Area Separation Wall supplied by Gold Bond Building Products, Charlotte, NC.)*

FIGURE 5.69
This mixed-use urban building is being constructed of cast-in-place concrete up to the second-floor level and of wood light framing above. The building code permits such *podium structures* to be larger in size than those constructed of wood exclusively. *(Photo by Joseph Iano.)*

range of building uses (Figure 5.69). Furthermore, its economies are such that most building owners will choose platform frame construction over more fire-resistant construction types if given the opportunity. And, despite their vulnerability to fire, the comprehensive life-safety requirements of modern building codes ensure that buildings of wood light frame construction are safe places for their occupants. A study published by the Canadian Wood Council of residential fire deaths in Canada showed that the rate of fire deaths in Type VA dwellings is about the same as the death rate in dwellings made of noncombustible types of construction.

UNIQUENESS OF WOOD LIGHT FRAME CONSTRUCTION

Wood light framing is popular because it is a flexible and economical way of constructing buildings (Figures 5.70 through 5.74). Its flexibility

FIGURE 5.70
The W. G. Low house, built in Bristol, Rhode Island, in 1887 to the design of architects McKim, Mead, and White, illustrates both the essential simplicity of wood light framing and the complexity of which it is capable. *(Photo Wayne Andrews.)*

stems from the ease with which carpenters using ordinary tools can create buildings of astonishing variety and complexity. Its economy can be attributed in part to the relatively unprocessed nature of the materials from which it is made, and in part to mass-market competition among suppliers of components and materials and local competition among small builders.

Platform framing is the one truly complete and open system of construction that we have. It incorporates structure, enclosure, thermal insulation, mechanical installations, and finishes into a single constructional concept. Thousands of products are made to fit it: competing brands of windows and doors; interior and exterior finish materials; electrical, plumbing, and heating products. For better or worse, it can be dressed up to look like a building of wood or of masonry in any architectural style from any era of history. Architects have failed to exhaust its formal possibilities, and engineers have failed to invent a new environmental control system that it cannot assimilate. Wood light frame construction can be used to construct the cheapest and most mundane buildings. Yet, one can look to the best examples of the Carpenter Gothic, Queen Anne, and Shingle-style buildings of the 19th century, or the Bay Region and Modern styles of more recent times, to realize that wood light framing also gives the designer the freedom to make a finely crafted building that nurtures life and elevates the spirit.

FIGURE 5.71

In the late 19th century, the sticklike qualities of wood light framing often found expression in the exterior ornamentation of houses. *(Photo by Edward Allen.)*

FIGURE 5.72
Light wood frame construction is used to build the vast majority of housing in North America. It remains to this day the most affordable and versatile construction system for buildings of this type. *(Project: New Holly Phase I, by Weinstein A|U Architects & Planners. Photo by Joseph Iano.)*

FIGURE 5.73
In this contemporary New England cottage, designer Dennis Wedlick has exploited the sculptural possibilities of platform framing. *(Photo: © Michael Moran.)*

FIGURE 5.74
The Thorncrown Chapel in Eureka Springs, Arkansas, designed by Fay Jones and
Associates, Architects, combines large areas of glass with special framing details
to create a richly inspiring space. To avoid damage to the sylvan site, all materials
were carried in by hand rather than on trucks. Thus, all framing was done with
nominal 2-inch (38-mm) lumber rather than heavy timbers. *(Photo by Christopher
Lark. Courtesy of American Wood Council.)*

CSI/CSC MasterFormat Sections for Wood Light Frame Construction	
06 10 00	ROUGH CARPENTRY
06 12 00	STRUCTURAL PANELS
06 14 00	TREATED WOOD FOUNDATIONS
06 16 00	SHEATHING
06 17 00	SHOP-FABRICATED STRUCTURAL WOOD
06 18 00	GLUED-LAMINATED CONSTRUCTION

KEY TERMS

wood light frame construction
balloon frame
joist
stud
rafter
fireblocking
platform frame
sheathing
rim board, band joist
subfloor
sole plate, bottom plate
top plate
ridge board
header
trimmer
sill
trimmer stud
o.c., on center
batter board
permanent wood foundation
framing plan
section detail
architectural floor plan
exterior elevation
building section

interior elevation
rough carpentry
foundation sill plate, mudsill
sill seal, sill gasket
termite shield
bridging
web blocking
jack stud
king stud
rough sill
cripple stud
let-in diagonal bracing
lateral force
hold-down
wracking
shear wall
braced wall, braced panel
collector, drag strut, drag tie
load path
squash block
ceiling joist
ridge beam
rafter tie
pitch
rise

run
framing square
hip
valley
hip rafter
valley rafter
pattern rafter
dormer
common rafter
ridge board
collar tie
valley jack rafter
hip jack rafter
birdsmouth cut
fascia
rake
plumb cut
level cut
lookout
fly rafter
advanced framing techniques,
    optimum value engineering
roof truss
floor truss
podium structure

REVIEW QUESTIONS

1. Draw a series of very simple section drawings to illustrate the procedure for erecting a platform frame building, starting with the foundation and continuing with the ground floor, the ground-floor walls; the second floor, the second-floor walls; and the roof. Do not show details of connections, but simply represent each plane of framing in your section drawing.

2. Draw from memory the standard detail sections for a two-story platform frame dwelling. *Hint:* The easiest way to draw a detail section is to draw the pieces in the order in which they are put in place during construction. If your simple drawings from Question 1 are correct, and if you follow this procedure, you will not find this question too difficult.

3. What are the differences between balloon framing and platform framing? What are the advantages and disadvantages of each? Why has platform framing become the method of choice?

4. Why is less fireblocking required in platform framing in comparison to balloon framing?

5. Why is a steel beam or glue-laminated wood beam preferred to a solid wood beam for supporting floor joists at the foundation level?

6. How is a platform frame building braced against wind and earthquake forces?

7. Light framing of wood is highly combustible. In what different ways does a typical building code take this fact into account?

EXERCISES

1. Visit a building site where a wood platform frame is being constructed. Compare the details that you see on the site with the ones shown in this chapter. Ask the carpenters about the procedures you see them using. When their details differ from the ones illustrated, make up your own mind about which is better and why.

2. Develop floor framing and roof framing plans for a building you are designing. Estimate the approximate sizes of the joists and rafters using the rules of thumb provided in this chapter.

3. Make thumbnail sketches of 20 or more different ways of covering an L-shaped building with combinations of sloping roofs. Start with the simple ones (a single shed, two intersecting sheds, two intersecting gables) and work up to the more elaborate ones. Note how the varying roof heights of some schemes could provide room for a partial second-story loft or for high spaces with clerestory windows. How many ways do you think there are of covering an L-shaped building with sloping roofs? Look around as you travel through areas with wood frame buildings, especially older areas, and see how many ways designers and framers have roofed simple buildings in the past. Build up a collection of sketches of ingenious combinations of sloping roof forms.

4. Build a scale model of a platform frame from basswood or pine, reproducing accurately all its details, as a means of becoming thoroughly familiar with them. Better yet, build a small frame building for someone at full scale (perhaps a toolshed, playhouse, or garage).

SELECTED REFERENCES

Allen, Edward, and Rob Thallon. *Fundamentals of Residential Construction.* Hoboken, NJ, John Wiley & Sons, Inc., 2011.

This book expands upon the chapters on residential-scale construction in the book you are now reading, giving full details of every aspect, including plumbing, mechanical and electrical systems, and landscaping.

American Forest & Paper Association. *Details for Conventional Wood Frame Construction.* Washington, DC, 2001.

This 55-page publication is available as a free download from the American Wood Council, at www.awc.org. It provides an excellent introduction to wood light framing methods and their use in residential construction, and includes design and construction guidelines and extensive illustrated details.

APA-The Engineered Wood Association. *I-Joist Construction Details: Performance Rated I-Joists in Floor and Roof Framing.* Tacoma, WA, 2004.

This 55-page publication is available as a free download from the APA-The Engineered Wood Association, at www.apawood.org. It provides extensive guidelines and illustrated details for wood light framing with engineered I-joists.

International Code Council, Inc. *International Residential Code.* Falls Church, VA, updated regularly.

This is the definitive legal guide for platform frame residential construction throughout most of the United States. It includes details of every aspect of construction in both wood light frame and light gauge steel.

Thallon, Rob. *Graphic Guide to Frame Construction.* Newtown, CT, Taunton Press, 2008.

Unsurpassed for clarity and usefulness, this is an encyclopedic collection of details for wood platform frame construction.

Woodward, G. E. *Woodward's Country Homes.* New York, 1989.

WEB SITES

Wood Light Frame Construction

Author's supplementary web site: **www.ianosbackfill.com/05_wood_light_frame_construction**

American Wood Council: **www.awc.org**

APA-The Engineered Wood Association: **www.apawood.org**

Canadian Wood Council: **www.cwc.ca**

Fine Homebuilding magazine: **www.finehomebuilding.com**

Journal of Light Construction magazine: **www.jlconline.com**

NAHB Toolbase Services: **www.toolbase.org**

Wood Design & Building magazine: **www.wood.ca**

EXTERIOR FINISHES FOR WOOD LIGHT FRAME CONSTRUCTION

6

- **Protection from the Weather**
  - Roofing Underlayment
  - Wall Moisture Barrier

- **Roofing**
  - Finishing the Eaves and Rakes
  - Roof Drainage
  - Roof Overhangs and Rain Protection
  - Ventilated Roofs
  - Unventilated Roofs
  - Roof Shingling

- **Windows and Doors**
  - Flashings
  - Installing Windows and Doors

  PAINTS AND COATINGS

- **Siding**
  - Board Siding
  - Plywood Siding

- Shingle Siding
- Metal and Plastic Siding
- Stucco
- Masonry Veneer
- Artificial Stone
- Fiber-Cement Panel Siding

- **Corner Boards and Exterior Trim**

- **Sealing Exterior Joints**

  SUSTAINABILITY OF PAINTS AND OTHER ARCHITECTURAL COATINGS

- **Exterior Painting, Finish Grading, and Landscaping**

- **Exterior Construction**

Architects Hartman–Cox clad this small church with an economical but attractive siding of plywood and vertical wood battens, all painted white. The roofs are of asphalt shingles and the windows are of wood. The rectangular windows are double hung. *(Photo by Robert Lautman. Courtesy of American Wood Council.)*

As the rough carpentry of a platform frame building nears completion, a logical sequence of exterior finishing operations begins. First, the eaves and rakes of the roof are finished, which permits the roof to be shingled so as to offer as much weather protection as possible to subsequent operations. When the roof has been completed, the windows and doors are installed (Figure 6.1). Then the siding is applied. The building is now "tight to the weather," allowing interior finishing work to take place safely sheltered from sun, rain, snow, and wind. The outside of the building is ready for exterior painting or staining. Finish grading, landscaping, and paving work may commence as electricians, plumbers, sheet metal workers, drywallers, finish carpenters, and flooring installers swarm about the interior of the building.

FIGURE 6.1
With roofing, skylights, doors, and windows installed, the interior of the structure can begin to dry out in preparation for interior finish work.

PROTECTION FROM THE WEATHER

Roofing Underlayment

In the interest of protecting the building structure from weather as quickly as possible, *roofing underlayment* is installed soon after the roof framing is completed and sheathed. The material most commonly used for this purpose is one or two layers of *building felt*, made of matted cellulose fibers saturated with asphalt, often referred to simply as "tarpaper." During this stage of construction, the roofing underlayment serves as a temporary barrier to rain, allowing the structure to begin drying out in preparation for interior work. Once the underlayment is covered by the final roofing, it continues to serve as a permanent backup to the roofing, reducing the chance that wind-driven rain or minor leakage can penetrate into the roof structure. Traditionally, building felt was designated as either 15-lb or 30-lb, referring to the weight of material used to cover 100 square feet (9.2 m²) of roof area. Contemporary standards specify roofing felt as either No. 15 or No. 30. Though still frequently referred to as 15-lb or 30-lb, today's felts actually weigh slightly less than these names suggest.

Roofing underlayment membranes made from materials such as polypropylene or polyethylene fabrics may be substituted for traditional building felt. These *synthetic roofing underlayments* tend to be lighter, more tear-resistant, less sensitive to prolonged exposure to sunlight, and more costly than building felt.

Wall Moisture Barrier

Later during construction, the sheathed walls of the structure are covered with a protective layer intended to repel water and reduce the leakage of air. This layer, variously referred to as a *moisture barrier, water-resistive barrier,* or *weather-resistive barrier,* must be applied before siding is installed. Whether it is applied before or after the installation of doors and windows is usually up to the builder. The traditional materials used are the same No. 15 felt used for roofing underlayment or an asphalt-saturated paper with similar water-resistive properties called *Grade D sheathing paper.*

Alternatives to traditional building felts and papers offer greater durability and the potential for increased energy efficiency by reducing air leakage through the building walls and roofs. More airtight *housewraps,* made of synthetic fibers, are manufactured in sheets as wide as 10 feet (3 m) in order to minimize seams (Figure 6.2). To further reduce air leakage, the seams that remain may be sealed with self-adhering tape provided by the housewrap manufacturer.

Walls and roofs may also be sheathed with OSB panels, with a resin-and-paper overlay applied directly to the outer face of the panels in the factory. Once these panels are fastened to the wall or roof framing, panel seams are sealed with self-adhering tape and no additional weather barrier paper or wrap is required. In addition to the labor saved in the elimination of a step, this system has the potential for better air leakage control in comparison to

FIGURE 6.2
Building felt covers most of the roof and housewrap completely covers the walls of this house under construction in preparation for the application of roofing and siding. The bottom few feet of the roof sheathing remain exposed, awaiting the application of a waterproof ice barrier material as described later in this chapter. Lightweight housewraps such as seen in this photograph have to a large extent replaced building felt or paper for protecting walls. *(Photo by Rob Thallon.)*

separately applied housewraps. Similar levels of performance can also be achieved with liquid-applied air/weather barrier coatings. These are applied to the wall sheathing in the field using a spray gun or roller, after the panels have been installed over the framing. (The control of air leakage is discussed further in Chapter 7.)

Like roofing underlayment, the wall moisture barrier provides temporary protection from rain and wind during construction and, once finish siding is installed, acts as a permanent secondary line of defense against water infiltration. Although it is important that the wall moisture barrier resist penetration of liquid water, it must (in most climates) nevertheless also allow water vapor (water in a gas state) to pass relatively easily. Especially during cold months, interior humidity that tends to diffuse outward through the wall assembly must not become trapped within the wall; rather, it must be allowed to continue to pass freely through to the exterior.

ROOFING

Finishing the Eaves and Rakes

Before a roof can be shingled, the *eaves* (horizontal roof edges) and *rakes* (sloping roof edges) must be completed. Several typical ways of doing this are shown in Figures 6.3, 6.4, and 6.5.

FIGURE 6.3
Two typical details for the rakes (sloping edges) of sloping roofs. The upper detail has no rake overhang, and the lower has an overhang supported on lookouts.

Vent spacer

Metal drip edge

Rafter

Lookout

Wood gutter

Wood fascia

Wood trim

Plywood soffit

Siding

Screened
ventilation strip

Vent spacer

2 × 4
(38 × 89 mm)
nailer

Wood edge strip

Trim

Screened
ventilation strip

Plywood soffit

Wood blocking
acts as vent
spacer

Aluminum drip
edge and gutter

Aluminum fascia

Aluminum soffit
with integral
ventilation
openings

Aluminum siding

FIGURE 6.4
Three ways, from among many, of finishing the eaves of a light frame wood building. The top detail has a fascia and gutter, both of wood. The gutter is spaced slightly away from the fascia on blocks or shims to prevent moisture from being trapped between them, where it could lead to decay. The designer may vary the width of the overhang, or may substitute a metal or plastic gutter for the wood one. The sloping line at the edge of the ceiling insulation indicates a vent spacer, as shown in Figure 6.6. The middle detail has no fascia or gutter; it works best for a steep roof with a sufficient overhang to drain water well away from the walls below. The bottom detail is finished entirely in aluminum. It shows wood blocking as an alternative to vent spacers for maintaining free ventilation through the attic.

When designing eave and rake details, several objectives should be kept in mind: The edges of the roof shingles should be positioned and supported in such a way that water flowing over them will drip free of the trim and siding below. Provision must be made to drain rainwater and snowmelt from the roof without damaging the structure below. In most roof assemblies, the eaves must be ventilated to allow free circulation of air beneath the roof sheathing. And siding, which is not installed until later, must fit easily against or into the eave and rake trim.

Roof Drainage

Gutters and *downspouts* (downspouts are also called *rain leaders*) are installed on the eaves of a sloping roof to remove rainwater and snowmelt without wetting the walls or causing splashing or erosion on the ground below. *External gutters* may be made of wood, plastic, aluminum, or other sheet metal and are fastened

FIGURE 6.5
An example of a badly done wood eave detail with an aluminum gutter and downspout. The carpentry at the intersection of the eave and the rake shows some poorly fitted joints and badly finished ends, and the soffit is not ventilated at all, which may lead to the formation of ice dams on the roof (see Figure 6.5) unless constructed as an unventilated roof, as explained in the text. *(Photo by Edward Allen.)*

to the outer edge of the roof eave (Figures 6.3 and 6.4). *Internal* or *concealed gutters*, which are recessed into the surface of the roof, are rare. They are difficult to waterproof, and if they do leak, they can cause more damage to the building structure and its finishes than can a leaky external gutter. Gutters are sloped toward the points at which downspouts drain away the collected water. On larger buildings, gutters and downspouts are sized using rainfall intensity data for the location in which the building is located and flow capacity formulas found in building and plumbing codes. At the bottom of each downspout, the water must be conducted away from the building to prevent soil erosion or basement flooding. A simple precast concrete *splash block* can

spread the water and direct it away from the foundation, or a system of underground piping can collect the water from the downspouts and conduct it to a storm sewer, a *dry well* (an underground seepage pit filled with coarse crushed stone), or a drainage ditch.

Where collection of rainwater from roofs is not required by the building code or other regulations, gutters may be omitted and their associated problems of clogging and ice buildup avoided. Such roofs should be designed with eaves of sufficient depth to protect the wall below from excessive wetting by water falling from the roof edge. To prevent soil erosion and mud spatter from the falling water, the drip line at ground level below should be provided with

a bed of crushed stone or other suitable surface material.

Roof Overhangs and Rain Protection

Roof eaves and overhangs play an important role in protecting buildings from the weather. Where ample overhangs are provided, the volume of rain that reaches the building wall is greatly reduced in comparison to walls without such protection, and the chance for water penetration into the walls is much less. The design of walls without overhangs should be approached with caution: These walls are more vulnerable to the effects of weather exposure, such as staining, leaking, decay, and premature deterioration of the windows, doors, and siding.

Ventilated Roofs

In cold climates, snow on a sloping roof tends to be melted by heat escaping through the roof from the heated space below. At the eave overhang, however, the shingles and gutter, which are not above heated space, can be much colder, and the snow-melt can refreeze and begin to build up layers of ice. If left unchecked, eventually an *ice dam* can form. Meltwater accumulating above the ice dam can then seep between the shingles, through the roof sheathing, and into the building, causing damage to walls and ceilings. Ice dams can be largely eliminated by keeping the entire roof cold enough so that snow will not melt. With a *ventilated roof*, this is accomplished by continuously passing outside air under the roof sheathing through vents at the eave and ridge. In buildings with an attic, the attic itself is ventilated and kept as cold as possible. Insulation is installed in the ceilings below to retain the heat in the building. Where there is no attic and the insulation is installed between the rafters, the spaces between the rafters are ventilated by means of a continuous airspace between the insulation and the roof sheathing (Figures 6.6 and 6.7).

UNVENTED AND UNINSULATED

Snow is melted by heat escaping from the heated space below

Snow melt refreezes over the cold eave to form an ice dam

Standing water runs around the shingles and into the building

Heated air

VENTED AND INSULATED

Cold air circulating under the roof sheathing prevents the roof from becoming warm and melting the snow

A vent spacer keeps the insulation from blocking the air passage

Vents at the eave and ridge allow free circulation of cold outside air

Snow   Cold air

Heated air

Insulation in the ceiling keeps the heat inside the building

FIGURE 6.6

Ice dams form because of inadequate insulation combined with a lack of ventilation, as shown in the top diagram. The lower two diagrams show how insulation, attic ventilation, and vent spacers are used to minimize the melting of snow on the roof. In areas prone to ice damming, building codes also require the installation of a strip of waterproof ice barrier material under the shingles in the area of the roof where ice dams are likely to form.

Continuous vent spacers may be required where the insulation is between the rafters

Heated air

Cold air

FIGURE 6.7
To maintain clear ventilation passages where thermal insulation materials come between the rafters, vent spacers of plastic foam or wood fiber are installed as shown here. The positioning of the blocking or vent spacers is shown in Figures 6.4 and 6.6. Vent spacers can be installed along the full slope of a roof to maintain free ventilation between the sheathing and the insulation where the interior ceiling finish material is applied to the bottom of the rafters. *(Proper Vent™ is a registered trademark of Poly-Foam Inc.)*

Soffit vents create the required ventilation openings at the eave; these usually take the form of a continuous slot covered with screening or a perforated aluminum strip made especially for the purpose (Figure 6.8). As an alternative to soffit vents, panels with hollow ventilation channels can be installed under the lowest several courses of roof shingles. A continuous slot running across the roof sheathing allows exterior air to circulate under the shingles, through the panel, and into the ventilation space under the roof sheathing. Ventilation openings at the ridge can be either *gable vents* just below the roof line in each of the end walls of the attic, or a continuous *ridge vent*, a screened cap that covers the ridge of the roof and draws air through gaps in the roof sheathing on either side of the ridge board (Figure 6.9). Building codes establish minimum area requirements for roof ventilation openings based on the floor area beneath the roof being ventilated.

In regions prone to ice damming, building codes also require an *ice barrier* to be installed along roof eaves. This may be several layers of roofing felt cemented together with asphalt, or more commonly, a self-adhered sheet of polymer-modified bitumen, referred to as *rubberized underlayment* or by one product of this type's proprietary trade name that has fallen into colloquial usage, *ice and water shield*. The sheet must extend from the lowest edge of the roof to a line that is 2 feet (600 mm) inside the insulated space of the building (Figure 6.10). The soft, sticky bitumen seals tightly around the shingle nails that are driven through it and prevents passage of water that may accumulate behind an ice dam. Alternatively, the lower portion of the sloped roof may be covered with sheet metal or other roofing material that is sufficiently waterproof to protect the roof from damage if ice damming occurs.

Roof ventilation serves other purposes as well. It keeps a roof cooler in summer by dissipating solar heat absorbed by the shingles. A lower roof temperature extends the life of the roofing materials and reduces the heat load on the building interior. For this reason, asphalt shingle manufacturers may require that their shingles be installed on ventilated roofs as part of their product warranty conditions. Roof ventilation also helps dissipate moisture that collects under the roof sheathing that may originate from leaks in the roofing or, in colder months, that may be generated by the condensation of interior humidity that comes in contact with the cold underside of the roof sheathing.

Unventilated Roofs

Unventilated roofs can be constructed with insulated foam panels applied over the roof sheathing or with spray foam or rigid foam insulation boards applied below the sheathing between the roof rafters (Figure 6.11). When unventilated roofs are used in cold climates, protection from ice damming depends on ice barriers and the installation of sufficient insulation to reduce heat flow through the roof assembly, thereby slowing the melting of snow on the roof. Condensation resistance in unventilated roofs depends on the inability of significant quantities of air or water vapor to pass through the air-impermeable foam insulation to the colder layers of the roof assembly.

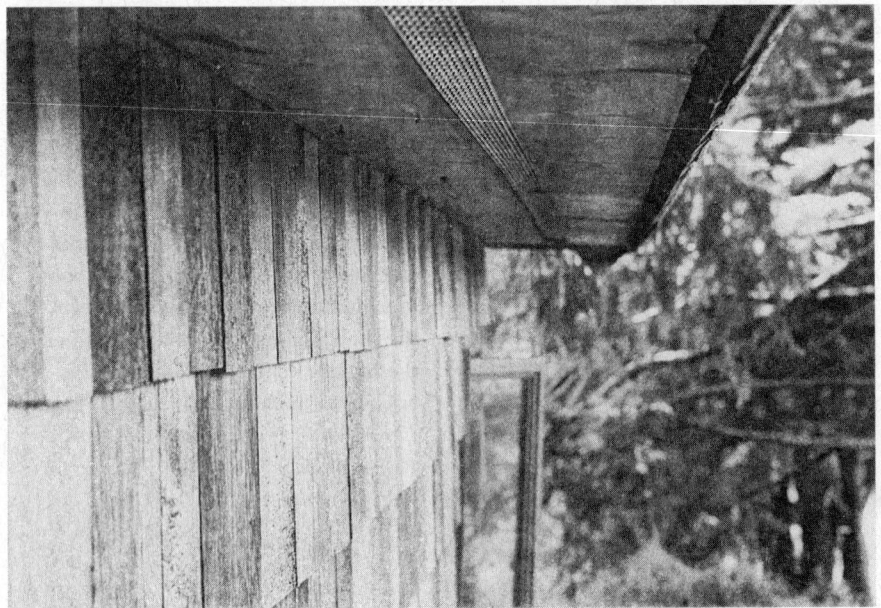

FIGURE 6.8
A continuous soffit vent made of perforated sheet aluminum permits generous airflow to all the rafter spaces but keeps out insects. The walls of this building are covered with cedar shingles. *(Photo by Edward Allen.)*

FIGURE 6.9
This building has both a louvered gable vent and a continuous ridge vent to discharge air that enters through the soffit vents. The gable vent is of wood and has an insect screen on the inside. The aluminum ridge vent is internally baffled to prevent snow or rain from entering even if blown by the wind. The roofing is a two-layered asphalt shingle designed to mimic the rough texture of a wood shake roof. *(Photo by Edward Allen.)*

FIGURE 6.10
In cold climates, building codes typically require that an ice barrier be installed beneath the lowest courses of shingles. This must extend at least 24 inches (600 mm) over the insulated portion of the inhabited space.

UNVENTED ROOF ASSEMBLY CONSTRUCTED WITH INSULATED PANELS

Insulated foam panel with OSB or plywood face

Wood roof decking

Rafter

UNVENTED ROOF ASSEMBLY CONSTRUCTED WITH SPRAY FOAM INSULATION

Plywood roof sheathing

Spray foam insulation

Rafter

FIGURE 6.11

Two ways to build an unventilated roof assembly. In the top diagram, insulated foam panels are installed on top of wood decking. The insulated panels have a plywood or OSB face that provides a surface to hold the nails or staples for attaching the roof underlayment and shingles. The panels themselves are secured with screws long enough to penetrate through the panel into the roof decking or rafters below. In the lower diagram, foam insulation is sprayed against the underside of the roof sheathing. Depending on climate conditions, the thickness of foam insulation required to control condensation may vary from as little as 1 inch to as much as 7 inches (25 to 178 mm). Other types of insulation may be added beneath the foam where necessary to achieve the total required insulation value of the roof.

With insulated panels, special care must be taken to seal the panel joints where air leakage is most likely to occur. With spray foam insulation applied under the sheathing, the foam must have low permeability to airflow and, in cold climates, low water vapor permeance as well. With rigid foam boards applied under the sheathing, edges of the boards must be carefully sealed to prevent the leakage of air around the boards. Where high roof temperatures are a concern (due to the lack of ventilation below the roof sheathing), shingles or other roof coverings with high solar reflectance may be considered, or insulated panels with integral ventilation channels may be used to create a ventilated assembly.

Advantages of unventilated roofing systems include simplicity of detailing, a thinner roof profile, less air leakage between the building interior and exterior, and no risk of windblown water entry into the roof assembly through poorly baffled ridge vents. However, the absence of ventilation also leaves this roof type potentially less able to dry out if moisture does find its way into the assembly. Where the potential for condensation is high (for example, in very cold climates or in roofs over high-humidity spaces), such assemblies should be thoroughly analyzed and carefully detailed to ensure that moisture conditions are adequately controlled.

Roof Shingling

Wood light frame buildings can be roofed with any of the materials described in Chapter 16. The least expensive, and most common, are *asphalt shingles*. These are applied either by the carpenters who have built the frame or by a roofing subcontractor (Figures 6.12, 6.1). Wood shingle and shake roofs, clay and concrete tile roofs, and architectural sheet metal roofs are also common. Low-slope roofing, either a built-up or a single-ply membrane, may also be applied to a light frame building. (See Chapter 16 for more information about these roofing types.)

WINDOWS AND DOORS

Flashings

Prior to the installation of windows and doors, *flashings* are installed to prevent water from seeping through gaps around the edges of these components. Flashings may be made of any of a number of corrosion-resistant metals, plastics, synthetic fabrics similar to housewraps, or modified bitumen sheets similar to the rubberized underlayment used for ice barriers on roofs. The choice of materials and the complexity of the detailing vary with the type of door or window frame, the severity of the exposure to wind and rain, and the detailing of the surrounding trim. The most important flashing in a wall opening is the sill flashing or *sill pan*. Any moisture that penetrates around the edges of the door or window frame tends to fall downward to the bottom of the wall opening. The sill flashing captures this water and ensures that it drains back to the exterior, preventing it from seeping further into the interior of the wall (Figures 6.13, 6.14, and 6.15).

FIGURE 6.12
Applying asphalt shingles over roofing felt. Shingles and other roof coverings are covered in detail in Chapter 16.
(Photo by Edward Allen.)

FIGURE 6.13

Steps in installing a vinyl-clad wood window in a wall covered with housewrap: *1.* With a single strip of self-adhering, flexible, waterproof sheet material, a U-shaped flashing is formed along the sill and up onto the side jambs. *2.* A bead of sealant is placed around the opening. *3.* The window unit is pushed into the opening from the outside. Its exterior flange is bedded in the sealant. *4.* One corner of the window unit is anchored with a nail through its flange and into the framing of the wall. With the aid of a level, a measuring tape, and thin wooden wedges, the unit is carefully squared as further nails are driven through the flanges. *5, 6.* Strips of self-adhering waterproof sheet are adhered to the window flanges and housewrap over the side jambs and head for airtightness and watertightness.

FIGURE 6.14
Window flashing in a wall exposed to heavy, wind-driven rain. The metal sill pan is made of soldered copper. Underneath the copper is a self-adhered bituminous flashing. Beneath that, and surrounding the entire opening, is a synthetic fabric flashing. Behind the fabric flashing is the wall moisture barrier paper. The window opening itself is temporarily protected with plastic that will be removed before the window is installed.

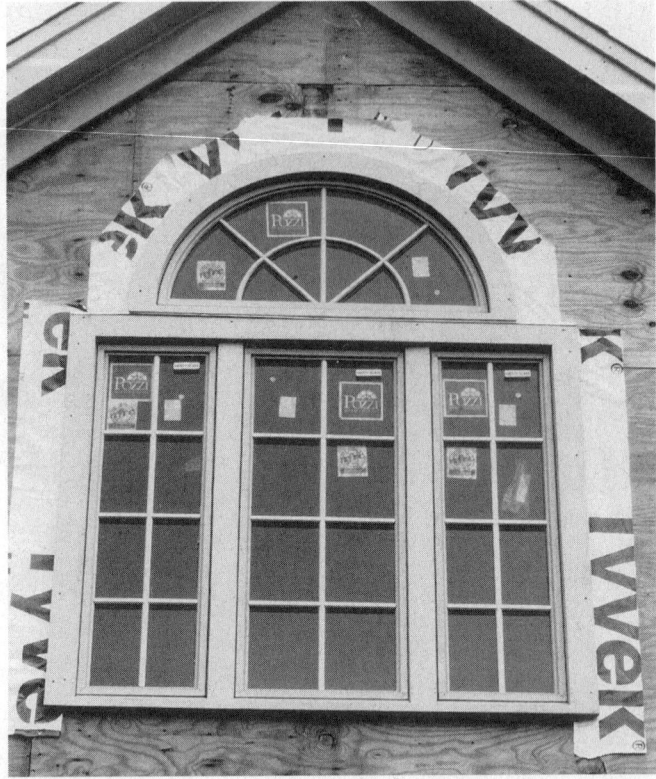

FIGURE 6.15
Synthetic fabric flashing and window unit installed into a wall opening. Housewrap has not yet been installed over the exterior sheathing. Sill flashing is missing from the installation. (*Photo by Edward Allen.*)

Installing Windows and Doors

Windows and doors are covered in detail in Chapter 18. *Nail-fin* or *flanged windows* are simple to install because they include a continuous flange around the perimeter through which nails are driven into the sheathing and studs (perpendicular to the wall plane) to fasten the units securely in place. The flange also provides a convenient surface for the application of self-adhering flashings. Most plastic-clad, aluminum, and solid plastic windows are manufactured as nail-fin types, as are some, but not all, wood windows. *Finless* or *flangeless windows* and most doors do not have nailing flanges. They are fastened through the sides of the window or door frame into the supporting studs (parallel to the wall plane) with long finish nails or screws. The fasteners are either covered by additional trim or they are countersunk, with the resulting holes later filled with putty and painted over. A typical entrance door and details for its installation are shown in Figures 6.16 and 6.17.

> **Everybody loves window seats, bay windows, and big windows with low sills and comfortable chairs drawn up to them. . . . A room where you feel truly comfortable will always contain some kind of window place.**
>
> **—Christopher Alexander et al., *A Pattern Language*, 1977**

FIGURE 6.16

A six-panel wood entrance door with flanking sidelights and a fanlight above. A number of elaborate traditional entrance designs such as this are available from stock for use in light frame buildings. (*Courtesy of Morgan Products, Ltd.*)

FIGURE 6.17

Details of an exterior wood door installation. The door opens toward the interior. The flashing in the head intercepts water running down the wall and directs it away from the door and frame. The sill flashing prevents water that penetrates around the door frame from leaking into the floor structure.

PAINTS AND COATINGS

Paints and other architectural coatings (stains, varnishes, lacquers, sealers) protect and beautify the surfaces of buildings. A good coating job begins with thorough surface preparation to make the surface (called the *substrate*) ready to receive the coating. The coating materials must be carefully chosen and skillfully applied using the proper tools and techniques. To finish the job, environmental conditions must be right for the drying or curing of the coating.

Material Ingredients

Most architectural coating materials are formulated with four basic types of ingredients: vehicles, solvents, pigments, and additives.

- The *vehicle* or *binder* provides adhesion to the substrate and forms a film over it; it is often referred to as a *film-former.*
- *Solvents* are volatile liquids used to improve the working properties of the paint or coating. The most common solvents used in coating materials are water and hydrocarbons, but turpentine, alcohols, ketones, esters, and ethers are also used.
- *Pigments* are finely divided solids that add color, opacity, and gloss control to the coating material. They also impart hardness, abrasion resistance, and weatherability to the coating.
- Additives modify various properties of the coating material. Driers, for example, are additives that hasten the curing of the coating. Other additives may relate to ease of application, resistance to fading, and other functions.

Solvent-Based Coatings and Water-Based Coatings

Coatings fall into two major groups, *solvent-based* and *water-based.* The most common solvents in solvent-based coatings are alkyd or polyurethane resins, or natural oils. These coatings cure by evaporation of the solvent, oxidation of the vehicle, or moisture curing from reaction of the vehicle with humidity in the air. Cleanup after painting is usually done with mineral spirits. Water-based coatings use water as the solvent. Most vehicles in water-based coatings are vinyl or acrylic resins. Cleanup is done with soap and water. In everyday use, solvent-based paints are most often referred to as "oil paints" or "alkyd paints" and the water-based paints as "latex paints."

Historically, water-based and solvent-based coatings each had uses for which they were preferred. Alkyd paints gave a smoother, harder surface than water-based paints

and were favored for interior painting, especially where a durable, high-gloss finish was desired. Alkyd primers also offered superior covering ability and reliable adhesion to problematic substrates. In comparison, latex paints generated fewer odors during application than alkyd paints, an important consideration for renovation work where building occupants could be exposed to paint fumes. Where used for exterior finishing, latex paints produced a more flexible coating able to stretch without breaking when subjected to expansion and contraction of the substrate with changes in temperature and humidity. Latex coatings are also more breathable, lessening the chance of moisture being trapped behind painted siding or trim.

However, many solvent-based coatings also emit significantly more *volatile organic compounds (VOCs)* than water-based coatings. As pollution regulations and sustainability-influenced indoor air quality standards have placed increasingly strict limits on VOC emissions, the use of solvent-based coatings has decreased markedly. In response, manufacturers have worked to improve the formulation of water-based coatings to the extent that, today, water-based coatings with low VOC or even no VOC emissions are available that can equal the performance of solvent-based coatings in almost all applications.

VOC emissions from solvent-based coatings can also be lowered by increasing the solids content and lowering the solvent content of the coating or using solvents that are exempt from low-VOC regulations. Also new to market are hybrid *alkyd-water emulsion coatings* in which alkyd resins are suspended in water. These products offer many of the advantages of traditional alkyds, such as ease of application, excellent coverage, and durability, but in low-VOC formulations that can be cleaned up with water.

Types of Paints and Coatings

The various types of architectural coatings can be defined by the relative proportions of vehicle, solvent, pigment, and additives in each.

Paints contain relatively high amounts of pigment. The highest pigment content is in flat paints, those that dry to a completely matte surface texture. Flat paints contain a relatively low proportion of film-forming vehicle.

Paints that produce glossier surfaces are referred to as *enamels.* A high-gloss enamel contains a very high proportion of vehicle and a relatively low proportion of pigment. The vehicle cures to form a hard, shiny film in which the pigment is fully submerged. A semigloss enamel has a somewhat lower proportion of vehicle, though still more than a flat paint.

Stains range from transparent stains to semitransparent and solid stains. Transparent stains are intended only to change the color of the substrate, usually wood and sometimes concrete. They contain little or no vehicle or pigment, a very high proportion of solvent, and a dye additive. Excess stain is wiped off with a rag a few minutes after application, leaving only the stain that has penetrated the substrate. Usually, a surface stained with a transparent stain is subsequently coated with a clear finish such as varnish to bring out the color and figure of the wood and produce a durable, easily cleaned surface.

Semitransparent stains have more pigment and vehicle than transparent ones. They are not wiped after application. They are intended for exterior application in two coats and do not require a clear topcoat. Also self-sufficient are the solid stains, which are usually water based. These contain much more pigment and vehicle than the other two types of stains and resemble a dilute paint more than they do a true stain. They are intended for exterior use.

The *clear coatings* are high in vehicle and solvent content and contain little or no pigment. Their purpose is to protect the substrate, make it easier to keep clean, and bring out its inherent beauty, whether it be wood, metal, stone, or brick. *Lacquers* are clear coatings that dry extremely rapidly by solvent evaporation. They are based on nitrocellulose or acrylics and are employed chiefly in factories and shops for rapid finishing of cabinets and millwork. A slower-drying clear coating is known as a *varnish*. Varnishes may be either solvent or water based. Most of them harden either by oxidation of an oil vehicle or by moisture curing. Varnishes are useful for on-site finishing. Varnishes and lacquers are available in gloss, semigloss, and flat formulations.

Shellac is a clear coating for interior use that is made from secretions of a particular Asian insect that have been dissolved in alcohol. Shellac dries rapidly and gives a very fine finish, but it is highly susceptible to damage by water or alcohol.

There are many finishes, intended primarily for furniture and indoor woodwork, that are based on simple formulations of natural oils and waxes. A mixture of boiled linseed oil and turpentine, rubbed into wood in many successive coats, gives a soft, water-resistant finish that is attractive to sight, smell, and touch. Waxes such as beeswax and carnauba can be rubbed over sealed (and sometimes unsealed) surfaces of wood and masonry to give a pleasingly lustrous finish. Finishes based on waxes and oils usually require periodic reapplication during service to maintain the character of the surface.

There are countless specialized coatings formulated for particular purposes. Intumescent coatings add fire resistance to steel (see Chapter 11). Industrial *high-performance coatings*, such as epoxies and urethanes, provide greater resistance to physical wear, chemical attack, and corrosion than are possible with ordinary paints. Asphaltic coatings may be applied to some roofing materials. Coatings based on portland cement may be applied over masonry or concrete.

Many of the newest and most durable architectural coatings are designed to be applied in the factory, where controlled environmental conditions and customized machines permit the use of many types of materials and techniques that would be difficult or impossible to use in the field. These include powder coatings, which are sprayed on dry and fused into continuous films by the application of heat; highly durable, multicomponent coating formulations such as fluoropolymer finishes (see Chapter 21), and many others.

Field Application of Architectural Coatings

No aspect of painting and finishing work is more important than surface preparation. Unless the substrate is clean, dry, smooth, and sound, no paint or clear coating will perform satisfactorily. Normal preparation of wood surfaces involves scraping and sanding to remove any previous coatings, patching and filling holes and cracks, and sanding the surface to make it smooth. Preparation of metals may involve cleaning with solvents to remove oil and grease; scraping and wire brushing to remove corrosion and mill scale; sandblasting if the corrosion and scale are tenacious; and, on some metals, chemical etching of the bare metal to improve the paint bond. New masonry, concrete, and plaster surfaces generally require some aging before coating to ensure that the chemical curing reactions within the materials themselves are complete and that excess water has evaporated from the material.

A number of materials are designed specifically to prepare a surface to receive paints or clear coatings. Paste fillers are used to fill the small pores in open-grained woods such as oak, walnut, and mahogany prior to finishing. Various patching and caulking compounds serve to fill larger holes in substrates. A *primer* is a pigmented coating especially formulated to make a surface more paintable. A wood primer, for example, improves the adhesion of paint to wood. It also hardens the surface fibers of the wood so that it can be sanded smooth after priming. Other primers are designed as first coats for various metals, masonry materials, plaster, and gypsum board. Wood

PAINTS AND COATINGS (CONTINUED)

trim and casings in high-quality work are back primed by applying primer to their back surfaces before they are installed; this helps equalize the rate of moisture change on both sides of the wood during periods of changing humidity, which reduces cupping and other distortions. A *sealer* is a thin, unpigmented liquid that can be thought of as a primer for a clear coating. It seals the pores in the substrate so that the clear coating will not be absorbed.

To prevent premature drying, surfaces to be painted should not be exposed to direct sunlight, extreme temperatures, or high-speed winds. The paint materials themselves should also be at normal room temperature.

Paint and other coatings may be applied by brush, roller, pad, or spray. Brushing is the slowest and most expensive method; it is best for detailed work and for applying many types of stains and varnishes. Spraying is the fastest and least expensive, but also the most difficult to control. Roller application is economical and effective for large expanses of flat surface. Many painters prefer to apply transparent stains to smooth surfaces by rubbing with a rag that has been saturated with stain.

A single coat of paint or varnish is usually insufficient to cover the substrate and build the required thickness of film over it. A typical requirement for a satisfactory paint coating is one coat of primer plus two coats of finish material. Two coats of varnish are generally required over raw wood. The surface is lightly sanded after the first coat has dried to produce a smooth surface to which the final coat is applied.

Different coatings require curing periods that range from minutes for lacquers to days for some paints. During this time, environmental conditions similar to those required for the application of the coating must be maintained to ensure that the coating cures properly.

Deterioration of Paints and Finishes

Coatings are the parts of a building that are exposed to the most wear and weathering, and they deteriorate with time, requiring recoating. The ultraviolet (UV) component of sunlight is particularly damaging, causing fading of paint colors and chemical decomposition of paint films. Clear coatings are especially susceptible to UV damage, often lasting no more than a year before discoloring and peeling, which is why they are generally avoided in exterior locations. Some clear coatings are manufactured with special UV-blocking ingredients so that they may be used on exterior surfaces. The other major force of

destruction for paints and other coatings is water. Most peeling of paint is caused by water getting behind the paint film and lifting it off. The most common sources of this water in wood sidings are lumber that is damp at the time it is coated, rainwater leakage at joints, and water vapor migrating from damp interior spaces to the outdoors during the winter. Good construction practices and proper design of air barriers and vapor retarders can minimize these problems.

Other major forces that cause deterioration of architectural coatings are oxygen, air pollutants, fungi, dirt, degradation of the substrate through rust or decay, and mechanical wear. Most exterior paints are designed to "chalk" slowly in response to these forces, allowing the rain to wash the surface clean at frequent intervals.

Typical Architectural Coating Systems

The accompanying table summarizes some typical specifications of coating systems for new surfaces of buildings. Especially where alkyd coatings are indicated, conformance of product VOC emissions to applicable limits should be verified.

Standards for Architectural Coating Systems

The Master Painters Institute (MPI) sets standards for paints and painting methods. Its manual of coating systems standardizes and simplifies the specification of complete coating systems, that is, combinations of surface preparation, primer, and topcoats appropriate for any given substrate and meeting a variety of performance levels.

Sustainability standards for paints and other architectural coatings continue to evolve. Green Seal's *GS-11 Standard for Paints and Coatings* sets performance standards for hiding power (opacity) and cleanability, limits VOC emissions, and restricts the inclusion of certain toxic or harmful ingredients in paints. Compliance with GS-11 is, at the time of this writing, recognized by LEED for credit for low-emitting architectural paints and coatings. MPI's *Green Performance Standards* establish criteria in the same categories of performance, emissions, and restricted ingredients, but in a manner that more closely coordinates with that organization's other product standards. Emissions limits set by the California Air Resources board are referenced in many sustainability guidelines, including the LEED rating system.

Substrate	Surface Preparation	Primer and Topcoats
Exterior		
Wood siding, trim, window, doors	Sand smooth, spot-prime knots and pitch streaks. Fill and sand surface blemishes after applying the prime coat	Exterior latex (or alkyd primer where required by difficult substrate conditions) and two coats of latex paint; or two coats of semitransparent or solid stain
Concrete masonry	Must be clean and dry, at least 30 days old	Block filler primer and two coats of latex paint
Concrete walls	Must be clean and dry, at least 30 days old	Masonry primer where required and two coats of latex paint
Stucco	Must be clean and at least 7 days old	Masonry primer where required and two coats of latex paint
Iron and steel	Remove rust, mill scale, oil, and grease	General-purpose latex or corrosion-resistant metal primer and two coats of latex or alkyd enamel; or two coats of direct-to-metal latex enamel
Aluminum	Clean with a solvent to remove oil, grease, and oxide	Latex metal primer and two coats of latex or alkyd enamel; or two coats of direct-to-metal latex enamel
Interior		
Plaster	Conventional plaster must be 30 days old (veneer plaster may be coated with latex paint immediately after hardening)	Interior latex primer and two coats of latex paint
Gypsum board	Must be clean, dry, and free from dust	Interior latex primer and one or two coats of latex paint
Wood doors, windows, and trim	Sand smooth. Fill and sand small surface blemishes after applying the prime coat. Sand lightly between coats. Remove sanding dust before coating	Interior latex or alkyd enamel primer and one or two coats of latex or alkyd enamel
Concrete masonry	Must be clean and dry, at least 30 days old	Masonry primer where required and two coats of latex paint
Hardwood floors	Fill surface blemishes and sand before coating. Sand lightly between coats	Oil stain if color change is desired, and two coats of oil or polyurethane varnish

SIDING

Exterior cladding materials applied to the walls of a wood light frame building are called *siding*. Many different types of materials may be used: wood boards with various profiles, applied either horizontally or vertically; plywood; wood shingles; metal or plastic materials; fiber-cement panels; brick or stone; and stucco.

Board Siding

Horizontally applied *board siding*, made of solid wood, wood composition board, or fiber-cement, is usually nailed through the wall moisture barrier and sheathing, into the studs, ensuring secure attachment. This procedure also allows horizontal sidings to be applied directly over insulating sheathing materials without requiring a *nail-base sheathing*, a material such as plywood or OSB that is itself dense enough to hold nails. *Siding nails*, whose heads are intermediate in size between those of common nails and finish nails, are used to give the best compromise between holding power and appearance when attaching horizontal siding. Siding nails should be hot-dip galvanized, or made of aluminum or stainless steel, to prevent corrosion and staining. Ring-shank nails are preferred because of their better resistance to pulling out as the siding boards shrink and swell with changes in moisture content. Nailing is done in such a way that the individual pieces of siding may expand and contract freely without damage (Figure 6.18). Horizontal sidings are butted tightly to corner boards and window and door trim, usually with a sealant material applied to the joint during assembly.

More economical horizontal siding boards made of various types of wood composition materials are also

| PLAIN BEVEL | RABBETED BEVEL | V SHIPLAP | COVE SHIPLAP | V-GROOVE TONGUE AND GROOVE | BOARD AND BATTEN |

FIGURE 6.18
Six types of wood siding from among many. The four *bevel* and *shiplap* sidings are designed to be applied in a horizontal orientation. *Tongue-and-groove* siding may be used either vertically or horizontally. *Board-and-batten* siding may only be applied vertically. The nailing pattern (shown with broken lines) for each type of siding is designed to allow for expansion and contraction of the boards. Nail penetration into the sheathing and framing should be a minimum of 1½ inches (38 mm) for a satisfactory attachment.

available. Some of these have experienced problems in service with excessive absorption of water, decomposition, or fungus growth, resulting in large class-action lawsuits and withdrawal of some such products from the market. Board sidings made from wood fiber in a portland cement binder have proven to perform well in service. They accept and hold paint well, do not decay or support fungus growth, and are dimensionally stable, highly resistant to fire, and very durable.

In North America, siding boards are frequently nailed tight to the wall moisture barrier and sheathing. Alternatively, horizontal siding may be nailed over vertical wood spacers called *furring strips,* usually made from preservative-treated 1 × 3s (19 × 63 mm) or similar lengths of preservative-treated plywood or plastic that are aligned over the studs (Figure 6.19). Such so-called *ventilated* or *rainscreen cladding* allows water that seeps through the siding to readily drain out of the assembly rather than soak further into the wall, permits more rapid drying of the siding should it become soaked, and enhances the wall's capability to dispel water vapor that could otherwise accumulate within the insulated

FIGURE 6.19
A ventilated or rainscreen siding application. Water that penetrates through joints or holes in the siding drains away before it reaches the sheathing. With vertical siding, horizontal wood furring strips are not practical, as they would block the drainage space. Instead, plastic furring strips with crosswise hollow channels may be used. The open channels in the furring allow water to pass through.

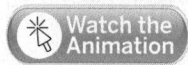

Sheathing

Air and moisture barrier

1x3 (19 x 63 mm) vertical spacers

3/4" (19 mm) airspace

Siding nailed to spacers

A folded strip of insect screening keeps insects out of the airspace and permits drainage

portions of the wall. With furred siding, special attention must be given to cornerboards and window and door casings to account for the overall thickness of the cladding. Alternatively, several manufacturers sell thin drainage mat materials, which can be sandwiched between the siding and the moisture barrier. These products provide some improvement in drainage and ventilation in comparison to conventional siding installations, but at lower cost and with less impact on the detailing of exterior trim in comparison to siding over vertical furring.

Vertically applied sidings (Figures 6.20 and 6.21) are nailed at the top and bottom plates of the wall

FIGURE 6.20
A carpenter applies V-groove tongue-and-groove redwood siding to an eave soffit, using a pneumatic nail gun. *(Photograph courtesy of Senco Products, Inc.)*

FIGURE 6.21
A completed installation of tongue-and-groove siding, vertically applied in the foreground of the picture and diagonally in the area seen behind the chairs. The lighter-colored streaks are sapwood in this mixed heartwood/sapwood grade of redwood. The windows are framed in dark-colored aluminum. *(Architect: Zinkhan/ Tobey. Photo by Barbeau Engh. Courtesy of California Redwood Association.)*

framing and at one or more intermediate horizontal lines of wood blocking installed between the studs.

Heartwood redwood, cypress, and cedar sidings may be left unfinished if desired to weather to various shades of gray. The bare wood will erode gradually over a period of decades and will eventually have to be replaced. Other woods must be either stained or painted to prevent weathering and decay. If these coatings are renewed faithfully at frequent intervals, the siding beneath will last indefinitely.

Plywood Siding

Plywood panel siding (Figure 6.22) is often chosen for its economy. The cost of the material per unit area of wall is usually somewhat less than for other siding materials, and labor costs tend to be relatively low because the large sheets of plywood are more quickly installed than equivalent areas of boards. In some cases, sheathing can be eliminated from the building (with the wall moisture barrier applied directly to the studs) if plywood is used for siding, yielding further cost savings. All plywood sidings must be painted or stained, even those made of decay-resistant heartwoods, because their veneers are too thin to withstand weather erosion for more than a few years. The most popular plywood sidings are those that are grooved to imitate board sidings and conceal the vertical joints between sheets.

The largest problem in using plywood sidings for multistory buildings is how to detail the horizontal panel joints between sheets. A *Z-flashing* of aluminum (Figure 6.23) is the usual solution, but remains clearly visible. The designer should include in the construction drawings a sheet layout for the plywood siding that organizes the horizontal joints in an acceptable manner. This will help avoid a random, unattractive pattern of joints on the face of the building.

FIGURE 6.22
Grooved plywood siding is used vertically on this commercial building by Roger Scott Group, Architects. The horizontal metal flashings between sheets of plywood are purposely emphasized here with a special projecting flashing detail that casts a dark shadow line. *(Courtesy of APA–The Engineered Wood Association.)*

plywood siding panel

2 × 4 (38 × 89 mm) nailer block

Aluminum Z-flashing

plywood siding panel

FIGURE 6.23
A detail of a simple Z-flashing, the device most commonly used to prevent water penetration at horizontal joints in plywood siding.

Shingle Siding

Wood shingles and *shakes* (Figures 6.24 through 6.30) require a nail-base sheathing material such as OSB or plywood. Either corrosion-resistant box nails or gun-driven staples may be used for attachment. Most shingles are of cedar or redwood heartwood and do not need to be coated with paint or stain unless such a finish is desired for cosmetic reasons.

The application of wood shingle siding is labor intensive, especially around corners and openings, where many shingles must be cut and fitted. Figures 6.26 and 6.27 show two different ways of turning corners with wood shingles. Several manufacturers produce shingle siding in panel form by stapling and/or gluing shingles to wood backing panels in a factory. A typical panel size is approximately 2 feet high and 8 feet long (600 × 2450 mm). Several different shingle application patterns are available in this form, along with prefabricated woven corner panels. Panelized shingles can be applied much more rapidly than individual shingles, which results in sharply lower on-site labor costs and, in most cases, a lower overall cost.

FIGURE 6.24
Applying wood shingle siding over asphalt-saturated felt building paper.
The corners are woven as illustrated in Figure 6.26. *(Photo by Edward Allen.)*

FIGURE 6.25
A detail of wood shingle siding at the sill of a wood platform frame building. The first course of shingles projects below the sheathing to form a drip, and is doubled so that all the open vertical joints between the shingles of the outside layer are backed up by the undercourse of shingles. Succeeding courses are single, but are laid so that each course covers the open joints in the course below.

FIGURE 6.26
Wood shingles can be woven at the corners to avoid corner boards. Each corner shingle must be carefully trimmed to the proper line with a block plane, which is time-consuming and relatively expensive, but the result is a more continuous, sculptural quality in the siding.

FIGURE 6.27
Corner boards save time when shingling walls and become a strong visual feature of the building. Notice in this and the preceding diagram how the joints and nail heads in each course of shingles are covered by the course above.

Metal and Plastic Siding

Painted wood sidings are apt to deteriorate unless they are carefully scraped and repainted every 3 to 6 years. *Aluminum* or *vinyl siding*, formed of prefinished sheets of aluminum or molded of vinyl plastic, are usually designed to imitate wood sidings and are generally guaranteed against needing repainting for long periods, typically 20 years (Figures 6.31 and 6.32). Such sidings do have their own problems, however, including the poor resistance of aluminum sidings to denting and the tendency of plastic sidings to crack and occasionally shatter on impact, especially in cold weather. Although plastic and aluminum sidings bear a superficial similarity to the wood sidings that they mimic, their details around openings and corners in the wall are sufficiently different that they are usually inappropriate for use in historic restoration projects.

FIGURE 6.28
Fancy cut wood shingles were often a featured aspect of shingle siding in the late 19th century. Notice the fish-scale shingles in the gable end, the serrated shingles at the lower edges of walls, and the sloping double-shingle course along the rakes. Corners are woven. *(Photo by Edward Allen.)*

FIGURE 6.29
Fancy cut wood shingles stained in contrasting colors are used here on a contemporary restaurant. *(Courtesy of Red Cedar Shingle & Handsplit Shake Bureau.)*

FIGURE 6.30
Both the roof and walls of this New England house by architect James Volney Righter are covered with wood shingles. Wall corners are woven. *(© Nick Wheeler/Wheeler Photographics.)*

ALUMINUM
SIDING

PVC (VINYL)
SIDING

FIGURE 6.31
Aluminum and vinyl sidings are both intended to imitate wood horizontal bevel siding. Their chief advantage in either case is low maintenance. Nails are completely concealed in both systems.

FIGURE 6.32
Retrofitting aluminum siding over insulating foam sheathing on an existing residence. Special aluminum pieces are provided for corner boards and window casings; each piece has a shallow edge channel to accept the cut ends of the siding. *(Photograph provided by The Dow Chemical Company.)*

Stucco

Stucco, or *portland cement plaster*, is a strong, durable, economical, fire-resistant cladding material. It is normally applied in three coats over wire lath, either by hand or by spray apparatus (Figures 6.33 and 6.34). Despite its monolithic appearance, stucco is a porous material and prone to the development of hairline shrinkage cracks. When it is used in locations exposed to heavy wind and rain, significant quantities of water can pass through the material. In such circumstances, stucco is best installed as a rain-screen cladding (Figure 6.19) or with a double-layer wall moisture barrier to provide good drainage and protect the wall construction from water.

Masonry Veneer

Light frame buildings can be faced with *masonry veneer*, a single wythe of brick or stone, in the manner shown in Figure 6.35. The corrugated metal ties

FIGURE 6.34
This small Los Angeles office building demonstrates the plasticity of form that is possible with stucco siding. Vertical and horizontal joints in the stucco minimize its tendency to shrink and crack as it dries out after curing. Although the building is supported from below on concrete columns and steel girders, it is actually a wood light frame building. (*Eric Owen Moss Architects. Photo © Tom Bonner '97.*)

FIGURE 6.33
Applying exterior stucco over woven wire lath, often referred to as "chicken wire." The workers to the right hold a hose that sprays the stucco mixture onto the wall, while the man at the left levels the surface of the stucco with a straightedge. The small rectangular opening at the base of the wall is a crawlspace vent. (*Photo compliments Keystone Steel & Wire Co.*)

Single wythe of brick or stone

Corrugated metal ties nailed to frame

Asphalt-saturated felt paper

Weep holes

FIGURE 6.35
A detail of masonry veneer facing for a platform frame building. The weep holes drain any moisture that might collect in the cavity between the masonry and the sheathing. The cavity should be at least 1 inch (25 mm) wide. A 2-inch (50-mm) cavity is better because it is easier to keep free of mortar droppings that can clog the weep holes.

Watch the Animation

prevent the masonry from falling away from the building while allowing for differential vertical movement between the masonry and the frame. Like stucco, brick veneer is a porous material. The cavity behind the brick creates a rainscreen-like system that provides a path of free drainage to protect the wood-framed wall from water penetration. Masonry materials and detailing are covered in Chapters 8 and 9.

Artificial Stone

Artificial stone or *manufactured masonry* is made from mixtures of cement, sand, other natural aggregates, and mineral pigments such as iron oxide. It is cast into any of a great variety of shapes, textures, and colors simulating the appearance of traditional brick and natural stone products but applied as a thin facing. Artificial stone units range in thickness from approximately 1 inch to 2⅝ inches (25–67 mm) and are typically applied over a base of metal lath and portland cement plaster in a manner similar to that of stucco.

Fiber-Cement Panel Siding

Fiber-cement panel siding is made of cement, sand, organic or inorganic fibers, and fillers manufactured into panels usually 4 feet × 8 feet (1219 mm × 2438 mm) in size and ¼ or ⁵⁄₁₆ inch (6 or 8 mm) thick (Figure 6.36). They are manufactured with a smooth face or with various textures, and can be painted in the field or provided prefinished. They are attached with nails or screws driven into the wall studs or other solid framing behind the moisture barrier and wall sheathing. Gaps between panels can be sealed with joint sealant or filled with preformed plastic or aluminum trim.

FIGURE 6.36

This wood-framed house illustrates the use of four siding materials. To the left, the chimney is clad in stucco. Note the horizontal expansion joints spaced at roughly 8 feet (2400 mm). The second story is clad in asphalt shingles. The area around the entry door is covered with fiber-cement panels. The panels were field-painted and fastened with stainless steel washer-head screws that remain visible and make a distinctive pattern. To the right of the fiber-cement panels is a concrete wall with no applied cladding. (*Architect: Studio Ectypos, www.studioectypos.com. Photo by Joseph Iano.*)

CORNER BOARDS AND EXTERIOR TRIM

Most siding materials require vertical corner boards as well as trim boards around windows and doors.

Roof edges require trim boards of various types, including fascias, soffits, frieze boards, and moldings, the exact patterns depending on the style and detailing of the building (Figure 6.37). For most sidings, the traditional choice of material for these exterior trim components is pine. For wood shingles, unpainted cedar or redwood may be preferred. Plastic and metal sidings are trimmed with special accessory strips of the same material. Fiber-cement siding may be trimmed either with wood or with

FIGURE 6.37
Painters finish the exterior of a house. The painter at the far left is standing on the top of a stepladder—a very unsafe practice.
(Courtesy of Energy Studies in Buildings Laboratory, Center for Housing Innovation, Department of Architecture, University of Oregon.)

fiber-cement boards made for the purpose.

A high appearance grade of pine is usually chosen for wood trim boards. Traditionally, the boards are back primed before installation, with the application of a prime coat of paint on the back side. This helps to reduce cupping and other distortions of the trim boards with changes of humidity. Fully preprimed trim boards are also available at most lumberyards.

Many historical styles require trim that is ornamentally shaped. The shaping generally takes place at a mill, so that the trim pieces come to the site ready to install.

As appearance grades of lumber have become more costly, substitutes have been introduced to the market, including boards made of high-density plastic foams, wood composites of various kinds, and lumber that has been cut into short pieces to eliminate knots, then reassembled end to end with finger joints and glue. These products tend to be more consistently straight and free of defects than natural, solid lumber. Some of them are not only produced as flat boards, but also molded to imitate traditional trim. They must be painted and are not suitable for transparent finishes. Many are also available preprimed or prefinished.

SEALING EXTERIOR JOINTS

After the completion of exterior carpentry work, exposed gaps between siding, trim, frames of doors and windows, and other exterior materials are sealed with *joint sealant*, traditionally called *caulk*, to protect against the entry of water. Exterior sealants must have good adhesion to the materials being sealed, remain permanently flexible, and be unaffected by exposure to sunlight. *Paintable sealants* can be installed before finish painting and then painted over to match adjacent surfaces. Other sealants, such as most silicones, cannot hold paint. These are selected from a range of available premixed colors and then applied after finish painting. Sealant materials and sealant joint design are discussed in detail in Chapter 19.

FIGURE 6.38
Careful detailing is evident in every aspect of the exterior finishes of this commercial building. Notice especially the cleanly detailed window casings, the purposeful use of both vertical boards and wood shingles for siding, and the neat junction between the sidewall shingles and the rake boards. *(Woo & Williams, Architects. Photographer: Richard Bonarrigo.)*

FIGURE 6.39
This house in a rainy climate is designed to shelter every window and door with a roof overhang. The base of the wall is sided with water-resistant cement board. Special attention to roof overhang proportions, exposed framing details, window muntin patterns, and railing details gives this house a uniquely appealing character. *(Photo by Rob Thallon.)*

SUSTAINABILITY OF PAINTS AND OTHER ARCHITECTURAL COATINGS

• Paints and other coatings can be significant emitters of VOCs, and thus require care in selection and specification. Water-based acrylic latex paints generally have lower VOC emissions than solvent-based paints.

• The quality and durability of low- and zero-VOC paints continue to improve.

• Unused paint can be reprocessed to make *recycled paint*, eliminating paint waste from the waste disposal stream.

• In addition to volatile organic compounds, paints and coatings can emit other toxic or unpleasant-smelling chemicals. Some manufacturers also report their products' emissions of such chemicals and offer products with low emissions in these categories.

• Paints that wear quickly and require frequent recoating may increase VOC emissions over the full life of a facility in comparison to others with higher emissions that are more durable and require less frequent recoating.

EXTERIOR PAINTING, FINISH GRADING, AND LANDSCAPING

The final steps in finishing the exterior of a light frame building are painting or staining of exposed wood surfaces; finish grading of the ground around the building; installation of paving for drives, walkways, and terraces; and seeding and planting of landscape materials. By the time these operations take place, interior finishing operations are also usually well underway, having begun as soon as the roofing, sheathing, windows, and doors were in place.

EXTERIOR CONSTRUCTION

Wood is widely used outdoors for porches, decks, stairs, stoops, and retaining walls (Figure 6.40). Decay-resistant heartwoods, wood that is pressure-treated with preservatives, and moisture-resistant wood/plastic composite planks are suitable for these exposed uses. If nondurable woods are used, they will soon decay at the joints, where water is trapped and held by capillary action. Fasteners must be hot-dip galvanized or stainless steel to avoid corrosion. Wood decking that is exposed to the weather should always have open,

FIGURE 6.40
Exterior carpentry for this residential addition includes ipe decking and painted pine railings and trim. The decking is spaced to allow water to drain through. *(Photograph by Michael Craig Moore, AIA.)*

spaced joints to allow for drainage of water through the deck and for expansion and contraction of the decking. Plastic composite planks are durable and attractive, although some are not as strong or stiff as wood, and may require more closely spaced joists for support.

CSI/CSC MasterFormat Sections for Exterior Finishes for Wood Light Frame Construction

04 20 00	**UNIT MASONRY**
04 21 13.13	Brick Veneer Masonry
04 73 00	**MANUFACTURED STONE MASONRY**
06 20 00	**FINISH CARPENTRY**
06 20 23	Exterior Finish Carpentry
06 53 00	**PLASTIC DECKING**
06 65 00	**PLASTIC SIMULATED WOOD TRIM**
07 31 00	**SHINGLES AND SHAKES**
07 31 13	Asphalt Shingles
07 31 29	Wood Shingles and Shakes
07 46 00	**SIDING**
07 46 16	Aluminum Siding
07 46 23	Wood Siding
07 46 29	Plywood Siding
07 46 33	Plastic Siding
07 46 46	Fiber-Cement Siding
07 25 00	**WEATHER BARRIERS**
07 62 00	**SHEET METAL FLASHING AND TRIM**
07 65 00	**FLEXIBLE FLASHING**
07 65 26	Self-Adhering Sheet Flashing
08 14 00	**WOOD DOORS**
08 50 00	**WINDOWS**
09 24 00	**CEMENT PLASTERING**
09 91 00	**PAINTING**
09 93 00	**STAINING AND TRANSPARENT FINISHING**

KEY TERMS

roofing underlayment
building felt
synthetic roofing underlayment
moisture barrier, water-resistive barrier, weather-resistive barrier
Grade D sheathing paper
housewrap
liquid-applied air/weather barrier coatings
eave
rake
gutter
downspout, rain leader
external gutter
internal gutter, concealed gutter

splash block
dry well
ice dam
ventilated roof
soffit vent
gable vent
ridge vent
ice barrier
rubberized underlayment, ice and water shield
unventilated roof
asphalt shingle
flashing
sill pan
nail-fin window, flanged window

finless window, flangeless window
substrate (for painting)
vehicle, binder, film-former
solvent
pigment
solvent-based coating
water-based coating
volatile organic compound (VOC)
alkyd-water emulsion coatings
paint
enamel
stain
clear coating
lacquer
varnish

shellac
high-performance coating
primer
sealer
siding
board siding
nail-base sheathing
siding nail
bevel siding
shiplap siding

tongue-and-groove siding
board-and-batten siding
furring strip
ventilated cladding, rainscreen
   cladding
plywood panel siding
Z-flashing
wood shingle
wood shake
aluminum siding

vinyl siding
stucco, portland cement plaster
masonry veneer
artificial stone, manufactured
   masonry
fiber-cement panel siding
joint sealant
caulk
paintable sealant
recycled paint

REVIEW QUESTIONS

1. In what order are exterior finishing operations carried out on a platform frame building, and why?

2. At what point in exterior finishing operations can interior finishing operations begin?

3. Which types of siding require a nail-base sheathing? What are some nail-base sheathing materials? Name some sheathing materials that cannot function as nail bases.

4. What are the reasons for the relative economy of plywood sidings? What special precautions are advisable when designing a building with plywood siding?

5. How does one make corners when siding a building with wood shingles?

6. Specify two alternative exterior coating systems for a building clad in wood bevel siding.

7. What are the usual reasons for premature paint failure on a wood-sided house?

EXERCISES

1. For a completed wood frame building, make a complete list of the materials used for exterior finishes and sketch a set of details of the eaves, rakes, corners, and windows. Are there ways in which each could be improved?

2. Visit a building materials supply store and look at all the alternative choices of sidings, windows, doors, trim lumber, and roofing. Study one or more systems of gutters and downspouts. Look at eave vents, gable vents, and ridge vents.

3. For a wood light frame building of your design, list precisely and completely the materials you would like to use for the exterior finishes. Sketch a set of typical details to show how these finishes should be applied to achieve the appearance you desire, with special attention to the roof edge details.

SELECTED REFERENCES

In addition to the references listed at the end of Chapter 5, the reader should acquire current product information from manufacturers of residential windows, doors, roofing, and siding materials. Many building supply retailers also distribute unified catalogs of windows, doors, and millwork that can be an invaluable part of the designer's reference shelf.

National Roofing Contractors Association. *NRCA Steep-Slope Roofing Manual.* Rosemont, IL, updated regularly.

This industry-standard reference provides guidelines and recommended details for asphalt shingle roofing and other types of steep-slope roofing.

The Taunton Press, Inc. *Fine Homebuilding.* Newtown, CT, published monthly; and *Journal of Light Construction*, Williston, VT, published monthly.

These two magazines are excellent references on all aspects of wood light frame construction, including topics related to exterior carpentry and finishing.

Williams, R. Sam. "Finishing of Wood." In Forest Products Laboratory, *Wood Handbook: Wood as an Engineering Material.* Madison, WI, 1999.

This article provides a comprehensive introduction to the topic of wood finishing. It can be viewed online for free on the Forest Product Laboratory's web site, www.fpl.fs.fed.us. The complete handbook is also available in printed form from various publishers.

Web Sites

Exterior Finishes for Wood Light Frame Construction

Author's supplementary web site: **www.ianosbackfill.com/06_exterior_finishes_for_wood_light_frame_construction**

Roofing

National Roofing Contractors Association: **www.nrca.net**

Windows and Doors

Anderson Windows & Doors: **www.andersonwindows.com**
Fleetwood Windows & Doors: **www.fleetwoodusa.com**
Pella Windows and Doors: **www.pella.com**
Quantum Windows & Doors: **www.quantumwindows.com**

Siding

Cedar Shake & Shingle Bureau: **www.cedarbureau.org**
James Hardie Fiber Cement Siding: **www.jameshardie.com**
Portland Cement Association—Stucco (Portland Cement Plaster): **www.cement.org/stucco**
Vinyl Siding Institute: **www.vinylsiding.org**
Western Red Cedar Lumber Association: **www.wrcla.org**

Exterior Painting

Benjamin Moore Paints: **www.benjaminmoore.com**
Cabot Stains: **www.cabotstain.com**
Green Seal: **www.greenseal.org**
Master Painters Institute (MPI): **www.paintinfo.com**
MPI Specify Green: **www.specifygreen.com**
Olympic Paints and Stains: **www.olympic.com**
Sherwin-Williams Coatings: **www.sherwin-williams.com**

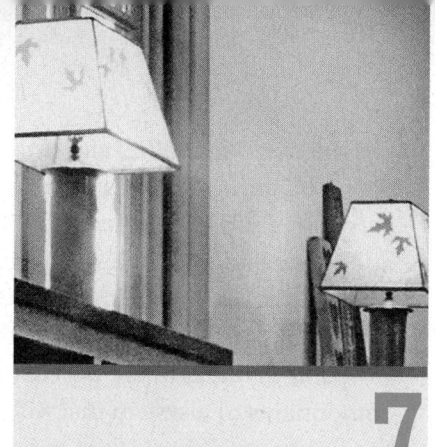

INTERIOR FINISHES FOR WOOD LIGHT FRAME CONSTRUCTION

- **Completing the Building Enclosure**
  Insulating the Building Frame
  Increasing Levels of Thermal Insulation
  Radiant Barriers
  Vapor Retarders

  SUSTAINABILITY OF INSULATION MATERIALS FOR WOOD LIGHT FRAME CONSTRUCTION

  Air Barriers
  Infiltration and Ventilation

- **Wall and Ceiling Finish**

- **Millwork and Finish Carpentry**

  PROPORTIONING FIREPLACES

Interior Doors
Window Casings and Baseboards
Cabinets
Finish Stairs
Miscellaneous Finish Carpentry

PROPORTIONING STAIRS

- **Flooring and Ceramic Tile Work**

- **Finishing Touches**

Architect Michael Craig Moore uses a rich palette of stone and wood contrasting with light-colored walls and ceiling surfaces. In the foreground is slate flooring. Stairs and wood floors are heartwood-only pine ("heart pine"). Wood cabinets and trim are vertical grain Douglas fir. Cabinet counters and the window seat top (middleground of the image) are a composite sheet material made from compressed paper and phenolic resin. Cabinet pulls are bronze. The exposed structural wood column, to the right in the image, is also Douglas fir. Wood surfaces are finished with clear polyurethane. Walls and ceilings are covered with gypsum wallboard finished with latex primer and **topcoats.** (*Photograph by Michael Craig Moore, AIA.*)

As the exterior roofing and siding of a platform frame building approach completion, the framing carpenters and roofers are joined by workers from additional building trades. Masons commence work on fireplaces and chimneys (Figures 7.1 and 7.2). Plumbers begin roughing in their piping (*roughing in* refers to the process of installing the components of a system that will not be visible in the finished building). First to be installed are the large *DWV (drain–waste–vent) pipes*, which drain by gravity and must therefore have first choice of space in the building; then the small *supply pipes*, which bring hot and cold water to the fixtures; and the gas piping (Figures 7.3, 7.4, 7.5, and 7.6). If the building will have central warm air heating and/or air conditioning, sheet metal workers install the furnace and *ductwork* (Figures 7.7, 7.8, and 7.9). If the building is to have a *hydronic (forced hot water) heating* system, the plumbers put in the boiler and rough in the heating pipes and *convectors* at this time (Figure 7.10). A special variety of hydronic heating is the *radiant heating* system that warms the building floors by circulating hot water through plastic piping built into the floors (Figure 7.11). The electricians are usually the last of the mechanical and electrical trades to complete their roughing in because their wires are flexible and can generally be routed around the pipes and ducts without difficulty (Figure 7.12). When the plumbers, sheet metal workers, and electricians have completed their rough work, which consists of everything except installing the plumbing fixtures, electrical outlets, and air registers and grills, inspectors from the local building department check each of the systems for compliance with the plumbing, electrical, and mechanical codes, as well as to ensure that framing has not been damaged during the installation of these other components.

Once these inspections have been passed, connections are made to external sources of water, gas, electricity, and communications services, and to a means of sewage disposal (either a sewer main or a septic tank and leaching field). Thermal insulation and a vapor retarder are added to the exterior ceilings and walls.

Now a new phase of construction, the interior finishing operations, begins, during which the

inside of the building undergoes a succession of radical transformations. The elaborate tangle of framing members, ducts, pipes, wires, and insulation rapidly disappears behind the finish wall and ceiling materials. The interior millwork—doors, finish stairs, railings, cabinets, shelves, closet interiors, and door and window casings—is installed. The finish flooring materials are installed as late in the process as possible to save them from damage by the passing armies of workers, and carpenters follow behind the flooring installers to add the baseboards that cover the last of the rough edges in the construction.

Finally, the building is visited by the painters who prime, paint, stain, varnish, and paper its interior surfaces. The plumbers, electricians, and sheet metal workers make brief return appearances on the heels of the painters to install the plumbing fixtures; the electrical receptacles, switches, and lighting fixtures; and the air grills and registers. At last, following a final round of inspections and a last-minute round of repairs and corrections to remedy lingering defects, the building is ready for occupancy.

FIGURE 7.1
Insulated metal flue systems (shown here) are often more economical than masonry chimneys for furnaces, boilers, water heaters, package fireplaces, and solid-fuel stoves. *(Photo courtesy of SELKIRK METALBESTOS.)*

FIGURE 7.2
A mason adds a section of clay flue liner to a chimney. The large flue is for a fireplace, and the three smaller flues are for a furnace and two woodburning stoves. *(Photo by Edward Allen.)*

FIGURE 7.3
A typical residential wastewater system. All fixtures drain to the building drain through sloping or vertical branch lines. The waste pipe is vented to the exterior at each fixture through a network of vent pipes, shown with broken lines.

FIGURE 7.4
A typical residential water supply system. Water enters the house through a buried line and branches into two parallel sets of distribution lines. One is for cold water. The other passes through the water heater and supplies hot water.

Shower/tub

Toilet Lavatory

Laundry
Lavatory

Toilet

Hose bib

Water heater

Kitchen sink

Hose bib

Cold water ————
Hot water – – – –

Pressurized underground water pipe from well or municipal system

FIGURE 7.5
This plumbing wall is framed with 2 × 8 (38 × 184 mm) studs to provide adequate width for the plethora of building services routed within it, including: cast iron and plastic DWV piping, insulated hot water and bare cold water copper supply piping, refrigerant piping for a distributed air conditioning system, communications wiring, and power wiring. *(Photo by Joseph Iano.)*

Vent through roof

Double stud wall with pipe space between

Joists headed off for water closet waste

2x6 (38x180 mm) stud wall

The water closet waste can be run below the basement ceiling

FIGURE 7.6

The plumber's work is easier and less expensive if the building is designed to easily accommodate piping. The "stacked" arrangement shown here, in which a second-floor bathroom and a back-to-back kitchen and bath on the first floor share the same vertical runs of pipe, is economical and easy to rough in, compared with plumbing that does not align vertically from one floor to the next. The double wall framing on the second floor allows plenty of space for the waste, vent, and supply pipes. The second-floor joists are located to provide a slot through which the pipes can pass at the base of the double wall, and the joists beneath the water closet (toilet) are headed off to house its waste pipe. The first floor shows an alternative type of wall framing using a single layer of deeper studs, which must be drilled to permit horizontal runs of pipe to pass through, as in Figure 7.5.

Second floor
return air grill

Return
air duct

First floor
return air grill

Heat
exchanger

Furnace

Supply trunk

Supply duct

Heat pump or
air conditioner

Register

FIGURE 7·7

A forced-air system in a two-story building with a basement. The furnace is in the basement. It burns gas or oil, or uses electric resistance heating to warm a heat exchanger that creates warm air. It blows the warm air through sheet metal supply ducts to registers in the floor near the exterior walls. The air returns to the furnace through a centrally located return air duct that has a return air grill near the ceiling on each floor. With the addition of a heat pump or air conditioner, this system can deliver cool, dehumidified air during the warm months.

FIGURE 7.8

The installation of this hot air furnace and air conditioning unit is almost complete, needing only electrical connections. The metal pipe running diagonally to the left carries the exhaust gases from the oil burner to the masonry chimney. The ductwork is insulated to prevent moisture from condensing on it during the cooling season and to prevent excessive losses of energy from the ducts. *(Photo by Edward Allen.)*

FIGURE 7.9
Ductwork and electrical wiring are installed conveniently through the openings in these floor trusses, making it easy to apply a finish ceiling if desired. The 2 × 6 that runs through the trusses in the center of the photograph is a bridging member that restrains the trusses from buckling. *(Courtesy of Trus Joist Corporation.)*

FIGURE 7.10
A hydronic heating system. The boiler burns gas or oil, or uses electric resistance heating to heat water. Pumps circulate the water through pipes that lead to convectors in various zones of the building. Inside each convector, a pipe heats closely spaced metal fins that warm the air in the room. The sheet metal convector covers are shown schematically and have been cut away in this drawing to reveal the metal fins.

Convector
Zone valves
Pump
Boiler

Return
manifold

Supply
manifold

Pump

Boiler

FIGURE 7.11

The radiant heating system in this two-story building delivers heat
through hydronic tubes that warm the floors. There are four zones,
two upstairs and two downstairs. Water is heated in a boiler and
pumped through a supply manifold, where thermostatically controlled
valves control the water flow to each of the zones.

FIGURE 7.12
The electrician begins work by nailing plastic fixture boxes to the framing of the building in the locations shown on the electrical plan. Then holes are drilled through the framing, and the plastic-sheathed cable, which houses two insulated copper conductors and an uninsulated ground wire, is pulled through the holes and into the boxes, where it is held by insulated staples driven into the wood. After the interior wall materials are in place, the electrician returns to connect outlets, lighting fixtures, and switches to the wires and the boxes and to affix cover plates to finish the installation. *(Photo by Edward Allen.)*

what it otherwise would be. Furthermore, they prevent the harmful accumulation of condensation within the exterior walls, roof, and other enclosure assemblies of the building.

Insulating the Building Frame

Thermal insulating materials resist the conduction of heat. *Thermal insulation* is added to virtually all buildings to limit winter heat loss and/or summer cooling loads. A material's resistance to the conduction of heat is measured as *thermal resistance*, abbreviated as *R-value*, or, in metric units, as *RSI-value* (or in many cases, also simply as R-value). A material with a higher R-value is a better insulator than one with a lower R-value. Figure 7.13 lists common thermal insulating materials used in wood light frame buildings and some of their characteristics. *Glass fiber batts* are the most popular, but all of those listed find use. For a more in-depth discussion of the role of thermal insulation in buildings, see Chapter 16.

Batt insulation comes in 16 in. or 24 in. (406 or 610 mm) widths. It is installed by inserting the insulation between framing members and is held in place either by friction or, if the insulation has a paper or foil facing, by stapling the facing to the framing (Figure 7.14). Loose fill insulation is blown into attic floors or, in insulation retrofits, into wall cavities through

COMPLETING THE BUILDING ENCLOSURE

The walls, roofs, and other surfaces of a building that separate the indoors from the outdoors may be referred to as the *building envelope* or *building enclosure*. The building enclosure controls the flow of heat, air, and moisture between the interior and exterior. Well-designed and carefully constructed enclosure assemblies help keep a building cooler in summer and warmer in winter by retarding the passage of heat and air through the exterior surfaces of the building. They help keep the occupants of the building more comfortable by moderating the temperatures of the interior surfaces of the building and reducing convective drafts. They reduce the building energy consumption for heating and cooling to a fraction of

Type	Materials	R-Value [a]	Combustibility	Advantages and Disadvantages
Batt	Glass fiber or mineral wool	3.2–4.3 _22–30_	Noncombustible; facings are combustible	Low in cost, easy to install
Batt	Treated cotton	3.0–3.7 _21–26_	Limited combustibility	High percentage of recycled materials content
Blanket	Aerogel	8–10 _55–69_	Noncombustible	Very high R-value, high cost
Spray fibers with binder	Glass and mineral fiber, treated cellulose	3.1–4.0 _22–28_	Same as above	Low in cost
Foamed in place, high-density	Polyurethane	5–7 _35–49_	Combustible, gives off toxic gasses when burned [b]	High R-value, seals against air leakage, vapor impermeable
Foamed in place, low-density	Polyicynene, polyurethane	3.6–4.0 _25–28_	Same as above	Seals against air leakage, vapor permeable
Rigid board	Expanded polystyrene (EPS) foam	3.6–4.2 _25–29_	Same as above	Low vapor permeance, can be used in contact with soil
Rigid board	Extruded polystyrene (XPS) foam	5 _35_	Same as above	Lower vapor permeance, higher compressive strength, and lower moisture absorption than EPS
Rigid board	Polyisocyanurate foam	5–6 _35–42_	Same as above	High R-value, compatible with asphaltic roof membranes, high cost
Semi-rigid board	Mineral fiber	3.6–4.2 _25–30_	Noncombustible	Moderate cost, vapor permeable
Rigid board	Wood or vegetable fiber	2.8 _19_	Combustible	Moderate cost, vapor permeable, high percentage rapidly renewable materials content
Vacuum insulation panel	Vacuum insulated fumed silica	28–39 _194–270_	Noncombustible	Very high insulation value, high cost, special care required during installation

[a] R-values are per unit of thickness, expressed first in U.S. customary units of ft²-hr-°F/Btu-in. and second in metric units (italicized) of m²-°K/W-m. For example, for an insulation material with R-values of 5 ft²-hr-°F/Btu-in. (35 m²-°K/W-m), 3 inches (76 mm) of the material would have a total insulation value of 5 ft²-hr-°F/Btu-in. × 3 in. = R-15 or 35 m²-°K/W-m × 0.076 m = RSI-2.7.

[b] Though all plastic foams are combustible, resistance to ignition and tendency for flame propagation vary among available products.

FIGURE 7.13
Commonly used thermal insulation materials. The R-values offer a direct means of comparing the relative effectiveness of the different types of equal thickness. For more information about insulation materials used with low-slope roofing, see Chapter 16.

holes drilled through the exterior siding. Spray fiber insulation is blown from a nozzle (Figure 7.15). A light spray of water activates a binder that adheres the insulation in place and prevents settlement. Foamed-in-place insulation usually consists of two components sprayed or injected into place, where they react chemically to expand and adhere to the surrounding surfaces (Figure 7.16). Rigid and semi-rigid board insulation are nailed in place over wood framing, or when used around foundations, adhered in place (Figure 5.12). Plastic foam insulation, both spray foam and rigid board, is combustible and releases toxic gasses when it burns. Where these materials face the building interior, most must be covered with gypsum wallboard or some other material that will delay the onset of combustion while occupants exit the building.

(a)

(b)

(c)

(d)

(e)

FIGURE 7.14

(*a*) Installing a polyethylene vapor retarder over glass fiber batt insulation using a staple hammer, which drives a staple each time it strikes a solid surface. The batts are unfaced and stay in place between the studs by friction. (*b*) Stapling paper-faced batts between roof trusses. (Better practice is to staple the paper facing to the face of the framing member rather than to its side as seen here.) (*c*) Placing unfaced batts between ceiling joists in an existing attic. Vent spacers should be used at the eaves (see Figure 6.7). (*d*) Working from below to insulate a floor over a crawlspace. Batts in this type of installation may be retained in place by pieces of stiff wire cut slightly longer than the distance between joists and sprung into place at frequent intervals below the insulation. (*e*) Insulating crawlspace walls with batts of insulation suspended from the sill. The header space between the joist ends has already been insulated.

(Courtesy of Owens-Corning Fiberglas Corporation.)

FIGURE 7.15
Blowing loose-fill glass fiber insulation into a ceiling below an attic. A vapor retarder was installed on the bottom side of the joists and then a gypsum board ceiling, which supports the insulation. Vent spacers were installed at the eaves to prevent the insulation from blocking eave vents. *(Courtesy of Owens-Corning Fiberglas Corporation.)*

Aerogel insulation consists of a silica gel embedded in a flexible fabric. It is available in blanket form, approximately ¼ inch (6 mm) thick, suitable for both interior and exterior applications. It has a very high insulation value and is very costly. Vacuum insulated panels are made from rigid fumed silica boards that have been evacuated and then sealed within an airtight, vapor-tight foil or polymer wrapping. They are also very expensive, but have the highest available insulation values per unit of thickness of any currently available building insulation product. However, vacuum insulation panels require careful handling during shipping and construction. If the panels are cut or punctured, they lose much of their insulation value. Vacuum insulated panels also gradually lose some portion of their insulation value over time, as air and moisture slowly seep into the panels. Because of their high costs, aerogel and vacuum insulation panels are best suited to applications where limiting dimensions make it impossible to meet required insulation values with other materials.

Increasing Levels of Thermal Insulation

The colder the climate, the greater the amounts of insulation required in the building enclosure to avoid wasteful energy loss. A wall framed with 2 × 4 (38 × 89 mm) studs and filled with glass fiber batt insulation can achieve a thermal resistance of R-13 to R-15 (RSI-2.3 to 2.6). According to current energy code standards, this is adequate only for a residential building in the very southernmost portions of the continental United States and Hawaii. To achieve higher insulation values, either the wall can be framed with 2 × 6 (38 × 140 mm) studs and insulated with thicker batts, or rigid or semi-rigid insulation boards can be added to one side or the other of a 2 × 4 insulated wall (Figure 7.17).

(*a*)

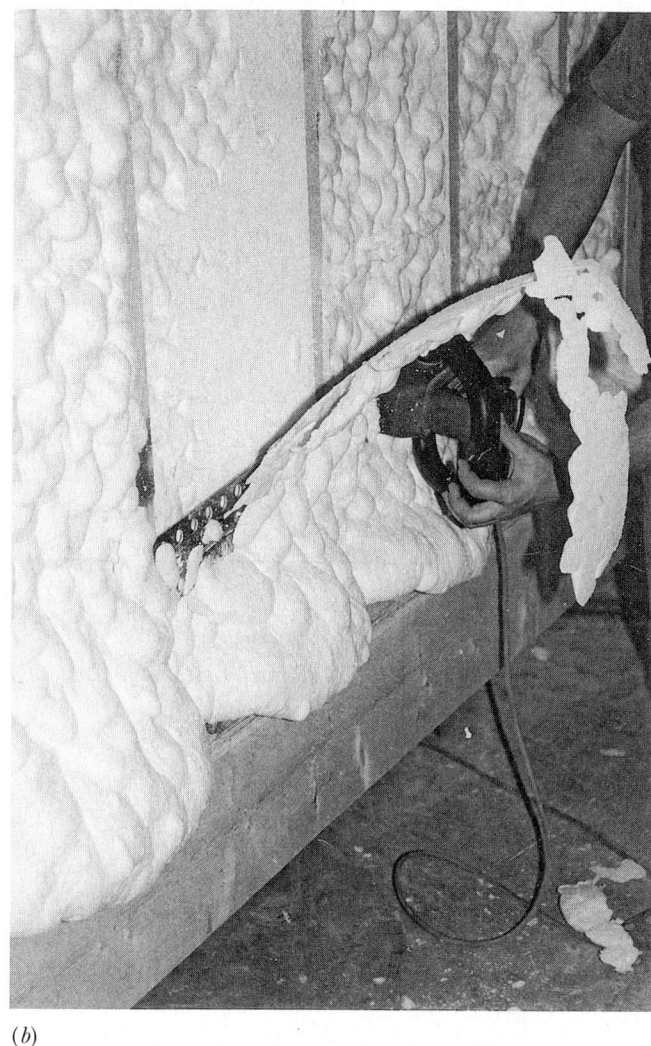

(*b*)

FIGURE 7.16

(*a*) **Spraying a low-density polyicynene foam insulation between studs of a wood light frame building. At the time of spraying, the components are dense liquids, but they react immediately with one another to produce a low-density foam, as has already occurred at the bottom of the cavity. The foam in the cavity to the right has already been trimmed off flush with the studs. (*b*) Trimming off excess foam with a special power saw.** (*Courtesy of Icynene, Inc., Toronto, Canada.*)

Of the two strategies illustrated in Figure 7.17, the insulation applied continuously across the wall in the right-hand example has the advantage of reducing heat loss caused by *thermal bridging*. Thermal bridging occurs wherever framing members interrupt the insulation, creating areas of wall with lower thermal resistance and reducing the insulating efficiency of the wall as a whole. Thus, even though the nominal R-value of the right-hand example in the figure is less than that of the one on the left, when the effects of thermal bridging are taken into account, the two are comparable in real-life energy performance. For residential buildings in colder climates, as well as for most commercial buildings, continuous insulation such as this is always required.

Walls may be insulated to levels greater than those illustrated in Figure 7.17 either to meet code requirements in colder climate zones, or, where desired, in excess of code requirements as a way to create an even more energy-efficient building.

For example, the thermal enclosure of a building designed to the *Passive House* (or *PassivHaus*) standard must be sufficiently efficient such that no major heating or cooling systems are required to maintain the building at comfortable temperatures. (The majority of the heat required to keep a Passive House building warm comes from the body heat of its occupants, waste heat from electrical equipment, and solar gains.) Figure 7.18 illustrates two possible ways to achieve high insulation values in a wood light frame

5½" (140 mm) batt insulation

1" (25 mm) rigid foam insulation

3½" (89 mm) batt insulation

spray foam insulation

2 × 6 STUDS WITH BATT INSULATION

2 × 4 STUDS WITH CONTINUOUS INSULATION AND BATTS

FIGURE 7.17
Left, a 2 × 6 (38 × 140 mm) framed wall with glass fiber batt insulation achieves an insulation value of approximately R-20 (RSI-3.5). *Right*, a 2 × 4 (38 × 89 mm) framed wall with batt insulation and 1 in. of extruded polystyrene foam applied continuously across the outside of the wall achieves an insulation value of approximately R-18 (RSI-3.2), but also reduces heat loss due to thermal bridging at solid framing members. Vapor retarders have not been shown in this figure or Figure 7.18. See the text for a discussion of vapor retarder types and when they are required.

⁑ Watch the Animation

Spray cellulose insulation

Spray foam insulation

(2) 3½" (89 mm) batts within stud cavities plus a continuous 2" (50 mm) batt

Spray foam insulation at platform band joists

2" (50 mm) rigid foam board insulation

3" (75 mm) semirigid mineral fiberboard insulation

5½" spray cellulose insulation

Plywood at window and door rough at openings provides a solid surface for supporting flashings and finishes

2 × 6 STUDS WITH CONTINUOUS INSULATION AND SPRAY-APPLIED CELLULOSE

DOUBLE-FRAMED SUPERINSULATED WALL

FIGURE 7.18
Two examples of wall assemblies that achieve insulation values exceeding current code requirements for any North American climate zone. *Left*, a 2 × 6 (38 × 140 mm) framed wall is insulated with spray-applied cellulose and a continuous, 3-in. (76-mm) layer of semi-rigid mineral wool insulation on the outside of the sheathing. This assembly achieves an insulation value of approximately R-32 (RSI-5.6). *Right*, a double-framed, superinsulated wall with a total of 9 in. (229 mm) of glass fiber batts and a continuous, 2-in. (51-mm) layer of rigid foam insulation. This wall attains a value of approximately R-42 (RSI-7.4). As an alternative, a wall such as this can be built with the inner frame loadbearing rather than the outer frame as shown here. By stopping framing for floors and roofs at the inner bearing wall, thermal bridging at these locations can be reduced further. However, this approach may also be more difficult and costly to construct.

wall. The *superinsulated* (very high insulation value) wall on the right has an insulation value in the range of what may be required in a Passive House project in a cold climate.

Heat loss due to thermal bridging across the wall cavity can also be reduced with modified framing techniques. Figures 7.19 and 7.20 illustrate details that reduce the area of solid framing at wall corners and headers over doors and window openings. Advanced framing techniques, discussed in Chapter 5, also significantly reduce thermal bridging in the wall frame.

Insulation continuity can also become compromised where ceiling insulation under a sloped roof must be compressed in the diminishing space between the roof sheathing and the top of the exterior wall, as can be seen in Figure 7.17. A raised-heel roof truss is one way to overcome this problem (Figure 7.21).

Minimum insulation values for other parts of the building enclosure, such as roofs, basement walls, and slabs on grade, are treated similarly, with lower insulation values permitted in milder climates and higher values required in climates that are more severe.

(a)

(b)

FIGURE 7.19

Insulated headers for window and door openings that minimize thermal bridging in a 2 × 6 (38 × 140 mm) framed wall, shown in section view: (*a*) At the time the header is constructed, rigid foam insulation is sandwiched between the two members. Plywood nailed across the bottom of the headers allows attachment of finishes. (*b*) A double header is installed flush with the exterior. Later, rigid foam insulation can be added on the interior side. A nailer provides attachment for wall finishes.

2 × 4 nailer

(a)

(b)

FIGURE 7.20

Insulated corner details that minimize thermal bridging in a 2 × 6 (38 × 140 mm) framed wall, shown in plan view: (*a*) Each wall frame ends with a full 2 × 6 stud. A single 2 × 4 nailer is added to accept fasteners from the interior wall finish. (*b*) For even greater thermal efficiency, one wall frame ends with a 2 × 4 (38 × 89 mm) stud flush with the interior surface.

Raised heel of truss provides space for full depth of insulation and free air passage

Blocking to keep insulation away from vents

Radiant Barriers

In warmer regions, *radiant barriers* may be used in roofs and walls to reduce the flow of heat into the building. These are thin sheets or panels faced with a bright metal foil that blocks the transmission of infrared (heat) radiation. Radiant barriers can be installed in various locations within the wall or roof. However, the reflective surface must face an airspace in order to function. In one configuration, air barriers are located close to the exterior side of the wall or roof assembly, in a ventilated airspace. The radiant barrier reflects the very high temperatures experienced by the outermost skin of the building and the ventilated space convectively exhausts accumulated heat. Alternatively, a radiant barrier may be attached to the underside of roof

framing in an attic. In this location, the barrier is equally effective at preventing the radiation of heat downward. A potential shortcoming of radiant barriers is their loss of effectiveness over time if the reflective surface becomes dulled by dust, grime, or corrosion.

Vapor Retarders

A *vapor retarder* (also called, less accurately, a *vapor barrier*) is a material placed on the warm side of thermal insulation to prevent water vapor (humidity) from diffusing into the insulation where it could condense into liquid. In the International Residential Code, vapor retarders are grouped accordingly to their relative resistance to water vapor diffusion. Class I vapor retarders are the most resistant; that is, they have the lowest

vapor permeance. Class II and III retarders are each increasingly permeable.

Where vapor retarders are used, they are normally placed on the interior (warm) side of the insulation. Polyethylene plastic sheet, a Class I vapor retarder, is stapled to the face of wood framing after the insulation has been installed in the framing cavities (Figure 7.14(*a*)). To further improve its performance (as well as to control air leakage), the edges of the sheets are lapped and taped or sealed, and penetrations at electrical outlets, plumbing, and other such elements are carefully sealed as well. Asphalt-coated kraft paper facing available with many batt insulation products is a Type II vapor retarder (Figure 7.14(*b*)). The insulation is installed with the paper side facing inward and the paper tabs that

SUSTAINABILITY OF INSULATION MATERIALS FOR WOOD LIGHT FRAME CONSTRUCTION

Building Insulation

Thermal insulation in walls, roofs, and other parts of the building enclosure is probably the most cost-effective, planet-saving material used in buildings.

• Insulation increases occupant comfort and reduces the energy consumed to heat and cool buildings to a fraction of what it would otherwise be.

• The cost to insulate buildings gets paid back through energy savings many times over the life of the building.

• Insulating buildings beyond code minimums can contribute to credits in LEED and other sustainability rating systems.

Natural Fiber Insulation Materials

Natural fiber alternatives to traditional petrochemical and mineral insulation materials include wood fiber, paper, wool, hemp, cotton, flax, and cork. These products can be manufactured as loose fill, batts, and rigid or semi-rigid boards. Insulation values are comparable to those of traditional glass and mineral fiber products.

• Natural fiber insulation materials typically are non-irritating and, as long as safe binders are used, do not have harmful emissions.

• Products made from wood and paper have high recycled materials content.

• Products derived from plants and animals require relatively little energy to manufacture and are often rapidly renewable. However, some of these products are transported great distances to market (for example, lamb's wool), which somewhat increases their total energy content.

• Products made from plants contribute to the sequestration of carbon that might otherwise be emitted into the

Earth's atmosphere as carbon dioxide gas, a major contributor to global warming.

• Nontraditional insulation products may not be readily available locally, and may not be extracted and processed locally.

Glass and Mineral Fiber Insulation

• The embodied energy of glass fiber insulation is approximately 13,000 BTU per pound (30 MJ/kg) and of mineral fiber insulation, 6300 BTU per pound (15 MJ/kg).

• The phenol-formaldehyde binder historically used in glass and mineral fiber batts can release low levels of formaldehyde gas, a recognized carcinogen. Formaldehyde-free alternatives are now readily available.

• Glass fiber insulation may contain up to 40 percent recycled material, and slag fiber insulation up to 70 percent. Rock wool insulation may include recycled slag fibers as well.

• Glass and mineral fiber insulation made with locally extracted and processed materials are available throughout many parts of North America.

• When airborne, glass and mineral fibers can irritate the skin and lungs. Installers must wear skin and breathing protection.

Cotton Fiber Insulation

• Cotton fiber insulation does not emit VOCs or chemical irritants.

• Cotton fiber insulation is made from up to 80 percent recycled content and up to 70 percent rapidly renewable resource (cotton).

extend beyond the edges of the insulation are stapled to the face of the adjacent framing members. Latex paint applied over conventional interior gypsum wallboard is an example of a Class III vapor retarder.

The importance of the vapor retarder in an insulated assembly increases with colder climates. But climate is not the only consideration. For example, the International Residential Code requires Type I or II vapor

retarders for buildings in roughly the upper half to two-thirds of the continental United States. However, where exterior walls are designed to be more tolerant of high rates of vapor diffusion without harmful condensation, more permeable Type III vapor retarders may be used. (Such walls may be constructed with ventilated exterior cladding, insulated sheathing, or relatively permeable exterior sheathing products.) In the National

Building Code of Canada, the insulated enclosure of small buildings must have vapor retarders comparable in performance to Class I or II.

In mixed or hot climates, the addition of vapor retarders where none is required or use of a vapor retarder with a much lower vapor permeance than needed should be approached with caution. This can lead to the inadvertent entrapment of water within the wall. For more

• No skin or breathing protection is required for installers.

Aerogel Insulation

• Aerogel is a silicon-based foam only recently introduced to the building industry as an insulating product. It has an embodied energy of approximately 23,000 BTU per pound (53 MJ/kg).

• Aerogel's very high insulating value makes it especially useful where limited space renders alternative insulation materials impractical.

• Skin and breathing protection are required for installers.

Cellulose Insulation

• The embodied energy of cellulose insulation is approximately 1000 BTU per pound (2.5 MJ/kg).

• Cellulose insulation is made from 85 percent recycled paper products, primarily newspaper.

• The binders used with cellulose insulation are simple glues that do not outgas.

• Cellulose insulation made with locally extracted and processed materials is available throughout many parts of North America.

• Installers must wear skin and breathing protection.

Lamb's Wool Insulation

• Lamb's wool has an insulation value comparable to that of other fibrous insulation products. Its manufactured embodied energy is very low. Even when the energy required to ship this product from its source (New Zealand) to North America is accounted for, overall energy content is still quite low: approximately 1.6 BTU per pound (690 MJ/kg).

• Lamb's wool is a rapidly renewable material.

• Lamb's wool does not emit VOCs or chemical irritants.

• No skin or breathing protection is required for installers.

Plastic Foam Insulation

• Plastic foam materials are made in large part from petroleum, a nonrenewable resource.

• Polystyrene foam has an embodied energy of about 38,000 BTU per pound (90 MJ/kg). Polyisocyanurate and spray-applied foam insulation have an embodied energy of about 30,000 BTU per pound (70 MJ/kg).

• Plastic foams have high insulation values, absorb little if any moisture, and are air impermeable, beneficial characteristics not readily available in many other insulating materials.

• Polystyrene foam can include recycled styrene and polystyrene in its manufacture.

• Extruded polystyrene, polyisocyanurate, and polyurethane foams were formerly manufactured using HCFC, an ozone-depleting blowing agent. All are now manufactured with chemicals less harmful to the atmosphere, although some may still retain higher global warming potential than alternative products due to the release over time of these blowing agents into the atmosphere.

• Expanded polystyrene foam (bead foam) is made using pentane gas as a blowing agent. This chemical does not contribute to global warming.

• Skin and breathing protection are required when installing spray foams.

---

information about vapor retarders in buildings, see Chapter 16.

Air Barriers

Air barriers control the leakage of air through the building enclosure. They significantly reduce building energy consumption and help protect enclosure assemblies from moisture condensation by restricting the infiltration of humid air.

Air barriers appear in some form in all insulated enclosures. Housewraps applied over exterior sheathing are one common way to incorporate an air barrier into wood light frame buildings. Or, when plastic sheeting used as a vapor retarder is carefully sealed against air leakage, it can function effectively as an air barrier. As an alternative to traditional building papers or housewraps, liquid air/weather barrier membranes can be applied to wall and roof sheathing; these provide the necessary moisture resistance and act as highly effective air barriers when combined with special tapes and sealants to seal around openings and penetrations in the wall.

The *airtight drywall approach (ADA)* relies on the gypsum wallboard (drywall) panels used to finish interior walls and ceilings to create an effective air barrier (Figure 7.22). Careful attention must be given

Gypsum wallboard

Airtight tape or caulk

Plywood subfloor

Solid framing

Airtight sill sealer

Concrete foundation wall

FIGURE 7.22
An air barrier system must form a continuous airtight boundary between the inside and outside of the building. In the ADA, the air barrier system is constructed of gypsum wallboard, wood framing members, and the concrete foundation wall (the shaded elements in the figure). Tapes and caulks (see the enlarged inset) are used to seal leakage paths between these components.

to details of construction with this method: Potential air leakage paths around the edges of gypsum panels and between abutting framing members must be sealed with compressible foam tape gaskets or joint sealants during the installation of these members; gypsum board is applied to all the interior surfaces of the outside walls before the interior partitions are framed, eliminating potential air leaks where these partitions intersect; and gaskets, sealants, or special airtight boxes are used to seal air leakage paths around electrical fixtures and other penetrations.

The ADA requires careful attention to joint sealing throughout rough framing and drywalling operations. A related system, *simple caulk and seal*, relies on the strategic application of joint sealants after framing and drywalling are complete to achieve much the same result as the ADA. Simple caulk and seal is easier to coordinate, because the joint sealing occurs separately from the work of other trades. However, it may be less successful at eliminating air leakage paths that become concealed and inaccessible before the sealing work takes place.

Many energy codes require *blower door testing* once the insulation and air barrier components of the building enclosure are fully in place. In this test, a large fan is temporarily installed into an exterior door opening and used to pressurize the entire building interior to a specified level. Calculations are then performed to determine the rate of air leakage through the complete building enclosure. If the leakage rate exceeds that permitted by code, the sources of leakage are sought out and additional testing is performed to bring the building into compliance. The permitted rate of air leakage varies, with buildings in colder regions requiring more airtight enclosures than those in milder ones. The role of the air barrier in the building enclosure is discussed in more detail in Chapter 19.

Infiltration and Ventilation

Reducing the flow of heat and air through the building enclosure reduces energy consumption. However, tightly constructed homes or apartment buildings may exchange so little air with the outdoors that if they are not adequately ventilated, indoor moisture, odors, and chemical pollutants can build up to intolerable or unhealthful levels. Opening a window to ventilate the dwelling is one way to introduce fresh air, but during heating or cooling seasons this wastes fuel. A better solution is a mechanical ventilation system designed to supply fresh air to interior spaces at a controlled and consistent rate. Such a ventilation system may be integrated with a forced-air heating or cooling system, or it may be dedicated solely to satisfying ventilation needs. For greater energy efficiency, a *heat recovery ventilator* or *energy recovery ventilator*—devices that recover heat from the air exhausted from the building and add it to the outside air that is drawn in—may be parts of such a system. Nonresidential buildings of wood light frame construction generally require higher rates of ventilation than residential structures, and here, too, low-infiltration construction and appropriately designed ventilation systems are important to maintaining

a healthful and comfortable interior environment while also reducing heating and cooling costs.

WALL AND CEILING FINISH

Gypsum-based plaster and drywall finishes have always been the most popular for walls and ceilings in wood frame buildings. Their advantages include substantially lower installed costs than any other types of finishes, adaptability to either painting or wallpapering, and, importantly, a degree of fire resistance that offers considerable protection to a combustible frame. *Three-coat gypsum plaster* applied to *wood strip lath* was the prevalent wall and ceiling finish system until the Second World War, when *gypsum board* (also called *drywall*) came into increasing use because of its lower material cost, more rapid installation, and utilization of less skilled labor. More recently, *veneer plaster* systems have been developed that offer surfaces of a quality and durability superior to those of gypsum board, often at comparable prices.

Plaster, veneer plaster, and gypsum board finishes are covered in detail in Chapter 23. Gypsum board remains the favored material for small builders who do all the interior finish work in a building themselves, because the skills and tools it requires are largely those of carpenters rather than plasterers. In geographic areas where there are plenty of skilled plasterers, veneer plaster captures a substantial share of the market. Almost everywhere in North America, there are subcontractors who specialize in gypsum board installation and finishing and who are able to finish the interior surfaces of larger projects, such as apartment buildings, retail stores, and rental office buildings, as well as individual houses, at highly competitive prices.

In most small buildings, all wall and ceiling surfaces are covered with plaster or gypsum board. Even wood

paneling should be applied over a gypsum board backup layer, for increased fire resistance. In buildings that require fire walls between dwelling units or fire separation walls between areas with different uses, a gypsum board wall of the required degree of fire resistance can be installed, eliminating the need to employ masons to put up a wall of brick or concrete masonry (Figures 5.67 and 5.68).

MILLWORK AND FINISH CARPENTRY

Millwork (so named because it is manufactured in a planing and molding mill) includes all the wood interior finish components of a building. Millwork is generally produced from much higher-quality wood than that used for framing: The softwoods used are those with fine, uniform grain structure and few defects, such as sugar pine, Ponderosa pine, or poplar. Flooring, stair treads, and millwork intended for transparent finish coatings such as varnish or shellac are customarily made of hardwoods such as red and white oak, cherry, mahogany, or walnut, or of similar quality hardwood plywoods.

Moldings are also produced from high-density plastic foams, parallel strand lumber, and medium-density fiberboard as a way of reducing costs and minimizing moisture expansion and contraction. All these materials must be painted.

The quality of millwork is regulated by the Architectural Woodwork Institute (AWI), which defines three grades: Economy, Custom, and Premium. *Economy Grade* cabinetry and millwork represents the minimum expectation of quality. *Custom Grade* provides a well-defined degree of control over the quality of materials, workmanship, and installation and is the most commonly specified grade. *Premium* is the most expensive grade, and is reserved for the very finest cabinets and millwork.

PROPORTIONING FIREPLACES

Ever since fireplaces were first developed in the Middle Ages, people have sought formulas for their construction to ensure that the smoke from a fire will go up the chimney and the heat will go into the room, rather than the reverse, which is too often the case. To this day, there is little scientific information on how fireplaces work and how to design them. What we do have are measurements taken from fireplaces that seem to work reasonably well. These have been correlated and arranged into a table of dimensions (Figures B and C) that enables designers to reproduce the critical features of these fireplaces as closely as possible.

Several general principles are clear: The chimney should be as tall as possible. The cross-sectional area of the *flue* should be about one-tenth the area of the front opening of the fireplace. A *damper* should be installed to close off the chimney when no fire is burning and to regulate the passage of air through the *firebox* when the fire is burning (Figure A). A *smoke shelf* above the damper reduces fireplace malfunctions caused by cold downdrafts in the chimney. Splayed sides and a sloping back in the firebox reduce smoking and throw more heat into the room.

Starting from these general principles, two schools of thought have developed concerning how a fireplace should be shaped and proportioned. Most fireplaces in North America are built to the conventional standards tabulated here. Many designers, however, favor the rules for fireplace construction that were formulated by Count Rumford in the 1790s. These produce a fireplace with a taller opening and a shallower firebox than the conventional fireplace. The intention of the Rumford design is to attain a higher efficiency by throwing more radiant heat from the fire-warmed bricks at the back of the firebox to the occupants of the building.

Building codes place a number of restrictions on fireplace and chimney construction. Typically, these call for a 2-in. (51-mm) clearance between the wood framing and the masonry of a chimney or fireplace, and clearances to combustible finish materials around the opening of the fireplace as shown in Figure A. Also specified by code are the minimum thicknesses of masonry around the firebox and flue, the minimum size of the flue, the minimum extension of the chimney above the roof, and steel reinforcing for the chimney. A combustion air inlet must be provided to bring air from the outdoors to the base of the fire. For ready reference in proportioning fireplaces, use the values in Figures B and C. In most cases, the designer need not detail the internal construction of the fireplace beyond the information given in these dimensions, because masons are well versed in the intricacies of assembling a fireplace.

There are several alternatives to the conventional masonry fireplace. One is a steel or ceramic fireplace liner that takes the place of the firebrick lining, damper, and smoke chamber. Many of these products have internal passages that draw air from the room, warm it with the heat of the fire, and return it to the room. The liner is set onto the underhearth and built into a masonry facing and chimney by the mason.

Another alternative is the "package" fireplace, a self-contained, fully insulated unit that needs no masonry whatsoever. It is usually set directly on the subfloor and fitted with an insulated chimney of prefabricated metal pipe. It may be faced with any desired ceramic or masonry materials. Many package fireplaces are made to burn gas rather than wood.

A third alternative is a freestanding metal stove that burns wood, coal, or other solid fuel. Stoves are available in hundreds of styles and sizes. Their principal advantage is that they provide more heat to the interior of the building per unit of fuel burned than a fireplace. A stove requires a noncombustible hearth and a fire-protected wall that are rather large. The designer should consult the local building code at an early stage of design to be sure that the room is big enough to hold a stove of the desired dimensions.

FIGURE A

A cutaway view of a conventional masonry fireplace. Concrete masonry is used wherever it will not show, to reduce labor and material costs, but if hollow concrete blocks are used, they must be filled solid with grout. The damper and ash doors are prefabricated units of cast iron, as is a combustion air inlet in the hearth (not shown here) that connects with a sheet metal duct to an outdoor air intake. The flue liners are made of fired clay and are highly resistant to heat. The flue from the furnace or boiler in the basement slopes as it passes the firebox so as to adjoin the fireplace flue and keep the chimney as small as possible.

Flue from furnace or boiler

Header

Clay flue liner

Brick or stone facing

Cast iron damper

Smoke shelf

Lintel

Tile or brick hearth

Firebrick lining

Firebrick underfire

Cast iron ash dump door

Reinforced concrete underhearth

Concrete block walls to footing

Ash cleanout door

Concrete fill

PROPORTIONING FIREPLACES (CONTINUED)

SECTION

ELEVATION

FIGURE B

The critical dimensions of a conventional masonry fireplace, keyed to the table in Figure C. Dimension *D*, the depth of the hearth, is commonly required to be 16 inches (405 mm) for fireplaces with openings up to 6 square feet (0.56 m²) and 20 inches (510 mm) if the opening is larger. The side extension of the hearth, *E*, is usually fixed at 8 inches (200 mm) for smaller fireplace openings and 12 inches (305 mm) for larger ones.

PLAN

FIGURE C

Recommended proportions for conventional masonry fireplaces, based largely on figures given in Ramsey and Sleeper, *Architectural Graphic Standards* (9th ed.), Hoboken, NJ, John Wiley & Sons, Inc., 1994.

Fireplace Opening						Flue Lining	
Height (*A*)	Width (*B*)	Depth (*C*)	Minimum Backwall Width (*F*)	Vertical Backwall Height (*G*)	Inclined Backwall Height (*H*)	Rectangular (outside dimensions)	Round (inside diameter)
24″	28″	16 to 18″	14″	14″	16″	8½ × 13″	10″
(610 mm)	(710 mm)	(405 to 455 mm)	(355 mm)	(355 mm)	(405 mm)	(216 × 330 mm)	(254 mm)
28 to 30″	30″	16 to 18″	16″	14″	18″	8½ × 13″	10″
(710 to 760 mm)		(405 to 455 mm)	(405 mm)	(355 mm)	(455 mm)	(216 × 330 mm)	(254 mm)
28 to 30″	36″	16 to 18″	22″	14″	18″	8½ × 13″	12″
(710 to 760 mm)	(915 mm)	(405 to 455 mm)	(560 mm)	(355 mm)	(455 mm)	(216 × 330 mm)	(305 mm)
28 to 30″	42″	16 to 18″	28″	14″	18″	13 × 13″	12″
(710 to 760 mm)	(1065 mm)	(405 to 455 mm)	(710 mm)	(355 mm)	(455 mm)	(330 × 330 mm)	(305 mm)
32″	48″	18 to 20″	32″	14″	24″	13 × 13″	15″
(815 mm)	(1220 mm)	(455 to 510 mm)	(815 mm)	(355 mm)	(610 mm)	(330 × 330 mm)	(381 mm)

Millwork is manufactured and delivered to the building site at a very low moisture content, typically about 10 percent, so it is important to protect it from moisture and high humidity before and during installation to avoid swelling and distortions. The humidity within the building is frequently high at the conclusion of plastering or gypsum board work. The framing lumber, concrete work, masonry mortar, plaster, drywall joint compound, and paint are still diffusing large amounts of excess moisture into the interior air. As much of this moisture as possible should be ventilated to the outdoors before *finish carpentry* (the installation of millwork) commences. Windows should be left open for a few days, and in cool or damp weather the building's heating system should be turned on to raise the interior air temperature and help drive off excess water. In hot, humid weather, the air conditioning system should be activated to dry the air.

Interior Doors

Figure 7.23 illustrates five doors that fall into three general categories: Z-brace, panel, and flush. Z-brace doors, mostly built on site, are used infrequently because they are subject to distortions and large amounts of moisture expansion and contraction in the broad surface of boards whose grain runs perpendicular to the width of the door. Panel doors were developed centuries ago to minimize dimensional changes and distortions caused by the seasonal changes in the moisture content of the wood. They are widely available in ready-made form from millwork dealers. Flush doors are smooth slabs with no surface features except the grain of the wood. They may be either solid core or hollow core. Solid-core doors consist of two veneered faces glued to a solid core of wood blocks or bonded wood chips (Figure 7.24). They are much heavier, stronger, and more resistant to the passage of sound than hollow-core doors and are also more expensive. In residential buildings, their use is usually confined to entrance doors, and because of their natural resistance to fire, doors between car garages and a residence. However, they are frequently installed throughout commercial and institutional buildings, where doors are subject to heavier use. *Hollow-core doors* have two thin plywood faces separated by an airspace.

The airspace is maintained by an interior grid of wood or paperboard spacers to which the veneers are bonded. Flush doors of either type are available in a variety of veneer species, the least expensive of which are intended to be painted.

For speed and economy of installation, most interior doors are furnished *prehung*, meaning that they have been hinged and fitted to frames at the mill. The carpenter on the site merely tilts the prehung door and frame unit up into the rough opening, plumbs it carefully with a spirit level, shims it with pairs of wood shingle wedges between the finish and rough jambs, and nails it to the studs with finish nails through the jambs (Figure 7.25). *Casings* are then nailed around the frame on both sides of the partition to close the ragged gap between the door frame and the wall finish (Figure 7.26). To save the labor of applying casings, door units can also be purchased with *split jambs* that enable the door to be cased at the mill. At the time of installation, each door unit is separated into halves, and the halves are installed from opposite sides of the partition to telescope snugly together before being nailed in place (Figure 7.27).

Z-BRACE

FOUR PANEL

SIX PANEL

FLUSH SOLID CORE

FLUSH HOLLOW CORE

FIGURE 7.23
Types of wood doors.

FIGURE 7.24
Edge details of three types of wood doors. The panel is loosely fitted to the stiles and rails in a panel door to allow for moisture expansion of the wood. The spacers and edge strips in hollow-core doors have ventilation holes to equalize air pressures inside and outside the door.

Panel

Stile or rail

PANEL DOOR

Face veneers

Wood or paperboard spacers

Wood block core

Edge strip

SOLID-CORE FLUSH DOOR

HOLLOW-CORE FLUSH DOOR

Top plate

Header

Hinge

Hinge

Hinge

Finish floor

Supporting stud

Door frame

Shingle wedges and finish nails attach the door frame to the studs

Latch strike

Sole plate

ELEVATION

FIGURE 7.25
Installing a door frame in a rough opening. The shingle wedges at each nailing point are paired in opposing directions to create a flat, precisely adjustable shim to support the frame.

(*a*)

(*b*)

Casing — Wood shingle wedges

Door

Casing

SECTION
THROUGH SPLIT
JAMB

FIGURE 7.27
A split-jamb interior door arrives on the construction site prehung and precased. The halves of the frame are separated and installed from opposite sides of the partition.

(*c*)

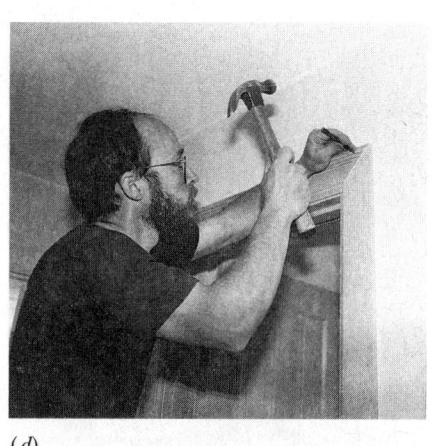

(*d*)

FIGURE 7.26
Casing a door frame. (*a*) The heads of the finish nails in the frame are recessed below the surface of the wood with a steel nail set. (*b*) The top piece of casing, mitered to join the vertical casings, is ready to install, and glue is spread on the edge of the frame. (*c*) The top casing is nailed into place. (*d*) The nails are set below the surface of the wood, ready for filling. (*Photos by Joseph Iano.*)

Window Casings and Baseboards

Windows are cased in much the same manner as doors (Figures 7.28 and 7.29). After the finish flooring is in place, baseboards (see Figures 7.36K–N and 24.31) are installed to cover the gap between the flooring and the wall finish and to protect the wall finish against damage by feet, furniture legs, and cleaning equipment.

When installing casings or baseboards, the carpenter recesses the heads of finish nails below the surface of the wood, traditionally using a hammer and a *nail set* (a hardened steel punch), or, as is more likely today, a powered finish nail gun. Later, the painters will fill these nail holes with a paste filler and sand the surface smooth after the filler has dried so that the holes will be invisible in the painted woodwork. For transparent wood finishes, nail holes are usually filled after the finishes have been applied, using wax-based fillers that are supplied in a range of colors to match the full range of wood species.

The nails in casings and baseboards must reach through the plaster or gypsum board to penetrate the framing members beneath in order to make a secure attachment. Eight- or ten-penny finish or casing nails are customarily used.

Cabinets

Cabinets for kitchens, bathrooms, bedrooms, workrooms, and other spaces may be either custom- or factory-fabricated. *Custom wood cabinets* are fabricated in specialty woodworking

FIGURE 7.28

Casing a window. (*a*) Marking the length of a casing. (*b*) Cutting the casing to length with a power miter saw. (*c*) Nailing the casing. (*d*) Coping the end of a molded edge of an apron with a coping saw so that the molding profile will terminate neatly at the end of the apron. (*e*) Planing the edge of the apron, which has been ripped (sawn lengthwise, parallel to the grain of the wood) from wider casing stock. (*f*) Applying glue to the apron. (*g*) Nailing the apron, which has been wedged temporarily in position with a stick. (*h*) The coped end of the apron, in place. (*i*) The cased window, ready for filling, sanding, and painting. *(Photos by Joseph Iano.)*

shops according to drawings and specifications prepared individually for each project. Like quality millwork, custom cabinets are constructed to AWI specifications, usually Premium or Custom Grade. Less expensive, *manufactured wood cabinets* are factory-manufactured to standard sizes and configurations. Both types of cabinets are usually delivered to the construction site fully finished.

On the construction site, cabinets are installed by shimming against wall and floor surfaces as necessary to make them level and screwing through the backs of the cabinet units into the wall studs (Figures 7.30 and 7.31). The tops are then attached with screws driven up from the cabinets beneath. Kitchen and bath *countertops* are cut out for built-in sinks and lavatories, which are subsequently installed by the plumber.

FIGURE 7.29
Simple but carefully detailed and skillfully crafted window casings in a restaurant. *(Woo & Williams, Architects. Photographer: Richard Bonarrigo.)*

FIGURE 7.30
Prepainted wood kitchen cabinets
installed before the installation of
shelves, drawers, doors, and countertops.
(Photo by Edward Allen.)

FIGURE 7.31
Custom, paneled wood cabinetry contrasts elegantly with contemporary kitchen fixtures and furnishings. The wall behind the
cabinets is finished with stone tile, and the floor with solid wood strip flooring. *(Design by Jane Lockhart; courtesy NKBA.)*

Finish Stairs

Finish stairs are either constructed in place (Figures 7.32 and 7.33) or shop built (Figures 7.34 and 7.35). Shop-built stairs tend to be less expensive to build and easier to install. Constructing stairs on site requires highly skilled carpenters but allows stairs to be built to any configuration and fitted more closely to framing or finish irregularities. Stair treads are usually made of wear-resistant hardwoods such as oak or maple. Risers and stringers may be made of any reasonably hard wood, such as oak, maple, or Douglas fir.

STAIR PLANS

FIGURE 7.32

Left: Stair terminology and clearances for wood frame residential construction. *Right*: Types of stairs. *1.* Straight run. *2.* L-shaped stair with landing. *3.* 180-degree turn with landing. *4.* L-shaped stair with winders (triangular treads). Winders are helpful in compressing a stair into a small space, but are perilously steep where they converge, and their treads become much too shallow for comfort and safety. Building codes limit their minimum dimensions and restrict their use to within dwelling units. *5.* L-shaped stair whose winders have an offset center. The offset center can increase the minimum tread dimension to within legal limits. *6.* A spiral stair (in reality a helix, not a spiral) consists entirely of winders and is generally illegal for any use but a secondary stair in a single-family residence. *7.* A spiral or circular stair with an open center of sufficient diameter can have its treads dimensioned to legal standards.

Handrail

Ramp

Baseboard

Landing

Balusters

Post

Tread

Return

Riser

Wall stringer

Open stringer

Rough stringer

FIGURE 7·33

Constructing a finished stair in place.
The joint between the riser and the open
stringer is a miter. The balusters, posts,
and handrail are purchased ready-made
from millwork suppliers and cut to fit.

Interactive
Exercise

Wedges are driven and glued into tapered grooves under treads and risers

Finish stringer

FIGURE 7.34

A shop-built stair. All the components are glued firmly together in the shop, and the stair is installed as a single piece.

Three openings are required in stair-cases; the first is the door thro' which one goes up to the staircase, which the less it is hid to them that enter into the house, so much the more it is to be commended. And it would please me much, if it was in a place, where before that one comes to it, the most beautiful part of the house was seen; because it makes the house (even tho' small) seem very large; but however, let it be manifest, and easily found. The second opening is the windows that are necessary to give light to the steps; they ought to be in the middle, and high, that the light may be spread equally every where alike. The third is the opening thro' which one enters into the floor above; this ought to lead us into ample, beautiful, and adorned places.

—Andrea Palladio, *The Four Books of Architecture*, 1570

Miscellaneous Finish Carpentry

Finish carpenters install dozens of miscellaneous items in the average building: closet shelves and poles, pantry shelving, bookshelves, wood paneling, chair rails, picture rails, ceiling moldings, mantelpieces, laundry chutes, folding attic stairs, access hatches, door hardware, weatherstripping, doorstops, and bath accessories (towel bars, paper holders, and so on). Many of these items are available ready-made from millwork and hardware suppliers (Figures 7.36 and 7.37), but others have to be crafted by the carpenter.

(a)

(b)

FIGURE 7.35
(a) A worker completes a highly customized curving stair in a shop. (b) Balusters, newel post, and handrail are finely finished in a historical style. (*Courtesy of Staircase & Millwork Co., Alpharetta, GA.*)

FIGURE 7.36

Some common molding patterns for wood interior trim. *A* and *I* are crowns, *C* is a bed, and *D* and *E* are coves. All are used to trim the junction of a ceiling and a wall. *F* is a quarter-round for general-purpose trimming of inside corners. Moldings *G–J* are used on walls. *G* is a picture molding, applied near the top of a wall so that framed pictures can be hung from it at any point on special metal hooks that fit over the rounded portion of the molding. *H* is a chair rail, installed around dining rooms to protect the walls from damage by the backs of chairs. *I* is a panel molding and *J* is a batten, both used in traditional paneled wainscoting. Baseboards (*K–N*) include three single-piece designs and one traditional design (*K*) using a separate cap molding and shoe in addition to a piece of S4S stock that is the baseboard itself (see also Figure 24.31). Notice the shallow groove, also called a "relieved back," in the single-piece baseboards and many other flat moldings on this page. This serves to reduce cupping forces on the piece and makes it easier to install even if it is slightly cupped. Designs *AA–FF* are standard casings for doors and windows. *GG*, *HH*, and *II* are handrail stock. *MM* is representative of a number of sizes of S4S material available to the finish carpenter for miscellaneous uses. *NN* is lattice stock, also used occasionally for flat trim. *OO* is square stock, used primarily for balusters. *PP* represents several available sizes of round stock for balusters, handrails, and closet poles. Wood moldings are furnished in either of two grades: N Grade, for transparent finishes, must be of a single piece. P Grade, for painting, may be finger-jointed or edge-glued from smaller pieces of wood. P Grade is less expensive because it can be made up of short sections of lower-grade lumber with the defects cut out. Once painted, it is indistinguishable from N Grade. The shapes shown here represent a fraction of the moldings that are generally available from stock. Custom molding patterns can be produced easily because the molding cutters used to produce them can be ground quickly to the desired profiles, working from the architect's drawings.

PROPORTIONING STAIRS

Building codes ensure the design of safe stairs through a number of dimensional requirements. The limitations for stairs as given in the International Building Code (IBC) are summarized in Figure A. The required width of an exit stairway is also calculated according to the number of occupants served by the stair, in accordance with formulas given in the code, and may be wider than the minimums indicated in this figure. A stair may also not rise more than 12 feet (3660 mm) between landings. Landings contribute to the safety of a stair by providing a moment's rest for the legs between flights of steps. (Architects also generally avoid designing flights of less than three risers because short flights, especially in public buildings, sometimes go unnoticed, leading to dangerous falls.) The width of the landing must equal the width of the stair. The length of a landing must also equal the width of the stair for stairs up to 48 inches (1219 mm) in width, but need not exceed this dimension for wider stairs.

Working within the dimensional limits of the IBC, combinations of tread and riser dimensions that are most comfortable underfoot can be found by using the proportional rule that twice the riser dimension added to the tread dimension should equal 24 to 25 inches (610 to 635 mm). This formula was derived in France two centuries ago from measurements of actual dimensions of comfortable stairs. Figure B gives an example of how this formula is used in designing a new stair, in this case for a single-family dwelling. Because the IBC does not allow variations that are greater than $3/8$ in. (9.5 mm) between successive treads or risers, the floor-to-floor dimension should be divided equally into risers to an accuracy of 0.01 inch or 1 mm to avoid cumulative errors. The framing square used by carpenters in the United States to lay out stair stringers has a scale of hundredths of an inch, and riser dimensions should be given in these units rather than fractions to achieve the necessary accuracy.

Monumental outdoor stairs, such as those that lead to entrances of public buildings, are designed with lower risers and deeper treads than indoor stairs. Many designers relax the proportions of the $2R + T$ formula a bit for outdoor stairs, raising the sum to 26 or 27 inches (660 or 685 mm), but it is best to make a full-scale mockup of a section of such a stair to be sure that it is comfortable underfoot.

	Minimum Width	Maximum Riser Height	Minimum Riser Height	Minimum Tread Depth	Minimum Headroom
Stair within a residence	36″ (915 mm)	7¾″ (197 mm)	none	10″ (254 mm)	6′ 8″ (2032 mm)
Nonresidential stair, serving an occupant load of less than 50	36″ (915 mm)	7″ (178 mm)	4″ (102 mm)	11″ (279 mm)	6′ 8″ (2032 mm)
Nonresidential stair, serving an occupant load of 50 or more	44″ (1118 mm)	7″ (178 mm)	4″ (102 mm)	11″ (279 mm)	6′ 8″ (2032 mm)

FIGURE A

Dimensional limitations for stairs as established by the IBC.

Procedure	U.S. Customary	Metric
1. Determine the height *(H)* from finish floor to finish floor.	$H = 9'4\frac{3}{8}''$, or $112.375''$	$H = 2854$ mm
2. Divide *H* by the approximate riser height desired—in this example, 7″ or 180 mm—and round off to obtain a trial number of risers for the stair.	Approximate riser height = 7″ $$\frac{H}{7''} = \frac{112.375''}{7''} = 16.05''$$ Try 16 risers	Approximate riser height = 180 mm $$\frac{H}{180 \text{ mm}} = \frac{2854 \text{ mm}}{180 \text{ mm}} = 15.9$$ Try 16 risers
3. Divide *H* by the trial number of risers to obtain an exact riser height *(R)*. Work to the nearest hundredth of a inch, or the nearest millimeter, to avoid any cumulative error that would result in one riser being substantially lower or higher than the rest. Check to make sure that this riser height falls within the limits set by the building code.	$$R = \frac{H}{16} = \frac{112.375''}{16} = 7.02''$$ $R = 7.02''$ $7.02'' < 7.75''$ maximum; riser OK	$$R = \frac{H}{16} = \frac{2854 \text{ mm}}{16} = 178 \text{ mm}$$ $R = 178$ mm 178 mm < 197 mm maximum; riser OK
4. Substitute this riser height into the formula given for proportioning treads and risers and solve for the tread depth. The depth can be rounded down somewhat if desired, as long as $2R+T > 24$. Check the tread depth against the code minimum.	$2R + T = 25''$ $2(7.02'') + T = 25''$ $T = 25'' - 14.04''$ T = 10.96″, say 10.9″ $10.9'' > 10''$ code minimum; tread OK	$2R + T = 635$ mm $2(178 \text{ mm}) + T = 635$ mm $T = 635$ mm $- 356$ mm T = 279 mm, say 275 mm 275 mm > 254 mm code minimum; tread OK
5. Summarize the results of these calculations. There is always one fewer tread than risers in a flight of stairs.	16 risers @ 7.02″ 15 treads @ 10.9″ Total run = (15)(10.9″) = 164.5″ = 13′7½″	16 risers @ 178 mm 15 treads @ 275 mm Total run = (15) (275 mm) = 4125 mm = 13′7½″
6. If desired, a steeper or shallower stair can be tried as an alternative by subtracting or adding one tread and one riser and recalculating dimensions. Reducing the number of treads and risers results in a steeper stair, but also shortens the total run of the stair significantly, which is helpful when designing a stair for a limited space. Adding risers and treads results in a stair that is less steep and occupies more space in the plan.	Try 15 risers: $$R = \frac{H}{15} = \frac{112.375''}{15} = 7.49''$$ $7.49'' < 7.75''$ code minimum; riser OK $2(7.49'') + T = 25''$ T = 10.02″, say 10.0″ $10'' = 10''$ code minimum; tread OK Summary: 15 risers @ 7.49″ 14 treads @ 10″ Total run = (14)(10″) = 11′8″ Subtracting one tread and riser shortens the stair by almost 2 feet.	Try 15 risers: $$R = \frac{H}{15} = \frac{2854 \text{ mm}}{15} = 190 \text{ mm}$$ 190 mm < 197 mm code minimum; riser OK $2(197 \text{ mm}) + T = 635$ mm $T = 255$ mm 255 mm > 254 code minimum; tread OK Summary: 15 risers @ 190 mm 14 treads @ 255 mm Total run = (14)(255 mm) = 3570 mm Subtracting one tread and riser shortens the stair by 555 mm.

FIGURE B

A sample calculation for proportioning a residential stair.

FIGURE 7·37
Fireplace mantels are available from specialized mills in a number of traditional and contemporary designs. Each mantel is furnished largely assembled, but is detailed in such a way that it can easily be adjusted to fit any fireplace within a wide range of sizes.
(Courtesy of Mantels of Yesteryear, P.O. Box 908, 70 West Tennessee Ave., McCaysville, GA 30555; www.mantelsofyesteryear.com; (706) 492-5534, (706) 492-3758 fax.)

FLOORING AND CERAMIC TILE WORK

Before finish flooring can be installed, the subfloor is scraped free of plaster droppings and swept thoroughly. *Underlayment panels* of C–C Plugged plywood or particleboard (in areas destined for resilient flooring materials or carpeting) are glued and nailed over the subfloor, their joints staggered with those in the subfloor to eliminate weak spots. The thicknesses of the underlayment panels are chosen to make the finished floor surfaces as nearly equal in level as possible at junctions between different flooring materials.

In multistory wood light frame commercial and apartment buildings, specially formulated lightweight poured *gypsum* or *portland cement underlayment* is often placed over the subfloor. This has a threefold function: It provides a smooth, level surface for finish floor materials; it furnishes additional fire resistance to the floor construction; and it reduces the transmission of sound through the floor to the apartment or office below. The gypsum or concrete is formulated with superplasticizer additives that make it *self-leveling* as it is applied (Figure 7.38). A minimum thickness is ¾ inch (19 mm). Poured underlayments are also used to level floors in older buildings, to add fire and sound resistance to precast concrete floors, and to embed plastic tubing or electric resistance wires for in-floor radiant heat.

Floor finishing operations require cleanliness and freedom from traffic, so members of other trades are banished from the area as the flooring materials are applied. Hardwood flooring is sanded level and smooth after installation, then vacuumed to remove the sanding dust. The finish coatings are applied in as dust-free an atmosphere as possible to avoid embedded specks. Resilient-flooring installers vacuum the underlayment meticulously so that particles of dirt will not become trapped beneath the thin flooring and cause small bumps in the surface. The finished floors are often covered with heavy sheet or thin board material to protect them during the final few days of construction activity. Carpet installation is less

sensitive to dust, and the installed carpets are less prone to damage than hardwood and resilient floorings, but temporary coverings are applied as necessary to protect the carpet from paint spills and water stains.

The application of ceramic tile to a portland cement plaster base coat over metal lath for a shower stall is illustrated in Figure 7.39, and finished ceramic tile work is

shown in Figure 7.40. Cementitious backer board may be used as a less costly substitute for a cement plaster base coat (Figure 24.28). Wood flooring installation is depicted in Figures 24.31 and 24.32. A finished hardwood floor is shown in Figures 7.41 and 24.33. The installation of ceramic tile and finish flooring materials is covered in more detail in Chapters 23 and 24.

FIGURE 7.38
Workers apply a gypsum underlayment to an office floor. The gypsum is pumped through a hose and distributed with a straightedge tool. Because the gypsum seals against the bottoms of the interior partitions, it can also help reduce sound transmission from one room to another. *(Copyright Gyp-Crete Corporation, Hamel, MN.)*

FIGURE 7.39
Installing sheets of ceramic tile in a shower stall. The base coat of portland cement plaster over metal lath has already been installed. Now the tilesetter applies a thin coat of tile adhesive to the base coat with a trowel and presses a sheet of tiles into it, taking care to align the tiles individually around the edges. A day or two later, after the adhesive has hardened sufficiently, the joints will be grouted to complete the installation. *(Photographs by Joseph Iano.)*

Interactive Exercise

FIGURE 7.40
Ceramic tile is used for the floor, countertops, and backsplash in this kitchen. The border was made by selectively substituting tiles of four different colors for the white tiles used for the field of the floor.
(Designer: Kevin Cordes. Courtesy of American Olean Tile.)

FIGURE 7.41
Varnished oak flooring, millwork, and casework. *(Woo & Williams, Architects. Photographer: Richard Bonarrigo.)*

FINISHING TOUCHES

When flooring and painting are finished, the plumbers install and activate the lavatories, water closets, tubs, sinks, and shower fixtures. Gas lines are connected to appliances and the main gas valve is opened. The electricians connect the wiring for the heating and air conditioning equipment and (if electric) water heater; mount the receptacles, switches, and lighting fixtures; and put metal or plastic cover plates on the switches and receptacles. The electrical circuits are energized and checked to be sure that they work. The smoke, heat, and carbon monoxide alarms required by most codes in residential structures, are also connected and tested by the electricians, along with any communications, entertainment, and security system wiring. The heating and air conditioning system is completed with the installation of air grills and registers, or with the mounting of metal convector covers, then turned on and tested. Painted surfaces that have been scuffed or marred are touched up, and last-minute problems are identified and corrected through cooperative effort by the contractors, the owner of the building, and the architect. The building inspector is called in for a final inspection and issuance of an occupancy permit. After a thorough cleaning, the building is ready for use.

CSI/CSC MasterFormat Sections for Interior Finishes for Wood Light Frame Construction

03 54 00	**CAST UNDERLAYMENT**
03 54 13	Gypsum Cement Underlayment
03 54 16	Hydraulic Cement Underlayment
06 20 00	**FINISH CARPENTRY**
06 20 23	Interior Finish Carpentry
06 22 00	**MILLWORK**
06 26 00	**BOARD PANELING**
06 40 00	**ARCHITECTURAL WOODWORK**
06 41 00	**ARCHITECTURAL WOOD CASEWORK**
06 42 00	**WOOD PANELING**
06 43 00	**WOOD STAIRS AND RAILINGS**
06 44 00	**ORNAMENTAL WOODWORK**
06 46 00	**WOOD TRIM**
07 21 00	**THERMAL INSULATION**
07 21 13	Board Insulation
07 21 16	Blanket Insulation
07 21 19	Foamed-In-Place Insulation
07 21 23	Loose-Fill Insulation
07 21 26	Blown Insulation
07 21 29	Sprayed Insulation
07 21 53	Reflective Insulation
07 26 00	**VAPOR RETARDERS**
07 27 00	**AIR BARRIERS**
08 14 00	**WOOD DOORS**
08 14 16	Flush Wood Doors
08 14 33	Stile and Rail Wood Doors
08 71 00	**DOOR HARDWARE**
09 26 00	**VENEER PLASTERING**
09 29 00	**GYPSUM BOARD**
09 30 00	**TILING**
09 64 00	**WOOD FLOORING**
09 65 00	**RESILIENT FLOORING**
09 68 00	**CARPETING**
09 91 00	**PAINTING**
09 91 23	Interior Painting
09 93 00	**STAINING AND TRANSPARENT FINISHING**
12 32 00	**MANUFACTURED WOOD CASEWORK**
12 36 00	**COUNTERTOPS**

KEY TERMS

roughing in
DWV (drain–waste–vent) pipe
supply pipe
ductwork
hydronic heating, forced hot water
   heating
convector
radiant heating
building envelope, building enclosure
thermal insulation
thermal resistance
R-value, RSI-value
glass fiber batt
cotton batt
aerogel
glass fiber fill
cellulose fill
foamed in place polyurethane
foamed in place polyicynene
rigid board expanded polystyrene
   (EPS) foam
rigid board extruded polystyrene
   (EXS) foam
rigid board polyisocyanurate foam
semi-rigid board mineral fiber
rigid board wood or vegetable fiber
vacuum insulation

thermal bridging
Passive House (PassivHaus) standard
superinsulated
radiant barrier
vapor retarder, vapor barrier
vapor permeance
air barrier
airtight drywall approach (ADA)
simple caulk and seal
blower door test
heat recovery ventilator, energy recovery
   ventilator
three-coat gypsum plaster
wood strip lath
gypsum board, drywall
veneer plaster
millwork
molding
AWI Economy Grade, Custom Grade,
   Premium Grade
flue
damper
firebox
smoke shelf
finish carpentry
Z-brace door
panel door

flush door
solid-core door
hollow-core door
panel
stile
rail
prehung door
casing
split jamb
baseboard
nail set
custom wood cabinet
manufactured wood cabinet
countertop
handrail
baluster
post
tread
return
riser
rough stringer
open stringer
wall stringer
underlayment panel
gypsum cement underlayment
portland cement underlayment
self-leveling

REVIEW QUESTIONS

1. List the sequence of operations required to complete the interior of a wood light frame building and explain the logic of the order in which these operations occur.

2. What are some alternative ways of insulating the walls of a wood light frame building to R-values beyond the range normally possible with ordinary 2 × 4 (38 × 89 mm) studs?

3. Why are plaster and gypsum board so popular as interior wall finishes in wood frame buildings? List as many reasons as possible.

4. What is the level of humidity in a building at the time installation of the interior wall finishes is completed? Why? What should be done about this and why?

5. Summarize the most important things to keep in mind when designing a stair.

6. Describe the function of thermal insulation. When and where is it used?

7. Describe the function of a vapor barrier. When and where is it used?

8. Describe the purpose of an air barrier. When and where is it used?

EXERCISES

1. Design and detail a fireplace for a building that you are designing, using the information provided in this chapter to work out the exact dimensions and the information in Chapter 8 to help in detailing the masonry.

2. Design and detail a stairway for a building that you are designing. Provide complete dimensions.

3. Visit a wood frame building that you admire. Make a list of the interior finish materials and components, including finishes and species of wood where possible. How does each material and component contribute to the overall feeling of the building? How do they relate to one another?

4. Make measured drawings of millwork details in an older building that you admire. Analyze each detail to discover its logic. What woods were used and how were they sawn? How were they finished?

SELECTED REFERENCES

Architectural Woodwork Institute. *AWI Quality Standards Illustrated*. Reston, VA, updated regularly.

Every detail of every grade of interior woodwork and cabinetry is illustrated and described in this thick volume.

Dietz, Albert G. H. *Dwelling House Construction* (5th ed.). Cambridge, MA, MIT Press, 1990.

This classic text has extensive chapters with clear illustrations concerning chimneys and fireplaces, insulation, wallboard, lath and plaster, and interior finish carpentry.

Lstiburek, Joseph. *Builder's Guide to Cold Climates*. Westford, MA, Building Science Corporation, 2006.

This guide, along with companion guides for other major climate zones, explains the roles of insulation, vapor retarders, and air barriers in the performance of the building enclosure, and provides guidelines for the construction of energy-efficient and weather-resistant homes.

Thallon, Rob. *Graphic Guide to Interior Details: For Builders and Designers*. Newtown, CT, Taunton Press, 2004.

Profusely illustrated, clearly written, and encyclopedic in scope, this book offers complete guidance on interior finishing of wood light frame buildings.

WEB SITES

Interior Finishes for Wood Light Frame Construction

Author's supplementary web site: **www.ianosbackfill.com/07_interior_finishes_for_wood_light_frame_construction**

Thermal Insulation and Vapor Retarder

Building Science Corporation: **www.buildingscience.com**
Dow Chemical rigid and spray foam insulation: **www.dow.com**
Gypsum Association: **www.gypsum.org**
Icynene Corporation spray -foam insulation: **www.icynene.com**
Owens Corning insulation products: **www.owenscorning.com**
Passive House Institute (International): **www.passiv.de/en/**
Passive House Institute: **www.passivehouse.us, www.passivehouse.ca, www.passivhaus.org.uk/, www.passiv.de/en/**

Wall and Ceiling Finish

USG gypsum products: **www.usg.com**

Millwork and Finish

Architectural Woodwork Institute: **www.awinet.org**
Hardwood Plywood & Veneer Association: **www.hpva.org**
Jeld-Wen Windows & Doors: **www.jeld-wen.com**
Window and Door Manufacturers Association: **www.wdma.org**

Proportioning Fireplaces

Buckley Rumford Company: **www.rumford.com**

Flooring and Ceramic Tile Work

American Olean Tile: **www.americanolean.com**
Tile Council of America: **www.tileusa.com**

8

BRICK MASONRY

- **History**

- **Mortar**
  Mortar Ingredients
  Mortar Mixes
  Lime Mortar
  Mortar Hydration
  Mortar Admixtures

  SUSTAINABILITY OF BRICK MASONRY

- **Brick Masonry**
  Brick Forming
  Firing Bricks
  Fly Ash Brick
  Brick Sizes
  Brick Classifications
  Choosing Bricks
  Laying Bricks
  Spanning Openings in Brick Walls
  Reinforced Brick Masonry

- **Masonry Wall Construction**

Flemish bond brickwork combines simply and directly with cut limestone lintels and sills in this townhouse on Boston's Beacon Hill. *(Photo JAMES AUSTIN, CAMBRIDGE (GB).)*

Masonry is the simplest of building techniques: The *mason* stacks pieces of material (bricks, stones, or concrete blocks, collectively called *masonry units*) atop one another to make walls. But masonry is also the richest and most varied of techniques, with its endless selection of colors, textures, and patterns. And because the pieces of which it is made are small, masonry can take any shape, from a planar wall to a sinuous surface that defies the distinction of roof from wall.

Masonry is the material of earth, taken from the earth and comfortably at home in foundations, pavings, and walls that grow directly from the earth. With modern techniques of reinforcing, however, masonry can rise many stories from the earth, and in the form of arches and vaults, masonry can take wing and fly across space.

The most ancient of our building techniques, masonry remains labor intensive, requiring the patient skills of experienced and meticulous artisans to achieve a satisfactory result. It has kept pace with the times and remains highly competitive technically and economically with other systems of structure and enclosure, the more so because one mason can produce in one operation a completely finished, insulated, loadbearing wall ready for use.

Masonry is durable. The designer can select masonry materials that are scarcely affected by water, air, or fire, ones with brilliant colors that will not fade, ones that will stand up to heavy wear and abuse, and make from them a building that will last for generations.

Masonry is a material of the small entrepreneur. One can set out to build a building of bricks with no more tools than a *trowel*, a shovel, a hammer, a measuring rule, a level, a square of scrap plywood, and a piece of string. Yet many masons can work together, aided by mechanized handling of materials, to put up projects as large as the human mind can conceive.

HISTORY

Masonry began spontaneously in the creation of low walls by stacking stones or pieces of caked mud taken from dried puddles. Mortar was originally the mud smeared into the joints of the rising wall to lend stability and weathertightness. Where stone lay readily at hand, it was preferred to bricks; where stone was unavailable, bricks were made from local clays and silts. Changes came with the passing millennia: People learned to quarry, cut, and dress stone with increasing precision. Fires built against mud brick walls brought a knowledge of the advantages of burned brick, leading to the invention of the kiln. Masons learned the simple art of turning limestone into lime, and lime mortar gradually replaced mud.

By the fourth millennium BC, the peoples of Mesopotamia were building palaces and temples of stone and sun-dried brick. In the

FIGURE 8.1

The Parthenon, constructed of marble, has stood on the Acropolis in Athens for more than 24 centuries. (*Photo JAMES AUSTIN, CAMBRIDGE (GB).*)

third millennium, the Egyptians erected the first of their stone temples and pyramids. In the last centuries prior to the birth of Christ, the Greeks perfected their temples of limestone and marble (Figure 8.1). Control of the Western world then passed to the Romans, who made the first large-scale use of masonry arches and roof vaults in their temples, basilicas, baths, palaces, and aqueducts.

Medieval civilizations in both Europe and the Islamic world brought masonry vaulting to a very high plane of development. The Islamic craftsmen built magnificent palaces, markets, and mosques of brick and often faced them with brightly glazed clay tiles. The Europeans directed their efforts toward fortresses and cathedrals of stone, culminating in the pointed vaults and flying buttresses of the great Gothic churches (Figures 8.2 and 8.3). In Central America, South America, and Asia, other civilizations were carrying on a simultaneous evolution of masonry techniques in cut stone.

During the Industrial Revolution in Europe and North America, machines were developed that quarried and worked stone, molded bricks, and sped the transportation of these heavy materials to the building site. Sophisticated mathematics were applied for the first time to the analysis of the structure of masonry arches and to the art of stonecutting. Portland cement mortar came into widespread use, enabling the construction of masonry buildings of greater strength and durability.

In the late 19th century, masonry began to lose its primacy among the materials of construction. The very tall buildings of the central cities required frames of iron or steel to replace the thick masonry bearing walls that had limited the heights to which one could build. Reinforced concrete, poured rapidly and economically into simple forms made of wood, began to replace brick and stone masonry in foundations and

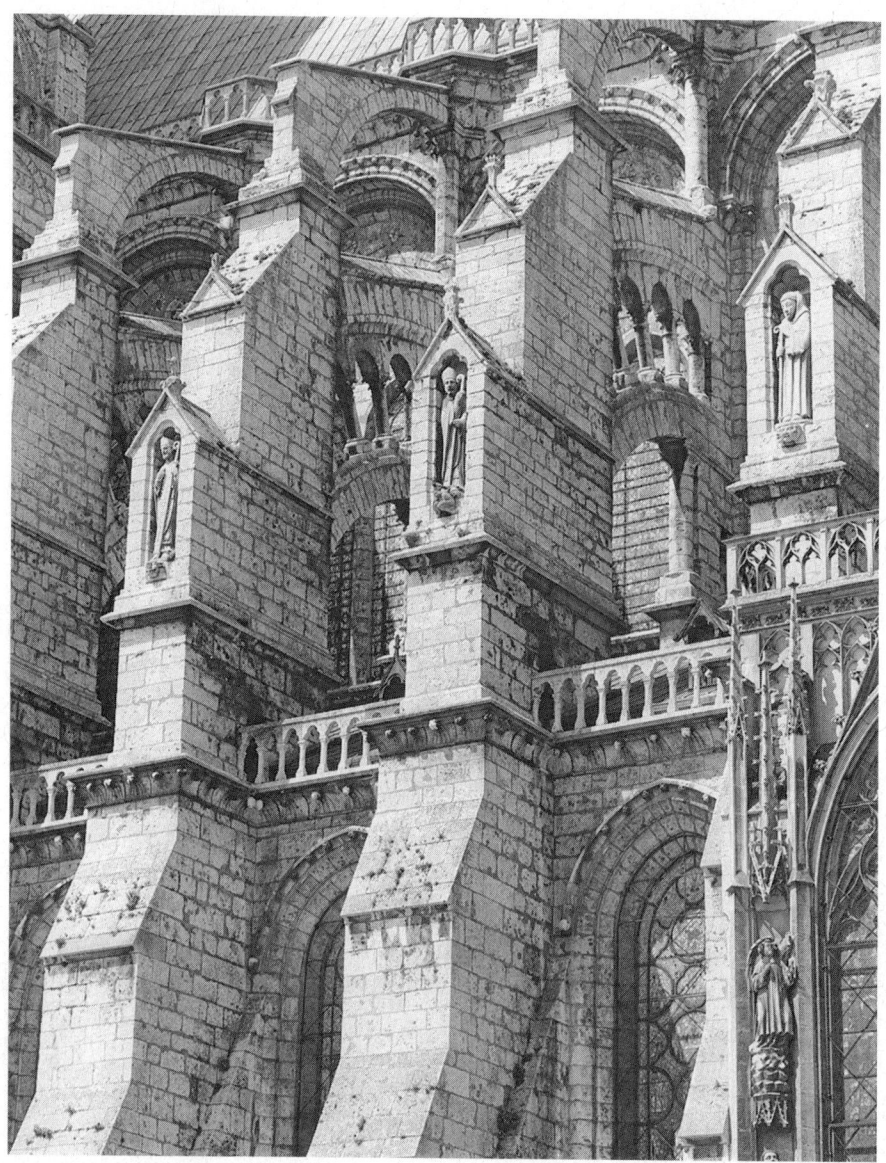

FIGURE 8.2

Construction in ashlar limestone of the magnificent Gothic cathedral at Chartres, France, was begun in AD 1194 and was not finished until several centuries later. Seen here are the flying buttresses that resist the lateral thrusts of the stone roof vaulting. *(Photo JAMES AUSTIN, CAMBRIDGE (GB).)*

walls. The heavy masonry vault was supplanted by lighter floor and roof structures of steel and concrete that were faster to erect.

The 19th-century invention of the hollow concrete block helped to avert the extinction of masonry as a craft. The concrete block was much cheaper than cut stone and required much less labor to lay than brick. It could be combined with brick or stone facings to make lower-cost walls that were still satisfactory in appearance. The brick cavity wall, an early-19th-century British invention, also contributed to the survival of masonry, for it produced a warmer, more watertight wall that was later to adapt easily to the introduction of thermal insulation when appropriate insulating materials became available in the mid-20th century.

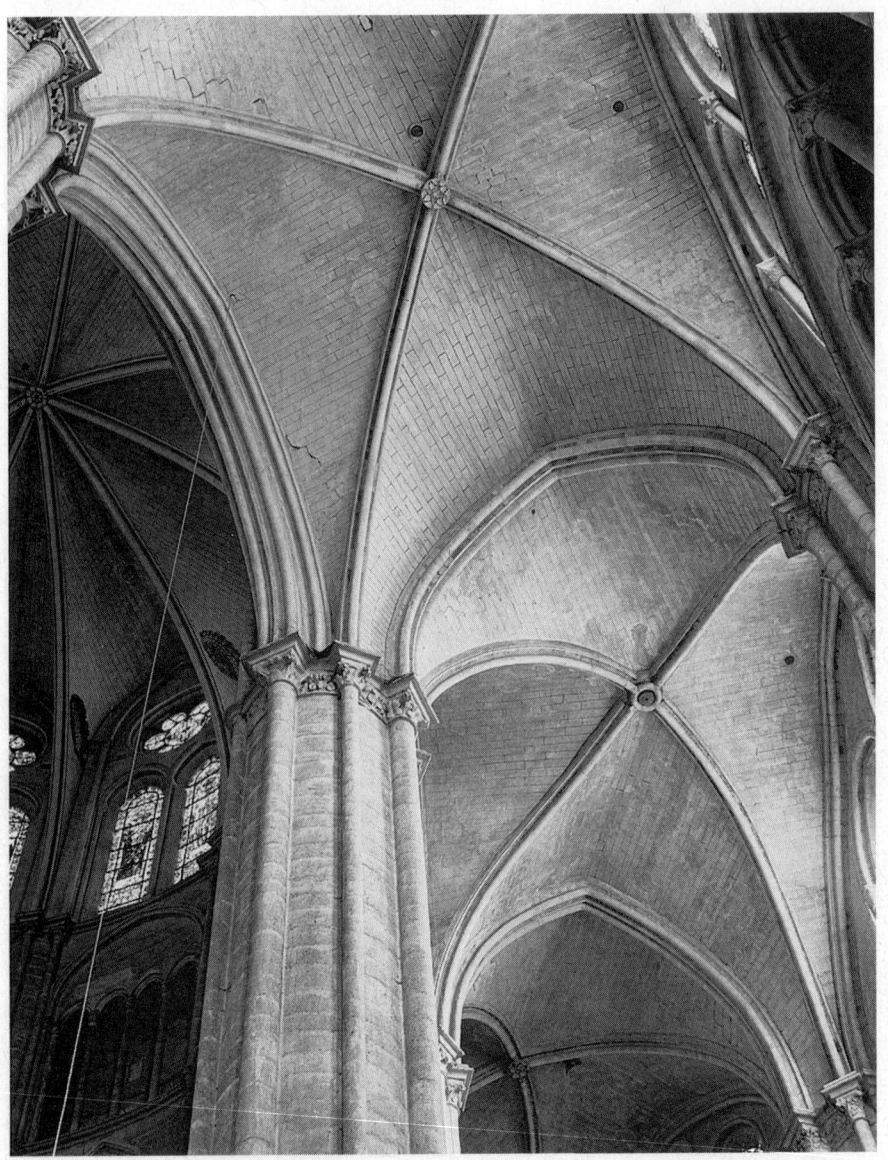

FIGURE 8.3
The Gothic cathedrals were roofed
with lofty vaults of stone blocks. The
ambulatory roof at Bourges (built
1195–1275) evidences the skill of the
medieval French masons in constructing
vaulting to cover even a curving
floor plan. *(Photo JAMES AUSTIN,
CAMBRIDGE (GB).)*

FIGURE 8.4
Despite the steady mechanization of
construction operations in general,
masonry construction in brick, concrete
block, and stone is still based on simple
tools and the highly skilled hands that
use them. *(Courtesy of International
Masonry Institute, Washington, D.C.)*

Other 20th-century contributions to masonry construction include the development of techniques for steel-reinforced masonry, high-strength mortars, masonry units (both bricks and concrete blocks) that are higher in structural strength, and masonry units of many types that reduce the amount of labor required for masonry construction.

...and the smothered incandescence of the kiln: in the fabulous heat, mineral and chemical treasure baking on mere clay, to issue in all the hues of the rainbow, all the shapes of imagination that never yield to time ... these great ovens would cast a spell upon me as I listened to the subdued roar deep within.

—Frank Lloyd Wright, *In the Nature of Materials*, 1942

If this book had been written as recently as 125 years ago, it would have had to devote little space to construction materials other than masonry and wood. Because the development of other construction materials occurred so late, most of the great works of architecture in the world, and many of the best-developed vernacular architectures, are built of masonry. We live amid a rich heritage of masonry buildings. There is scarcely a town in the world that is without a number of beautiful examples from which the serious student of masonry architecture can learn.

MORTAR

Mortar is as vital a part of masonry as the masonry units themselves. Mortar cushions the masonry units, giving them full bearing against one another despite their surface irregularities. Mortar seals between the units to keep water and wind from penetrating; it adheres the units to one another to bond them into a monolithic structural unit; and, inevitably, it is important to the appearance of the finished masonry wall.

Mortar Ingredients

The most characteristic type of mortar is *cement–lime mortar*, made of portland cement, hydrated lime, an inert aggregate, and water. The *aggregate*, sand, must be clean and must be screened to eliminate particles that are too coarse or too fine; ASTM specification C144 establishes standards for mortar sand. The *portland cement* is the bonding agent in the mortar. Its composition and manufacture are described in more detail in Chapter 13 of this book. (Only portland cement Types I, II, and III are recommended for use in masonry mortars.) Mortar made only with portland cement, however, is "harsh," meaning that it does not flow well on the trowel or under the brick, so lime is added to impart smoothness and workability. *Lime* is produced by burning limestone or seashells (calcium carbonate) in a kiln to drive off carbon dioxide and leave *quicklime* (calcium oxide). The quicklime is then slaked by adding water that chemically combines with it to form calcium hydroxide, called *slaked lime* or *hydrated lime*. The slaking process, which releases large quantities of heat, is usually carried out in the factory. The hydrated lime is subsequently dried, ground, and bagged for shipment. ASTM specification C207 governs the production of lime. Water is also an important ingredient in mortar because it is chemically involved in the curing of the cement and lime. Water used in mortar should be clean and free of acids, alkalis, and organic matter. Water that is potable is generally considered suitable for use in mortar.

Blended hydraulic cements, ASTM C595, are combinations of portland cement with other cementitious materials, such as blast furnace slag, that may be used in place of ordinary portland cement alone in the cement–lime mortar mix.

Masonry cements and mortar cements are prepackaged cements that do not require the addition of lime by the mason on the job site. Their main advantages are convenience, consistency (since they are premixed), and good workability. *Masonry cements* are proprietary formulations that may contain portland cement or blended hydraulic cements, lime, additional plasticizing ingredients, and other additives. Formulations vary from one manufacturer to another, but all must comply with ASTM C91. To achieve a workability equivalent to that of conventional cement–lime mortars, masonry cement mortars are formulated with *air-entraining admixtures* that result in a higher air content in the cured mortar than cement–lime mortar. This reduces the bond strength between the mortar and the masonry unit to about half that of conventional mortar, which means that the flexural and shear strength of the wall is reduced and the wall is more permeable to water. For these reasons, masonry cements should not be specified for masonry work that requires high strength and low permeability.

Mortar cements are also blends of portland cement, lime, and other additives. However, they are formulated according to ASTM C1329, with limits on air entrainment that enable them to meet bond strength requirements comparable to those of cement–lime mortars. Structural codes treat mortars made with mortar cement as equivalent to traditional cement–lime mortars.

Cements are available in a range of colors. The most common color is light gray, about the same color as ordinary concrete block. Cements are also available in white, as well as in a range of darker grays, all achieved by controlling the ingredients used in the manufacture of the cement itself. In the final mortar mix, a much broader range of colors can be produced either by adding pigments to the mortar at the time of mixing or by purchasing dry mortar mix that has been custom colored at the factory. Packaged mortar mixes can be obtained in shades ranging from pure white to pure black, including all the colors of the spectrum.

Because mortar makes up a considerable fraction of the exposed surface area of a brick wall, typically about 20 percent, the color of the mortar is extremely important in the appearance of a brick wall and is almost as important in the appearance of stone or concrete masonry walls. Small sample walls are often constructed before a major building goes under construction to view and compare different color combinations of brick and mortar and make a final selection.

Mortar Mixes

Mortars mixed with portland cement, blended hydraulic cement, masonry cement, or mortar cement are all specified according to ASTM C270. Four basic *mortar types*, distinguished primarily by differences in strength, are defined:

• *Type N mortar* is a general-purpose mortar with a balance of good bonding capabilities and good workability. It is recommended for exterior veneers, nonloadbearing exterior walls, parapets, chimneys, and interior loadbearing walls.

• *Type S mortar* has a higher flexural bond strength than Type N mortar. It is recommended for exterior reinforced masonry, exterior loadbearing masonry walls, and veneers and walls subject to high wind forces or high seismic loads.

• *Type O mortar* is a low-strength mortar recommended mainly for interior nonloadbearing masonry and historic restoration work.

• *Type M mortar* is a high-strength mortar with less workability than Type S or N mortars. It is recommended for masonry construction below grade, masonry subject to high lateral or compressive loads, or masonry exposed to severe frost action.

Because lower-strength mortars are more workable than higher-strength mortars, as a general rule, the lowest-strength mortar that meets project requirements should be chosen. (*Workability* is a general characterization of wet mortar's usability: a workable mortar is smooth and plastic, easy to spread with a trowel, and adheres well to vertical surfaces.) The majority of mortar for masonry work used in new construction is either Type N or Type S. As a memory aid, the letters used to designate mortar types, in order of decreasing strength, come from taking every other letter in the phrase MaSoN wOrK. (Type K mortar is a very-low-strength mortar used in historic preservation work that is no longer part of the ASTM C270 specification.)

According to ASTM C270, mortar mixes may be specified in one of two ways: either by *proportion specification*, in which the quantities of ingredients used to prepare the mix are specified, or by *property specification*, in which the compressive strength and other properties of the hardened mortar as determined by laboratory testing are defined. Proportion specification is the simpler (no laboratory testing is required) and more common method. On large jobs, however, property specification gives the mason more flexibility in the choice of mortar ingredients and can result in an overall cost savings, even after the costs of laboratory testing are considered. These two methods are summarized in Figures 8.5 and 8.6, respectively.

Lime Mortar

Modern masonry mortars are made with *hydraulic cements*, that is, cements that cure by chemical reaction with water, a process called "hydration" and discussed in more detail in the next subsection. Until the late 19th and early 20th centuries, however, mortar was made without portland cement, and the lime itself was the bonding agent. Such traditional *lime mortars*, made from a mix of lime, sand, and water, continue to find use principally in the restoration of historic structures. Unlike modern hydraulic cements, lime is a *nonhydraulic cement*, and mortars made with lime as the sole cementing ingredient cure through a reaction with carbon dioxide in the atmosphere. This process, called *carbonation*, occurs gradually and may continue for many years. Such mortars remain at least partially water soluble and retain some ability to self-heal in the event of hairline cracking caused by movement within the wall. By adding other cementitious materials, lime mortars with greater degrees of hydraulic properties can also be formulated.

Mortar Hydration

Hydraulic cement mortars cure by *hydration*, not by drying: In a complex set of chemical reactions, they take up water and combine it with the constituents of the cement and lime to create a dense, strong crystalline structure that binds the sand particles and masonry units together. Once hydraulic cements harden, they become water insoluble.

Mortar that has been mixed but not yet used can become too stiff for use, either by drying out or by commencing its hydration. If the mortar was mixed less than 90 minutes prior to its stiffening, it has merely dried and the mason can safely *retemper* it with water to make it workable again. If the unused mortar is more than 2½ hours old, it must be discarded because it has begun to hydrate. Adding water beyond this time will result in a mortar with reduced final strength. On large masonry projects, an *extended-life admixture* is sometimes

Mortar Type	Parts by Volume of Portland Cement or Blended Hydraulic Cement	Parts by Volume of Mortar Cement or Masonry Cement	Part by Volume of Hydrated Lime	Aggregate, Measured in a Damp, Loose Condition
M	1		¼	Not less than 2¼ and not more than 3 times the sum of the volumes of cement and lime materials used
M	1	1 (Type N)		
M		1 (Type M)		
S	1		Over ¼ to ½	
S	½	1 (Type N)		
S		1 (Type S)		
N	1		over ½ to 1¼	
N		1 (Type N)		
O	1		over 1¼ to 2½	
O		1 (Type N)		

FIGURE 8.5

Mortar types by proportion specification. The general principle is that the greater the proportion of cement to lime, the greater the compressive strength of the mortar. For each of the four mortar types, the first example is a cement–lime mortar mix consisting of portland cement or blended hydraulic cement, lime, and aggregate. The second and sometimes third examples for each type are mortars made with mortar cement or masonry cement. Note that no added lime is required for these mixes; the necessary lime and other plasticizers are included in the prepackaged ingredients. Note also that a Type M mortar can be made with a Type N cement by increasing the total cement content of the mix with the addition of portland cement or blended hydraulic cement on the job site. In the same way, Type S mortar can be made with Type N prepackaged mortar and added cement.

Mortar Type	Minimum Average Compressive Strength at 28 days
M	2500 psi (17.2 MPa)
S	1800 psi (12.4 MPa)
N	750 psi (5.2 MPa)
O	350 psi (2.4 MPa)

FIGURE 8.6

Minimum compressive strength for mortar types by property specification. Not shown here, but also included in the ASTM C270 specification, are requirements for maximum air content, minimum water retention (a factor affecting the bond strength of the mortar), and volume ratio of aggregate to cement and lime. When mortar is specified by property, the mason is free to use any mix that meets the specified strength and other requirements as demonstrated through laboratory testing.

included in the mortar. This allows the mortar to be mixed in large batches and kept for as long as 72 hours before it must be discarded. When working with highly absorptive brick, the mortar may have to be protected from premature drying that could occur when water is drawn out of the mortar by the bricks. In this circumstance, the bricks are slightly dampened just before laying.

Mortar Admixtures

ASTM C1384 defines other *admixtures* that may be added to mortar to adjust various of its properties, either immediately after it is mixed and in a workable state, or once it has hardened. *Set accelerators* help keep mortar from freezing when masonry work takes place in very cold conditions, and *set retarders* help keep mortar from hydrating too quickly in very hot water. *Integral water repellents* help make masonry walls more watertight and reduce *efflorescence* (staining that forms on the surface of a wall when excess moisture carries minerals to the wall surface). Other admixtures may be used to improve workability before setting and bond strength in the finished work.

SUSTAINABILITY OF BRICK MASONRY

Brick Masonry Materials

• Mortar is made of minerals that are generally abundant in the earth. Portland cement and lime are energy-intensive to produce. (For more information about the sustainability of cement production, see Chapter 13.)

• Clay and shale, the raw materials for bricks, are plentiful. They are usually obtained from open pits, with the attendant disruption of drainage, vegetation, and wildlife habitat.

• Clay brick can include recycled brick dust, postindustrial wastes such as fly ash, and a variety of other waste products in their manufacture.

Brick Manufacturing

• Brick manufacturing plants are usually located close to the sources of their raw materials.

• Brick manufacturing produces few waste materials. Unfired clay is easily recycled into the production process. Fired bricks that are unusable are ground up and recycled into the production process or used as landscaping material.

• Brick manufacturing requires relatively large amounts of water. Water that does not evaporate can be reused many times. Little if any water need be discharged as waste.

• Because of the energy used in its firing, brick is a relatively energy-intensive product. Its embodied energy may range from about 1000 to 4000 BTU per pound (2.3–9.3 MJ/kg).

• The most common energy source for brick kilns is natural gas, although oil and coal are also used. Firing of clay masonry produces fluorine and chlorine emissions. Other types of air pollution can result from improperly regulated kilns.

• Brick made from fly ash, a waste product of coal-fired power generation, and sand need not be kiln-fired. Therefore, the energy required for fly ash brick manufacture is only about 15 percent of that needed for clay brick. The use of fly ash gives this brick type a high recycled materials content and contributes to the diversion of this waste material from landfills or storage ponds. However, although fly ash has been used as an ingredient in a variety of building materials for many decades, the possible long-term environmental impacts of introducing this hazardous material into the building materials product stream continues to be a point of some disagreement.

• Most bricks are sold for use in regional markets close to their point of manufacture. This reduces the energy required for shipping and makes much brick eligible for credit as a regional material.

Brick Masonry Construction

• Relatively small amounts of waste are generated on a construction site during brick masonry work, including partial bricks, unsatisfactory bricks, and unused mortar. These wastes generally go into landfills or are buried on site.

• Sealers applied to brick masonry to provide water repellency and protection from staining are potential sources of emissions. Solvent-based sealers generally have higher emissions than water-based products.

Brick Masonry Buildings

• Brick masonry is not normally associated with any indoor air quality problems, although in rare circumstances it can be a source of radon gas.

• The thermal mass effect of brick masonry can be a useful component of energy-saving heating and cooling strategies such as solar heating and nighttime cooling.

• Brick masonry is a durable form of construction that requires relatively little maintenance and can last a very long time.

• Construction with brick masonry can reduce reliance on paint finishes, a source of volatile organic compounds.

• Brick masonry is resistant to moisture damage and mold growth.

• When a brick building is demolished, sound bricks may be cleaned of mortar and reused (once their physical properties have been verified as adequate for the new use). Brick waste can be crushed and used for landscaping. Brick and mortar waste can also be used as on-site fill. Much such waste, however, is disposed of offsite in landfills.

BRICK MASONRY

Among the masonry materials, brick is special in two respects: fire resistance and size. As a product of fire, it is the most resistant to building fires of any masonry unit type. Its size may account for much of the love that many people instinctively feel for brick: A traditional brick is shaped and dimensioned to fit the human hand. Hand-sized bricks are less likely to crack during drying or firing than larger bricks, and they are easy for the mason to manipulate. This small unit size makes brickwork very flexible in adapting to small-scale geometries and patterns and gives a pleasing scale and texture to a brick wall or floor.

Brick Forming

Because of their weight and bulk, which make them expensive to ship over long distances, *bricks* are produced by a large number of relatively small, widely dispersed factories from a variety of local clays and shales. The raw material is dug from pits, crushed, ground, and screened to reduce it to a fine consistency. It is then tempered with water to produce a plastic clay ready for forming into bricks.

Three major methods are used today for forming bricks: the soft mud process, the dry-press process, and the stiff mud process. The oldest is the *soft mud process*, in which a relatively moist clay (20–30 percent water) is pressed into simple rectangular molds, either by hand or with the aid of molding machines (Figure 8.7). To keep the sticky clay from adhering to the molds, the molds may be dipped in water immediately before being filled, producing bricks with a relatively smooth, dense surface that are known as *water-struck bricks*. If the wet mold is dusted with sand just before forming the brick, *sand-struck* or *sand-mold bricks* are produced, with a matte-textured surface.

The *dry-press process* is used for clays that shrink excessively during drying. Clay mixed with a minimum of water (up to 10 percent) is pressed into steel molds by a machine working at a very high pressure.

The high-production *stiff mud process* is the least costly and most widely used today. Clay containing 12 to 15 percent water is passed through a vacuum to remove any pockets of air, then extruded through a rectangular die (Figures 8.8 and 8.9). As the clay leaves the die, textures or thin mixtures of colored clays may be applied to its surface as desired. The rectangular column of moist clay is pushed by the pressure of extrusion across a cutting table, where automatic cutter wires slice it into bricks.

Bricks produced by the stiff mud process are highly uniform in dimension and shape. Where more

FIGURE 8.7

A simple wooden mold produces seven water-struck bricks at a time. *(Photo by Edward Allen.)*

FIGURE 8.8

A column of clay emerges from the die in the stiff mud process of molding bricks. *(Courtesy of Brick Institute of America.)*

FIGURE 8.9
Rotating groups of parallel wires cut the column of clay into individual bricks, ready for drying and firing. (*Courtesy of Brick Institute of America.*)

variation in appearance among units is desired, the bricks can be *tumbled* prior to firing. Tumbling softens edges and corners, and introduces a greater individuality in appearance among units.

After molding by any of these three processes, the bricks are dried for 1 or 2 days in a low-temperature dryer kiln. They are then ready for transformation into their final form by a process known as *firing* or *burning*.

Firing Bricks

Before the advent of modern kilns, bricks were most often fired by stacking them in a loose array called a *clamp*, covering the clamp with earth or clay, building a wood fire under the clamp, and maintaining the fire for a period of several days. After cooling, the clamp was disassembled and the bricks were sorted according to the degree of burning each had experienced. Bricks adjacent to the fire (*clinker bricks*) were often over-burned and distorted, making them unattractive and therefore unsuitable for use in exposed brickwork. Bricks in a zone of the clamp near the fire were fully burned but undistorted, suitable for exterior facing bricks with a high degree of resistance to weather. Bricks farther from the fire were softer and were set aside for use as backup bricks, while some bricks from around the perimeter of the clamp were not burned sufficiently for any purpose and were discarded. In the days before mechanized transportation, bricks for a building were often produced from clay obtained from the building site and were burned in clamps adjacent to the work.

(*a*)

(b)

(c)

FIGURE 8.10

Three stages in the firing of water-struck bricks in a small factory. (*a*) Bricks stacked on a kiln car ready for firing. The open passages between the bricks allow the hot kiln gases to penetrate to the interior of the stack. The bed of the kiln car is made of a refractory material that is unaffected by the heat of the kiln. The rails on which the car runs are recessed in the floor. (*b*) The cars of bricks are rolled into the far end of this gas-fired periodic tunnel kiln. When firing has been completed, the large door in the near end of the kiln is opened and the cars of bricks are rolled out on the rails that can be seen at the lower right of the picture. (*c*) After the fired bricks have been sorted, they are strapped into these "cubes" for shipping. (*Photo by Edward Allen.*)

Today, bricks are usually burned in either a periodic kiln or a continuous tunnel kiln. The *periodic kiln* is a fixed structure that is loaded with bricks, fired, cooled, and unloaded (Figure 8.10). For higher productivity, bricks are passed continuously through a long *tunnel kiln* on special railcars to emerge at the far end fully burned. In either type of kiln, the first stages of burning are *water smoking* and *dehydration*, which drive off the remaining water from the clay. The next stages are *oxidation* and *vitrification*, during which the temperature rises to 1800 to 2400 degrees Fahrenheit (1000 to 1300°C) and the clay is transformed into a ceramic material. This may be followed by a stage called *flashing*, in which the fire is regulated to create a reducing atmosphere in the kiln that develops color variations in the bricks. Finally, the bricks are cooled under controlled conditions to achieve the desired color and avoid thermal cracking. The cooled bricks are inspected, sorted, and packaged for shipment. The entire process of firing takes 40 to 150 hours and is monitored continuously to maintain product quality. Considerable shrinkage takes place in the bricks during drying and firing; this must be taken into account when designing the molds for the bricks. The higher the temperature is, the greater the shrinkage and the darker the color of the brick. Bricks are often used in a mixed range of colors, with the darker bricks inevitably being smaller than the lighter bricks. Even in bricks of uniform color, some size variation is to be expected, and bricks, in general, are subject to a certain amount of distortion from the firing process.

The color of a brick depends on the chemical composition of the clay or shale and the temperature and chemistry of the fire in the kiln. Higher temperatures, as noted in the previous paragraph, produce darker bricks. The iron that is prevalent in most clays turns red in an oxidizing fire and purple in a reducing fire.

Other chemical elements interact in a similar way to the kiln atmosphere to make still other colors. For bright colors, the faces of the bricks can be glazed like pottery, either during the normal firing or in an additional firing.

Fly Ash Brick

Fly ash bricks are made from fly ash, a waste product from coal-fired power generation, and sand. The fly ash acts as the binder in the bricks and the sand as the aggregate. These two ingredients are mixed with water and smaller amounts of other ingredients, compacted into molds, and steam cured. The resulting masonry units are equal in performance and similar in appearance to traditional clay bricks.

In comparison to the tradition kiln brick firing process, steam curing of fly ash bricks requires significantly less energy. As a result, the embodied energy and global warming potential of fly ash bricks is significantly less than that of clay masonry.

Brick Sizes

There is no truly standard brick. The nearest thing in the United States is the *modular brick*, dimensioned to construct walls in modules of 4 inches (101 mm) horizontally and 8 inches (203 mm) vertically, but the modular brick has not found ready acceptance in some parts of the country and traditional sizes persist on a regional basis. Figure 8.11 shows the brick sizes that represent about 90 percent of all the bricks used in the United States. In practice, the designer, when selecting bricks for a building, usually views actual samples before completing the drawings for the building and dimensions the drawings in accordance with the size of the particular bricks selected (see Figure 8.22). For most bricks in the traditional range of sizes, three courses of bricks plus the accompanying three mortar joints add up to a height of 8 inches (203 mm). Length dimensions must be calculated specifically for the brick selected and must include the thicknesses of the mortar joints.

The use of larger bricks can lead to substantial economies in construction. A *Utility brick*, for example, has the same face height as a standard modular brick, but because it is longer, its in-the-wall cost per square foot is about 25 percent lower and the compressive strength of the wall is about 25 percent higher because of the smaller proportion of mortar. The designer should also consider, however, that a wall built with larger bricks can deceive the viewer regarding the scale of the building.

Custom shapes and sizes of brick are often required for buildings with special details, ornamentation, or unusual geometries (Figures 8.12 and 8.13). These are readily produced by most brick manufacturers if sufficient lead time is given.

Name	Width	Length	Height
Modular	3½″ or 3⅝″ (90 mm)	7½″ or 7⅝″ (190 mm)	2¼″ (57 mm)
Standard	3½″ or 3⅝″ (90 mm)	8″ (200 mm)	2¼″ (57 mm)
Engineer Modular	3½″ or 3⅝″ (90 mm)	7½″ or 7⅝″ (190 mm)	2¾″ to 2¹³⁄₁₆″ (70 mm)
Engineer Standard	3½″ or 3⅝″ (90 mm)	8″ (200 mm)	2¾″ (70 mm)
Closure Modular	3½″ or 3⅝″ (90 mm)	7½″ or 7⅝″ (190 mm)	3½″ or 3⅝″ (90 mm)
Closure Standard	3½″ or 3⅝″ (90 mm)	8″ (200 mm)	3⅝″ (90 mm)
Roman	3½″ or 3⅝″ (90 mm)	11½″ or 11⅝″ (290 mm)	1½″ to 1⅝″ (40 mm)
Norman	3½″ or 3⅝″ (90 mm)	11½″ or 11⅝″ (290 mm)	2¼″ (57 mm)
Engineer Norman	3½″ or 3⅝″ (90 mm)	11½″ or 11⅝″ (290 mm)	2¾″ to 2¹³⁄₁₆″ (70 mm)
Utility	3½″ or 3⅝″ (90 mm)	11½″ or 11⅝″ (290 mm)	3½″ or 3⅝″ (90 mm)
King Size	3″ (75 mm)	9⅝″ (240 mm)	2⅝″ or 2¾″ (70 mm)
Queen Size	3″ (75 mm)	7⅝″ or 8″ (190 mm)	2¾″ (70 mm)

FIGURE 8.11

Dimensions of bricks commonly used in North America, as established by the Brick Industry Association. This list gives an idea of the diversity of sizes and shapes available, and of the difficulty of generalizing about brick dimensions. Modular bricks are dimensioned so that three courses plus mortar joints add up to a vertical dimension of 8 inches (203 mm), and one brick length plus mortar joint has a horizontal dimension of 8 inches (203 mm). The alternative dimensions of each brick are calculated for ⅜″ (9.5-mm) and ½″ (12.7-mm) mortar joint thicknesses.

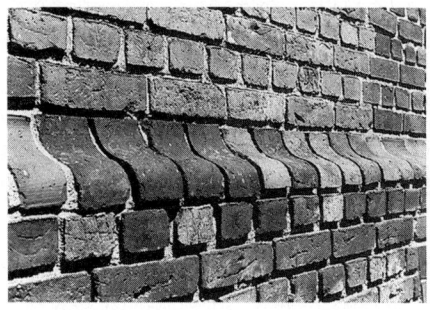

FIGURE 8.12
Bricks may be custom molded to perform particular functions. This rowlock water table course in an English bond wall was molded to an ogee curve.
(Photo by Edward Allen.)

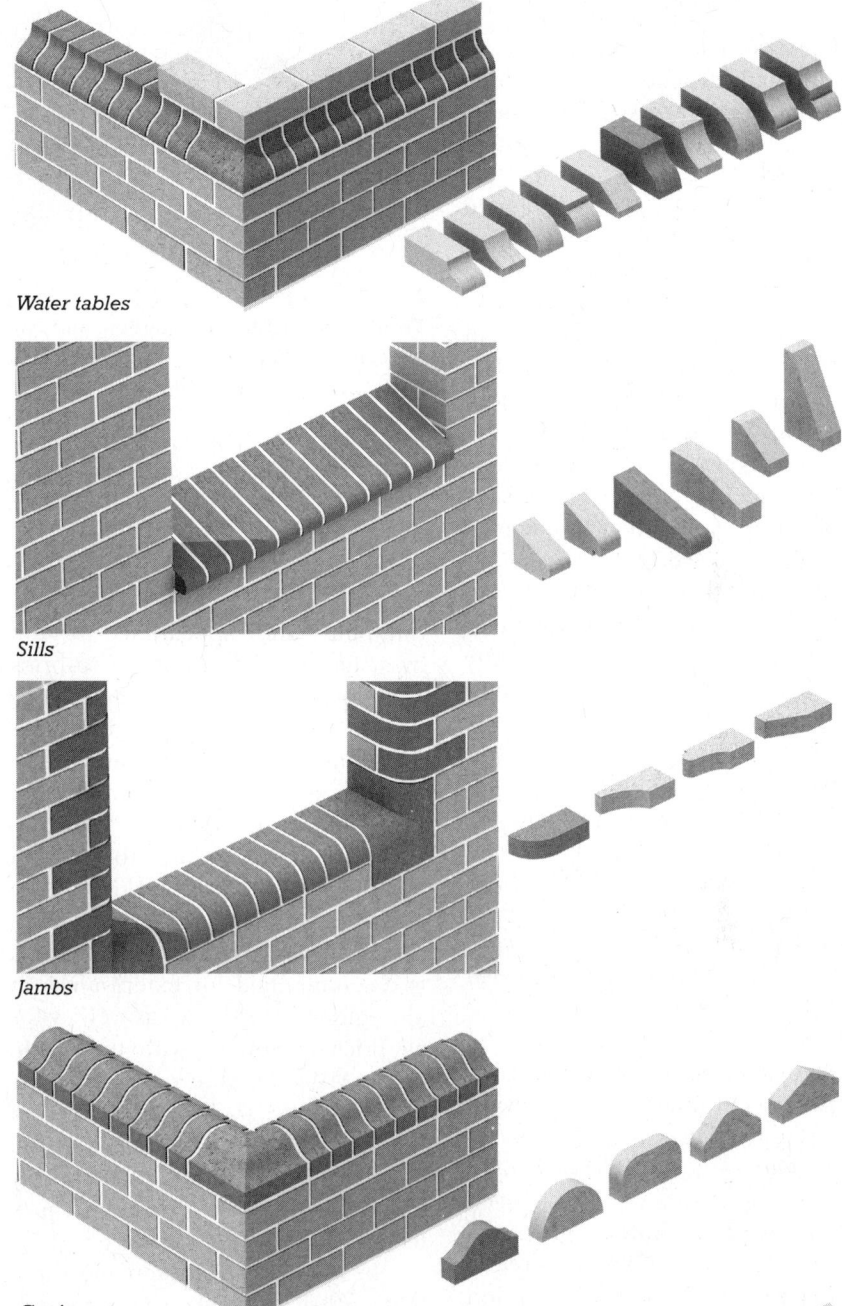

Water tables

Sills

Jambs

Copings

Angles and radials

FIGURE 8.13
Some commonly used custom brick shapes. Notice that each water table, jamb, and coping brick shape requires special inside and outside corner bricks as well as the basic rowlock or header brick. The angle bricks are needed to make neat corners in walls that meet at other than right angles. The hinge brick at the extreme lower right of the drawing is a traditional shape that can be used to make a corner at any desired angle. Radial bricks produce a smoothly curved wall surface of any specified radius.

Common shapes not pictured here include voussoirs for any desired shape and size of arch and rounded-edge bricks for stair treads.

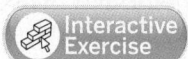
Interactive Exercise

Brick Classifications

The bricks most commonly used in building construction are classified as facing brick (ASTM C216), building brick (ASTM C62), or hollow brick (ASTM C652). *Facing bricks* (also called *face brick*) are intended for both structural and nonstructural uses where appearance is important. *Building bricks* are used where appearance does not matter, such as in backup wythes of masonry that will be concealed in the finished work. Both facing bricks and building bricks are specified as *solid units*. Solid units may, in fact, be genuinely solid, or despite their name, they may be *cored* or *frogged* (Figure 8.14), as long as any plane measured parallel to the bearing surface of the brick is at least 75 percent solid. By reducing the volume and thickness of the clay, cores and frogs permit more even drying and firing of bricks, reduce fuel costs for firing, reduce shipping costs, and create bricks that are lighter and easier to handle. (Where genuinely solid bricks are required, specifications should call for 100 percent solid, uncored, unfrogged brick.) *Hollow bricks*, defined according to ASTM C652, may be up to 60 percent void and are used primarily to enable the insertion and grouting of steel reinforcing bars in single wythes of brickwork (Figure 8.14).

Paving bricks (ASTM C902), used for the paving of walks, drives, and patios, must conform to special requirements not only for freeze-thaw resistance, but water absorption and abrasion resistance as well. *Firebricks* (ASTM C64) are used for the lining of fireplaces or furnaces. These are made from special *fireclays* which produce bricks with refractory qualities (resistance to very high temperatures). Firebricks are laid in very thin joints of *fireclay mortar.*

Choosing Bricks

In addition to molding process, color, size, and other aesthetic qualities,

FIGURE 8.14
From left to right: cored, hollow, and frogged bricks. Cored and frogged bricks are considered solid, as long as they remain at least 75 percent solid. A hollow brick may be up to 60 percent void.

durability and appearance uniformity may also be important to brick selection (Figure 8.15). *Brick grade* establishes minimum requirements for compressive strength and water absorption. The overall durability of the brick and its resistance to weathering can then be related to a map of weathering indices derived from data on winter rainfall and freeze-thaw cycles (Figure 8.16). Grade SW brick is recommended for exterior use in all regions on the map as well as for all brick in contact with the earth. Grade MW brick is recommended for use only above grade, in areas labeled on the map as negligible weathering. Grade NW brick should be used only indoors or in exterior locations protected from the weather.

In addition to influencing durability, a brick's compressive strength is also of obvious importance when used in the construction of loadbearing walls and piers. Minimum compressive strengths for building bricks and face bricks range, depending on the grade, from 1500 to 3000 pounds per square inch (psi) (10 to 21 MPa). However, higher-strength bricks are readily available. A compressive strength of 10,000 psi (69 MPa) is not uncommon for bricks used in structural masonry. In high-strength applications, brick strength may exceed

20,000 psi (138 MPa). The strength of constructed brickwork depends not only on the strength of the bricks but also on the strength of the mortar and, when steel reinforced, on the strength and quantity of reinforcing.

Brick type defines limits on the variation in size, distortion in shape, and *chippage* (extent of physical damage to face or visible corners) among brick units (Figure 8.15). Type FBS is considered a general-purpose face brick and is the most common type sold. Type FBX bricks have more stringent limits on appearance characteristics and are intended for masonry work with very thin joints or for bond patterns demanding very close dimensional tolerances. Type FBA bricks are characterized by significant variations in size and shape, as is typical of handmade brick or brick intentionally manufactured for such effect.

Laying Bricks

Figure 8.17 shows the basic vocabulary of bricklaying. Bricks are laid in various positions for visual reasons, structural reasons, or both. The simplest brick wall is a single *wythe* of *stretcher courses*. For walls two or more wythes thick, *headers* are used to bond the wythes together into a structural unit. *Rowlock courses* are often used for caps

Durability of Facing Brick, Building Brick and Hollow Brick	
Grade SW	Any weathering region, all brick in contact with the earth
Grade MW	Above-ground brick only, in regions indicated in Figure 8.16 as *Negligible weathering* only
Grade NW	Sheltered or indoor locations only
Appearance Uniformity of Facing Brick	
Type FBX	Least variation in size per unit, least distortion in shape, minimum chippage
Type FBS	General-purpose face brick, with wider variation in shape and size, greater chippage
Type FBA	Great nonuniformity in size and shape permitted, as defined by manufacturer

FIGURE 8.15

Brick grade and type. Brick grades classify bricks according to their durability and resistance to freeze-thaw action. The grades listed here apply to facing brick, building brick, and hollow brick. Paving brick is graded similarly, but using the designations, in decreasing order of weathering resistance, SX, MX, and NX. Brick types classify bricks according to their uniformity of size and shape. The types listed here apply only to facing brick. Hollow brick is manufactured in types designated HBX, HBS, HBB, and HBA, in decreasing order of uniformity, and paving brick is manufactured in types PX, PS, and PA. Building bricks, which are not visible in the finished construction, are not classified for appearance.

FIGURE 8.16

Weathering regions of the United States, as determined by winter rainfall and freezing cycles. *(Courtesy of Brick Institute of America.)*

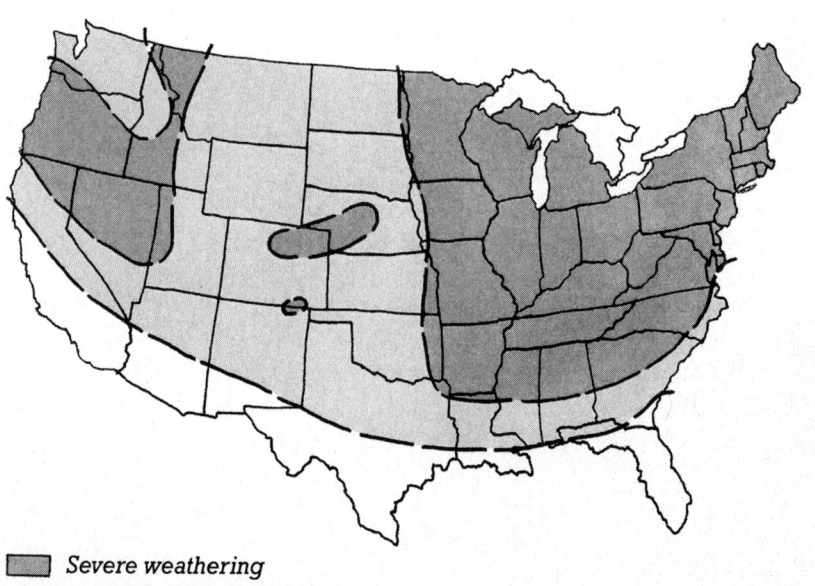

■ *Severe weathering*
□ *Moderate weathering*
□ *Negligible weathering*

FIGURE 8.17
Basic brickwork terminology.

A course is a horizontal layer of bricks or other masonry units

A wythe is a vertical layer of masonry units, one unit thick

Collar joint

Bed joint

Head joint

A stretcher is a brick laid with its face parallel to the wall and its long dimension horizontal

A header is a brick laid so as to bond two wythes together

A soldier is a brick laid on its end with its face parallel to the wall

A rowlock is a brick laid on its face with its end visible in the wall face on garden walls and for sloping sills under windows, although such caps and sills are not durable in severe climates. Architects frequently employ *soldier courses* for visual emphasis in such locations as window lintels or tops of walls.

The problem of bonding multiple wythes of brick has been solved in many ways in different regions of the world, often resulting in surface patterns that are particularly pleasing to the eye. Figures 8.18 and 8.19 show some *structural bonds* for brickwork,

among which *common bond*, *Flemish bond*, and *English bond* are the most popular. On the exterior of buildings, the *cavity wall*, with its single outside wythe, offers the designer little excuse to use anything but *running bond*. Inside a building, safely out of the weather, one may use solid brick walls in any desired bond. For fireplaces and other very small brick constructions, however, it is often difficult to create a long enough stretch of unbroken wall to justify the use of bonded brickwork.

FIGURE 8.18

Frequently used structural bonds for brick walls. Partial closer bricks are necessary at the corners to make the header courses come out even while avoiding alignments of head joints in successive courses. The mason cuts the closers to length with a mason's hammer or diamond saw.

³/₄ brick closers

Running Bond consists entirerly of stretchers

Common Bond (also known as Amercian Bond) has a header course every sixth course. Notice how the head joints are aligned between the header and the stretcher courses

Queen closer

English Bond alternates courses of headers and stretchers

³/₄ brick closer

Flemish Bond alternates headers and stretchers in each course

(a)

(b)

(c)

(d)

(e)

(f)

FIGURE 8.19

Photographs of some brick bonds. (*a*) Running bond, (*b*) common bond, and (*c*) English garden wall bond with Flemish header courses. In the right column, (*d*) English bond, (*e*) Flemish bond, and (*f*) monk bond, which is a Flemish bond with two stretchers instead of one between headers. The running bond example shown here is from the late 18th century. Its extremely thin joints require mortar made with very fine sand. Notice in the common bond wall (dating from the 1920s in this case) that the header course began to fall out of alignment with the stretcher courses, so the mason inserted a partial stretcher to make up the difference; such small variations in workmanship account for some of the visual appeal of brick walls. Flemish header courses, such as those used in the English garden wall bond, are often used with bricks whose length, including mortar joint, is substantially more than twice their width; the Flemish header course avoids the thick joints between headers that would otherwise result. The Flemish bond example is modern and is composed of modular sand-mold bricks. The monk bond shown here has unusually thick bed joints, approximately ¾″ (19 mm) high; these joints are difficult for the mason to lay unless the consistency of the mortar is very closely controlled. (*Photos by Edward Allen.*)

The process of bricklaying is summarized in Figures 8.20 and 8.21. While conceptually simple, bricklaying requires both extreme care and considerable experience to produce a satisfactory result, especially where a number of bricklayers working side by side must produce identical work on a major structure. Yet speed is essential to the economy of masonry construction. The work of a skilled mason is impressive both for its speed and for its quality. This level of expertise takes time and hard work to acquire, which is why the apprenticeship period for brickmasons is both long and demanding.

The laying of *leads* (pronounced "leeds") is relatively labor intensive. A mason's rule or a *story pole* that is marked with the course heights is used to establish accurate course heights in the leads. The work is checked frequently with a spirit level to ensure that surfaces are flat and plumb and courses are level. When the leads have been completed, a mason's line (a heavy string) is stretched between the leads, using L-shaped *line blocks* at each end to locate the end of the line precisely at the top of each course of bricks.

The laying of the infill bricks between the leads is much faster and easier because the mason needs only a trowel in one hand and a brick in the other to *lay to the line* and create a perfect wall. It follows that leads are expensive compared to the wall surfaces between them, so where economy is important, the designer should seek to minimize the number of corners in a brick structure.

Bricks may be cut as needed, either with sharp, well-directed blows of the chisel-pointed end of a mason's hammer or, for greater accuracy and more intricate shapes, with a power saw that utilizes a water-cooled diamond blade (see Figure 9.25). Cutting of bricks slows the process of bricklaying considerably, however, and ordinary brick walls should be dimensioned to minimize cutting (Figure 8.22).

I remember the masons on my first house. I was not much older than the apprentice whom I found choking back tears of frustration with the clumsiness of his work and the rebukes of his boss.

Nearly thirty years later, we still collaborate on sometimes difficult masonry walls, fireplaces and paving patterns. I work with bricklayers . . . whose years of learning their craft paralleled my years of trying to understand my profession. We are friends and we talk about our work like pilgrims on a journey to the same destination.

—Henry Klein, architect, at his acceptance of the Louis Sullivan Award for Architecture, 1981, quoted in *Blueprints Magazine*, 1985.

1. The construction of a brick wall begins with the laying of leads. The leads establish the wall planes and course heights

Line

2. the bricks between the leads are laid to a line, a heavy string stretched taut between line blocks at each lead

Line block

3. When the entire wall is laid to the level of the top of the leads, a second set of leads is laid, and the entire process is repeated as many times as necessary

FIGURE 8.20
The procedure for building brick walls. This example is a single wythe of running bond.

(a)

(b)

(c)

(d)

(e)

(f)

FIGURE 8.21

Laying a brick wall. (*a*) The first course of bricks for a lead is bedded in mortar, following a line marked on the foundation. (*b, c, d*) As each lead is built higher, the mason uses a spirit level to make sure that each course is level, straight, plumb, and in the same plane as the rest of the lead. A mason's rule or a story pole is also used to check the heights of the courses. (*e*) A finished lead. (*f*) A mason lays brick to a line stretched between two leads. When laying to a line, there is no need to use a level or rule. (*Courtesy of International Masonry Institute, Washington, D.C.*)

Mortar joints can vary in thickness from about ¼ inch (6 mm) to more than ½ inch (13 mm). Thin joints work only when the bricks are identical to one another within very small tolerances and the mortar is made with a fine sand. Very thick joints require a stiff mortar that is difficult to work with. Mortar joints are usually standardized at ⅜ inch (10 mm), which is easy for the mason and allows for considerable distortion and unevenness in the bricks. One-half inch (13-mm) joints are also common.

The joints in brickwork are *tooled* an hour or two after laying as the mortar begins to harden, to give a neat appearance and to compact the mortar into a profile that meets the visual and weather-resistive requirements of the wall (Figures 8.23 and 8.24). Outdoors, the *vee joint* and *concave joint* shed water and resist freeze-thaw damage better than the others. Indoors, a *raked* or *stripped joint* can be used if desired to accentuate the pattern of bricks in the wall and deemphasize the mortar.

After joint tooling, the face of the brick wall is swept with a soft brush to remove the dry crumbs of mortar left by the tooling process. If the mason has worked cleanly, the wall is now finished, but most brick walls are later given a final cleaning by scrubbing with *muriatic acid* (hydrochloric acid) or other cleaning agent and rinsing with water to remove mortar stains from the faces of the bricks. Light-colored bricks can be stained by acids and should be cleaned by other means.

FIGURE 8.22

Dimensions for brick buildings are worked out in advance by the architect based on the actual dimensions of the bricks and mortar joints to be used in the building. Bricks and mortar joints are carefully counted and converted to numerical dimensions for each portion of the wall.

FIGURE 8.23

(*a*) Tooling horizontal joints to a concave profile. (*b*) Tooling vertical joints to a concave profile. The excess mortar squeezed out of the joints by the tooling process will be swept off with a brush, leaving a finished wall. (*c*) Raking joints with a common nail held in a skate-wheel joint raker. The head of the nail digs out the mortar to a preset depth. (*Courtesy of Brick Institute of America.*)

Weathered joint

Concave joint

Vee joint

Flush joint

Raked joint

Stripped joint

Struck joint

(a)

(b)

(c)

FIGURE 8.24
Joint tooling profiles for brickwork. The concave joint and vee joint are the only ones suitable for outdoor use in severe climates.

Spanning Openings in Brick Walls

Brick walls must be supported above openings for windows and doors. *Lintels* of reinforced concrete, reinforced brick, or steel angles (Figures 8.25 and 8.26) are all equally satisfactory from a technical standpoint. The near invisibility of the steel lintel is a source of delight to some designers but dissatisfies those who prefer that a building visually express its means of support. Wood is no longer used for lintels because of its tendency to burn, to decay, and to shrink and allow the masonry above to settle and crack.

The *corbel* is an ancient structural device of limited spanning capability, one that may be used for small openings in brick walls, for beam brackets, and for ornament (Figures 8.27, 8.28, and 8.29). A good rule of thumb for designing corbels is that the projection of each course should not exceed half the course height; this results in a corbel angle of about 60 degrees to the horizontal and minimizes flexural stress in the bricks.

FIGURE 8.25
Three types of lintels for spanning openings in brick walls. The double angle steel lintel (*top*) is scarcely visible in the finished wall. The reinforced brick lintel (*center*) works in the same manner as a reinforced concrete beam and gives no visible clues as to what supports the bricks over the opening. The precast reinforced concrete lintel (*bottom*) is clearly visible. For short spans, cut stone lintels without reinforcing can be used in the same manner as concrete lintels.

FIGURE 8.26
Because of corbelling and arching action in the bricks, a lintel is considered to carry only the triangular area of brickwork indicated by the shaded portion of this drawing. The broken line indicates a concealed steel angle lintel.

FIGURE 8.27
Corbelling has many uses in masonry construction. It is used in this example both to span a door opening and to create a bracket to support a beam.

FIGURE 8.28
Corbelling creates a transition from the cylindrical tower to a hexagonal roof. Cut limestone is used for window sills, lintels, arch intersections, and grotesquely carved rainwater spouts. The building is the Gothic cathedral in Albi, France. *(Photo by Edward Allen.)*

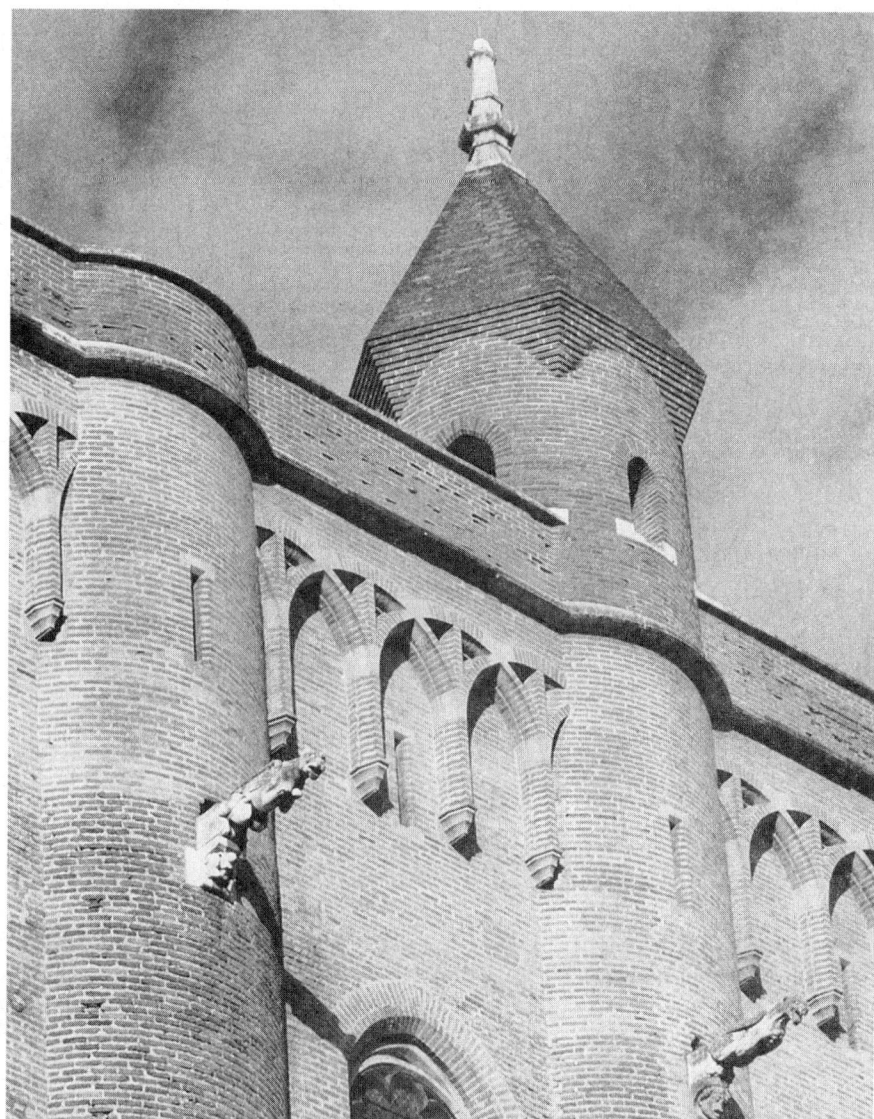

FIGURE 8.29
All the skills of the 19th-century mason were called into play to create the corbels and arches of this brick cornice in Boston's Back Bay. *(Photo by Edward Allen.)*

SEGMENTAL

3 COURSE
2 COURSE
ROWLOCK COURSE
SPRING LINE

TUDOR

BRICK
STONE
SPRING LINE

ROMAN

LAY OUT FULL BRICK PLUS JOINT ON PERIMETER
RADIUS
STONES EQUAL

PARABOLIC

SPRING LINE MAJOR ARCH
ALTERNATING ROWLOCK AND SOLDIER COURSES
SPRING LINE MINOR ARCH

JACK

SKEWBACK -½" PER FT. OF SPAN FOR EACH 4" OF ARCH DEPTH
ALL JOINTS ARE UNIFORM CAMBER -⅛" PER FT. OF SPAN
EQ EQ
STONE SKEWBACK
STONE JOINT ¼"

ELLIPTICAL

FULL BRICK WIDTH HERE
MINOR AXIS
MAJOR AXIS
SPRING LINE

GOTHIC

CENTERS ALWAYS ON SPRING LINE

NOTE: Stone joints may be handled in a variety of ways. This is one illustration.

ARCH TERMINOLOGY

RISE (F)
ARCH AXIS
CROWN
EXTRADOS
DEPTH (D)
SKEWBACK
SOFFIT
RISE (R)
INTRADOS
ABUTMENT
SPRING LINE (MINOR ARCH)
SPRING LINE (MAJOR ARCH)
SPAN (S)
SPAN (L)

FIGURE 8.30

Arch forms and arch terminology in brick and cut stone. The *spandrel* is the area of wall that is bounded by the extrados of the arch. *(Reprinted by permission of John Wiley & Sons, Inc., from Ramsey/Sleeper,* Architectural Graphic Standards *(7th ed.), Robert T. Packard, A.I.A., ed., copyright © 1981 by John Wiley & Sons, Inc.)*

The brick *arch* is a structural form so widely used and so powerful, both structurally and symbolically, that entire books have been devoted to it (Figure 8.30). Given a *centering* of wood or steel (Figure 8.31), a mason can lay a brick arch rapidly, although the *spandrel*, the area of flat wall that adjoins the arch, is time-consuming to construct because many of its bricks must be cut to fit. In an arch of *gauged brick*, each brick is rubbed

(*a*)

(*b*)   (*c*)

FIGURE 8.31

(*a*) Two rough brick arches under construction, each on its wooden centering. (*b*) The brick locations were marked on the centering in advance to be sure that no partial bricks or unusual mortar joint thicknesses will be required to close the arch. This was done by laying the centering on its side on the floor and placing bricks around it, adjusting their positions by trial and error to achieve a uniform spacing. Then the location of each brick was marked with pencil on the curved surface of the centering. (*c*) The brick arches whose construction is illustrated in the previous two photographs span a fireplace room that is roofed with a brick barrel vault. The firebox is lined with firebrick and the floor is finished with quarry tiles. *(Photo by Edward Allen.)*

to the required wedge shape on an abrasive stone, which is laborious and expensive. The *rough arch*, which depends on wedge-shaped mortar joints for its curvature, is therefore much more usual in today's buildings (Figures 8.32 and 8.33). A number of brick manufacturers will mold to order sets of tapered bricks for arches of any shape and span.

An arch translated along a line perpendicular to its plane produces a *barrel vault*. An arch rotated about its vertical centerline becomes a *dome*. From various intersections of these two basic roof shapes comes the infinite vocabulary of vaulted masonry construction. Brick vaults and domes, if their lateral thrusts are sufficiently tied or *buttressed*, are

strong, stable forms. In parts of the world where labor is inexpensive, they continue to be built on an every-day basis (Figures 8.34 and 8.46). In North America and most of Europe, where labor is more costly, they have been replaced almost entirely by less expensive, more compact spanning elements such as beams and slabs of wood, steel, or concrete.

FIGURE 8.32

A deep semicircular brick arch frames a view of a brick arcade and the harbor beyond at the Federal Courthouse in Boston, Massachusetts. *(Architects: Pei, Cobb, Freed and Partners. Photo © 1998 Steve Rosenthal.)*

FIGURE 8.33

A rough *jack arch* (also called a *flat arch*) in a wall of Flemish bond brickwork. *(Photo by Edward Allen.)*

(a)

(b)

FIGURE 8.34

(a) Masons in Mauritania, drawing on thousands of years of experience in masonry vaulting, build a dome for a patient room of a new hospital. The masonry is self-supporting throughout the process of construction; only a simple radius guide is used to maintain a constant diameter. The dome is double, with an airspace between to insulate the room from the sun's heat. (b) The walls are buttressed with stack bond brick headers to resist the outward thrust of the domes. (*Courtesy of ADAUA, Geneva, Switzerland.*)

Reinforced Brick Masonry

Reinforced brick masonry (RBM) is analogous to reinforced concrete construction. The same deformed steel reinforcing bars used in concrete are placed in thickened collar joints to strengthen a brick wall or lintel. A reinforced brick wall (Figure 8.35) is created by constructing two wythes of brick 2 to 4 inches (50–100 mm) apart, placing the reinforcing steel in the cavity, and filling the cavity with *grout*. Grout is a mixture of portland cement, aggregate, and water. ASTM C476 specifies the proportions and qualities of grout for use in filling masonry loadbearing walls. It is important that grout be fluid enough to flow readily into the narrow cavity and fill it completely. The excess water in the grout that is required to achieve this fluidity is quickly absorbed by the bricks and does not detract from the eventual strength of the grout, as it would from concrete poured into formwork. Highly flowable *self-consolidating grout*, analogous to self-consolidating concrete as described in Chapter 13, can also be used.

There are two methods for grouting reinforced brick walls: low lift and high lift. In *low-lift grouting*, the masonry is constructed to a height not greater than 4 feet (1200 mm) before grouting, taking care to keep the cavity free of mortar squeeze-out and droppings, which might interfere with the placement of the reinforcing and grout. The vertical reinforcing bars are inserted into the cavity and are left projecting at least 30 bar diameters above the top of the brickwork to transfer their loads to the steel in the next lift. The cavity is then filled with grout to within 1½ inches (38 mm) of the top, and the process is repeated for the next lift.

In *high-lift grouting*, the wall is grouted a story at a time. The cleanliness of the cavity is ensured by temporarily omitting some of the bricks in the lowest course of masonry to create *cleanout holes*. As the bricklaying progresses, the cavity is flushed periodically from above with water to drive debris down and out through the cleanouts. To resist the hydrostatic pressure of the wet grout, the wythes are held together by galvanized steel wire *ties* laid into the bed joints and across the cavity, usually at intervals of 24 inches (600 mm) horizontally and 16 inches (400 mm) vertically. After the cleanouts have been filled with bricks and the mortar has cured for at least 3 days, the reinforcing bars are placed and grout is pumped into the cavity from above in increments not more than 4 feet (1200 mm) high. To minimize pressure on the brickwork, each increment is allowed to harden for an hour or so before the next increment is poured above it.

The low-lift method is generally easier for small work where the grout is poured by hand. Where grout-pumping equipment must be rented, the high-lift method is preferred because it minimizes rental costs.

Although unreinforced brick walls were once adequate for many structural purposes, RBM walls are much stronger against vertical loads, flexural loads from wind or earth pressure, and shear loads. With RBM, it is possible to build bearing wall buildings to heights formerly possible only with steel and concrete frames, and to do so with surprisingly thin walls (Figures 8.36 and 8.37). RBM is also used for brick piers, which are analogous to concrete columns, and, less commonly, for structural lintels (Figure 8.25), beams, slabs, and retaining walls.

Reinforced brickwork may also be created at a smaller scale by inserting reinforcing bars and grout into the cores of hollow bricks. This technique is especially useful for single-family residential construction and for single-wythe prefabricated curtain wall panels (Chapter 20).

FIGURE 8.35
A reinforced brick loadbearing wall is built by installing steel reinforcing bars in a thickened collar joint, then filling the joint with portland cement grout. The cleanout holes shown here are used in the high-lift method of grouting.

Grout

Wire ties

Reinforcing bars

Bricks are left out of the bottom course at intervals to create cleanout holes, then inserted before grouting

FIGURE 8.36
Twelve-inch (300-mm) reinforced brick walls of constant thickness bear the concrete floor and roof structures of a hotel. *(Photo by Edward Allen.)*

MASONRY WALL CONSTRUCTION

Bricks, stones, and concrete masonry units are mixed and matched in both loadbearing and nonloadbearing walls of many types. Stone and concrete masonry are discussed in Chapter 9, and the more important types of masonry wall constructions are presented in Chapter 10. Figures 8.38 through 8.46 illustrate just some of the richness and stylistic diversity possible with brick masonry construction.

FIGURE 8.37

Because loads in a building accumulate from top to bottom, the unreinforced brick walls of the 16-story Monadnock Building, built in Chicago in 1891, are 18 inches (460 mm) thick at the top and 6 feet (1830 mm) thick at the base of the building. *(Nicholas Janberg, www.structurae.de)*

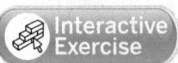

FIGURE 8.38
Ornamental corbelled brickwork in an 18th-century New England chimney. Step flashings of lead sheet waterproof the junction between the chimney and the wood shingles of the roof. *(Photo by Edward Allen.)*

FIGURE 8.39
In the gardens he designed at the University of Virginia, Thomas Jefferson used unreinforced serpentine walls of brick that are only a single wythe thick. The shape of the wall makes it extremely resistant to overturning despite its thinness. *(Photo Wayne Andrews.)*

FIGURE 8.40
Cylindrical bays of brick with stone lintels front these Boston rowhouses.
(Photo by Edward Allen.)

FIGURE 8.41
Bricks were laid diagonally in two of the courses to create this mouse-tooth pattern. The window is spanned with a segmental arch of cut limestone. *(Photo by Edward Allen.)*

FIGURE 8.42
Quoins originated long ago as cut stone blocks used to form strong corners on walls of weak masonry materials such as mud bricks or round fieldstones. In more recent times, quoins (pronounced "coins") have been used largely for decorative purposes. At the left, cut limestone quoins and a limestone *water table* dress up a common bond brick wall. The mortar joints between quoins are finished in a protruding beaded profile to emphasize the pattern of the stones. At the right, brick quoins are used to make a graceful termination of a concrete masonry wall at a garage door opening. Notice that three brick courses perfectly match one block course. *(Photos by Edward Allen.)*

FIGURE 8.43
Louis Sullivan's National Farmers' Bank in Owatonna, Minnesota, completed in 1908, rises from a red sandstone base. Enormous rowlock brick arches span the windows in the two street facades. Bands of glazed terra-cotta ornament in rich blues, greens, and browns outline the walls, and a flaring cornice of corbelled brick and terra cotta caps the building. *(Photo Wayne Andrews.)*

FIGURE 8.44
Architects Pei, Cobb, Freed employed brick half-domes to mark the entrances to courtrooms in the Boston Federal Courthouse. *(Photo © 1998 Steve Rosenthal.)*

FIGURE 8.45
Frank Lloyd Wright used long, flat Roman bricks and cut limestone wall copings to emphasize the horizontality of the Robie House, built in Chicago in 1906. *(Photo by Mildred Mead. Courtesy of Chicago Historical Society, IChi-14191.)*

FIGURE 8.46
During the second half of the 20th century, Uruguayan engineer Eladio Dieste constructed hundreds of industrial buildings with long-span roof vaults of reinforced clay masonry, either bricks or, like the example shown here, hollow clay tiles. The span of this roof is about 100 feet (30 m). (*Left*) The vaults are long strips with clerestory windows between for daylighting. (*Right*) Each strip is S-shaped in cross section. This stiffens the vault to prevent buckling and also provides the openings for the windows. *(Photo by Edward Allen.)*

CSI/CSC MasterFormat Sections for Brick Masonry	
04 05 00	**COMMON WORK RESULTS FOR MASONRY**
04 05 13	Masonry Mortaring
04 05 16	Masonry Grouting
04 05 19.26	Masonry Reinforcing Bars
04 21 00	**CLAY UNIT MASONRY**
04 21 13	Brick Masonry
04 54 00	**REFRACTORY MASONRY**

KEY TERMS

mason
masonry unit
trowel
mortar
cement–lime mortar
aggregate
portland cement
lime
quicklime
slaked lime, hydrated lime
blended hydraulic cement
masonry cement
air-entraining admixture
mortar cement
mortar type
Type N mortar
Type S mortar
Type O mortar
Type M mortar
workability
proportion specification
property specification
hydraulic cement
lime mortar
nonhydraulic cement
carbonation
hydration
retemper
extended-life admixture
(mortar) admixture
set accelerator
set retarder
integral water repellent
efflorescence
brick
soft mud process
water-struck brick
sand-struck brick, sand-mold brick
dry-press process
stiff mud process

tumbled brick
firing, burning
clamp
clinker brick
periodic kiln
tunnel kiln
water smoking
dehydration
oxidation
vitrification
flashing
fly ash brick
modular brick
Utility brick
facing brick, face brick
building brick
solid unit
cored unit
frogged unit
hollow brick
paving brick
firebrick
fireclay
fireclay mortar
brick grade
brick type
chippage
wythe
stretcher course
header
rowlock course
soldier course
course
bed joint
head joint
collar joint
stretcher
header
soldier
rowlock

structural bond
common bond
Flemish bond
English bond
cavity wall
running bond
lead
story pole
line block
lay to the line
tooled joint
weathered joint
vee joint
concave joint
flush joint
raked joint
stripped joint
struck joint
muriatic acid
lintel
corbel
arch
centering
spandrel
gauged brick
rough arch
jack arch, flat arch
barrel vault
dome
buttress
reinforced brick masonry (RBM)
grout
self-consolidating grout
low-lift grouting
high-lift grouting
cleanout hole
tie
quoin
water table

REVIEW QUESTIONS

1. How many syllables are in the word *masonry*? (Hint: There cannot be more syllables in a word than there are vowels. Many people, even masons and building professionals, mispronounce this word.)

2. What are the most common types of masonry units?

3. What are the molding processes used in manufacturing bricks? How do they differ from one another?

4. List the functions of mortar.

5. What are the ingredients of mortar? What is the function of each ingredient?

6. Why are mortar joints tooled? Which tooling profiles are suitable for a brick wall in a severe climate?

7. What is the function of a structural brick bond such as common or Flemish bond? Draw the three brick bonds from memory.

EXERCISES

1. What is the exact height of a brick wall that is 44 courses high when 3 courses of brick plus their 3 mortar joints are 8 inches (203.2 mm) high?

2. What are the inside dimensions of a window opening in a wall of modular bricks with ⅜-inch (9.5-mm) mortar joints if the opening is 6 bricks wide and 29 courses high?

3. Obtain sand, hydrated lime, several hundred bricks, and basic bricklaying tools from a masonry supply house. Arrange for a mason to help everyone in your class learn a bit of bricklaying technique. Use lime mortar (hydrated lime, sand, and water), which hardens so slowly that it can be retempered with water and used again and again for many weeks. Lay small walls in several different structural bonds. Make simple wooden centering and construct an arch. Construct a dome about 4 feet (1.2 m) in diameter without using centering, as is done in Figure 8.34. Dismantle what you build at the end of each day, scrape the bricks clean, stack them neatly for reuse, and retemper the mortar with water, covering it with a sheet of plastic to keep it from drying out before it is used again.

4. Design a brick fireplace for a house that you are designing. Select the size and color of brick and the color of mortar. Proportion the fireplace according to the guidelines in Chapter 7. Dimension the fireplace so that it uses only full and half bricks. Draw every brick and every mortar joint in each view of the fireplace. Use rowlocks, soldiers, corbels, and arches as desired for visual effect. How will you span the fireplace opening?

SELECTED REFERENCES

Beall, Christine. *Masonry Design and Detailing for Architects, Engineers, and Builders* (6th ed.). New York, McGraw-Hill, 2012.

This 500-page book is an excellent general design reference on brick, stone, and concrete masonry.

Brick Industry Association. *BIA Technical Notes of Brick Construction*. McLean, VA, various dates.

This collection of more than 50 bulletins is available in ring-binder format or as free downloads from the Brick Industry Association's web site, www.bia.org. It includes up-to-date information on every aspect of bricks and brick masonry.

Brick Industry Association. *Principles of Brick Masonry*. Reston, VA, BIA, 1989.

Despite its age, this 70-page booklet still presents a complete curriculum in the fundamentals of clay masonry construction for the student of building construction.

WEB SITES

Brick Masonry

Author's supplementary web site: **www.ianosbackfill.com/08_brick_masonry**

Brick Industry Association: **www.bia.org**

General Shale Brick: **www.generalshale.com**

Glen-Gery Brick: **www.glengerybrick.com**

International Masonry Institute: **www.imiweb.org**

STONE AND CONCRETE MASONRY

- **Stone Masonry**
  Types of Building Stone
  Quarrying and Milling of Stone
  Selecting Stone for Buildings
  Stone Masonry

  SUSTAINABILITY IN STONE AND CONCRETE MASONRY

- **Concrete Masonry**
  Manufacture of Concrete Masonry Units

Laying Concrete Masonry
Dry-Stacked Unit Masonry
Decorative Concrete Masonry Units
The Economy and Utility of Concrete Masonry Construction

- **Other Types of Masonry Units**

- **Masonry Wall Construction**

Stone offers a wide range of expressive possibilities to the architect. In this detail of a 19th-century church, the columns to the left are made of polished granite and rest on bases of carved limestone. The limestone blocks to the right, squared and dressed by hand, have rough-pointed faces and tooth-axed edges. *(Architects: Cummings and Sears. Photo by Edward Allen.)*

Stone masonry and concrete masonry are similar in concept to brick masonry. Both involve the stacking of masonry units in the same mortar that is used for brick masonry. However, there are important differences: Whereas bricks are molded to shape, building stone must be wrested from quarries in rough blocks, then cut and carved to the shapes desired. We can control the physical and visual properties of bricks to some extent, but we cannot control the properties of stone, so we must learn to select from the bountiful assortment provided by the Earth the type and color that we want and to work with it as nature provides it to us. Concrete masonry units, like bricks, are molded to shape and size, and their properties can be closely controlled. Most concrete masonry units, however, are much larger than bricks, and, like stone, they require slightly different techniques for laying.

STONE MASONRY

Types of Building Stone

Building stone is obtained by taking rock from the earth and reducing it to the required shapes and sizes for construction. It is a natural, richly diverse material that can vary greatly in its chemistry, structure, properties, and appearance. Geologically, stone can be classified into three types according to how it was formed:

- *Igneous rock* is rock that was deposited in a molten state.
- *Sedimentary rock* is rock that was deposited by the action of water and wind.
- *Metamorphic rock* was formerly either igneous or sedimentary rock. Subsequently, its properties were transformed by heat and pressure.

For commercial purposes, ASTM C119 classifies stone used in building construction into six groups: Granite, Limestone, Quartz-Based Stone, Slate, Marble, and Other.

Granite Group

Granite is the igneous rock most commonly quarried for construction in North America. It is a mosaic of mineral crystals, principally feldspar and quartz (silica), and can be obtained in a range of colors that includes gray, black, pink, red, brown, buff, and green. Granite is nonporous, hard, strong, and durable, and is the most nearly permanent of building stones, suitable for use in contact with the ground or in locations where it is exposed to severe weathering. Its surface can be finished in any of a number of textures, including a mirror-like polish. In North America, it is quarried chiefly in the East and the upper Midwest. Various granites are also imported from abroad, chiefly from Brazil, China, India, and Italy. Domestic granites are classified according to whether they are fine-grained, medium-grained, or coarse-grained. Requirements for granite dimension stone are defined in ASTM C615.

Basalt, like granite, is a very dense and durable igneous rock. It is usually found only in a dark gray color, and is one of a group of stones that may be collectively referred to as "black granites." It is generally used in the form of rubble and is seldom machined.

FIGURE 9.1

Austin Hall at Harvard University (1881–1884), designed by Henry Hobson Richardson, is a virtuoso performance in stone masonry. Notice the intricate carving of the yellow Ohio sandstone capitals and arch components. The spandrels above the arches are a mosaic of two colors of Longmeadow sandstone blocks. The depth of the arches is intentionally exaggerated to impart a feeling of massiveness to the wall at the entrance to the building. (*Photo by Steve Rosenthal.*)

FIGURE 9.2
The loadbearing walls of the Cistercian Abbey Church in Irving, Texas, are made of 427 rough blocks of Big Spring, Texas, limestone, each 2 × 3 × 6 feet (0.6 × 0.9 × 1.8 m) and weighing about 5000 pounds (2300 kg). The stones were brought directly from the quarry to the site without milling; drill holes are visible in many of the stones. Each stone was bedded in Type S mortar 1 inch (25 mm) thick to allow for irregularities in its horizontal faces. Darker stones are grouped in bands for visual effect. The columns were turned from limestone. Cunningham Architects, designing for a Catholic order that originated 900 years ago in Europe, wanted to build a church that would last for another 900 years. *(Photo by James F. Wilson.)*

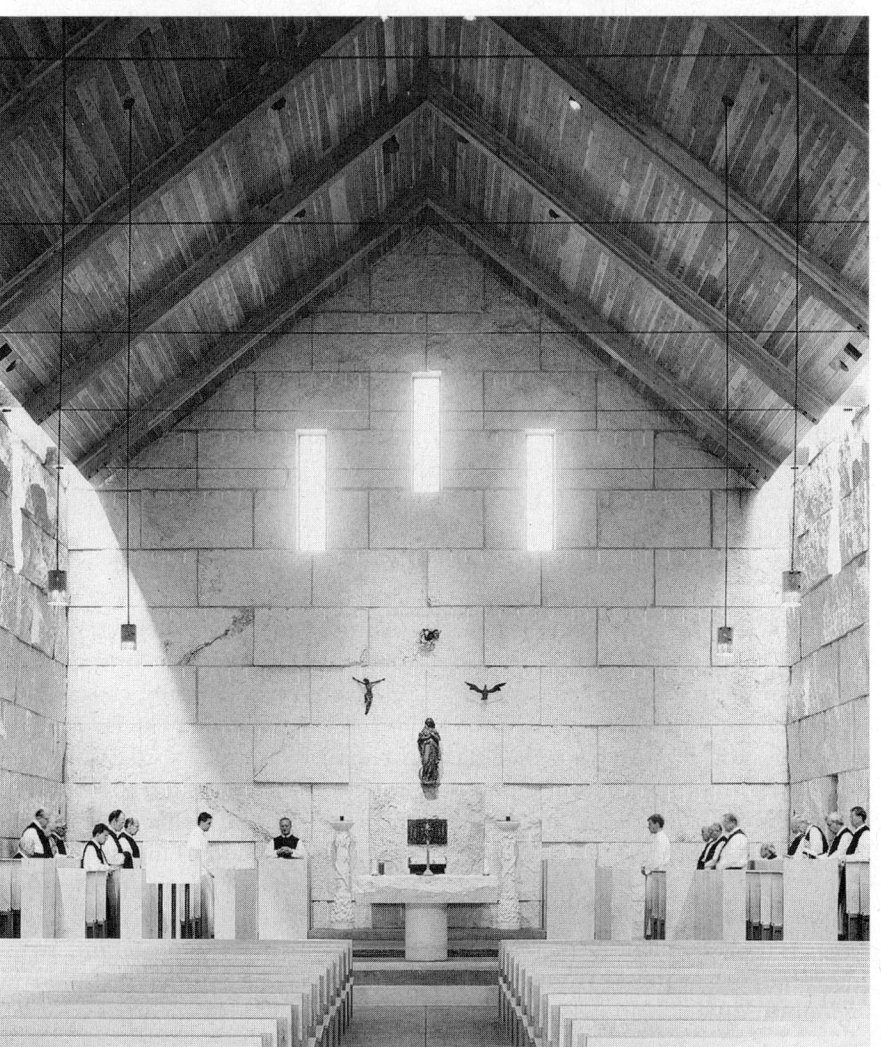

FIGURE 9.3
The heavy timber roof of the Cistercian Abbey Church, flanked by continuous skylights, appears to float above the simple stone volume of the nave. *(Photo by James F. Wilson.)*

Limestone Group

Limestone is one of the two principal sedimentary rock types used in construction. It may be found in a strongly stratified form or in deposits that are more homogeneous in structure (*freestone*). It is quarried throughout North America, with major quarries for large-dimension stone found in Missouri and Indiana. France, Germany, Italy, Spain, Portugal, and Croatia are major suppliers of imported limestone. Requirements for limestone dimension stone are specified in ASTM C568.

Limestone may be composed either of calcium carbonate (*oolitic limestone*) or of a mixture of calcium and magnesium carbonates (*dolomitic limestone*). Both types were formed long ago from the skeletons or shells of marine organisms. Colors range from almost white through gray and buff to iron oxide red. Limestone is porous and contains considerable groundwater, or *quarry sap*, when quarried. While still saturated with quarry sap, most limestones are easy to work but are susceptible to frost damage. After seasoning in the air to evaporate the quarry sap, the stone becomes harder and resistant to frost damage. Some dense limestones can be polished (and may be classified as marbles), but most are produced with varying degrees of surface texture.

According to ASTM C568, limestones are classified as I Low-Density, II Medium-Density, or III High-Density. Stones in higher density classifications generally are stronger and less porous than those with lower densities. All three classifications are suitable for building applications. The Indiana Limestone Institute classifies limestone into two colors, buff and gray, and four grades. Select grade is the finest-grained and has the fewest natural flaws. Standard and rustic grades are progressively coarser-grained with more natural flaws. The fourth grade, variegated, consists of an unselected mixture of the first three grades.

Quartz-Based Dimension Stone Group

Sandstone is the second major sedimentary rock type used in building construction. Like limestone, it may be found in either a strongly stratified form or as more homogeneous freestone. Sandstone was formed in ancient times from deposits of quartz sand (silicon dioxide). Its color and physical properties vary significantly with the material that cements the sand particles, which may consist of silica, carbonates of lime, or iron oxide. Sandstone is quarried principally in New York, Ohio, and Pennsylvania. Two of its more familiar forms are *brownstone*, widely used in wall construction, and *bluestone*, a highly stratified, durable stone especially suitable for paving and wall copings. It is quarried in the northeastern United States. Sandstone will not accept a high polish. Requirements for quartz-based dimension stone are specified in ASTM C616.

Slate Group

Slate is one of the two metamorphic stone groups utilized in building construction. It is formed from clay. It is a dense, hard stone with closely spaced planes of cleavage, along which it is easily split into sheets, making it useful for paving stones, roof shingles, and thin wall facings. It is quarried in Vermont, Virginia, New York, Pennsylvania, and parts of Canada, in a variety of colors including black, gray, purple, blue, green, and red. China and India are the largest foreign suppliers of slate to North America, with Italy, the United Kingdom, and Spain also providing significant quantities. Requirements for slate dimension stone are specified in ASTM C629.

Marble Group

Marble is the second of the major metamorphic rock groups. In its true geologic form, it is a recrystallized form of limestone. It is easily carved and polished and occurs in white, black, and nearly every other color, often with beautiful patterns of veining. The marbles used in North America come chiefly from Alabama, Tennessee, Vermont, Georgia, Missouri, and Canada. Marbles are also imported to this continent from many parts of the world, but primarily Turkey, Lebanon, Mexico, Italy, China, and France. The properties and appearance of marble vary greatly, depending on the chemistry of the original limestone from which it was formed and, even more so, on the processes by which it was metamorphosed. Requirements for marble dimension stone are specified in ASTM C503.

The Marble Institute of America has established a four-step grading system for marbles, in which Group A includes sound marbles and stone with uniform and favorable working qualities. Groups B through D have progressively less favorable working qualities and more natural faults that may require sticking (cementing together) and waxing (filling of voids with cements, shellac, or other materials). Many of the marbles most prized for their color and figure belong to Group D. The Marble Group also includes other stones that can take a high polish but are not true marbles, such as dense limestones called limestone marble, onyx marble, serpentine marble, and others.

Other Group

The ASTM C119 Other Group includes a variety of less frequently used building stones. *Travertine* is a relatively rare, partially crystallized, and richly patterned calcite (having a chemistry similar to that of limestone) rock deposited by ancient springs. It is marblelike in its physical qualities. Requirements for travertine dimension stone are specified in ASTM C1527. Alabaster, greenstone, schist, serpentine, and soapstone are other stones included in this group.

Quarrying and Milling of Stone

The construction industry uses stone in many different forms. *Fieldstone* is rough building stone obtained from riverbeds and rock-strewn fields. *Rubble* consists of irregular quarried fragments that have at least one good face to expose in a wall. *Dimension stone* is stone that has been quarried and cut into rectangular shapes; large slabs are often referred to as *cut stone*, and small rectangular blocks are called *ashlar*. *Flagstone* consists of thin slabs of stone, either rectangular or irregular in outline, that are used for flooring and paving. Crushed and broken stone are useful in site work as freely draining fill material, as base layers under concrete slabs and pavings, as surfacing materials, and as aggregates in concrete and asphalt. Stone dust and powder are used in landscaping for walks, drives, and mulch.

The maximum sizes and minimum thicknesses of sheets of cut stone vary from one type of stone to another. Granite, the strongest stone, may be utilized in sheets as thin as ⅜ inch (9.5 mm) in some applications. Marble is generally cut no thinner than ¾ inch (19 mm). However, in most applications, somewhat greater thicknesses than these are advisable. Limestone, the weakest building stone, is never cut thinner than 2 inches (51 mm), and 3 inches (76 mm) is the preferred thickness for conventionally set stone. In a 6-inch (152-mm) thickness, limestone may be handled in sheets as large as 5 × 18 feet (1.5 × 5.5 m). A Group A marble has a maximum recommended sheet size of 5 × 7 feet (1.5 × 2.1 m).

Dimension stone, whether granite, limestone, sandstone, or marble, is cut from the quarry in blocks. The methods for doing this vary somewhat with the type of stone, and quarrying technology continues to evolve. The most advanced machines for cutting marble and limestone in the quarry are chain saws and belt saws equipped with diamond blades (Figures 9.4 and

9.5). Granite is much harder than other stones, so it is quarried either by drilling and blasting or by the use of a *jet burner* that combusts fuel oil with compressed air. The hot flame at the end of the jet burner lance induces local thermal stress and *spalling* (cracking or flaking of the surface) in the granite. With repeated passes of the lance, the operator gradually

excavates a deep, narrow trench to isolate each block of granite from the bedrock. Certain types of granite are quarried successfully with a diamond wire technique, in which a long, taut loop of diamond-studded wire is run at high speed against the stone to make the cut. (See the Chapter 14 sidebar, "Cutting Concrete, Stone, and Masonry," for more information

FIGURE 9.4
A typical procedure for quarrying limestone. (*A*) A diamond belt saw divides the limestone bedrock into long cuts, each about 50 feet (15 m) long and 12 feet (3.6 m) high. Multiple horizontal drill holes create a plane of weakness under the outermost cut. (*B*) Rubber air bags are inflated in the saw kerf to "turn the cut," breaking it free of the bedrock and tipping it over onto a prepared bed of stone chips that cushions its fall. (*C*) Quarry workers use sledgehammers to drive steel wedges into shallow drilled holes to split the cut into blocks. A front-end loader removes the blocks and stockpiles them, ready to be trucked to the mill.

(a)

(b)

FIGURE 9.5

(a) The long blade of a diamond belt saw, only the shank of which is visible here, cuts limestone full depth in one pass, using water to lubricate the saw and flush away the stone dust. The saw advances automatically on its portable rails at a maximum rate of about 2½ inches (65 mm) per minute. A chain saw is similar to the belt saw shown here, but it uses a chain of linked, rigid teeth, whereas the belt saw uses a narrow, flexible belt of steel-reinforced polyethylene with diamond cutting segments. (b) After the cut has been turned, quarry workers split it into transportable blocks. The man at the top of the photograph is laying out the splitting pattern with a measuring tape and straightedge, working from a list of the sizes of blocks that the mill requires for the specific jobs it is working on. At the right, a worker drills shallow holes in the stone. He is followed by a second worker who places steel wedges in the holes and a third who drives the wedges until the cut splits. Each split takes just several minutes to accomplish. (*Photos by Edward Allen.*)

about cutting concrete, stone, and masonry).

When preparing cut stone for a building, the stone producer works from the architect's drawings to make a set of shop drawings that show the shape and dimension of each individual stone in the building. After these drawings have been checked by the architect, they are used to guide the work of the mill in producing the stones. Rough blocks of stone are selected in the yard, brought into the mill, and sawed into slabs. The slabs may be sawed into smaller pieces, edged, planed flat or to a molding profile, turned on a lathe, or carved, as required. Automated equipment is often used to cut and carve repetitive pieces. After the desired surface finish has been created, holes for *lewises* and anchors are drilled as needed (Figures 9.6, 9.7, and 9.11). Each finished piece of stone is marked to correspond to its position in the building as indicated on the shop drawings before being shipped to the construction site.

Selecting Stone for Buildings

The selection of building stone is complicated by several factors. Common names for stone do not necessarily correspond to their geologic origins, mineral composition, or physical properties. For example, some stones labeled commercially as marble are devoid of carbonates, the essential mineral group in true marbles, and they differ markedly in their dimensional stability. Commercially applied names for stone often differ regionally as well. In North America, for example, the name "basalt" is applied to a granite-like igneous rock, whereas elsewhere around the globe this term may be used to describe several varieties of sedimentary sandstones or siltstones. Even within one region, the physical characteristics of a particular stone type can vary, sometimes significantly, from one quarry to another, or even within single batches of stone extracted from the same locale.

> His powers continually increased, and he invented ways of hauling the stones up to the very top, where the workmen were obliged to stay all day, once they were up there. Filippo had wineshops and eating places arranged in the cupola to save the long trip down at noon. . . . He supervised the making of the bricks, lifting them out of the ovens with his own hands. He examined the stones for flaws and hastily cut model shapes with his pocket knife in a turnip or in wood to direct the men. . . .
>
> Giorgio Vasari, writing of Filippo Brunelleschi (1377–1446), the architect of the great masonry dome of the Cathedral of Santa Maria del Fiore in Florence, in *The Lives of the Artists*, 1550

When stone is obtained from an established source, its properties and fitness for use can usually be extrapolated from past experience. When a particular stone has a history of successful performance as exterior cladding, for example, it is generally safe to assume that new supplies of this stone from the same source will continue to be suitable for this type of application. However, when stone comes from an unproven source or will be used in a new way, it should be tested in the laboratory to determine its physical properties and verify its suitability for the proposed use. Stone intended for masonry may undergo *petrographic analysis* (microscopic examination of the stone's mineral content and structure), as well as testing for water absorption, density, compressive strength, dimensional stability, frost resistance, and resistance to attack from salt or other chemicals. Stone panels used as external cladding, as discussed in Chapter 20, may also be tested for bending strength, modulus of rupture, thermal expansion and contraction, and anchor capacity (the load capability of the proposed metal anchor system). Examples of physical property requirements for some common building stone types are listed in Figure 9.8.

The stone industry is international in scope and operation. Architects and building owners have tended to select stone primarily on the basis of appearance, durability, and cost, often with minimal regard to national origin. The most technically advanced machinery for stoneworking, used all over the world, is designed and manufactured in Italy and Germany. A number of Italian companies have earned reputations for cutting and finishing stone to a very high standard at a reasonable cost. As a result of these factors and the low cost of ocean freight relative to the value of stone, it is possible to choose, for example, a granite that is quarried in Finland, to have the quarry blocks shipped to Italy for cutting and finishing, and to have the finished stone shipped to a North American port, where it is transferred to railcars or trucks for delivery to the building site. Because of the unique character of many domestic stones, both U.S. and Canadian quarries and mills also ship millions of tons of stone to foreign countries each year, in addition to the even larger amount that is quarried, milled, and erected at home. With a heightened awareness of sustainability factors, however, there is an increasing emphasis on the use of local stone and local stone mills.

FIGURE 9.6

Stone milling operations, showing a composite of techniques for working granite and limestone. (*a*) An overhead crane lifts a rough block of granite from the storage yard to transport it into the plant. The average block in this yard weighs 60 to 80 tons (55 to 75 metric tons). (*b*) Two reciprocating gang saws slice blocks of limestone into slabs. The saw at the right has just completed its cuts, while the one at the left is just beginning. The intervals between the parallel blades are set by the saw operator to produce the desired thicknesses of slabs. Water cools the diamond blades of the saws and flushes away the stone dust. (*c*) A slab of granite is ground to produce a flat surface prior to polishing operations. (*d*) If a textured thermal finish is desired on a granite slab, a propane–oxygen torch is passed across the slab under controlled conditions to cause small chips to explode off the surface. (*e*) A layout specialist, working from shop drawings for a specific building, marks a polished slab of granite for cutting. (*f*) The granite slab is cut into finished pieces with a large diamond circular saw that is capable of cutting 7 feet (2.1 m) per minute at a depth of 3 inches (76 mm).

(*g*)

(*h*)

(*i*)

(*j*)

(*k*)

(*l*)

(*g*) Small pneumatic chisels with carbide-tipped bits are used for special details in granite. (*h*) Hand polishers are used to finish edges of granite that cannot be finished by automatic machinery. (*i*) Cylindrical components of limestone are turned on a lathe. (*j*) A cylindrical column veneer is ground to its true radius. (*k*) Linear shapes of Indiana limestone, a relatively soft stone, can be formed by a profiled silicon carbide blade in a planer. The piece of stone is clamped to the reciprocating bed, which passes it back and forth beneath the blade. The blade's pressure and depth are controlled by the operator's left hand pressing on the wooden lever. Here the planer is making a stepped profile for a cornice. (*l*) Working freehand with a vibrating pneumatic chisel, a carver finishes a column capital of Indiana limestone. Different shapes and sizes of interchangeable chisel bits rest in the curl of the capital. *(Photos* a, c–h, *and* j *courtesy of Cold Spring Granite Co.; photo* i *courtesy Indiana Limestone Institute; photos* b, k, *and* l *by Edward Allen.)*

FIGURE 9.7
This 9-ton (8-metric-ton) Corinthian column capital was carved from a single 30-ton (27-metric-ton) block of Indiana limestone. Rough cutting took 400 hours and carving another 500. Eight of these capitals were manufactured for a new portico on an existing church. *(Architect: I. M. Pei & Partners. Courtesy Indiana Limestone Institute.)*

	Maximum Water Absorption by Weight	Minimum Density	Minimum Compressive Strength	Minimum Modulus of Rupture	Minimum Flexural Strength
Granite ASTM C615	0.40%	160 lb/ft³ *2560 kg/m³*	19,000 psi *131 MPa*	1500 psi *10.3 MPa*	1200 psi *8.27 MPa*
Limestone ASTM C568	3–12%	110–160 lb/ft³ *1760–2560 kg/m³*	1800–8000 psi *12.4–55.2 MPa*	400–1000 psi *2.76–6.89 MPa*	
Sandstone ASTM C616	1–8%	125–160 lb/ft³ *2000–2560 kg/m³*	4000–20,000 psi *27.6–138 MPa*	350–2000 psi *2.41–13.8 MPa*	
Marble ASTM C503	0.20%	144–162 lb/ft³ *2310–2600 kg/m³*	7500 psi *51.7 MPa*	1000 psi *6.89 MPa*	1000 psi *6.89 psi*

FIGURE 9.8

Minimum property requirements for some common stone types, according to ASTM standards. Moisture absorption is a good indicator of stone durability. The less absorptive the stone, generally the less susceptible it is to freeze-thaw damage or chemical deterioration. Higher moisture levels within stone can also cause accelerated corrosion of metal anchors or reinforcing within the constructed masonry assembly. Stone density is also a good general measure of durability, correlating with higher strength and lower absorption. Compressive strength is especially important for stone used in loadbearing walls. Modulus of rupture is a measure of a stone's resistance to shear and tension forces, a property particularly relevant to the performance of metal anchors used to attach stone to the building. Flexural strength is important in determining the wind load resistance of relatively thin stone panels.

Stone Masonry

Stone is used in two fundamentally different ways in buildings: It may be laid in mortar, much like bricks or concrete blocks, to make walls, arches, and vaults, a method of construction referred to as *stone masonry*. Or it may be mechanically attached to the structural frame or walls of a building as a facing, called *stone cladding*. This chapter deals with stone masonry laid in mortar; the detailing and installation of mechanically attached stone cladding are covered in Chapters 20 and 23.

Two simple distinctions are useful in classifying patterns of stone masonry (Figures 9.9 and 9.10):

• Rubble masonry is composed of unsquared pieces of stone, whereas ashlar is made up of squared pieces.

• *Coursed stone masonry* has continuous horizontal joint lines, whereas *uncoursed* or *random stone masonry* does not.

Rubble can take many forms, from rounded, river-washed stones to broken pieces from a quarry. It may be either coursed or uncoursed. Ashlar masonry may be coursed or uncoursed and may be made up of blocks that are all the same size or several different sizes. The terms are obviously very general in their meaning, and the reader will find some variation in their use even among people experienced in the field of stone masonry.

Rubble stonework is laid very much like brickwork, except that the irregular shapes and sizes of the stones require the mason to select each stone carefully to fit the available space and, occasionally, to trim a stone with a mason's hammer or chisel. Ashlar stonework, though also similar in many ways to brickwork, frequently uses stone pieces that are too heavy to lift manually, and therefore relies on hoisting equipment to assist the mason in positioning the stones. This, in turn, requires a means of attaching the hoisting ropes to the sides or top of the stone block so as not to interfere with the mortar joint, and several types of devices are commonly used for this purpose (Figure 9.11). Both ashlar and rubble are usually laid with the *quarry bed* or *grain* of the stone running in the horizontal direction, because stone is both stronger and more weather resistant in this orientation.

Stone masonry is frequently combined with concrete masonry to reduce costs. The stone is used where it can be seen, but concrete masonry, which is less expensive, quicker to erect, and more easily reinforced, is used for concealed parts of the wall. One example of such techniques, an exterior stone veneer with a concrete masonry backup wall, is illustrated in Figure 9.12.

Random Rubble

Coursed Rubble

Random Ashlar

Coursed Ashlar

FIGURE 9.9
Rubble and ashlar stone masonry, coursed and random.

(a)   *(b)*

FIGURE 9.10
Random granite rubble masonry (*left*) and random ashlar limestone (*right*). *(Photos by Edward Allen.)*

Lewis Pins

Box Lewis

FIGURE 9.11
Lewises permit the lifting and placing of blocks of building stone without interfering with the bed joints of mortar.

Strap anchors are inserted into slots in the edges of the stone facing blocks

This type of bracket allows for both horizontal and vertical adjustment of the anchor location

Weep holes

Continuous flashing

FIGURE 9.12
A conventional method of anchoring blocks of cut stone facing to a concrete masonry backup wall. The metal strap anchors, which are in direct contact with the stone, should be made only of highly corrosion-resistant stainless steel. (When steel corrodes, it expands. Corrosion of the anchors where they engage with the stone could quickly cause damage to the stone itself.) The brackets attached to the concrete masonry backup wall may be stainless steel or less expensive galvanized steel. They are built into the concrete masonry backup wall as it rises. The airspace, or cavity, between the stone and concrete masonry creates a drainage plane that conducts water that penetrates the stone facing to the bottom of the cavity. The waterproof flashing at the base of the cavity forces the water to exit the wall back to the exterior through small drains in the facing called weep holes. For more information about cavity wall construction, see Chapter 10. For methods of attaching larger panels of stone to a building, see Chapter 20.

SUSTAINABILITY IN STONE AND CONCRETE MASONRY

Stone and Concrete Masonry Materials

• Stone is a plentiful but finite resource. It is usually obtained from open pits, with the attendant disruption of drainage, vegetation, and wildlife habitat.

• The detrimental impacts of stone quarrying can long outlive the buildings for which the stone was extracted.

• Quarry reclamation practices, such as revegetation, land reshaping, and habitat restoration, can mitigate some of the adverse environmental impacts of stone quarrying and convert exhausted quarry sites to other beneficial uses.

• Concrete used in the manufacture of masonry units may include recycled materials such as fly ash, crushed glass, slag, and other postindustrial wastes. For more information regarding the sustainability of concrete, see Chapter 13.

• Mortar used for stone and concrete masonry is made from minerals that are generally abundant in the earth. However, portland cement and lime are energy-intensive products to manufacture. For more information about the sustainability of cement production, see Chapter 13.

Stone and Concrete Masonry Processing and Manufacturing

• Stone is heavy. It is expensive and energy intensive to transport. Stone may originate from local quarries or from sources in many places around the world. Fabrication may take place close to the source of the stone, close to the building site, or in some other remote location. Where uniquely sourced stones are desired or where specialized fabrication processes or skills are required, shipping over long distances may be required.

• The cutting, shaping, and polishing operations that take place during stone fabrication use large quantities of water that becomes contaminated with stone residue, lubricants, and abrasives. Water filtration and recycling systems can prevent contaminants from entering the wastewater stream and minimize water consumption.

• As much as one-half of quarried stone may become waste during fabrication. Depending on the type of stone, waste may be crushed and used as fill material on construction sites or as aggregate in concrete or asphalt. Stone with a strong color or other unique appearance qualities may be processed into aggregate for use in the manufacture of terrazzo, architectural concrete masonry units, or synthetic stone products. Much stone waste, however, is disposed of as landfill.

• The embodied energy of building stone can vary significantly with the source of the stone, fabrication processes, and distances and methods of shipping. Stone that is easily quarried and fabricated, and that is used locally, may have an embodied energy of as little as 300 to 400 BTU per pound (0.7 to 0.9 MJ/kg). In contrast, stone that requires more effort and energy to extract and fabricate, and that is transported over long distances before arriving at the building site, may have an embodied energy 10 or even 20 times greater.

• Most concrete masonry units are manufactured in regional plants relatively close to their final end-use destinations.

• The use of lightweight concrete masonry reduces transportation-related costs and energy consumption.

Mortar joints in stone masonry are frequently *raked* (excavated) to a depth ranging from ½ inch to 1 inch (13–25 mm) or more after setting of the stones. This prevents uneven settling of stones or spalling of the stone edges that can occur when mortar at the face of the wall dries out and hardens faster than mortar deeper in the joint. After the remaining mortar has fully cured, the masons return to *point* the wall by filling the joints out to the face with *pointing mortar* and tooling them to the desired profile.

The primary function of the pointing mortar is to form a good weather seal at the face of the stone. For this reason, low-strength mortar with good workability and adhesion characteristics, such as Type O or N mortar, is used. To ensure that the pointing mortar does not itself lead to concentrated stresses and possible spalling at the face of the masonry, it should never be of higher strength than the mortar deeper in the joint. (See Chapter 8 for a discussion of mortar types.) Using pointing mortar also affords the opportunity to use a mortar of different color from the setting mortar. Alternatively, raked joints can be pointed with elastomeric joint sealant. High-quality sealant that does not stain the stone and has the appropriate elastic capabilities should be used. Joint sealant material types and their application are described more fully in Chapter 19.

Some building stones, especially limestone and marble, deteriorate rapidly in the presence of acids. This restricts their outdoor use in regions

- The embodied energy of concrete masonry units is slightly higher than that of the concrete from which they are made, due to the additional energy consumed in the curing of the units. Ordinary concrete masonry units have an embodied energy of approximately 250 BTU per pound (0.6 MJ/kg).

Stone and Concrete Masonry Construction

- Relatively small amounts of waste are generated on a construction site during stone and concrete masonry construction, including, for example, stone cutoffs, partial blocks, and unused mortar. These wastes generally go into landfills or are buried on the site.

- Sealers applied to stone and concrete masonry to provide water repellency and protection from staining are potential sources of emissions. Solvent-based sealers generally emit more air pollutants than water-based products.

Stone and Concrete Masonry Buildings

- Stone and concrete masonry are not normally associated with indoor air quality problems. In rare instances, stone aggregate in concrete or stone used in stone masonry has been found to be a source of radon gas emissions.

- The thermal mass effect of stone and concrete masonry can be a useful component of fuel-saving heating and cooling strategies such as solar heating and nighttime cooling.

- Stone and concrete masonry are dense materials that can effectively reduce sound transmission between adjacent spaces.

- Stone and concrete masonry construction are noncombustible. Lightweight concrete masonry units are especially effective for construction of fire-resistance rated assemblies.

- Lightweight concrete masonry units have greater thermal resistance than more dense concrete units, stone, or brick.

- Construction with stone or concrete masonry can reduce reliance on paint finishes, a source of volatile organic compounds.

- Stone and concrete masonry are durable forms of construction that require relatively little maintenance and can last a very long time.

- Stone and concrete masonry are resistant to moisture damage and mold growth.

- When a building with stone or concrete masonry is demolished, the stone or masonry units can be crushed and recycled for use as on-site fill or as aggregate for paving. Some building stone can be salvaged for new construction.

Concrete Masonry Sitework

- Concrete masonry permeable pavers can facilitate on-site capture of stormwater.

- Light-colored concrete pavers can lessen urban heat island effects.

- Interlocking concrete masonry units used in earth retaining walls are easily disassembled and reused.

where the air is heavily polluted, and it also prevents their being cleaned with weak acid, as is often done with bricks. Exceptional care is taken during construction to keep stonework clean: Nonstaining mortars are used, high standards of workmanship are enforced, and the work is kept covered as much as possible. Flashings in stone masonry must be of plastic or nonstaining metal. Stonework normally may be cleaned only with mild soap, water, and a soft brush. Once the stonework is clean and dry, it may be treated with a *clear sealer* or *water repellent*, a breathable and virtually invisible coating that reduces water absorption by the stonework, helping to protect it from staining and weathering.

From thousands of years of experience in building with stone, we have inherited a rich tradition of styles and techniques (Figures 9.13, 9.14, 9.15, 9.16, 9.17, 9.18, 9.19, and 9.20). This tradition is all the richer for its regional variations that are nurtured by the locally abundant stones: warm, creamy Jerusalem limestone in Israel; pristine, white Pentelic marble in Greece; stern gray granite and gray-streaked white Vermont marble in the northeastern United States; joyously colorful marbles and wormy travertine in Italy; golden limestone and gray Aswan granite in Egypt; red and brown sandstones in New York; cool, carveable limestone in France; gray-black basalt and granite in Japan. Though the stone industry is now global in character, its greatest glories are often regional and local.

doric **ionic** **corinthian**

FIGURE 9.13

Proportioning rules for the classical orders of architecture. *(Courtesy Indiana Limestone Institute.)*

Brick Relieving Arch

detail of tracery
Showing Dowel Connections

section f-f

detail of
cusp 'g'

section d-d

Dotted Line Shows
section e-e

Building Line

Floor

section a-a

exterior elevation

interior elevation

plan b-b

plan c-c

6'-10"

FIGURE 9.14

Details of Gothic window framing and tracery in limestone. In the lower portion of section a-a, note the 8-inch-deep facing stones that act as bonding units with the interior masonry wythes. This type of composite construction is discussed further in Chapter 10. *(Courtesy Indiana Limestone Institute.)*

FIGURE 9.15
Details for a limestone stair and balustrade in the classical manner. The balusters are turned in a lathe, as shown in **Figure 9.6i**. *(Courtesy Indiana Limestone Institute.)*

FIGURE 9.16
The Marshall Field Wholesale Store, built in 1885 in Chicago, rested on a two-story base of red granite. Its upper walls were built of red sandstone, and its interior was framed with heavy timber. *(Architect: H. H. Richardson. Photo by John McCarthy. Courtesy of Chicago Historical Society, IChi-01688.)*

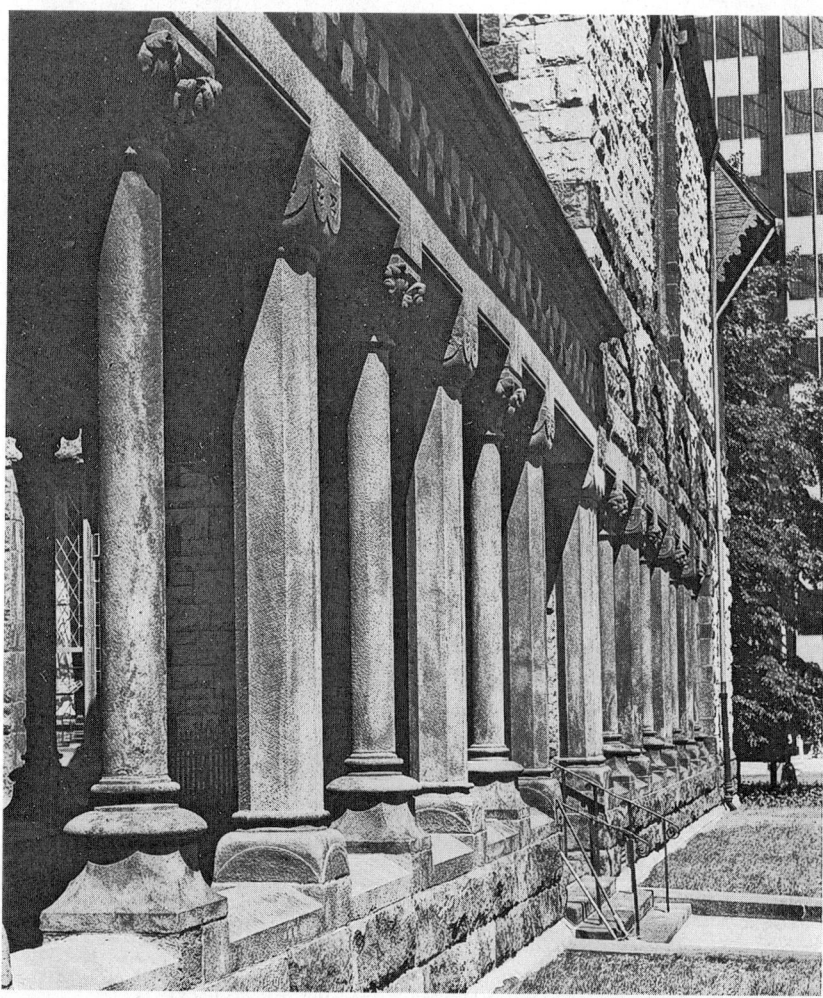

FIGURE 9.17
Granite and sandstone details from the rectory of H. H. Richardson's Trinity Church in Boston, 1872–1877. *(Photo by Edward Allen.)*

(a) (b) (c)

(d) (e) (f)

FIGURE 9.18

Cut stone detailing. (*a*) Random ashlar of broken face granite on H. H. Richardson's Trinity Church in Boston (1872–1877) still shows the drill holes of its quarrying. (*b*) Dressed stonework in and around a window of the same church contrasts with the rough ashlar of the wall. (*c*) Stonework on a 19th-century apartment building grows more refined as it distances itself from the ground and meets the brick walls above. (*d*) The base of the Boston Public Library (1888–1895, McKim, Mead, and White, Architects) is constructed of pink granite, with strongly rusticated blocks between the windows. (*e*) A college chapel is simply detailed in limestone. (*f*) A contemporary college library is clad in bands of limestone. (*Architects: Shepley, Bulfinch, Richardson and Abbott. Photos by Edward Allen.*)

FIGURE 9.19
Ashlar limestone loadbearing walls at the Wesleyan University Center for the Arts, designed by architects Kevin Roche, John Dinkeloo, and Associates. *(Courtesy Indiana Limestone Institute.)*

FIGURE 9.20
The Suzallo Library, University of Washington, built in 1926, combines richly detailed limestone, brick, and terra cotta in its ornate exterior walls. *(Photo by Joseph Iano.)*

CONCRETE MASONRY

Concrete masonry units (CMUs) are manufactured in three basic forms: larger hollow units that are commonly referred to as *concrete blocks*, solid bricks, and, less commonly, larger solid units.

Manufacture of Concrete Masonry Units

Concrete masonry units are manufactured by vibrating a stiff concrete mixture into metal molds, then immediately turning out the wet blocks or bricks onto a rack so that the mold can be reused at the rate of 1000 or more units per hour. The racks of concrete masonry units are cured at an accelerated rate by subjecting them to steam, either at atmospheric pressure or, for faster curing, at higher pressure (Figure 9.21). After steam curing, the units are bundled on wooden pallets for shipping to the construction site.

Concrete masonry units are made in a variety of sizes and shapes (Figures 9.22 and 9.23). They are also made with different densities of concrete, some of which use cinders, pumice, blast furnace slag, or expanded lightweight aggregates rather than crushed stone or gravel. Many colors and surface textures are available. Special shapes are relatively easy to produce if a sufficient number of units will be produced to amortize the expense of the mold. The major ASTM standards under which concrete masonry units are manufactured are C90 for loadbearing units, C129 for nonloadbearing units, and C55 for concrete bricks.

ASTM C90 establishes three weights of loadbearing concrete masonry units, as shown in Figure 9.24. Although all three weight classifications require the same minimum compressive strength, heavier blocks are denser and typically have greater compressive strength than lighter blocks. Heavier blocks are also less expensive to manufacture,

4" block (100 mm)

6" block (150 mm)

8" block (200 mm)

10" block (250 mm)

12" block (300 mm)

FIGURE 9.21
A forklift truck loads newly molded concrete masonry units into an autoclave for steam curing. *(Reprinted with permission of the Portland Cement Association from* Design and Control of Concrete Mixtures, *12th edition; PHOTOS: Portland Cement Association, Skokie, IL.)*

FIGURE 9.22
American standard concrete blocks and half-blocks. Each full block is nominally 8 inches (200 mm) high and 16 inches (400 mm) long.

Channel bond beam

Low-web bond beam

Solid unit

Capping unit

A-block

H-block

Header unit

Control joint unit

Both ends plain

Regular stretcher

One end plain

Single bullnose

Steel sash unit

4-inch-high stretcher

Cap or paving unit

concrete brick

FIGURE 9.23
Other concrete masonry shapes. Concrete bricks are interchangeable with modular clay bricks. Header units accept the tails of a course of headers from a brick facing. The use of control joint units is illustrated in Figure 10.22. Bond beam units have space for horizontal reinforcing bars and grout and are used to tie a wall together horizontally. They are also used for reinforced block lintels. A-blocks are used to build walls with vertical reinforcing bars grouted into the cores in situations where there is insufficient space to lift the blocks over the tops of the projecting bars; one such situation is a concrete masonry backup wall that is built within the frame of a building, as seen in Figure 20.1.

ASTM C90 Weight Classification	Density of Concrete (dry)	Typical Weights of Individual Units
Normal weight (sometimes also referred to as Heavyweight)	125 pcf (2000 kg/m³) or more	33–39 lb (15–18 kg)
Medium weight	From 105 pcf to less than 125 pcf (1680–2000 kg/m³)	28–32 lb (13–15 kg)
Lightweight	Less than 105 pcf (1680 kg/m³)	20–27 lb (9–12 kg)

FIGURE 9.24
Specified densities and typical weights of hollow concrete masonry units. Note the possibly confusing naming of normal and medium weight blocks: normal weight blocks are heavier than medium weight.

FIGURE 9.25
Concrete blocks and bricks can be cut very accurately with a water-cooled, diamond-bladed saw. For rougher sorts of cuts, a few skillful blows from the mason's hammer will suffice. (*Reprinted with permission of the Portland Cement Association from* Design and Control of Concrete Mixtures, *12th edition; PHOTOS: Portland Cement Association, Skokie, IL.*)

absorb moisture less readily, have better resistance to sound transmission, and are more resistant to abuse. But their greater weight also makes heavier blocks more expensive to ship and more labor intensive and expensive for masons to lay in comparison to lighter-weight blocks. The greater density of heavier blocks also gives them lower thermal resistance and lower fire resistance. Medium-weight blocks are probably the most common weight classification specified, although the availability of blocks of different weights varies with regional differences in building practices.

Although innumerable special sizes, shapes, and patterns are available, the dimensions of American concrete masonry units are based on an 8-inch (203-mm) cubic module. The most common block is nominally 8 × 8 × 16 inches (203 × 203 × 406 mm). The actual size of the block is 7⅝ × 7⅝ × 15⅝ inches (194 × 194 × 397 mm), which allows for a mortar joint ⅜ inch (9 mm) thick. This standard block is designed to be

lifted and laid conveniently with two hands (compared to a brick, which is designed to be laid with one). Its double-cube proportions work well for running bond stretchers, for headers, and for corners.

Although concrete masonry units can be cut with a diamond-bladed power saw (Figure 9.25), it is more economical and produces better results if the designer lays out buildings of concrete masonry in dimensional units that correspond to the module of the block (Figure 9.26). Nominal 4-inch, 6-inch, and 12-inch (102-mm, 152-mm, and 305-mm) block thicknesses are also common, as is a solid concrete brick that is identical in size and proportion to a modular clay brick. A handy feature of the standard 8-inch (203-mm) block height is that it corresponds exactly to three courses of ordinary clay or concrete brickwork, or two courses of oversize bricks, making it easy to interweave blockwork and brickwork in composite walls.

FIGURE 9.26
Modular dimensioning of concrete masonry construction. Concrete masonry buildings should be dimensioned to use uncut blocks except under special circumstances.

Interactive Exercise

Laying Concrete Masonry

The accompanying photographic sequence (Figure 9.27) illustrates the technique for laying concrete block walls. The mortar is identical to that used in brick walls, but in most walls only the face shells of the block are mortared, with the webs left unsupported (Figure 9.27c, j, and k).

Concrete masonry is frequently reinforced with steel to increase its loadbearing capacity, or resistance to seismic forces and cracking. *Horizontal reinforcing* may take the form of

(a)

(b)

(c)

(d)

FIGURE 9.27

Laying a concrete masonry wall. (a) A bed of mortar is spread on the footing. (b) The first course of blocks for a lead is laid in the mortar. Mortar for the head joint is applied to the end of each block with the trowel before the block is laid. (c) The lead is built higher. Mortar is normally applied only to the face shells of the block and not to the webs. (d) As each new course is started on the lead, its height is meticulously checked with either a folding rule or, as shown here, a story pole marked with the height of each course.

(continued)

joint reinforcing, welded grids of small-diameter steel rods that are laid into the bed joints at set vertical intervals (Figure 9.28). For greater strength, bond beam blocks (Figure 9.23) or special blocks with channeled webs (Figure 9.29) allow heavier reinforcing bars to be placed horizontally. The horizontal bars may be embedded in grout before the next course of masonry is laid. In this case, the grout is confined within the cores of the reinforced course by a strip of metal mesh laid in the bed joint beneath the course. Or, the bars may

(e)

(f)

(g)

(h)

FIGURE 9.27 *continued*

(*e, f*) Each new course is also checked with a spirit level to be sure that it is level and plumb. Time expended in making sure that the leads are accurate is amply repaid in the accuracy of the wall and the speed with which blocks can be laid between the leads. (*g*) The joints of the lead are tooled to a concave profile. (*h*) A soft brush removes mortar crumbs after tooling.

(*i*)

(*j*)

(*k*)

(*l*)

(*i*) A mason's line is stretched taut between the leads on line blocks. (*j*) The courses of blocks between the leads are laid rapidly by aligning each block with the stretched line; no story pole or spirit level is necessary. The mason has laid bed joint mortar and "buttered" the head joints for a number of blocks. (*k*) The last block to be installed in each course of infill blocks, the closer, must be inserted between blocks that have already been laid. The bed and head joints of the already laid blocks are buttered. (*l*) Both ends of the closer blocks are also buttered with mortar, and the block is lowered carefully into position. Some touching up of the head joint mortar is often necessary. (*Reprinted with permission of the Portland Cement Association from* Design and Control of Concrete Mixtures, *12th edition; PHOTOS: Portland Cement Association, Skokie, IL.*)

be grouted later, at the same time as grouting for vertical bars.

Vertical block cores are easily reinforced by inserting bars and grouting, using either the low-lift or high-lift technique, as described in Chapter 8. In most cases only those cores that contain reinforcing bars are grouted, but sometimes all the vertical cores are filled, whether or not they contain reinforcing, for added strength (Figure 9.30).

Lintels for concrete block walls may be made of steel angles, combinations of rolled steel shapes, reinforced concrete, or bond beam blocks with grouted horizontal reinforcing (Figure 9.31).

FIGURE 9.28
Concrete masonry walls subject to only moderate stresses are reinforced horizontally with steel joint reinforcing. This type of reinforcing is thin enough to fit into the ordinary mortar bed joint between courses. Horizontal joint reinforcing is available in a truss pattern, as illustrated here, or in a rectilinear ladder pattern. Both are equally satisfactory. Vertical reinforcing is done with ordinary mild steel reinforcing bars grouted into block cores.

FIGURE 9.29
These specially formed masonry units are grooved to allow the insertion of horizontal reinforcing bars. The course of block will then be grouted to embed the bars. *(Courtesy of G. R. Ivany and Associates, Inc.)*

FIGURE 9.30
Grout is deposited in the cores of a reinforced concrete masonry wall using a grout pump and hose. *(Reprinted with permission of the Portland Cement Association from* Design and Control of Concrete Mixtures, *12th edition; PHOTOS: Portland Cement Association, Skokie, IL.)*

Dry-Stacked Unit Masonry

Dry-stacked and *surface-bonded* masonry walls are constructed by stacking concrete masonry units in a running bond directly upon one another without the application of mortar. Where leveling is required, metal or high-density plastic shims are used. Once the wall stack is complete, a thin layer of a special cement plaster containing short reinforcing fibers of alkali-resistant glass is applied to each side. After the plaster coating has cured, this surface coating joins the blocks securely to one another both in tension and compression. The plaster coating also serves as a finish with an appearance resembling stucco. Surface-bonded masonry unit construction and materials are governed by ASTM C949. It can be performed with relatively unskilled labor, making it an attractive alternative for low-rise masonry construction where more skilled masonry trades are not readily available.

At least one manufacturer is developing a system of dry-stacked cast masonry units that are internally reinforced rather than surface bonded. Face shells of the masonry units are made of architectural precast concrete with tongue-and-groove edges so that units interlock as they are stacked. The webs of the units are made of metal or plastic. Units are stacked over a preassembled cage of horizontal and vertical reinforcing bars. After stacking, the entire void space between faces is filled with cast-in-place concrete. Like surface-bonded walls, internally reinforced dry-stacked masonry can be built more quickly, by less skilled crews. Additionally, because the masonry units are manufactured with architectural quality faces, no additional finishing materials or operations are required.

Decorative Concrete Masonry Units

Decorative (or *architectural*) *concrete masonry units* are easily and economically manufactured in an unending variety of surface patterns, textures, and colors intended for exposed use in exterior and interior walls. A few such units are diagrammed in Figure 9.33, and some of the resulting surface textures are depicted in Figures 9.34, 9.35, 9.36, and 9.37. Mold

FIGURE 9.31

Lintels for openings in concrete masonry walls. At the top, a steel lintel for a broad opening is made up of a wide-flange section welded to a plate. Steel angle lintels are used for narrower openings. In the middle, a reinforced block lintel is composed of bond beam units. At the bottom, a precast reinforced concrete lintel is seen.

FIGURE 9.32

A small house by architects Clark and Menefee is cleanly detailed in concrete masonry. A reinforced masonry retaining wall helps to link the building visually to the site.
(Photo by Jim Rounsevell.)

Scored-face unit

Ribbed-face unit

Ribbed-face unit

Fluted-face unit

Ribbed split-face unit

Angular-face unit

FIGURE 9.33
Some decorative concrete masonry units, representative of literally hundreds of designs currently in production. The scored-face unit, if the slot in the face is filled with mortar and tooled, produces a wall that looks as if it were made entirely of halfblocks. The ribbed split-face unit is produced by casting "Siamese twin" blocks that are joined at the ribs, then shearing them apart.

FIGURE 9.34
A facade of split-face concrete masonry. *(Architects: Paderewski, Dean, Albrecht & Stevenson. Courtesy of National Concrete Masonry Association.)*

(a)

(b)

(c)

(e)

(d)

FIGURE 9.35
Some walls of decorative concrete masonry. (*a*) Split block. (*b*) Slump block, which is molded from relatively wet concrete and allowed to sag slightly after molding and before curing. (*c*) Split blocks of varying sizes laid in a random ashlar pattern. (*d*) Ribbed split-face blocks. (*e*) Striated blocks. (*Courtesy of National Concrete Masonry Association.*)

FIGURE 9.36
Split-face blocks are used indoors in this high-school auditorium. (*Architects: Marcel Breuer and Associates. Courtesy of National Concrete Masonry Association.*)

FIGURE 9.37
High-rise apartments constructed with reinforced bearing walls of fluted-face units.
Specially cast blocks were used to produce the curved balcony fronts. *(Architect: Paul*
Rudolph. Courtesy of National Concrete Masonry Association.)

costs for producing special units are low when spread across the number of units required for a medium- to large-sized building. Many of the textured concrete masonry units that are now considered standard originated as special designs created by architects for particular buildings.

The Economy and Utility of Concrete Masonry Construction

Concrete masonry is a versatile building material, and walls built from it are usually more economical than comparable ones made of brick or stone masonry: The concrete blocks themselves are cheaper on a volumetric basis and are made into a wall much more quickly because of their larger size (a single standard concrete block occupies the same volume as 12 modular bricks). Concrete blocks can be produced to required degrees of strength, and because their hollow cores allow for the easy insertion of reinforcing steel and grout, they are widely used in bearing wall construction. In one recent year, nearly 8 billion concrete blocks were manufactured in North America alone.

However, concrete masonry is relatively porous, and single-wythe exterior concrete masonry walls are prone to leaking when exposed to heavy rains. Except in dry climates, such walls should be coated on the outside with masonry paint, stucco, or breathable elastomeric coatings, or the masonry units and mortar should include integral water-repellent additives. For this reason, concrete masonry in exterior wall construction is more often used as the backup wythe in a cavity wall with brick or stone facing, as described in Chapter 10.

OTHER TYPES OF MASONRY UNITS

Bricks, stones, and concrete blocks are the most commonly used types of masonry units. In the past, hollow tiles of cast gypsum or fired clay were often used for partition construction (Figure 23.38). Both have been supplanted in the United States by concrete blocks, though hollow clay tiles are still widely used in other parts of the world. *Structural glazed facing tiles* of clay remain in use, especially for partitions, where their durable, easily cleaned surfaces are advantageous, as in public corridors, toilet rooms, institutional kitchens, locker and shower rooms, and industrial plants (Figure 23.39).

Structural terra cotta—glazed or unglazed molded decorative units of fired clay—was widely used until the mid-20th century and is often seen on the facades of late-19th-century masonry buildings in the United States (Figure 8.43). Terra cotta had almost disappeared from the U.S. market a few decades ago, but restoration work on older buildings created a demand for it and generated a renewed desire to use it in new buildings, rescuing the last manufacturers from the brink of obsolescence. Today, terra cotta

FIGURE 9.38

Exterior use of glass blocks in the wall of a stairwell. *(Architect: Gwathmey/Siegel. Courtesy of Pittsburgh Corning Corp.)*

masonry, in both historical and contemporary shapes is readily available.

Glass blocks are available in many textures and in clear, heat-absorbing, and reflective glass (Figure 9.38). Glass blocks are nonabsorbent. When glass masonry walls are constructed, the mortar stiffens more slowly than it does with more absorbent units of clay or concrete, so temporary spacers are inserted between units to maintain proper spacing until the mortar sets up.

Autoclaved aerated concrete (AAC), manufactured and used in Europe for many years, has found limited application in North America. Its ingredients are sand, lime, water, and a small amount of aluminum powder. These materials are reacted with steam to produce a relatively lightweight, aerated concrete that consists primarily of calcium silicate hydrates. AAC is available in solid blocks that are laid in mortar like other concrete masonry units. It is also made into unreinforced wall panels and steel-reinforced lintels and floor/ceiling panels. Because of its trapped gas bubbles, which are created by the reaction of the aluminum powder with the lime, the density of AAC is similar to that of wood, and it is easily sawed, drilled, and shaped. It has moderately good thermal insulating properties. It is not nearly as strong as normal-density concrete, but it is sufficiently strong to serve in loadbearing walls, floors, and roofs in low-rise construction. Walls of AAC are too porous to be left exposed; they are usually stuccoed on the exterior and plastered on the interior.

MASONRY WALL CONSTRUCTION

Bricks, stones, and concrete masonry units are mixed and matched in both loadbearing and nonloadbearing walls of many types. The more important of these wall constructions are presented in the next chapter.

FIGURE 9.39

Dark granite, simply and beautifully detailed, creates a feeling of quiet dignity for the main stair of the Federal Courthouse in Boston. The floor is made of terrazzo with matching and contrasting colors of stone chips. *(Photo © Steve Rosenthal.)*

CSI/CSC MasterFormat Sections for Stone and Concrete Masonry

04 05 00	**COMMON WORK RESULTS FOR MASONRY**
04 05 16	Masonry Grouting
04 05 16.26	Masonry Reinforcing Bars
04 05 19.13	Continuous Joint Reinforcing
04 05 19.26	Masonry Reinforcing Bars
04 21 00	**CLAY UNIT MASONRY**
04 21 23	Structural Clay Tile Masonry
04 21 26	Glazed Structural Clay Tile Masonry
04 21 29	Terra Cotta Masonry
04 22 00	**CONCRETE UNIT MASONRY**
04 22 00.16	Surface-Bonded Concrete Unit Masonry
04 22 23	Architectural Concrete Unit Masonry
04 22 26	Autoclaved Aerated Concrete Unit Masonry
04 43 00	**STONE MASONRY**
04 43 13	Stone Masonry Veneer

KEY TERMS

building stone
igneous rock
sedimentary rock
metamorphic rock
granite
basalt
limestone
freestone
oolitic limestone
dolomitic limestone
quarry sap
sandstone
brownstone
bluestone
slate
marble
travertine

fieldstone
rubble
dimension stone
cut stone
ashlar
flagstone
jet burner
spalling
lewis
petrographic analysis
stone masonry
stone cladding
coursed stone masonry
uncoursed stone masonry, random stone
    masonry
quarry bed, grain
raked mortar joint

pointed mortar joint
pointing mortar
clear sealer, water repellent
concrete masonry unit (CMU)
concrete block
horizontal reinforcing
joint reinforcing
dry-stacked concrete masonry
surface-bonded concrete masonry
decorative concrete masonry unit,
    architectural concrete masonry unit
structural glazed facing tile
structural terra cotta
glass block
autoclaved aerated concrete (AAC)

REVIEW QUESTIONS

1. What are the major types of stone used in construction? How do their properties differ?

2. What sequence of operations would be used to produce rectangular slabs of polished marble from a large quarry block?

3. In what ways is the laying of stone masonry different from the laying of bricks?

4. What are the advantages of concrete masonry units over other types of masonry units?

5. How long is a wall that is made up of 22 concrete masonry units, each nominally 8 × 8 × 16 inches (203 × 203 × 406 mm), joined with ⅜-inch (10-mm) mortar joints?

6. How may horizontal and vertical steel reinforcement be introduced into a CMU wall?

EXERCISES

1. Design a stone masonry façade for a downtown bank building that is 32 feet (9.8 m) wide and 2 stories high. Draw all the joints between stones on the elevation. Draw a detail section to show how the stones are attached to a concrete masonry backup wall.

2. Design a masonry gateway for one of the entrances to a college campus with which you are familiar. Choose whatever type of masonry you feel is appropriate, and make the fullest possible use of the decorative and structural potentials of the material. Show as much detail of the masonry on your drawings as you can.

3. Visit a masonry supply company and view all the types of concrete masonry units that are available. What is the function of each?

4. Design a simple concrete masonry house that a student could build for his or her own use, keeping the floor area to 400 square feet (37 m²).

5. Design a decorative CMU that could be used to build a richly textured wall.

SELECTED REFERENCES

Indiana Limestone Institute of America, Inc. *ILI Handbook.* Bedford, IN, updated regularly.

This is the definitive reference on the use of limestone in building. It includes the history and provenance of Indiana limestone, recommended standards and details for its use, and architectural case histories.

Marble Institute of America. *Dimensional Stone Design Manual.* Cleveland, OH, updated regularly.

Standards, specifications, and details for dimensional stonework are included in this large looseleaf volume. Though all stone types are included, the emphasis is on marble.

National Concrete Masonry Association. *Annotated Design and Construction Details for Concrete Masonry.* Herndon, VA, 2003.

A treasury of typical concrete masonry details.

National Concrete Masonry Association. *TEK Manual for Concrete Masonry Design and Construction.* Herndon, VA, updated regularly.

This comprehensive collection of technical bulletins covers every major aspect of concrete masonry products and design. The complete TEK bulletin series is also available free online, through links from the NCMA web site, www.ncma.org.

Panarese, W., S. Kosmatka, and F. Randall. *Concrete Masonry Handbook for Architects, Engineers, and Builders* (6th ed.). Skokie, IL, Portland Cement Association, 2008.

This is a clearly written, beautifully illustrated guide to every aspect of concrete masonry.

WEB SITES

Stone and Concrete Masonry

Author's supplementary web site: **www.ianosbackfill.com/09_stone_and_concrete_masonry**

Building Stone Institute: **www.buildingstoneinstitute.org**

Bybee Stone Company: **www.bybeestone.com**

Genuine Stone: **www.genuinestone.com**

Indiana Limestone Institute: **www.iliai.com**

Marble Institute of America: **www.marble-institute.com**

National Concrete Masonry Association: **www.ncma.org**

Portland Cement Association: **www.cement.org**

Trenwyth Architectural Masonry Units: **www.trenwyth.com**

10

MASONRY WALL CONSTRUCTION

- **Types of Masonry Walls**
  Composite Masonry Walls
  Masonry Cavity Walls
  Masonry Loadbearing Walls

  FOR PRELIMINARY DESIGN OF A
  LOADBEARING MASONRY STRUCTURE

- **Spanning Systems for Masonry Bearing Wall Construction**
  Ordinary Joisted Construction
  Heavy Timber or Mill Construction
  Steel and Concrete Decks with
  Masonry Bearing Walls

- **Detailing Masonry Walls**
  Cavity Drainage and Flashings
  Thermal Insulation

- **Some Special Problems of Masonry Construction**
  Expansion and Contraction

  MOVEMENT JOINTS IN BUILDINGS

  Efflorescence
  Mortar Joint Deterioration
  Moisture Resistance of Masonry
  Cold- and Hot-Weather Construction

- **Masonry Paving**

- **Masonry and the Building Codes**

- **Uniqueness of Masonry**

America's rich fabric of 19th-century downtown buildings is made up largely of buildings whose construction is defined by building codes as ordinary construction: exterior masonry loadbearing walls with floors and roof spanned by wood joists. *(Photo by Edward Allen.)*

The world's most enduring buildings are constructed of *solid masonry*. Such buildings' thick, monolithic walls can stand for centuries, their great mass providing strength, durability, and moderation of the flow of heat and moisture between indoors and outdoors. Traditional masonry construction methods survive to the present day in such buildings as the Washington National Cathedral (Washington, D.C.), constructed primarily of solid limestone masonry. However, most contemporary construction combines the essential materials of masonry—brick, stone, concrete block, and mortar—with more modern materials and technological innovations to create walls that are thinner, lighter in weight, faster to erect, and better able to control the flow of heat and moisture through the wall.

TYPES OF MASONRY WALLS

Composite Masonry Walls

Solid *composite masonry walls* use different masonry materials in different parts of the wall construction to achieve an optimum balance between appearance, longevity, and economy in the completed wall. In traditional examples of this wall type, face units of terra cotta, stone, or facing brick, chosen for their appearance and durability, are combined with interior wythes of more utilitarian builder's brick (Figure 10.1). The face wythes are bonded to the interior mass of the wall with some combination of header units (see, for example, *section a-a* in Figure 9.14) and iron straps or ties. In contemporary construction, backup wythes are constructed of concrete masonry and wythes are bonded with steel ties or reinforcing (Figure 10.2*D, F, G*). As with any

FIGURE 10.2

Masonry ties and joint reinforcing. These are only a few examples from among dozens of types available. Corrugated ties (*A*) have little strength or stiffness and are recommended only for anchoring low-rise veneer to wood framed backup walls. Z-ties (*B*) are stiffer than corrugated ties. Adjustable ties (*C*) accommodate irregularities in course height between wythes. Adjustable stone ties (*D*) have excessive play, unless the collar joint is filled with mortar. The ladder and truss ties (*E–G*) combine horizontal joint reinforcement with the masonry tie. Truss-type horizontal reinforcing (*G*) can develop composite action between wythes, resulting in a stiffer wall. It is suitable for solid walls but should not be used with cavity walls, where differential rates of expansion and contraction between the wythes can lead to bowing of the wall. (Cavity wall construction is illustrated here. In the case of composite construction, the space between wythes would be filled with mortar.)

FIGURE 10.1

Nineteenth-century solid composite masonry, consisting of facing brick and stone, with builder's brick backup wythes. Traditional masonry construction relies on its compressive strength and mass for strength. However, the lack of tensile reinforcing makes this type of construction vulnerable to seismic forces. This building was nearly destroyed in Seattle's 2001 Nisqually earthquake and subsequently underwent lengthy and extensive reconstruction. Steel crossbracing, part of an internal structural framework inserted for added lateral force resistance, can be seen through the center street-level window. Efflorescence (the white, powderlike deposits just below the parapet corbelling) is evidence of moisture in the masonry and is discussed later in this chapter. (*Photo by Joseph Iano.*)

Nailed to wood stud backup

A. Corrugated Tie

B. Z-Tie

C. Adjustable Tie

D. Adjustable Stone Tie

E. Two-Wire Ladder Tie

F. Ladder Loop Tie

G. Three-Wire Truss Tie

H. Dovetail Anchors for Concrete Backup

I. Steel Column Anchor

solid masonry construction, the space between wythes is filled solid with mortar. Composite masonry walls can be designed to be either loadbearing or nonloadbearing.

In a composite masonry wall, the separate wythes are intimately bonded and behave as a single mass. When designing such walls, care must be taken that differences in the rates of expansion and contraction between the different masonry materials, due to changes in temperature or moisture, are not so great as to cause unintended stresses within the wall. Otherwise, bowing or cracking of the wall could occur. In loadbearing composite walls, the different strengths and elasticities of the materials used must also be taken into account to ensure the proper placement of reinforcing and for calculating the deformations that the wall will experience under load.

Masonry Cavity Walls

Every masonry wall is porous to some degree. Solid masonry construction relies on the wall's mass to absorb and then gradually release moisture that enters the wall. A *cavity wall* interposes a hollow vertical space within the wall to intercept water that penetrates the outer wythe. Only masonry ties, made of corrosion-resistant galvanized or stainless steel, span the cavity and connect the face wythe to the backup wythes (Figures 10.2 and 10.3). When water penetrates the outer wythe and reaches the cavity, it has no place to go but down. At the bottom of the cavity, water is caught by a thin, impervious membrane called a *flashing* and drained through *weep holes* back to the exterior (Figure 10.4).

To further protect against water penetration, a water-repellent coating

dampproofing, may be applied to the cavity side of the backup wythe of the wall. In the event that water does bridge the cavity, this coating discourages seepage further toward the building interior. Membranes or coatings that can also function as air barriers may be applied to this surface as well, as illustrated in Figures 20.1 and 20.3. To control the flow of heat though a cavity wall, rigid foam plastic or semirigid mineral fiber insulation boards may also be inserted into the cavity (Figure 10.3*d–f*).

During construction, the cavity must be kept free of mortar droppings, brick chips, and other debris, which can clog the weep holes or form bridges that transport water across the cavity to the backup wythe. This requires good construction practices and careful inspection. When a masonry unit is pushed into position in a bed joint of mortar, some mortar is squeezed out of the joint on each side. Normally, the mason cuts off this extruded mortar with a trowel and removes it from the cavity, but inevitably some of the mortar falls into the cavity. Eventually, enough mortar may accumulate at the bottom of the cavity to block drainage holes. One solution to this problem is to place a strip of wood on the steel ties in the cavity to catch such mortar droppings, then to pull out this strip and scrape the mortar off it before placing the next row of ties. However, this method is an added inconvenience for the mason and is prone to spilling mortar as the wood strip is lifted from the cavity. An alternative is for the mason to bevel the bed joint with the trowel, making it thinner at the cavity face and thicker at the outside face, before placing the masonry units in it. This minimizes the squeezing-out of mortar into the cavity.

FIGURE 10.3

Photographs of joint reinforcing and ties. (*a*) Ladder reinforcing for a single wythe of concrete masonry. (*b*) Truss reinforcing. (*c*) The double rods of seismic ladder reinforcing offer greater strength than the single rods of normal ladder reinforcing. (*d*) Ladder-type side rod reinforcing reaches across an insulated cavity to reinforce the face wythe of brick masonry and tie it to the backup wythe of concrete masonry. (*e*) Truss-type eye-and-pintle tie reinforcing allows for some adjustment in course heights. (*f*) Ladder-type seismic tab tie reinforcing allows the construction of the face wythe to lag behind that of the concrete masonry backup wythe, and still provides reinforcement and positive tying of the face wythe to the backup.

(Courtesy of Dur-O-Wal, Inc.)

(a)

(b)

(c)

(d)

(e)

(f)

Regardless of precautions taken during construction of the wall, the cavity should be inspected before it is closed to ensure that weep holes have not been obstructed by mortar droppings or other debris. It was once common practice to deposit a layer of small-diameter gravel at the base of the cavity to maintain drainage. However, this has been found to have little if any positive effect. Today, where clogging by mortar droppings is a concern, *cavity drainage materials* in the form of various proprietary free-draining, woven, or matted products are used. These materials are inserted within the cavity, either at the bottom of the cavity or continuously throughout its full height, to catch mortar droppings so that they do not obstruct drainage (Figure 10.5). Such products, though helpful in keeping weep holes open, may still permit mortar droppings to accumulate, creating bridges that can transfer heat and moisture across the cavity. For this reason, they should not be considered a substitute for good construction practices aimed at keeping the cavity as free of mortar droppings as possible.

In a loadbearing cavity wall, the backup wythe normally carries the structural loads while the outer wythe serves as a nonstructural facing or *veneer.* (Less commonly, both wythes can participate in carrying structural loads.) In nonloadbearing walls, the backup wythe does not carry structural gravity loads but still provides support to the veneer through the metal ties, anchoring the veneer in place and bracing it against wind and seismic forces. Additional examples of cavity wall construction can be found elsewhere in this text: Brick veneer curtain wall with concrete masonry backup is illustrated in Figures 20.1 through 20.3 and is discussed in detail in the accompanying captions and text. Brick veneer with a steel stud backup wall is shown in Figure 20.5 and in the case study, "From Concept to Reality: Seattle University Law School," in Chapter 20. Brick

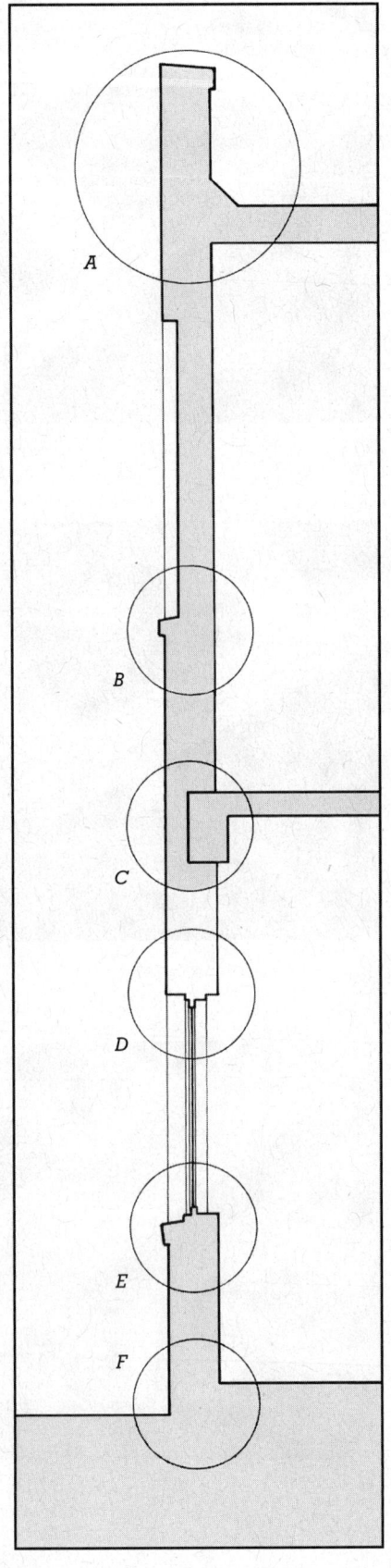

FIGURE 10.4

Typical flashings in masonry walls. Flashings are required at the top of the wall and anywhere the cavity is obstructed horizontally. The letters on the full-height wall section on this page refer to the large-scale details on the facing page. The flashings in a parapet (*A*) may be thought of as external flashings because they do not drain the cavity, and they do not usually have weep holes associated with them. The flashing shown for a recess or projection in *B* is for a composite masonry wall. All other details illustrate cavity wall construction. The reglet used in the second shelf angle detail, *C*, is a length of formed metal or molded plastic that is attached to the inside of the formwork before the concrete is poured. The end dam shown in the center drawing prevents water trapped by the flashing from draining back into the cavity rather than out through the weep holes.

veneer with wood light framing is illustrated in Figure 6.35. An illustration of exterior stone facing and cavity wall is shown in Figure 9.12.

Masonry Loadbearing Walls

Walls constructed of brick, stone, or concrete masonry can be used to support roof and floor structures of virtually any material. Because these *masonry loadbearing walls* (usually called simply *bearing walls*) do double duty by serving also as exterior walls and interior partitions, they are often an economical system of construction compared to others, such as wood, steel, or concrete, that require separate systems to fill in between structural parts. Figure 10.6 lists common ranges of compressive strengths and densities for concrete masonry, bricks, and various types of stone used in bearing wall construction. In practice, the working compressive stresses used in structural

A. Roof Edge (two alternatives)

Weep holes

D. Lintel

B. Recess or Projection

End Dam in Sill Flashing (lintel flashing similar)

Weep holes

E. Sill

Weep holes

Soft joint

Reglet

Weep holes

Soft joint

C. Shelf Angle (two alternatives)

Weep holes

F. Base of Wall

calculations are significantly lower than the ultimate strengths listed in the table, accounting for the lower strength of the mortar between units and to provide appropriate factors of safety.

Reinforced Masonry Walls

Loadbearing masonry walls may be built with or without reinforcing. However, unreinforced masonry walls are not as strong, and are particularly vulnerable to lateral forces from earthquake, heavy winds, or earth pressures (when used below grade). In the past, unreinforced walls were used in North America to support buildings as tall as 16 stories (Figure 8.37). Today, however, most contemporary loadbearing masonry construction is reinforced with steel or other materials that add tensile strength and ductility. Reinforced masonry loadbearing walls can also be constructed thinner than comparable unreinforced walls, resulting (especially in taller buildings) in savings in materials, construction labor, and floor area taken up by the wall.

The engineering design of masonry loadbearing walls is governed by *Building Code Requirements for Masonry Structures, ACI 530/ASCE 5/ TMS 402*, a standard jointly prepared by the American Concrete Institute, the American Society of Civil Engineers, and The Masonry Society. This document establishes requirements for masonry materials (masonry units, reinforcing materials, mortar, metal ties and accessories, and grout) and methods of construction. It also sets forth the engineering procedures by which the strength and stiffness of masonry structural elements are calculated. Methods of reinforcing brick and concrete masonry walls have been described in Chapters 8 and 9, and are further illustrated in this chapter in Figures 10.2, 10.3, 10.7, 10.8, 10.9, 10.10, and 10.17. Reinforcing requirements for relatively short, lightly loaded bearing walls often can be read from standard design tables. However, for larger or more heavily loaded

FIGURE 10.5

One manufacturer's proprietary cavity drainage material. Strips of tangled matting cut with a keystone pattern are inserted into the bottom of the wall cavity, where they prevent accumulations of mortar from obstructing weep holes. (© Copyright 2007, *MortarNet® USA, Ltd. All rights reserved.*)

	Ultimate Compressive Strength	Density
Concrete masonry units	1700–6000 psi	75–145 lb/ft³
	12–41 MPa	*1200–2320 kg/m³*
Bricks	2000–20,000 psi	100–140 lb/ft³
	14–140 MPa	*1600–2240 kg/m³*
Limestone	1800–33,000 psi	110–170 lb/ft³
	12–230 MPa	*1760–2720 kg/m³*
Sandstone	4000–35,000 psi	125–165 lb/ft³
	28–240 MPa	*2000–2640 kg/m³*
Marble	7500–27,000 psi	144–170 lb/ft³
	50–190 MPa	*2310–2720 kg/m³*
Granite	19,000–45,000 psi	160–170 lb/ft³
	130–310 MPa	*2560–2720 kg/m³*

FIGURE 10.6

Common ranges of ultimate strength and density for bricks, concrete blocks, and building stone.

FIGURE 10.7
One system for posttensioning a concrete masonry wall. Short sections of round, threaded high-strength steel bar are joined with threaded couplers as the wall rises. At the base of the wall, the bar is anchored to a threaded insert that has been epoxied into a hole drilled in the concrete foundation. At the top, the bar passes through a steel plate. When the nut at the top end of the bar is tightened, the masonry wall is placed in compression. A load indicator washer tells when the bar is applying sufficient compression to the wall. Load indicator washers are shown in Figure 11.17.

Posttensioned bar puts masonry in compression

Nut

Washer

Posttensioning plate

Threaded couplers allow splicing of posttensioning bars

Alignment plates keep bars centered in walls

Posttensioning bar cast into foundation

walls, the number, locations, and sizes of the steel reinforcing bars must be determined individually for each condition by a structural engineer.

Posttensioned Masonry Walls

Masonry walls may be posttensioned utilizing high-strength steel threaded rods or flexible tendons rather than ordinary, untensioned vertical reinforcing bars (Figure 10.7). The posttensioning elements are anchored into the foundation and run vertically through the masonry wall, either in a cavity between wythes or in the cores of the masonry units. After the wall has been completed and the mortar has cured, each rod or tendon is tensioned (stretched very tightly) and anchored in its taut condition to a horizontal steel plate at the top of the wall. Posttensioning rods are threaded so that each may be tensioned by tightening a nut against the steel plate. Tendons are stretched by a special hydraulic jack and then anchored with the aid of a steel chuck that grips the wires of which the tendon is made. In either case, this tensioning of the reinforcing places the entire wall under a vertical compressive prestress that is considerably higher than would be created by the weights of the masonry and the floors and roofs that it supports. The effect of posttensioning is to strengthen the wall against forces that would normally induce tension in the wall, such as from wind or seismic loads. This allows the use of thinner walls with fewer grouted cores, which saves material and labor. The concept of posttensioning is discussed and illustrated in more detail in Chapter 13.

FOR PRELIMINARY DESIGN OF A LOADBEARING MASONRY STRUCTURE

• To estimate the size of a reinforced brick masonry column, add up the total roof and floor area supported by the column. A 12-inch (300-mm) square column can support up to about 1500 square feet (140 m^2) of area, a 16-inch (400-mm) column 3000 square feet (280 m^2), a 20-inch (500-mm) column 5000 square feet (465 m^2), and a 24-inch (600-mm) column 7000 square feet (650 m^2).

• To estimate the size of a reinforced concrete masonry column, add up the total roof and floor area supported by the column. A 12-inch (300-mm) square column can support up to about 1000 square feet (95 m^2) of area, a 16-inch (400-mm) column 2000 square feet (185 m^2), a 20-inch (500-mm) column 3000 square feet (280 m^2), and a 24-inch (600-mm) column 4000 square feet (370 m^2).

• To estimate the thickness of a reinforced brick masonry loadbearing wall, add up the total width of floor and roof decks that contribute load to a 1-foot (305-mm) length of wall. An 8-inch (200-mm) wall can support up to approximately 600 feet (180 m) of deck and a 12-inch (300-mm) wall about 1000 feet (300 m).

• To estimate the thickness of a reinforced concrete masonry loadbearing wall, add up the total width of floor and roof decks that contribute load to a 1-foot (305-mm) length of wall. An 8-inch (200-mm) wall can support up to approximately 400 feet (120 m) of deck, a 12-inch (300-mm) wall 700 feet (215 m).

These approximations are valid only for purposes of preliminary building layout and must not be used to select final member sizes. They apply to the normal range of building occupancies, such as residential, office, commercial, and institutional buildings, and parking garages.

For manufacturing and storage buildings, use somewhat larger members. For more comprehensive information on preliminary selection and layout of a structural system and sizing of structural members, see Edward Allen and Joseph Iano, *The Architect's Studio Companion* (5th ed.), Hoboken, NJ, John Wiley & Sons, Inc., 2010.

SPANNING SYSTEMS FOR MASONRY BEARING WALL CONSTRUCTION

Ordinary Joisted Construction

So-called *Ordinary construction*, in which the floors and roof are framed with wood joists and rafters and supported at the perimeter on masonry walls, is the fabric of which American center cities largely were built in the 19th century. It still finds use today in a small percentage of new buildings and is listed as Type III construction in the building code table in Figure 1.3. Ordinary construction is essentially balloon framing (Figure 5.2) in which the outer walls of wood are replaced with masonry bearing walls. Balloon framing of the interior loadbearing partitions is used instead of platform framing because it minimizes the sloping of floors that might be caused by wood shrinkage along the interior lines of support. Figure 10.8 shows the essential features of Ordinary construction. Notice two very important details: the firecut ends of the joists and the metal anchors used to tie the wood framing and the masonry wall together.

Heavy Timber or Mill Construction

Heavy Timber or *Mill construction*, listed as Type IV-HT in the code table (Figure 1.3), like Ordinary construction, combines masonry exterior walls with a wood frame interior. However, it uses heavy timbers rather than light joists, rafters, and studs, and thick timber decking rather than thin wood panel sheathing and subflooring. Because heavy timbers are slower to catch fire and burn than nominal 2-inch (38-mm) framing members, somewhat larger floor areas and greater building heights are permitted with Mill construction than with Ordinary construction. Mill construction is discussed in detail in Chapter 4 and a contemporary example is illustrated in Figure 4.10.

Steel and Concrete Decks with Masonry Bearing Walls

Spanning systems of structural steel, sitecast concrete, and precast concrete are frequently used in combination with masonry bearing walls. Figures 10.9 and 10.10 show representative details of two of these combinations. Depending on the degree of fire resistance of the spanning elements, these constructions may be classified under Type I or Type II in the table in Figure 1.3.

FIGURE 10.8

Ordinary construction, shown here with a cavity wall of brick with a concrete masonry loadbearing wythe. Thermal insulation is installed within the CMU and between wood furring strips on the interior side. The interior of a building of Ordinary construction is framed with balloon framing of wood studs and joists.

At the perimeter of the building, the joists are supported by the masonry wall. The rowlock window sill detailed here is similar to the one shown in Figure 10.15.

EXTERIOR WALL

Shingles

Plywood roof sheathing

Gutter

Attic ventilation strip in soffit

Brick facing

Cavity

Bituminous coating on face of concrete block

Vertical reinforcing bars in grouted cores

8' concrete block backup wall with insulating foam inserts in cores

A continuous flashing and weep holes drain the cavity at window heads

Steel lintel

Wood sash

Wood sash

Rowlock sill

A continuous flashing and weep holes drain any leakage through the sill

Wire ties and joint reinforcing

A continuous flashing and weep holes drain the cavity at the bottom

INTERIOR BEARING WALL

Wood rafters and ceiling joists

Insulation

The wood top plate is anchored to the masonry with long threaded rods and steel anchor plates

Bond beam blocks with horizontal reinforcing bars at top of wall

Wood flooring and subfloor

Wood joists with firecut ends

Plaster ceiling

A metal strap tie anchors the joist to the masonry

Bond beam blocks with horizontal reinforcing bars at joist bearings

Wood casing

Wood stool and apron

Plaster

Insulation between furring strips

Wood flooring and subfloor

Wood joists with firecut ends

Metal tie strap

Bond beam blocks with horizontal reinforcing bars at joist bearings

Each stud is a single full-height piece from basement to attic, to minimize vertical wood shrinkage

Wood fireblocking closes off internal cavities in the framing at each floor to slow the spread of fire

1" (19 mm) ledgers notched into the studs support the floor joists

Roof membrane

Gravel stop

Wood

Tapered rigid insulation boards

Corrugated steel roof decking

Metal fascia and ventilated soffit

Textured concrete block facing

Concrete block backup with insulated cores

Foam rubber filler gasket closes between the steel deck and the masonry wall

A lower chord extension on the open web steel roof joists supports the edge of a suspended ceiling

Glass fiber batt insulation and vapor retarder

Vertical reinforcing bars in grouted cores

Wire ties and joint reinforcing

Cavity

Continuous flashing and weep holes

Rigid foam insulation with metal Z-furring strips

Gypsum board

Sheet vinyl floor and vinyl base

Corrugated steel floor decking and concrete topping

Open-web steel floor joists

Each joist is welded to a steel plate anchored to the bond beam

A reinforced concrete block bond beam ties the wall together at each floor level and at the roof

Joint reinforcing

Interior steel column and beam

Vertical reinforcing bars in grouted cores

FIGURE 10.9

An example of a concrete masonry exterior bearing wall with a roof and floor of steel open-web joists and corrugated steel decking. Insulation could also be inserted into the wall cavity, as an alternative to either of the strategies shown here. With suitable fire protection for the steel structure, this type of construction could be built many stories high.

Metal coping/counterflashing

Sealant

Lightweight concrete roof fill

Stone ballast

Polystyrene foam insulation

Roof membrane

Precast concrete hollow-core slabs with plugged cores

Plaster

Reinforced concrete block wall

Plastic foam insulation adhered to the wall

Acrylic stucco finish reinforced with glass fiber mesh

Welded wire fabric temperature reinforcing in poured concrete topping

Reinforcing bars from the walls into the slab topping create structural continuity

Precast concrete hollow-core slabs with plugged cores

All precast concrete slabs rest on neoprene bearing strips

A reinforced concrete block bond beam ties the wall together horizontally at each floor

Reinforced concrete block wall

Plaster

Vinyl base

Carpet

Continuous strip footing

Concrete slab on grade

Reinforced concrete foundation wall

2'

500 mm

1'

250 mm

0

0

12"

FIGURE 10.10

An example of concrete block exterior and interior bearing walls with precast concrete hollow-core slabs spanning the roof and floors. This system can be used for multistory buildings. Full wall reinforcing is shown, with EIFS applied to the outside of the building. Details of this exterior finishing system are further discussed later in this chapter.

DETAILING MASONRY WALLS

Cavity Drainage and Flashings

The minimum recommended separation between wythes of a cavity wall is 2 inches (50 mm). This provides sufficient space for the masons to keep the cavity free of mortar obstructions while the wall is being built. Where insulation boards are inserted into the cavity, the remaining clear space between the face of the insulation and the back of the outer wythe should not be reduced to less than 1 inch (25 mm).

Weep holes should be installed not more than 24 inches (600 mm) apart horizontally in brick and 32 inches (800 mm) apart in concrete masonry. Weeps should lie immediately on top of the flashing in the wall to keep the bottom of the cavity as dry as possible. The recommended method for creating weeps in brick masonry is to leave open head joints between units, making the weep the full width of the joint and at least 2 inches (50 mm) tall. The same method can also be used on concrete masonry. To prevent insects from taking up residence within the cavity, stainless steel wool or plastic or metal accessories with built-in insect screening may be inserted into the open joints (Figure 10.11). Weeps can also be formed with plastic tubes laid in the head joints as they are mortared. Traditionally, weeps were frequently formed with a short piece of rope laid in the mortar joint. This was later pulled out after the mortar had set (Figure 10.15). The minimum diameter for a weep hole is ³⁄₁₆ inch (5 mm). However, where smaller weeps are used, they should be spaced more closely.

Two general types of flashing are used in masonry construction: *External flashings* prevent moisture from penetrating into the masonry wall at its exposed top or where it intersects the roof. *Internal flashings* (also known as *concealed* or *through-wall flashings*) catch water that has penetrated into the wall and drain it through weep holes back to the exterior.

The external flashing at the intersection of a flat roof and a wall parapet is usually constructed in two overlapping parts, a *base flashing* and a *counterflashing* or *cap flashing*. This makes installation easier and allows for movement between the wall and roof components. The base flashing is formed by turning the roof membrane up for a height of at least 8 inches (200 mm). The metal counterflashing is embedded in the masonry wall above the base flashing and extends downward, lapping over the base flashing (Figures 10.4*A*, 10.10, 16.26). Counterflashings are often made with two interlocking pieces, allowing for easy removal when the roof membrane must be replaced. External flashings are also used to protect horizontal surfaces of the masonry wall that are especially vulnerable to water intrusion, such as the tops of masonry parapets (Figure 10.12).

FIGURE 10.11

An open head joint weep, with a plastic insert to prevent the entry of insects. The thin black line underneath the bottom-most mortar joint is the edge of an internal flashing made of modified asphalt. It is being squeezed out of the joint by the weight of the masonry above, pressing on this soft material. A better design would add a sheet metal edge that would extend out from the face of the joint and form a more effective drip edge.

FIGURE 10.12

Metal coping protection for the top of a masonry parapet. The sheet metal pieces are joined with standing seams that add stiffness to this unusually broad wall top.

Internal flashings are installed by masons as they construct the wall. We have already discussed the flashing that is placed at the bottom of the wall cavity. Additional flashings are required wherever the cavity is interrupted: for example, at heads of windows and doors, window sills, shelf angles, and spandrel beams (Figures 10.4B–F, and 10.13). Where an internal flashing crosses the wall cavity, it should be turned up 6 to 9 inches (150–225 mm) at the back of the cavity and penetrate the inner wythe at least 2 inches (50 mm). In this manner, water draining down the cavity is intercepted and directed toward the wall exterior. If the cavity is backed up by a concrete beam or wall, the flashing may terminate in a *reglet*, a horizontal slot formed in the face of the concrete, as shown on the right-hand side of Figure 10.4(C). At the outside face of the wall, flashings should extend beyond the wall face and turn down at an angle so that water draining from the flashing drips free of the wall, rather than being drawn under the flashing by capillary action and returning into the wall (Figure 10.14). It is a

FIGURE 10.13
Three internal metal flashings protect a masonry parapet from water intrusion through any of its exposed surfaces. Uppermost, a flashing directly under the cast concrete coping units intercepts water that infiltrates through the units or the joints between them. The middle flashing intercepts water that infiltrates the back of the parapet. Where it extends out the back of the wall, it will connect to an external counterflashing and the roof membrane will be turned up underneath the counterflashing. Lower down, a flashing spanning the cavity intercepts water that infiltrates the outer face of the wall. *(Courtesy of Copper Development Association Inc.)*

FIGURE 10.14
Internal metal flashings extend beyond the face of the wall under a rowlock brick window sill and soldier course window head. In the window head, weeps formed with plastic tubes are discernible at the bottom of every third brick joint. Weeps in the window sill flashing are hidden in shadow. Vertical expansion joints, aligned with either side of the window openings, are sealant joints with sand embedded into the surface of the sealant to simulate the appearance of mortar. Joints such as these as discussed later in this chapter.

common but insecure practice to terminate flashings just inside the face of the wall in order to hide the flashing; this allows water to be reabsorbed by the wall underneath the flashing (Figure 10.11).

Flashings may be made of sheet metal, modified asphalt membranes, plastic, rubber, or composite sheets. Sheet metal flashings are the most durable and the most expensive. Copper and stainless steel are best. Galvanized steel eventually rusts and disintegrates; it is acceptable for external flashings where it can easily be replaced when needed, but it is not recommended for internal flashings. Aluminum is unsuitable for flashings in masonry walls because it reacts chemically with mortar. Asphaltic flashings are made of polymer-modified asphalts laminated to plastic backings. Most are manufactured with preapplied adhesive on one side, for which reason they are frequently referred to as *self-adhered flashings*. In comparison to metal flashings, they can be more easily shaped and sealed at corners and laps. They are often used in combination with sheet metal. The sheet metal can support the flexible asphaltic membrane where it spans the wall cavity and can extend beyond the outer face of the wall to form the recommended drip. (Asphaltic flashings are too flexible to form a projecting drip, and this material cannot be permanently exposed to sunlight.) Composite (laminated) flashings combine layers of various materials and are intermediate in price. Most consist of a heavy foil of copper or lead laminated with polyester film, glass fiber mesh, bitumen-coated fabric, or waterproofed kraft paper. Many composite flashings are durable. Flashings of plastic and rubber are the least expensive. Some are durable, but others deteriorate too rapidly for use in masonry walls. The designer should specify only those types that have been proven suitable for the purpose.

Flashings within a wall are extremely difficult to replace if they fail in service. Even the most expensive flashing materials cost only a very small fraction of the total price of a masonry wall, so there is little reason to use cheap ones in a misguided effort to save money.

Careful supervision is necessary to ensure that flashings are properly installed. At corners and other junctions where separate pieces of flashing material meet, the pieces should be lapped and soldered or sealed with a suitable mastic. Head and sill flashings should extend beyond the jambs of the opening and terminate in folded *end dams* (Figures 10.4 and 10.15a). The end dams ensure that water trapped by the flashing drains back to the exterior of the wall rather than spilling off the end of the flashing within the cavity. End dams should also be formed wherever else a flashing is interrupted, such as at exposed columns or expansion and control joints.

Cavity wall vents, similar to weeps but located close to the top of the wall cavity, may also be part of a cavity wall design. Along with the weeps low in the cavity, these promote convective air movement through the cavity as an aid to drying. For walls subject to strong, wind-driven rain, vents may be part of pressure-equalized rainscreen design, as discussed in Chapter 19.

Thermal Insulation

A solid masonry wall is a good conductor of heat, which is another way to say that it is a poor insulator. In many hot, dry climates, the capacity of an uninsulated masonry wall to store heat and slow its passage keeps the inside of the building cool during the hot day and warm during the cold night. But in climates with sustained cold or hot seasons, measures must be taken to improve the thermal resistance of masonry walls. The introduction of an empty cavity into a wall improves its thermal insulating properties considerably, but not to a level fully sufficient for cold climates.

Masonry walls can be insulated on the outside face, within the wall, and/or on the inside face. Insulation on the outside face is usually accomplished by means of an *exterior insulation and finish system (EIFS)*, which consists of panels of plastic foam that are attached to the masonry and covered with a thin, continuous layer of polymeric stucco reinforced with glass fiber mesh. The appearance is that of a stucco building. The masonry is completely concealed, and can be of inexpensive materials and unrefined workmanship. EIFS is frequently used for insulating existing masonry buildings in cases where the exterior appearance of the masonry need not be retained. An advantage of EIFS is that the masonry is protected from temperature extremes and can function effectively to stabilize the interior temperature of the building. Disadvantages are that the thin stucco coatings are usually not very resistant to denting or penetration damage and that the EIFS is combustible. Furthermore, some EIFSs have no internal drainage or flashings, so damage and lapses in workmanship, especially around edges and openings, can lead to substantial moisture leakage into the walls and the building. Figure 10.10 details a building insulated in this manner, and photographs of EIFS installation and a completed building are shown in Figures 20.23 and 20.24.

Insulation within the wall can take several forms. If the cavity in a wall is made sufficiently wide, the masons can insert slabs of plastic foam insulation against the inside wythe of masonry as the wall is built (Figure 10.3d–f). The hollow cores of a concrete block wall can be filled

(a)

FIGURE 10.15
Flashing a rowlock brick window sill
with composite flashing sheet. (*a*) The
flashing sheet is turned up to form a
dam at the end. The projecting portion
should be made of sheet metal. (*b*) In
this traditional method of forming weep
holes, pieces of chord are laid in the bed
joint beneath the sloping rowlock sill
bricks. These will be pulled out after the
mortar has stiffened. (*c*) The finished
sill. (*Courtesy of Brick Institute of America.*)

(b)

(c)

FIGURE 10.16

Insulating the cores in a concrete block wall with a dry fill insulation. The insulation is noncombustible, inorganic, nonsettling, and treated to repel water that might be present in the block cores from condensation or leakage. *(Courtesy of W. R. Grace & Co.)*

with loose granular insulation (Figure 10.16) or with special molded-to-fit liners of foam plastic (Figure 10.17). Insulating the cores of concrete blocks does not slow the passage of heat through the webs of the blocks or through portions of the masonry that have been reinforced and fully grouted. Therefore, it is most effective when coupled with an unbroken layer of insulation in the cavity or on one face of the wall.

The inside surface of a masonry wall can be insulated by attaching wood or metal *furring strips* to the wall with masonry nails or powder-driven metal fasteners (Figures 10.8, 10.9, 10.18, and 23.3). Furring strips may be of any desired depth to allow for the necessary thickness of fibrous or foam insulation. The gypsum wallboard or other interior finish material is then fastened to the furring strips. Furring strips can also solve another chronic problem of masonry construction by creating a space in which electrical wiring and plumbing can easily be concealed.

FIGURE 10.18
A worker installs foam plastic insulation on the interior of a concrete masonry wall, using a system of steel furring strips in which only isolated clips, visible in this photograph, contact the masonry wall to minimize thermal conduction through the metal. The furring strips serve as a base to which interior finish panels such as gypsum board can be attached. *(Courtesy of W. R. Grace & Co.)*

FIGURE 10.17
A proprietary concrete masonry unit and insulation system. The webs of the blocks are minimal in area, to reduce thermal bridging where they pass through the foam. The specially shaped polystyrene inserts provide insulation even in parts of the wall with grouted reinforcing. Grouting has not yet occurred in this photograph. *(Courtesy of Korfil, Inc., P. O. Box 1000, West Brookfield, MA 01585.)*

SOME SPECIAL PROBLEMS OF MASONRY CONSTRUCTION

Expansion and Contraction

Masonry walls expand and contract slightly in response to changes in both temperature and moisture content. Thermal movement is relatively easy to quantify (Figure 10.19). Moisture movement is more difficult: New clay masonry units tend to absorb water and expand under normal atmospheric conditions. New concrete masonry units usually shrink somewhat as they give off excess water following manufacture. Expansion and shrinkage in masonry materials are small compared to the moisture movement in wood or the thermal movement in plastics or aluminum. Still, these changes must be taken into account in the design of the building by providing joints of the appropriate types to prevent excessive buildup of forces that could crack or spall the masonry. (See the accompanying sidebar, "Movement Joints in Buildings.")

Three different kinds of movement joints are used in masonry. Expansion joints are breaks in the material that can close or open to accommodate expansion and contraction due to changes in moisture content, temperature, or other effects. Control joints are intentionally created planes of weakness that can open to accommodate shrinkage in masonry surfaces, usually during initial curing. Isolation joints are placed at junctions between masonry and other materials, or between new masonry and old masonry, to accommodate differences in movement between these materials. Figures 10.20 and 10.21 illustrate the placement of movement joints in masonry walls. An example of expansion joints aligned with window openings can be seen in Figure 10.14. Movement joints are also critical in masonry facings applied over multistory structural frames of steel or concrete, to prevent fracture of the masonry when the frame deflects under load, as discussed in Chapter 20.

Joint reinforcing must be interrupted at movement joints so that it does not restrain the opening or closing of the joint. To prevent out-of-plane displacements of the wall, various kinds of vertically interlocking details are often used, as seen in Figure 10.22. Most movement joints in masonry walls are closed with flexible sealants to prevent air and water from passing through.

> **Masonry is a massive material, taking forms permitted by the law of gravity. Our vocabulary of masonry forms was developed in buildings which became essays about gravity—great weights piled high, buttresses braced against the thrust of arched vaulting. Long after the internal steel frame relieved the need for such forms, they still retain meaning for us through their historical references and in their familiarity. The basic masonry forms have become symbols.**
>
> **Michael Shellenbarger, *Landmarks: A Tradition of Portland Masonry Architecture*, 1984**

	Coefficient of Linear Thermal Expansion	
Material	in./in-°F	mm/mm-°C
Clay or shale brick masonry	3.6×10^{-6}	6.5×10^{-6}
Lightweight concrete masonry	4.3×10^{-6}	7.7×10^{-6}
Limestone	4.4×10^{-6}	7.9×10^{-6}
Granite	4.7×10^{-6}	8.5×10^{-6}
Normal weight concrete masonry	5.2×10^{-6}	9.4×10^{-6}
Marble	7.3×10^{-6}	13.1×10^{-6}
Normal weight concrete	5.5×10^{-6}	9.9×10^{-6}
Clay or shale brick masonry	3.6×10^{-6}	6.5×10^{-6}
Structural steel	6.5×10^{-6}	11.7×10^{-6}

FIGURE 10.19
Average linear coefficients of thermal expansion for some masonry materials. (See the Appendix for additional such data.)

FIGURE 10.21
Expansion joints in masonry walls should also be located at discontinuities in the wall, where cracks due to expansion and contraction are most likely to occur.

Unbroken wall length should not exceed 125' (40 m) for masonry, or 200' (60 m) for steel or concrete

Where walls change direction, locate expansion joint next to corners

FIGURE 10.20

Expansion joints in masonry walls should be located near changes of direction in the walls.

Change in thickness

Change in height

Columns or pilasters

Opening

MOVEMENT JOINTS IN BUILDINGS

Buildings and their materials are constantly in motion. Many of these motions are cyclical and never-ending. Some are caused by temperature changes: All materials shrink as they grow cooler and expand as they become warmer, each doing so at its own characteristic rate (see Figure 10.19 and the table of coefficients of thermal expansion in the Appendix). Some are caused by changes in moisture content: Most porous materials grow larger when wetted by water or humid air and smaller when they dry out, also at rates that vary from one material to another. These cyclical motions caused by temperature and moisture can occur on a seasonal basis (warm and moist in summer, cold and dry in winter), and they can also occur in much shorter cycles (warm days, cool nights; warm when the sun is shining on a surface, cool when a cloud covers the sun). Some motions are caused by structural deflections under load, such as the slight sagging of beams, girders, joists, and slabs. These motions can be very long term for dead loads and for floors supporting stored materials in a warehouse. Deflections can be medium term for snow on a roof and very short term for walls resisting gusting winds. Some motions are

one-time phenomena: Concrete and stucco shrink as they cure and dry out, whereas gypsum plaster expands upon curing. Clay bricks expand slightly over time as they absorb atmospheric moisture. Concrete columns shorten slightly, and concrete beams and slabs sag a bit due to plastic creep of the material during the first several years of a building's life; then they stabilize. Posttensioned slabs and beams grow measurably shorter as they take up the compression induced by the stretched steel tendons. Soil compresses under the pressure of the foundations of a new building and then, in most cases, stops moving.

Chemical processes can cause movement in building components: If a steel reinforcing bar rusts, it expands, cracking the concrete around it. Solvent-release sealants shrink as they cure. Some plastics shrink and crack upon prolonged exposure to sunlight. Motion can also be caused by the freezing expansion of water, as happens in the upward heaving of insufficiently deep footings during a cold winter, or in the spalling of concrete and masonry surfaces exposed to wetting and freezing.

Most of these motions are small in magnitude, but they occur in buildings of every size and every material.

If they are ignored in design and construction, they can cause aesthetically unattractive damage to finishes or more serious distortions and failures of building components.

We accommodate these inevitable movements in buildings in two ways: In some cases, we strengthen a material to enable it to resist the stress that will be caused by an anticipated movement, as we do in adding shrinkage–temperature steel to a concrete slab. In most cases, however, we install movement joints in the fabric of the building that are designed to allow the movements to occur without damage. We locate these joints in places where we anticipate maximum potential distress from expected movements. And we place them at regular intervals in large surfaces and assemblies to relieve cumulative stresses. A building that is not provided with suitable movement joints will make its own joints by cracking and spalling at points of maximum stress, creating a situation that is unsightly at best and sometimes dangerous or catastrophic.

Types of Building Joints

Figure A defines two broad classifications of building joints, movement

FIGURE A

and nonmovement. *Nonmovement joints* include most types of joints that are used to connect pieces of material in a building: joints such as the nailed connections in the wooden frame of a house, mortar joints between masonry units, welded and bolted connections in a steel frame, and joints between pours of concrete. A nonmovement joint can be made to move only by overloading the joint, as in the pulling-apart of a nailed connection, the slipping of steel members in a bolted connection, or the cracking of a weld, mortar joint, or concrete slab.

Movement joints are of many different kinds. What they have in common is a designed-in ability to adjust to expected amounts of motion without distress.

- The simplest movement joints are *working joints*, which are designed into various building materials and created in the normal process of assembling a building. An excellent example is the ordinary shingled roof, which is made up of small units of material that are applied in an overlapping pattern so that small amounts of thermal or moisture movement in the underlying roof structure or in the shingles themselves can be tolerated without distress (Figure 16.37). Other examples are wood bevel siding (Figure 6.18), which is nailed in such a way that moisture expansion and contraction are provided for; the metal clips and pans from which a sheet metal roof is assembled (Figure 16.49), which slip as necessary to allow for thermal movement; and most types of sealant joints and glazing joints (Chapters 17 and 19).

- *Structure/enclosure joints* separate structural from nonstructural elements so that they will act independently. A simple example is the deflection track sometimes provided at the top of a metal-framed interior partition to allow the structure above to deflect without imparting loads into the nonbearing wall framing (Figure 23.2b). Another important example of a structure/enclosure joint is the soft joint that is placed just beneath a shelf angle that supports a masonry veneer (Figures 10.4C and 20.3); like the joint at the top of a partition, it prevents a nonstructural element (in this case, a brick facing) from being subjected to a structural load for which it is not designed. Many other cladding attachment details shown in Chapters 19–21 are designed to allow the structural frame of the building and the exterior skin to move independently of one another, and these attachments are always associated with soft sealant or open joints in the skin panels.

- *Isolation joints* separate materials or assemblies that must be able to move independently without interfering with one another. For example, when an existing building is altered or enlarged, isolation joints may be inserted between old and new construction to allow normal, one-time movements to take place in the new materials without disturbing the original construction. Or, where a concrete slab on grade abuts columns or walls, isolation joints are provided to prevent unintentional restraint of the slab and that could lead to cracking. The last drawing in Figure 10.22, Figure 14.3c, and the joint filler strip between the concrete slab and concrete wall shown in Figure 5.6 are all examples of such joints. (Structure/enclosure joints can also be considered one particular type of isolation joint.)

- *Control joints* are deliberately created lines of weakness along which cracking will occur as a surface of brittle material shrinks, relieving the stresses that would otherwise cause uncontrolled or random cracking. The regularly spaced grooves across concrete sidewalks are control joints; they channel the cracking tendency of the pavement into an orderly pattern of straight lines rather than jagged, irregular cracks. Control joints for plaster are shown in Figure 23.11.

- *Expansion joints* are open seams that can close and open to allow expansion and contraction to occur in adjacent areas of material. Expansion joints in brick walls permit the bricks to expand slightly under moist conditions (Figure 10.22). Expansion joints in aluminum curtain wall mullions (Figure 21.15) allow the elements of the wall to increase in size when warmed by sunlight to shorten when they cool.

- Control joints and expansion joints should be located at geometric discontinuities such as corners, changes in the height or width of a surface, or openings. In long or large surfaces, they should also be spaced at intervals that will relieve the expected stresses in the material before those stresses rise to levels that can cause damage.

- *Building separation joints* divide a large or geometrically complex building mass into smaller, discrete structures that can move independently of one another. Building separation joints can be classified into three types: volume-change joints, settlement joints, and seismic separation joints.

- Large-scale effects of expansion and contraction caused by temperature and moisture are relieved by *volume-change joints*. These are generally placed at horizontal or vertical discontinuities in the massing of the building, where cracking would be most likely to occur (Figure B). They are also located at intervals of

MOVEMENT JOINTS IN BUILDINGS (*CONTINUED*)

150 to 200 feet (40 to 60 m) in very long buildings, the exact dimension depending on the nature of the materials and the rate at which dimensional changes occur.

• *Settlement joints* are designed to avoid distress caused by different rates of anticipated foundation settlement between different portions of a building, as between a high-rise tower and a connected low-rise wing, or between portions of a building that bear on different soils or have different types of foundations.

• *Seismic separation joints* are used to divide a geometrically complex building into smaller units that can move independently of one another during an earthquake. (Buildings in seismic zones should also be detailed with structure/enclosure joints that permit the frame of the building to deform during an earthquake without damaging brittle cladding or partition elements.)

Building separation joints are created by constructing independent structures on either side of the plane of the joint, sometimes with entirely separate foundations, columns, and slabs (Figure B). Each of these independent structures is small enough and compact enough in its geometry that it is reasonable to believe that it will move as a unit in response to the forces that are expected to act upon it.

Detailing Movement Joints

The first imperative in detailing a movement joint is to determine what types of movement the joint must

FIGURE B

150' to 200' (40 to 60 m)

accommodate. This is not always simple. Often the same joint is called upon to perform simultaneously in several of the ways that are outlined here—as a volume-change joint, a settlement joint, and a seismic joint, for example. A joint in a composite masonry wall may serve as both an expansion joint for the brick facing and a control joint for the concrete masonry backup. Once the functions of a joint have been determined, the expected character and magnitude of motion can be estimated with the aid of standard technical reference works, and the joint can then be designed accordingly.

It is important that any structural materials that would restrict movement be discontinued at a movement joint. Reinforcing bars or welded wire fabric are often cut (in whole or in part) at control joints in masonry and concrete. Expanded metal lath is interrupted at control joints in plaster or stucco. The primary loadbearing frame of a building is interrupted at building separation joints. At the same time, it is often important to detail a movement joint so that it will maintain a critical alignment of one sort or another. Figure 10.22 shows several expansion and control joints that use interlocking masonry units or hard rubber gaskets to avoid out-of-plane movement across the joint. In concrete slabs, smooth, greased, closely spaced steel dowels may be inserted across an expansion joint; these permit the joint to move while ensuring that the slab will remain at the same level on both sides of the joint. The curtain wall mullion in

Figure 21.15 allows for movement along one axis while maintaining alignment along the two other axes.

Joints must often be designed to stop the passage of heat, air, water, light, sound, and fire. Some must carry traffic, as in the case of joints in floors or bridge pavements. All must be durable and maintainable, while simultaneously adjusting to movement and maintaining an acceptable appearance. Each joint must be detailed to allow the expected direction and extent of movement: Some joints will have to operate only in a push-pull manner, while others are expected to accommodate a shearing motion, or even a twisting motion as well. The exterior joint closure is usually obtained by means of a bellows of metal or synthetic rubber (inset detail, Figure B). Some typical interior joint closures are shown in Figure 22.4. A perusal of manufacturers' catalogs classified under Section 07 95 00, Expansion Control, of the CSI/CSC MasterFormat system will reveal hundreds of different joint closure devices for every purpose.

Every designer of buildings must develop a sure sense of where movement joints are needed in buildings and the knowledge of how to design them. This is neither quick nor easy to do, for the topic is large and complex, and authoritative reference material is widely scattered. Numerous buildings are built each year by designers who have not acquired this intuition. Many of these buildings are filled with cracks even before they have been completed. This brief essay and the related illustrations

throughout the book are intended to create an awareness of the problem of movement in buildings and to establish a framework that the reader can fill in with more detailed information over time.

References

In addition to the references listed at the ends of the chapters of this book, many of which treat movement problems, a good general reference is *Cracks, Movements and Joints in Buildings,* Ottawa, National Research Council of Canada (NRCC 15477). For a practical summary, see also Edward Allen and Patrick Rand, *Architectural Detailing: Function Constructibility, Aesthetics,* Hoboken, NJ, John Wiley & Sons, Inc., 2007.

FIGURE 10.22

Some ways of allowing for movement in masonry construction. Notice that in many of these details the masonry units interlock to prevent out-of-plane movement of the walls. Detailing of sealant joints is discussed in Chapter 19.

Efflorescence

Efflorescence is a fluffy crystalline powder, usually white, that sometimes appears on the surface of a wall of brick, stone, or concrete masonry (Figure 10.1). It consists of one or more water-soluble salts that were originally present either in the masonry units or in the mortar. These were brought to the surface and deposited there by water that seeped into the masonry, dissolved the salts, and then migrated to the surface and evaporated. Efflorescence can usually be avoided by choosing masonry units that have been shown by laboratory testing not to contain water-soluble salts, by using clean ingredients in the mortar, and by minimizing water intrusion into the masonry construction. Most types of efflorescence form soon after the completion of construction and are removed with water and a brush. Although efflorescence is likely to reappear after such a washing, it will normally diminish and finally disappear with time as the salt is gradually leached out of the wall. Efflorescence that forms for the first time after a period of years is an indication that water has only recently begun to enter the wall, and is best controlled by investigating and correcting the source of leakage.

Mortar Joint Deterioration

Mortar joints are the weakest link in most masonry walls. Water running down a wall tends to accumulate in the joints, where cycles of freezing and thawing weather can gradually *spall* (split off flakes of) the mortar in an accelerating process of destruction that eventually creates water leaks and loosens the masonry units. To forestall this process as long as possible, a suitably weather-resistant mortar formulation must be used, and joints must be well filled and tightly compacted with a concave or vee tooling at the time the masonry is laid. Even with these precautions, a masonry wall in a severe climate will show substantial joint deterioration after many years of weathering and may require *repointing*: a process of raking and cutting out the defective mortar and replacing it with fresh mortar. Repointing may also sometimes be referred to as *tuckpointing*, although this term more properly refers to a special treatment of mortar joints with contrasting mortar colors.

Moisture Resistance of Masonry

Most masonry materials, including mortar, are porous and can conduct water from one side of the wall to the other. Water can also enter a wall through small cracks that may develop between masonry units and the mortar that surrounds them. Managing water penetration of masonry begins by specifying appropriate types of masonry units, mortar, and joint tooling. Cavity wall construction should be considered rather than solid wall construction. Properly designed external and internal flashings should be provided. The construction process should be supervised to ensure that mortar joints are free of voids, flashings and weeps are properly installed, and cavities are kept clean. Masonry walls should be protected against excessive wetting to the extent practical through proper roof drainage and sheltering roof overhangs. Beyond these measures, consideration may also be given to coating the wall with stucco, paint, or *clear water repellent*. It is important that any exterior coating be permeable to water vapor to avoid blistering and rupture of the coating from outward vapor migration. Masonry primer/ sealers and paints based on portland cement fill the pores of the wall without obstructing the outward passage of water vapor. Other latex-based or elastomeric coatings that are suitably breathable and intended for application over masonry may also be used. When exterior walls are constructed of solid masonry, *integral water repellents* should be added to the mortar, and if concrete masonry units are used, to the concrete from which these are made as well.

Below grade, masonry should first be parged (plastered on the outside) with two coats of Type M mortar to a total thickness of ½ inch (13 mm) to seal cracks and pores. After the parging has cured and dried, it can be coated with a bituminous dampproofing compound or, if a truly watertight wall is required below grade, it can be covered with a waterproofing layer, as discussed in Chapter 2.

Cold- and Hot-Weather Construction

Mortar cannot be allowed to freeze before it has cured; otherwise, its strength and watertightness may be seriously damaged. In cold climates, possible special precautions include keeping masonry units and sand dry, protecting them from freezing temperatures prior to use, warming the mixing water (and sometimes the sand as well) to produce mortar at an optimum temperature for workability and curing, using a Type III (high early strength) cement to accelerate the curing of the mortar, and mixing the mortar in smaller quantities so that it does not cool excessively before it is used. The masons' workstations should be protected from wind with temporary enclosures. They should also be heated if temperatures inside the enclosures do not remain above freezing. The finished masonry must be protected against freezing for at least 2 to 3 days after it is laid, and the tops of walls should be protected from rain and snow.

In hot weather, mortar may dry excessively before it cures. Some types of masonry units may have to be dampened before laying so that they do not absorb too much water from the mortar. It is also helpful to keep the masonry units and mortar ingredients, as well as the masons' workstations, in the shade.

MASONRY PAVING

Brick, concrete, and stone masonry can all be used as paving materials for both pedestrian and vehicular surfaces. Units may be laid dry over a thin layer of sand and a gravel base, or they may be set in mortar or asphalt over a concrete slab (Figure 10.23).

Permeable or *porous unit paving* systems allow rainwater to pass through the paving system and infiltrate into the soil beneath, rather than flow over the surface of the paving and into a municipal stormwater collection system. Pavements can be made permeable by spacing conventional paving units to leave sand or planted soil in between, or by manufacturing pavers with sufficient openings or permeability for water to freely flow through the pavers themselves.

FIGURE 10.23
Some masonry paving patterns in brick and granite. All six of these examples are laid without mortar in a bed of sand. (*Photos by Edward Allen.*)

Interactive Exercise

MASONRY AND THE BUILDING CODES

Because masonry construction is noncombustible, its use is permitted in any code-defined construction type, examples of which have been discussed in earlier sections of this chapter. Walls made of masonry are also effective barriers to the passage of fire, making them suitable for use as fire walls and other types of fire-resistance-rated separations between spaces within a building or between adjacent buildings. Figure 10.24 gives rule-of-thumb values for fire resistance ratings for brick and concrete masonry walls of various thicknesses.

Due to their mass, masonry walls are also effective in limiting the transmission of sound from one space to another. Sound Transmission Class (STC) ratings for brick and concrete masonry walls, also listed in Figure 10.24, allow comparison of the acoustical properties of these walls with those of other partition systems discussed in Chapter 23.

UNIQUENESS OF MASONRY

Masonry is often chosen as a construction material for its association in people's minds with beautiful buildings and architectural styles of the past, and with the qualities of permanence and solidity. It is often selected for its unique colors, textures, and patterns; for its fire resistance; and for its easy compliance with building code requirements for noncombustibility and fire resistance. Masonry is often chosen, too, because it is economical. Although masonry construction is labor intensive, it can create a high-performance, long-lasting structure and enclosure in a single operation by a single trade, bypassing the difficulties that are frequently encountered in managing the numerous trades and subcontractors needed to erect a comparable building of other materials.

Masonry construction, like wood light frame construction, is carried out with small, relatively inexpensive tools and machines on the construction site. Unlike steel and concrete construction, it does not require (except in the case of ashlar stonework) a large and expensively equipped shop to fabricate the major materials prior to erection. It shares with sitecast concrete construction the long construction schedule that requires special precautions, and can encounter delays during periods of very hot, very cold, or very wet weather. In general, however, it does not require an extensive period of preparation and fabrication in advance of the beginning of construction, because it uses standardized units and materials that are put into final form as they are placed in the building.

From the beginning of human civilization, masonry has been the medium from which we have created our most nearly permanent, most carefully crafted, most highly prized buildings. It has given us the massiveness of the Egyptian pyramids, the inspirational elegance of the Parthenon, and the light-filled loftiness of the great European cathedrals, as well as the reassuring coziness of the fireplace, the brick cottage, and the walled garden. Masonry can express our highest aspirations and our deepest yearnings for rootedness in the earth. It reflects both the tiny scale of the human hand and the boundless power of that hand to create.

Wall Type	Fire Resistance (hours)	STC
4″ (100-mm) brick	1	45
6″ (150-mm) brick	2	51
8″ (200-mm) brick	2–4	52
10″, 12″ (250-mm, 300-mm) brick	4	59
4″ CMU	½–1	43–47[a]
6″ CMU	1	44–51[a]
8″ CMU	1–2	45–55[a]
10″ CMU	3–4	46–59[a]

[a] STC ratings for concrete masonry assume paint or plaster finishes both sides.

FIGURE 10.24
Approximate fire resistance ratings and Sound Transmission Class (STC) ratings for some types of masonry partitions. Fire resistance of masonry construction varies with the density of the masonry units (less dense units conduct heat more slowly and can achieve higher ratings), the total mass of the wall (greater mass absorbs more heat with less rise in temperature and can achieve a higher rating), and other factors, such as whether the wall is solid or has a cavity, whether combustible members are framed into the wall, and the presence of applied finishes, such as plaster or gypsum board, that can contribute to fire resistance. STC ratings also vary with wall density, mass, and the presence of finish coatings.

Material	Working Strength in Tension[a]	Working Strength in Compression[1]	Density	Modulus of Elasticity
Wood (framing lumber)	300–1000 psi	600–2000 psi	30 pcf	1,000,000–1,900,000 psi
	2.1–6.9 MPa	4.1–14 MPa	480 kg/m^3	6900–13,000 MPa
Brick masonry (including mortar, unreinforced)	0	250–1300 psi	120 pcf	700,000–3,700,000 psi
		1.7–9.0 MPa	1900 kg/m^3	4800–25,000 MPa
Structural steel	24,000–43,000 psi	24,000–43,000 psi	490 pcf	29,000,000 psi
	170–300 MPa	170–300 MPa	7800 kg/m^3	200,000 MPa
Concrete (unreinforced)	0	1000–4000 psi	145 pcf	3,000,000–4,500,000 psi
		6.9–28 MPa	2300 kg/m^3	21,000–31,000 MPa

[a]Allowable stress or approximate maximum stress under normal working conditions.

FIGURE 10.25

Comparative physical properties for four common structural materials: wood, brick masonry (shaded row), steel, and concrete. Brick masonry itself has no useful tensile strength, but its compressive strength is considerable, and when combined with steel reinforcing, it can be used for a wide range of structural types. The ranges of values of strength and stiffness reflect variations among available bricks and mortar types.

FIGURE 10.26

A detail of the porch of H. H. Richardson's First Baptist Church, Boston, built in 1871. (*Photo by Edward Allen.*)

CSI/CSC MasterFormat Sections for Masonry Wall Construction

04 05 00	**COMMON WORK RESULTS FOR MASONRY**
04 05 19	Masonry Anchorage and Reinforcing
04 05 23.13	Masonry Control and Expansion Joints
04 05 23.16	Masonry Embedded Flashing
04 50 23.19	Masonry Cavity Drainage, Weep Holes, and Vents
04 27 00	**MULTIPLE-WYTHE UNIT MASONRY**
04 27 13	Composite Unit Masonry
04 27 23	Cavity Wall Unit Masonry
04 42 00	**STONE MASONRY**
07 11 00	**DAMPPROOFING**
07 11 13	Bituminous Dampproofing
07 11 16	Cementitious Dampproofing
07 19 00	**WATER REPELLENTS**
07 24 00	**EXTERIOR INSULATION AND FINISH SYSTEMS**
07 62 00	**SHEET METAL FLASHING AND TRIM**
07 65 00	**FLEXIBLE FLASHING**
09 97 00	**SPECIAL COATINGS**
09 97 23	Concrete and Masonry Coatings
32 14 00	**UNIT PAVING**
32 14 13	Precast Concrete Unit Paving
32 14 16	Brick Unit Paving
32 14 40	Stone Paving
32 14 43	Porous Unit Paving

KEY TERMS

solid masonry
composite masonry wall
cavity wall
flashing
weep hole
dampproofing
cavity drainage material
veneer
masonry loadbearing wall, bearing wall
Ordinary construction
Heavy Timber construction, Mill construction
external flashing

internal flashing, concealed flashing, through-wall flashing
base flashing
counterflashing, cap flashing
reglet
self-adhered flashing
end dam
cavity wall vent
exterior insulation and finish system (EIFS)
furring strip
nonmovement joint
movement joint
working joint

structure/enclosure joint
isolation joint
control joint
expansion joint
building separation joint
volume-change joint
settlement joint
seismic separation joint
efflorescence
spall
repointing, tuckpointing
clear water repellent
integral water repellent
permeable unit paving, porous unit paving

REVIEW QUESTIONS

1. Describe how a cavity wall works and sketch its major construction features. What aspects of cavity wall construction are most critical to its success in preventing water leakage?

2. Where should flashings be installed in (or on) a masonry wall? What is the function of the flashing in each of these locations?

3. Where should weep holes be provided? Describe the function of a weep hole and indicate several ways in which it may be constructed.

4. What are the differences between Ordinary construction and Mill construction? What features of each are related to fire resistance?

5. What types of movement joints are required in a concrete or brick masonry wall? Where should these joints be located?

6. What are some ways of insulating masonry walls?

7. Why is balloon framing used rather than platform framing in Ordinary construction?

8. What precautions should be taken when constructing masonry walls in Minneapolis in the winter?

EXERCISES

1. What are the allowable height and floor area for a restaurant (Occupancy Group A-2) in a building of Mill construction (Type IV)? How do these figures change if unprotected Ordinary construction (Type IIIB) is used instead? (Assume that sprinklers are not required.) What if unprotected steel joists are substituted for the wood joists? Or precast concrete plank floors with a 2-hour fire rating?

2. Examine the masonry walls of some new buildings in your area. Where have movement joints been placed in these walls? What type of joint is each? Why is it placed where it is? Do you agree with this placement?

3. Look at weep holes and flashings in these same buildings. How are they detailed? Can you improve on these details?

4. Design a masonry gateway for one of the entrances to a college campus with which you are familiar. Choose whatever type of masonry you feel is appropriate, and make the fullest possible use of the decorative and structural potentials of the material.

SELECTED REFERENCES

See also the references listed in Chapters 8 and 9.

Masonry Society. *Masonry Designers' Guide.* Boulder, CO, updated regularly.

This comprehensive design guide is based on Building Code Requirements for Masonry Structures, ACI 530/ASCE 5/ TMS 402 (discussed in this chapter) and the related Specifications for Masonry Structures. It also includes extensive commentary, illustrations, and design examples relating to the design and construction of masonry structures.

WEB SITES

Author's supplementary web site: **www.ianosbackfill.com/10_masonry_loadbearing_wall_construction**
Dur-O-Wall: **www.dur-o-wal.com**
Grace Construction Products: **www.na.graceconstruction.com** (select "Masonry Products")
Hyload Flashings: **www.hyloadflashing.com**
The Masonry Society: **www.masonrysociety.org**
Mortar Net: **www.mortarnet.com**
Thermadrain: **www.thermadrain.com**

11

STEEL FRAME CONSTRUCTION

- **History**

- **The Material Steel**
  Steel

  FOR PRELIMINARY DESIGN OF A
  STEEL STRUCTURE

  Steel Alloys
  Production of Structural Shapes
  Cast Steel
  Cold-Worked Steel
  Open-Web Steel Joists

- **Joining Steel Members**
  Rivets
  Bolts
  Welding

- **Details of Steel Framing**
  Typical Connections
  Stabilizing the Building Frame
  Shear Connections and Moment
    Connections

- **The Construction Process**
  The Fabricator
  The Erector

  Floor and Roof Decking
  Architectural Structural Steel

- **Fire Protection of Steel
  Framing**

- **Longer Spans and High-
  Capacity Columns in Steel**
  Improved Beams
  Trusses
  Arches
  Tensile Structures

  FABRIC STRUCTURES

  Composite Columns

- **Industrialized Systems in
  Steel**

  SUSTAINABILITY IN STEEL FRAME
  CONSTRUCTION

- **Steel and the Building Codes**

- **Uniqueness of Steel**

Ironworkers place open-web steel joists on a frame of steel wide-flange beams as a crane lowers bundles of joists from above. *(Photo by Balthazar Korab. Courtesy Vulcraft Division of Nucor.)*

421

Steel, strong and stiff, is a material of slender towers and soaring spans. Precise and predictable, light in proportion to its strength, it is also well suited to rapid construction, highly repetitive building frames, and architectural details that satisfy the eye with a clean, precise elegance. Among the metals, it is uniquely plentiful and inexpensive. If its weaknesses—a tendency to corrode in certain environments and a loss of strength during severe building fires—are held in check by intelligent construction measures, it offers the designer possibilities that exist with no other material.

HISTORY

Prior to the beginning of the 19th century, metals had little structural role in buildings except in connecting devices. The Greeks and Romans used hidden cramps of bronze to join blocks of stone, and architects of the Renaissance countered the thrust of masonry vaults with wrought iron chains and rods. The first all-metal structure, a cast iron bridge, was built in the late 18th century in England and still stands across the Severn River more than two centuries after its construction. Cast iron, produced from iron ore in a blast furnace, and wrought iron, iron that has been purified by beating it repeatedly with a hammer, were used increasingly for framing industrial buildings in Europe and North America in the first half of the 19th century, but their usefulness was limited by the unpredictable brittleness of cast iron and the relatively high cost of wrought iron.

Until that time, steel had been a rare and expensive material,

FIGURE 11.1
Landscape architect Joseph Paxton designed the Crystal Palace, an exposition hall of cast iron and glass, which was built in London in 1851. (*Bettmann Archive.*)

FIGURE 11.2
Allied Bank Plaza, designed by Architects Skidmore, Owings, and Merrill. (*Permission of American Institute of Steel Construction.*)

produced only in small batches for applications such as weapons and cutlery. Plentiful, inexpensive steel first became available in the 1850s with the introduction of the *Bessemer process*, in which air was blown into a vessel of molten iron to burn out the impurities. By this means, a large batch of iron could be made into steel in about 20 minutes, and the structural properties of the resulting metal were vastly superior to those of cast iron. Another economical steelmaking process, the *open-hearth method*, was developed in Europe in 1868 and was soon adopted in America. By 1889, when the Eiffel Tower was built of wrought iron in Paris (Figure 11.3), several steel frame skyscrapers had already been erected in the United States (Figure 11.4). A new material of construction had been born.

THE MATERIAL STEEL

Steel

Steel is any of a range of alloys of iron that contain less than roughly 2 percent carbon. Ordinary structural steel, called *mild steel*, contains less than three-tenths of 1 percent carbon, plus traces of beneficial elements such as manganese and silicon, and of detrimental impurities such as phosphorus, sulfur, oxygen, and nitrogen. In contrast, ordinary *cast iron* contains 2 to 4 percent carbon and greater quantities of impurities than steel, while *wrought iron* contains even less carbon than most steel alloys. Carbon content is a crucial determinant of the properties of any *ferrous* (iron-based) *metal*. Too much carbon makes a hard but brittle metal (like cast iron), while too little produces a malleable, weaker material (like wrought iron). Thus, mild steel is iron whose properties have been optimized for structural purposes by controlling the amounts of carbon and other elements in the metal.

> **The gap between stone and steel-and-glass was as great as that in the evolutionary order between the crustaceans and the vertebrates.**
>
> —Lewis Mumford, *The Brown Decades*, New York, Dover Publications, 1955, pp. 130–131

The process of converting *iron ore* (oxides of iron extracted from the earth in mineral form) to steel begins with the *smelting* of the ore into cast iron. This is done in a *blast furnace*, which is charged with alternating layers of iron ore, *coke* (coal whose volatile constituents have been distilled out, leaving only carbon), and crushed limestone. The coke is burned by large quantities of air

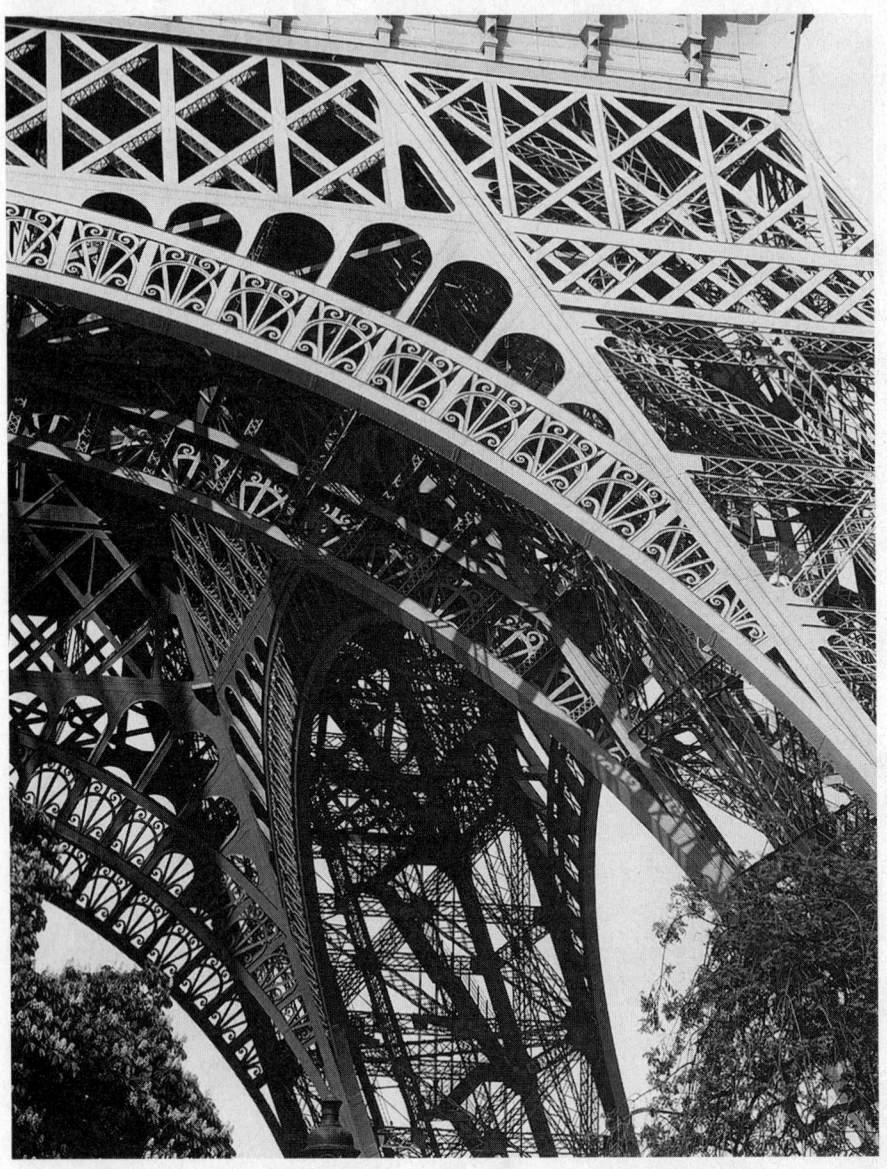

FIGURE 11.3
Engineer Gustave Eiffel's magnificent tower of wrought iron was constructed in Paris from 1887 to 1889. *(Photo by JAMES AUSTIN, CAMBRIDGE (GB).)*

forced into the bottom of the furnace. This produces carbon monoxide, which reacts with the ore, reducing it to elemental iron. The limestone forms a *slag* containing various impurities that floats on top of the molten metal. But large amounts of carbon and other elements inevitably remain incorporated into the iron as well. The molten iron is drawn off at the bottom of the furnace and held in a liquid state for processing into steel (Figure 11.5).

Today, roughly 40 percent of North American steel is made using the *basic oxygen process* (Figure 11.6). In this method, a hollow, water-cooled lance is lowered into a container of molten iron produced from ore (as above) along with recycled steel scrap. A stream of pure oxygen at very high pressure is blown from the lance into the metal to burn off the excess carbon and impurities. A flux of lime and fluorspar is added to the metal to react with other

impurities, particularly phosphorus, and forms a slag that is discarded. New metallic elements may be added to the container at the end of the process to adjust the composition of the steel as desired: Manganese gives resistance to abrasion and impact, molybdenum gives strength, vanadium imparts strength and toughness, and nickel and chromium give corrosion resistance, toughness, and stiffness. The entire process, which takes less than an hour from start to finish, is performed with the aid of careful sampling and analysis techniques to ensure the finished quality of the steel. Steel made by the basic oxygen process contains roughly 25 to 35 percent recycled material.

The greater share of structural steel produced on this continent is made from virtually 100 percent scrap steel in so-called *mini-mills*, utilizing *electric arc furnaces*. These mills are miniature only in comparison to the conventional mills that they have replaced; they are housed in enormous buildings and roll structural shapes of any size and shape. The scrap from which structural steel is made comes mostly from defunct automobiles, one mini-mill alone consuming 300,000 junk cars in an average year. Through careful metallurgical testing and control, these are recycled into top-quality steel.

Regardless of the particular steel-making process, finished steel is cast continuously into *beam blanks* or *blooms*, very thick approximations of the desired final shape, which are then rolled into final form, as described later in this chapter.

FIGURE 11.4

The Home Insurance Company Building, designed by William LeBaron Jenney and built in Chicago in 1884, was among the earliest true skyscrapers. The steel framing was fireproofed with masonry, and the exterior masonry facings were supported on the steel frame. (*Photo by Wm. T. Barnum. Courtesy of Chicago Historical Society ICHi-18293.*)

FIGURE 11.5

The steelmaking process, from iron ore to structural shapes. Notice particularly the steps in the evolution of a wide-flange shape as it progresses through the various stands in the rolling mill. Today, most structural steel in the United States is made from steel scrap in electric furnaces. *(Adapted from "Steelmaking Flowlines," by permission of the American Iron & Steel Institute.)*

For Preliminary Design of a Steel Structure

- Estimate the depth of **corrugated steel roof decking** at 1/40 of its span. Standard depths are 1, 1½, 2, and 4 inches (25, 38, 50, and 100 mm).

- Estimate the overall depth of **corrugated steel floor decking plus concrete topping** at 1/24 of its span. Typical overall depths range from 2½ to 7 inches (65 to 180 mm).

- Estimate the depth of **open-web steel joists** at 1/20 of their span for heavily loaded floors or widely spaced joists and at 1/24 of their span for roofs, lightly loaded floors, or closely spaced joists. The spacing of joists depends on the spanning capability of the decking material. Typical joist spacings range from 2 to 10 feet (0.6–3.0 m). Standard joist depths are given elsewhere in this chapter.

- Estimate the depth of **steel beams** at 1/20 of their span and the depth of **steel girders** at 1/15 of their span. The width of a beam or girder is usually 1/3 to 1/2 of its depth. For composite beams and girders, use the same ratios but apply them to the overall depth of the beam or girder, including the floor deck and concrete topping. Standard depths of steel wide-flange shapes are given elsewhere in this chapter.

- Estimate the depth of **triangular steel roof trusses** at 1/4 to 1/5 of their span. For rectangular trusses, the depth is typically 1/8 to 1/12 of their span.

- To estimate the size of a **steel column**, add up the total roof and floor area supported by the column. A W8 column can support up to about 4000 square feet (370 m²) and a W14 column 30,000 square feet (2800 m²). Very heavy W14 shapes, which are substantially larger than 14 inches (355 mm) in dimension, can support up to 50,000–100,000 square feet (4600–9300 m²). Steel column shapes are usually square or nearly square in proportion.

These approximations are valid only for purposes of preliminary building layout and must not be used to select final member sizes. They apply to the normal range of building occupancies such as residential, office, commercial, and institutional buildings and parking garages. For manufacturing and storage buildings, use somewhat larger members.

For more comprehensive information on preliminary selection and layout of a structural system and sizing of structural members, see Edward Allen and Joseph Iano, *The Architect's Studio Companion* (5th ed.), Hoboken, NJ, John Wiley & Sons, Inc., 2010.

Wide Flange Shape

Intermediate Stand

Edging Stand

Roll Changing Rig

Finishing Stand

Cooling Bed

Rotary Straightener

Hot Saw

Cold Saw

Shipping

FIGURE 11.6
Molten iron is poured into a crucible to begin its conversion to steel in the basic oxygen process. *(Courtesy of U.S. Steel Corp.)*

paint or another protective coating. Though used mostly for highway and bridge construction where it reduces maintenance costs, weathering steel also finds occasional use in buildings, where the deep, warm hue of the oxide coating can be exploited as an aesthetic feature. With the addition of nickel and chromium to steel, various grades of *stainless steel* (ASTM A240 and A276) with even greater corrosion resistance—and costing significantly more than conventional structural steel—can be produced. Steel can also be protected from corrosion by galvanizing, the application of a zinc coating, which is discussed further in Chapter 12.

In some cases, the properties of steel are adjusted after the final shapes are rolled, using various thermo-mechanical processes. For example, immediately after rolling, ASTM A913 steel is subjected to a process of *quenching* (rapid cooling) and then *tempering* (partial reheating) to give the steel an optimized balance of strength, toughness, and weldability characteristics.

Production of Structural Shapes

In the *structural mill* or *breakdown mill*, the beam blank is reheated as necessary and then passed through a succession of rollers that squeeze the metal into progressively more refined approximations of the desired shape and size (Figure 11.7). The finished shape exits from the last set of rollers as a continuous length that is cut into shorter segments by a *hot saw* (Figure 11.8). These segments are cooled on a *cooling bed* (Figure 11.9),

Steel Alloys

By adjusting the mix of metallic elements used in the production of steel, its strength and other properties can be manipulated. Mild structural steel, known by its ASTM designation of A36, was for decades the predominant steel type used in building frames. But today's mini-mills, using scrap as their primary raw material, routinely produce stronger, less expensive *high-strength, low-alloy steels*, such as those designated ASTM A992 or ASTM A572. ASTM A992 steel is the preferred steel type for standard wide-flange structural shapes; ASTM A36 steel, or, where higher strength is needed, ASTM A572 steel, are specified for angles, channels, plates, and bars.

Where steel without any protective finish will remain exposed to exterior conditions in the completed construction, *weathering steel* (ASTM A588 for structural shapes and A606 for thin sheet) may be specified. This steel alloy develops a tenacious oxide coating when exposed to the atmosphere that, once formed, protects against further corrosion and eliminates the need for

FIGURE 11.7
A glowing steel wide-flange shape emerges from the rolls of the finishing stand of the rolling mill. *(Photo by Mike Engestrom. Courtesy of Nucor-Yamato Steel Company.)*

FIGURE 11.8
A hot saw cuts pieces of wide-flange stock from a continuous length that has just emerged from the finishing stand in the background. Workers in the booth control the process. *(Courtesy of U.S. Steel Corp.)*

FIGURE 11.9
Wide-flange shapes are inspected for quality on the cooling bed. *(Photo by Mike Engestrom. Courtesy of Nucor-Yamato Steel Company.)*

and then a *roller straightener* corrects any residual crookedness. Finally, each piece is cut to length and labeled with its shape designation and the number of the batch of steel from which it was produced. Later, when the piece is shipped to a fabricator, it will be accompanied by a certificate that gives the chemical analysis of that particular batch, as evidence that the steel meets standard structural specifications.

The roller spacings in the structural mill are adjustable; by varying the spacings between rollers, a number of different shapes with the same nominal dimensions can be produced (Figure 11.10). This provides the architect and the structural engineer with a finely graduated selection of shapes from which to select each structural member in a building, thereby avoiding waste through the specification of shapes that are larger than required.

Wide-flange shapes are used for most beams and columns, having

FIGURE 11.10
Examples of the standard shapes of structural steel. Where two shapes are superimposed, they illustrate different weights of the same section, produced by varying the spacing of the rollers in the structural mill. Structural steel shapes and their general requirements are defined in ASTM A6. *Bars* **are round, rectangular, and hexagonal solid shapes generally not greater than 8 inches (203 mm) in any cross-sectional dimension. Wider solid shapes are called** *plate* **or** *sheet,* **depending on their thickness in relation to their width. Plate is thicker than sheet.**

many decades ago superseded the older *American Standard (I-beam)* shapes. American Standard shapes are less efficient structurally than wide flanges. The roller arrangement that produces them is incapable of increasing the thickness of the outer flanges without also adding steel to the middle web (as can be seen in Figure 11.10), where it does little to increase the load-carrying capacity of the member. Wide-flange shapes are available in a vast range of sizes and weights. The smallest available depth in the United States is nominally 4 inches (100 mm), and the largest is 44 inches (1117 mm). Weights per linear foot of member range from 9 to 730 pounds (13–1080 kg/m), the latter for a nominal 14-inch

(360-mm) shape with flanges nearly 5 inches (130 mm) thick. Some producers construct heavier wide-flange sections by welding together flange and web plates rather than rolling, a procedure that is also used for producing very deep, long-span plate girders (Figure 11.79).

Structural shapes are identified using a standard nomenclature, in two variations. For example, a MC 10 × 8.4 is a channel shape (MC), nominally 10 inches deep (10), and weighing 8.4 pounds per foot (8.4). The same form is used for wide-flange American Standard, and tee shapes. For angles and HSS members, the shape designation is followed by numerals indicating depth, width, and thickness, in that order (Figure 11.11).

Wide flanges are manufactured in two basic proportions: tall and narrow for beams and squarish for columns and foundation piles. The accepted nomenclature for wide-flange shapes begins with the letter W, followed by the nominal depth of the shape in inches, a multiplication sign, and the weight of the shape in pounds per foot. Thus, a W12 × 26 is a wide-flange shape nominally 12 inches (305 mm) deep that weighs 26 pounds per foot of length (38.5 kg/m). More information about this shape is contained in a table of dimensions and properties in the *Manual of Steel Construction* published by the American Institute of Steel Construction (Figure 11.12): Its actual depth is 12.22 inches (310.4 mm), and its

Shape	Example Designation	Explanation	Range of Available Sizes
Wide-flange	W21 × 83	Nominal depth × weight	4″–44″ *102 mm–1118 mm*
American Standard beam	S18 × 70	Nominal depth × weight	3″–24″ *76 mm–610 mm*
Channel	MC10 × 36	Nominal depth × weight	3″–18″ *76 mm–457 mm*
American Standard channel	C6 × 13	Nominal depth × weight	3″–15″ *76 mm–381 mm*
Structural tee	WT13.5 × 47	Nominal depth × weight	WTs are split from wide-flange shapes. See the WF beam sizes listed above and divide by 2 for available WT depths
Angle	L4 × 3 × ⅜	Length of each leg followed by thickness	Legs: 2″–8″ *(51–203 mm)* Thickness: ⅛″–1⅛″ *(3–29 mm)*
HSS Square, Rectangular, or Elliptical	HSS10 × 8 × ½	Nominal depth and width followed by wall thickness	Depth and width: 1″–48″ *(25–1219 mm)* Wall thickness: ⅛″–⅝″ *(3–16 mm)*
HSS Round	HSS8 × ½	Nominal diameter followed by wall thickness	Diameter: 1.66″–20″ *(42–508 mm)* Wall thickness: 0.109″–0.625″ *(2.8–16 mm)*

FIGURE 11.11
Commonly used steel shapes and sizes.

W SHAPES
Properties

Designation	Nominal Wt per Ft (Lb)	Compact Section Criteria bf/2tf	Compact Section Criteria F_y' (Ksi)	Compact Section Criteria d/tw	Compact Section Criteria F_y''' (Ksi)	rT (In.)	d/Af	X-X I (In.⁴)	X-X S (In.³)	X-X r (In.)	Y-Y I (In.⁴)	Y-Y S (In.³)	Y-Y r (In.)	Torsional constant J (In.⁴)	Zx (In.³)	Zy (In.³)
W 12×336	336	2.3	—	9.5	—	3.71	0.43	4060	483	6.41	1190	177	3.47	243	603	274
×305	305	2.4	—	10.0	—	3.67	0.46	3550	435	6.29	1050	159	3.42	185	537	244
×279	279	2.7	—	10.4	—	3.64	0.49	3110	393	6.16	937	143	3.38	143	481	220
×252	252	2.9	—	11.0	—	3.59	0.53	2720	353	6.06	828	127	3.34	108	428	196
×230	230	3.1	—	11.7	—	3.56	0.56	2420	321	5.97	742	115	3.31	83.8	386	177
×210	210	3.4	—	12.5	—	3.53	0.61	2140	292	5.89	664	104	3.28	64.7	348	159
×190	190	3.7	—	13.6	—	3.50	0.65	1890	263	5.82	589	93.0	3.25	48.8	311	143
×170	170	4.0	—	14.6	—	3.47	0.72	1650	235	5.74	517	82.3	3.22	35.6	275	126
×152	152	4.5	—	15.8	—	3.44	0.79	1430	209	5.66	454	72.8	3.19	25.8	243	111
×136	136	5.0	—	17.0	—	3.41	0.87	1240	186	5.58	398	64.2	3.16	18.5	214	98.0
×120	120	5.6	—	18.5	—	3.38	0.96	1070	163	5.51	345	56.0	3.13	12.9	186	85.4
×106	106	6.2	—	21.1	—	3.36	1.07	933	145	5.47	301	49.3	3.11	9.13	164	75.1
×96	96	6.8	—	23.1	—	3.34	1.16	833	131	5.44	270	44.4	3.09	6.86	147	67.5
×87	87	7.5	—	24.3	—	3.32	1.28	740	118	5.38	241	39.7	3.07	5.10	132	60.4
×79	79	8.2	62.6	26.3	—	3.31	1.39	662	107	5.34	216	35.8	3.05	3.84	119	54.4
×72	72	9.0	52.3	28.5	—	3.29	1.52	597	97.4	5.31	195	32.8	3.04	2.93	108	49.2
×65	65	9.9	43.0	31.1	—	3.28	1.67	533	87.9	5.28	174	29.1	3.02	2.18	96.8	44.1
W 12×58	58	7.8	—	33.9	57.6	3.72	1.90	475	78.0	5.28	107	21.4	2.51	2.10	86.4	32.5
×53	53	8.7	55.9	35.0	54.1	2.71	2.10	425	70.6	5.23	95.8	19.2	2.41	1.58	77.9	29.1
W 12×50	50	6.3	—	32.9	60.9	2.17	2.36	394	64.7	5.18	56.3	13.9	1.96	1.78	72.4	21.4
×45	45	7.0	—	36.0	51.0	2.15	2.61	350	58.1	5.15	50.0	12.4	1.94	1.31	64.7	19.0
×40	40	7.8	—	40.5	40.3	2.14	2.90	310	51.9	5.13	44.1	11.0	1.93	0.95	57.5	16.8
W 12×35	35	6.3	—	41.7	38.0	2.74	3.66	285	45.6	5.25	24.5	7.47	1.54	0.74	51.2	11.5
×30	30	7.4	—	47.5	29.3	2.73	4.30	238	38.6	5.21	20.3	6.24	1.52	0.46	43.1	9.56
×26	26	8.5	57.9	53.1	23.4	2.72	4.95	204	33.4	5.17	17.3	5.34	1.51	0.30	37.2	8.17
W 12×22	22	4.7	—	47.3	29.5	1.02	7.19	156	25.4	4.91	4.66	2.31	0.847	0.29	29.3	3.66
×19	19	5.7	—	51.7	24.7	1.00	8.67	130	21.3	4.82	3.76	1.88	0.822	0.18	24.7	2.98
×16	16	7.5	—	54.5	22.2	0.96	11.3	103	17.1	4.67	2.82	1.41	0.773	0.10	20.1	2.26
×14	14	8.8	54.3	59.6	18.6	0.95	13.3	88.6	14.9	4.62	2.36	1.19	0.753	0.07	17.4	1.90

W SHAPES
Dimensions

Designation	Area A (In.²)	Depth d (In.)	Depth d	Web Thickness tw (In.)	Web tw	Web tw/2 (In.)	Flange Width bf (In.)	Flange Width	Flange Thickness tf (In.)	Flange Thickness	Distance T (In.)	Distance k (In.)	Distance k1 (In.)
W 12×336	98.8	16.82	16 7/8	1.775	1 3/4	7/8	13.385	13 3/8	2.955	2 15/16	9 1/2	3 11/16	1 1/2
×305	89.6	16.32	16 3/8	1.625	1 5/8	13/16	13.235	13 1/4	2.705	2 11/16	9 1/2	3 7/16	1 7/16
×279	81.9	15.85	15 7/8	1.530	1 1/2	3/4	13.140	13 1/8	2.470	2 1/2	9 1/2	3 3/16	1 3/8
×252	74.1	15.41	15 3/8	1.395	1 3/8	11/16	13.005	13	2.250	2 1/4	9 1/2	2 15/16	1 5/16
×230	67.7	15.05	15	1.285	1 5/16	11/16	12.895	12 7/8	2.070	2 1/16	9 1/2	2 3/4	1 1/4
×210	61.8	14.71	14 3/4	1.180	1 3/16	5/8	12.790	12 3/4	1.900	1 7/8	9 1/2	2 5/8	1 1/4
×190	55.8	14.38	14 3/8	1.060	1 1/16	9/16	12.670	12 5/8	1.735	1 3/4	9 1/2	2 7/16	1 3/16
×170	50.0	14.03	14	0.960	15/16	1/2	12.570	12 5/8	1.560	1 9/16	9 1/2	2 1/4	1 1/8
×152	44.7	13.71	13 3/4	0.870	7/8	7/16	12.480	12 1/2	1.400	1 3/8	9 1/2	2 1/8	1 1/16
×136	39.9	13.41	13 3/8	0.790	13/16	7/16	12.400	12 3/8	1.250	1 1/4	9 1/2	1 15/16	1
×120	35.3	13.12	13 1/8	0.710	11/16	3/8	12.320	12 3/8	1.105	1 1/8	9 1/2	1 13/16	15/16
×106	31.2	12.89	12 7/8	0.610	5/8	5/16	12.220	12 1/4	0.990	1	9 1/2	1 11/16	7/8
×96	28.2	12.71	12 3/4	0.550	9/16	5/16	12.160	12 1/8	0.900	7/8	9 1/2	1 5/8	7/8
×87	25.6	12.53	12 1/2	0.515	1/2	1/4	12.125	12 1/8	0.810	13/16	9 1/2	1 1/2	7/8
×79	23.2	12.38	12 3/8	0.470	1/2	1/4	12.080	12 1/8	0.735	3/4	9 1/2	1 7/16	7/8
×72	21.1	12.25	12 1/4	0.430	7/16	1/4	12.040	12	0.670	11/16	9 1/2	1 3/8	13/16
×65	19.1	12.12	12 1/8	0.390	3/8	3/16	12.000	12	0.605	5/8	9 1/2	1 5/16	13/16
W 12×58	17.0	12.19	12 1/4	0.360	3/8	3/16	10.010	10	0.640	5/8	9 1/2	1 3/8	13/16
×53	15.6	12.06	12	0.345	3/8	3/16	9.995	10	0.575	9/16	9 1/2	1 1/4	13/16
W 12×50	14.7	12.19	12 1/4	0.370	3/8	3/16	8.080	8 1/8	0.640	5/8	9 1/2	1 3/8	13/16
×45	13.2	12.06	12	0.335	5/16	3/16	8.045	8	0.575	9/16	9 1/2	1 1/4	13/16
×40	11.8	11.94	12	0.295	5/16	3/16	8.005	8	0.515	1/2	9 1/2	1 1/4	3/4
W 12×35	10.3	12.50	12 1/2	0.300	5/16	3/16	6.560	6 1/2	0.520	1/2	10 1/2	1	9/16
×30	8.79	12.34	12 3/8	0.260	1/4	1/8	6.520	6 1/2	0.440	7/16	10 1/2	15/16	1/2
×26	7.65	12.22	12 1/4	0.230	1/4	1/8	6.490	6 1/2	0.380	3/8	10 1/2	7/8	1/2
W 12×22	6.48	12.31	12 1/4	0.260	1/4	1/8	4.030	4	0.425	7/16	10 1/2	7/8	1/2
×19	5.57	12.16	12 1/8	0.235	1/4	1/8	4.005	4	0.350	3/8	10 1/2	13/16	1/2
×16	4.71	11.99	12	0.220	1/4	1/8	3.990	4	0.265	1/4	10 1/2	3/4	1/2
×14	4.16	11.91	11 7/8	0.200	3/16	1/8	3.970	4	0.225	1/4	10 1/2	11/16	1/2

FIGURE 11.12

A portion of the table of dimensions and properties of wide-flange shapes from the *Manual of Steel Construction* of the American Institute of Steel Construction. One inch equals 25.4 mm. *(Permission of American Institute of Steel Construction.)*

flanges are 6.49 inches (164.9 mm) wide. These proportions indicate that the shape is intended mainly for use as a beam or girder and not as a column or foundation pile. Reading across the table column by column, the designer can learn everything there

is to know about this section, from its thicknesses and the radii of its fillets to various quantities that are useful in computing its structural behavior. At the upper end of the portion of the table dealing with 12-inch (305-mm)-wide-flanges, we find shapes weighing up to 336 pounds per foot (501 kg/m), with actual depths of almost 17 inches (432 mm). These heavier shapes have flanges nearly as wide as the shapes are deep, suggesting that they are intended for use as columns. U.S. producers manufacture steel shapes only in conventional units of measurement, inches and pounds. In other parts of the world, a standard range of metric sizes is used. The United States has adopted a soft conversion to metric sizes, merely tabulating metric dimensions for shapes that are produced in conventional units.

Steel *angles* (Figure 11.13) are extremely versatile. They

ANGLES
Equal legs and unequal legs
Properties for designing

Size and Thickness	k	Weight Per Foot	Area	AXIS X-X				AXIS Y-Y				AXIS Z-Z		
				I	S	r	y	I	S	r	x	r	Tan
In.	In.	Lb.	In.²	In.⁴	In.³	In.	In.	In.⁴	In.³	In.	In.	In.	α
L 4 x3 x ⅝	1¹/₁₆	13.6	3.98	6.03	2.30	1.23	1.37	2.87	1.35	0.849	0.871	0.637	0.534
½	¹⁵/₁₆	11.1	3.25	5.05	1.89	1.25	1.33	2.42	1.12	0.864	0.827	0.639	0.543
⁷/₁₆	⅞	9.8	2.87	4.52	1.68	1.25	1.30	2.18	0.992	0.871	0.804	0.641	0.547
⅜	¹³/₁₈	8.5	2.48	3.96	1.46	1.26	1.28	1.92	0.866	0.879	0.782	0.644	0.551
⁵/₁₆	¾	7.2	2.09	3.38	1.23	1.27	1.26	1.65	0.734	0.887	0.759	0.647	0.554
¼	¹¹/₁₆	5.8	1.69	2.77	1.00	1.28	1.24	1.36	0.599	0.896	0.736	0.651	0.558
L 3½x3½ x ½	⅞	11.1	3.25	3.64	1.49	1.06	1.06	3.64	1.49	1.06	1.06	0.683	1.000
⁷/₁₆	¹³/₁₆	9.8	2.87	3.26	1.32	1.07	1.04	3.26	1.32	1.07	1.04	0.684	1.000
⅜	¾	8.5	2.48	2.87	1.15	1.07	1.01	2.87	1.15	1.07	1.01	0.687	1.000
⁵/₆	¹¹/₁₆	7.2	2.09	2.45	0.976	1.08	0.990	2.45	0.976	1.08	0.990	0.690	1.000
¼	⅝	5.8	1.69	2.01	0.794	1.09	0.968	2.01	0.794	1.09	0.968	0.694	1.000
L 3½x3 x ½	¹⁵/₁₆	10.2	3.00	3.45	1.45	1.07	1.13	2.33	1.10	0.881	0.875	0.621	0.714
⁷/₁₆	⅞	9.1	2.65	3.10	1.29	1.08	1.10	2.09	0.975	0.889	0.853	0.622	0.718
⅜	¹³/₁₆	7.9	2.30	2.72	1.13	1.09	1.08	1.85	0.851	0.897	0.830	0.625	0.721
⁵/₁₆	¾	6.6	1.93	2.33	0.954	1.10	1.06	1.58	0.722	0.905	0.808	0.627	0.724
¼	¹¹/₁₆	5.4	1.56	1.91	0.776	1.11	1.04	1.30	0.589	0.914	0.785	0.631	0.727
L 3½x3½ x ½	¹⁵/₁₆	9.4	2.75	3.24	1.41	1.09	1.20	1.36	0.760	0.704	0.705	0.534	0.486
⁷/₁₆	⅞	8.3	2.43	2.91	1.26	1.09	1.18	1.23	0.677	0.711	0.682	0.535	0.491
⅜	¹³/₁₆	7.2	2.11	2.56	1.09	1.10	1.16	1.09	0.592	0.719	0.660	0.537	0.496
⁵/₁₆	¾	6.1	1.78	2.19	0.927	1.11	1.14	0.939	0.504	0.727	0.637	0.540	0.501
¼	¹¹/₁₆	4.9	1.44	1.80	0.755	1.12	1.11	0.777	0.412	0.735	0.614	0.544	0.506
L 3 x3 x ½	¹³/₁₆	9.4	2.75	2.22	1.07	0.898	0.932	2.22	1.07	0.898	0.932	0.584	1.000
⁷/₁₆	¾	8.3	2.43	1.99	0.954	0.905	0.910	1.99	0.954	0.905	0.910	0.585	1.000
⅜	¹¹/₁₆	7.2	2.11	1.76	0.833	0.913	0.888	1.76	0.833	0.913	0.888	0.587	1.000
⁵/₁₆	⅝	6.1	1.78	1.51	0.707	0.922	0.865	1.51	0.707	0.922	0.865	0.589	1.000
¼	⁹/₁₆	4.9	1.44	1.24	0.577	0.930	0.842	1.24	0.577	0.930	0.842	0.592	1.000
³/₁₆	½	3.71	1.09	0.962	0.441	0.939	0.820	0.962	0.441	0.939	0.820	0.596	1.000

Angles In shaded rows may not be readily available. Availability is subject to rolling accumulation and geographical location, and should be checked with material suppliers.

FIGURE 11.13

A portion of the table of dimensions and properties of angle shapes from the *Manual of Steel Construction* of the American Institute of Steel Construction. One inch equals 24.5 mm. *(Permission of American Institute of Steel Construction.)*

can be used for very short beams supporting small loads and are frequently found playing this role as lintels spanning door and window openings in masonry construction. In steel frame buildings, their primary role is in connecting wide-flange beams, girders, and columns, as we will see shortly. They also find use as diagonal braces in steel frames and as members of steel trusses, where they are paired back to back to connect conveniently to flat *gusset plates* at the joints of the truss (Figure 11.82). *Channel* sections are also used as truss members and bracing, and for short beams, lintels, and stringers in steel stairs. *Tees*, plates, bars, and sheets all have their various roles in a steel frame building, as shown in the diagrams that accompany this text.

Cast Steel

Although the vast majority of structural steel is produced as rolled shapes, structural shapes can also be produced as *cast steel*, that is, by pouring molten steel into molds and allowing the steel to cool. Although steel castings are, pound for pound, more expensive than rolled steel shapes, they have other advantages: Because cast parts are produced in small quantities, they can economically utilize specialized steel alloys selected on the basis of a part's unique requirements. Because they are cast in discrete molds rather than formed through a continuous rolling process, cast steel parts can be nonuniform in section, they can readily incorporate curves or complex geometries, and their shapes can be carefully tailored to the particular requirements of the part. Cast steel is especially well suited for the production of custom-shaped connections for steel structures that are stronger, lighter, and more

attractive than possible with those fabricated from conventional rolled steel.

Cold-Worked Steel

Steel can be *cold-worked* or *cold-formed* (rolled or bent) in a cold state (at room temperature). Cold working causes the steel to gain considerable strength through a realignment of its crystalline structure. Light gauge (thin) steel sheet is formed into C-shaped sections to make short-span framing members that are frequently used to frame partitions and exterior walls of larger buildings and floor structures of smaller buildings (see Chapter 12). Steel sheet stock is also rolled into corrugated configurations utilized as floor and roof decking in steel-framed structures (Figures 11.58 through 11.63).

Heavier sheet or plate stock may be cold-formed into square, rectangular, round, and elliptical hollow shapes that are then welded along the longitudinal seam to form *hollow structural sections* (HSSs) (Figures 11.11, 11.14, and 11.67). Also called structural tubing, they are often used for columns and for members of welded steel trusses and space trusses (Figure 11.85). Their hollow shape makes them especially suitable for members that are subjected to torsional (twisting) stresses or to buckling associated with compressive loads.

The normal range of wide-flange shapes is too large to be cold-rolled, but cold rolling is used to produce small-section steel rods and steel components for open-web joists, where the higher strength can be utilized to good advantage. Steel is also cold-drawn through dies to produce the very-high-strength wires used in wire ropes, bridge cables, and concrete prestressing strands.

Open-Web Steel Joists

Among the many structural steel products fabricated from hot- and cold-rolled shapes, the most common is the *open-web steel joist* (OWSJ), a mass-produced truss used in closely spaced arrays to support floor and roof decks (Figure 11.14). According to Steel Joist Institute (SJI) specifications, open-web joists are produced in four series: K series joists are for spans up to 60 feet (18 m) and range in depth from 8 to 30 inches (200–760 mm). LH series joists are designated as Long-span and can span as far as 96 feet (29 m). Their depths range from 18 to 48 inches (460–1220 mm). The DLH Deep Longspan series of open-web joists (for roofs only) are 52 to 72 inches deep (1320–1830 mm) and can span up to 144 feet (44 m). Most buildings that use open-web joists utilize K series joists that are less than 2 feet (600 mm) deep to achieve spans of up to 40 feet (13 m). The spacings between joists commonly range from 2 to 10 feet (0.6–3 m), depending on the magnitude of the applied loads and the spanning capability of the decking. Often, wider joist spacings are frequently more economical: Fewer joists and connections are needed and the thicker slabs required help to dampen floor vibrations.

CJ composite joists are rated for composite floor construction, explained later in this chapter. *Joist girders* are prefabricated steel trusses designed to carry heavy loads, particularly bays of steel joists. They range in depth from 20 to 72 inches (500–1800 mm). They can be used instead of wide-flange beams and girders in roof and floor structures where their greater depth is not objectionable. Open-web joists and joist girders are invariably made of high-strength steel.

FIGURE 11.14
Open-web steel floor and roof joists are supported on wide-flange beams and square
HSS columns. The heavy, diagonally braced frames in the far right corner of the
image are part of the lateral-force resisting system of the structure, explained later in
this chapter. They will be incorporated into the enclosure around an exit stair to be
installed in that location. *(Photo by Joseph Iano.)*

Joining Steel Members

Rivets

Steel shapes can be joined into a building frame with any of three fastening techniques—rivets, bolts, or welds—and by combinations of these. A *rivet* is a steel fastener consisting of a cylindrical body and a formed head. It is brought to a white heat in a forge, inserted while hot through holes in the members to be joined, and hot-worked with a pneumatic hammer to produce a second head opposite the first (Figure 11.15). As the rivet cools, it shrinks, clamping the joined pieces together and forming a tight joint. Riveting was for many decades the predominant fastening technique used in steel frame buildings, but it has been completely replaced in contemporary construction by the less labor-intensive techniques of bolting and welding.

FIGURE 11.15
How riveted connections are made.
(*A*) A hot steel rivet is inserted through holes in the two members to be joined.
(*B, C*) Its head is placed in the cup-shaped depression of a heavy, hand-held hammer. A pneumatic hammer drives a rivet set repeatedly against the body of the rivet to form the second head.
(*D*) The rivet shrinks as it cools, drawing the members tightly together.

Bolts

Bolts used in steel frame construction may be either *high-strength bolts* (ASTM A325 and A490), which are heat treated during manufacturing to develop their greater strength, or lower-strength *carbon steel bolts* (ASTM A307). In contemporary steel frame construction, bolted structural connections rely almost exclusively on high-strength bolts. Carbon steel bolts (also called unfinished or common bolts) find only limited use, such as in the fastening of minor framing elements or temporary connections.

The manner in which a bolted structural steel connection derives its strength depends on how the bolts are installed. In a *bearing-type connection*, bolts need only be installed to a *snug-tight* condition. In this case, movement between the joined members is resisted by the bolts themselves as the sides of the bolt holes in the connected members bear against the bodies of the bolts. In a *slip-critical* (or *friction-type*) *connection*, bolts are *preloaded* (tightened during installation) to such an extent that friction between the adjoining faces of the steel members (the *faying surfaces*) resists movement between the members. Under normal load conditions, bolts in bearing-type connections are stressed primarily in shear, while those in slip-critical connections are stressed in tension.

When a bearing-type connection is first loaded, a slight slippage of the joint occurs as the sides of the bolt holes in the joined members achieve full bearing against the bodies of the bolts. In contrast, a slip-critical connection will reach its full design capacity with virtually no initial slippage. For this reason, only slip-critical connections are used where the small changes in alignment that can occur with bearing-type connections could be detrimental to the performance of a structure. For example, column splices and beam-to-column connections in tall buildings must be designed as slip-critical, as must

FIGURE 11.16
An ironworker tightens high-strength bolts with a pneumatic impact wrench. *(Courtesy of Bethlehem Steel Corporation.)*

FIGURE 11.17
Top: **An untightened high-strength bolt with a load indicator washer under the head.** *Bottom:* **The bolt and washer after tightening; notice that the protrusions on the load indicator washer have flattened.**

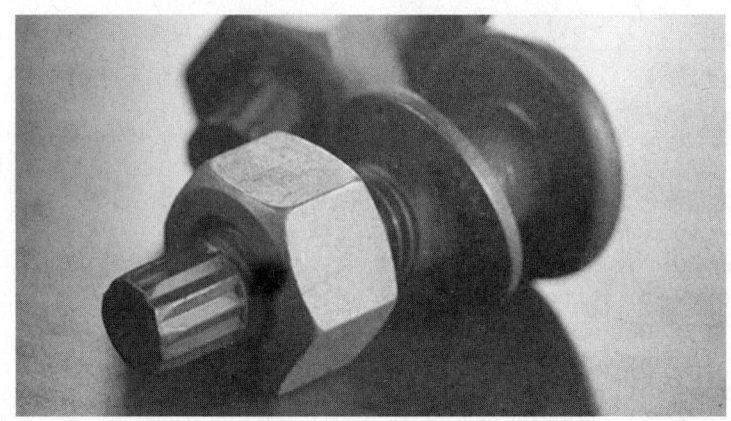

FIGURE 11.18
A splined tension control bolt. *(Courtesy of LeJeune Bolt Company.)*

connections that experience load reversals.

In a typical connection, bolts are inserted into holes ¹⁄₁₆ inch (2 mm) larger than the diameter of the bolt. Depending on a variety of factors, hardened steel washers may be inserted under one or both ends of the fastener. Washers are required with slotted or oversized holes to ensure that the bolt head and nut

have adequate contact with the surfaces of the joined members. When installing preloaded bolts, washers may be required to prevent damage, such as *galling* (tearing), to the surfaces of the joined members. Many bolt tension verification methods, discussed later in this section, also require washers under at least one end of the bolt to ensure consistent tensioning results.

A

B

C

FIGURE 11.19
Tightening a tension control bolt.
(A) The wrench holds both the nut and the splined body of the bolt and turns them against one another to tighten the bolt. (B) When the required torque has been achieved, the splined end twists off in the wrench. (C) A plunger inside the wrench discharges the splined end into a container. *(Courtesy of LeJeune Bolt Company.)*

Bolts are usually tightened using a pneumatic or electric *impact wrench* (Figure 11.16). In a bearing-type connection, the amount of tension in the bolts is not critical. In a slip-critical connection, bolts must be tightened reliably to at least 70 percent of their ultimate tensile strength.

A major problem in the assembly of slip-critical connections is how to verify that the necessary tension has been achieved in each bolt. This can be accomplished in any of several ways. In the *turn-of-nut method,* each bolt is tightened snug, then turned a specified additional fraction of a turn. Depending on bolt length, bolt alloy, and other factors, the additional tightening required will range from one-third of a turn to a full turn.

In another method, a *load indicator washer,* also called a *direct tension indicator (DTI),* is placed under the head or nut of the bolt. As the bolt is tightened, protrusions on the washer are progressively flattened in proportion to the tension in the bolt (Figure 11.17). Inspection for proper bolt tension then becomes a relatively simple matter of inserting a feeler gauge to determine whether the washer has flattened sufficiently to indicate the required tension. One washer manufacturer has made inspection even easier by attaching tiny dye capsules to the washer; when the protrusions on the washer flatten sufficiently, the capsules squirt a highly visible dye onto the surface of the washer.

Less frequently used to verify bolt tension is the *calibrated wrench method,* in which a special torque control wrench is used to tighten the bolts. The torque setting of the wrench is carefully calibrated for the particular size and type of fasteners being installed so as to achieve the required bolt tension. A washer under the turned end of the bolt minimizes friction and ensures a consistent correlation between the tightening force applied and the tension achieved in the bolt.

Yet another method of bolt tension verification employs *tension control bolts.* These have protruding splined ends that extend beyond the threaded portion of the body of the bolt (Figure 11.18). The nut is tightened by a special power-driven *shear wrench* that grips both the nut and the splined end simultaneously, turning one against the other. The splined end is formed in such a way that when the required torque has been reached, the end twists off (Figure 11.19). Verification of adequate bolt tensioning in installed bolts is then done by visually checking for the absence of splines. Another advantage of this fastener type is that it is installed by a single worker, unlike conventional bolts, which require a second worker with a wrench to prevent the other end of the bolt assembly from turning during tightening.

An alternative to the high-strength bolt is the *lockpin and collar fastener* or *swedge bolt,* a boltlike steel pin with annular rings that relies on a steel collar in lieu of a conventional nut to hold the pin. The swedge bolt is installed using a special power tool to hold the pin under high tension while cold forming (swaging, a crimpinglike action) the collar around its end to complete the connection. As the installation process is completed, the tail of the lockpin breaks off, furnishing visual evidence that the necessary tension has been achieved in the fastener. Like the tension control bolt, the swedge bolt can be installed by a single worker.

Welding

Welding offers a unique and valuable capability to the structural designer: It can join the members of a steel frame as if they were a monolithic whole. Welded connections, properly designed and executed, are stronger than the members they join. Although it is possible to achieve this same performance with high-strength bolted connections, such connections are often cumbersome compared to equivalent welded joints. Bolting, however, has its own advantages: It is quick and easy for field connections that need only resist shearing forces, and it can be accomplished under conditions of adverse weather or difficult physical access that would make welding impossible. Often welding and bolting are combined in the same connections to take advantage of the unique qualities of each: Welding may be used in the fabricator's shop for its inherent economies and in the field for its structural continuity, whereas bolting is often employed in the simpler field connections and to hold connections in alignment prior to welding. The choice of bolting, welding, or combinations of the two is often dictated by the designer, but it may also be influenced by considerations such as the fabricator's and erector's equipment and expertise, availability of electric power, climate, and location.

Electric arc welding is conceptually simple. An electrical potential is established between the steel pieces to be joined and a metal *electrode* held either by a machine or by a person. When the electrode is held close to the seam between the steel members, a continuous electric arc is established that generates sufficient heat to melt both a localized area of the steel members and the tip of the electrode (Figure 11.20). The molten steel from the electrode merges with that of the members to form a single puddle. The electrode is drawn slowly along the seam, leaving behind a continuous bead of metal that cools and solidifies to form a continuous

FIGURE 11.20
Close-up diagram of the electric arc welding process.

connection between the members. For small members, a single pass of the electrode may suffice to make the connection. For larger members, a number of passes are made so as to build up a weld of the required depth.

In practice, welding is a complex science. The metallurgy of the structural steel and the welding electrodes must be carefully coordinated. Voltage, amperage, and polarity of the electric current are selected to achieve the right heat and penetration for the weld. Air must be kept away from the electric arc to prevent rapid oxidation of the liquid steel; this is accomplished in simple welding processes either by a thick coating on the electrode that melts to create a liquid and gaseous shield around the arc or by a core of vaporizing flux in a tubular steel electrode. It may also be done by means of a continuous flow of inert gas around the arc, or with a dry flux that is heaped over the end of the electrode as it moves across the work.

The required thickness and length of each weld are calculated by the designer to match them to the forces to be transmitted between the members, and are indicated on fabrication drawings using standardized *weld symbols* (Figure 11.21). For deep welds, such as the full-penetration welds shown in Figure 11.22, the edges of the members are beveled to create a groove that permits access of the electrode to the full thickness of the piece. Small strips of steel called *backup bars* or *backing bars* are welded beneath the groove before the actual weld is begun to prevent the molten metal from flowing out the bottom of the groove. The weld then is deposited in a number of passes of the electrode until the groove is fully filled. In some cases, *runoff bars* are required at the ends of a groove weld to facilitate the formation of a full thickness of weld metal at the edges of the member (Figure 11.45).

Workers who do structural welding are methodically trained and periodically tested to ensure that they have the required level of skill and knowledge. When an important weld is completed, it is inspected to make sure that it is of the required size and quality; this often involves sophisticated magnetic particle, dye penetrant, ultrasonic, or radiographic testing procedures that search for hidden voids and flaws within each weld.

Welds in structural connections that may be subjected to very high stresses during a seismic event and that are critical to maintaining the stability of the building structure are termed *demand-critical welds.* They are subject to their own special design, materials, installation, and inspection requirements to ensure their reliability under these conditions.

FIGURE 11.21
Standard weld symbols, as used on steel connection detail drawings.

THE ARROW

The reference line carries the descriptive symbols

The arrow points to the weld

THE BASIC SYMBOLS

The basic weld symbol is located on either side of the reference line as follows:

Symbols on the top of the reference line refer to welds on the side of the joint opposite the arrow

Symbols on the bottom of the reference line refer to welds on the same side of the joint as the arrow

The basic symbols are

BACK	FILLET	PLUG or SLOT	GROOVE or BUTT						
			SQUARE	V	BEVEL	U	J	FLARE V	FLARE BEVEL
⌒	◁	▭	‖	∨	∨	∪	⊍	∨	∨

SUPPLEMENTARY SYMBOLS

FIELD WELD
This weld must be done in the field during erection. Other welds are done earlier in the fabricator's shop

WELD ALL AROUND
This indicates that the weld should be carried fully around the perimeter of the joining pieces

BACKUP BAR
As indicated in the example, a backup bar to support the first pass of the weld must be placed on the side of the joint opposite the arrow

SPACER
Small metal spacers are used to maintain a gap between the pieces to be joined, prior to welding

A sharp bend near the end of the arrow indicates that the arrowhead is pointing toward the grooved side of a bevel or J-grooved joint

*V-Groove Weld
With Backup Bar*

*Single-Bevel
Groove Weld
With Backup Bar*

Fillet Weld

V-Groove Weld

*Double-Bevel
Groove Weld*

*Double Fillet
Weld*

Puddle Weld

*Partial-Penetration
Single-Bevel
Groove Weld*

Fillet Weld

FIGURE 11.22

Typical welds used in steel frame construction. Fillet welds are the most economical because they require no advance preparation of the joint, but full-penetration groove welds are stronger. The standard symbols used here are explained in Figure 11.21.

DETAILS OF STEEL FRAMING

Typical Connections

Most steel frame connections use angles, plates, or tees as transitional elements between the members being connected. A simple bolted beam-to-column-flange connection requires two angles and a number of bolts (Figures 11.23, 11.24, 11.25, and 11.26). The angles are cut to length, and the holes are made in all the components prior to assembly. The angles are usually bolted to the web of the beam in the fabricator's shop. The bolts through the flange of the column are added as the beam is erected on the construction site. This type of connection, which joins only the web of the beam, but not the flanges, to the column is known as a *shear connection*. It is capable of transmitting vertical forces (*shear*) from a beam to a column. However, because it does not connect the beam flanges to the column, it is of no value in transmitting bending forces (*bending moment*) from one to the other.

To produce a *moment connection*, one capable of transmitting bending forces between a beam and column, it is necessary to connect the beam flanges strongly across the joint, most commonly by means of *full-penetration groove welds* (Figures 11.27 and 11.28). If the column flanges are insufficiently strong to accept the forces transmitted from the beam flanges, *stiffener plates* must be installed inside the flanges of the column to better distribute these forces into the body of the column. Though much less common, it is also possible to design moment-transmitting connections that rely solely on bolting.

FIGURE 11.23
A generic steel building frame. The letters are keyed to the connection details in the figures that follow.

FIGURE 11.24
Exploded and assembled views of a bolted beam-to-column-flange connection, *A* on the frame shown in Figure 11.23. The size of the angles and the number and size of the bolts are determined by the magnitude of the load that the connection must transmit from the beam to the column.

Interactive Exercise

Shop bolts hold the connecting angles to the beam

Field bolts connect the beam to the column

An end clearance allows for slight inaccuracies, and makes it easier to lower the beam into place during erection

FIGURE 11.25
Two elevation views of the bolted beam-to-column-flange connection shown in Figure 11.24. This is a shear connection (AISC simple connection) and not a moment connection, because the flanges of the beam are not rigidly connected to the column. This type of shear connection, in which the beam is connected to the column by angles, plates, or tees fastened to the web of the beam, is also called a *framed connection*. Alternatively, shear connections can be seated, as illustrated in Figure 11.31.

FIGURE 11.26
A pictorial view of a framed, bolted beam-to-column-flange shear connection.

FIGURE 11.27
A welded moment connection (AISC Fully Restrained) for joining a beam to a column flange. This is the type of connection that would be used instead of the shear connection at location *A* in Figure 11.23 if a moment connection were required. The bolts hold the beam in place for welding and also provide shear resistance. Small rectangular backup bars are welded beneath the end of each beam flange to prevent the welding arc from burning through. A clearance hole is cut from the top of the beam web to permit the backup bar to pass through. A similar clearance hole at the bottom of the beam web allows the bottom flange to be welded entirely from above for greater convenience. The groove welds develop the full strength of the flanges of the beam, allowing the connection to transmit moments between the beam and the column. If the column flanges are not stiff enough to accept the moments from the beam, stiffener plates are welded between the column flanges, as shown here. The flanges of the beam are cut to a dog-bone configuration to create a zone of the beam that is slightly weaker in bending than the welded connection itself. During a violent earthquake, the beam will deform permanently in this zone while protecting the welded connection against a more severe failure.

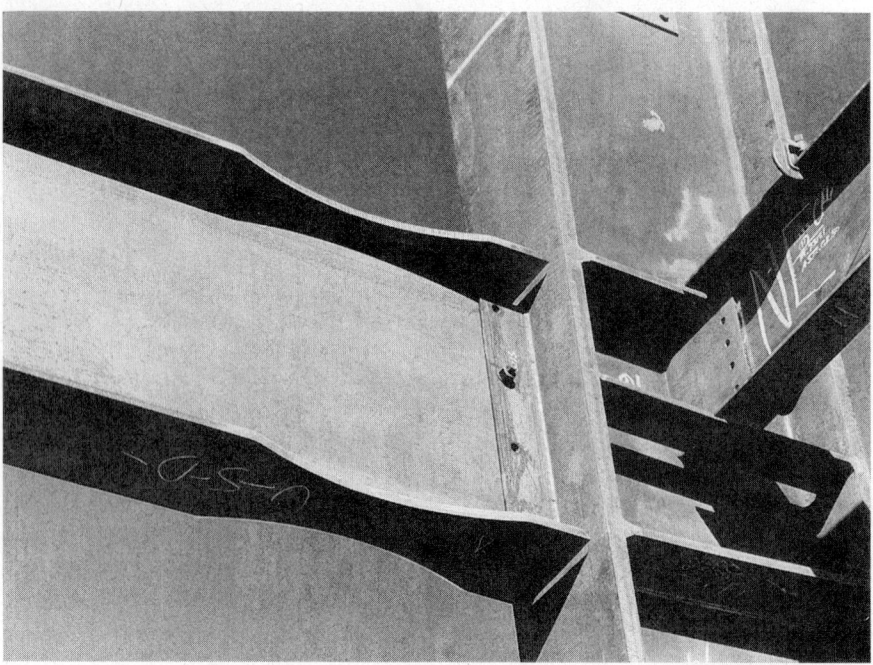

FIGURE 11.28
Photograph of a moment connection similar to the one shown in Figure 11.27. The beam has just been bolted to a *shear tab* that is welded to the column. Next, backup bars will be welded to the column just under the beam flanges, after which the flanges will be welded to the column.
(Permission of American Institute of Steel Construction.)

Stabilizing the Building Frame

To understand the roles of shear and moment connections in the building structure, it is necessary to understand the means by which large buildings are made stable against the lateral forces of wind and earthquake. Three types of stabilizing elements are used: braced frames, shear walls, and moment-resisting frames (Figure 11.29). A *braced frame* uses *diagonal bracing* to create stable triangular configurations within the otherwise naturally unstable rectilinear geometry of the frame. The connections between beams and columns within a braced frame need not transmit moments (bending forces); they can behave like pins or hinges, which is another way of saying that they can be shear connections such as the one in Figure 11.26. (Though it may not be readily apparent, this type of connection is capable of the small rotations necessary for it to behave essentially as if it were hinged.)

A special case of the braced frame is the *eccentrically braced frame* in which the ends of diagonal braces are offset some distance from each other where they connect to horizontal members (Figure 11.29). These connections introduce greater energy-absorbing capacity or resilience into the building frame, important for resistance to severe earthquake stresses. Like conventional braced frames, eccentrically braced frames can rely exclusively on shear connections between beams and columns. (Another way to add energy-absorbing capacity into a building frame is with the introduction of dedicated *damping* devices. One type of such damping, buckling restrained bracing, is shown in Figure 11.76.)

Shear walls are very stiff walls made of steel, concrete, or reinforced concrete masonry. They act in much the same way as the diagonal bracing within a braced frame structure and, like the braced frame, permit the use of shear connections between beams and columns in the remainder of the frame.

Braced Frame

Eccentrically Braced Frame

Moment-Resisting Frame

Shear Walls

FIGURE 11.29
Elevation views of the basic means for imparting lateral stability to a structural frame. Connections made with dots are shear connections, and solid intersections indicate moment connections. The braced frame (*top*) is illustrated here with *Chevron* (or *inverted V*) bracing. Crossbracing, in which paired diagonals run from opposite corners of the braced bay, are also common. Eccentric bracing (*second from top*) is one way to increase a building frame's capacity to absorb seismic energy. The connections in the moment-resisting frame (*third from top*), called "moment connections," are sufficiently resistant to rotation to stabilize the structure against lateral forces without bracing or shear walls. Moment-resisting frames are also sometimes called "rigid frames," but this name can be misleading. Although the connections in such frames are relatively rigid, whole frames that rely solely on such connections for their lateral stability are usually less stiff than those stabilized with bracing or shear walls. Shear walls (*bottom*) perform similarly to braced framing.

Moment-resisting frames have neither diagonal bracing nor shear walls. They rely instead on stronger and stiffer moment connections between beams and columns to provide lateral stability. Depending on the configuration of the structure and the magnitude of the forces involved, not all of the connections in the frame must necessarily be moment connections. Because these connections are more difficult to make, they are used only to the extent required, with the remainder of the frame relying on simpler, less costly shear connections.

In tall buildings, these stabilizing elements must be arranged in a manner that is structurally efficient as well as conducive to good building planning (Figure 11.30). The most common arrangement is the *rigid core*. The core, the central area containing elevators, stairs, mechanical chases, and washrooms, is structured as a stiff tower, using diagonal bracing

Rigid Core

Rigid Perimeter

FIGURE 11.30
Rigid core versus rigid perimeter.

or shear walls. (Moment connections are generally not efficient enough to stiffen a rigid core.) The remainder of the building frame may then be constructed with shear connections, relying on the core for its lateral stability. Where additional resistance to lateral forces or greater stiffness is required, beam-to-column moment connections may be introduced into some portions of the building frame as well. Figure 1.11 illustrates an example of a tall building structure stabilized by rigid concrete shear wall cores.

A second common arrangement of stabilizing elements in very tall buildings is the *rigid perimeter*. In this case, diagonal bracing, shear walls, or (less frequently) beam-to-column moment connections are incorporated into the outer walls of the building frame and the entire interior structure is assembled with simpler shear connections. Tall building structures relying on rigid perimeters, also sometimes called *tube structures*, can be more efficient than rigid core structures, since the stabilizing elements are spread further apart and act over a great area in plan.

In summary, shear connections between beams and columns are sufficient to transmit vertical loads through the building frame, but they are not, on their own, capable of providing adequate resistance to lateral forces. Lateral force resistance may be provided by the introduction of diagonal bracing, shear walls, beam-to-column moment connections, or some combination of these elements into the frame. Braced frames and shear walls tend to be more structurally efficient than moment frames and are usually the preferred choices for stabilizing structures that are very tall or otherwise subject to high wind or earthquake forces.

Shear Connections and Moment Connections

The American Institute of Steel Construction (AISC) defines three types of beam-to-column connections, classified according to their moment-resisting capability. *Fully-Restrained (FR) moment connections* are sufficiently rigid that the geometric angles between members will remain virtually unchanged under normal loading. *Partially-Restrained (PR) moment connections* are not as rigid as FR connections, but nonetheless possess a dependable and predictable moment-resisting capacity that can be used to stabilize a building frame. FR and PR moment connections are also sometimes referred to as rigid and semirigid connections, respectively. Both connection types can be used to construct moment-resisting building frames. *Simple connections*, otherwise known as shear connections, are considered to be capable of unrestrained rotation under normal loading conditions and to have negligible moment-resisting capacity. Buildings framed solely with simple connections must depend on diagonal bracing or shear walls for lateral stability.

A series of simple, fully bolted shear (AISC simple) connections are shown as a beginning basis for understanding steel connection details (beginning with Figure 11.24). These are interspersed with a corresponding series of welded moment (AISC Fully- and Partially-Restrained) connections (Figures 11.27, 11.28, 11.32, and 11.35). Welding is also widely used for making shear connections, two examples of which are shown in Figures 11.33 and 11.34. A series of column connections is illustrated in Figures 11.37, 11.38, 11.39, and 11.40.

In practice, there are a number of different ways of making any of these connections, using various kinds of connecting elements and different combinations of bolting and welding. The object is to choose a method of stabilization and designs for individual connections that will result in the greatest possible economy of construction for the building as a whole. For standard joint conditions in simple structures, the choice of which connection to use may be left to the fabricator, who has firsthand

FIGURE 11.31
A *seated* beam-to-column-web connection, location *B* on the frame in Figure 11.23. Although the beam flanges are connected to the column by a seat angle below and a stabilizing angle above, this is a shear (AISC simple) connection, not a moment connection, because the two bolts are incapable of developing the full strength of the beam flange. This seated connection is used rather than a framed connection, as illustrated in Figure 11.26, to connect to a column web because there is usually insufficient space between the column flanges to insert a power wrench to tighten all the bolts in a framed connection.

FIGURE 11.32
A welded beam-to-column-web moment (AISC Fully-Restrained) connection, used at location *B* on Figure 11.23 when a rigid connection is required. A vertical shear tab, welded to the web of the column at its centerline, serves to receive bolts that join the column to the beam web and hold the beam in place during welding. The horizontal stiffener plates that are welded inside the column flanges are thicker than the beam flanges and extend out beyond the column flanges to reduce concentrations of stress at the welds.

FIGURE 11.33
A single-tab shear (AISC simple frame) connection is an economical alternative to the connection shown in Figure 11.26 when the load on the connection is relatively light. A single connector plate is welded to the column in the shop, and the beam is bolted to it on the construction site.

FIGURE 11.34
Shear (AISC simple frame) connections may also be made entirely by welding. The angles are welded to the beam in the shop. Bolts through the angles hold the beam in place while it is welded to the column. The angles are not welded to the column along their top and bottom edges. This permits the angles to flex slightly to allow the beam to rotate away from the column as it bends.

FIGURE 11.35
A welded/bolted *end plate* beam–column connection. As shown, this is a semi-rigid (AISC Partially-Restrained) connection. With more bolts, this can become a rigid, AISC Fully-Restrained connection, and could be used to support a short cantilever beam such as the one at location *C* in Figure 11.23. The plate is welded to the end of the beam in the shop and bolted to the column on the building site.

FIGURE 11.36
A coped beam–girder shear (AISC simple) connection, used at location *D* in Figure 11.23. A girder is a beam that supports other beams. This connection may also be made with single tabs rather than angles if the load is not too great.
The top flanges of the beams are cut away (coped) so that the tops of the beams and the girder are all level with one another, ready to receive the floor or roof decking. Bending moments at the ends of a beam are normally so small that the flanges may be coped without compromising the strength of the beam.

FIGURE 11.37
A bolted column–column connection for columns that are the same size. The plates are bolted to the lower section of the column in the shop and to the upper section on the site. All column connections are made at waist height above the floor, location *E* in Figure 11.23.

FIGURE 11.38
Column sizes diminish as the building rises, requiring frequent use of shim plates at connections to make up for differences in flange thicknesses.

FIGURE 11.39
Column connections may be welded rather than bolted. The connector plate is welded to the lower column section in the fabricator's shop. The hole in the connector plate is used to attach a lifting line during erection. The bolts hold the column sections in alignment, while the flanges are connected in the field with partial-penetration welds in bevel grooves. Partial-penetration welding allows one column to rest on the other prior to welding.

FIGURE 11.40
A welded butt plate connection is used where a column changes from one nominal size of wide flange to another. The thick butt plate, which is welded to the lower column section in the shop, transfers the load from one section of column to the other. The partial-penetration weld at the base of the upper column is made on the site.

knowledge of the safest, most erectable methods that will utilize the company's labor and equipment most efficiently. For more complex structures or for unique joining conditions, the structural engineer may dictate a specific connection detail.

THE CONSTRUCTION PROCESS

A steel building frame begins as a rough sketch on the drafting board of an architect or engineer (Figure 11.41). As the building design process progresses, the sketch evolves through many stages of drawings and calculations to become a finished set of structural drawings. These show accurate column locations, the shapes and sizes of all the members of the frame, and all the loads of the members, but they do not give the exact length to which each member must be cut to mate with the members it joins, and they do not give details of the more routine connections of the frame. These are left to be worked out by a subsequent recipient of the drawings, the *fabricator.*

The Fabricator

The fabricator's job is to deliver to the construction site steel components that are ready to be assembled without further processing. This work begins with the preparation in the fabricator's shop of detailed drawings that show exactly how each piece will be made and what its precise dimensions will be. The fabricator designs connections to transmit the loads indicated by the engineer's drawings. Within the limits of accepted engineering practice, the fabricator is free to design the connections to be made as economically as possible, using various combinations of welding and bolting that best suit available equipment and expertise. The fabricator also prepares drawings to show the general contractor exactly where and how to install foundation anchor bolts to connect to the columns of the building and to guide the erector in assembling the steel frame on the building site. When completed, the fabricator's *shop drawings* are submitted to the engineer and the architect for review and approval to be sure that they conform exactly to the intentions of the design team. Meanwhile, the fabricator places an order with a producer of steel for the stock from which the structural steel members will be fabricated. (The major beams, girders, and columns are usually ordered cut to exact length by the mill.) When the approved shop drawings, with corrections and comments, are returned to the fabricator by the design team, revisions are made as necessary, and full-size templates of cardboard or wood are prepared as required to assist the shop workers in laying out the various connections on the actual pieces of steel.

Plates, angles, and tees for connections are brought into the shop and cut to size and shape with gas-fueled cutting torches, power shears, and saws. With the aid of the templates, bolt hole locations are marked. If the plates and angles are not unusually thick, the holes may be made rapidly and economically with a punching machine. In very thick stock, or in pieces that will not fit conveniently into the punching machines, holes are drilled rather than punched.

Pieces of steel stock for the beams, girders, and columns are brought into the fabricator's shop with an overhead traveling crane or conveyor system. Each piece is stenciled or

FIGURE 11.41
A typical framing plan for a multistory steel-framed building, showing size designations for beams and girders. Notice how this frame requires beam-to-column-flange connections where the W30 girders meet the columns, beam-to-column-web connections where the W27 beams meet the columns, and coped beam–girder connections where the W18 beams meet the W30 girders. The small squares in the middle of the building are openings for elevators, stairways, and mechanical shafts. An architect's or engineer's framing plan would also give dimensions between centers of columns, and would indicate the magnitudes of the loads that each joint must transfer, to enable the fabricator to design each connection.

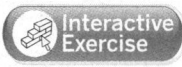

painted with a code that tells which building it is intended for and exactly where it will go in the building. With the aid of the shop drawings, each piece is measured and marked for its exact length and for the locations of all holes, stiffeners, connectors, and other details. Cutting to length, for those members not already cut to length at the mill, is done with a power saw or a flame cutting torch. The ends of column sections that must bear fully on the base plates or on one another are squared and are made perfectly flat by sawing, milling, or facing. In cases where the columns will be welded to one another, and for beams and girders that are to be welded, the ends of the flanges are beveled as necessary. Beam flanges are *coped* as required. Bolt holes are punched or drilled (Figure 11.42). *Plasma* (high-temperature ionized gas) *cutting* and *laser cutting* are also finding increased use in steel fabrication. Both of these types of cutting devices can be driven by machines that allow the fully automated cutting and shaping of parts from digitally prepared models.

Where called for, beams and girders are *cambered* (curved slightly in an upward direction) so that they will deflect into a straight line under load. Cambering may be accomplished by a hydraulic ram that bends the beam enough to force a permanent deformation. Steel shapes can also be bent to a smooth radius with a large machine that passes the shape through three rollers that flex it sufficiently to impart a permanent curvature (Figure 11.43). An older, much more costly means of cambering involves heating local areas of one flange of the member with a large oxyacetylene torch. As each area is heated to a cherry-red color, the metal softens, expands, and deforms to make a slight bulge in the width and thickness of the flange because the surrounding steel, which is cool, prevents the heated flange from lengthening. As the heated flange cools, the metal contracts, pulling

FIGURE 11.42
Punching bolt holes in a wide-flange beam. *(Courtesy of W. A. Whitney Corp., An Esterline Company.)*

FIGURE 11.43
Rectangular hollow structural sections for this frame were bent into curves by the fabricator. Wide-flange shapes can also be bent. *(Architect & Photographer: Eliot W. Goldstein, AIA, The Goldstein Partnership.)*

FIGURE 11.44
Welders attach connector plates to an exceptionally heavy column section in a fabricator's shop. The twin channels bolted to the end of the column will be used to attach a lifting line for erection, after which they will be removed and reused. *(Courtesy of U.S. Steel Corp.)*

FIGURE 11.45
Machine-welding plates together to form a box column. The torches to the left preheat the metal to help avoid thermal distortions in the column. Mounds of powdered flux around the electrodes in the center indicate that this is the submerged arc process of welding. The small steel plates tacked onto the corners of the column at the extreme left are runoff bars, which are used to allow the welding machine to go past the end of the column to make a complete weld. These will be cut off as soon as welding is complete. *(Courtesy of U.S. Steel Corp.)*

the member into a slight bend at that point. By repeating this process at several points along the beam, a camber of the desired shape and magnitude is produced.

As a last step in fabricating beams, girders, and columns, stiffener plates are arc-welded to each piece as required, and connecting plates, angles, and tees are welded or bolted at the appropriate locations (Figure 11.44). As much connecting as possible is done in the shop, where tools are handy and access is easy. This saves time and money during erection, when tools and working conditions are less optimal and total costs per worker-hour are higher.

Plate girders, built-up columns, trusses, and other large components are assembled in the shop in units as large as can practically be transported to the construction site, whether by truck, railway, or barge. Intricate assemblies such as large trusses are usually preassembled in their entirety in the shop, to be sure that they will go together smoothly in the field, then broken down again into transportable components.

As the members are completed, each is straightened, cleaned, and prime painted as necessary, and inspected for quality and for conformance to the job specifications and shop drawings. The members are then taken from the shop to the fabricator's yard by crane, conveyor, trolley, or forklift, where they are organized in stacks according to the order in which they will be needed on the building site.

As an alternative to the traditional shop drawing process just described, a structural engineer may use three-dimensional modeling software to design the steel connections and supply digital data to the fabricator to automatically guide the fabricator's equipment. While this method requires the engineer to assume more responsibility for the final design of the steel connection details, it also can shorten the time required for steel to arrive on site and can

improve the coordination between the structural steel and other building systems.

The Erector

Where the fabricator's job ends, the *erector's* job begins. Some companies both fabricate and erect, but more often the two operations are done by separate companies. The erector is responsible for assembling into a frame on the building site the steel components furnished by the fabricator. The erector's workers, by tradition, are called *ironworkers*.

Erecting the First Tier

Erection of a multistory steel building frame starts with assembly of the first two-story *tier* of framing. Lifting of the steel components may begin with either some type of mobile crane or (on larger projects) fixed tower cranes (Figure 11.48). In accordance with the erection drawings prepared by the fabricator, the columns for the first tier, usually furnished in sections two stories high, are picked up from organized piles on the site or directly from the trucks they are delivered on and lowered carefully over the anchor bolts and onto the foundation, where the ironworkers bolt them down.

There are 175,000 ironworkers in this country . . . and apart from our silhouettes ant-size atop a new bridge or skyscraper, we are pretty much invisible.

—Mike Cherry, *On High Steel: The Education of an Ironworker*, New York, Quadrangle/The New York Times Book Co., 1974, p. xiii

---

Foundation details for steel columns vary (Figure 11.46). Steel *baseplates*, which distribute the concentrated loads of the steel columns across a larger area of the concrete

foundation, are shop-welded to all but the largest columns. The foundations and anchor bolts were put in place previously by the general contractor, following the plan prepared by the fabricator. The contractor may, if requested, provide thin steel *leveling plates* that are set perfectly level at the proper height on a bed of *grout* atop each concrete foundation. The baseplate of the column rests upon the leveling plate and is held down with the protruding anchor bolts. Alternatively, especially for larger baseplates with four anchor bolts, the leveling plate is omitted. The column is supported at the proper elevation on stacks of steel shims inserted between the baseplate and the foundation, or on leveling nuts placed beneath the baseplate on the anchor bolts. After the first tier of framing is plumbed up (as described later in this section), the baseplates are grouted and the anchor bolts tightened. For very large, heavy columns, baseplates are shipped independently of the columns (Figure 11.47). Each is leveled in place with shims, wedges, or shop-attached leveling screws, then grouted prior to column placement.

After the first tier of columns has been erected, the beams and girders for the first two stories are bolted in place (Figures 11.49 through 11.54). First, a *raising gang*, working with a crane, positions the components and inserts enough bolts to hold them together temporarily. A gang of bolters follows behind, inserting bolts

FIGURE 11.46
Three typical column base details.
Upper left: A small column with a welded baseplate set on a steel leveling plate.
Upper right: A larger column with a welded baseplate set on leveling nuts.
Below: A heavy column field-welded to a loose baseplate that has been previously leveled and grouted.

A thin steel plate is leveled on a bed of grout prior to erection of the column

Leveling nuts on the anchor bolts support the baseplate and column before grouting

For very large columns, the heavy baseplates are installed separately in advance of the columns

A pair of angles and two bolts support the column before it is welded to the baseplate

Holes through the baseplate may be provided on each side of the column as a way of introducing grout under the middle of the baseplate

Three leveling screws support the plate before grouting

in all the holes and partially tightening them. The two-story tier of framing is then *plumbed up* (straightened and squared) with diagonal cables and turnbuckles while checking the alignment with plumb bobs, transits, or laser levels. When the tier is plumb, connections are tightened, baseplates are grouted if necessary, welds are made, and permanent diagonal braces, if called for, are rigidly attached. Ironworkers scramble back and forth, up and down on the columns and beams, protected from falling by safety harnesses that are connected to steel cable lifelines (Figure 11.49).

At the top of each tier, a temporary working surface of 2- or 3-inch (50- or 75-mm) wood planks or corrugated steel decking may be laid over the steel framing. Similar platforms will be placed every second story as the frame rises, unless safety nets are used instead or the permanent floor decking is installed as erection progresses. The platforms protect workers on lower levels of the building from falling objects. They also furnish a convenient working surface for tools, materials, and derricks. Column splices are made at waist level above this platform, both as a matter of convenience and as a way of avoiding conflict between the column splices and the beam-to-column connections. Columns are generally fabricated in two-story

lengths, a transportable size that also corresponds to the two-story spacing of the plank surfaces.

Erecting the Upper Tiers

Erection of the second tier proceeds much like that of the first. Two-story column sections are hoisted into position and connected by splice plates to the first tier of columns. The beams and columns for the two floors are set, the tier is plumbed and tightened

up, and another layer of planks, decking, or safety netting is installed.

If the building is not too tall, the mobile crane will do the lifting for the entire building. For a taller building, tower cranes are used (Figure 11.48). The tower crane builds itself an independent tower as the building rises, either alongside the building or within an elevator shaft or a vertical space temporarily left open in the frame.

FIGURE 11.47
Ironworkers guide the placement of a very heavy column fabricated by welding together two rolled wide-flange sections and two thick steel plates. It will be bolted to its baseplate through the holes in the small plate welded between the flanges on either side. (*Courtesy of Bethlehem Steel Corporation.*)

*Tower Crane With
Hammerhead Boom*

Mobile Crane

*Tower Crane With
Luffing Boom*

FIGURE 11.48

Two common types of *tower cranes* and a mobile crane. The *luffing-boom crane* is useful in congested situations where the wide arc of the *hammerhead boom crane* could be limited by obstructions. Tower cranes can either be supported on their own foundations, or, in the case of internal cranes, they may also be supported by the frame of the building. With very tall buildings, tower cranes climb as the building rises by means of self-contained hydraulic jacks.

FIGURE 11.49

An ironworker clips his body harness to a safety line as he moves around a column. *(Photograph by James Digby. Courtesy of L.P.R. Construction Company.)*

FIGURE 11.50
A tower crane lowers a series of beams to ironworkers. The worker to the left uses a tagline to maneuver the beam into the proper orientation. *(Photograph by James Digby. Courtesy of L.P.R. Construction Company.)*

As each piece of steel is lowered toward its final position in the frame, it is guided by an ironworker who holds a rope called a *tagline*, the other end of which is attached to the piece. Other ironworkers in the raising gang guide the piece by hand as soon as they can reach it, until its bolt holes align with those in the mating pieces (Figure 11.52). Sometimes crowbars or hammers must be used to pry, wedge, or drive components until they fit properly, and bolt holes may, on occasion, have to be reamed larger to admit bolts through slightly misaligned pieces. When an approximate alignment has been achieved, tapered steel *drift pins* from the ironworker's tool belt are shoved into enough bolt holes to hold the pieces together until a few bolts can be inserted. The bolters follow behind the raising gang, filling the remaining holes with bolts from leather carrying baskets and tightening them first with hand wrenches and then with impact wrenches. Field-welded connections are initially held in alignment with bolts, then welded when the frame is plumb.

The last beam is placed at the top of the building with a degree of ceremony appropriate to the magnitude of the building. At the very least, a small evergreen tree, a national flag, or both are attached to the beam before it is lifted (Figures 11.55 and 11.56). For major buildings, assorted dignitaries are likely to be invited to a building-site *topping-out* party that includes music and refreshments. After the party, work goes on as usual, for although the frame is complete, the building is not. Roofing, cladding, and finishing operations will continue for many months.

FIGURE 11.51
Connecting a beam to a column. *(Courtesy of Bethlehem Steel Corporation.)*

FIGURE 11.52
Ironworkers attach a girder to a box column. Each worker carries two combination wrench–drift pin tools in a holster on his belt and inserts the tapered drift pins into each connection to hold it until a few bolts can be added. Bundles of corrugated steel decking are ready to be opened and distributed over the beams to make a floor deck. *(Courtesy of Bethlehem Steel Corporation.)*

FIGURE 11.53
Bolting heavy joist girders to a column.
(Courtesy Vulcraft Division of Nucor.)

FIGURE 11.54
Welding open-web steel joists to a wide-flange beam. *(Courtesy Vulcraft Division of Nucor.)*

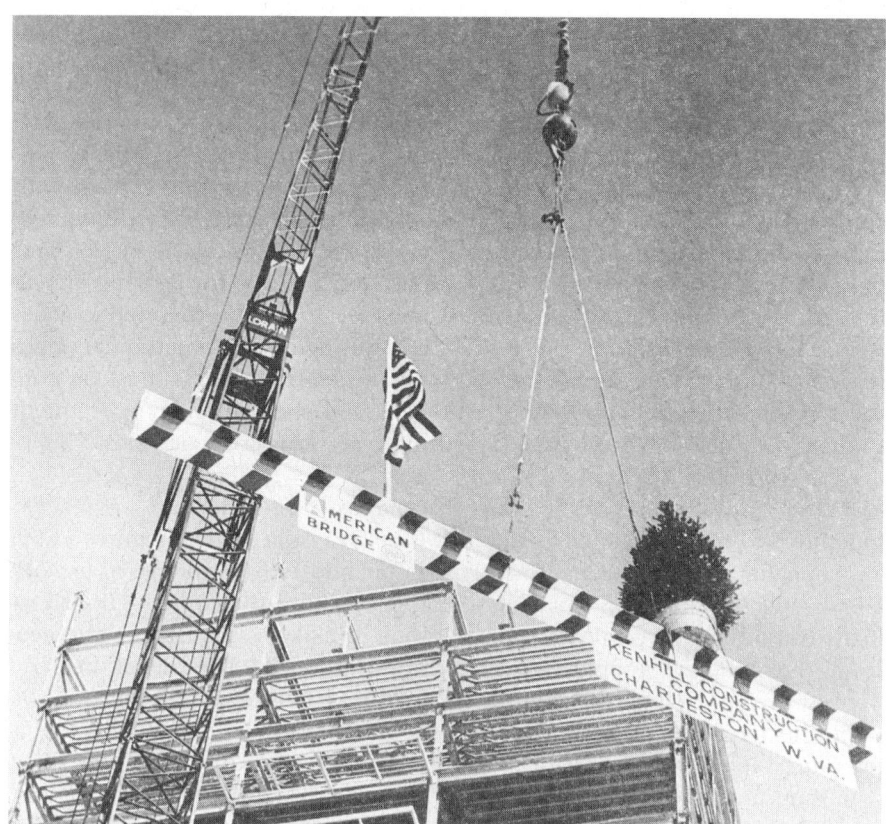

FIGURE 11.55
Topping out: The last beam in a steel frame is special. *(Courtesy of U.S. Steel Corp.)*

FIGURE 11.56
A 10-story steel frame nears completion. The lower floors have already been decked with corrugated steel decking. *(Courtesy of Vulcraft Division of Nucor.)*

If nobody plumbed-up, all the tall buildings in our cities would lean crazily into each other, their elevators would scrape and bang against the shaft walls, and the glaziers would have to redress all the windowglass into parallelograms. . . . What leaned an inch west on the thirty-second floor is sucked back east on the thirty-fourth, and a column that refused to quit leaning south on forty-six can generally be brought over on forty-eight, and by the time the whole job is up the top is directly over the bottom.

—Mike Cherry, *On High Steel: The Education of an Ironworker*, New York, Quadrangle/The New York Times Book Co., 1974, pp. 110–111

Floor and Roof Decking

If plank decks are used during erection of the frame, they must be replaced with permanent floor and roof decks of incombustible materials. In early steel frame buildings, shallow arches of brick or tile were often built between the beams, tied with steel tension rods, and filled over with concrete to produce level surfaces (Figure 11.57). These were heavier than the metal deck systems commonly used today and required larger framing members to carry their weight. They were also much more labor intensive.

Metal Decking

Metal decking at its simplest is a thin sheet of steel that has been corrugated to increase its stiffness. The spanning capability of the deck is determined mainly by the thickness of the sheet from which it is made and the depth and spacing of the corrugations. It also depends on whether the decking sheets are single or cellular. Single corrugated sheets are commonly used for *roof decking*, where concentrated loads are not expected to be great and deflection criteria are not as stringent as in floors. *Form deck* is used as permanent formwork for concrete floor decks, with the reinforced concrete slab supported by the steel decking until the slab can support itself and its live loads (Figures 11.58, 11.59, and 11.60). *Cellular decking* is manufactured by welding together two sheets, one corrugated and one flat. It can be made sufficiently stiff to support normal floor loads without structural assistance from the concrete topping that is poured over it. Cellular decking can offer the important side benefit of providing spaces for electrical and communications wiring, as illustrated in Chapter 24. *Vented metal decking* is manufactured with small, slotted openings comprising roughly 1 percent of the overall decking area. It is sometimes used where impermeable roofing membranes are installed over lightweight insulating concrete fills to allow excess moisture in the concrete to escape downward through the decking, rather than become trapped under the roofing membrane.

Metal decking is usually *puddle welded* to the joists, beams, and girders by melting through the decking to the supporting members below with a welding electrode. Self-drilling, self-tapping screws or powder-driven pins may also be used for decking attachment. If the deck is required to act as a diaphragm, the longitudinal edges of the decking panels must be connected to one another at frequent intervals with screws or welds.

Composite Construction

Composite metal decking (Figures 11.59, 11.60, and 11.61) is designed to work together with the concrete floor topping to make a stiff, lightweight, economical deck. The metal decking serves as tensile reinforcing for the concrete, to which it bonds by means of special rib patterns in the sheet metal or by small steel rods or wire fabric welded to the tops of the corrugations.

Composite construction is often carried a step beyond the decking to include the beams of the floor. Before the concrete is poured over the metal deck, *shear studs* are welded every few inches to the top of each beam, using a special

A concrete fill over the arches creates a level floor

Steel tie rods through the beam webs resist the thrust of the arches

Special bricks or tiles protect the beam flange against fire

FIGURE 11.57
Tile or brick arch flooring is found in many older steel frame buildings.

FIGURE 11.58
Three types of corrugated metal decking can be seen in this image. The roof decking above has smaller, more closely spaced corrugations than the composite floor decking on the level below. A single sheet of decking with even deeper, more widely spaced corrugations has been laid across the framing in the foreground, but not yet fastened in place. *(Photo by Joseph Iano.)*

Form Deck   *Composite Deck*   *Roof Deck*

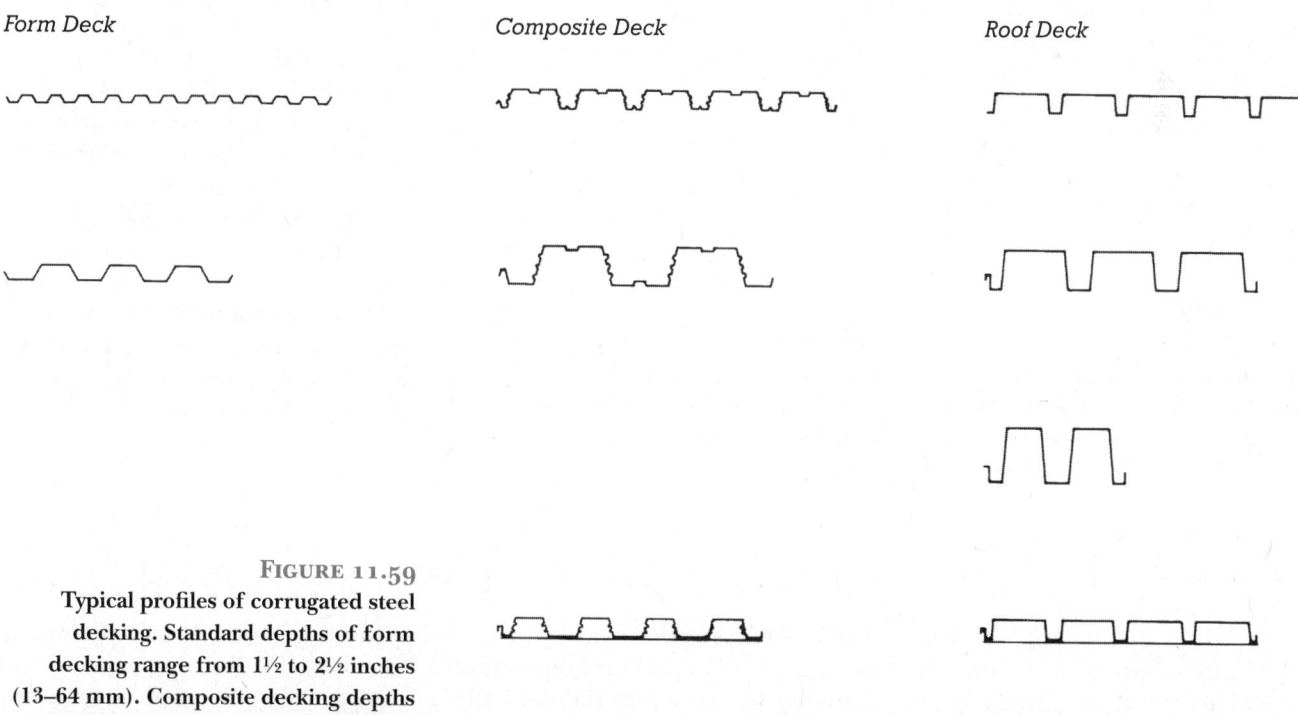

FIGURE 11.59
Typical profiles of corrugated steel decking. Standard depths of form decking range from 1½ to 2½ inches (13–64 mm). Composite decking depths lie between 1½ and 3 inches (38 and 76 mm). Roof decking is available in depths of 1½ to 7 inches (38–178 mm). The bottom two examples of composite and roof decking are cellular.

FIGURE 11.60
Samples of corrugated steel decking. The second sample from the bottom achieves composite action by keying of the concrete topping to the deformations in the decking. The bottom sample has a closed end, which is used at the perimeter of the building to prevent the concrete topping from escaping during pouring. *(Courtesy of Wheeling Corrugating Company Division, Wheeling-Pittsburgh Steel Corp.)*

FIGURE 11.61
Composite decking acts as steel reinforcing for the concrete topping installed over it. The top example bonds to the concrete with deformed ribs and the middle example with welded steel rods. The bottom type makes an attractive ceiling texture if left exposed and furnishes dovetail channels for the insertion of special fastening devices to hang ductwork, piping, conduits, and machinery from the ceiling.

electric welding gun (Figures 11.62 and 11.63). It would be more economical to attach the shear studs in the shop rather than in the field, but the danger that ironworkers would trip on the studs during the erection process delays their installation until the steel decking is in place. The purpose of the studs is to create a strong shear connection between the concrete slab and the steel beam. A strip of the slab can then be assumed to act together with the top flange of the steel shape to resist compressive forces. The result of the composite action of the two materials is a steel member whose loadbearing capacity has been greatly enhanced at relatively low cost by taking advantage of the unused strength of the concrete topping that must be present in the construction anyway. The payoff is a stiffer, lighter, less expensive frame.

FIGURE 11.62
Composite beam construction.

Shear studs connect the concrete deck to the steel beam

FIGURE 11.63
Pouring a concrete fill on a steel roof deck, using a concrete pump to deliver the concrete from the street below to the point of the pour. Shear studs are plainly visible over the lines of the beams below. The welded wire reinforcing strengthens the concrete against cracking. *(Courtesy of Schwing America, Inc.)*

Concrete Decks

Concrete floor and roof slabs are often used in steel building frames instead of metal decking and concrete fill. Concrete may be poured in place over removable plywood forms, or it may be erected in the form of precast concrete planks lifted into place much like the metallic elements of the building (Figure 11.64). Precast concrete decks are relatively light in weight and quick to erect, even under weather conditions that would delay the pouring of concrete. Usually, a thin, poured-in-place concrete topping is added later during construction, to produce a smooth floor.

Roof Decking

For roofs of low steel-framed buildings, many different types of decks are available. Corrugated metal may be used, with or without a concrete fill; many types of rigid insulation boards are capable of spanning the corrugations to make a flat surface for the roof membrane. Some corrugated decks are finished with a weather-resistant coating that allows them to serve as the water-resistive surface of the roof. Many different kinds of insulating deck boards are produced of wood fibers or glass fibers bonded with portland cement, gypsum, or organic binders. Many of these insulating boards are designed as permanent formwork for reinforced, poured slabs of gypsum or lightweight concrete. In this application, they are

FIGURE 11.64
A tower crane installs precast concrete hollow-core planks for floor decks in an apartment building. Precast concrete is also used for exterior cladding of the building. The steel framing is a design known as the staggered truss system, in which floors are supported by full story-height steel trusses at alternate levels of the building. The trusses are later enclosed with interior partitions. *(Courtesy of Blakeslee Prestress, Inc.)*

usually supported on steel *subpurlins* (Figures 11.65 and 11.66). Heavy timber decking, or even wood joists and plywood sheathing, are also used over steel framing in situations where building codes permit combustible materials.

Corrugated steel sheets are also often used for siding of industrial buildings, where they are supported on *girts*, which are horizontal Z-shapes or channels that span between the outside columns of the building (see Figure 11.80).

Architectural Structural Steel

Where structural steel members will remain exposed in the finished building and a high standard of appearance quality is desired, steel may be specified as *architecturally exposed structural steel (AESS)*. AESS specifications may include special requirements for dressing and finishing of welds, closer tolerances in connections between members, removal of marks made on the steel during fabrication, application of high-quality finishes, and other considerations (Figure 11.67).

FIGURE 11.65
Welding truss-tee subpurlins to open-web steel joists for a roof deck. *(Courtesy of Keystone Steel & Wire Co..)*

FIGURE 11.66
Installing insulating formboard over truss-tee subpurlins. A light wire reinforcing mesh and poured gypsum fill will be installed over the formboards. The trussed top edges of the subpurlins become embedded in the gypsum slab to form a composite deck. *(Courtesy of Keystone Steel & Wire Co..)*

FIGURE 11.67
Architectural steel tree-like columns, fabricated from HSS round members, support curved rolled structural sections in this rail transportation station. *(Photo by Joe Iano.)*

FIRE PROTECTION OF STEEL FRAMING

Normal building fires are not hot enough to melt steel, but are often able to weaken it sufficiently to cause structural failure (Figures 11.68 and 11.69). For this reason, building codes generally limit the use of exposed steel framing to buildings of one to five stories, where escape in case of fire is rapid. For taller buildings, it is necessary to protect the steel frame from heat long enough for the building to be fully evacuated and the fire extinguished or allowed to burn out on its own.

Fireproofing ("fire protection" being a more accurate term) of steel framing was originally done by encasing steel beams and columns in brick masonry (Figure 11.57) or poured concrete (Figure 11.70A). These heavy encasements were effective, absorbing heat into their great mass and dissipating some of it through dehydration of the mortar and concrete, but their weight added considerably to the load that the steel frame

FIGURE 11.68
An exposed steel structure following a prolonged fire in the highly combustible contents of a warehouse. *(Courtesy of National Fire Protection Association, Quincy, MA.)*

Yield Strength of ASTM A 36 Structural Steel at Elevated Temperatures

FIGURE 11.69
The relationship between temperature and strength in structural steel. *(Courtesy of American Iron & Steel Institute.)*

had to bear. This added, in turn, to the weight and cost of the frame. The search for lighter-weight fireproofing led first to thin enclosures of metal lath and plaster around the steel members. These derive their effectiveness from the large amounts of heat needed to dehydrate the water of crystallization from the gypsum plaster. Plasters based on lightweight aggregates such as vermiculite instead of sand have come into use to further reduce the weight and increase the thermal insulating properties of the plaster (Figures 11.70*B*, 11.71*B* and *E*, and 11.72).

Today's designers can also choose from a group of fireproofing techniques that are lighter still. Plaster fireproofing has largely been replaced by beam and column enclosures made of boards or slabs of gypsum or other fire-resistive materials. These are fastened mechanically around the steel shapes, and in the case of gypsum board fireproofing, they can also serve as the finished surface on the interior of the building (Figures 11.70*C*, 11.71*C*, 11.73, and 11.74).

Where the fireproofing material need not serve as a finished surface, *spray-applied fire-resistive materials*

A

B

C

D

E

F

FIGURE 11.70
Some methods for fireproofing steel columns. (*A*) Encasement in reinforced concrete.
(*B*) Enclosure in metal lath and plaster. (*C*) Enclosure in multiple layers of gypsum
board. (*D*) Spray-on fireproofing. (*E*) Loose insulating fill inside a sheet metal
enclosure. (*F*) Water-filled box column made of a wide-flange shape with added steel
plates.

(*SFRM*), commonly referred to as
spray-applied fireproofing, have become
the most prevalent type. These
generally consist of either a fiber and
a binder or a cementitious mixture,
and are sprayed over the steel to the
required thickness (Figures 11.70*D*,
11.71*D*, 11.75, and 11.76). These
products are available in weights of

about 12 to 40 pounds per cubic foot
(190–640 kg/m³). The lighter materials are fragile and must be covered
with finish materials. The denser
materials are generally more durable.
All spray-applied materials act primarily by insulating the steel from
high temperatures for long periods
of time. They are usually the least

expensive form of steel fireproofing. Spray-applied fireproofing is
most commonly applied in the field
after the steel has been erected and
the connections between members
completed. Less frequently, it may
be applied in the fabrication shop,
where controlled environmental conditions and easier access to the steel

FIGURE 11.71
Some methods for fireproofing steel beams and girders. (*A*) Encasement in reinforced concrete. (*B*) Enclosure in metal lath and plaster. (*C*) Rigid slab fireproofing. (*D*) Spray-on fireproofing. (*E*) Suspended plaster ceiling. (*F*) Flame-shielded exterior spandrel girder with spray-on fireproofing inside.

members can result in faster application and more consistent quality.

When terrorists flew airliners into Manhattan's World Trade Center towers in 2001, the fireproofing on steel structural components of the buildings was dislodged by the impact of the airplanes. This left these structures vulnerable to the heat of the fires that followed the crashes and is believed to be a primary cause of the eventual collapse of both buildings. In response to this failure, the International Building Code now requires higher bond strengths for spray-applied fireproofing used in buildings with occupied floors greater than 75 feet (23 m) above the level of firefighter access.

Intumescent mastics and *intumescent paints* are thin coatings that allow steel structural elements to remain exposed to view in situations of low to moderate fire risk. They expand when exposed to fire to form a thick, stable char that insulates the steel from the heat of the fire for varying lengths of time, depending on the thickness of the coating. Most intumescent coatings are available in an assortment of colors. They can also serve as a base coat under ordinary paints if another color is desired.

FIGURE 11.72
Lath-and-plaster fireproofing around a steel beam. *(Courtesy of United States Gypsum Company.)*

FIGURE 11.73
Gypsum board fireproofing around a steel column. The gypsum board layers are screwed to the four cold-formed steel C-channels at the corners of the column, and finished with steel corner bead and drywall compound on the corners. *(Courtesy of United States Gypsum Company.)*

FIGURE 11.74
Attaching slab fireproofing made of mineral fiber to a steel column, using welded attachments.
(Courtesy of United States Gypsum Company.)

FIGURE 11.75
Applying spray-on fireproofing to a steel beam, using a gauge to measure the depth. *(Courtesy of W. R. Grace & Co.)*

FIGURE 11.76
In this photograph, spray-applied fireproofing appears as the light-colored, lumpy coating on steel members. Note that the diagonal bracing remains unprotected. Even in a building of Type I-A construction with the highest possible fire-protection requirements, structural members that serve only to resist lateral forces (but not gravity loads) are not required to be fireproofed. The braces in this photograph are *buckling-restrained braces*, designed to provide both lateral stability and structural damping. Within the brace, a relatively slender steel core is encased in grout and surrounded by the square steel casings visible in the photograph. In an extreme seismic event, the inner core experiences repeated plastic elongation and compression, absorbing dynamic energy imparted into the building structure and lessening the structure's lateral movements. Other types of damping devices rely on the controlled flow of viscous fluids within hydraulic cylinders, friction between tightly squeezed steel plates, and other clever mechanisms. *(Photo by Joseph Iano.)*

A rather specialized technique, applicable only to steel box or tube columns exposed on the exterior of buildings, is to fill the columns with water and antifreeze (Figure 11.70*F*). Heat applied to a region of a column by a fire is dissipated throughout the column by convection in the liquid filling.

Mathematical and computer-based techniques allow the calculation of temperatures that will be reached by steel members in various situations during a fire. These allow the designer to experiment with a variety of ways of protecting the members, for example, metal flame shields that allow the component to be left exposed on the exterior of a building (Figure 11.71*F*).

LONGER SPANS AND HIGH-CAPACITY COLUMNS IN STEEL

Standard wide-flange beams are suitable for the range of structural spans normally encountered in offices, schools, hospitals, apartment buildings, hotels, retail stores, warehouses, and other buildings in which columns may be brought to earth at intervals without obstructing the activities that take place within. For many other types of buildings—athletic buildings, certain types of industrial buildings, aircraft hangars, auditoriums, theaters, religious buildings, transportation terminals—longer spans are required than can be accomplished with wide-flange beams. A rich assortment of longer-span structural devices is available in steel for these uses.

Improved Beams

The *castellated beam* (Figures 11.77 and 11.78) is produced by flame-cutting the web of a wide-flange section along a zigzag or arced path, then reassembling the beam by

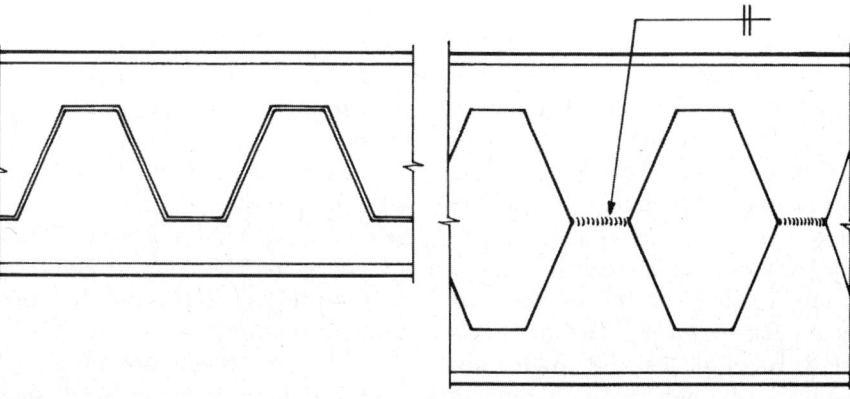

FIGURE 11.77
Manufacture of a castellated beam.

FIGURE 11.78
A long-span floor system is framed with castellated beams in this four-story office building. *(Photo by Joseph Iano.)*

a rectangular space (Figure 11.80) or arrayed around a vertical axis to cover a circular area. Their structural action lies midway between that of a rectilinear frame and that of an arch. Like an arch, they may sometimes require steel tie rods at the base to resist lateral thrust, in which case these rods are usually concealed within the floor slab.

Castellated beams, plate girders, and rigid steel frames share the characteristic that because they are long, slender elements, they frequently must be braced laterally by purlins, girts, decking, or diagonal bracing to prevent them from buckling.

Trusses

Steel trusses are triangular arrangements of steel members that are generally deeper and lighter than improved beams and can span correspondingly longer distances. They can be designed to carry light or heavy loads. Earlier in this chapter, open-web joists and joist girders were presented. These are standardized trusses made from light members. They can span further than conventional steel framing systems, while still remaining less expensive than longer-spanning custom-made trusses.

Custom-designed and fabricated roof trusses for light loadings are traditionally made up of steel tee or paired-angle top and bottom *chords* with paired-angle internal members. The angles within each pair are spaced just far enough apart to leave space between for the steel gusset plate connectors that join them to the other members of the truss. They may be either welded or bolted to

welding its two halves point to point, thus increasing its depth without increasing its weight. This greatly augments the spanning potential of the beam, provided that the super-imposed loads are not exceptionally heavy. The ubiquitous openings in castellated beams can also facilitate the routing of various building services within the floor/ceiling space.

For long-span beams tailored to any loading condition, *plate girders* are custom-designed and fabricated. Steel plates and angles are assembled by bolting or welding in such a way as to put the steel exactly where it is needed: The flanges are

often made thicker in the middle of the span where bending forces are higher, more web stiffeners are provided near the ends where web stresses are high, and areas around the supports are specially reinforced. Almost any depth can be manufactured as needed, and very long spans are possible, even under heavy loads (Figure 11.79). These members are often tapered, having greater depth where the bending moment is largest.

Rigid steel frames are efficiently produced by welding together steel wide-flange sections or plate girders. They may be set up in a row to roof

FIGURE 11.79
Erecting a welded steel plate girder.
Notice how the girder is custom made
with cutouts for the passage of pipes and
ductwork. The section being erected is
115 feet (35 m) long, 13 feet (4 m) deep,
and weighs 192,000 pounds (87,000 kg).
(Courtesy of Bethlehem Steel Corporation.)

FIGURE 11.80
The steel rigid frames of this industrial building carry steel purlins that will support
the roof deck and girts to support the wall cladding. The depth of each frame varies
with the magnitude of the bending forces and is greatest at the eave connections,
where these forces are at a maximum. *(Courtesy of Metal Building Manufacturers
Association.)*

the gusset plates (Figures 11.81 and 11.82).

Trusses for heavier loadings, such as for very-long-span roofs or the lateral transfer of column loads in a tall building where columns from floors above are interrupted by a large, open space lower in the building, can be made from virtually any type of larger section structural members (Figure 11.83). Where long-span trusses remain exposed in the finished building, their design and detailing can make a significant contribution to the architecture of the space (Figure 11.84).

FIGURE 11.82

Architect Suzane Reatig structured the roof of a Washington, D.C., church with trusses made of steel angles. The ribs of the roof decking add a strong texture to the ceiling. *(Photo by Robert Lautman. From* Detailing for Steel Construction, *Chicago, AISC, 1983. By permission of American Institute of Steel Construction.)*

TRUSS - 110T3

FIGURE 11.81

A fabricator's shop drawing of a welded steel roof truss made of tees and paired-angle diagonals. *(From* Detailing for Steel Construction, *Chicago, AISC, 1983. By permission of American Institute of Steel Construction.)*

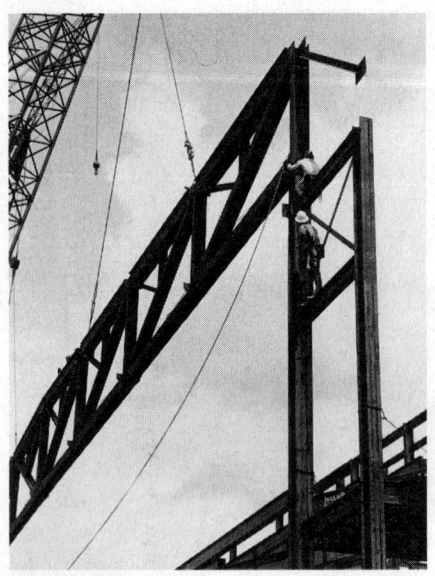

FIGURE 11.83
Ironworkers seat the end of a heavy roof truss made of wide-flange sections. *(Permission of American Institute of Steel Construction.)*

The *staggered truss system* (Figure 11.64) relies on story-high steel trusses, staggered on alternating floors of a building structure, and spanning between perimeter columns, to support the floors and roof. Typically, the trusses are concealed within partitions, leaving the remaining floor areas column free. The depth of the floor structure in a staggered truss building is only as much as the thickness of the decking system itself, resulting in overall floor-to-floor heights of as little as 8 ft 8 in.

(2.6 m). An additional advantage is that the ground-floor level can be made completely free of columns or other interior structural members. (The first floor above grade is supported by trusses spanning between it and the next level above grade.)

A steel *space truss* (more popularly called a *space frame*) is a truss made three dimensional (Figures 11.85 and 11.86). It carries its load by bending along both of its axes, much like a two-way concrete slab (Chapter 13). It must be supported by columns that are spaced more or less equally in both directions.

Arches

Short-span steel *arches* can be made by bending standard wide-flange shapes or by joining sections of straight members. When the truss form is extruded or rotated, vaults and domes can also be made. Heavier arches are one of the longest-span structural forms available to the building designer (Figure 11.87). Lateral thrusts are

FIGURE 11.84
Steel lenticular trusses supporting the roof of the Shanghai Pudong Airport passenger terminal. The truss upper chords are concealed above the finish ceiling. The steel cable lower chords and struts connecting the top and bottom chords create a striking visual pattern within the space. *(Photo by Joseph Iano.)*

FIGURE 11.85
Assembling a space truss. *(Courtesy of Unistrut Space-Frame Systems, GTE Products Corp.)*

FIGURE 11.86
A space truss carries the roof of a ferry terminal. *(Architects: Braccia/DeBrer/Heglund. Structural engineer: Kaiser Engineers. Photo by Barbeau Engh. Permission of American Institute of Steel Construction.)*

FIGURE 11.87

A steel roof arch at Century Link Field, Seattle, spans more than 700 ft (215 m). The thrust of the arch is resisted by high-tensile-strength cables that are just visible within the roof trusswork underneath the arch. *(Photo by Joseph Iano.)*

produced at the base of any arch and must be resisted by the foundations, or with a tie rod or other structural member that connects opposite ends of the arch. One of the longest single-span roof structures in the world is the retractable roof for Cowboys Stadium in Arlington, Texas. The roof structure is supported on a pair of steel box trusses, spanning 1225 ft (373 m), each 35 feet (11 m) deep and weighing 3255 tons (2955 metric tons).

Tensile Structures

Fabrics and cables made of cold-drawn, high-tensile-strength steel wire are the material for a fascinating variety of tentlike roofs that can span from short to very long distances. With *anticlastic* (saddle-shaped) *curvature, cable stays,* or other means of restraining the cable net, hanging roofs are fully rigid against wind uplift and flutter. For smaller spans, fabrics can do most of the work, supported by steel cables along the edges and at points of maximum stress, as presented in the accompanying sidebar (Figure 11.88).

The spider web is a good inspiration for steel construction.

Frank Lloyd Wright, "In the Cause of Architecture: The Logic of the Plan," *Architectural Record* **(January 1928)**

FIGURE 11.88

The Olympic Stadium roof in Munich, Germany, is made of steel cables and transparent acrylic plastic panels. For scale, notice the worker seen through the roof at the upper left. (*Architects: Frei Otto, Ewald Bubner, and Benisch and Partner. Courtesy of Institute for Lightweight Structures, Stuttgart.*)

FABRIC STRUCTURES

Fabric structures are not new: People have constructed tents since the earliest days of human civilization. During the past several decades, however, new, durable fabrics and computerized methods for finding form and forces have helped to create a new construction type: a permanent, rigid, stable fabric structure that will last for 20 years or more.

Types of Fabric Structures

Fabric structures are either tensile or pneumatic (Figure A). A *tensile fabric structure* is a membrane supported by masts or other rigid structural elements such as frames or arches. The membrane usually consists of a woven textile fabric and is generally reinforced with steel cables along the main lines of stress. The fabric and cables transmit external loads to the rigid supports and ground anchors by means of tensile forces.

Pneumatic structures depend on air pressure for their stability and their capacity to carry snow and wind loads. The most common type of pneumatic structure is the *air-supported structure*, in which an airtight fabric, usually reinforced with steel cables, is held up by pressurizing the air in the inhabited space below it. The fabric and cables in an air-supported structure are stressed in tension.

Fabrics for Permanent Structures

Nearly all fabric structures are made of woven cloth that has been coated with a synthetic material. The cloth provides structural strength to resist the tensile forces in the structure, and the coating makes the fabric airtight and water resistant. The most widely used fabric is polyester cloth that has been laminated or coated with polyvinyl chloride (vinyl, PVC). Two other frequently used, longer-lasting fabrics are based on glass fiber cloth: One is coated with polytetrafluorethylene (PTFE, the most common brand of which is known as Teflon); the other is coated with silicone. The polyester/PVC fabric is the most economical of the three, but it does not meet U.S. building code requirements for a noncombustible material and is used predominantly for smaller structures. The glass/PTFE and glass/silicone fabrics are classified as noncombustible. Though it is more expensive, the glass/PTFE fabric remains clean longer than the other two. All three fabrics are highly resistant to such forces of deterioration as ultraviolet light, oxidation, and fungi.

Fabrics may be white, colored, or imprinted with patterns or graphics. A fabric may be totally opaque, obstructing all passage of light, or it may be translucent, allowing a controlled percentage of light to pass through.

Though a single layer of fabric has little resistance to the flow of heat, a properly designed fabric structure can achieve substantial energy savings over conventional enclosures through selective use of translucency and reflectivity. Translucency can be used to provide natural illumination, gather solar heat in the winter, and cool the space at night in the summer. A highly reflective fabric can reduce solar heat gain and conserve artificial illumination. With the addition of a second layer, a fabric liner that is suspended about a foot (300 mm) below the structural fabric, the thermal resistance of the structure can be improved. An acoustic inner liner can help to control internal sound reflection, which is especially important in air-supported structures, which tend to focus sound. Tensile structures, because of their anticlastic curvature, tend to disperse sound rather than focus it.

FIGURE A

Simple tensile and pneumatic (air-supported) fabric structures. *(Sketch by Edward Allen.)*

A.

B.

FIGURE B
One of the world's largest roof structures covers the Haj Terminal in Jeddah, Saudi Arabia, an airport building that is used to facilitate the travel of vast numbers of Muslim faithful during a short period of annual pilgrimage. The roof is made up of radial shapes. *(Architects: Skidmore, Owings & Merrill. Roof designer and structural engineer: Geiger Berger Associates. Photos for Figures B–G courtesy of the photographer, Horst Berger.)*

FIGURE C
The fabric of the Haj Terminal is PTFE-coated glass fiber cloth.

FIGURE E
The San Diego Convention Center's roof is raised at the middle on rigid steel struts that rest on cables supported by the concrete frame. *(Architects: Arthur Erickson Associates. Roof designer and structural engineer: Horst Berger.)*

FIGURE D
The roof canopy of the San Diego Convention Center is supported at the perimeter on the concrete frame of the building below.

FABRIC STRUCTURES *(CONTINUED)*

FIGURE F
The form of the Denver International Airport roof combines saddle and radial shapes to echo the forms of the surrounding mountains. *(Architects: C. W. Fentriss, J. H. Bradburn & Associates. Roof designer and structural engineer: Severud Associates; Horst Berger, Principal Consultant.)*

FIGURE G
Like the petals of a giant flower, tensile structures arranged in a huge circle shade the grandstand of King Fahd Stadium in Riyadh, Saudi Arabia. The masts are 190 feet (58 m) high. *(Architects: Ian Fraser, John Roberts & Partners. Roof designer and structural engineer: Geiger Berger Associates.)*

Tensile Structures

Tensile structures are stabilized by anticlastic curvature and prestress. Anticlastic curvature means that the fabric is curved simultaneously in two opposite directions. Two basic geometries may be used to create anticlastic curvature: One is the saddle shape (Figures D–E and G), the other the radial tent (Figures B and C). It is from combinations and variations of these geometries, as in Figure F, that all tensile structures are shaped.

Prestress is the introduction of permanent tension into the fabric in two opposing directions. Without anticlastic curvature and prestress, the fabric would flutter in the wind and destroy itself within a short time. The amount of curvature and the amount of prestressing force must both be sufficient to maintain stability under expected wind and snow conditions. If the curvature is too flat or if the prestressing tension is too low, excessive deflection or flutter will occur.

The design of a tensile structure usually begins by experimenting with simple physical models. These often are made of pantyhose material or stretch fabric, either of which is easily stretched and manipulated. After a general shape has been established with the model, a computer is used to find the exact equilibrium shape, determine the stresses in the fabric and supporting members under wind and snow loadings, and generate cutting patterns for the fabric. The design process is referred to as *form finding*, because a tensile structure cannot be made to take any arbitrary shape. Just as a hanging chain will always take a form that places its links in equilibrium with one another, a tensile structure must take a form that maintains proportionate amounts of tension in all parts of the fabric under all expected loading conditions. The designer's task is to find such a form.

A good design for a tensile structure employs short *masts* to minimize buckling problems. The fabric generally cannot come to a peak at the mast, but must terminate in a cable ring that is attached to the mast in order to avoid high tensile stresses in the fabric. The perimeter edges of the fabric usually terminate along curving steel cables. To make a stable structure, these cables must have adequate curvature and must be anchored to foundations that offer firm resistance to uplift forces. The fabric may be attached to the cables by sleeves sewn into the edges of the fabric or clamps that grip the fabric and pull it toward the cable.

Air-Supported Structures

Air-supported structures are pressurized by the fans that are used to heat, cool, and ventilate the building. The required air pressures are so low that they are scarcely discernible by people entering or leaving the building, but they are high enough (5–10 pounds per square foot, or 0.25–0.50 kPa) to prevent ordinary swinging doors from opening. For this reason, revolving doors, whose operation is unaffected by internal pressure and that maintain a continual seal against loss of air, are usually used for access.

The fabric of an air-supported structure is prestressed by its internal air pressure to prevent flutter. For low-profile roof shapes, a cable net is employed to resist the high forces that result from the flat curvature. The fabric spans between the cables. The fabric and cables pull up on the foundations with a total force that is equal to the internal air pressure multiplied by the area of ground covered by the roof. The supporting elements and foundations must be designed to resist this force.

Wind causes suction forces to occur on many areas of an air-supported structure, which results in additional tension in the fabric and cables. The downward forces from wind or snow load on an air-supported structure must be resisted directly by the internal air pressure pushing outward against them. In geographic areas where snow loads are larger than acceptable internal pressures, snow must be removed from the roof. Failure to do so has led to unplanned deflations of several air-supported roofs.

In theory, air-supported structures are not limited in span. In practice, flutter and perimeter uplift forces restrict their span to a few hundred meters, but this is sufficient to house entire football stadiums. For safety, the outer edges of most air-supported roofs terminate at a level that is well above the floor level within. Thus, if the roof deflates because of fan failure, inadequate snow removal, or air leakage, the roof fabric will hang in suspension at a height well above the floor of the building (Figure H).

FIGURE H

Most air-supported structures are designed so that if air pressure fails, the membrane will hang at a safe level above the heads of the occupants. *(Sketches by Edward Allen.)*

Composite Columns

Columns that combine the strength of structural steel shapes and sitecast concrete have been used in buildings for many years. One type of composite column surrounds a steel wide-flange column with sitecast reinforced concrete. Another type consists of a steel pipe that is filled with concrete. In a third type, a wide-flange column is inserted within the pipe before the concrete is added to create a higher loadbearing capacity.

Contemporary design of tall buildings frequently makes buildings use large steel pipe columns filled with very-high-strength concrete to carry a major portion of both vertical and lateral loads. These columns enable reductions of as much as 50 percent in the overall quantity of steel required for the building (Figure 11.89). In one example, in a 720-foot (200-m) office tower, four 10-foot-diameter (3-m) pipe columns filled with 19,000-psi (131-MPa) concrete carry 40 percent of the building's gravity loads and a large proportion of the wind loads. There is no reinforcing or other steel inside the pipes except at certain connections that carry very heavy loads. The potential advantages of composite columns in tall buildings include reduced steel usage, greater rigidity of the building against wind forces, and simplified beam–column connections.

INDUSTRIALIZED SYSTEMS IN STEEL

Steel adapts well to industrialized systems of construction. The two most successful and most economical prefabrication systems in the United States are probably the manufactured home (often referred to as a mobile home) and the package industrial building. The manufactured home, built largely of wood, is made possible by a rigid undercarriage (chassis) welded together from rolled steel shapes. The package building is most commonly based on a structure of welded steel rigid frames supporting an enclosure of corrugated metal sheets. The manufactured home is founded on steel because of steel's matchless stiffness and strength. The package building depends on steel for these qualities, for the repeatable precision with which components can be produced, and for the ease with which the relatively light steel components can be transported and assembled. It is but a short step from the usual process of steel fabrication and erection to the serial production of repetitive building components.

SUSTAINABILITY IN STEEL FRAME CONSTRUCTION

Manufacture

• The raw materials for steel are iron ore, coal, limestone, air, and water. The ore, coal, and limestone are minerals whose mining and quarrying cause disruption of land and loss of wildlife habitat, often coupled with pollution of streams and rivers. Coal, limestone, and low-grade iron ore are plentiful, but high-grade iron ore has been depleted in many areas of the earth.

• The steel industry continues to achieve reductions in the pollution of air, water, and soil caused by its manufacturing processes.

• Supplies of some alloying metals, such as manganese, chromium, and nickel, are becoming depleted.

• The manufacture of a ton of steel from iron ore by the basic oxygen process consumes 3170 pounds (1440 kg) of ore, 300 pounds (140 kg) of limestone, 900 pounds (410 kg) of coke (made from coal), 80 pounds (36 kg) of oxygen, and 2575 pounds (1170 kg) of air. In the process, 4550 pounds (2070 kg) of gaseous emissions are given off, and 600 pounds (270 kg) of slag and 50 pounds (23 kg) of dust are generated. Further emissions emanate from the process of converting coal to coke.

• The embodied energy of steel produced from ore by the basic oxygen process is about 14,000 BTU per pound (33 MJ/kg). In modern facilities, scrap steel is typically added as an ingredient during this process, resulting in recycled materials content of 25 to 35 percent.

• Today, most structural steel in North America is made from recycled scrap by the electric arc furnace process; its embodied energy is approximately 4000 BTU per pound (9.3 MJ/kg), less than one-third that of steel made from ore. The recycled materials content of steel made by this process is 90 percent or higher.

• Since 1990, the North American steel industry has reduced energy consumption and carbon dioxide emissions per ton of steel produced by roughly 25 and 35 percent, respectively.

• Worldwide, the recycled content of steel averages roughly 35 to 45 percent.

• In North America, virtually all hot-rolled structural steel shapes are manufactured by the electric arc furnace process. Steel plate and sheet, used in the manufacture, for example, of light gauge steel members, decking, and hollow structural sections, may be produced by either the electric arc furnace or basic oxygen processes.

• Ninety-five percent or more of all structural steel used in North American building construction is eventually recycled or reused, which is a very high rate. In the United States alone, roughly 80 million tons of steel are recycled annually.

• Scrap used in the production of structural steel in minimills usually comes from sources within approximately 300 miles (500 km) of the mill. When the steel produced in such mills is then used for the construction of buildings not too far from the mill, the steel may be eligible for credit as a regionally extracted, processed, and produced material. This is most likely for the commonly used steel alloys that are produced in the greatest number of mills. However, some less commonly produced steel alloys are only available from a limited number of mills or, in some cases, are produced solely overseas.

Construction

• Steel fabrication and erection are relatively clean, efficient processes, although the paints and oils used on steel members can cause air pollution.

• Steel frames are lighter in weight than concrete frames that would do the same job. This means that a steel building generally has smaller foundations and requires less excavation work.

• Some spray-on fireproofing materials can pollute the air with stray fibers.

In Service

• Steel framing, if protected from water and fire, will last for many generations with little or no maintenance.

• Steel exposed to weather must be repainted periodically unless it is galvanized, given a long-lasting polymer coating, or made of more expensive stainless steel.

• Steel framing members in building walls and roofs should be thermally broken or insulated in such a way that they do not conduct heat between indoors and outdoors.

• When a steel building frame is demolished, its material is almost always recycled.

• Steel seldom causes indoor air quality problems, although surface oils and protective coatings sometimes outgas and cause occupant discomfort.

FIGURE 11.89

A core structure of eight large composite columns, each a concrete-filled pipe 7½ feet (2.3 m) in diameter, carries the majority of gravity and wind loads in this 44-story Seattle office building. The perimeter of the building is supported by smaller-diameter composite pipe columns. (*Courtesy of Skilling Ward Magnusson Barkshire, Inc., Seattle, WA, Consulting Structural & Civil Engineers.*)

STEEL AND THE BUILDING CODES

Steel frame construction appears in the typical building code tables in Figures 1.3 and 1.4 as six different construction types, I-A through III-B, the exact classification depending on the degree of fireproofing treatment applied to the various members of the frame. With a high degree of fireproofing, especially on members supporting more than one floor, unlimited building heights and areas are permitted for most occupancy groups. With no fireproofing of steel members whatsoever, building heights and areas are limited, but many occupancy groups can easily meet these restrictions.

UNIQUENESS OF STEEL

Among the common structural materials for fire-resistant construction—masonry, concrete, and steel—steel alone has useful tensile strength, which, along with compressive strength, it possesses in great abundance (Figure 11.90). A relatively small amount of steel can do a structural job that would take a much greater amount of another material. Thus, steel, the densest structural material, is also the one that produces the lightest structures and those that span the greatest distances.

The infrastructure needed to bring steel shapes to a building site—the mines, the mills, the fabricators, and the scrap metal industry—is vast and complex. An elaborate sequence of advance planning and preparatory activities is required for making a steel building frame. Once on the site, however, a steel frame goes together quickly, and with relatively few tools, in an erection process that is rivaled for speed and all-weather reliability only by certain precast concrete systems. With proper design and planning, steel can frame almost any shape of building, including irregular angles and curves. Ultimately, of course, structural steel produces only a frame. Unlike masonry or concrete, it does not lend itself easily to forming a total building enclosure except in certain industrial applications. This is of little consequence, however, because steel mates easily with glass, masonry, and panel systems of enclosure and because steel does its own job, that of carrying loads high and wide with apparent ease, so very well.

Material	Working Strength in Tension[a]	Working Strength in Compression[a]	Density	Modulus of Elasticity
Wood (framing lumber)	300–1000 psi 2.1–6.9 MPa	600–2000 psi 4.1–14 MPa	30 pcf 480 kg/m³	1,000,000–1,900,000 psi 6900–13,000 MPa
Brick masonry (including mortar, unreinforced)	0	250–1300 psi 1.7–9.0 MPa	120 pcf 1900 kg/m³	700,000–3,700,000 psi 4800–25,000 MPa
Structural steel	24,000–43,000 psi 170–300 MPa	24,000–43,000 psi 170–300 MPa	490 pcf 7800 kg/m³	29,000,000 psi 200,000 MPa
Concrete (unreinforced)	0	1000–12,000 psi 6.9–83 MPa	145 pcf 2300 kg/m³	3,000,000–4,500,000 psi 21,000–31,000 MPa

[a] Allowable stress or approximate maximum stress under normal working conditions.

FIGURE 11.90
Comparative physical properties for four common structural materials: wood, brick masonry, steel (shaded row), and concrete. Steel is many times stronger and stiffer than these other structural materials. The ranges of values of strength and stiffness reflect differences among steel alloys.

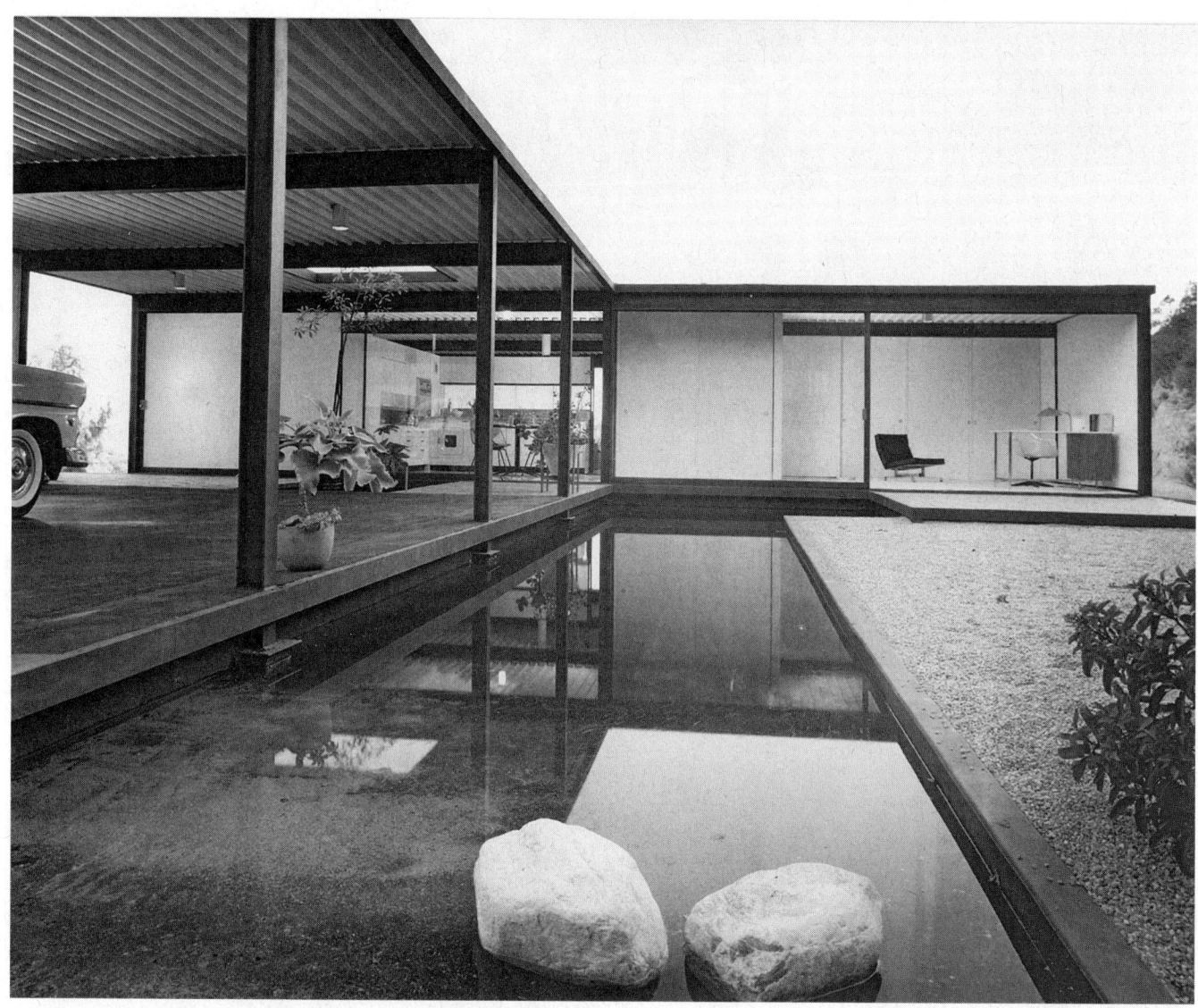

FIGURE 11.91
This elegantly detailed house in southern
California is an early example of the use
of structural steel at the residential scale.
(Architect: Pierre Koenig, FAIA. Photo: Julius
Shulman, Hon AIA.)

FIGURE 11.92
Architect Peter Waldman utilized steel pipe columns, wide-flange beams, open-web steel joists, and corrugated steel roof decking for his own house in Charlottesville, Virginia. *(Photo by Maxwell MacKenzie.)*

FIGURE 11.93
The Chicago Police Training Center expresses elegantly the logic and simplicity of a straightforward steel frame. *(Architect: Jerome R. Butler, Jr. Engineer: Louis Koncza. Photograph by Lautman Photography. Permission of American Institute of Steel Construction.)*

FIGURE 11.94
The United Airlines Terminal at Chicago's O'Hare Airport is a high-tech wonderland of steel framing and fritted glass. *(Architect: Murphy-Jahn. Photo by Edward Allen.)*

FIGURE 11.95
Chicago is famous for its role in the development of the steel frame skyscraper (see Figure 11.4). One of the tallest in the United States is the Sears Tower, seen in the foreground of this photograph. *(Architect and engineer: Skidmore, Owings and Merrill. Photo by Chicago Convention and Tourism Bureau, Inc. Permission of the American Institute of Steel Construction.)*

CSI/CSC MasterFormat Sections for Steel Frame Construction		
05 12 00	STRUCTURAL STEEL FRAMING	
05 12 13	Architecturally Exposed Structural Steel Framing	
05 12 19	Buckling Restrained Braces	
05 16 00	STRUCTURAL CABLING	
05 21 00	STEEL JOIST FRAMING	
05 21 19	Open-Web Steel Joist Framing	
05 31 00	STEEL DECKING	
05 31 13	Steel Floor Decking	
05 31 23	Steel Roof Decking	
05 31 33	Steel Form Decking	
05 36 00	COMPOSITE METAL DECKING	
05 56 00	METAL CASTINGS	
07 81 00	APPLIED FIREPROOFING	
07 81 16	Cementitious Fireproofing	
07 81 23	Intumescent Fireproofing	
07 81 29	Mineral-Fiber Cementitious Fireproofing	
07 81 33	Mineral-Fiber Fireproofing	
07 82 00	BOARD FIREPROOFING	
09 96 00	HIGH-PERFORMANCE COATINGS	
09 96 46	Intumescent Painting	

KEY TERMS

Bessemer process
open-hearth method
steel
mild steel
cast iron
wrought iron
ferrous metal
iron ore
smelting
blast furnace
coke
slag
basic oxygen process
mini-mill
electric arc furnace
beam blank, bloom
high-strength, low-alloy steel
weathering steel
stainless steel
quenching
tempering
structural mill, breakdown mill
hot saw

cooling bed
roller straightener
(steel) bar
(steel) plate
(steel) sheet
wide-flange shape
American Standard shape, I-beam
angle
gusset plate
channel
tee
cast steel
cold-worked steel, cold-formed steel
hollow structural section (HSS)
open-web steel joist (OWSJ)
joist girder
rivet
high-strength bolt
carbon steel bolt
bearing-type connection
snug-tight
slip-critical connection, friction-type
  connection

(bolt) preloading
faying surface
galling
impact wrench
turn-of-nut method
load indicator washer, direct tension
  indicator (DTI)
calibrated wrench method
tension control bolt
shear wrench
lockpin and collar fastener, swedge bolt
electric arc welding
electrode
weld symbol
backup bar, backing bar
runoff bar
demand-critical weld
shear connection
shear
bending moment
framed connection
moment connection
full-penetration groove weld

stiffener plate
shear tab
braced frame
diagonal bracing
Chevron bracing, inverted V bracing
crossbracing
eccentrically braced frame
damping
shear wall
moment-resisting frame
rigid core
rigid perimeter, tube structure
Fully-Restrained moment connection
Partially-Restrained moment connection
Simple connection
seated connection
end plate connection
fabricator
shop drawing
coped flange
plasma cutting
laser cutting
camber
erector
ironworker

tier
baseplate
leveling plate
grout
raising gang
plumbing up
tower crane
luffing-boom crane
hammerhead boom crane
tagline
drift pin
topping out
metal decking
roof decking
form deck
cellular decking
vented metal decking
puddle weld
composite metal decking
shear stud
subpurlin
girt
architecturally exposed structural steel (AESS)

fireproofing
spray-applied fire-resistive materials (SFRM), spray-applied fireproofing
buckling-restrained bracing
intumescent mastic
intumescent paint
castellated beam
plate girder
rigid steel frame
steel truss
chord
staggered truss system
space truss, space frame
arch
anticlastic curvature
cable stay
tensile fabric structure
pneumatic structure
air-supported structure
prestress
form finding
mast

REVIEW QUESTIONS

1. What is the difference between iron and steel? What is the difference between wrought iron and cast iron?

2. By weight, what is the major raw material used in the making of cast iron?

3. How are steel structural shapes produced? How are the weights and thicknesses of a shape changed?

4. How does the work of the fabricator differ from that of the erector?

5. Explain the designation W21 × 68.

6. How can you tell a shear connection from a moment connection? What is the role of each?

7. Why might a beam be coped?

8. What is the advantage of composite construction?

9. Explain the advantages and disadvantages of a steel building structure with respect to fire. How can the disadvantages be overcome?

10. List three different structural systems in steel that might be suitable for the roof of an athletic fieldhouse.

EXERCISES

1. For a simple multistory office building of your design:

a. Draw a steel framing plan for a typical floor.

b. Draw an elevation or section showing a suitable method of making the building stable against lateral forces—wind and earthquake.

c. Make a preliminary determination of the approximate sizes of the decking, beams, and girders.

d. Sketch details of the typical connections in the frame, using actual dimensions from the *Manual of Steel Construction* (see references for this chapter) for the member size you have determined and work to scale.

2. Select a method of fireproofing, and sketch typical column and beam fireproofing details for the building in Exercise 1.

3. What fire-resistance ratings in hours are required for the following elements of a steel-framed department store, three stories in height, unsprinklered, with 21,000 square feet of area per floor? (The

necessary information is found in Figures 1.2 and 1.3.)

a. Lower-floor columns

b. Floor beams

c. Roof beams

d. Interior nonbearing walls and partitions

4. Find a steel building frame under construction. Observe the connections carefully and figure out why each is detailed as it is. If possible, arrange to talk with the structural engineer of the building to discuss the design of the frame.

SELECTED REFERENCES

Ambrose, James, and Patrick Tripeny. *Simplified Design of Steel Structures* (8th ed.). Hoboken, NJ, John Wiley & Sons, Inc., 2007.

This is an excellent introduction to the calculations needed for designing and specifying steel beams, columns, and connections.

American Institute of Steel Construction, Inc. *Steel Construction Manual*. Chicago, updated regularly.

This is the bible of the steel construction industry in the United States. It contains detailed tables of the dimensions and properties of all standard rolled steel sections, data on standard connections, and specifications and code information.

American Iron and Steel Institute. *Designing Fire Protection for Steel Beams*. Washington, DC, Author, 2002.

The problem of fireproofing steel building elements is discussed, and a range of fireproofing details is illustrated in these concise booklets.

American Iron and Steel Institute. *Specification for Structural Steel Buildings*. Washington, DC, Author, 2010.

This specification, included in the *Steel Construction Manual*, can also be viewed for free on the American Institute of Steel Construction's web site, www.aisc.org.

Kulak, Geoffrey L., John W. Fisher, and John H. A. Struik. *Guide to Design Criteria for Bolted and Riveted Joints* (2nd ed.). Chicago, AISC, 2001.

This 300-plus-page guide provides detailed engineering guidelines for the design of bolted and riveted steel connections. This guide can be viewed for free on the Research Council on Structural Connections web site, www. boltcouncil.org.

Steel Joist Institute. *Catalog of Standard Specifications and Load and Weight Tables for Steel Joists and Joist Girders*. Myrtle Beach, SC, updated regularly.

Load tables, sizes, and specifications for open-web joists are given in this booklet.

WEB SITES

Steel Frame Construction

Author's supplementary web site: **www.ianosbackfill.com/11_steel_frame_construction**

The Material Steel

American Institute of Steel Construction (AISC): **www.aisc.org**

American Iron and Steel Institute: **www.steel.org**

Nucor Steel: **www.nucor.com**

Nucor–Vulcraft Group: **www.vulcraft.com**

Research Council on Structural Connections: **www.boltcouncil.org**

Steel Joist Institute (SJI): **www.steeljoist.org**

Steel Recycling Institute: **www.recycle-steel.org**

Longer Spans in Steel

Birdair Tensioned Membrane and Lightweight Structures: **www.birdair.com**

## 12

LIGHT GAUGE STEEL FRAME CONSTRUCTION

- **The Concept of Light Gauge Steel Construction**

  SUSTAINABILITY IN LIGHT GAUGE STEEL FRAMING

- **Light Gauge Steel Framing**

- **Other Uses of Light Gauge Steel Framing**

  FOR PRELIMINARY DESIGN OF A LIGHT GAUGE STEEL FRAME STRUCTURE

- **Insulating Light Gauge Steel Frame Structures**

- **Advantages and Disadvantages of Steel Framing**

- **Light Gauge Steel Framing and the Building Codes**

- **Finishes for Light Gauge Steel Framing**

  METALS IN ARCHITECTURE

  FROM CONCEPT TO REALITY
  Camera Obscura at Mitchell Park, Greenport, New York

Driving self-drilling, self-tapping screws with electric screw guns, framers add diagonal bracing straps to a wall frame made from light gauge steel studs and runner channels.
(Photographs courtesy of United States Gypsum Company.)

To manufacture the members used in *light gauge steel* frame construction, sheet steel is fed from continuous coils through machines at room temperature that cold-work the metal (see Chapter 11) and fold it into efficient structural shapes, producing linear members that are stiff and strong. Thus, these members are referred to as *cold-formed metal framing*, to differentiate them from the much heavier hot-rolled shapes that are used in structural steel framing. The term "light gauge" refers to the relative thinness (gauge) of the steel sheet from which the members are made.

THE CONCEPT OF LIGHT GAUGE STEEL CONSTRUCTION

Light gauge steel construction is the noncombustible equivalent of wood light frame construction. The external dimensions of the standard sizes of light gauge members correspond closely to the dimensions of the standard sizes of nominal 2-inch (38-mm) framing lumber. These steel members are used in framing as closely spaced studs, joists, and rafters in much the same way as wood light frame members are used, and a light gauge steel frame building may be sheathed, insulated, wired, and finished inside and out in the same manner as a wood light frame building.

The steel used in light gauge members is manufactured to ASTM standard A1003 and is metallic-coated with zinc or aluminum-zinc alloy to protect against corrosion. The thickness of the metallic coating can be varied, depending on the severity of the environment in which the members will be placed.

For wall, floor, and roof framing, the steel is formed into *stud or joist sections*, also called *C-sections* (Figure 12.1). The webs of these members may be punched at the factory to provide holes at 2-foot (600-mm) intervals; these allow wiring, piping, and bracing to pass through without the necessity of cutting holes in members on the construction site (Figures 12.11, 12.15). *Track sections* are used for top and bottom plates. They are slightly oversized so that stud or joist members can nest into them. *Channel sections* and *furring channels* are used for lighter bracing and framing tasks. Other, more specialized shapes are also available.

The strength and stiffness of a member depend on its shape and size, as well as the strength and thickness of sheet metal used in its manufacture. The range of metal thicknesses available for both load-bearing and nonbearing members are listed in Figure 12.2. Steel with yield strengths of both 33,000 and 50,000 psi (230 or 350 MPa) is used.

Light gauge steel members are identified using a standard nomenclature, similar to that used with hot-rolled steel structural shapes. For example, a 600S162-54 member is a 6.00-inch-deep ("600") stud or joist section ("S"), 1.625 inch wide ("162"), made from 54-mil-thick sheet metal ("54"). Standard designations and the range of sizes available for common member types are listed in Figure 12.3.

FIGURE 12.1
Three typical light gauge steel framing member shapes, from left to right: studs or joists, channels, and tracks.

SUSTAINABILITY IN LIGHT GAUGE STEEL FRAMING

The sustainability issues raised in Chapter 11 also apply here. In addition:

• The embodied energy of light gauge steel framing members is about 7000 BTU per pound (15 MJ/kg).

• The postconsumer recycled content of light gauge steel framing members varies from about 25 to 45 percent, depending on the manufacturing process.

• The largest issue concerning the sustainability of light gauge steel construction is the high thermal conductivity of the framing members. If a building framed with light gauge steel members is constructed as if it were framed with wood, and insulated only in the framing cavities, it will lose heat in winter at a rate as much as double that of the equivalent wood structure. However, this limitation can be overcome with proper attention to insulating techniques and metal component detailing, as discussed later in this chapter.

Nominal Gauge	Minimum Thickness of Steel Sheet	
	Loadbearing Members	**Nonloadbearing Members**
10	0.118″ (3.00 mm)	
12	0.097″ (2.45 mm)	
14	0.068″ (1.72 mm)	0.068″ (1.72 mm)
16	0.054″ (1.37 mm)	0.054″ (1.37 mm)
18	0.043″ (1.09 mm)	0.043″ (1.09 mm)
20	0.033″ (0.84 mm)	0.030″ (0.75 mm)
22		0.027″ (0.68 mm)
25		0.018″ (0.45 mm)

FIGURE 12.2
Minimum thicknesses of base sheet metal (not including the metallic coating) for light gauge steel framing members. Shaded figures indicate the most commonly used. Traditional *gauge* designations are also shown. Note how lower gauge numbers correspond to greater metal thickness. Though still used in common parlance, gauge numbers are not recommended for specification purposes due to lack of standards for their translation into actual metal thickness. Metal thickness is also sometimes expressed in mils, or thousandths of an inch. For example, 0.033 inch equals 33 mils.

Cold-Formed Member	Example Designation	Available Depths	Available Widths
Stud or Joist	600S162-54	1.625″, 2.5″, 3.5″, 3.625″, 4″, 5.5″, and 6″ to 16″ *41, 64, 90, 92, 102, 140, and 152–406 mm*	1.25″–2.5″ *32 mm–64 mm*
Track	362T125-30	same as above	1.25″–2″ *32 mm–51 mm*
Channel	250U050-54	0.75″–2.5″ *19 mm–64 mm*	0.5″ *13 mm*
Furring	087F125-43	0.875″–1.25″ *22 mm–32 mm*	1.25″–2.5″ *32 mm–64 mm*

FIGURE 12.3
Standard light gauge steel shapes and sizes.

END CLIPS

WEB STIFFENER

FOUNDATION CLIP

V-BRACING

JOIST HANGER

FLAT STRAP BRACING

$1^{1}/_{2}"$ COLD ROLLED CHANNEL

FIGURE 12.4
Standard accessories for light gauge steel framing. End clips are used to join members that meet at right angles. Foundation clips attach the ground-floor platform to anchor bolts embedded in the foundation. Joist hangers connect joists to headers and trimmers around openings. The web stiffener is a two-piece assembly that is inserted inside a joist and screwed to its vertical web to help transmit wall loads vertically through the joist. The remaining accessories are used for bracing.

Some manufacturers produce proprietary, improved member types. For example, the cross-sectional profile of a stud or joist may be altered to enhance its strength and stiffness. Or, studs may be passed through steel rollers that produce a dense array of dimples in the formed metal. The additional cold working of the metal that occurs during this process and the finished patterning of the surface produce a stud with greater strength and stiffness. In addition, a wide variety of angles, straps, plates, channels, and other shapes are used in the joining and bracing of members (Figure 12.4).

Like wood light members, light gauge steel shapes are furnished in standard lengths. They are cut to length on the construction job site, with power saws, or special electric or hand shears, depending on their thickness. For large projects, members may be delivered precut to the required lengths.

Light gauge steel members are usually joined with *self-drilling screws*, which drill their own holes and form helical threads in the holes as they are installed. Driven rapidly by hand-held electric or pneumatic tools, these screws are plated with cadmium or zinc to resist corrosion, and they are available in an assortment of diameters and lengths to suit a full range of connection needs. Welding is often employed to assemble panels of light gauge steel framing that are prefabricated in a factory, and is sometimes used on the building site where stronger connections are needed. Other fastening techniques include hand-held clinching devices that join members without screws or welds and pneumatically driven pins that penetrate the members and hold by friction.

LIGHT GAUGE STEEL FRAMING

The sequence of construction for a building constructed with light gauge steel members is analogous to that described in Chapter 5 for wood light framing (Figures 12.5 through 12.12). Framing proceeds

Anchor clip
Steel joist roof rafter
Steel joist soffit framing

B. EAVE

Steel joist roof rafter
End clip
Ridge beam—nested steel joists

A. RIDGE

Stud
Runner
Continuous bead of adhesive
Web stiffener
Closure channel
C-runner

C. JOIST BEARING AT UPPER FLOOR

FIGURE 12.5

Typical light gauge framing details. Each detail is keyed by letter to a circle on the central whole-building diagram included as part of this figure, to show its location in the frame. (*A*) A pair of nested joists makes a boxlike ridge beam. (*B*) Anchor clips are sandwiched between the ceiling joists and rafters to hold the roof framing down to the wall. (*C*) A web stiffener helps transmit vertical forces from each stud through the end of the joist to the stud in the floor below. Mastic adhesive cushions the joint between the subfloor and the steel framing. (*D*) Foundation clips anchor the entire frame to the foundation. (*E*) At interior joist bearings, joists are overlapped back to back and a web stiffener is inserted.

(continued)

Runner—fasten through plywood into closure
Plywood subfloor
Web stiffener
Grout and shim as required
Foundation clip

D. JOIST BEARING AT FOUNDATION

Web stiffener
Steel joists
Steel stud or beam

E. INTERIOR JOIST BEARING

Rafter

Ceiling joists

Steel stud

H. GABLE END FRAMING

DETAIL KEY

Closure channel
or joist section

End tabs

1½" × 20-
gauge bracing
strap

G. JOIST PARALLEL TO END WALL

Closure channel
or joist section

F. JOIST PARALLEL TO FOUNDATION

FIGURE 12.5 *continued*

(*F, G*) Short crosspieces brace the last joist at the end of
the building and help transmit stud forces through to the
wall below. (*H*) Like all these details, the gable end framing
is directly analogous to the corresponding detail for a wood
light frame building as shown in Chapter 5.

Interactive
Exercise

Opening

Joist hanger

Double joist header
(nested)

Steel joist framing
into header

Double joist
trimmer (nested)

FIGURE 12.6
Headers and trimmers for floor openings
are doubled and nested to create a
strong, stable box member. Only one
vertical flange of the joist hanger is
attached to the joist; the other flange
would be used instead if the web of the
joist were oriented to the left rather than
the right.

Steel gusset plate

Runner channel

Lintel—2 steel joists

Steel stud

FIGURE 12.7
A typical window or door head detail. The header is made of
two joists placed with their open sides together. The top plate
of the wall, which is a runner channel, continues over the
top of the header. Another runner channel is cut and folded
at each end to frame the top of the opening. Short studs are
inserted between this channel and the header to maintain the
rhythm of the studs in the wall.

FIGURE 12.8
Diagonal strap braces stabilize upper-floor wall framing for an apartment building. A special tool is used to draw the metal strapping tight as it is installed. The studs are also braced against buckling with horizontal strapping across both faces of the studs at approximately their mid-height. (*Photograph courtesy of United States Gypsum Company.*)

platform-fashion: The ground floor is framed with steel joists. Mastic adhesive is applied to the upper edges of the joists, and structural wood panels are laid down and screwed to the upper flanges of the joists. Next, steel studs are laid out on the completed subfloor and assembled to make wall frames. These frames are sheathed, tilted up, attached to the subfloor, and braced. Upper-floor platforms and walls proceed, in sequence, in the same manner. Finally, the ceiling and roof are framed in much the same way. Prefabricated trusses of light gauge steel members that are screwed or welded together are often used to frame ceilings and roofs (Figures 12.13 and 12.14). It is possible, in fact, to frame any building with light gauge steel members that can be framed with nominal 2-inch (38-mm) wood members.

Openings in floors and walls are framed analogously to openings in wood light frame construction, with doubled members around each opening and headers over doors and windows (Figures 12.6, 12.7, 12.10). Joist hangers and right-angle clips of sheet steel are used to join members around openings. Light gauge members are designed so that they can be *nested* to form a tubular configuration that is especially strong and stiff when used for a ridge board or header (Figures 12.5A and 12.6).

Because light gauge steel members are more prone than their wood counterparts to twisting or buckling under load, more attention must be paid to their bracing and bridging. The studs in tall walls are generally braced at 4-foot (1200-mm) intervals, either with steel straps screwed to the faces of the studs or with

1½-inch (38-mm) channels passed through the punched openings in the studs and welded or screwed to an angle clip at each stud (Figures 12.8 and 12.17). Floor joists are bridged with joist blocking between the joists or steel straps screwed to their top and bottom edges. In locations where large vertical forces must pass through floor joists (as occurs where loadbearing studs sit on the edge of a floor platform), steel *web stiffeners* are screwed to the thin webs of the joists to prevent them from buckling (Figure 12.5C, D, E). Lateral bracing for walls consists of diagonal steel straps screwed to the studs (Figure 12.8). Subflooring, wall sheathing, roof sheathing, and interior finish materials also all contribute significantly to resistance to buckling, twisting, and lateral wind and earthquake loads.

FIGURE 12.9
Ceiling joists in place for an apartment building. A brick veneer cladding has already been added to the ground floor. *(Photograph courtesy of United States Gypsum Company.)*

FIGURE 12.10
Close-up view of a window header connection. Because a supporting stud has been inserted under the end of the header, a large gusset plate such as the one shown in Figure 12.7 is not required. *(Photograph(s) courtesy of Unimast Incorporated—www.unimast.com.)*

FIGURE 12.11
Flexible metal conduit runs through prepunched openings in metal wall studs. The junction box is supported on metal bracket spanning between studs. *(Photo by Joseph Iano.)*

Where a light gauge steel frame building is constructed to be fully noncombustible, wood structural panels cannot be used for subfloors or wall sheathing. Instead, floors may be constructed of corrugated steel decking with concrete topping or precast concrete. Exterior walls are sheathed with *gypsum sheathing panels,* which are similar to gypsum wallboard but made more resistant to moisture and weather with glass mat faces and specially formulated water-resistant gypsum cores (Figure 12.16).

FIGURE 12.12
A detail of eave framing. *(Photograph(s) courtesy of Unimast Incorporated—www. unimast.com.)*

FIGURE 12.13
A worker tightens the last screws to complete a connection in a light gauge steel roof truss. The truss members are held in alignment during assembly by a simple jig made of plywood and blocks of framing lumber. *(Photograph(s) courtesy of Unimast Incorporated—www. unimast.com.)*

FIGURE 12.14
Installing steel roof trusses. *(Photograph(s) courtesy of Unimast Incorporated—www.unimast.com.)*

FIGURE 12.15
Light gauge steel joists supporting corrugated, composite metal decking. Note the preformed openings in the joists, for easy routing of piping and other services within the depth of the framing. *(Photo by Joseph Iano.)*

FIGURE 12.16
Gypsum sheathing panels being applied over metal stud framing on a residential noncombustible building. *(Photo by Joseph Iano.)*

OTHER USES OF LIGHT GAUGE STEEL FRAMING

Light gauge steel members are used to construct many components of fire-resistant buildings whose structures are made of structural steel, concrete, or masonry. These components include interior walls and partitions (Chapters 23); suspended ceilings (Chapter 24); and fascias, parapets, and backup walls for many kinds of exterior cladding materials (Chapters 19 and 20; Figures 12.17, 12.18). Light gauge steel members used for framing interior partitions and other nonloadbearing applications are specified as *nonstructural metal framing*, as distinct from cold-formed metal framing, the latter term being reserved for light gauge steel members used in structural applications and exterior wall cladding systems (even though both types of members are, in fact, cold-formed). The different specifications reflect the different requirements to which these two types of framing must be designed and constructed.

Light gauge steel studs can be combined with concrete to produce thin, but relatively stiff, wall panel systems. Both loadbearing and nonloadbearing panels can be made that are suitable for use in residential and light commercial buildings. A variety of production methods are possible that involve casting an approximately 2-inch (50-mm)-thick concrete facing onto a framework of steel studs. The concrete may be sitecast on the building site or precast in a factory. The concrete-to-steel bond may be created by stud anchors, sheet metal shear strips, welded wire reinforcing, or expanded metal that is welded or screwed to the studs and then becomes embedded in the concrete. In loadbearing applications, the concrete panels provide shear resistance while the steel studs provide most of the resistance to gravity and wind loads acting perpendicular to the face of the panel.

FIGURE 12.17
Light gauge steel stud infill between concrete and structural steel will support the exterior cladding. Note the horizontal rows of steel strap bracing.
(Photo by Joseph Iano.)

FIGURE 12.18
Light gauge steel stud framing forms the exterior enclosure for this building structured with posttensioned concrete.
(Photo by Joseph Iano.)

FOR PRELIMINARY DESIGN OF A LIGHT GAUGE STEEL FRAME STRUCTURE

- Estimate the depth of rafters on the basis of the horizontal (not slope) distance from the outside wall of the building to the ridge board in a gable or hip roof and the horizontal distance between supports in a shed roof. Estimate the depth of a rafter at ¹⁄₂₄ of this span, rounded up to the nearest 2-inch (50-mm) dimension.

- The depth of light gauge steel **roof trusses** is usually based on the desired roof pitch. A typical depth is one-quarter of the width of the building, which corresponds to a 6:12 pitch.

- Estimate the depth of light gauge steel **floor joists** as ¹⁄₂₀ of the span, rounded up to the nearest 2-inch (50-mm) dimension.

- For **loadbearing studs**, add up the total width of floor and roof slabs that contribute load to the stud wall. A 3⅝-inch (92-mm) or 4-inch (102-mm) stud wall can support a combined width of approximately 60 feet (18 m), and a 6-inch (152-mm) or 8-inch (203-mm) stud wall can support a combined width of approximately 150 feet (45 m).

- For **exterior cladding backup walls**, estimate that a 3⅝-inch (92-mm) stud may be used to a maximum height of 12 feet (3.7 m), a 6-inch (150-mm) stud to 19 feet (5.8 m), and an 8-inch (100-mm) stud to 30 feet (9.1 m). For brittle cladding materials such as brick masonry, select a stud that is 2 inches (50 mm) deeper than these numbers would indicate.

All framing members are usually spaced at 24 inches (600 mm) o.c.

These approximations are valid only for purposes of preliminary building layout and must not be used to select final member sizes. They apply to the normal range of building occupancies, such as residential, office, commercial, and institutional buildings.

For manufacturing and storage buildings, use somewhat larger members. For more comprehensive information on preliminary selection and layout of structural members, see Edward Allen and Joseph Iano, *The Architect's Studio Companion* (5th ed.), Hoboken, NJ, John Wiley & Sons, Inc., 2012.

In situations where noncombustibility is not a requirement, metal and wood light framing are sometimes mixed in the same building. Some builders find it economical to use wood to frame exterior walls, floors, and roof, with steel framing for interior partitions. Sometimes all walls, interior and exterior, are framed with steel, and floors are framed with wood. Steel trusses made of light gauge members may be applied over wood frame walls. In such mixed uses, special care must be taken in the details to ensure that wood shrinkage will not create unforeseen stresses or damage to finish materials. Steel framing also may be used in lieu of wood where the risk of damage from termites is high.

INSULATING LIGHT GAUGE STEEL FRAME STRUCTURES

The thermal conductivity of light gauge steel framing members is much higher than that of wood, and any other primary structural material as well. In cold regions, light gauge steel framing must be detailed and insulated to minimize *thermal bridging* through the steel members. Without such measures, the thermal performance of the wall or roof is greatly reduced, energy losses increase substantially, and moisture condensation within the framing cavity or on interior building surfaces may occur, with attendant damage to materials,

growth of mold and mildew, and discoloration of surface finishes.

As a first step, a continuous layer of insulation rigid foam, semirigid mineral fiber, or other insulation type that can span across the stud cavities, must be placed on one side or the other of framing members. The minimum amount of such insulation varies, with colder climates requiring higher insulation values than warmer ones. For an example of continuous insulation applied to the outside of a (wood) light framed structure, see Figure 7.17—the techniques are the same for a light gauge steel frame. The preferred location for continuous insulation in a metal-framed wall or roof is on the exterior (cold) side of the metal components, as shown in Figure 7.17 (right). Placing continuous insulation on the interior, warm side of the framing exposes the metal components to lower temperatures, increasing the risk of condensation within the framing cavity.

Additionally, where steel members may extend from inside to outside the building enclosure, such as at the roof eave of a simple house frame (Figures 12.5B, 12.12), steel members must be broken, insulating spacers must be inserted between inside and outside steel, or some other method must be used to prevent thermal bridging at such locations.

ADVANTAGES AND DISADVANTAGES OF STEEL FRAMING

Light gauge steel framing shares most of the advantages of wood light framing: It is versatile and flexible; requires only simple, inexpensive tools; furnishes internal cavities for utilities and thermal insulation; and accepts an extremely wide range of exterior and interior finish materials. Additionally, steel framing may be used in buildings for which non-combustible construction is required by the building code, thus extending its use to larger buildings and those whose uses require a higher degree of resistance to fire.

Steel framing members are significantly lighter in weight than the wood members to which they are structurally equivalent, an advantage

that is often enhanced by spacing steel studs, joists, and rafters at 24 inches (600 mm) o.c. rather than 16 inches (400 mm) o.c. Light gauge steel joists and rafters can span slightly longer distances than nominal 2-inch (50-mm) wood members of the same depth. Steel members tend to be straighter and more uniform than wood members (Figure 12.19), and they are much more stable dimensionally because they are unaffected by changing humidity. Although they may corrode if exposed to moisture over an extended period of time, steel framing members cannot fall victim to termites or decay.

Compared to walls and partitions of masonry construction, equivalent walls and partitions framed with steel studs are much lighter in weight, easier to insulate, and accept electrical wiring and pipes for plumbing and heating much more readily. Steel framing, because it is a dry process, may be carried out during wet or cold weather conditions that would make masonry construction difficult. Masonry walls tend to be much stiffer and more resistant to the passage of sound than steel-framed walls, however.

LIGHT GAUGE STEEL FRAMING AND THE BUILDING CODES

Although light gauge steel framing members will not burn, they will lose their structural strength and stiffness rapidly if exposed to the heat of fire.

They must therefore be protected from fire in accordance with building code requirements. With suitable protection provided by gypsum sheathing and gypsum wallboard or plaster, light gauge steel construction may be classified as either Type I or Type II Construction in the building code table shown in Figure 1.3, enabling its use for a wide range of building types and sizes.

In its *International Residential Code for One- and Two-Family Dwellings*, the International Code Council has incorporated *prescriptive requirements* for steel-framed residential construction. In many cases, these requirements, with their structural tables and standard details, allow builders to design and construct light gauge steel-framed houses without having to employ an engineer or architect, just as they are able to do with wood light frame construction.

FINISHES FOR LIGHT GAUGE STEEL FRAMING

Any exterior or interior finish material that is used in wood light frame construction may be applied to light gauge steel frame construction. Whereas finish materials are often fastened to a wood frame with nails, only screws may be used with a steel frame. Wood trim components are applied with special finish screws, analogous to finish nails, which have very small heads.

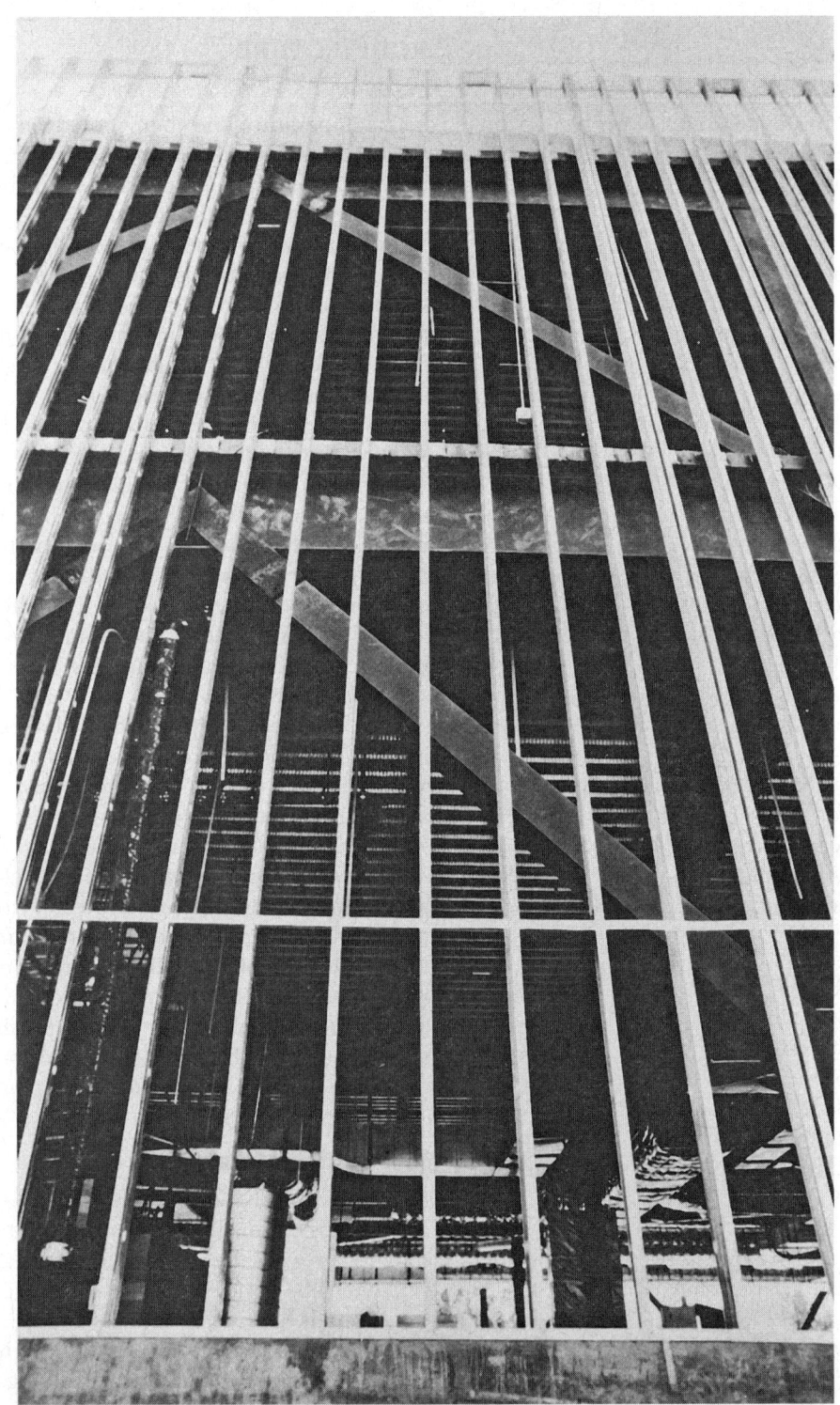

FIGURE 12.19
The straightness of steel studs is apparent in these tall stud walls enclosing a building framed with structural steel. *(Photograph(s) courtesy of Unimast Incorporated—www.unimast.com.)*

METALS IN ARCHITECTURE

Metals are dense, lustrous materials that are highly conductive of heat and electricity. They are generally *ductile*, meaning that they can be hammered thin or drawn into wires. They can be liquefied by heating and will resolidify as they cool. Most metals corrode by oxidation. Metals include the strongest building materials presently in common use, although stronger materials based on carbon or aramid fibers are beginning to appear more frequently in building construction applications.

Metals are usually found in nature in the form of oxide ores. These ores are refined by processes that involve heat and reactant materials or, in the case of aluminum, electrolysis.

In the building construction industry, metals may be classified broadly as either ferrous, meaning that they consist primarily of iron, or nonferrous, that is, all other metals. Because iron ore is an abundant mineral and is relatively easy to refine, ferrous metals tend to be much less expensive than nonferrous ones. The ferrous metals are also the strongest, but most have a tendency to rust. Nonferrous metals in general are considerably more expensive on a volumetric basis than ferrous metals, but unlike ferrous metals, most of them form thin, tenacious oxide layers that protect from further corrosion under normal atmospheric conditions. This makes many of the nonferrous metals valuable for finish components of buildings. Many of the nonferrous metals are also easy to work and attractive to the eye.

Modifying the Properties of Metals

A metal is seldom used in its chemically pure state. Instead, it is mixed with other elements, primarily other metals, to modify its properties for a particular purpose. Such mixtures are called *alloys*. An alloy that combines copper with a small amount of tin is known as bronze. A very small, closely controlled amount of carbon mixed with iron makes steel. In both of these examples, the alloy is stronger and harder than the elemental metal that is its primary ingredient. Various alloys of iron (different steel alloys, to be more specific) are discussed in Chapter 11. Some of these have higher strengths and some form self-protecting oxide layers because of the influence of the alloying elements they contain. Similarly, there are many alloys that consist primarily of aluminum; some are soft and easy to form, others are very hard and springy, still others are very strong, and so on.

The properties of many metals can also be changed by *heat treatment*. Steel that is *quenched*, that is, heated red-hot and then plunged in cold water, becomes much harder but brittle. Steel can be *tempered* by heating it to a moderate degree and cooling it more slowly, making it both hard and strong. Steel that is brought to a very

high temperature and then cooled very slowly, a process called *annealing*, will become softer, easier to work, and less brittle. Many aluminum alloys can also be heat-treated to modify their characteristics.

Cold working is another way of changing the properties of a metal. When steel is beaten or rolled thinner at room temperature, its crystalline structure is altered in a way that makes it much stronger, though also somewhat more brittle. The highest-strength metals used in construction are steel wires and cables used to prestress concrete. Their high strength (about four times that of normal structural steel) is the result of drawing the metal through smaller and smaller orifices to produce the wire, a process that subjects the metal to a high degree of cold working. Cold-rolled steel shapes with substantially higher strengths than hot-rolled structural steel are used as reinforcing and as components of open-web joists. The effects of cold working are easily reversed by annealing. Hot rolling, which is, in effect, a self-annealing process, does not increase the strength of metal.

To change the appearance of metal or to protect it from oxidation, it can be coated with a thin layer of another metal. Steel is often *galvanized* by coating it with zinc to protect against corrosion, as described later in this sidebar. *Electroplating* is widely used to coat metals such as chromium and cadmium onto steel to improve the appearance of the steel and protect it from oxidation. An electrolytic process is used to *anodize* aluminum, adding a thin oxide layer of controlled color and consistency to the surface of the metal. To protect them and enhance their appearance, metals are frequently finished with nonmetallic coatings such as paints, lacquers, organic coatings, porcelain enamel, and thermosetting powders.

Fabricating Metals

Metals can be shaped in many different ways. *Casting* is the process of pouring molten metal into a shaped mold; the metal retains the shape of the mold as it cools. *Rolling*, which may be done either hot or cold, forms the metal by squeezing it between a series of shaped rollers. *Extrusion* is the process of squeezing heated but not molten metal through a shaped die to produce a long metal piece with a shaped profile matching the cutout in the die. *Forging* involves heating a piece of metal until it becomes soft, then beating it into shape. Forging was originally done by hand with a blacksmith's forge, hammer, and anvil. Now most forging is done with powerful hydraulic machinery that forces the metal into shaped dies. *Stamping* is the process of squeezing sheet metal between two matching dies to give it a desired shape or texture. *Drawing* produces wires by pulling a metal rod through a series of progressively smaller orifices in hardened steel plates until the desired diameter is reached. These forming processes have varying effects on the strength of the resulting material: Cold drawing and cold rolling will harden and strengthen many metals. Forging imparts a grain orientation to the metal that closely follows the shape of the piece for improved structural performance. Casting tends to produce somewhat weaker metal than most other forming processes, but it is useful for making elaborate shapes (like lavatory faucets) that could not be manufactured economically in any other way. Recent developments in steel casting enable the production of castings that are as strong as rolled steel shapes.

Metals can also be shaped by *machining*, which is a process of cutting unwanted material from a piece of metal to produce the desired shape. Among the most common machining operations is *milling*, in which a rotating cutting wheel is used to cut metal from a workpiece. To produce cylindrical shapes, a piece of metal is rotated against a stationary cutting tool in a *lathe*. Holes are produced by *drilling*, which is usually carried out either in a *drill press* or a lathe. Screw threads may be produced in a hole by the use of a helical cutting tool called a *tap*, and the external threads on a steel rod are cut with a *die*. (The threads on mass-produced screws and bolts are formed at high speed by special rolling machines.) Grinding and polishing machines are used to create and finish flat surfaces. Sawing, shearing, and punching operations, described in Chapter 11, are also common methods of shaping metal components.

An economical method of cutting steel of almost any thickness is with a *flame cutting torch* that combines a slender, high-temperature gas flame with a jet of pure oxygen to burn away the metal. *Plasma cutting* with a tiny supersonic jet of superheated gas that blows away the metal can give more precise cuts at thicknesses of up to 2 inches (50 mm), and *laser cutting* gives high-quality results in thin metal plates.

Sheet metal is fabricated with its own particular set of tools. Shears are used to cut metal sheets. Folds are made on large machines called *brakes* (the results of which are called *brake metal*).

Joining Metal Components

Metal components may be joined either mechanically or by fusion. Most mechanical fastenings require drilled or punched holes for the insertion of screws, bolts, or rivets. Some small-diameter screws that are used with thin metal components are shaped and hardened so that they are capable of drilling and tapping as they are driven. Many sheet metal components, especially roofing sheet and ductwork, are joined primarily with interlocking, folded connections.

High-temperature fusion connections are made by *welding*, in which a gas flame or electric arc melts the metal on both sides of the joint and allows it to flow together with additional molten metal from a welding rod or consumable electrode. *Brazing* and *soldering* are lower-temperature processes in which the parent metal is not melted. Instead, a different metal with a lower melting point (bronze or brass in the case of brazing and a tin alloy in the case of soldering) is melted into the joint and bonds to the pieces that it joins. A fully welded connection is generally as strong as the pieces it connects. A soldered connection is not as strong, but it is easy to make and works well for connecting copper plumbing pipes and sheet metal roofing. As an alternative to welding or soldering, adhesives are occasionally used to join metals in certain nonstructural applications.

Common Metals Used in Building Construction

The ferrous metals include cast iron, wrought iron, steel, and stainless steel. **Cast iron** contains relatively large amounts of carbon and impurities. It is the most *brittle* (vulnerable to sudden failure) ferrous metal. **Wrought iron** is produced by hammering semi-molten iron to produce a metal with long fibers of iron interleaved with long fibers of slag. It has very low iron content, making it stronger in tension and much less brittle than cast iron. Both cast iron and wrought iron found significant use in early metal structures. With the introduction of economical steelmaking processes, though, the roles of both of these earlier metals were largely taken over by steel. Even the ornamental metalwork that we refer to today as wrought iron is frequently made of mild steel. **Steel** is discussed in some detail in Chapter 11, and its many uses are noted throughout this book. In general, all these ferrous metals are very strong, relatively inexpensive, easy to form and machine, and must be protected from corrosion.

Stainless steel, made by alloying steel with other metals, primarily chromium and nickel, forms a self-protecting oxide coating that makes it highly resistant to corrosion. It is harder to form and machine than mild steel and is more costly. It is available in attractive finishes that range from matte textures to a mirror polish. Stainless steel is frequently used in the manufacture of fasteners, roofing and flashing sheet, hardware, railings, and other ornamental metal items.

Stainless steel is available in different alloys distinguished, most importantly, by their level of corrosion resistance. *Type 304 stainless steel* is the type most commonly specified and provides adequate corrosion resistance for most applications. Type 304 stainless steel may also be referred to as Type 18-8, the two numbers referring to the percentages of chromium and nickel, respectively, in this alloy. *Type 316 stainless steel*, with higher nickel content and the addition of small amounts of molybdenum, is more corrosion resistant than Type 304. It is specified for use in marine environments where salt-laden air can lead to the accelerated corrosion of less resistant stainless steel alloys. *Type 410 stainless steel* has a lower chromium content and is less corrosion resistant than the 300 series alloys. However, this alloy also has a different metallic crystal structure that, unlike the 300 series alloys, allows it to be hardened through heat treatment. Self-drilling, self-tapping stainless steel fasteners, whose threads must be tough enough to cut through structural steel or concrete, are frequently made of hardened Type 410 stainless steel.

Aluminum (spelled and pronounced "aluminium" in the British Commonwealth) is the nonferrous metal most often used in construction. Its density is about one-third that of steel and it has moderate to high strength and stiffness, depending on which of a multitude of alloys is selected. It can be hardened by cold working, and some alloys can be heat-treated for increased strength. It can be hot- or cold-rolled, cast, forged, drawn, and stamped, and is particularly well adapted to extrusion (see Chapter 21). Aluminum is self-protecting from corrosion, easy to machine, and has thermal and electrical conductivities that are almost as high as those of copper. It is easily made into thin foils that find wide use in thermal-insulating and vapor-retarding materials. With a mirror finish, aluminum in foil or sheet form reflects more heat and light than any other architectural material. Typical uses of aluminum in buildings include roofing and flashing sheet, ductwork, curtain wall components, window and door frames, grills, ornamental railings, siding, hardware, electrical wiring, and protective coatings for other metals, chiefly steel. Aluminum powder is used in metallic paints, and aluminum oxide is used as an abrasive in sandpaper and grinding wheels.

Copper and copper alloys are widely used in construction. Copper is slightly more dense than steel and is bright orange-red in color. When it oxidizes, it forms a self-protecting coating that ranges in color from blue-green to black, depending on the contaminants in the local atmosphere. Copper is moderately strong and can be made stronger by alloying or cold working, but it is not amenable to heat treatment. It is ductile and easy to fabricate. It has the highest thermal and electrical conductivity of any metal used in construction. It may be formed by casting, drawing, extrusion, and hot- or cold-rolling. The primary uses of copper in buildings are roofing and flashing sheet, piping and tubing, and wiring for electricity and communications. Copper is an alloying element in certain corrosion-resistant steels, and copper salts are used as wood preservatives.

Copper is the primary constituent of two versatile alloys, bronze and brass. **Bronze** is a reddish-gold metal that traditionally consists of 90 percent copper and 10 percent tin. Today, however, the term "bronze" is applied to a wide range of alloys that may also incorporate such metals as aluminum, silicon, manganese, nickel, and zinc. These various bronzes are found in buildings in the form of statuary, bells, ornamental metalwork, door and cabinet hardware, and weatherstripping. **Brass** is formulated of copper and zinc plus small amounts of other metals. It is usually lighter in color than bronze, more of a straw yellow, but in contemporary usage the line between brasses and bronzes has become rather indistinct, and the various brasses occur in a wide range of colors, depending on the formulation. Brass, like bronze, is resistant to corrosion. It can be polished to a high luster. It is widely used in hinges and doorknobs, weatherstripping, ornamental metalwork, screws, bolts, nuts, and plumbing faucets (where it is usually plated with chromium). On a volumetric basis, brass, bronze, and copper are expensive metals, but they are often the most economical materials for applications that require their unique combination of functional and visual properties. For greater economy, they are frequently plated electrolytically onto steel for such uses as door hinges and locksets.

Zinc is a blue-white metal that is low in strength, relatively brittle, and moderately hard. Zinc alloy sheet is used for roofing and flashing. Alloys of zinc are also used for casting small hardware parts such as doorknobs, cabinet pulls and hinges, bathroom accessories, and components of electrical fixtures. These *die castings*, which are usually electroplated with another metal such as chromium for appearance, are not especially strong, but they are economical and they can be very finely detailed.

The most important use of zinc in construction is for galvanizing, the application of a zinc coating to prevent steel from rusting. The zinc itself forms a self-protecting gray oxide coating, and even if the zinc is accidentally scratched through to the steel beneath, the zinc interacts electrochemically with the exposed steel to continue to protect the steel from corrosion—a phenomenon called *galvanic protection. Hot-dip galvanizing*, in which the steel parts are submerged in a molten zinc bath to produce a thick coating, is the most durable form of galvanizing. Much less durable is the thin coating produced by *electrogalvanizing*. Threaded steel fasteners and other small parts may be *mechanically galvanized*, in which zinc is fused to the steel at room temperature in a tumbler that contains zinc dust, impact media (such as ball bearings, for example), and other materials. Mechanical galvanizing produces a coating that is especially uniform and consistent in thickness. Steel sheet for architectural roofing is also frequently coated with an aluminum-zinc alloy coating. The aluminum provides a superior protective oxide coating, and the zinc provides galvanic protection if the coating becomes damaged and the base steel exposed. (For a more detailed discussion of galvanic action, see Chapter 16.)

Tin is a soft, ductile, silvery metal that forms a self-protecting oxide layer. The ubiquitous tin can is actually made of sheet steel with an internal corrosion-resistant coating of tin. Tin is found in buildings primarily as a constituent of terne metal, an alloy of 80 percent lead and 20 percent tin that was used in the past as a corrosion-resistant coating for steel or stainless steel roofing sheet. Today, zinc-tin alloy coated steel and stainless steel sheets are available for use as roofing metals that are close in appearance and durability to traditional terne metal.

Chromium is a very hard metal that can be polished to a brilliant mirror finish. It does not corrode in air. It is often electroplated onto other metals for use in ornamental metalwork, bathroom and kitchen accessories, door hardware, and plumbing and lighting fixtures. It is also a major alloying ingredient in stainless steel and many other metals, to which it imparts hardness, strength, and corrosion resistance. Chromium compounds are used as colored pigments in paints and ceramic glazes.

Magnesium is a strong, remarkably lightweight metal (less than one-quarter the density of steel) that is much used in aircraft but remains too costly for general use in buildings. It is found on the construction site as a material for various lightweight tools and as an alloying element that increases the strength and corrosion resistance of aluminum.

Titanium is also low in density, about half the weight of steel, very strong, and one of the most corrosion resistant of all metals. It is a constituent in many alloys, and its oxide has replaced lead oxide in paint pigments. Though expensive, titanium has recently begun to appear on the building construction site in the form of architectural sheet metal for siding and roofing.

CSI/CSC MasterFormat Sections for Light Gauge Steel Frame Construction	
05 40 00	**COLD-FORMED METAL FRAMING**
05 41 00	**STRUCTURAL METAL STUD FRAMING**
05 42 00	**COLD-FORMED METAL JOIST FRAMING**
05 44 00	**COLD-FORMED METAL TRUSSES**
09 22 00	**SUPPORTS FOR PLASTER AND GYPSUM BOARD**
09 22 16	Non-Structural Metal Framing

KEY TERMS

light gauge steel	temper	die
cold-formed metal framing	anneal	flame cutting torch
stud or joist section, C-section	cold working	plasma cutting
track section	galvanize	laser cutting
channel section	electroplating	sheet metal brake
furring channel	anodize	brake metal
gauge	casting	welding
self-drilling screw	rolling	brazing
nested member	extrusion	soldering
web stiffener	forging	brittle
gypsum sheathing panel	stamping	Types 304, 316, 410 stainless steel
nonstructural metal framing	drawing	die casting
thermal bridging	machining	galvanic protection
prescriptive requirement	milling	hot-dip galvanizing
ductile	lathe	electrogalvanizing
alloy	drilling	mechanical galvanizing
heat treatment	drill press	
quench	tap	

REVIEW QUESTIONS

1. How are light gauge steel framing members manufactured?

2. How do the details for a house framed with light gauge steel members differ from those for a similar house with wood platform framing?

3. What special precautions should you take when detailing a steel-framed building to avoid excessive conduction of heat through the framing members?

4. If a building framed with light gauge steel members must be totally noncombustible, what materials would you use for subflooring and wall sheathing?

5. What is the advantage of a prescriptive building code for light gauge steel framing?

6. Compare the advantages and disadvantages of wood light frame construction and light gauge steel frame construction.

EXERCISES

1. Convert a set of details for a wood light frame house to light gauge steel framing.

2. Visit a construction site where light gauge steel studs are being installed. Grasp an installed stud that has not yet been sheathed, at chest height, and twist it clockwise and counterclockwise. How resistant is the stud to twisting? How is this resistance increased as the building is completed?

3. On this same construction site, make sketches of how electrical wiring, electric fixture boxes, and pipes are installed in metal framing.

SELECTED REFERENCES

American Iron and Steel Institute. *AISI Cold-Formed Steel Design Manual.* 1996, Chicago.

This engineering reference work contains structural design tables and procedures for light gauge steel framing.

International Code Council. *International Residential Code for One- and Two-Family Dwellings.* Falls Church, VA, Author, 2002.

This code incorporates full design information and other code provisions, applicable throughout most of the United States, for light gauge steel frame residential construction.

WEB SITES

Light Gauge Steel Frame Construction

Author's supplementary web site: **www.ianosbackfill.com/12_light_gauge_steel_frame_construction**

Dietrich Metal Framing: **www.dietrichindustries.com**

Steel Framing Alliance: **www.steelframing.org**

Steel Stud Manufacturers Association (SSMA): **www.ssma.com**

United States Gypsum: **www.usg.com**

Wei-Wen Yu Center for Cold-Formed Steel Structures (CCFSS): **www.ccfssonline.org**

FROM CONCEPT TO REALITY

PROJECT: Camera Obscura at Mitchell Park, Greenport, New York
ARCHITECT: SHoP/Sharples Holden Pasquarelli

The *camera obscura* is an ancient device: a room-sized projector used to display views of the room's surroundings within the camera, where these images may be viewed by the camera's occupants. In undertaking the Camera Obscura at Mitchell Park, SHoP Studio accepted the nostalgic theme of the client's program, and added to it its own interests in developing cutting-edge design and construction methods (Figures A–D).

SHoP designed and documented the Camera Obscura entirely in the form of a three-dimensional digital model. Beyond facilitating the project's unconventional geometry, the

FIGURE A
Section and elevation.

use of digital modeling created significant opportunities for changing the way in which this project would be built and altering the architect's contribution to that process.

For example, as a consequence of the digital model, much of the traditional construction-phase shop drawing preparation process was stood on its head in this project. Instead of the fabricator preparing drawings for review by the architect/ engineer team, the model created by SHoP for the project design was used to generate templates that were supplied by the architect in digital form to the fabricator. The fabricator used these templates to drive automated machinery that transformed raw materials stock to cut, formed, and drilled components. Pieces were delivered to the construction site individually prelabeled, ready for assembly in the final structure.

FIGURE B
Cutting templates derived from the digital model.

TEMPLATE KEY - INNER SHEATHING (SKATELITE PRO)
SCALE: 1/4"=1'-0"

5'X12' SKATELITE PRO TM PANELS
1/4" THICK, BLACK FINISH (TYP.)

E | INNER SHEATHING - SKATELITE PRO

NOTE: FULL SCALE PAPER TEMPLATES OF ALL SKATELITE SHEATHING PANELS TO BE PROVIDED BY ARCHITECT FOR FABRICATION.

FIGURE C

Individually sized and shaped aluminum fins.

The digital building model also allowed SHoP to explore the possibilities of customization beyond what is practical with more conventional design methods. In the Camera Obscura, many of the building pieces were unique in shape and were intended for use in a single predetermined location within the building. If this proposition were undertaken using conventional construction methods, it would imply significant cost premiums. By capitalizing on the descriptive capabilities of the model coupled with automated fabrication, the costs to produce these items and to organize their assembly can be made competitive with traditional construction.

SHoP also used the digital building model to generate construction drawings that communicated how the building would be assembled in the field. For example, exploded assembly diagrams were used to study and illustrate the sequences in which systems were constructed. Cutting patterns were organized to minimize cutting time and material waste. Templates were plotted full size on paper and delivered to the building site to assist with construction layout.

SHoP's interest in creatively exploring the means of construction carried with it additional responsibilities. Because SHoP provided templates for forming various components, it assumed greater responsibility for ensuring that these components would fit properly when assembled in the field. As a consequence, SHoP worked closely with fabricators and suppliers to educate themselves regarding both the potential capabilities and the limitations of the materials with which they designed. In some instances, material properties, such as the practical bend radii of metals of various gauges, were built into the parameters of the digital model itself. Full-size mockups could be constructed on site to verify assembly concepts and tolerances prior to fabricating the bulk of the project's components. And as with any design firm committed to improving its professional capabilities, the lessons SHoP learned from completed work are conscientiously applied to new projects.

SHoP's goal was to connect the tools of design with the techniques of construction. Note that all images shown here are taken from actual construction drawings, for a project awarded through a competitive, public bid process. With the innovative application of new design tools and a willingness to challenge the conventional professional boundaries, SHoP aims to open up new architectural possibilities. These efforts are still new, and their full potential is perhaps not yet realized. Yet they already demonstrate how the exploration of materials and techniques of construction can be an integral part of a creative design practice.

Special thanks to SHoP/Sharples Holden Pasquarelli, and William Sharples, Principal, for assistance with the preparation of this case study.

Figure D
Assembly diagram.

FLAT SEAM ZINC ROOF - SEE ROOF PLAN, SHEET A323.

PLASTIC DRAINGE SHEET (VENTILATION SHEET).

HIGH TEMPERATURE ICE & WATER SHIELD.

3/4" THICK PLYWOOD ROOF AND PARAPET SHEATHING - SCREW THROUGH 1/4" THICK SKATELITE PRO INTO 2X6 IPE ROOF JOIST BELOW, STAGGER SCREWS AS REQUIRED.

1/4" THICK BLACK SKATELITE PRO - ALL SEAM LINES TO MATCH AT ROOF JOISTS; SCREW INTO 2×6 IPE ROOF JOISTS AT 6" O.C., TYP. DOUBLE 2×4 HEADER AS SHOWN.

ALLUMINUM "SAWTOOTH" IPE PLANKING GUIDE AND GUSSET ASSEMBLY - SEE SHEET A340.

IPE PLANKING MOUNTING HINGE - SEE SHEET A340.

MILLED 2×6 IPE PLANKING - WEATHER SCREEN - SEE SHEETS A300, A301, A323, A340.

IPE 2×4

IPE 2X6 ROOF JOIST, SCREWED COPLANAR WITH IPE 2×4, TYP. - SEE DETAIL, SHEET A330.

TO STEEL COMPRESSION RING / BEAM ASSEMBLY

TO STEEL COMPRESSION RING / BEAM ASSEMBLY

IPE 3X6 COMPRESSION CHORD (WHERE OCCURS).

STEEL TENSION RING ASSEMBLY - SEE SHEET A334; FASTEN TO IPE 3X6 COMPRESSION CHORDS WITH STAINLESS STEEL 1/2" DIA. THRU-BOLTS.

ALUMINUM FIN / SHELF SUPPORT - SEE SHEETS A335 & A336.

ALUMINUM FIN SHELF ASSEMBLY - SEE SHEETS A335 & A336.

REVOLVING METAL WALL ASSEMBLY, SEE SHEET A334.

SCREW THROUGH EPDM SEALANT BED FOR VERTICAL PENETRATIONS AS REQUIRED.

FULLY ADHERED EPDM.

HIGH TEMPERATURE ICE & WATER SHIELD.

1/2" PLYWOOD SHEATHING - FULLY GLUED AND SCREWED INTO 1/2" PLYWOOD SHEATHING; 3/4" SCREWS.

1/2" PLYWOOD SHEATHING - SCREW THROUGH 1/4" THICK SKATELITE PRO INTO IPE 2X4; STAGGER SCREWS AS REQUIRED.

1/4" SKATELITE PRO SHEATHING, SCREW INTO IPE 2×4 - ALL SEAM LINES TO MATCH AT IPE 2X4s AND ALUMINUM SHELVES - SEE SHEET A339 FOR TEMPLATES.

SHEATHING COMPOSITE

HILTI ANCHOR BOLT, TYPE

1/4"THICK STEEL COVER PLATE - FLUSH MOUNTED INTO FINISHED CONCRETE FLOOR

1 1/2" DIAMETER TRANSFER BALLS MOUNTED ON STEEL ANGLE.

STEEL MOUNTING PLATES PROVIDED AS REQUIRED FOR LEVEL INSTALLATION OF TRANSFER BALLS / REVOLVING METAL WALL ASSEMBLY.

1/2" DIAMETER STAINLESS STEEL THRU-BOLT.

STEEL SILL-PLATE ASSEMBLY - SEE SHEETS A335 & A336.

CONCRETE FOUNDATION / PAD - SEE SHEETS A310 & A 320.

2 / A338   EXPLODED AXONOMETRIC - ASSEMBLY DETAILS
NOT TO SCALE

13

CONCRETE CONSTRUCTION

- **History**

- **Cement and Concrete**
  Cement

  SUSTAINABILITY IN CONCRETE CONSTRUCTION

  Aggregates and Water
  Supplementary Cementitious Materials
  Admixtures

- **Making and Placing Concrete**
  Proportioning Concrete Mixes
  Handling and Placing Concrete
  Curing Concrete

- **Formwork**

- **Reinforcing**
  The Concept of Reinforcing

Steel Bars for Concrete Reinforcement
Fabrication and Erection of Reinforcing Bars
Reinforcing a Simple Concrete Beam
Reinforcing a Continuous Concrete Beam
Reinforcing Structural Concrete Slabs
Two-Way Slab Action
Reinforcing Concrete Columns
Fibrous Reinforcing

- **Concrete Creep**

- **Prestressing**
  Pretensioning
  Posttensioning

- **ACI 301**

- **Innovations in Concrete Construction**

The Chapel of St. Ignatius, Seattle University, designed by architect Steven Holl. Holl creates a striking composition of precisely detailed tilt-up concrete wall panels and architectural sheet metal roofing. *(Photo by Joseph Iano.)*

Concrete is the universal material of construction. According to the World Business Council for Sustainable Development, concrete is, after water, the most widely used material on earth. The raw ingredients for its manufacture are readily available in almost every part of the globe, and concrete can be made into buildings with tools ranging from a primitive shovel to a computerized precasting plant. Concrete does not rot or burn; it is relatively low in cost; and it can be used for every building purpose, from lowly pavings to sturdy structural frames to handsome exterior claddings and interior finishes.

But concrete is the only major structural material commonly manufactured on site, it has no form of its own, and it has no useful tensile strength. Before its limitless architectural potential can be realized, the designer and builder must learn to produce concrete of consistent and satisfactory quality, to combine concrete skillfully with steel reinforcing to bring out the best structural characteristics of each material, and to mold and shape it to forms appropriate to its qualities and to our building needs.

HISTORY

The ancient Romans, while quarrying limestone for mortar, accidentally discovered a silica- and alumina-bearing mineral on the slopes of Mount Vesuvius that, when mixed with limestone and burned, produced a cement that exhibited a unique property: When mixed with water and sand, it produced a mortar that could harden underwater as well as in the air. In fact, it was stronger when it hardened underwater. This mortar was also harder, stronger, much more adhesive, and cured much more quickly than the ordinary lime mortar to which they were accustomed. In time, it not only became the preferred mortar for use in all their building projects, but it also began to alter the character of Roman construction. Masonry of stone or brick came to be used to build only the surface layers of piers, walls, and vaults, and the hollow interiors were filled entirely with large volumes of the new type of mortar (Figure 13.2). We now know that this mortar contained

all the essential ingredients of modern portland cement and that the Romans were the inventors of concrete construction.

Knowledge of concrete construction was lost with the fall of the Roman Empire, not to be regained until the latter part of the 18th century, when a number of English inventors began experimenting with both natural and artificially produced cements. Joseph Aspdin, in 1824, patented an artificial cement that he named *portland cement*, after English Portland limestone, whose durability as a building stone was legendary. His cement was soon in great demand, and the name "portland," for the cementitious component of concrete, remains in use today.

Reinforced concrete, in which steel bars are embedded to resist tensile forces, was developed in the 1850s by several people simultaneously. Among them were the Frenchman J. L. Lambot, who built several reinforced concrete boats in Paris in 1854, and an American, Thaddeus Hyatt, who made and tested a number of reinforced concrete beams. But the combination of steel and concrete did not come into widespread

FIGURE 13.1

At the time concrete is placed, it has no form of its own. This bucket of fresh, wet concrete was filled on the ground by a transit-mix truck and hoisted to the top of the building by a crane. The worker at the right has opened the valve in the bottom of the bucket to discharge the concrete. *(Reprinted with permission of the Portland Cement Association from* Design and Control of Concrete Mixtures, *12th edition; PHOTOS: Portland Cement Association, Skokie, IL.)*

use until a French gardener, Joseph Monier, obtained a patent for reinforced concrete flower pots in 1867 and went on to build concrete water tanks and bridges of the new material. By the end of the 19th century, engineering design methods had been developed for structures of reinforced concrete and a number of major structures had been built. By this time, the earliest experiments in prestressing (placing the reinforcing steel under tension before the structure supports a load) had also been carried out, although it remained for Eugene Freyssinet in the 1920s to establish a scientific basis for the design of prestressed concrete structures.

CEMENT AND CONCRETE

Concrete is a rocklike material produced by mixing coarse and fine *aggregates*, portland cement, and water and allowing the mixture to harden. *Coarse aggregate* is normally gravel or crushed stone, and *fine aggregate* is sand. Portland cement, hereafter referred to simply as cement, is a fine gray powder. During the hardening, or *curing*, of concrete, the cement combines chemically with water to form strong crystals that bind the aggregates together, a process called *hydration*. During this process, considerable heat, called *heat of hydration*, is given off. In addition, especially as excess water evaporates from the concrete, the concrete shrinks slightly, a phenomenon referred to as *drying shrinkage*. The curing process, and the gradual increase in the strength of the concrete that occurs with it, does not end abruptly unless it is artificially interrupted. Rather, it tapers off over long periods of time, though, for practical purposes, concrete is normally considered cured and at full design strength after 28 days.

In properly formulated concrete, the majority of the volume consists of coarse and fine aggregate, proportioned and graded so that the fine particles fill the spaces between the coarse ones (Figure 13.3). Each particle is completely coated with a paste of cement and water that bonds it fully to the surrounding particles.

Cement

Portland cement may be manufactured from any of a number of raw materials, provided that they are combined to yield the necessary amounts of lime, iron, silica, and alumina. Lime is commonly furnished by limestone, marble, marl (a sedimentary rock), or seashells. Iron, silica, and alumina

FIGURE 13.2
Hadrian's Villa, a large palace built near Rome between AD 125 and 135, used unreinforced concrete extensively for structures such as this dome. *(Photo by Edward Allen.)*

FIGURE 13.3
Photograph of a polished cross section of hardened concrete, showing the close packing of coarse and fine aggregates and the complete coating of every particle with cement paste. *(Reprinted with permission of the Portland Cement Association from* Design and Control of Concrete Mixtures, *12th edition; PHOTOS: Portland Cement Association, Skokie, IL.)*

may be provided by clay or shale. The exact ingredients depend on what is readily available, and the recipe varies from one geographic region to another, often including slag or flue dust from iron furnaces, chalk, sand, ore washings, bauxite, and other minerals. To make portland cement, the selected constituents are crushed, ground, proportioned, and blended. Then they are conducted through a long, rotating kiln at temperatures of 2600 to 3000 degrees Fahrenheit (1400–1650°C) to produce *clinker* (Figures 13.4 and 13.5). After cooling, the clinker is pulverized to a powder finer than flour. Usually at this stage a small amount of gypsum is added to act as a retardant during the later concrete curing process. This finished powder, portland cement, is either packaged in bags or shipped in bulk. In the United States, a standard bag of cement contains 1 cubic foot (0.09 m^2) of volume and weighs 94 pounds (43 kg).

The quality of portland cement is established by ASTM C150, which identifies eight different types:

Type I	Normal
Type IA	Normal, air entraining
Type II	Moderate resistance to sulfate attack
Type IIA	Moderate sulfate resistance, air entraining
Type III	High early strength
Type IIIA	High early strength, air entraining
Type IV	Low heat of hydration
Type V	High resistance to sulfate attack

Type I cement is used for most purposes in construction. Types II and V are used where the concrete will be in contact with water that has a high concentration of sulfates. Type III hardens more quickly than the other types and is employed in situations where a reduced curing period is desired (as may be the case in cold weather), in the precasting of concrete structural elements, or when the construction schedule must be accelerated. Type IV is used in massive structures such as

STEPS IN THE MANUFACTURE OF PORTLAND CEMENT

STONE IS FIRST REDUCED TO 5-IN. SIZE, THEN 3/4-IN., AND STORED

BURNING CHANGES RAW MIX CHEMICALLY INTO CEMENT CLINKER

FIGURE 13.4

A rotary kiln manufacturing cement clinker. The blended raw ingredients for cement are introduced into the kiln's higher end, top right in this photograph. As the kiln rotates, the ingredients slowly tumble downward toward the lower, far end. Heated air is introduced into the kiln through the large round ducts clustered around the kiln's lower end. (*Photo by Joseph Iano.*)

dams, where the heat emitted during curing may raise the temperature of the concrete to damaging levels. Most cement that is manufactured in North America is designated as Type I/II,

meaning it meets the requirements of both these types.

ASTM C150 also permits the addition of limestone to the finished portland cement (distinct from its use as a raw ingredient in the manufacture of clinker). This relatively recent change results in reductions in raw material consumption, energy consumption, carbon dioxide emissions, and cement dust generation during cement manufacturing.

Air-entraining cements contain ingredients that cause microscopic air bubbles to form in the concrete during mixing (Figure 13.6). These bubbles, which usually comprise 2 to 8 percent

⊢—⊣ 0.01 in

FIGURE 13.6

A photomicrograph of a small section of air-entrained concrete shows the bubbles of entrained air (0.01 inch equals 0.25 mm). (*Reprinted with permission of the Portland Cement Association from* Design and Control of Concrete Mixtures, *12th edition; PHOTOS: Portland Cement Association, Skokie, IL.*)

FIGURE 13.5

Steps in the manufacture of portland cement. (*Reprinted with permission of the Portland Cement Association from* Design and Control of Concrete Mixtures, *12th edition; PHOTOS: Portland Cement Association, Skokie, IL.*)

RAW MATERIALS ARE GROUND TO POWDER AND BLENDED

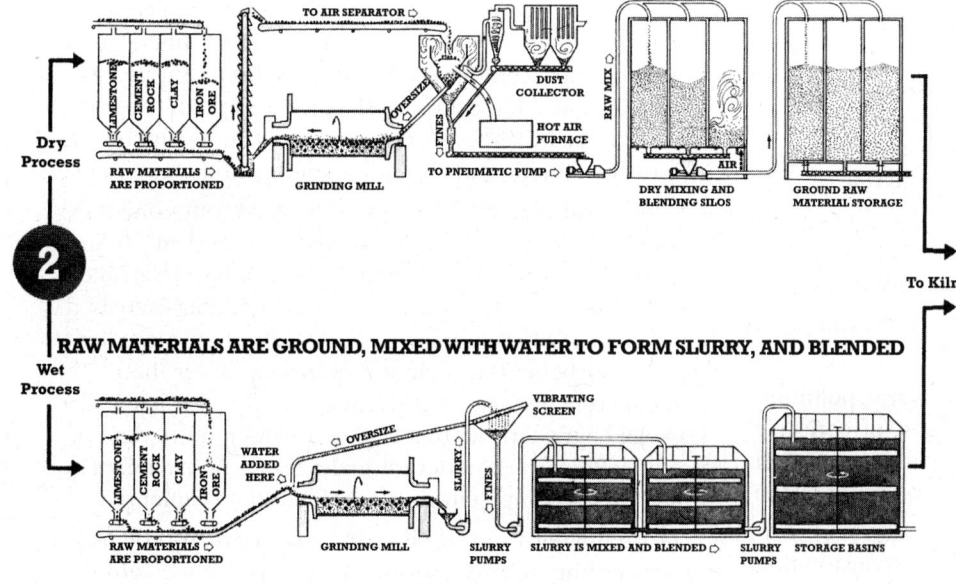

RAW MATERIALS ARE GROUND, MIXED WITH WATER TO FORM SLURRY, AND BLENDED

CLINKER WITH GYPSUM ADDED IS GROUND INTO PORTLAND CEMENT AND SHIPPED

of the volume of the finished concrete, improve workability during placement of the concrete and, more importantly, greatly increase the resistance of the finished concrete to damage caused by repeated cycles of freezing and thawing. Air-entrained concrete is commonly used for pavings and exposed architectural concrete in cold climates. With appropriate adjustments in the formulation of the mix, air-entrained concrete can achieve the same structural strength as normal concrete.

White portland cement is produced by controlling the quantities of certain minerals, such as oxides of iron and manganese, found in the ingredients of cement, that contribute to cement's usual gray color. White portland cement is used for architectural applications to produce concrete that is lighter and more uniform in color or, when combined with other coloring agents, to enhance the appearance of integrally colored concrete.

SUSTAINABILITY IN CONCRETE CONSTRUCTION

• Worldwide each year, the making of concrete consumes 1.6 billion tons (1.5 billion metric tons) of portland cement, 10 billion tons (9 billion metric tons) of sand and rock, and 1 billion tons (0.9 billion metric tons) of water, making the concrete industry the largest user of natural resources in the world.

• The quarrying of the raw materials for concrete in open pits can result in soil erosion, pollutant runoff, habitat loss, and ugly scars on the landscape.

• Concrete construction also uses large quantities of other materials—wood, wood panel products, steel, aluminum, plastics—for formwork and reinforcing.

• The total energy embodied in a pound of concrete varies, especially with the design strength. This is because higher-strength concrete relies on a greater proportion of portland cement in its mix, and the energy required to produce portland cement is very high in comparison to concrete's other ingredients. For average-strength concrete, the embodied energy ranges from about 200 to 300 BTU per pound (0.5–0.7 MJ/kg).

• There are various useful approaches to increasing the sustainability of concrete construction:

  • Use waste materials from other industries, such as fly ash from power plants, slag from iron furnaces, copper slag, foundry sand, mill scale, sandblasting grit, and others, as components of cement and concrete.

  • Use concrete made from locally extracted materials and local processing plants to reduce the transportation of construction materials over long distances.

  • Minimize the use of materials for formwork and reinforcing.

  • Reduce energy consumption, waste, and pollutant emissions from every step of the process of concrete construction, from quarrying of raw materials through the eventual demolition of a concrete building at the end of its useful life.

  • In regions where the quality of the construction materials is low, improve the quality of concrete so that concrete buildings will last longer, thus reducing the demand for concrete and the need to dispose of demolition waste.

Portland Cement

• The production of portland cement is by far the largest user of energy in the concrete construction process, accounting for about 85 percent of the total energy required. Portland cement production also accounts for roughly 5 percent of all carbon dioxide gas generated by human activities worldwide and about 1.5 percent of such emissions in North America.

• For every ton of cement clinker produced, almost a ton of carbon dioxide, a greenhouse gas, is released into the atmosphere.

• Since 1970, the North American cement industry has reduced the amount of energy expended in cement production by one-third, and the industry continues to work toward further reductions.

• The manufacture of cement produces large amounts of air pollutants and dust. In the past 35 years, the emission of particulates from cement production has been reduced by more than 90 percent.

• The cement industry is committed to reducing greenhouse gas emissions per ton of product by 10 percent from 1990 levels by the year 2020. According to the Portland Cement Association, over concrete's lifetime, it reabsorbs roughly half of the carbon dioxide released during the original cement manufacturing process.

• The amount of portland cement used as an ingredient in concrete, and as a consequence, the energy required to produce the concrete, can be substantially reduced by the addition of certain industrial waste materials with cementing properties to the concrete mix. Substituting such supplementary cementitious materials, including fly ash, silica fume, and blast furnace slag, for up to half the portland cement in the concrete, can result in as much as a one-third reduction in embodied energy of the concrete.

• When added to concrete, fly ash is most commonly substituted for portland cement at rates of between 15 and 25 percent. Mixes with even higher replacement rates, called *high-volume-fly-ash (HVFA) concrete*, are also finding increased acceptance. Concrete mixed with fly ash as an ingredient gains other benefits as well: It requires less water than normal concrete, its heat of hydration is lower, and it shrinks less—all characteristics that lead to a denser, more durable product. Research is underway to develop concrete mixes in which fly ash completely replaces all portland cement.

• Waste materials from other industries can also be used as cementing agents; wood ash and rice-husk ash are two examples. Used motor oil and used rubber vehicle tires can be employed as fuel in cement kilns. While consuming waste products from other industries, a cement manufacturing plant can, if efficiently operated, generate virtually no solid waste itself.

• Technologies currently under development aim to significantly reduce carbon dioxide emissions in comparison to current manufacturing methods. By injecting carbon dioxide into concrete during its manufacture, it may be possible to create a stronger and harder material, but with a 20 percent reduction in net carbon dioxide emissions. Carbon dioxide emissions from aluminosilicate cement manufacturing are as much as 80 to 90 percent less than those of ordinary portland cement. Cement made from magnesium produces a material that is actually carbon negative, absorbing significantly more carbon over its lifetime than is emitted during its manufacture.

Aggregates and Water

• Sand and crushed stone come from abundant sources in many parts of the world, but high-quality aggregates are becoming scarce in some countries.

• In rare instances, aggregate in concrete has been found to be a source of radon gas. Concrete itself is not associated with indoor air quality problems.

• Waste materials such as crushed, recycled glass, used foundry sand, and crushed, recycled concrete can substitute for a portion of the conventional aggregates in concrete.

• Water of a quality suitable for concrete is scarce in many developing countries. Concretes that use less water by using superplasticizers, air entrainment, and fly ash could be helpful.

Wastes

• A significant percentage of fresh concrete is not used because the truck that delivers it to the building site contains more than is needed for the job. This concrete is often dumped on the site, where it hardens and is later removed and taken to a landfill for disposal. An empty transit-mix truck must be washed out after transporting each batch, which produces a substantial volume of water that contains portland cement particles, admixtures, and aggregates. These wastes can be recovered and recycled as aggregates and mixing water, but more concrete suppliers need to implement schemes for doing this.

Formwork

• Formwork components that can be reused many times have a clear advantage over single-use forms, which represent a large waste of construction material.

• Form release compounds and curing compounds should be chosen for low volatile organic compound content and biodegradability.

• Insulating concrete forms eliminate most temporary formwork and produce concrete walls with high thermal insulating values.

Reinforcing

• In North America, reinforcing bars are made almost entirely from recycled steel scrap, primarily junked automobiles. This reduces resource depletion and energy consumption significantly.

Concrete Buildings

• Considering only the materials production phase, concrete has a relatively high global warming potential in comparison to other major structural materials. However, once occupied, concrete buildings can be designed to operate more efficiently, for example by exploiting the energy saving potential of thermal mass. Comparative studies indicate that, over the complete material life cycle, the global warming potential of concrete buildings is at least as good as that of buildings constructed of wood or steel.

Demolition and Recycling

• When a concrete building is demolished, its reinforcing steel can be recycled.

• In many if not most cases, fragments of demolished concrete can be crushed, sorted, and used as aggregates for new concrete. At present, however, most demolished concrete is buried on the site, used to fill other sites, or dumped in a landfill.

Green Uses of Concrete

• Pervious concrete can be used to make porous pavings that allow stormwater to filter into the ground, helping to recharge aquifers and reduce stormwater runoff.

• Concrete is a durable material that can be used to construct buildings that are long-lasting and suitable for adaptation and reuse, thereby reducing the environmental impacts of building demolition and new construction.

• In brownfield development, concrete fill materials can be used to stabilize soils and reduce leachate concentrations.

• Where structured parking garages (often constructed of concrete) replace surface parking, open space is preserved.

• Concrete's thermal mass can be exploited to reduce building heating and cooling costs by storing excess heat during overheated periods of the day or week and releasing it back to the interior of the building during underheated periods.

• Lighter-colored concrete paving reflects more solar radiation than darker asphalt paving, leading to lower paving surface temperatures and reduced urban heat island effects.

• Interior concrete slabs made with white concrete can improve illumination, visibility, and worker safety within interior spaces without the expense or added energy consumption of extra light fixtures or increasing the light output from existing fixtures. White concrete is made with white cement and white aggregates.

• Photocatalytic agents can be added to concrete used in the construction of roads and buildings. In the presence of sunlight, the concrete chemically breaks down carbon monoxide, nitrogen oxide, benzene, and other air pollutants.

Aggregates and Water

Because aggregates make up roughly three-quarters of the volume of concrete, the strength of a concrete is heavily dependent on the quality of its aggregates. Aggregates for concrete must be strong, clean, resistant to freeze-thaw deterioration, chemically stable, and properly graded for size distribution. An aggregate that is dusty or muddy will contaminate the cement paste with inert particles that weaken it. An aggregate that contains any of a number of chemicals, from sea salt to organic compounds, can cause problems ranging from corrosion of reinforcing steel to retardation of the curing process and weakening of the concrete. A number of standard ASTM laboratory tests are used to assess the various qualities of aggregates (Figure 13.7).

Size distribution of aggregate particles is important because a range of sizes must be included and properly proportioned in the concrete mix to achieve close packing of the particles. A concrete aggregate is graded for size using a standard assortment of sieves with diminishing mesh spacings, then weighing the percentage of material that passes through each sieve, in the same manner as described in Chapter 2. Aggregates must also be sized so that the largest particle in the mix is small enough to pass easily between reinforcing bars and fit easily into the formwork. In general, the aggregate should be no larger than three-fourths of the clear spacing between bars or one-third the depth of a slab. For very thin slabs and toppings, a ⅜-inch (9-mm) maximum aggregate diameter is often specified. A ¾-inch or 1½-inch (19-mm or 38-mm) maximum size is common for much slab and structural work. Aggregate with diameters up to 6 inches (150 mm) are used in dams and other massive structures.

Lightweight aggregates are used for various special types of concrete. *Structural lightweight aggregates* are made from minerals such as shale. The shale is crushed to the desired particle sizes, then heated in an oven to a temperature at which it becomes plastic in consistency. The small amount of water that occurs naturally in the shale turns to steam and expands the softened particles like popcorn. Concrete made from this *expanded shale aggregate* has a density about 25 percent less than that of normal concrete, yet it is nearly as strong. Nonstructural lightweight concretes are made for use in insulating roof toppings that have densities only one-fourth to one-sixth that of normal concrete. The aggregates in these concretes are usually expanded mica (*vermiculite*) or expanded volcanic glass (*perlite*), both produced by processes much like that used to make expanded shale. However, both of these aggregates are much less dense than expanded shale, and the density of the concretes in which they are used is further reduced by admixtures that entrain large amounts of air during mixing.

ASTM standard C1602 defines the requirements for mixing water for concrete. Generally, *water* must be free of harmful substances, especially organic material, clay, and salts such as chlorides and sulfates. Water that is suitable for drinking has traditionally been considered suitable for making concrete.

Supplementary Cementitious Materials

Various mineral products, called *supplementary cementitious materials (SCMs)*, may be added to concrete mixtures as a substitute for some portion of the portland cement to achieve a range of benefits. Supplementary cementitious materials are classified as either pozzolans or hydraulic cements.

Pozzolans are materials that react with the calcium hydroxide in wet concrete to form cementing compounds. They include:

• *Fly ash*, a fine powder that is a waste product from coal-fired power plants, increases concrete strength, decreases permeability, increases sulfate resistance, reduces temperature rise during curing, reduces the amount of mixing water needed, and improves pumpability and workability of concrete. Fly ash also reduces concrete drying shrinkage.

• *Silica fume*, also known as *microsilica*, is a powder that is approximately 100 times finer than portland cement, consisting mostly of silicon dioxide. It is a byproduct of electronic semiconductor chip manufacturing. When added to a concrete mix, it produces extremely high-strength concrete that also has very low permeability.

• *Natural pozzolans*, mostly derived from shales or clays, are used for purposes such as reducing the internal temperature of curing concrete, reducing the reactivity of concrete with aggregates containing sulfates, or improving the workability of concrete. *High-reactivity metakaolin* is a unique white-colored natural pozzolan that enhances the brilliance of white or colored concrete while also improving the material's workability, strength, and density. These characteristics make it especially well suited as an ingredient in exposed architectural concrete applications where appearance and finish quality are critical.

FIGURE 13.7
Taking a sample of coarse aggregate from a crusher yard for testing. *(Reprinted with permission of the Portland Cement Association from* Design and Control of Concrete Mixtures, *12th edition; PHOTOS: Portland Cement Association, Skokie, IL.)*

Blast furnace slag (also called *slag cement*), a byproduct of iron manufacture, is a *hydraulic cement*, meaning that, like portland cement, it reacts directly with water to form a cementitious compound. It may be added to concrete mixes to improve workability, increase strength, reduce permeability, reduce temperature rise during curing, and improve sulfate resistance.

Supplementary cementitious materials may be added to portland cement during the cement manufacturing process, in which case the resulting product is called a *blended hydraulic cement* (ASTM C595), or they may be added to the concrete mix at the batch plant. The use of SCMs also enhances the sustainability of concrete by reducing reliance on more energy-intensive portland cement and, in many cases, by making productive use of waste products from other industrial manufacturing processes. Half or more of the concrete produced in North America includes some supplementary cementitious materials in its mix.

Admixtures

Ingredients other than cement and other cementitious materials, aggregates, and water, broadly referred to as *admixtures*, are often added to concrete to alter its properties in various ways:

• *Air-entraining admixtures* increase the workability of the wet concrete, reduce freeze-thaw damage, and, when used in larger amounts, create very lightweight nonstructural concretes with thermal insulating properties.

• *Water-reducing admixtures* allow a reduction in the amount of mixing water while retaining the same workability, which results in a higher-strength concrete.

• *High-range water-reducing admixtures*, also known as *superplasticizers*, are organic compounds that transform a stiff concrete mix into one that flows freely into the forms. They are used either to facilitate placement of concrete under difficult circumstances or to reduce the water content of a concrete mix so as to increase its strength.

• *Accelerating admixtures* cause concrete to cure more rapidly, and *retarding admixtures* slow its curing to allow more time for working with the wet concrete.

• *Workability agents* improve the plasticity of wet concrete to make it easier to place in forms and finish. They include pozzolans and air-entraining admixtures, along with certain fly ashes and organic compounds.

• *Shrinkage-reducing admixtures* reduce drying shrinkage and the cracking that results therefrom.

• *Corrosion inhibitors* are used to reduce rusting of reinforcing steel in structures that are exposed to road deicing salts or other corrosion-causing chemicals.

• *Freeze protection admixtures* allow concrete to cure satisfactorily at temperatures as low as 20 degrees Fahrenheit (7°C).

• *Extended set-control admixtures* may be used to delay the curing reaction in concrete for any period up to several days. They include two components: The stabilizer component, added at the time of initial mixing, defers the onset of curing indefinitely; the activator component, added when desired, reinitiates the curing process.

• *Coloring agents* are dyes and pigments used to alter and control the color of concrete for building components whose appearance is important.

MAKING AND PLACING CONCRETE

The quality of cured concrete is measured by any of several criteria, depending on its end use. For structural columns, beams, and slabs, compressive strength and stiffness are important. For pavings and floor slabs, flatness, surface smoothness, and abrasion resistance are also important. For pavings and exterior concrete walls, a high degree of weather resistance is required. Watertightness is important in concrete tanks, dams, and walls. Regardless of the criterion to which one is working, however, the rules for making high-quality concrete are much the same: Use clean, sound ingredients; mix them in the correct proportions; handle the wet concrete properly; and cure the concrete carefully under controlled conditions.

Proportioning Concrete Mixes

The starting point of any mix design is to establish the desired workability characteristics of the wet concrete, the desired physical properties of the cured concrete, and the acceptable cost of the concrete, keeping in mind that there is no need to spend money to make concrete better than it has to be for a given application. Concretes with ultimate compressive strengths as low as 2000 psi (13.8 MPa) are satisfactory for some foundation elements. At the other extreme, concretes with ultimate compressive strengths of 20,000 psi (140 MPa) and more, produced with the aid of silica fume, fly ash, and superplasticizer admixtures, are used in columns of high-rise buildings. Acceptable workability is achievable at any of these strength levels.

Given a proper gradation of satisfactory aggregates, the strength of cured concrete is primarily dependent on the amount of cement in the mix and on the *water–cement (w-c) ratio*. Although a proportion of water is required as a reactant in the curing of concrete, much more must be added to give the wet concrete the necessary fluidity and plasticity for placing and finishing. The extra water eventually evaporates from the concrete, leaving microscopic voids that reduce the strength and surface qualities of the concrete (Figure 13.8). For common concrete applications, water–cement ratios range from about 0.45 to 0.60 by weight, meaning that the weight of the water in the mix does not exceed 45 to 60 percent of the weight of the cement. Relatively

COMPRESSIVE STRENGTH

FIGURE 13.8

The effect of the water–cement ratio on the strength of concrete. "A-E concrete" refers to air-entrained concrete. (*Reprinted with permission of the Portland Cement Association from* Design and Control of Concrete Mixtures, *12th edition; PHOTOS: Portland Cement Association, Skokie, IL.*)

FIGURE 13.9

Charging a transit-mix truck with measured quantities of cement, aggregates, admixtures, and water at a central batch plant. (*Reprinted with permission of the Portland Cement Association from* Design and Control of Concrete Mixtures, *12th edition; PHOTOS: Portland Cement Association, Skokie, IL.*)

high water–cement ratios are often favored by concrete workers, because they produce a fluid mixture that is easy to place in the forms, but such a mix may be deficient in strength and surface qualities. Lower water–cement ratios make concrete that is denser and stronger and that shrinks less during curing. However, without air-entraining or water-reducing admixtures, such a mix may not flow easily into the forms or may finish poorly. It is important that concrete be formulated with the right quantity of water for each situation, enough to ensure workability but not enough to adversely affect the properties of the cured material.

Most concrete in North America is proportioned at central batch plants, using laboratory equipment and engineering knowledge to produce concrete of the proper quality for each project. The concrete is *transit mixed* en route in a rotating drum on the back of a truck so that it is ready to pour by the time it reaches the job site (Figures 13.9 and 13.10).

FIGURE 13.10

A transit-mix truck discharges its concrete, which was mixed en route in the rotating drum, into a truck-mounted concrete pump, which forces it through a hose to the point in the building at which it is being poured. (*Reprinted with permission of the Portland Cement Association from* Design and Control of Concrete Mixtures, *12th edition; PHOTOS: Portland Cement Association, Skokie, IL.*)

For very small jobs, concrete may be mixed at the job site, either in a small power-driven mixing drum or on a flat surface with shovels. For these small jobs, where the quality of the finished concrete generally does not have to be precisely controlled, proportioning is usually done by rule of thumb. Typically, the dry ingredients are measured volumetrically, using a shovel as a measuring device, in proportions such as one shovel of cement to two of sand to three of gravel, with enough water to make a wet concrete that is neither soupy nor stiff.

Each load of transit-mixed concrete is delivered with a certificate from the batch plant that lists its ingredients and their proportions. As a further check on quality, a *slump test* may be performed at the time of pouring to determine if the desired degree of workability has been achieved without making the concrete too wet (Figure 13.11). For structural concrete, standard test cylinders are also poured from each

FIGURE 13.12

Inserting a standard concrete test cylinder into a structural testing machine, where it will be crushed to determine its strength. *(Reprinted with permission of the Portland Cement Association from* Design and Control of Concrete Mixtures, *12th edition; PHOTOS: Portland Cement Association, Skokie, IL.)*

FIGURE 13.11

Measuring concrete slump. The hollow metal cone is filled with concrete and tamped with the rod according to a standard procedure. The cone is carefully lifted off, allowing the wet concrete to sag, or slump, under its own weight. The slump is then measured in the manner shown *(Reprinted with permission of the Portland Cement Association from* Design and Control of Concrete Mixtures, *12th edition; PHOTOS: Portland Cement Association, Skokie, IL.)*

truckload. Within 48 hours of pouring, the cylinders are taken to a testing laboratory, cured for a specified period under standard conditions, and tested for compressive strength (Figure 13.12). If the laboratory results are not up to the required standard, test cores are drilled from the actual members made from the questionable batch of concrete. If the strength of these core samples is also deficient, the contractor may be required to cut out the defective concrete and replace it. Frequently, test cylinders are also cast and cured on the construction site under the same conditions as the concrete in the forms; these may then be tested to determine when the concrete is strong enough to allow removal of forms and temporary supports.

Handling and Placing Concrete

Freshly mixed concrete is not a liquid but a *slurry*, a semistable mixture of solids suspended in liquid. If it is vibrated excessively, moved horizontally for long distances in the forms, or dropped through constrained spaces, it tends to segregate, which means that the coarse aggregate works its way to the bottom of the form and the water and cement paste rise toward the top. The result is concrete of nonuniform and generally unsatisfactory properties. *Segregation* is prevented by depositing the concrete, fresh from the mixer, as close to its final position as possible.

When large quantities of concrete must be placed over a large area or a great height, truck-mounted pumps with extendable booms deliver concrete to its final position (Figure 13.13). Usually, the concrete is pumped through a flexible hose. Or, fixed, rigid pipes may be temporarily installed to carry concrete over unusually long distances. The concrete mixture is designed so that it will not clog the line when it is put under pressure by the pump. Concrete can be pumped to astonishing heights and horizontal distances: For the Burj Khalifa tower in Dubai, concrete was pumped more than 1970 vertical feet (600 m).

For lesser distance and heights, concrete can be conveyed on portable conveyor belt systems. Crane-mounted buckets (Figure 13.1), powered buggies, or even hand wheelbarrows may also be used. The method selected will depend on the scale of the job and the accessibility of the work to the truck delivering the concrete. To prevent segregation, concrete dropped a distance of more than 3 to 5 feet (1 m or so) should be allowed to fall freely, without obstruction, or it should be deposited through *dropchutes* that control its fall. Concrete should also not be pushed over horizontal distances from one location to another once it has been deposited.

FIGURE 13.13
A transit mixer feeds fresh concrete directly to a truck-mounted pump that delivers the concrete to an upper floor level. The end of the boom is radio controlled by an operator standing close to where the concrete is being deposited. (*Photo by Joseph Iano.*)

Once placed, concrete must be *consolidated* to eliminate trapped air and to completely fill the space around the reinforcing bars and in all corners of the formwork. This may be done by repeatedly thrusting a rod, spade, or immersion-type vibrator into the concrete at closely spaced intervals throughout the formwork. Excessive agitation of the concrete must be avoided, however, or segregation will occur.

Self-consolidating concrete (SCC), a concrete that fills forms completely without requiring vibration or any other method of consolidation, is a more recent development. It is formulated with more fine aggregates than coarse ones, a reversal of the usual proportions; it also includes special superplasticizing admixtures and, in some cases, other viscosity-modifying agents. The result is a concrete that flows freely, yet does not allow its coarse aggregate to sink to the bottom of the mix. Self-consolidating concrete may be used where forms are crowded with steel reinforcing, making consolidation of stiffer conventional concrete problematic. The consistent surface characteristics and crisp edges produced by self-consolidating concrete make it well suited to the production of high-finish-quality architectural concrete. By eliminating the separate consolidation step and allowing more rapid placement, self-consolidating concrete can improve productivity in precast concrete and large-volume sitecast concrete operations. However, formwork costs for self-consolidating concrete may be higher than those for conventional concrete, as the greater fluid pressures exerted by the freely flowing material require stiffer and stronger forms.

Curing Concrete

Because concrete cures by hydration, the chemical bonding of the water and cement, and not by simple drying, it is essential that it be kept moist until its required strength is achieved. The curing reaction takes place over a very long period of time, but concrete is commonly designed on the basis of the strength that it reaches after 28 days. If it is allowed to dry out at any point during this time period, the strength of the resulting concrete will be reduced, and its surface hardness and durability can be adversely affected (Figure 13.14). Concrete cast in formwork is protected from dehydration on most surfaces by the formwork, but the top surfaces must be kept moist by repeatedly spraying or flooding with water, by covering with moisture-resistant sheets of paper or film, or by spraying on a curing compound that seals the surface of the concrete against loss of moisture. These measures are even more important for concrete slabs, whose large exposed surface areas make them especially susceptible to

premature drying. This is a particular danger when slabs are poured in hot or windy weather, which can cause the surface of the pour to dry out and crack even before the concrete begins to cure. Temporary windbreaks may be erected, shade may be provided, evaporation retarders may be added to the concrete, and frequent fogging of the air directly over the surface of the slab with a fine spray of water may be required until the slab is hard enough to be finished and covered or sprayed with curing compound.

At low temperatures, the curing reaction in concrete proceeds much more slowly. If concrete reaches subfreezing temperatures while curing, the reaction stops completely until the temperature of the concrete rises above the freezing mark. To achieve full strength, it is important that the concrete be protected from very low temperatures or freezing until it is adequately cured. If freshly poured

concrete is covered and insulated, its heat of hydration may be sufficient to maintain an adequate temperature in the concrete even at fairly low air temperatures. Under more severe conditions, the ingredients of the concrete may have to be heated before mixing, and both a temporary enclosure and a temporary source of heat may have to be provided during placing and curing.

In very hot weather, the hydration reaction is greatly accelerated, and concrete may begin curing before there is time to place and finish it. This tendency can be controlled by using cool ingredients and, under extreme conditions, by replacing some of the mixing water with an equal quantity of crushed ice, making sure that the ice has melted fully and the concrete has been thoroughly mixed before placing. Another method of cooling concrete is to bubble liquid nitrogen through the mixture at the batch plant.

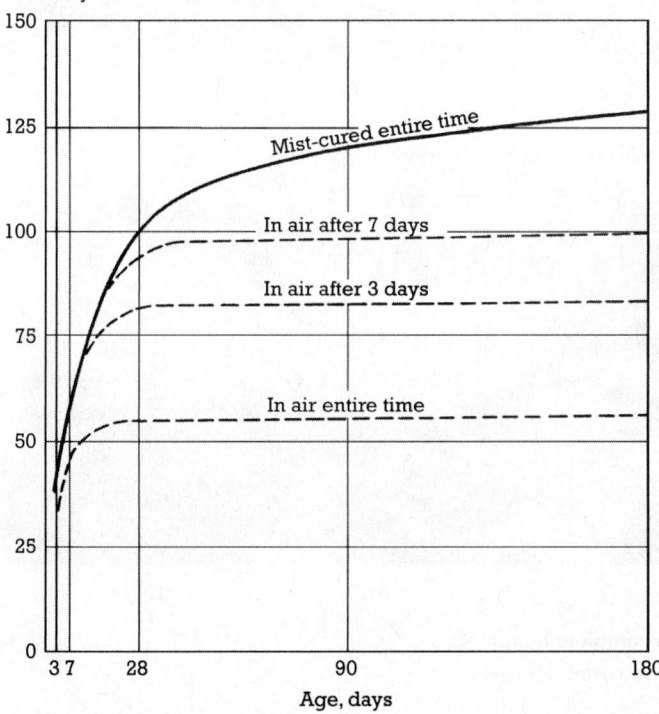

FIGURE 13.14
The growth of compressive strength in concrete over time. Moist-cured concrete is still gaining strength after 6 months, whereas air-dried concrete virtually stops gaining strength altogether. (*Reprinted with permission of the Portland Cement Association from* Design and Control of Concrete Mixtures, *12th edition; PHOTOS: Portland Cement Association, Skokie, IL.*)

FORMWORK

Because concrete is put in place as a shapeless slurry with no physical strength, it must be shaped and supported by *formwork* until it has cured sufficiently to support itself. Formwork is usually made of braced panels of wood, metal, or plastic. It is constructed as a negative of the shape intended for the concrete. Formwork for a beam or slab serves as a temporary working surface during the construction process and as the temporary means of support for reinforcing bars. Formwork must be strong enough to support the considerable weight and fluid pressure of wet concrete without deflection, which often requires temporary supports that are major structures in themselves.

During curing, the formwork helps to retain the necessary water of hydration in the concrete. When curing is complete, the formwork must pull away cleanly from the concrete surfaces without damage either to the concrete or to the formwork, which is usually used repeatedly as a construction project progresses. This means that the formwork should have no reentrant corners that will trap or be trapped by the concrete. Any element of formwork that must be withdrawn directly from a location in which it is surrounded on four or more surfaces by concrete, such as a joist pan (Figures 14.24 and 14.25), must be tapered. Formwork surfaces that are in contact with concrete are also usually coated with a *form release compound,* an oil, wax, or plastic, that prevents adhesion of the concrete to the form.

The quality of the concrete surfaces can be no better than the quality of the forms in which they are cast, and the requirements for surface quality and structural strength of formwork are rigorous. Top-grade wooden boards and resin-overlaid plywood panels are frequently used to achieve high-quality surfaces. The ties and temporary framing members that support the boards or plywood are spaced closely to avoid bulging of the forms under the high pressure of the wet concrete.

In a sense, formwork constitutes an entire temporary building that must be erected and then demolished in order to produce a second, permanent building of concrete (Figure 13.15). In fact, the cost of conventional formwork accounts for a major portion—often one-half or more—of

FIGURE 13.15
Casting concrete on the building site requires the construction of a complete temporary structure that will be removed once the concrete has been placed and cured. *(Photo by Joseph Iano.)*

the overall cost of a concrete building frame. This cost is one of the factors that has led to the development of *precasting*, a process in which concrete is cast into reusable forms at an industrial plant. Rigid, fully cured structural units from the plant are then transported to the job site, where they are hoisted into place and connected much as if they were structural steel shapes. The alternative to precasting, and the more usual way of building with concrete, is *sitecasting*, also called *cast-in-place construction*, in which concrete is poured into forms that are erected on the job site. In Chapters 14 and 15, formwork is shown for both sitecast and precast concrete.

REINFORCING

The Concept of Reinforcing

Concrete has no useful tensile strength (Figure 13.16). Historically, its structural uses were limited until the concept of steel reinforcing was

developed. The compatibility of steel and concrete is a fortuitous accident. If the two materials had grossly different coefficients of thermal expansion, a reinforced concrete structure would tear itself apart during seasonal cycles of temperature variation. If the two materials were chemically incompatible, the steel would corrode or the concrete would be degraded. If concrete did not adhere to steel, a very different and more expensive configuration of reinforcing would be necessary. Concrete and steel, however, change dimension at nearly the same rate in response to temperature changes; steel is protected from corrosion by the alkaline chemistry of concrete; and concrete bonds strongly to steel, providing a convenient means of adapting brittle concrete to structural elements that must resist not only compression, but tension, shear, and bending as well.

The basic theory of *reinforced concrete* is extremely simple: Put the reinforcing steel where there are tensile (stretching) forces in a structural

member, and let the concrete resist the compression (squeezing) forces. This accounts fairly precisely for the location of most of the reinforcing steel that is used in a concrete structure. However, there are some important exceptions: Steel is used to resist a share of the compression in concrete columns and in beams whose height must be reduced for architectural reasons. It is used as column ties, discussed later in this chapter, to prevent buckling of vertical reinforcing in columns. It is used to resist cracking that might otherwise be caused by curing shrinkage, and by thermal expansion and contraction in slabs and walls.

Steel Bars for Concrete Reinforcement

Steel reinforcing bars (rebar) for concrete construction are hot-rolled in much the same way as structural shapes. They are round in cross section and *deformed* with surface ribs for better bonding to concrete

Material	Working Strength in Tension[a]	Working Strength in Compression[a]	Density	Modulus of Elasticity
Wood (framing lumber)	300–1000 psi 2.1–6.9 MPa	600–2000 psi 4.1–14 MPa	30 pcf 480 kg/m^3	1,000,000–1,900,000 psi 6900–13,000 MPa
Brick masonry (including mortar, unreinforced)	0	250–1300 psi 1.7–9.0 MPa	120 pcf 1900 kg/m^3	700,000–3,700,000 psi 4800–25,000 MPa
Structural steel	24,000–43,000 psi 170–300 MPa	24,000–43,000 psi 170–300 MPa	490 pcf 7800 kg/m^3	29,000,000 psi 200,000 MPa
Concrete (unreinforced)	0	1000–12,000 psi 6.9–83 MPa	145 pcf 2300 kg/m^3	3,000,000–4,500,000 psi 21,000–31,000 MPa

[a] Allowable stress or approximate maximum stress under normal working conditions.

FIGURE 3.16
Comparative physical properties of four common structural materials: wood, brick masonry, steel, and concrete (shaded row). Concrete, like masonry, has no useful tensile strength, but its compressive strength is considerable, and when combined with steel reinforcing, it can be used for every type of structure. The ranges of values in strength and stiffness reflect variations in concrete mix properties. (Specially formulated concretes are capable of substantially higher strengths than those listed in this table.)

FIGURE 13.17
Glowing strands of steel are reduced to reinforcing bars as they snake their way through a rolling mill. *(Courtesy of Bethlehem Steel Corporation.)*

FIGURE 13.18
The deformations rolled onto the surface of a reinforcing bar help it to bond tightly to concrete. *(Photo by Edward Allen.)*

(Figures 13.17 and 13.18). At the end of the rolling line in the mill, the bars are cut to a standard length (commonly 60 feet, or 18.3 m, in the United States), bundled, and shipped to local fabricating shops.

Reinforcing bars are rolled in standard diameters. In the United States, bars are specified by a number that corresponds to eighths of an inch (3.2 mm) of bar diameter (Figure 13.19). For example, a number 6 reinforcing bar is ⅝ or ¾ inch (19 mm) in diameter, and a number 8 is ⅝ or 1 inch (25.4 mm) in diameter. Bars larger than number 8 vary slightly from these nominal diameters in order to correspond to convenient cross-sectional areas. For the increasing volume of work in the United States that is carried out in SI units, a soft conversion of units is used: The bars are exactly the same, but a different numbering system is used, corresponding roughly to the diameter of each bar in millimeters. This avoids the expensive process of converting rolling mills to produce a slightly different set of bar sizes. In most countries other than the United States, a hard metric range of reinforcing bar sizes is standard (Figure 13.20).

In selecting reinforcing bars for a given beam or column, the structural engineer knows from calculations the required cross-sectional area of steel that is needed in a given location. This area may be achieved with a larger number of smaller bars, or a smaller number of larger bars, in any of several combinations. The final bar arrangement is based on the physical space available in the concrete member, the required depth of concrete that must cover the reinforcing, the clear spacing required between bars to allow passage of the concrete aggregate, and the sizes and number

Bar Size		Nominal Dimensions					
		Diameter		Cross-Sectional Area		Weight	
American	Metric	in.	mm	in.²	mm²	lb/ft	kg/m
#3	#10	0.375	9.5	0.11	71	0.376	0.560
#4	#13	0.500	12.7	0.20	129	0.668	0.944
#5	#16	0.625	15.9	0.31	199	1.043	1.552
#6	#19	0.750	19.1	0.44	284	1.502	2.235
#7	#22	0.875	22.2	0.60	387	2.044	3.042
#8	#25	1.000	25.4	0.79	510	2.670	3.973
#9	#29	1.128	28.7	1.00	645	3.400	5.060
#10	#32	1.270	32.3	1.27	819	4.303	6.404
#11	#36	1.410	35.8	1.56	1006	5.313	7.907
#14	#43	1.693	43.0	2.25	1452	7.650	11.38
#18	#57	2.257	57.3	4.00	2581	13.60	20.24

FIGURE 13.19
American standard sizes of reinforcing bars. These sizes were originally established in conventional units of inches and square inches. More recently, "soft metric" designations have also been given to the bars without changing their sizes. Notice that the size designations of the bars in both systems of measurement correspond very closely to the rule-of-thumb values of ⅛ inch or 1 mm per bar size number.

Size Designation	Nominal Mass, kg/m	Nominal Dimensions	
		Diameter, mm	Cross-Sectional Area, mm²
10M	0.785	11.3	100
15M	1.570	16.0	200
20M	2.355	19.5	300
25M	3.925	25.2	500
30M	5.495	29.9	700
35M	7.850	35.7	1000
45M	11.775	43.7	1500
55M	19.625	56.4	2500

FIGURE 13.20
These hard metric reinforcing bar sizes are used in most countries of the world.

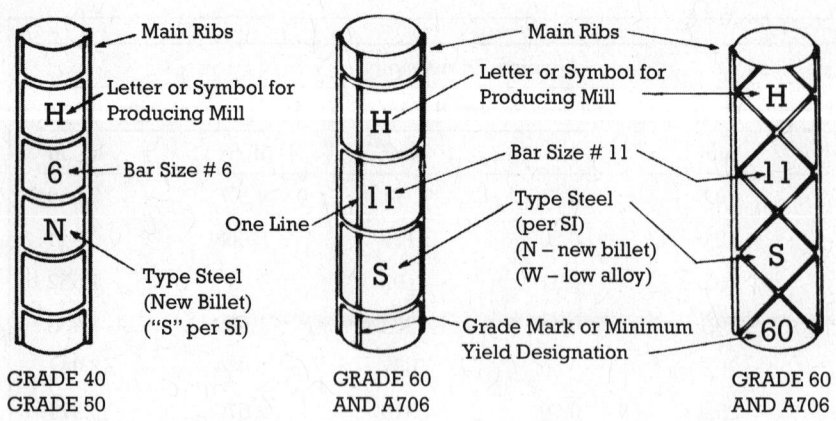

GRADE 40
GRADE 50

GRADE 60
AND A706

GRADE 60
AND A706

FIGURE 13.21
Reinforcing bars are manufactured with identification marks, denoting the mill that produced the bars, bar size, type of steel, and steel grade. Steel grade is indicated either with a number, such as "60" for Grade 60 steel, or with short bars (grade lines), in which no bars indicates Grade 40 or 50 steel, one bar indicates Grade 60 steel, and two bars indicates grade 75 steel. *(Courtesy of Concrete Reinforcing Steel Institute.)*

of bars that will be most convenient to fabricate and install.

Most reinforcing bars are manufactured according to ASTM standard A615 and are available in grades 40, 60, and 75, corresponding to steel with yield strengths of 40,000, 60,000, and 75,000 psi (280, 420, and 520 MPa), respectively (Figure 13.21). Grade 60 is generally the most economical and readily available of the three, although grade 75 is finding increasing use in column reinforcing. Reinforcing bars conforming to ASTM A706, made with low-alloy steel meeting special ductility requirements, are used where concrete structures must meet special seismic design criteria or where extensive welding of reinforcing is required. In structures with especially heavy reinforcing requirements, reinforcing bars conforming to ASTM A1035, with strengths as high as 120,000 psi (830 MPa), may be used. With such high-strength bars, bar size may be reduced and the spacing between the bars may be increased in comparison to designs with conventional-strength reinforcing. This reduces *rebar congestion*, making it easier to place and consolidate the concrete around the reinforcing.

Reinforcing bars in concrete structures that are exposed to salts such as deicing salts or those in seawater are prone to rust. Galvanized reinforcing bars and epoxy-coated reinforcing bars are often used in marine structures, highway structures, and parking garages to resist this corrosion. Stainless steel bars, stainless-steel-clad bars, zinc-and-polymer-coated bars, and proprietary corrosion-resistant alloy bars are newer types of corrosion-resistant reinforcing. Still in the experimental stage or newest to market are nonmetallic reinforcing bars made from high-strength fibers of carbon, aramid, or glass embedded in a polymeric matrix.

As an alternative to conventional reinforcing bars, reinforcing steel for slabs is produced in sheets or rolls of *welded wire reinforcing (WWR)*, also called *welded wire fabric (WWF)*, a grid of wires or round bars spaced 2 to 12 inches (50–300 mm) apart (Figure 13.22). The lighter styles of welded wire fabric resemble cattle fencing and are used to reinforce concrete slabs on grade and certain precast concrete elements. The heavier styles find use in concrete walls and structural slabs. The principal advantage of welded wire fabric over individual bars is economy of labor in placing the reinforcing, especially where a large number of small bars can be replaced by a single sheet of material. The size and spacing of the wires or bars for a particular application are specified by the structural engineer or architect of the building.

Fabrication and Erection of Reinforcing Bars

The fabrication of reinforcing steel for a concrete construction project is analogous to the fabrication of steel shapes for a steel frame building (Chapter 11). The fabricator receives the engineering drawings for the building from the contractor and prepares shop drawings for the reinforcing bars. After the shop drawings have been checked by the engineer or architect, the fabricator sets to work cutting the reinforcing bar stock to length, making the necessary bends (Figure 13.23) and tying the fabricated bars into bundles that are tagged to indicate their destination in the building. The bundles are shipped to the building site as needed. There they are broken down, lifted by hand or hoisted by crane, and wired (or occasionally welded) together in the forms to await pouring of the concrete. The wire has a temporary function only, which is to hold the reinforcement in position until the concrete has cured. Any transfer of load from one reinforcing bar to another in the completed building is done by the concrete. Where two bars must be spliced, they are overlapped a specified number of bar diameters (typically 30), and the loads are transferred from one to the other by the surrounding concrete.

Wire Size Number		Nominal Diameter, in.	Nominal Weight lb/lin ft	Area in Sq. In. per Ft. of Width for Various Spacings						
				Center-Center Spacing						
Smooth	Deformed			2″	3″	4″	6″	8″	10″	12″
W20	D20	0.505	0.680	1.20	0.80	0.60	0.40	0.30	0.24	0.20
W18	D18	0.479	0.612	1.08	0.72	0.54	0.36	0.27	0.216	0.18
W16	D16	0.451	0.544	0.96	0.64	0.48	0.32	0.24	0.192	0.16
W14	D14	0.422	0.476	0.84	0.56	0.42	0.28	0.21	0.168	0.14
W12	D12	0.391	0.408	0.72	0.48	0.36	0.24	0.18	0.144	0.12
W11	D11	0.374	0.374	0.66	0.44	0.33	0.22	0.165	0.132	0.11
W10.5		0.366	0.357	0.63	0.42	0.315	0.21	0.157	0.126	0.105
W10	D10	0.357	0.340	0.60	0.40	0.30	0.20	0.15	0.12	0.10
W9.5		0.348	0.323	0.57	0.38	0.285	0.19	0.142	0.114	0.095
W9	D9	0.338	0.306	0.54	0.36	0.27	0.18	0.135	0.108	0.09
W8.5		0.329	0.289	0.51	0.34	0.255	0.17	0.127	0.102	0.085
W8	D8	0.319	0.272	0.48	0.32	0.24	0.16	0.12	0.096	0.08
W7.5		0.309	0.255	0.45	0.30	0.225	0.15	0.112	0.09	0.075
W7	D7	0.299	0.238	0.42	0.28	0.21	0.14	0.105	0.084	0.07
W6.5		0.288	0.221	0.39	0.26	0.195	0.13	0.097	0.078	0.065
W6	D6	0.276	0.204	0.36	0.24	0.18	0.12	0.09	0.072	0.06
W5.5		0.265	0.187	0.33	0.22	0.165	0.11	0.082	0.066	0.055
W5	D5	0.252	0.170	0.30	0.20	0.15	0.10	0.075	0.06	0.05
W4.5		0.239	0.153	0.27	0.18	0.135	0.09	0.067	0.054	0.045
W4	D4	0.226	0.136	0.24	0.16	0.12	0.08	0.06	0.048	0.04
W3.5		0.211	0.119	0.21	0.14	0.105	0.07	0.052	0.042	0.035
W3		0.195	0.102	0.18	0.12	0.09	0.06	0.045	0.036	0.03
W2.9		0.192	0.099	0.174	0.116	0.087	0.058	0.043	0.035	0.029
W2.5		0.178	0.085	0.15	0.10	0.075	0.05	0.037	0.03	0.025
W2.1		0.162	0.070	0.126	0.084	0.063	0.042	0.031	0.025	0.021
W2		0.160	0.068	0.12	0.08	0.06	0.04	0.03	0.024	0.02
W1.5		0.138	0.051	0.09	0.06	0.045	0.03	0.022	0.018	0.015
W1.4		0.134	0.048	0.084	0.056	0.042	0.028	0.021	0.017	0.014

FIGURE 13.22

Standard configurations of welded wire reinforcing. The heaviest "wires" are more than ½ inch (13 mm) in diameter, making them suitable for structural slab reinforcing. Welded wire reinforcing is specified by first indicating the spacing of the wires and then the wire types. For example, the designation 6 × 12-W12 × W5 indicates welded wire reinforcing with W12 longitudinal wires spaced at 6 inches (150 mm) and W5 transverse wires spaced at 12 inches (300 mm). *(Courtesy of Concrete Reinforcing Steel Institute.)*

STANDARD HOOKS

All specific sizes recommended by CRSI below meet minimum requirements of ACI 318

RECOMMENDED END HOOKS

All Grades

D=Finished bend diameter

Bar Size	180° HOOKS			90° HOOKS
	D	A or G	J	A or G
# 3	2¼	5	3	6
# 4	3	6	4	8
# 5	3¾	7	5	10
# 6	4½	8	6	1-0
# 7	5¼	10	7	1-2
# 8	6	11	8	1-4
# 9	9½	1-3	11¾	1-7
# 10	10¾	1-5	1-1¼	1-10
# 11	12	1-7	1-2¾	2-0
# 14	18¼	2-3	1-9¾	2-7
# 18	24	3-0	2-4½	3-5

STIRRUP AND TIE HOOKS

STIRRUPS
(TIES SIMILAR)

STIRRUP AND TIE HOOK DIMENSIONS
Grades 40-50-60 ksi

Bar Size	D (in.)	90° Hook	135° Hook	
		Hook A or G	Hook A or G	H Approx.
#3	1¼	4	4	2½
#4	2	4½	4½	3
#5	2½	6	5½	3¾
#6	4½	1-0	7¾	4½
#7	5¼	1-2	9	5¼
#8	6	1-4	10¼	6

135° SEISMIC STIRRUP/TIE HOOKS

135° SEISMIC STIRRUP/TIE
HOOK DIMENSIONS
Grades 40-50-60 ksi

Bar Size	D (in.)	135° Hook	
		Hook A or G	H Approx.
#3	1½	5	3½
#4	2	6½	4½
#5	2½	8	5½
#6	4½	10¾	6½
#7	5¼	1-0½	7¾
#8	6	1-2¼	9

FIGURE 13.23

The bending of reinforcing bars is done according to precise standards in a fabricator's shop. *(Courtesy of Concrete Reinforcing Steel Institute.)*

The one common exception occurs in heavily reinforced columns where there is insufficient space to overlap the bars; there they are often spliced end to end rather than overlapped, and loads are transferred through welds or sleevelike mechanical splicing devices (Figure 13.24).

The combined action of concrete and steel in reinforced concrete structural elements is such that the reinforcing steel is usually loaded axially in tension or compression, and occasionally in shear, but never in bending. The bending stiffness of the reinforcing bars themselves is of no consequence in imparting strength to the concrete.

Reinforcing a Simple Concrete Beam

In an ideal, simply supported beam under uniform loading, compressive (squeezing) forces follow a set of arch-like curves that create a maximum stress in the top of the beam at midspan, with progressively lower compressive stresses toward either end. A mirrored set of curves correspond to paths of tensile (stretching) force, with stresses again reaching a maximum at the middle of the span (Figure 13.25). In an ideally reinforced concrete beam, steel reinforcing bars would be bent to follow these lines of tension, and the bunching of the bars at midspan would serve to resist the higher stresses at that point. It is difficult, however, to bend bars into these curves and to support the curved bars adequately in the formwork, so a simpler rectilinear arrangement of reinforcing steel is substituted.

This arrangement consists of a set of bottom bars and stirrups. The *bottom bars* are placed horizontally near the bottom of the beam, leaving a specified amount of concrete below and to the sides of the rods as *cover* (Figure 13.26). The concrete cover provides a full embedment for the reinforcing bars and protects them against fire and corrosion. The bars are most heavily stressed at the

Figure 13.24
Some mechanical devices for splicing reinforcing bars. From left to right: A lapped, wedged connection, used primarily to connect new bars to old ones when adding to an existing structure. A welded connector, very strong and tough. A grouted sleeve connector for joining precast concrete components: One bar is threaded and screwed into a collar at one end of the sleeve, and the other bar is inserted into the remainder of the sleeve and held there with injected grout. A threaded sleeve, with both bars threaded and screwed into the ends of the sleeve. A simple clamping sleeve that serves to align compression bars in a column. A flanged coupler for splicing bars at the face of a concrete wall or beam: The coupler is screwed onto the threaded end of one bar, and its flange is nailed to the inside face of the formwork. After the formwork has been stripped, the other bar is threaded and screwed through a hole in the flange and into the coupler. *(Photo courtesy of ERICO, Inc.)*

(a)

Concrete cover protects the steel against fire and corrosion

Bar spacings must be large enough that coarse aggregate can pass through easily

FIGURE 13.26
A cross section of a rectangular concrete beam showing cover and bar spacing.

(b)

FIGURE 13.25
(a) The directions of force in a *simply supported beam* (supported only at its ends) under uniform loading. The archlike lines represent compression, and the cablelike lines represent tension. Near the ends of the beam, the lines of strongest tensile force move upward diagonally through the beam. (b) Steel reinforcing for the same simply supported beam. The concrete resists compressive forces. The horizontal bars near the bottom of the beam resist the tensile forces. The vertical stirrups resist the lesser diagonal tensile forces near the ends of the beam.

Interactive Exercise

midpoint of the beam span, with progressively smaller amounts of stress toward each of the supports. The differences in stress are dissipated from the bars into the concrete by means of *bond* forces, the adhesive forces between the concrete and the steel, aided by the ribs on the surface of the bars. At the ends of the beam, some stress remains in the steel, but there is no further length of concrete into which the stress can be dissipated. This problem is solved by bending the ends of the bars into *hooks*, which are semicircular bends of standard dimensions.

The bottom bars do the heavy tensile work in the beam, but some lesser tensile forces occur in a diagonal orientation near the ends of the beam. These are resisted by a series of *stirrups* (Figure 13.25). The stirrups may be either open *U-stirrups*, as shown in Figure 13.25, or *closed*

FIGURE 13.27
A two-piece plastic bar support, called a "tower chair," supports a steel reinforcing bar for a structural concrete slab. To the left of the chair, a small concrete brick supporting a second bar in a position closer to the bottom of the slab is also partially visible. *(Photo by Joseph Iano.)*

stirrup-ties, which are full rectangular loops of steel that wrap all the way around the longitudinal bars. U-stirrups are less expensive to make and install and are sufficient for many situations, but stirrup-ties are required in beams that will be subjected to torsional (twisting) forces or to high compressive forces in the top or bottom bars. In either case, the stirrups furnish vertical tensile reinforcing to resist the cracking forces that run diagonally across them. A more efficient use of steel would be to use diagonal stirrups oriented in the same direction as the diagonal tensile forces, but they would be difficult to install.

When the simple beam of our example is formed, the bottom steel is supported at the correct cover height by *chairs* made of heavy steel wire or plastic (Figures 13.27 and 13.28). In a broad beam or slab, bars are supported by long chairs called *bolsters*. Chairs and bolsters remain in the concrete after pouring, even though their work is finished, because there is no way to get them out. In outdoor concrete

work, the feet of the chairs and bolsters sometimes rust where they come in contact with the face of the beam or slab unless plastic or plastic-capped steel chairs are used. Where reinforced concrete is poured in direct contact with the soil, concrete bricks or small pieces of concrete may be used to support the bars instead of chairs, to avoid the possibility of rust forming under the feet of the metal chairs and spreading up into the reinforcing bars.

The stirrups in the simple beam that we have been examining are supported by wiring them to the bottom bars and by tying their tops to horizontal #3 top bars (the smallest standard size) that have no function in the beam other than to keep the stirrups upright and properly spaced until the concrete has been poured and cured.

Reinforcing a Continuous Concrete Beam

Most sitecast concrete beams are not of this simple type, because concrete lends itself most easily to one-piece

structural frames with a high degree of continuity from one beam span to the next. In a continuous beam, the bottom of the beam is in tension at midspan, and the top of the beam is in tension at points of support (such as at girders, columns, or walls). This means that top bars must be provided over the supports, and bottom bars in midspan, along with the usual stirrups, as illustrated in Figure 13.29.

Reinforcing Structural Concrete Slabs

A concrete slab that spans across parallel beams or walls (*one-way action*) is, in effect, a very wide beam. The reinforcing pattern for such a slab is similar to the reinforcing pattern in a beam, but with a larger number of smaller top and bottom bars distributed evenly across the width of the slab. Because the slab is wide, it has a large cross-sectional area of concrete that can usually resist the relatively weak diagonal tension forces near its supports without the aid of stirrups.

SYMBOL	BAR SUPPORT ILLUSTRATION	BAR SUPPORT ILLUSTRATION PLASTIC CAPPED OR DIPPED	TYPE OF SUPPORT	SIZES
SB	5"	CAPPED 5"	Slab Bolster	¾, 1, 1½, and 2 inch heights in 5 ft. and 10 ft. lengths
SBU	5"		Slab Bolster Upper	Same as SB
BB	2½" 2½"	CAPPED 2½" 2½"	Beam Bolster	1, 1½, 2, over 2" to 5" heights in increments of ¼" in lengths of 5 ft.
BBU	2½" 2½"		Beam Bolster Upper	Same as BB
BC		DIPPED	Individual Bar Chair	¾, 1, 1½, and 1¾" heights
JC		DIPPED DIPPED	Joist Chair	4, 5, and 6 inch widths and ¾, 1 and 1½ inch heights
HC		CAPPED	Individual High Chair	2 to 15 inch heights in increments of ¼ inch
HCM			High Chair for Metal Deck	2 to 15 inch heights in increments of ¼ in.
CHC	8"	CAPPED 8"	Continuous High Chair	Same as HC in 5 foot and 10 foot lengths
CHCU	8"		Continuous High Chair Upper	Same as CHC
CHCM			Continuous High Chair for Metal Deck	Up to 5 inch heights in increments of ¼ in.
JCU	TOP OF SLAB #4 or ½" θ ¾" MIN HEIGHT 14"	TOP OF SLAB #4 or ½" θ ¾" MIN HEIGHT 14" DIPPED	Joist Chair Upper	14" span. Heights –1" thru 3½" vary in ¼" increments

FIGURE 13.28
Chairs and bolsters for supporting reinforcing bars in beams and slabs. Bolsters and continuous chairs are made in long lengths for use in slabs. Chairs support only one or two bars each. *(Courtesy of Concrete Reinforcing Steel Institute.)*

One-way slabs must be provided with *shrinkage–temperature steel*, a set of small-diameter reinforcing bars set at right angles to, and on top of, the primary reinforcing in the slab. Their function is to prevent cracks from forming parallel to the primary reinforcing because of concrete shrinkage, temperature-induced stresses, or miscellaneous forces that may occur in the building (Figure 13.30).

Two-Way Slab Action

A structural economy mostly unique to concrete frames is realized through the use of *two-way action* in floor and roof slabs. Two-way slabs, which work best for bays that are square or nearly square, are reinforced equally in both directions and share the bending forces equally between the two directions. In comparison to equivalent one-way slabs, two-way slabs can be somewhat shallower, use less reinforcing steel, and cost less. Figure 13.31 illustrates the concept of two-way action. Several different two-way concrete framing systems will be shown in detail in Chapter 14.

A continuous beam bends downward in the middle portion of each span, and upward in the portion around the supports

Top bars are required in the zone of upward bending

A right angle bend is used in place of a hook where there is sufficient space for it in the concrete

Bottom bars are heaviest in the center portion of the span, and lighter through the columns, where some bottom tensile forces can occur during strong winds or earthquakes

Stirrups are sometimes spaced more closely near columns, where the diagonal tension forces are highest

FIGURE 13.29
Reinforcing for a *continuous beam* that is supported across several spans. The upper diagram shows in exaggerated form the shape taken by a continuous beam under uniform loading; the broken line is the centerline of the beam. The lower diagram shows the arrangement of bottom steel, top steel, and stirrups conventionally used in this beam. The bottom bars are usually placed on the same level, but they are shown on two levels in this diagram to demonstrate the way in which some of the bottom steel is discontinued in the zones near the columns. There is a simple rule of thumb for determining where the bending steel must be placed in a beam: Draw an exaggerated diagram of the shape the beam will take under load, as in the top drawing of this illustration, and put the bars as close as possible to the convex edges.

A concrete slab supported by a number of beams bends in the same pattern as a concrete beam supported by a number of columns

The spandrel beam twists slightly from the bending forces in the slab at the face of the building

Concrete girder

Concrete beams

The thickness of the slab in these drawings is exaggerated in order to show the reinforcing better

Top steel for beam

Shrinkage-temperature bars reinforce against cracks parallel to the main reinforcing bars

The entire thickness of the concrete acts as a part of the beams

The stirrups have been omitted from the beams in this drawing for the sake of clarity

FIGURE 13.30

Reinforcing for a one-way concrete slab. The reinforcing is similar to that for a continuous beam, except that stirrups are not usually required in the slab, and shrinkage–temperature bars must be added in the perpendicular direction. The slab does not sit on the beams; rather, the concrete around the top of a beam is part of both the beam and the slab. A concrete beam in this situation is considered to be a T-shaped member, with a portion of the slab acting together with the stem of the beam, resulting in greater structural efficiency and reduced beam depth.

The country . . . near Taliesin, my home and workshop, is the bed of an ancient glacier drift. Vast busy gravel pits abound there, exposing heaps of yellow aggregate once and still everywhere near, sleeping beneath the green fields. Great heaps, clean and golden, are always waiting there in the sun. And I never pass . . . without an emotion, a vision of the long dust-whitened stretches of the cement mills grinding to impalpable fineness the magic powder that would "set" my vision all to shape; I wish both mill and gravel endlessly subject to my will. . . . Materials! What a resource.

—Frank Lloyd Wright, in *Architectural Record*, October 1928

Reinforcing Concrete Columns

Columns contain two types of reinforcing: *Vertical bars* (also called *column bars*) are large-diameter bars that share the compressive loads with the concrete, resist tensile stresses that may occur in columns, and impart ductility to the column (important for resistance to earthquake forces). *Ties* of small-diameter steel bars wrapped around the vertical bars help to prevent them from buckling under load: Inward buckling is prevented by the concrete core of the

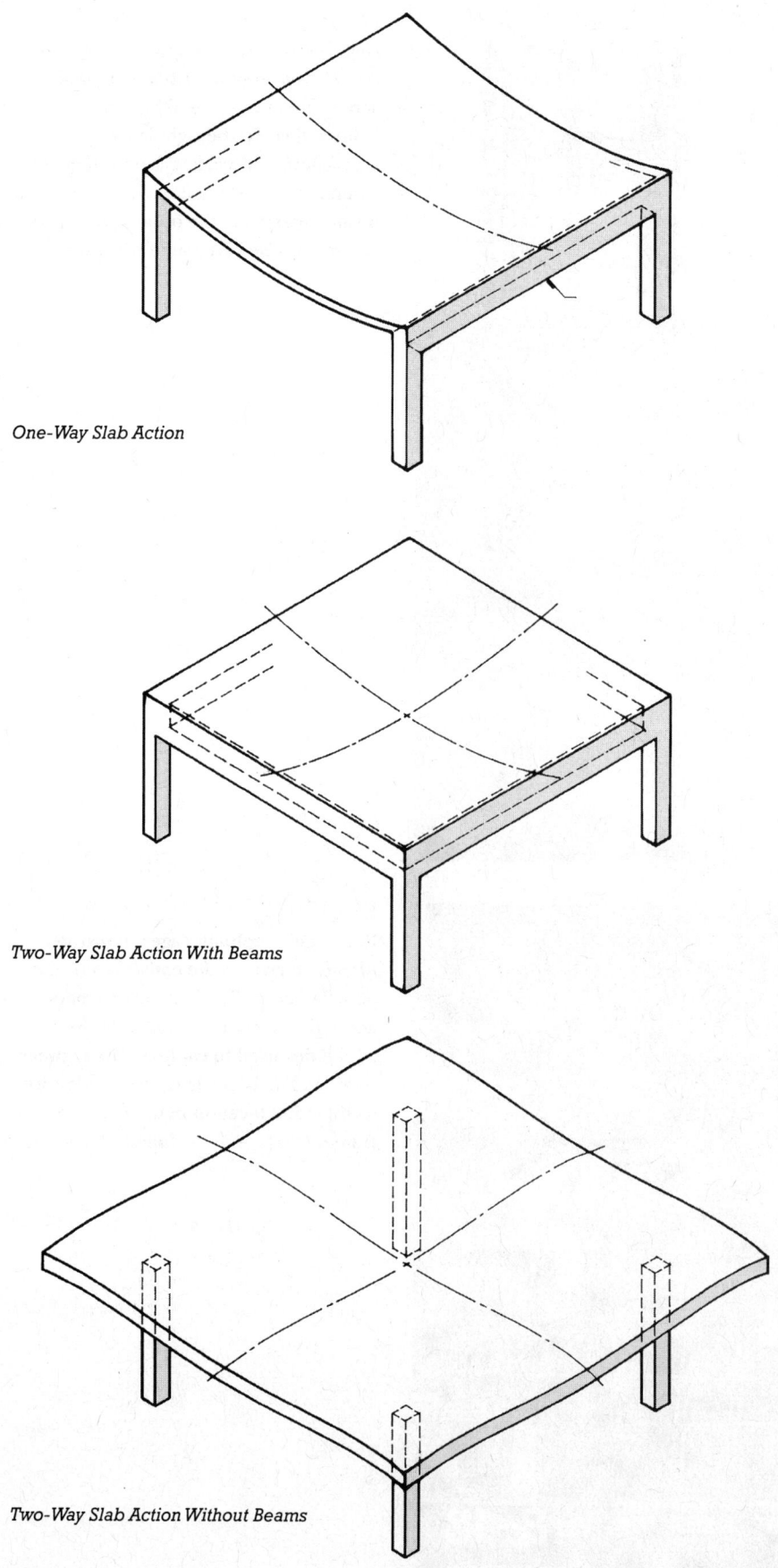

One-Way Slab Action

Two-Way Slab Action With Beams

Two-Way Slab Action Without Beams

FIGURE 13.31
One-way and two-way slab action, with deflections greatly exaggerated.

column and outward buckling by the ties (Figure 13.32). The vertical bars may be arranged in either circular or rectangular patterns. The ties may be either of two types: column spirals or column ties.

Column spirals are shipped to the construction site as tight coils of rod that are expanded accordion fashion to the required spacing and wired to the vertical bars. The vertical bars must be in a circular (or four-square) arrangement so as to make contact with the spiral. The outer form of the column, though, can be circular or rectangular. By overlapping spirals, even long lines of vertical bars in a reinforced wall can be tied with spirals.

Column ties are discrete, closely spaced hoops, individually wired in place. They are used mostly to tie rectangular arrangements of vertical bars. A single column tie can effectively restrain four vertical bars, one at each bent corner of the tie. For columns with more than four bars, additional ties are added to provide restraint for the bars between corners (Figure 13.33).

A circular arrangement of vertical bars is often more economical than a rectangular one because it avoids the need for added ties to capture bars between corners. However, column ties are generally more economical than spirals, so even columns with circular bar arrangements may be tied with discrete, circular ties. However, spiral-tied columns are more effective at resisting extreme dynamic loads, such as occur during earthquakes. To minimize labor costs, column reinforcing may be wired together while lying on the ground, and the finished column cage lifted into final position by crane (Figure 13.34). Or, column cages may be assembled in the reinforcing fabricator's shop, and

FIGURE 13.32
Reinforcing for concrete columns. To the left is a column with a rectangular arrangement of vertical bars and column ties. To the right is a circular arrangement of vertical bars with a column spiral. Regardless of the bar and tie arrangements, the outer form of the column can be cast round or square.

FIGURE 13.33
Rectangular column cages, fabricated off-site, arrive on the construction site on a flatbed trailer. The vertical bars are tied with column ties. Note the added ties used to capture bars between corners. Each cage is tagged to identify its intended location in the building frame. *(Photo by Joseph Iano.)*

FIGURE 13.34
A large reinforcing cage being assembled while resting on its side. Once complete, it will be lifted into final position, where it will form part of the central core structure of a tall building. Note the large rectangular ties. The worker standing on top of the reinforcing cage is installing shorter ties between pairs of column bars. Additional ties are stacked toward the back, right-hand side of the top of the cage (*Photo by Joseph Iano.*)

delivered ready for installation to the construction site (Figure 13.33).

Fibrous Reinforcing

Fibrous reinforcing is composed of short fibers of glass, steel, or polypropylene that are added to the concrete mix. *Microfiber reinforcing* is added in relatively low dosages and is intended to reduce *plastic shrinkage cracking*, which frequently occurs while the concrete is still in a plastic state, during the earliest stages of curing. Microfiber reinforcing makes little if any contribution to the properties of fully cured concrete. *Macrofiber reinforcing*, usually of polypropylene or a steel-polypropylene blend, not only protects against plastic shrinkage but also resists long-term cracking due to drying shrinkage and thermal stresses. Macrofiber reinforcing is added to concrete at dosages 10 or 20 times greater than those of microfiber reinforcing and, in some cases, fully replaces the usual shrinkage–temperature steel in concrete slabs. Macrofiber reinforcing can also improve concrete's resistance to impact, abrasion, and shock. Glass fibers are also added to concrete to produce glass-fiber-reinforced concrete (GFRC), which is used for cladding panels (see Chapter 20).

CONCRETE CREEP

In addition to plastic and drying shrinkage, concrete is subject to long-term *creep*. When placed under sustained compressive stress from its own weight, the weight of other permanent building components, or the force of prestressing (as described later in this chapter), concrete will gradually and permanently shorten over a period of months or years. In some circumstances, this dimensional change is of sufficient magnitude that it must be accounted for in the design and detailing of other parts. For example, when a brick veneer cladding system is supported on a concrete building frame, the shrinkage of the concrete combined with other factors affecting movement of the masonry require that horizontal movement joints be designed into the cladding system to accommodate the differential movement between the cladding and the supporting structure. If these joints are not provided, or if they are too narrow and unable to accommodate the extent of movement, the cladding system can fail as it becomes compressed, in part, by the shortening of the concrete structure. As a rule of thumb, sitecast concrete building frames can be expected to shorten in height under the influence of their own weight and other dead loads at the rate of 1/16 inch for every 10 feet (½ mm per meter) of building height.

PRESTRESSING

When a beam supports a load, the compression side of the beam is squeezed slightly and the tension side is stretched. In a reinforced concrete beam, the stretching tendency is resisted by the reinforcing steel but not by the concrete. When the steel elongates under tension, the concrete around it forms cracks that run from the edge of the beam to the horizontal plane in the beam, above which compressive forces occur. This cracking is even visible to the unaided eye in reinforced concrete beams that are loaded to (or beyond) their full capacity. In effect, more than half of

Distribution of stresses across the profile of the beam

Compression in concrete
Tension in steel

In a reinforced concrete beam, less than half the concrete is in compression, and cracks will appear in the bottom of the beam under full load

Compression in concrete

When a concrete beam is prestressed, all the concrete acts in compression. The off-center location of the prestressing steel causes a camber in the beam

Prestress Bending stress

Under loading, the prestressed beam becomes flatter, but all the concrete still acts in compression, and no cracks appear

FIGURE 13.35
The rationale for prestressing concrete. In addition to the elimination of cracks in the prestressed beam, the structural action is more efficient than that of a reinforced beam. Therefore, the prestressed beam uses less material. The small diagrams to the right indicate the distribution of stresses across the vertical cross section of each of the beams at midspan.

easier and cheaper to transport. For this reason, structural precast concrete used for slabs, beams, and girders (and in some cases columns as well) is usually prestressed.

In practice, ordinary reinforcing bars are not sufficiently strong to serve as prestressing steel. Prestressing is practical only with very high-strength steel strands that are manufactured for the purpose. These are made of cold-drawn steel wires that are formed into small-diameter cables.

Pretensioning

Prestressing is accomplished in two different ways. *Pretensioning* is used with precast concrete members: High-strength steel strands are stretched tightly between abutments in a precasting plant before the concrete is cast. The concrete member (or, more commonly, a series of concrete members arranged end to end) is then cast around the stretched steel (Figure 15.8). The curing concrete adheres to the strands along their entire length. After the concrete has cured to a specified minimum

the concrete in the beam is doing no useful work except to hold the steel in position and protect it from fire and corrosion (Figure 13.35, top).

If the reinforcing bars could be stretched to a high tension before the beam is loaded and then released against the concrete that surrounds them, they would place the concrete in compression. If a load were subsequently put on the beam, the tension in the stretched steel would increase further, and the compression in the concrete surrounding the steel would diminish. If the initial tension or *prestress* in the steel bars

were of sufficient magnitude, however, the surrounding concrete would never be subjected to tension, and no cracking would occur. Furthermore, the beam would be capable of carrying a greater load with the same amounts of concrete and steel than if it were merely reinforced in the conventional manner. This is the rationale for *prestressed concrete.* Prestressed members, particularly those designed to work in bending, contain less concrete than reinforced members of equivalent strength. Their lighter weight also pays off by making precast, prestressed concrete members compressive strength, the strands are cut off at either end of each member. This releases the external tension on the steel, allowing it to recoil slightly, which squeezes all of the concrete of the member into compression. If, as is usually the case, the steel is placed as close as possible to the tension side of the member, the member takes on a decided *camber* (lengthwise arching) at the time the steel strands are cut (Figure 13.36). Much or all of this camber disappears later when the member is subjected to loads in a building.

The strong abutments needed to hold the tensioned strands prior to the pouring of concrete are very expensive to construct except in a single fixed location where many concrete members can be created within the same set of abutments. For this reason, pretensioning is useful only for concrete members cast in precasting plants.

Posttensioning

Unlike pretensioning, *posttensioning* is done almost exclusively in place on the building site. High-strength steel strands (called *tendons*) are covered with a steel or plastic tube to prevent them from bonding to the concrete and are not tensioned until the concrete is poured and has achieved adequate strength. Each tendon is anchored to a steel plate embedded in one end of the beam or slab. A hydraulic jack is inserted between the other end of the tendon and a similar steel plate in the other end of the member. The jack applies a large tensile force to the tendon while compressing the concrete with an equal but opposite force that is applied through the plate. The stretched tendon is anchored to the plate at the second end of the member before the jack is removed (Figures 13.37, 13.38, and 13.39). For very long members, the tendons are jacked from both

ends to be sure that frictional losses in the tubes do not prevent uniform tensioning.

The net effect of posttensioning is the same as that of pretensioning. The difference is that in posttensioning, abutments are not needed because the member itself provides the opposing force needed to tension the steel. When the posttensioning process is complete, the tendons may be left unbonded, or, if they are in a steel tube, they may be bonded by injecting grout to fill the space between the tendons and the tube. Bonded construction is common in bridges and other heavy structures, but most posttensioning in buildings is done with unbonded tendons. These are made up of cold-drawn steel wires and are roughly ½ inch (13 mm) in diameter (Figures 13.39 and 13.40). The tendon is coated with a lubricant and covered with a plastic sheath at the factory.

1. The first step in pretensioning is to stretch the steel prestressing strands tightly across the casting bed

2. Concrete is cast around the stretched strands and cured. The concrete bonds to the strands

3. When the strands are cut the concrete goes into compression and the beam takes on a camber

FIGURE 13.36
Pretensioning. Photographs of pretensioned steel strands for a beam are shown in Chapter 15.

1. In posttensioning, the concrete is not allowed to bond to the steel strands during curing

2. After the concrete has cured, the strands are tensioned with a hydraulic jack and anchored, to the ends of the beam. If the strands are draped, as shown here, higher structural efficiency is possible than with straight strands

FIGURE 13.37
Posttensioning, using draped strands to more nearly approximate the flow path of tensile forces in the beam.

FIGURE 13.38
Posttensioning draped tendons in a large concrete beam with a hydraulic jack. Each tendon consists of a number of individual high-strength steel strands. The bent bars projecting from the top of the beam will be embedded in the concrete slab that the beam will support to allow them to act together as a composite structure. *(Reprinted with permission of the Portland Cement Association from* Design and Control of Concrete Mixtures, *12th edition; PHOTOS: Portland Cement Association, Skokie, IL.)*

FIGURE 13.39
Most beams and slabs in buildings are posttensioned with plastic-sheathed, unbonded tendons. The pump and hydraulic jack (also called a ram) are small and portable. *(Courtesy of Constructive Services, Inc., Dedham, MA.)*

Even higher structural efficiencies are possible in a prestressed beam or slab if the steel strands are placed to follow as closely as possible the lines of tensile force that are diagrammed in Figure 13.25. In a posttensioned beam or slab, this is done by using chairs of varying heights to support the tendons along a curving line that closely follows the center of the tensile forces in the member. Such a tendon is referred to as being *draped* (Figure 13.41). Draping is impractical in pretensioned members because the tendons would have to be pulled down at many points along their length. But pretensioned strands can be *harped*, that is, pulled up and down in the formwork to make a downward-pointing or flattened V shape in each member that approximates very roughly the shape of a draped tendon (Figures 13.42 and 15.9).

Because it is always highly compressed by prestressing force, the concrete in a prestressed member is subject to creep. The steel strands also stretch slightly over time and

FIGURE 13.40

End anchorage for a posttensioning tendon. The steel anchor plate and plastic pocket former are cast into the edge of the concrete slab. After the concrete has cured, the formwork and pocket former are removed, exposing the anchor plate. The two conical wedges are inserted around the tendon and into the hole in the anchor plate. The ram presses against the wedges and draws the tendon through until the gauge on the pump indicates that the required tension has been reached. When the ram is withdrawn, the wedges are drawn into the conical hole, grip the tendon, and maintain the tension. After all the tendons have been tensioned, the excess lengths of tendon are cut off and the pockets grouted flush with the edge of the slab. *(Photo by Edward Allen.)*

FIGURE 13.41
Draped posttensioning tendons. As explained in Chapter 14, banded tendons run from lower left to upper right and distributed tendons run in the perpendicular orientation. *(Photo by Joseph Iano.)*

lose some of their prestressing force. Initial prestressing forces must be increased slightly above their optimal values to make up for these long-term changes. Further increases in initial tension are needed to accommodate the slight curing shrinkage that takes place in concrete, small, short-term movements caused by elastic shortening of the concrete during structural loading, and frictional losses and initial slippage or set of the strand anchors in posttensioned members.

Succeeding chapters will discuss prestressed concrete, both pretensioned and posttensioned, in greater detail, showing its application to various standard precast and cast-in-place systems of construction.

Straight pretensioning strands

Depressed pretensioning strands

Harped pretensioning strands

FIGURE 13.42
Shaping pretensioning strands to improve structural efficiency. Examples of depressed and harped strands are shown in Chapter 15.

ACI 301

In North America, concrete structures are built according to the requirements of *ACI 301: Specifications for Structural Concrete for Buildings,* a publication of the American Concrete Institute. This comprehensive document covers every aspect of concrete work: formwork, reinforcement, chairs and bolsters, concrete mixtures, handling and placing of concrete, lightweight concrete, prestressing, and the use of concrete in exposed architectural surfaces. It is a standard that is familiar to architects, engineers, contractors, and building inspectors, and it furnishes the common basis for everyone who works in designing and constructing a concrete building.

INNOVATIONS IN CONCRETE CONSTRUCTION

The basic materials, concrete and steel, are constantly undergoing research and development, leading to higher allowable strengths and decreased weight of structures.

Structural lightweight concretes are being used more widely to reduce dead loads. Shrinkage-compensating cements have been developed for use in concrete structures that cannot be allowed to shrink during curing. Self-consolidating concrete is being used to improve productivity and reduce labor costs. Improved coatings for reinforcing steel, steel alloys with greater corrosion resistance, and nonmetallic reinforcing are being used to extend the life of structures exposed to the weather, marine environments, and road salts.

Concretes with compressive strengths as high as 30,000 psi (200 MPa) and that exhibit even usable tensile strength are called *ultra-high performance concretes (UHPCs)*. They are formulated from portland cement, silica fume, silica or quartz flour (extremely finely ground silica or quartz), fine sand, high-range water reducer, water, and steel fibrous reinforcing. There is no large aggregate. The resulting concrete is characterized by a dense packing of the fine aggregate and finely ground mineral powders within the cement matrix, resulting in concrete that is stronger, less permeable, and more durable than

conventional high-strength mixes. The addition of steel macrofiber reinforcing imparts tensile strength and ductility (toughness). Ultra-high performance concrete can be used in the casting of remarkably thin architectural elements. Precast exterior cladding panels and arched canopies, for example, both no thicker than ¾ inch (20 mm) and with no conventional reinforcing, have been produced from this material (Figures 13.43 and 13.44). Ultra-high performance concrete is suitable for structural applications where very high load capacity, low weight, and high durability are required. By reducing the volume of concrete required in such applications, its use also reduces greenhouse gas emissions resulting from the manufacture of concrete in comparison to the use of conventional, lower-strength concretes.

Pervious concrete is made with gap- or uniformly graded aggregate so that 15 to 30 percent of the finished concrete consists of open voids. Pervious concrete paving allows rainwater runoff to pass directly through the paving into the soil below, rather than flow across its surface. Pervious paving reduces water volumes directed to municipal stormwater systems and

FIGURE 13.43
The Shawnessy LRT Station in Calgary, Alberta, designed by Stantec Architecture, Ltd. The canopy shells were cast from ultra-high-performance concrete and are only ¾ inch (19 mm) thick. *(Photograph by Tucker Photography. Courtesy of Lafarge North America, Inc.)*

FIGURE 13.44
Casting of a concrete shell for the Shawnessy LRT Station shown in Figure 13.43. The shells were injection molded, a technique in which the concrete is cast in a fully enclosed mold rather than in a conventional open-sided form. In this photograph, the two halves of a mold have been separated, and the shell is being lifted with the aid of a temporary frame. *(Photograph by Tucker Photography. Courtesy of Lafarge North America, Inc.)*

lessens the contamination of nearby bodies of water by surface runoff carrying oils and other contaminants.

Light-transmitting concrete is made from precast concrete blocks or panels with embedded optic fibers or fabrics that allow light to pass through while retaining the strength and durability of the concrete. This recently formulated material is finding application in nonstructural partitions, countertops, and other architectural elements.

Under development is the possibility of self-healing concrete, in which small bacteria- and nutrient-containing granules are added to the concrete mix. When micro-cracks occur and water intrudes into the hardened concrete, the bacteria are activated and produce calcium carbonate (limestone), sealing the crack and protecting the steel reinforcing from corrosion.

CSI/CSC MasterFormat Sections for Concrete Construction	
03 11 00	**CONCRETE FORMING**
03 21 00	**REINFORCEMENT BARS**
03 21 11	Plain Steel Reinforcement Bars
03 21 13	Galvanized Steel Reinforcement Bars
03 21 16	Epoxy-Coated Steel Reinforcement Bars
03 21 19	Stainless Steel Reinforcement Bars
03 21 21	Composite Reinforcement Bars
03 22 00	**FABRIC AND GRID REINFORCING**
03 23 00	**STRESSED TENDON REINFORCING**
03 24 00	**FIBROUS REINFORING**
03 31 00	**STRUCTURAL CONCRETE**
03 31 16	Lightweight Structural Concrete
03 31 19	Shrinkage-Compensating Structural Concrete
03 31 23	High-Performance Structural Concrete
03 31 24	Ultra High-Performance Structural Concrete
03 38 00	**POST-TENSIONED CONCRETE**
03 40 00	**PRECAST CONCRETE**

KEY TERMS

portland cement
concrete
aggregate
coarse aggregate
fine aggregate
curing
hydration
heat of hydration
drying shrinkage
clinker
air-entraining cement
white portland cement
high-volume-fly-ash concrete (HVFA concrete)
lightweight aggregate
structural lightweight aggregate
expanded shale aggregate
vermiculite
perlite
water
supplementary cementitious material (SCM)
pozzolan
fly ash
silica fume, microsilica
natural pozzolan
high-reactivity metakaolin

blast furnace slag, slag cement
hydraulic cement
blended hydraulic cement
(concrete) admixture
air-entraining admixture
water-reducing admixture
high-range water-reducing admixture, superplasticizer
accelerating admixture
retarding admixture
workability agent
shrinkage-reducing admixture
corrosion inhibitor
freeze protection admixture
extended set-control admixture
coloring agent
water–cement ratio, w-c ratio
transit-mixed concrete
slump test
slurry
segregation
dropchute
consolidation
self-consolidating concrete (SCC)
formwork
form release compound
precasting

sitecasting, cast-in-place construction
reinforced concrete
steel reinforcing bar
deformed reinforcing bar
rebar congestion
welded wire reinforcing (WWR), welded wire fabric (WWF)
simply supported beam
bottom bar
cover
bond
hook
stirrup
U-stirrup
closed stirrup-tie
chair
bolster
continuous beam
one-way action
shrinkage–temperature steel
two-way action
vertical bar, column bar
tie
column spiral
column tie
fibrous reinforcing
microfiber reinforcing

plastic shrinkage cracking
macrofiber reinforcing
creep
prestress
prestressed concrete
pretensioning

camber
posttensioning
tendon
draped tendon
harped tendon
ACI 301

ultra-high performance concrete (UHPC)
pervious concrete
light-transmitting concrete

REVIEW QUESTIONS

1. What is the difference between cement and concrete?

2. List the conditions that must be met to make a satisfactory concrete mix.

3. List the precautions that should be taken to cure concrete properly. How do these change in very hot, very windy, and very cold weather?

4. What problems are likely to occur if concrete has too low a slump? Too high a slump? How can the slump be increased without increasing the water content of the concrete mixture?

5. Explain how steel reinforcing bars work in concrete.

6. Explain the role of stirrups in beams.

7. Explain the role of ties in columns.

8. What does shrinkage temperature steel do? Where is it used?

9. Explain the differences between reinforcing and prestressing and the relative advantages and disadvantages of each.

10. Under what circumstances would you use pretensioning, and under what circumstances would you use posttensioning?

11. Explain the advantages of using higher-strength reinforcing bars in concrete that requires very heavy reinforcing.

EXERCISES

1. Design a simple concrete mixture. Mix it and pour some test cylinders for several water–cement ratios. Cure and test the cylinders. Plot a graph of concrete strength versus water–cement ratio.

2. Sketch from memory the pattern of reinforcing for a continuous concrete beam. Add notes to explain the function of each feature of the reinforcing.

3. Design, form, reinforce, and cast a small concrete beam, perhaps 6 to 12 feet (2–4 m) long. Get help from a teacher or professional, if necessary, in designing the beam.

4. Visit a construction site where concrete work is being done. Examine the forms, reinforcing, and concrete work. Observe how concrete is brought to the site, transported, placed, compacted, and finished. How is the concrete supported after it has been poured? For how long?

SELECTED REFERENCES

American Concrete Institute. *Manual of Standard Practice.* Schaumburg, IL, Author, updated regularly.

A compendium of standards for all aspects of reinforced concrete design, mixing, pouring, curing, finishing, and maintenance. Includes ACI 301.

Concrete Reinforcing Steel Institute. *Manual of Standard Practice.* Schaumburg, IL, Author, updated regularly.

Specifications for reinforcing, steel, welded wire fabric, bar supports, detailing, fabrication, and installation are standardized in this booklet.

Mehta, P. Kumar, and J. M. Monteiro Paulo. *Concrete: Microstructure, Properties, and Materials* (3rd ed.). New York, McGraw-Hill, 2005.

For the reader who wishes to explore further the science and mechanics of concrete, this text provides an in-depth treatment of concrete materials, formulation, and behavior.

Portland Cement Association. *Design and Control of Concrete Mixtures* (15th ed.). Skokie, IL, Author, 2011.

The 372 pages of this book summarize clearly and succinctly, with many explanatory photographs and tables, the state of current practice in making, placing, finishing, and curing concrete.

WEB SITES

Concrete Construction

Author's supplementary web site: **www.ianosbackfill.com/13_concrete_construction**

American Concrete Institute International: **www.concrete.org**

Concrete Sustainability Hub: **web.mit.edu/cshub/**

Portland Cement Association: **www.cement.org**

Reinforcing

Concrete Reinforcing Steel Institute: **www.crsi.org**

14

SITECAST CONCRETE FRAMING SYSTEMS

- **Casting a Concrete Slab on Grade**
  - Pouring and Finishing
  - Controlling Cracking

- **Casting a Concrete Wall**
  - Insulating Concrete Forms

- **Casting a Concrete Column**

- **One-Way Floor and Roof Framing Systems**
  - One-Way Solid Slab
  - One-Way Concrete Joist (Ribbed Slab)
  - Wide-Module Concrete Joist

- **Two-Way Floor and Roof Framing Systems**
  - Two-Way Flat Slab and Two-Way Flat Plate
  - Two-Way Waffle Slab

- **Other Uses of Sitecast Concrete**

- **Sitecast Posttensioned Framing Systems**

- **Selecting a Sitecast Concrete Framing System**

- **Innovations in Sitecast Concrete Construction**

  FOR PRELIMINARY DESIGN OF A SITECAST CONCRETE STRUCTURE

- **Architectural Concrete**

  CUTTING CONCRETE, STONE, AND MASONRY

- **Longer Spans in Sitecast Concrete**

- **Designing Economical Sitecast Concrete Buildings**

- **Sitecast Concrete and the Building Codes**

- **Uniqueness of Sitecast Concrete**

Kane Hall, University of Washington, 1969. Architects Walker and McGough made bold use of sitecast concrete construction. *(Photo by Joseph Iano.)*

Concrete that is cast into forms on the building site offers almost unlimited possibilities to the designer. Any shape that can be formed can be cast, with a limitless selection of surface textures, and books on modern architecture are filled with graphic examples of the realization of this extravagant promise. Certain types of concrete elements cannot be precast, but can only be cast on the site: foundation caissons and spread footings, slabs on grade, structural elements too large or too heavy to transport from a precasting plant, elements so irregular or special in form as to rule out precasting, slab toppings over precast floor and roof elements, and many types of structures with two-way slab action or full structural continuity from one member to another. In many cases where sitecast concrete could be replaced with precast, sitecast remains the method of choice simply because of its more massive, monolithic architectural character.

Sitecast concrete structures tend to be heavier than most other types of structures, which can lead to the selection of a precast concrete or structural steel frame instead if foundation loadings are critical. Sitecast buildings are also relatively slow to construct because each level of the building must be formed, reinforced, poured, cured, and stripped of formwork before construction of another level can begin. In effect, each element of a sitecast concrete building is manufactured in place, often under variable weather conditions, whereas the majority of the work on steel or precast concrete buildings is done in factories and shops where worker access, tooling, materials-handling equipment, and environmental conditions are generally superior to those on the job site. The technology of sitecasting has evolved rapidly, however, in response to its own inherent limitations, with streamlined methods of materials handling, systems of reusable formwork that can be erected or taken down almost instantaneously, extensive prefabrication of reinforcing elements, and mechanization of finishing operations. The rapid pace of this evolution has kept sitecast concrete among the construction techniques most favored by building owners, architects, and engineers.

FIGURE 14.1

Unity Temple in Oak Park, Illinois, was constructed by architect Frank Lloyd Wright in 1906. Its structure and exterior surfaces were cast in concrete, making it one of the earliest buildings in the United States to be built primarily of this material. *(Photo by John McCarthy. Courtesy of Chicago Historical Society ICHi-18291.)*

Casting a Concrete Slab on Grade

A concrete *slab on grade* is a level surface of concrete that lies directly on the ground. Slabs on grade are used for roads, sidewalks, patios, airport runways, and basements or ground floors of buildings. A slab-on-grade floor usually experiences little structural stress except a direct transmission of compression between its superimposed loads and the ground beneath, so it furnishes a simple example of the operations involved in the sitecasting of concrete (Figure 14.2).

To prepare for the placement of a slab on grade, the unstable topsoil is scraped away to expose the subsoil beneath. If the exposed subsoil is too soft, it is compacted or replaced with more stable material. Next, a layer of 1½-inch-diameter (38-mm) crushed stone at least 4 inches (100 mm) deep, sometimes referred to as a *capillary break*, is compacted over the subsoil. This acts as a drainage layer to keep moisture away from the underside of the slab. Where the slab is not being cast within surrounding walls, a simple edge form—a strip of wood or metal fastened to stakes driven into the ground—is constructed around the perimeter of the area to be poured and is coated with a form release compound to prevent the concrete from sticking (Figure 14.3*a, b*). The top edge of the form is carefully leveled; when the slab is poured, this edge will be used to guide the screed that levels the slab surface. Where walls surround the slab to be poured, an isolation joint is creating by inserting a compressible filler material (Figure 14.3*c*). The thickness of the slab may range from 3 inches (100 mm) for a residential floor to 6 or 8 inches (150 or 200 mm) for an industrial floor. For interior floor slabs on grade, a *moisture barrier* (also called a *vapor retarder*), such as a heavy plastic sheet, is laid over the crushed stone to further protect the slab from moisture in the ground (Figure 14.3*b*).

For some interior floor slab construction, a layer of fine crushed stone or sand, 2 to 4 inches (50 to 100 mm) thick, may be placed over the vapor retarder before concrete is placed; this layer of fine aggregate protects the vapor retarder sheet from damage. As concrete is placed, this layer also absorbs excess water from the concrete. This can help to prevent curling (warping) of the slab that can occur when the top of the slab loses moisture more rapidly than the bottom, a concern especially for some types of industrial floors. However, this practice also runs the risk of creating a reservoir for the storage of water under the slab that can lead to moisture-related problems after the slab is in service. With a better recognition of this risk and the advent of tougher vapor retarder sheets that are less vulnerable to damage, the use

Concrete

Welded wire reinforcing

Vapor retarder

Crushed stone
drainage layer

Edge form

Brace

FIGURE 14.2

The construction of a concrete slab on grade. Notice how the welded wire reinforcing is overlapped where two sheets of fabric meet.

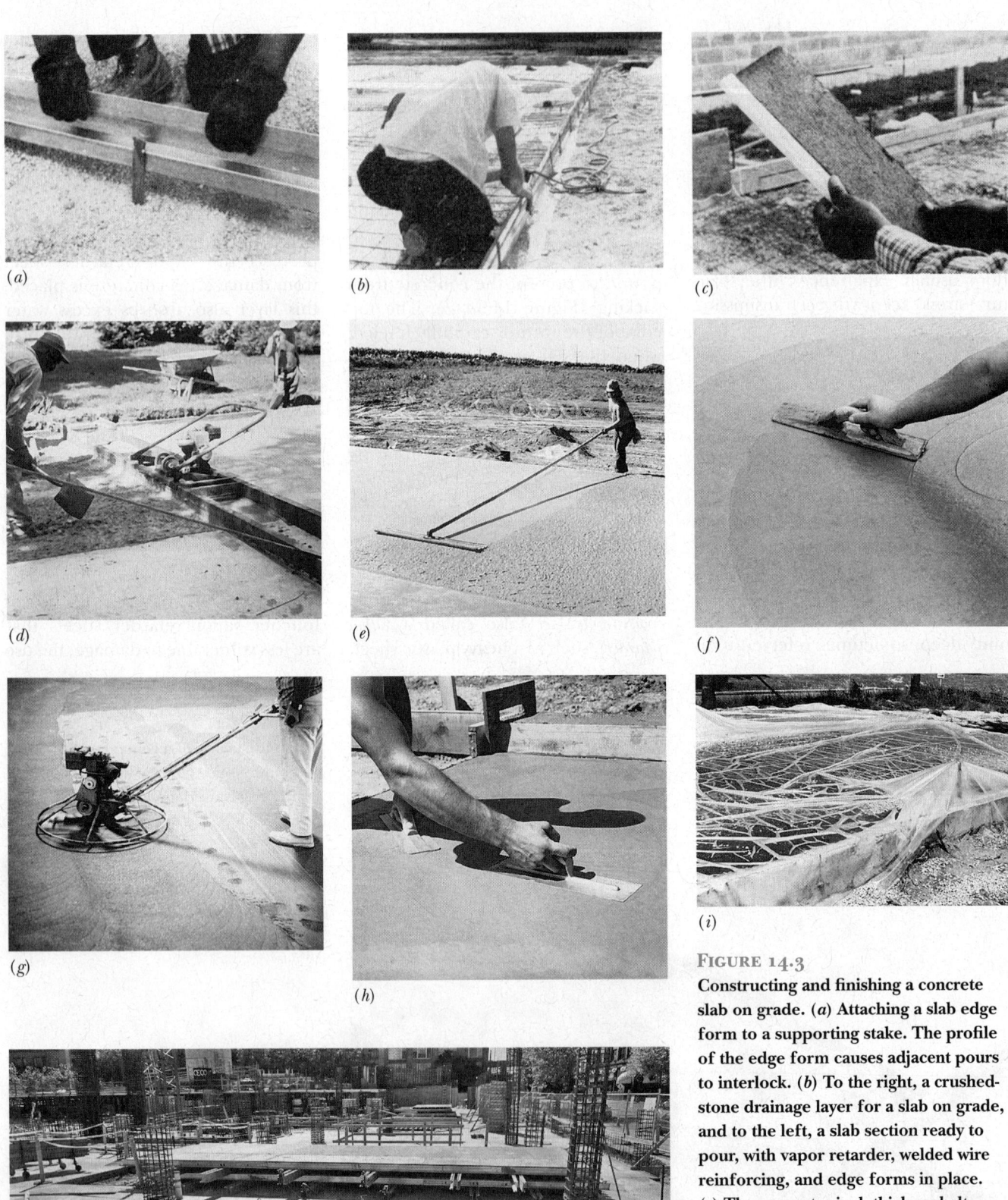

(a)

(b)

(c)

(d)

(e)

(f)

(g)

(h)

(i)

(j)

FIGURE 14.3

Constructing and finishing a concrete slab on grade. (*a*) Attaching a slab edge form to a supporting stake. The profile of the edge form causes adjacent pours to interlock. (*b*) To the right, a crushed-stone drainage layer for a slab on grade, and to the left, a slab section ready to pour, with vapor retarder, welded wire reinforcing, and edge forms in place. (*c*) Three-quarter-inch-thick asphalt-impregnated fiberboard used to form an isolation joint where slab edges abut walls or other vertical surfaces. The plastic cap is removed immediately after slab finishing to leave a clean slot for the later insertion of an elastomeric joint sealant.

of an aggregate layer over the vapor retarder has become less prevalent.

A reinforcing mesh of welded wire reinforcing, cut to a size just a bit smaller than the dimensions of the slab, is laid over the moisture barrier or crushed stone. The reinforcing most commonly used for lightly loaded slabs, such as those in houses, is 6 × 6-W1.4 × W1.4, which has a wire spacing of 6 inches (150 mm) in each direction and a wire diameter of 0.135 inch (3.43 mm). For slabs in factories, warehouses, and airports, a fabric made of heavier wires or a grid of reinforcing bars may be used instead. The grid of wires or bars helps protect the slab against cracking caused by concrete shrinkage during curing, temperature stresses, concentrated loads, frost heaving, or settlement of the ground beneath. Fibrous, macrofiber reinforcing, discussed in Chapter 13, can also be used in addition to or in place of wire reinforcing for the same purposes.

(d) Striking off the surface of a concrete slab on grade just after pouring, using a motorized screed. The motor vibrates the screed from end to end to work the wet concrete into a level surface. (e) A bull float is used for initial floating, to flatten and consolidate the slab surface immediately after screeding. (f) Floating, here performed by hand, brings cement paste to the surface and produces a smoother, denser finish. (g) For larger slabs, floating can be done by machine. (h) Steel troweling after floating produces a dense, hard, smooth surface. (i) Damp curing of the slab, here using a polyethylene plastic sheet cover. (j) This slab was cured with a moisture-retaining liquid compound that also provides a protective coating, allowing the next stages of work to proceed sooner on the finished slab. (Photos a, b, c courtesy of Vulcan Metal; d, e, f, g, h reprinted with permission of the Portland Cement Association from Design and Control of Concrete Mixtures, 12th edition; PHOTOS: Portland Cement Association, Skokie, IL; j by Joseph Iano.)

Pouring and Finishing

Casting of the slab commences with the placing of wet concrete in the formwork, using any of the methods described in Chapter 13. The concrete is spread by workers using shovels or rakes until the form is full. Then the same tools, or an immersion vibrator, are used to agitate the concrete slightly, especially around the edges, to eliminate air pockets. Unless the welded wire fabric reinforcing was set on chairs or other supports, the concrete masons reach into the wet concrete with metal hooks and raise this reinforcing to approximately the midheight of the slab. This locates the reinforcing optimally to resist tension caused by forces acting either from above or below the slab.

The first operation in finishing the slab is to *strike off* or *screed* the concrete by drawing a stiff plank of wood or metal across the top edges of the formwork to achieve a level concrete surface (Figure 14.3d). This is done with an end-to-end sawing motion of the plank that avoids tearing the projecting pieces of coarse aggregate from the surface of the wet concrete. A bulge of concrete is maintained in front of the screed as it progresses across the slab, so that when a low point is encountered, concrete from the bulge will flow in to fill it.

Immediately after striking off the concrete, the slab receives its initial *floating*. This step is usually performed by hand, using flat-surfaced tools, typically 4 to 10 feet (1.2 to 3 m) in length, called *bull floats* or *darbies* (Figure 14.3e). These are drawn across the concrete to flatten and consolidate its surface. After this initial floating, the top of the slab is level but still rather rough. If a concrete topping will later be poured over the slab, or if a floor finish of terrazzo, stone, brick, or quarry tile will be applied, the slab may be left to cure without further finishing.

If a smoother surface is desired, additional finishing operations proceed after a period of time during which the concrete begins to stiffen and the watery sheen, called *bleed water*, evaporates from the surface of the slab. First, specially shaped hand tools may be used to form neatly rounded edges around the perimeter of the slab and control joints in the interior. Next, the slab is floated a second time to further consolidate its surface. At this stage, small slabs may be floated by hand (Figure 14.3f), but for larger slabs, rotary power floats are used (Figure 14.3g). The working surfaces of floats are made of wood or of metal with a slightly rough surface. As the float is drawn across the surface, it gently vibrates the concrete and brings cement paste to the surface, where it is smoothed over the coarse aggregate and into low spots by the float. If too much floating is done, however, an excess of paste and free water rises to the surface and forms puddles, making it almost impossible to get a good finish. Experience on the part of the mason is essential to floating, as it is to all slab finishing operations, to know just when to begin each operation and just when to stop. The floated slab has a lightly textured surface that is appropriate for outdoor walks and pavings without further finishing.

For a completely smooth, dense surface, the slab must also be *troweled*. This is done immediately after the second floating, either by hand, using a smooth, rectangular *steel trowel* (Figure 14.3h), or with a *rotary power trowel*. If the concrete mason cannot reach all areas of the slab from around the edges, squares of plywood called *knee boards*, two per mason, are placed on the surface of the concrete. These distribute the mason's weight sufficiently that he or she can kneel on the surface without making indentations. Any marks left by the knee boards are removed by the trowel as the mason works backward across the surface from one edge to the other. If a nonslip surface is required, a stiff-bristled janitor's pushbroom may be drawn across the surface of the slab after troweling to produce a striated texture called a *broom finish.*

Where a concrete slab must meet narrow floor flatness limits, it may be restraightened after each floating or troweling operation. *Restraightening* is performed with a rectangular flat-bottomed *straightedge,* roughly 10 feet (3 m) in length, which is drawn across the concrete slab surface to remove minor undulations produced during floating or troweling.

Shake-on hardeners are sometimes sprinkled over the surface of a slab between the screeding and floating operations. These dry powders react with the concrete to form a very hard, durable surface for such heavy-wear applications as warehouses and factories.

When the finishing operations have been completed, the slab should be cured under damp conditions for at least a week; otherwise, its surface may crack or become dusty from premature drying. Damp curing may be accomplished by covering the slab with an absorbent material such as sawdust, earth, sand, straw, or burlap and maintaining the material in a damp condition for the required length of time. An impervious sheet of plastic or waterproof paper may be placed over the slab soon after troweling to prevent the escape of moisture from the concrete (Figure 14.3*i*). Or, the concrete surface may be sprayed with one or more applications of a liquid *curing compound,* which forms an almost invisible moisture barrier membrane over the slab surface (Figure 14.3*j*).

No concrete floor is perfectly flat. The normal finishing process produces a surface that undulates, usually imperceptibly, between low and high areas that go unnoticed in everyday use. Traditionally, the flatness of concrete slabs is specified as the maximum gap size, typically in the range of ⅛ to ⅜ inch (3 to 10 mm), permitted under a 10-foot (3-m) straightedge placed anywhere on the floor. Industrial warehouses that use high-rise forklift trucks, however, require floors whose flatness is controlled to within narrower tolerances. These *superflat floors,* as well as other floors where close control over flatness is desired, are specified according to a more complicated system of indexes, called *F-numbers,* that correspond to the degrees of flatness (waviness) and levelness (conformity to a horizontal plane) required, and are produced using special finishing equipment and techniques. Because of its extreme accuracy, a laser-guided automatic straightedging machine (Figure 14.4) is often used in the creation of superflat floors. This device produces a slab surface that is flat and level to within consistently small tolerances, and does so at a very rapid rate.

Controlling Cracking

Because concrete slabs on grade are relatively thin in relation to their horizontal dimensions and normally only lightly reinforced, they are prone to cracking. The stresses that cause cracking may originate from the shrinkage as the concrete cures, thermal expansion and contraction, or differential movement between the slab and abutting building elements. If such cracks are allowed to occur randomly, they can be unsightly and can compromise the functionality of the slab.

The most common method of controlling cracking in concrete slabs on grade is to introduce an organized system of joints into the slab that allow cracking to occur in a neat and visually acceptable pattern. *Control joints,* also called *contraction joints,* are intentionally weakened sections created through the concrete slab where the tensile forces caused by concrete drying shrinkage are relieved. They are usually formed as grooves that

FIGURE 14.4

Guided by a laser beam, the motorized straightedging device on this machine can strike off 240 square feet (22 m^2) of slab surface per minute to an extremely exacting standard of flatness. The worker to the right smoothes the surface with a bull float.

(Photo by Wironen, Inc. Courtesy of the Laser Screed Co., Inc., New Ipswich, NH.)

extend at least one-quarter of the depth of the slab. They are made either by running a special trowel along a straightedge while the concrete is still plastic or by sawing partially through the concrete shortly after it begins to harden using a diamond or abrasive saw blade in a power circular saw. To provide a further inducement for cracks to occur at control joint locations rather than elsewhere, reinforcing in the slab may be partially discontinued where it crosses these joints. Control joint spacing recommendations vary with the thickness of the slab and the shrinkage rate of the concrete. For normal concrete slabs 4 to 8 inches (100–200 mm) thick, joint spacings from 11 feet 6 inches to 17 feet 6 inches (3.6–5.3 m) are recommended, with thinner slabs requiring closer joint spacing than thicker ones. Control joints should be arranged in perpendicular directions such that they create panels roughly square in proportion.

Reinforced concrete made "pilotis" possible. The house is in the air, away from the ground; the garden runs under the house, and it is also above the house, on the roof. . . . Reinforced concrete is the means which makes it possible to build all of one material. . . . Reinforced concrete brings the free plan into the house! Floors no longer have to stand simply one on top of the other. They are free. . . . Reinforced concrete revolutionizes the history of the window. Windows can run from one end of the facade to the other. . . .

—Le Corbusier and P. Jeanneret, *Oeuvre Complète 1910–1929*, 1956.

Isolation joints, sometimes called *expansion joints*, are formed by casting full-depth preformed joint materials, typically ⅜ to ¾ inch (10–20 mm) in width, into the slab (Figure 14.3c), completely separating the slab from adjacent elements. Isolation joints allow freedom of movement of the slab with respect to other building parts or other portions of the slab—movements that may occur due to thermal expansion and contraction, structural loading, or differential settlement. Isolation joints are commonly provided where the edge of a concrete slab abuts walls or curbs, as well as around columns or loadbearing walls that pass through the slab within its perimeter. Isolation joints are also used to divide large or irregularly shaped slabs into smaller, more simply shaped areas that are less prone to stress accumulation.

Concrete itself can be manipulated to reduce cracking: Shrinkage-reducing chemical admixtures and some supplementary cementitious materials, such as fly ash, reduce drying shrinkage. Lowering the water–cement ratio of the concrete mix reduces drying shrinkage and results in finished concrete that is stronger and more crack resistant, though at increased cost. Specially formulated *shrinkage-compensating cements* can completely nullify drying shrinkage, allowing the casting of large slabs on grade that are entirely free of contraction joints. The amount of reinforcing in the slab can be increased or fibrous reinforcing can be added to the concrete mix to enhance the slab's resistance to tensile forces. Protecting a freshly poured concrete slab from premature drying during the damp curing process reduces cracking while the concrete hardens and ensures that the concrete attains its full design strength.

A slab on grade may also be posttensioned, using level tendons in both directions at the midheight of the slab in place of conventional reinforcing. Posttensioning places the entire slab under sufficient compression to

have it remain free from tensile stress under any anticipated loading condition. Posttensioning eliminates the need for control joints, makes floors more resistant to concentrated loads, and often permits the use of a thinner slab. It is especially effective for slabs over unstable or inconsistent soils and for superflat floors.

CASTING A CONCRETE WALL

A reinforced concrete wall at ground level usually rests on a poured concrete strip footing (Figures 14.5, 14.6, and 14.7). The footing is formed and poured much like a concrete slab on grade. Its cross-sectional dimensions and its reinforcing, if any, are determined by the structural engineer. A *key*, a groove that will form a mechanical connection to the wall, is sometimes formed in the top of the footing with strips of wood that are temporarily embedded in the wet concrete. In addition, vertical *dowels* consisting of steel reinforcing bars are usually installed in the footing; these will later be overlapped with the bars in the walls to form a structural connection. After pouring, the top of the footing is screeded; no further finishing operations are required. The footing is left to cure for at least a day before the wall forms are erected.

The wall reinforcing, either in one vertical layer of horizontal and vertical bars at the center of the wall or two layers near the faces of the wall, as specified by the structural engineer, is installed next, with the bars wired to one another at the intersections. The vertical bars are overlapped with the dowels projecting from the footing. L-shaped horizontal bars are installed at wall corners to maintain structural continuity between the two portions of the wall. If the top of the wall will connect to a concrete floor or another wall above, rods are left projecting from the formwork. These will be embedded in the later pour of concrete to form a continuous connection.

Wall forms may be custom-built of lumber and plywood for each job, but it is more usual to employ reusable prefabricated panels. The panels for one side of the form are coated with a form release compound, set on the footing, aligned, and braced. The *form ties*, which are small-diameter steel rods specially shaped to hold the formwork together under the pressure of the wet concrete, are inserted through holes provided in the formwork panels and secured to the back of the form by devices supplied with the ties. Both ties and fasteners vary in detail from one manufacturer to another and the type of wall being constructed. A typical example is shown in Figure 14.8. The ties will pass straight through the concrete wall from one side to the other, with all but the ends remaining embedded

Wall reinforcing

Dowels

Footing

Formwork panels

Studs

Walers

Bracing

Form ties

FIGURE 14.5

The process of casting a concrete wall. (a) Vertical reinforcing bars are wired to the dowels that project from the footing, and horizontal bars are wired to the vertical bars. (b) The formwork is erected. Sheets of plywood form the faces of the concrete. They are supported by vertical wood studs. The studs are supported against the pressure of the wet concrete by horizontal walers. The walers are supported by steel rod ties that pass through holes in the plywood to the walers on the other side. The ties also act as spreaders to maintain a spacing between the plywood walls that is equal to the desired thickness of the wall. Diagonal braces keep the whole assembly plumb and straight. (c) After the concrete has been poured, consolidated, and cured, the wedges that secure the walers to the form ties are driven out, the formwork is pulled off the concrete, and the projecting ends of the form ties are broken off.

Interactive Exercise

FIGURE 14.6
Protected against falling by a safety harness, a worker stands on the reinforcing bars for a concrete wall to wire another horizontal bar in position. *(Photo courtesy of DBI/SALA, Red Wing, MN.)*

—*One or two layers of vertical and horizontal bars reinforce the wall*

—*Ties are required in walls that bear heavy loads*

Dowels and a key unite the footing with the wall—

FIGURE 14.7
Section through a reinforced concrete wall, with two layers of horizontal and vertical reinforcing bars for greater strength.

permanently in the wall after it is poured.

When the ties are in place and the reinforcing has been inspected, the formwork for the second side of the wall is erected, the *walers* and *braces* are added, and the forms are inspected to be sure that they are straight, plumb, correctly aligned, and adequately tied and braced. A surveyor's transit or laser leveling device is used to establish the exact height to which the concrete will be poured, and this height is marked all around the inside of the forms (Figures 14.9 and 14.10). Pouring may then proceed.

Concrete is brought to the site, test cylinders are made, and a slump

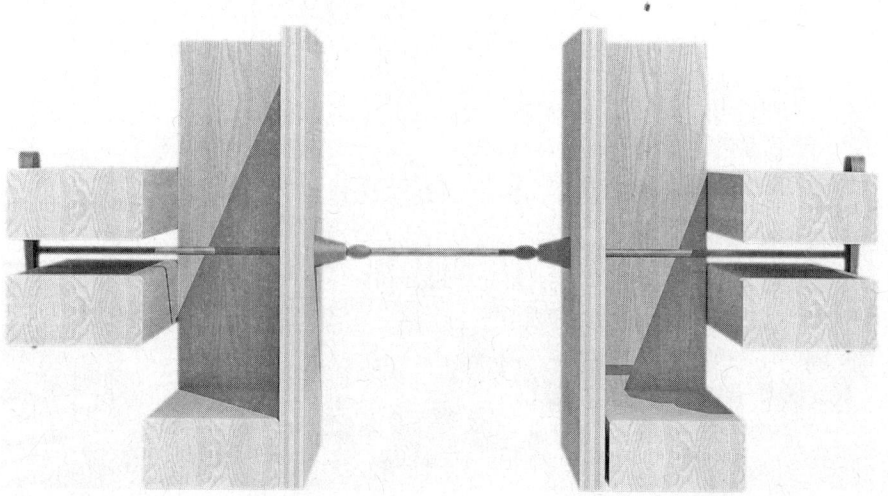

FIGURE 14.8
Detail of a typical form tie assembly. Plastic cones just inside the faces of the form hold the faces in position. Special hardware at the tie ends clamps against the walers. After the forms have been stripped, the cones will be removed from the concrete and the ties snapped off inside the voids left by the cones. The conical holes may be left open, filled with mortar, or plugged with conical plastic plugs. *(Photographs courtesy of Richmond Screw Anchor Co., Inc., 7214 Burns Street, Fort Worth, TX 76118.)*

FIGURE 14.9
Wall formwork, similar to that diagrammed in Figure 14.5, has been erected and bracing is being added. In the background the wall reinforcing is still exposed. Bent reinforcing bars projecting from the top of the formwork will connect to a concrete slab yet to be poured. In the background, bars project vertically, for added wall construction to continue above. (*Photo by Joseph Iano.*)

FIGURE 14.10
Heavy-strength wall formwork for a 40-ft (12-m)-plus-tall concrete wall. (*Photo by Joseph Iano.*)

FIGURE 14.11
Consolidating wet concrete after pouring, using a mechanical vibrator immersed in the concrete. *(Reprinted with permission of the Portland Cement Association from* Design and Control of Concrete Mixtures, *12th edition; PHOTOS: Portland Cement Association, Skokie, IL.)*

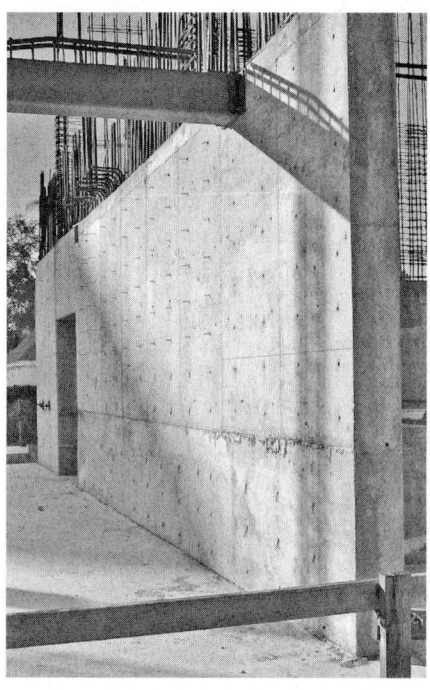

FIGURE 14.12
A concrete wall with formwork stripped. In the foreground and along the bottom of the wall, the form tie ends have been snapped off. Elsewhere, they still protrude from the wall. *(Photo by Joseph Iano.)*

test is performed to check for the proper pouring consistency. Workers standing on planks at the tops of the forms deposit the concrete into the forms, consolidating it with a vibrator to eliminate air pockets (Figure 14.11). When the form has been filled and consolidated up to the level that was marked inside the formwork, hand floats are used to smooth and level the top of the wall. The top of the form is then covered with a plastic sheet or canvas, and the wall is left to cure.

After a few days of curing, the bracing and walers are taken down, the connectors are removed from the ends of the form ties, and the formwork is *stripped* from the wall (Figure 14.12). This leaves the wall bristling with projecting form tie ends. These are twisted off with heavy pliers, and the *form tie holes* that they leave in the surfaces of the wall are carefully filled with grout. If required, major defects in the wall surface, such as caused by poor

consolidation of the concrete, can be repaired at this time. The wall is now complete.

Insulating Concrete Forms

Insulating concrete forms (ICFs) serve to contain the concrete when the wall is poured but, unlike conventional formwork, also become a permanent part of the wall, for which they act as thermal insulation (Figure 14.13). The forms are manufactured in differing configurations. Most are some kind of interlocking hollow blocks of polystyrene foam. Other manufacturers produce planks or panels of foam with plastic ties that join them to make wall forms. Whatever the configuration, these systems weigh so little and are so accurately made that they can be assembled almost as easily and quickly as a child's plastic building blocks. The tops of the blocks are channeled in such a way that horizontal reinforcing bars may be laid in the top of

each course and vertical bars may be inserted into the vertical cores. The wall forms must be braced to prevent them from moving during the concrete pour. Concrete is usually deposited in the cores of the foam blocks from the hose of a pump. The full height of a wall cannot be cast in one operation because the pressure of so great a depth of wet concrete would blow out the sides of the fragile blocks. The normal procedure is to deposit the concrete in several *lifts* of limited height, working all the way around the structure with each lift so that by the time the second lift is begun, the first lift has had an hour or two to harden somewhat, relieving the pressure at the bottom of the forms. Interior and exterior finish materials must be applied to the foam plastic faces to protect them from sunlight, mechanical damage, and fire. The thermal insulating value of the finished wall is usually in the range of R-17 to R-22 (RSI-3.0 to RSI-3.9).

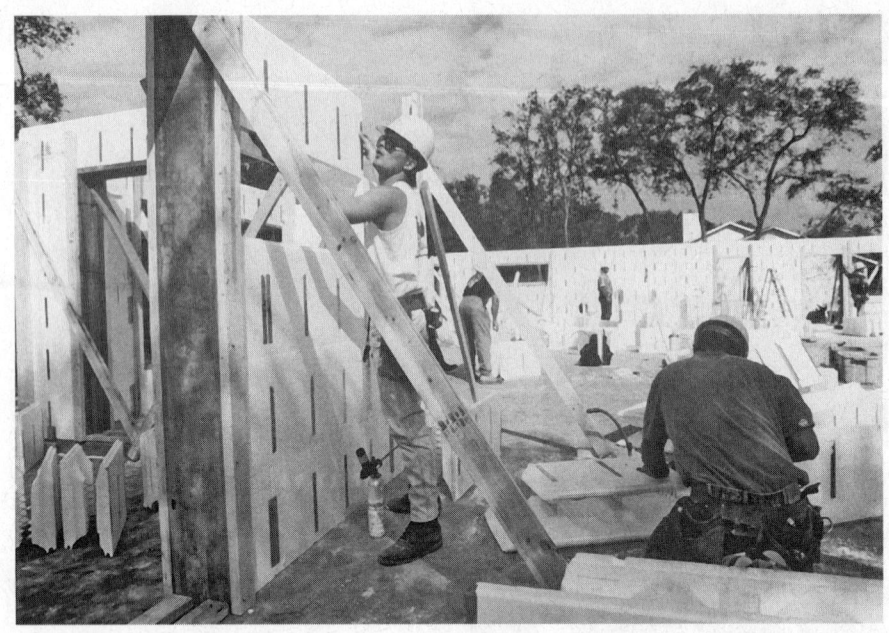

(a)

FIGURE 14.13
Insulating concrete forms are
manufactured as interlocking blocks or
panels. In the system illustrated here,
the inner and outer halves of the blocks
are tied together by steel mesh webs
that connect to sheet metal strips on the
outer surfaces. These strips later serve
to receive screws that fasten interior
and exterior finish materials to the wall.
(a) Workers stack the blocks to form
the exterior walls of a house. Openings
for doors and windows are formed with
dimension lumber. Blocks can be cut to
length with a simple hand saw. (b) This
sample wall, from which some of the
blocks have been removed, shows that
the completed wall contains a continuous
core of reinforced concrete with thermal
insulation inside and out. (Courtesy:
American Polysteel Forms.)

(b)

Casting a Concrete Column

A column is formed and cast much like a wall, with a few important differences. The footing is usually an isolated column footing, a pile cap, or a caisson rather than a strip footing (Figure 14.14). The dowels are sized and spaced in the footing to match the vertical bars in the column. The cage of column reinforcing is assembled with wire ties and hoisted into place over the dowels. If space is too tight in the region where the vertical bars and dowels overlap, the bars may be spliced end to end with welds or mechanical connectors instead (Figure 13.24). The column form may be a rectangular box of plywood or composite panels, a cylindrical steel or plastic tube bolted together in halves so that it can later be removed, or a waxed cardboard tube that is stripped after curing by unwinding the layers of paper that make up the tube (Figures 14.15, 14.16, and 14.54). Unless a rectangular column is very broad and wall-like,

(a)

Figure 14.14
(a) A column footing almost ready for pouring but lacking dowels. The reinforcing bars are supported on pieces of concrete brick. (b) Column footings poured with projecting dowels to connect to both round and rectangular columns.
(*Photos by Edward Allen.*)

(b)

FIGURE 14.15
In the foreground, a square column form is tied with pairs of
L-shaped steel brackets. In the background, a worker braces
a round column form made of sheet steel. *(Courtesy of the Ceco
Corporation, Oakbrook Terrace, IL.)*

FIGURE 14.16
Round columns may also be formed with single-use heavy
cardboard tubes. Notice the density of the steel shoring
structure that is being erected to support the slab form,
which will carry a very heavy load of wet concrete. *(Courtesy
of Sonoco Products Company.)*

form ties through the concrete are
not required. The vertical bars project from the top of the column to
overlap or splice to the bars in the
column for the story above, or they
are bent over at right angles to splice
into the roof structure. Where vertical bars overlap, the tops of the bars
from the column below are offset
(bent inward) by one bar diameter to
avoid interference.

ONE-WAY FLOOR AND ROOF FRAMING SYSTEMS

One-Way Solid Slab

A *one-way solid slab* (Figures 14.17,
14.18, and 14.19) spans across lines of
support furnished by walls or beams.
The walls and columns are poured
before the formwork for the slab is
erected, but the forms for the girders

and beams are nearly always built continuously with those for the slab, and
girders, beams, and slab are poured
simultaneously as a single piece.

The girder and beam forms are
erected first, then the slab forms.
The forms are supported on temporary joists and beams of metal
or wood, and the temporary beams
are supported on temporary *shores*
(adjustable-length columns). The

weight of uncured concrete that must be supported is enormous, and the temporary beams and shoring must be both strong and closely spaced. Formwork is, in fact, designed by a contractor's structural engineers as carefully as it would be if it were a permanent building, because a structural

failure in formwork is an intolerable risk to workers and property.

Edges of concrete structural elements are beveled or rounded by inserting shaped strips of wood or plastic into the corners of the formwork to produce the desired profile. This is done because sharp edges of

concrete often break off during form stripping to leave a ragged edge that is almost impossible to patch. In service, sharp edges are easily damaged by, and are potentially damaging to, people, furniture, and vehicles.

A form release compound is applied to all formwork surfaces

FIGURE 14.17

Plan and larger-scale section of a typical one-way solid slab system. For the sake of clarity, the girder and beam reinforcing are not shown in the plan, and the girder and column reinforcing are left out of the section. The slabs span between the beams, the beams are supported by the girders, and the girders rest on the columns.

that will be in contact with concrete. Then, in accordance with reinforcing diagrams and schedules prepared by the structural engineer, the girder and beam reinforcing—bottom bars, top bars, and stirrups—is installed in the forms, supported on chairs and bolsters to maintain the required cover of concrete. Next, the slab reinforcing—bottom bars, top bars, and shrinkage–temperature bars—is placed on bolsters. After the reinforcing and formwork have been inspected, the girders, beams, and slab are poured in a single operation, with sample cylinders being made for later testing to be sure that the concrete meets its specified strength.

One-way slab depths are typically 4 to 10 inches (100–250 mm), depending on the span and loading intensity. The top of the slab is finished in the same manner as a slab on grade, usually to a steel trowel finish, and the slab is sealed or covered for damp curing. The only components left projecting above the slab surface at

FIGURE 14.18
Isometric view of a one-way solid slab system under construction. The slab, beams, and girders are created in a single pour.

FIGURE 14.19
An example of a beam–column connection in a one-way solid slab structure, with the slab reinforcing omitted for clarity. Notice how the column bars are spliced by overlapping them just above floor level. The bars from the column below are offset at the top so that they lie just inside the bars of the column above at the splice. Structural continuity between the beam and column is established by running the top bars from the beam into the column. U-stirrups are shown in the beam; closed stirrup ties, shown in the inset detail, are often used instead.

this stage are the offset column bars, which are now ready to overlap with, or splice to, the column bars for the floor above.

When the slab and beams have attained enough strength to support themselves safely, the formwork is stripped and the slabs and beams are *reshored* with vertical props to

relieve them of loads until they have reached full strength, which will take several more weeks. Meanwhile, the formwork and the remainder of the shoring are cleaned and moved up a level above the slab and beams just poured, where the cycle of forming, reinforcing, pouring, and stripping is repeated (Figure 14.20).

Ordinarily, the most efficient and economical concrete beam is one whose depth is twice or three times its breadth. One-way solid slabs are often supported, however, by beams that are several times as broad as they are deep. These are called *slab bands* (Figure 14.21). Banded slab construction offers two kinds of

FIGURE 14.20
Reshoring supports the first slab above grade, which was recently stripped. Above that, formwork for the next level concrete slab has been installed. The columnar supports for reshoring and formwork look similar. But reshoring is installed after the slab formwork has been removed and each column makes direct contact with the underside of the concrete above. The supports for formwork contact the underside of the formwork. *(Photo by Joseph Iano.)*

FIGURE 14.21
Banded slab construction. Compare the depth and breadth of the slab band in the center of the building to those of the conventional concrete beams around the perimeter. *(Reprinted with permission of the Portland Cement Association from* Design and Control of Concrete Mixtures, *12th edition; PHOTOS: Portland Cement Association, Skokie, IL.)*

FIGURE 14.22
This helical ramp is a special application of one-way solid slab construction. The formwork is made of overlaid plywood for a smooth surface finish. *(Courtesy American Plywood Association.)*

economy: The width of the slab band reduces the span between the bands, which can result in a reduced thickness for the slab and consequent savings of concrete and reinforcing steel. Also, the reduced depth of the slab band compared to a more conventionally proportioned beam allows reduction of the story height of the building, with attendant economies in columns, cladding, partitions, and vertical runs of piping and ductwork.

One-Way Concrete Joist (Ribbed Slab)

As one-way solid slab spans increase, a progressively thicker slab is required. Beyond a certain span, the slab becomes so thick that its own weight becomes an excessive burden, unless a substantial portion of the nonworking concrete in the lower part of the slab can be eliminated to lighten the load. This is the rationale for the *one-way concrete joist system* (Figures 14.23 through 14.26), also called a *ribbed slab*. The bottom steel is concentrated in spaced ribs or joists. The thin slab

FIGURE 14.23

Plan and larger-scale section of a typical one-way concrete joist system. For the sake of clarity, no reinforcing is shown in the plan, and the column reinforcing is not shown on the section. All bottom and top reinforcing occurs in the ribs, and all shrinkage–temperature bars are placed in the slab.

FIGURE 14.24
Standard steel form dimensions for one-way concrete joist construction. (One inch equals 25.4 mm.) *(Courtesy of the Ceco Corporation, Oakbrook Terrace, IL.)*

TAPERED ENDFORM

16" FOR 20" STD. WIDTHS
25" FOR 30" STD. WIDTHS

3'-0"

SLOPE 1″ IN 12″

STIFFENER

8″ 10″ 12″ 14″ 16″ 20″

7⁄8″ 7⁄8″
30″

Filler widths (10″ and 15″) are available for filling nonstandard spaces only.

SLOPE 1″ IN 12″

STIFFENER

8″ 10″ 12″

7⁄8″ 7⁄8″
20″

FIGURE 14.25
Reinforcing being placed for a one-way concrete joist floor. Electrical conduits and boxes have been put in place, and welded wire fabric is being installed as shrinkage–temperature reinforcing. Both the tapered end pans and the square endcaps for the midspan distribution rib are clearly visible. *(Courtesy of the Ceco Corporation, Oakbrook Terrace, IL.)*

FIGURE 14.26
A one-way concrete joist system after stripping of the formwork, showing broadened joist ends at the lower edge of the photograph and a distribution rib in the foreground. The dangling wires are hangers for a suspended finish ceiling. *(Reprinted with permission of the Portland Cement Association from* Design and Control of Concrete Mixtures, *12th edition; PHOTOS: Portland Cement Association, Skokie, IL.)*

that spans across the top of the joists is reinforced only by shrinkage-temperature bars. There is little concrete in this system that is not working, with the result that a one-way concrete joist system can efficiently and economically span considerably longer distances than a one-way solid slab. Each joist is reinforced as a small beam, except that stirrups are not usually used, because of the restricted space in the narrow joist. Instead, the ends of the joists are broadened sufficiently that the concrete itself can resist the diagonal tension forces.

The joists are formed with metal or plastic *pans* supported on a temporary plywood deck. Pans are available in two standard widths, 20 inches (508 mm) and 30 inches (762 mm), and in depths ranging up to 20 inches (508 mm), as shown in Figure 14.24. The sides of the pans taper from bottom to top, to allow them to drop easily from the hardened concrete. The joist width can be varied by placing the rows of pans closer together or farther apart. The bottom of each joist is formed by the wood deck on which the pans are placed. The joist ends are broadened with standard end pans whose width tapers. A *distribution rib* is sometimes formed across the joists at midspan to distribute concentrated loads to more than one joist. After application of a form release compound, the beam and joist reinforcing are placed, the shrinkage–temperature bars are laid crosswise on bolsters over the pans, and the entire system is poured and finished (Figures 14.25 and 14.26).

One-way concrete joists are usually supported on *joist bands*, which are broad beams that are only as deep as the joists. Although a deeper beam would be more efficient structurally, a joist band is formed by the same plywood deck that supports the pans, which eliminates expensive beam formwork entirely and produces a simpler underside of slab profile with a uniform floor-to-ceiling height throughout.

Wide-Module Concrete Joist

When fire-resistance requirements of the building code dictate a slab thickness of 4½ inches (115 mm) or more, the slab is capable of spanning a much greater distance than the normal space between joists in a one-way concrete joist system. This has led to the development of the *wide-module concrete joist system*, also called the *skip-joist system*, in which the joists are placed 4 to 6 feet (1220–1830 mm) apart. The name "skip-joist" arose from the original practice of achieving this wider spacing by laying strips of wood over alternate joist cavities in conventional joist pan formwork to block out the concrete. Pans are now specially produced for wide-module construction (Figures 14.27 and 14.28).

Because wide-module joists must each carry about double the weight carried by conventionally spaced joists, stirrups are required near the ends of each joist. If conventional U-stirrups are used, they must be installed on a diagonal as seen from above, in order to fit into the narrow joist, or single-leg stirrups may be used instead.

FIGURE 14.27
Formwork for a wide-module concrete joist system. These pans have been placed over a flat plywood deck, which will also serve to form the bottoms of the joist band beams. *(Courtesy of the Ceco Corporation, Oakbrook Terrace, IL.)*

FIGURE 14.28
The underside of the finished wide-module joists, joist bands, and slab. *(Courtesy of the Ceco Corporation, Oakbrook Terrace, IL.)*

Two-Way Floor and Roof Framing Systems

Two-Way Flat Slab and Two-Way Flat Plate

Where columns supporting a slab can be arranged in bays that are square or nearly square in proportion, two-way concrete framing systems that are more economical than one-way systems can be used. A *two-way solid slab* system is rarely seen, though it is occasionally used for very heavily loaded industrial floors; in such a system, the slab is supported by a grid of beams running in both directions over the columns. Most two-way floor and roof framing systems, however, even for heavy loadings, are made without beams. The slab is reinforced in such a way that the varying stresses in the different zones of the slab are accommodated within a uniform thickness of concrete. This simplifies formwork construction and reinforcing bar patterns considerably.

The *two-way flat slab* (Figure 14.29), a system suited to heavily loaded buildings such as storage and industrial buildings, illustrates this concept. The formwork is completely flat except for

Column and drop panel

½ column strip

Middle strip

Column strip

Middle strip

Column strip

Edge panel

Interior panel

½ column strip

Middle strip

Column strip

Middle strip

Column strip

Edge panel

Interior panel

PLAN VIEW

TYPICAL COLUMN STRIP SECTION, EITHER DIRECTION

TYPICAL MIDDLE STRIP SECTION, EITHER DIRECTION

Edge panel

Interior panel

FIGURE 14.29
Plan and larger-scale section of a typical two-way flat slab system. The reinforcing pattern consists of column strips and middle strips, with each strip changing pattern slightly around the perimeter of the building to accommodate the different bending forces that occur in the edge panels. The system shown uses only drop panels without mushroom capitals. The reinforcing in a two-way flat plate system is essentially identical to this example; the only difference is that the flat plate has no drop panels.

Bottom bars ———————
Top bars ‑ ‑ ‑ ‑ ‑ ‑ ‑ ‑

a thickening of the concrete to resist the high shear forces around the top of each column. Traditionally, this thickening was accomplished with both a funnel-shaped *mushroom capital* and a square *drop panel*, but today the capital is usually eliminated to reduce the formwork cost, leaving a drop panel to do the work alone (Figure 14.30). Typical depths for the slab itself range from 6 to 12 inches (150 to 300 mm).

Reinforcing for a two-way slab is laid in both directions in half-bay-wide strips of two fundamental types: *Column strips* are designed to carry the higher bending forces encountered in the zones of the slab that cross the columns. *Middle strips* have a lighter reinforcing pattern. Shrinkage–temperature steel is not needed in two-way systems because the concrete is already reinforced in both directions to resist bending. The drop panel and capital (if any) have no additional reinforcing beyond that provided by the column strip; the greater thickness of concrete furnishes the required shear resistance.

In more lightly loaded buildings, such as hotels, hospitals, dormitories, and apartment buildings, the slab need not be thickened at all over the columns. This makes the formwork extremely simple and even allows some columns to be moved off the grid somewhat if it will improve the arrangement of the floor plan. The completely flat ceilings of this system allow room partitions to be placed anywhere with equal ease. Because there are no beams or girders, only a thin slab, the story heights of the building may be kept to an absolute minimum, which reduces the costs of exterior cladding and other systems. Typical slab depths for such a *two-way flat plate* system range from 5 to 12 inches (125–305 mm) (Figure 14.31).

The zones along the exterior edges of both the two-way flat slab system and the two-way flat plate system

require special attention. To take full advantage of structural continuity, the slabs should be cantilevered beyond the last row of columns a distance equal to about 30 percent of the interior span. If such a cantilever is impossible, additional reinforcing must be added to the slab edges to carry the higher stresses that will result.

Because a two-way flat plate has no drop panel, it requires additional reinforcing in the slab at the top of each column to resist the high shear stresses that occur in this region. This can be accomplished with added conventional steel reinforcing bars or a more compact, proprietary system of shear studs (Figure 14.32).

Two-Way Waffle Slab

The *waffle slab*, or *two-way concrete joist system* (Figure 14.33), is the two-way equivalent of the one-way concrete joist system. Metal or plastic pans called *domes* are used as formwork to eliminate nonworking concrete from the slab, allowing considerably longer spans than are feasible with the two-way flat plate. The standard domes form joists on 36-inch (914-mm), 48-inch (1219-mm), and 60-inch (1524-mm) centers, in a variety of depths up to 24 inches (610 mm) (Figures 14.34 and 14.35). Solid concrete *heads* are created around the tops of the columns by leaving the domes out of the formwork in these areas. A head serves the same function as a drop panel in the two-way flat slab system. If a waffle slab cannot be cantilevered at the perimeter of the building, a solid perimeter beam must also be provided. The waffle slab system is suited to longer-span, heavily loaded applications, and its coffered underside presents rich architectural opportunities as an exposed ceiling (Figure 14.36). However, the complexity of waffle slab formwork also makes it the most expensive of sitecast framing systems.

FIGURE 14.30
Column capitals for two-way concrete framing systems. For slabs bearing heavy loads, shear stresses around the column are reduced by means of a mushroom capital and drop panel or a drop panel alone. For lighter loads, no thickening of the slab is required.

FIGURE 14.31
Flat plate construction for a high-rise apartment building. Columns located off of a regular grid are readily accommodated by this system. Note the cantilevered slab edges, very thin floor slabs, and minimal floor-to-floor heights. Reshoring is in place on two of the upper floors. A climbing system of formwork and protection from the weather tops out the structure.

FIGURE 14.32
The high shear forces around the top of the column require either added conventional reinforcing or a system such as the Studrails shown here. The remainder of the slab reinforcing has not yet been installed. *(U.S. and Canada patents Nos. 4406103 and 1085642, respectively. Licensee: Deha, represented by Decon, 105C Atsion Rd., P.O. Box 1575, Medford, NJ 08055-6675 and 35 Devon Road, Bramton, Ontario L6T 5B6. US: 1-800-527-7245.)*

PLAN VIEW

TYPICAL COLUMN STRIP SECTION, EITHER DIRECTION

TYPICAL MIDDLE STRIP SECTION, EITHER DIRECTION

FIGURE 14.33
Plan and larger-scale section of a typical two-way concrete joist system, also known as a waffle slab. For the sake of clarity, no reinforcing is shown on the plan drawing, and the section does not show the welded wire fabric that is spread over the entire form before pouring.

2´-0˝ MODULE
(19˝ × 19˝ Dome System)

3´-0˝ MODULE
(30˝ × 30˝ Dome System)

FIGURE 14.34
Some examples of steel dome forms for two-way concrete joist construction. (One inch equals 25.4 mm.) *(Courtesy of the Ceco Corporation.)*

FIGURE 14.35
Steel domes being placed on a temporary plywood deck to form a two-way concrete joist floor. Pans are omitted around columns to form solid concrete heads.

(Courtesy of the Ceco Corporation.)

FIGURE 14.36
The underside of a two-way concrete joist floor. Notice how the joists are cantilevered beyond the column line for maximum structural efficiency. (*Courtesy of the Ceco Corporation.*)

OTHER USES OF SITECAST CONCRETE

Sitecast concrete finds a variety of other uses in the construction of most buildings, such as site pavings and curbs, raised pads for mechanical equipment, toppings over metal or precast concrete decking, stairs, and other uses.

Concrete is frequently used to fill the pans of a metal stair. Also, stairs may be constructed wholly of concrete, as illustrated in Figure 14.37. A concrete stair may be thought of as an inclined one-way solid slab. The underside of the formwork is planar. The top is built with riser forms, which are usually inclined to provide greater toe space and make the stair more comfortable to users. The concrete is poured in one operation, and the treads are tooled to a steel trowel finish.

SITECAST POSTTENSIONED FRAMING SYSTEMS

Posttensioning can be applied to any of the sitecast concrete framing systems. It is used in beams, girders, and slabs, both one-way and two-way, to reduce member sizes, reduce deflection, and extend spanning capability.

Two-way flat plate structures are very commonly posttensioned, especially when spans are long or restrictions on the height of the building require minimal slab depths. The tendon layout, however, is quite different from the conventional reinforcing layout shown in Figure 14.29. Instead of being placed identically in both directions, the *draped tendons* are evenly distributed in one direction and banded closely together over the line of columns in the other direction (Figures 14.38 and 14.39). This arrangement functions better structurally in posttensioned slabs because it balances the maximum upward force from the banded tendons against the maximum downward force from the distributed tendons. It is also much easier to install than distributed, draped tendons running in both directions. If the structural bay is square, the same number of tendons is used in each direction. The prestressing force from the banded tendons becomes evenly distributed throughout the width of the slab within a short distance of the end anchorages because of action of the concrete.

As with any prestressed concrete framing system, both short-term and long-term losses of prestressing force must be anticipated. The short-term losses in posttensioning are caused by elastic shortening of the concrete, friction between the tendons and the concrete, and initial movements (*set*) in the anchorages. The long-term losses are caused by concrete shrinkage, concrete creep, and steel relaxation. The structural engineer calculates the total of these expected losses and specifies an additional amount of initial posttensioning force to compensate for them.

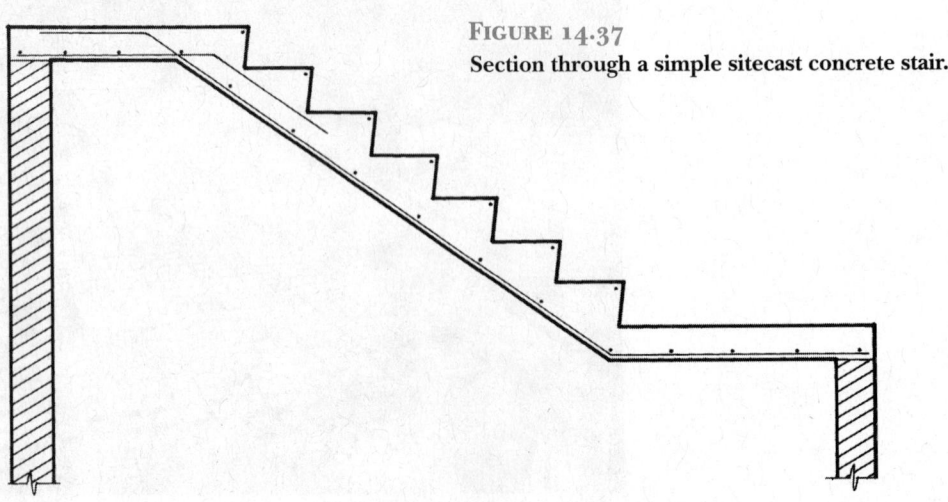

FIGURE 14.37
Section through a simple sitecast concrete stair.

PLAN VIEW

TYPICAL SECTION PARALLEL TO DISTRIBUTED TENDONS (A-A)

TYPICAL SECTION PARALLEL TO BANDED TENDONS (B-B)

FIGURE 14.38

A plan and two larger-scale sections of the tendon layout in a two-way flat plate floor with banded posttensioning. The number of tendons running in each of the two directions is identical, but those in one direction are concentrated into bands that run over the tops of the columns. The draping of the tendons is evident in the two section drawings. Building codes require that at least two distributed tendons run directly over each column to help reinforce against shear failure of the slab in this region. In addition to the tendons, conventional steel reinforcing is used around the columns and in midspan, but this has been omitted from these drawings for the sake of clarity.

FIGURE 14.39
Banded tendons run directly through the concrete column of this flat plate floor. A substantial amount of conventional reinforcing is used here for shear reinforcing. Notice the end anchorage plates nailed to the vertical surface of the formwork at the upper right; see also Figure 13.39. *(Courtesy of Post-Tensioning Institute.)*

SELECTING A SITECAST CONCRETE FRAMING SYSTEM

Preliminary factors to be considered in the selection of a sitecast concrete framing system for a building include the following (Figures 14.40 and 14.41):

1. Are the bays of the building square or nearly square? If so, a two-way system will probably be more economical than a one-way system.

2. How long are the spans? Spans up to 25 or 30 feet (7.6 or 9.1 m) are usually accomplished most economically with a two-way flat plate or flat slab system because of the relative simplicity of the formwork. For longer spans, a one-way joist system may be a good choice. Posttensioning significantly extends the economical span range of any of these systems.

3. How heavy are the loads? Heavy industrial loadings are borne better by thicker slabs and larger beams than

they are by light joist construction. Ordinary commercial, institutional, and residential loadings are carried easily by flat plate or joist systems.

4. Will there be a finish ceiling beneath the slab? If not, flat plate and one-way slab construction have smooth, paintable undersides that can serve as ceilings.

5. Does the lateral stability of the building against wind and seismic loads have to be provided by the rigidity of the concrete frame? Flat plate floors may not be sufficiently rigid for this purpose, which would favor a one-way system with its deeper beam-to-column connections.

INNOVATIONS IN SITECAST CONCRETE CONSTRUCTION

The development of sitecast concrete construction continues along several lines. The basic materials, concrete and steel, continue to undergo inno-

vations, as described in Chapter 13. The continuing evolution of high-strength, high-stiffness concrete, along with improvements in concrete forming systems and concrete pumping technology, have enabled sitecast concrete construction to remain economically competitive with structural steel for buildings of virtually any type or size. One of the world's tallest buildings, the Burj Khalifa in Dubai, United Arab Emirates, is constructed for most of its height as a steel-reinforced sitecast concrete structure. Self-consolidating concrete with strengths as high as 14,500 psi (100 MPa) was used in its construction.

Formwork generally accounts for more than half the cost of conventional sitecast concrete construction. Efforts to reduce this cost have led to many innovations, including new types of formwork panels that are especially smooth, durable, and easy to clean after they have been stripped. These can be reused dozens of times before they wear out.

FIGURE 14.40
The one-way sitecast concrete framing systems. (*a*) One-way solid slab with beams and girders. (*b*) One-way solid slab with slab bands. (*c*) One-way concrete joist system (rib slab) with joist bands. (*d*) Wide-module joist system with joist bands. (*Illustrations by Edward Allen.*)

FIGURE 14.41
The two-way sitecast concrete framing
systems. (*a*) Two-way solid slab.
(*b*) Two-way flat slab with drop panels
and mushroom capitals. (*c*) Two-way flat
plate. (*d*) Two-way concrete joist system
(waffle slab). (*Drawings by Edward Allen.*)

(*a*)

(*b*)

(*c*)

(*d*)

FOR PRELIMINARY DESIGN OF A SITECAST CONCRETE STRUCTURE

- Estimate the depth of a **one-way solid slab** at $\frac{1}{22}$ of its span if it is conventionally reinforced or $\frac{1}{40}$ of its span if it is posttensioned. Depths range typically from 4 to 10 inches (100–250 mm).

- Estimate the total depth of a **one-way concrete joist system** or **wide-module system** at $\frac{1}{18}$ of its span if it is conventionally reinforced or $\frac{1}{36}$ of its span if it is posttensioned. To arrive at the total depth, a slab thickness of 3 to 4½ inches (75–115 mm) must be added to the depth of the pan that is selected.

- Estimate the depth of **concrete beams** at $\frac{1}{16}$ of their span if they are conventionally reinforced or $\frac{1}{24}$ of their span if they are posttensioned. For concrete girders, use ratios of $\frac{1}{12}$ and $\frac{1}{20}$, respectively.

- Estimate the depth of **two-way flat plates** and **flat slabs** at $\frac{1}{30}$ of their span if they are conventionally reinforced or $\frac{1}{45}$ of their span if they are posttensioned. Typical depths are 5 to 12 inches (125–305 mm). The minimum column size for a flat plate is approximately twice the depth of the slab. The width of a drop panel for a flat slab is usually one-third of the span, and the projection of the drop panel below the slab is about one-half the thickness of the slab.

- Estimate the depth of a **waffle slab** at $\frac{1}{24}$ of its span if it is conventionally reinforced or $\frac{1}{35}$ of its span if it is posttensioned. To arrive at the total depth, a slab thickness of 3 to 4½ inches (75–115 mm) must be added to the depth of the dome that is selected.

- To estimate the size of a **concrete column** of normal height, add up the total roof and floor area supported by the column. A 12-inch (300-mm) column can support up to about 2000 square feet (190 m²) of area, a 16-inch (400-mm) column 4000 square feet (370 m²), a 20-inch (500-mm) column 6000 square feet (560 m²), a 24-inch (600-mm) column 9000 square feet (840 m²), and a 28-inch (700-mm) column 10,500 square feet (980 m²). These sizes are greatly influenced by the strength of the concrete used and the ratio of reinforcing steel to concrete. Columns are usually round or square.

- To estimate the thickness of a **concrete loadbearing wall**, add up the total width of floor and roof slabs that contribute load to the wall. An 8-inch (200-mm) wall can support approximately 1200 feet (370 m) of slab, a 10-inch (250-mm) wall 1500 feet (460 m), a 12-inch (300-mm) wall 1700 feet (520 m), and a 16-inch (400-mm) wall 2200 feet (670 m). These thicknesses are greatly influenced by the strength of the concrete used and the ratio of reinforcing steel to concrete.

These approximations are valid only for purposes of preliminary building layout and must not be used to select final member sizes. They apply to the normal range of building occupancies, such as residential, office, commercial, and institutional buildings, and parking garages. For manufacturing and storage buildings, use somewhat larger members.

For more comprehensive information on preliminary selection and layout of a structural system and sizing of structural members, see Edward Allen and Joseph Iano, *The Architect's Studio Companion* (5th ed.), Hoboken, NJ, John Wiley & Sons, Inc., 2012.

Lift-slab construction, used chiefly with two-way flat plate structures, virtually eliminates formwork. The floor and roof slabs of a building are cast in a stack on the ground. Then hydraulic jacks are used to lift the slabs up the columns to their final elevations, where they are welded in place using special cast-in-place steel slab collars (Figure 14.42).

Ganged forms for wall construction are large units made up of a number of panels that are supported by the same set of walers. These are handled by cranes and are often more economical than conventional small panels that are maneuvered by hand. For floor slabs that are cast in place, *flying formwork* is fabricated in large sections that are supported on deep metal trusses. The sections are moved from one floor to the next by crane, eliminating much of the labor usually expended on stripping and reerecting formwork (Figure 14.43).

Slip forming is used for tall-walled structures such as elevator shafts, stairwells, and storage silos. A ring of formwork is pulled steadily upward by jacks supported on the vertical reinforcing bars or previously cast sections of the concrete, while workers add concrete and horizontal reinforcing in a continuous process. Manufacturers of concrete formwork have developed more sophisticated systems of self-climbing formwork that offer many advantages over conventional slip forming (Figure 14.44).

In *tilt-up construction*, reinforced concrete wall panels are cast lying down over a previously poured slab that serves as a level, smooth work surface. When the wall panels have cured to sufficient strength, they are tilted up into a vertical orientation and hoisted into position by a crane, then grouted together. The elimination of most of the usual wall formwork results in formwork costs that are typically less than 5 percent of the cost of the overall system, making tilt-up construction economically favorable for single-story buildings.

FIGURE 14.42
Lift-slab construction in progress. The paired steel rods silhouetted against the sky to the right are part of the lifting jacks seen at the tops of the columns. In North America, this form of construction is used infrequently, in part due to a history of past construction accidents. *(Reprinted with permission of the Portland Cement Association from* Design and Control of Concrete Mixtures, *12th edition; PHOTOS: Portland Cement Association, Skokie, IL.)*

FIGURE 14.43
Flying formwork for a one-way concrete joist system being moved from one floor to the next in preparation for pouring. Stiff metal trusses allow a large area of formwork to be handled by a crane as a single piece. *(Courtesy of Molded Fiber Glass (MFG) Concrete Forms Company.)*

Although most tilt-up panel construction is for walls no taller than 45 feet (13.7 m), walls approaching heights as great as 100 feet (30 m) are feasible (Figures 14.45 and 14.57). Tilt-up wall panels can also be cast as sandwich panels with integral rigid plastic insulation to create panels with greater thermal efficiency.

Shotcrete (pneumatically placed concrete) is sprayed into place from the nozzle of a hose by a stream of compressed air. Because of its very low slump, even walls with vertical sides can be placed with little in the way of conventional formwork, though some kind of solid surface to spray against is required. Shotcrete is used for foundation walls, stabilization of steep slopes, repairing damaged concrete on the faces of beams and

FIGURE 14.44
A proprietary system of self-climbing formwork is being used to form these sitecast concrete elevator shafts for a tall building. The top level is a working surface from which reinforcing bars are handled and the concrete is poured. The outer panels of the formwork are mounted on overhead tracks just beneath the top level. The panels can be rolled back to the outside of the perimeter walkway after each pour, allowing workers to clean the formwork and install the reinforcing for the next pour. The entire two-story apparatus raises itself a story at a time with built-in hydraulic jacks. *(Courtesy of Patent Scaffolding Company, Fort Lee, NJ.)*

FIGURE 14.45
Tilt-up construction. The exterior wall panels were reinforced and cast flat on the floor slab. Using special lifting rings that were cast into the panels and a lifting harness that exerts equal force on each of the lifting rings, a crane tilts up each panel and places it upright on a strip foundation at the perimeter of the building. Each erected panel is braced temporarily with diagonal steel struts until the roof structure has been completed. *(Reprinted with permission of the Portland Cement Association from* Design and Control of Concrete Mixtures, *12th edition; PHOTOS: Portland Cement Association, Skokie, IL.)*

columns, seismic retrofits, and the production of free-form structures such as swimming pools and playground structures.

Even greater savings in formwork costs can be realized by casting concrete in reusable molds in a precasting plant, or by combining sitecast concrete with precast forming elements, both of which are the subject of Chapter 15.

Advances in reinforcement for sitecast concrete, other than the adoption of posttensioning, include a move to higher-strength steels and a trend toward increased prefabrication of reinforcing bars prior to installation in the forms. With developments in welding and fabricating machinery, the concept of welded wire reinforcing is expanding beyond the familiar grid of heavy wire to include complete cages of column reinforcing and entire bays of slab reinforcing. As lighter-weight reinforcing bars based on carbon fiber and aramid fiber come into more common use, other new forms of reinforcing are likely to evolve.

ARCHITECTURAL CONCRETE

If we were to train ourselves to draw as we build, from the bottom up ... stopping our pencil to make a mark at the joints of pouring or erecting, ornament would grow out of our love for the expression of method.

—Louis I. Kahn, quoted in Vincent Scully, Jr., *Louis I. Kahn*, 1962

Concrete that is intended as exposed interior or exterior surfaces, and is specified with highly prescribed finish characteristics, is known as *architectural concrete*. Most

formed concrete surfaces, although structurally sound, have many blemishes and irregularities. As a result, a vast amount of thought and effort has been expended to develop handsome surface finishes for architectural concrete (Figures 14.46 through 14.49). *Exposed aggregate finishes* involve the scrubbing and hosing of concrete surfaces shortly after the initial set of the concrete, to remove the cement paste from the surface and reveal the

FIGURE 14.46
Exposed wall surfaces of sitecast concrete. Narrow boards were used to form the walls, and form tie locations were carefully worked out in advance. *(Architect: Eduardo Catalano. Photo by Erik Leigh Simmons. Courtesy of the architect.)*

aggregate. This process is often aided by chemicals that retard the set of the cement paste; these are either sprayed on the surface of a slab or used as a coating inside formwork. Because concrete can take on almost any texture that can be imparted to the surface of formwork, much work has gone into developing formwork surfaces of wood, wood panel products, metal, plastic, and rubber to produce textures that range from almost glassy smooth to ribbed, veined, board textured, and corrugated. After partial curing, other steps can be taken to change the texture of concrete, including sandblasting, rubbing with abrasive stones, grinding smooth, and hammering with various types of flat, pointed, or toothed masonry hammers. Many types of pigments, dyes, paints, and sealers can be used to add color or gloss to concrete surfaces and to give protection against weather, dirt, and wear.

Exposed wall surfaces of concrete require special attention from the designer and contractor (Figure 14.48). The chairs and bolsters selected must not create rust spots on exterior concrete surfaces. Form tie locations in exposed concrete walls should be patterned to harmonize with the layout of the walls themselves, and the holes left in the concrete surfaces by snapped-off ties must be patched or plugged securely to prevent rusting through. Joints between pours can be concealed gracefully with recesses called *rustication strips* in the face of the concrete. The formulation of the concrete must be closely controlled for color consistency from one batch to the next. In cold climates, air entrainment is advisable to prevent freeze-thaw damage of exterior wall surfaces.

FIGURE 14.47
Exposed wall surfaces of concrete, sandblasted to expose the aggregate. Note the regular spacing of the form tie holes, which was worked out by the architect as an integral feature of the building design. *(Architect: Eduardo Catalano. Photo by Gordon H. Schenck, Jr. Courtesy of the architect.)*

REINFORCED CONCRETE NORMS

Eduardo Catalano - Architect • Deborah Forsman - Structural Engineer

(a) (b) (c)

(d) (e) (f)

FIGURE 14.48

Standards specified by the architect to ensure satisfactory visual quality in the exposed concrete walls of the buildings illustrated in Figures 14.46 and 14.47.

(Courtesy of Eduardo Catalano, Architect.)

FIGURE 14.49

Close-up photographs of some surface textures for exposed concrete walls. (*a*) Concrete cast against overlaid plywood to obtain a very smooth surface shows a crazing pattern of hairline cracks after 10 years of service. (*b*) The boat-shaped patches and rotary-sliced grain figure of A-veneered plywood formwork are mirrored faithfully in this surface. A neatly plugged form tie hole is seen at the upper left, and several lines of overspill from a higher pour have dribbled over the surface. (*c*) This exposed aggregate surface was obtained by coating the formwork with a curing retarder and scrubbing the surface of the concrete with water and a stiff brush after stripping the formwork. (*d*) The bush-hammered surface of this concrete column is framed by a smoothly formed edge. (*e, f*) Architect Paul Rudolph developed the techniques of casting concrete walls against ribbed formwork, then bush-hammering the ribs to produce a very heavily textured, deeply shadowed surface. In (*f*), the ribbed wall surface is contrasted to a board-formed slab edge, with a recessed rustication strip between. *(Photos by Edward Allen.)*

CUTTING CONCRETE, STONE, AND MASONRY

It is often necessary to cut hard materials, both in the course of obtaining and processing construction materials and during the construction process itself. The quarrying and milling of stone require many cutting operations. Precast concrete elements are frequently cut to length in the factory. Masonry units often have to be cut on the construction site, and masonry walls sometimes require the cutting of fastener holes and utility openings. Concrete cutting has become an industry in itself because of the need for utility openings, fastener holes, control joints, and surface grinding and texturing. Cutting and drilling are required to create new openings and remove unwanted construction during the renovation of masonry and concrete buildings. Core drilling is used to obtain laboratory test specimens of concrete, masonry, and stone. Finally, cutting is sometimes necessary to remove incorrect work and to perform building demolition operations.

In preindustrial times, hard materials were cut with hand tools, such as steel saws that employed an abrasive slurry of sand and water beneath the blade, and hardened steel drills and chisels that were driven with a heavy hammer. Wedges and explosives in drilled holes were used to split off large blocks of material. These techniques and mechanized variations of them are still used to some extent, but today, diamond cutting tools are used for the bulk of the tough cutting chores in the construction industry.

Diamond tools are expensive in first cost, but they cut much more rapidly than other types of tools, they cut more cleanly, and they last much longer, so they are usually more economical overall. Furthermore, diamond tools can sometimes do things that conventional tools cannot, such as precision sawing marble and granite into very thin sheets for floor and wall facings. Diamonds cut hard materials efficiently because they are the hardest known material. Most of the industrial diamonds that go into cutting tools are synthetic. They are produced by subjecting graphite and a catalyst to extreme heat and pressure, then sorting and grading the small diamonds that result. Although some natural diamonds are still used in industry, synthetic diamonds are preferred for most tasks because of their more consistent behavior in use.

To manufacture a cutting tool, the diamonds are first embedded in a metallic bonding matrix and the mixture is formed into small cutting segments. The choice of diamonds and the exact composition of the bonding matrix are governed by the type of material to be cut. The cutting segments are brazed to steel cutting tools—circular saw blades, gangsaw blades, core drill cylinders—and the cutting tools are mounted in the machines that drive them. Some tools are designed to cut dry, but most are used with a spray of water that cools the blade and washes away the waste material. The accompanying illustrations (Figures A–J) show a variety of machines that use diamond cutting tools. Another example is shown in Figure 9.25. Diamonds are also made into grinding wheels, which are used for everything from sharpening tungsten carbide tools to flattening out-of-level concrete floors and polishing granite.

Cutting tools based on materials other than diamonds are still common on the construction job site. Tungsten carbide is used for the tips of small-diameter masonry and concrete drills and for the teeth of woodworking saws. Low-cost circular saw blades composed of abrasives such as silicon carbide are useful for occasional cutting of

Magnified Detail of Saw Blade Segment

FIGURE A

A *diamond saw* blade is made up of cutting segments brazed to a steel blade core. Each segment consists of diamond crystals embedded in a metallic bonding matrix. The diamonds in the cutting segment fracture chips from the material being cut. In doing so, each diamond gradually becomes chipped and worn and finally falls out of the matrix altogether. The bonding matrix wears at a corresponding rate, exposing new diamonds to take over for those that have fallen out.

Figure B

A hand-held pneumatically powered diamond circular saw cuts excess length from a concrete pile. *(Courtesy of Sinco Products, Inc.)*

Figure C

A diesel-powered diamond circular saw slices a clean joint at the edge of a concrete pavement. *(Copyright © 2013 Husqvarna AB (Publ). All rights reserved.)*

Figure D

Cutting a new opening in a masonry wall with a diamond circular saw. *(Copyright © 2013 Husqvarna AB (Publ). All rights reserved.)*

Figure E

This remote-controlled wall saw is powered by a water-cooled electric motor and can make cuts up to 29 in. (730 mm) deep. *(Copyright © 2013 Husqvarna AB (Publ). All rights reserved.)*

CUTTING CONCRETE, STONE, AND MASONRY (CONTINUED)

FIGURE F

A hydraulically driven chain saw with diamond teeth cuts a concrete masonry wall. (*C-150 Hydracutter, photo courtesy of Reimann & George Construction, Buffalo, NY.*)

FIGURE G

A diamond core drill is used for cutting round holes. (*Photo courtesy of Sprague & Henwood, Inc.—Scranton, PA, USA.*)

FIGURE H

Using a technique called stitch drilling, a core drill cuts an opening in a very thick concrete wall. (*Courtesy of GE Superabrasives.*)

metals, concrete, and masonry, but the cutting action is slow and the blades wear very quickly. Less precise tools such as the pneumatic jackhammer, hydraulic splitters, and the traditional sledgehammer also have their uses. For large-volume precision cutting and drilling, however, there is no substitute for the industrial diamond.

FIGURE I
Sawed and drilled openings for utility lines in a concrete floor slab. *(Courtesy of GE Superabrasives.)*

FIGURE J
Wire saws are capable of cutting depths and thicknesses of material that cannot be cut with any other kind of tool. Holes are first drilled at the corners of the cut, and the diamond cutting wire is strung through the holes and around the wheels of the sawing equipment. The saw maintains a constant tension on the wire as it is pulled at high speed through the material. The wire gradually cuts its way out, leaving smooth planar surfaces. The wire is actually a steel cable on which steel beads with diamonds embedded in them are strung. *(© 1988 Cutting Technologies, Inc. (Cincinnati, Ohio). All rights reserved.)*

Basically there are two approaches to the problem of producing a good surface finish on concrete. One is to remove the cement that is the cause of the blemishes and expose the aggregate. The other is to superimpose a pattern or profile that draws attention from the blemishes.

—Henry Cowan, *Science and Building: Structural and Environmental Design in the Nineteenth and Twentieth Centuries,* Hoboken, NJ, John Wiley & Sons, Inc., 1978, p. 283

LONGER SPANS IN SITECAST CONCRETE

The ancient Romans built unreinforced concrete vaults and domes as roofs for temples, baths, palaces, and basilicas (Figure 13.2). Impressive spans were constructed, including a dome over the Pantheon in Rome, still standing, that approaches 150 feet (45 m) in diameter. Today, the arch, dome, and vault remain favorite devices for spanning long distances in concrete because of concrete's suitability to structural forms that work entirely in compression (Figures 14.50, 14.51, and 14.52). Through folding or scalloping of vaulted forms, or through the use of warped geometries such as the hyperbolic paraboloid, the required resistance to buckling can be achieved with a surprisingly thin layer of concrete, often proportionally thinner than the shell of an egg (Figure 14.52*b*).

Long-span beams and trusses are possible in concrete, including post-tensioned beams and girders and reinforced deep girders analogous to steel plate girders and rigid frames. Concrete trusses and space frames are not common, but are built from time to time. By definition, a truss includes strong tensile forces as well as compressive forces and is heavily dependent on steel reinforcing or prestressing.

Barrel shells and *folded plates* (Figures 14.51, 14.52*c*, and 14.53) derive their stiffness and strength from the folding or scalloping of a thin concrete plate to increase its rigidity and structural depth without adding material. Each of these forms depends on reinforcing or

FIGURE 14.50
The same wooden centering was used four times to form this concrete arch bridge. *(Courtesy of Gang-Nail Systems, Inc.)*

Rigid Frame

Vierendeel Truss

Space Frame

Arch

Dome

Hyperbolic Paraboloid Shells

Barrel Shell

Folded Plate

FIGURE 14.51

Examples of eight types of longer-span structures in concrete. Each is a special case of an infinite variety of forms. All can be sitecast, but the rigid frame, space frame, and Vierendeel truss are more likely to be precast for most applications.

(a)

FIGURE 14.52
Three concrete shell structures by 20th-century masters of concrete engineering. (*a*) **A domed sports arena by Pier Luigi Nervi.** (*b*) **A lakeside restaurant of hyperbolic paraboloid shells by Felix Candela.** (*c*) **A racetrack grandstand roofed with cantilevered concrete barrel shells by Eduardo Torroja.** (*Drawings by Edward Allen.*)

(b)

(c)

FIGURE 14·53
Flying formwork is removed from a bay of a folded plate concrete roof for an air terminal. *(Architects: Thorshov and Cerny. Courtesy American Plywood Association.)*

posttensioning to resist the tensile forces that it may experience.

DESIGNING ECONOMICAL SITECAST CONCRETE BUILDINGS

The cost of a concrete building frame can be broken down into the costs of the concrete, the reinforcing steel, and the formwork. Of the three, the cost of concrete is usually the least significant in North America and the cost of formwork is the most significant. Accordingly, simplification and standardization of formwork are the first requirements for an economical concrete frame. Repetitive, identical column spacings and bay sizes allow the same formwork to be used

again and again without alterations. Flat plate construction is often the most economical, simply because its formwork is so straightforward. Joist band construction is usually more economical than joist construction that uses beams proportioned more efficiently for their structural requirements, because enough is saved on formwork costs to more than compensate for the added concrete and reinforcing steel in the beams. This same reasoning applies if column and beam dimensions are standardized throughout the building, even though loads may vary; the amount of reinforcing and the strengths of the concrete and reinforcing steel can be adjusted to meet the varying structural requirements (Figure 14.54).

SITECAST CONCRETE AND THE BUILDING CODES

Concrete structures are inherently fire resistant. When fire attacks concrete, the water of hydration is gradually driven out and the concrete loses strength, but this deterioration is slow because considerable heat is needed to raise the temperature of the mass of concrete to the point where dehydration begins, and a large additional quantity of heat is required to vaporize the water. The steel reinforcing bars or prestressing strands are buried beneath a concrete cover that protects them for an extended period of time. Except under unusual circumstances, such as a prolonged fire

FIGURE 14.54
By keeping sizes of columns, beams, and other formed elements as consistent as possible, significant savings can be achieved in the cost of formwork for sitecast concrete construction. Here, a column form, having been stripped from a recently cast column, is being moved to a new location for reuse. *(Photo by Joseph Iano.)*

fueled by stored petroleum products, concrete structures usually survive fires with only cosmetic damage and are repaired with relative ease.

Concrete structures with adequate cover over reinforcing and adequate slab thicknesses are classified as Type I buildings under the International Building Code. Slab thickness requirements for the several construction types are complex, depending on the type of aggregate used in the concrete and whether or not a given structural member is restrained from movement by surrounding construction. Fire resistance requirements for the highest construction types can be met in joist and waffle systems either by increasing their slab thickness beyond what is structurally necessary or by applying fireproofing

materials to the lower surfaces of the floor structures.

Sitecast concrete buildings have rigid joints and in many cases require no additional structural elements to achieve the necessary resistance to wind and seismic forces. More restrictive seismic design provisions in the building codes have, however, increased the attention paid by structural engineers to column ties and beam stirrups, particularly in the zones where beams and columns meet, to be sure that vertical bars in columns and horizontal bars in beams are adequately restrained against the greater forces that can occur in these zones under seismic loadings. The joints between flat plate floors and columns may not be sufficiently rigid to brace a building of more than

modest height, unless drop panels or beams are added to stiffen the slab-to-column junction.

UNIQUENESS OF SITECAST CONCRETE

Concrete is a shapeless material that must be given form by the designer. For economy, the designer can adopt a standard system of concrete framing. For excitement, one can invent new shapes and textures. Some have pursued its sculptural possibilities, others its surface patterns and textures, still others its structural logic. From each of these routes have come masterpieces—Le Corbusier's chapel at Ronchamp (Figure 14.59); Wright's Unity Temple (Figure 14.1);

FIGURE 14.55
Concrete work nears the 1475-foot (450-m) summit of the twin Petronas Towers in Kuala Lumpur, Malaysia, which were, at the time of their construction, the world's tallest buildings. Each tower is supported by a perimeter ring of 16 cylindrical concrete columns and a central core structure, also made of concrete. The columns vary in diameter from 8 feet (2400 mm) at the base of the building to 4 feet (1200 mm) at the top. For speed of construction, the floors were framed with steel and composite metal decking. Concrete with strengths as high as 11,600 psi (80 MPa) was used in the columns. The architect was Cesar Pelli & Associates, Inc. The structural engineers were Thornton-Tomasetti and Rahnill Bersekutu Sdn Bhd. The U.S. partner in the joint venture team that constructed the towers was J. A. Jones Construction Co. of Charlotte, North Carolina. (*Photograph by Uwe Hausen, J. A. Jones, Inc.*)

and the elegant structures of Torroja, Candela, and Nervi, examples of which are sketched in Figure 14.52. Many of these masterpieces, especially from the latter three designers, were also constructed with impressive economy. Sitecast concrete can do almost anything, be almost anything, at almost any scale, and in any type of building. It is a potent architectural material, and therefore a material both of spectacular architectural achievements and dismal architectural failures. A material that is so malleable demands skill and restraint from those who would build with it, and a material so commonplace requires imagination if it is to rise above the mundane.

FIGURE 14.56
The plastered surfaces of Frank Lloyd Wright's Guggenheim Museum (1943–1956) cover a helical ramp of cast-in-place concrete. (*Photo by Wayne Andrews.*)

FIGURE 14.57
The Chapel of St. Ignatius, Seattle
University, designed by architect Steven
Holl, is a tilt-up concrete structure. *(Photo
by Joseph Iano.)*

FIGURE 14.58
**A sitecast concrete house in Lincoln,
Massachusetts.** *(Architects: Mary Otis
Stevens and Thomas F. McNulty.)*

FIGURE 14.59
Le Corbusier's most sculptural building in his favorite material, concrete: the chapel of Notre Dame de Haut at Ronchamp, France (1950–1955). (*Drawing by Edward Allen.*)

FIGURE 14.60
The TWA Terminal at John F. Kennedy Airport, New York, 1956–1962. (*Architect: Eero Saarinen. Photo by Wayne Andrews.*)

CSI/CSC MasterFormat Sections for Sitecast Concrete Framing Systems

03 31 00	**STRUCTURAL CONCRETE**
03 33 00	**ARCHITECTURAL CONCRETE**
03 35 00	**CONCRETE FINISHING**
03 37 00	**SPECIALTY PLACED CONCRETE**
03 37 14	Shotcrete
03 38 00	**POST-TENSIONED CONCRETE**
03 39 00	**CONCRETE CURING**
03 47 00	**SITE-CAST CONCRETE**
03 47 13	Tilt-Up Concrete
03 47 16	Lift-Slab Concrete
03 80 00	**CONCRETE CUTTING AND BORING**

KEY TERMS

slab on grade
capillary break
moisture barrier, vapor retarder
strike off, screed
floating
bull float
darby
bleed water
troweled slab
steel trowel
rotary power trowel
knee board
broom finish
restraightening
straightedge
shake-on hardener
curing compound
superflat (concrete) floor
F-number
control joint, contraction joint
isolation joint, expansion joint
shrinkage-compensating cement

key
dowel
form tie
waler
brace
stripped (formwork), formwork stripping
form tie hole
insulating concrete form (ICF)
lift
one-way solid slab
shore
reshoring
slab band
one-way concrete joist system, ribbed slab
pan
distribution rib
joist band
wide-module concrete joist system, skip-joist system
two-way solid slab
two-way flat slab
mushroom capital

drop panel
column strip
middle strip
two-way flat plate
waffle slab, two-way concrete joist system
dome
head
draped tendon
set
lift-slab construction
ganged form
flying formwork
slip forming
tilt-up construction
shotcrete, pneumatically placed concrete
architectural concrete
exposed aggregate finish
rustication strip
diamond saw
barrel shell
folded plate

REVIEW QUESTIONS

1. Draw from memory a detail of a typical slab on grade and list the steps in its production.

2. Why can't the surface of a slab on grade be finished in one operation, instead of waiting for hours before final finishing?

3. What are control joints and isolation joints? Explain the purpose and typical locations for each in a concrete slab.

4. List the steps that are followed in forming and pouring a concrete wall.

5. Distinguish one-way concrete framing systems from two-way systems. Are steel and wood framing systems one-way or two-way? Is one-way construction more efficient structurally than two-way construction?

6. List the common one-way and two-way concrete framing systems and indicate the possibilities and limitations of each.

7. Why posttension a concrete structure rather than merely reinforce it?

EXERCISES

1. Propose a suitable reinforced concrete framing system for each of the following buildings and determine an approximate thickness for each:

  a. An apartment building with a column spacing of about 16 feet (5 m) in each direction

  b. A newsprint warehouse, column spacing 20 feet by 22 feet (6 m × 6.6 m)

  c. An elementary school, column spacing 24 feet by 32 feet (7.3 m × 9.75 m)

  d. A museum, column spacing 36 feet by 36 feet (11 m × 11 m)

  e. A hotel where overall building height must be minimized in order to build as many stories as possible within a municipal height limit

2. Look at several sitecast concrete buildings. Determine the type of framing system used in each and explain why you think it was selected. If possible, talk to the designers of the building and find out if you were right.

3. Observe a concrete building under construction. What is its framing system? Why? What types of forms are used for its columns, beams, and slabs? In what form are the reinforcing bars delivered to the site? How is the concrete mixed? How is it raised and deposited into the forms? How is it consolidated in the forms? How is it cured? Are samples taken for testing? How soon after pouring are the forms stripped? Are the forms reused? How long are the shores kept in place? Keep a diary of your observations over a period of a month or more.

SELECTED REFERENCES

See also the Selected References listed in Chapter 13.

American Concrete Institute. *ACI 303R: Guide to Cast-in-Place Architectural Concrete Practice.* Farmington Hills, MI, updated regularly.

This is a comprehensive handbook on how to produce attractive surfaces in concrete.

American Concrete Institute. *ACI 318: Building Code Requirements for Structural Concrete and Commentary.* Farmington Hills, MI, updated regularly.

This booklet establishes the basis for the engineering design and construction of reinforced concrete structures in the United States.

Concrete Reinforcing Steel Institute. *CRSI Design Handbook.* Schaumburg, IL, Author, updated regularly.

Structural engineers working in concrete use this handbook, which is based on the ACI Code, as their major reference. It contains examples of engineering calculation methods and hundreds of pages of tables of standard designs for reinforced concrete structural elements.

Concrete Reinforcing Steel Institute. *Placing Reinforcing Bars.* Schaumburg, IL, Author, updated regularly.

Written as a handbook for those engaged in the business of fabricating and placing reinforcing steel, this small volume is clearly written and beautifully illustrated with diagrams and photographs of reinforcing for all the common concrete framing systems.

Hurd, M. K. *Formwork for Concrete* (7th ed.). Farmington Hills, MI, American Concrete Institute, 2005.

Profusely illustrated, this book is the bible on formwork design and construction for sitecast concrete.

Post-Tensioning Institute. *Post-Tensioning Manual.* Phoenix, AZ, Author, updated regularly.

This heavily illustrated volume is both an excellent introduction to posttensioning for the beginner and a basic engineering manual for the expert.

WEB SITES

Sitecast Concrete Framing Systems

Author's supplementary web site: **www.ianosbackfill.com/14_sitecast_concrete_framing_systems**
Dayton Superior formwork accessories: **www.daytonsuperior.com**
Dywidag-Systems International posttensioning systems: **www.dsiamerica.com**
Molded Fiber Glass Construction Products Co. formwork: **www.mfgcp.com**

15

PRECAST CONCRETE FRAMING SYSTEMS

- **Precast, Prestressed Concrete Structural Elements**
  Precast Concrete Slabs
  Precast Concrete Beams, Girders, and Columns

  FOR PRELIMINARY DESIGN OF A PRECAST CONCRETE STRUCTURE

  Precast Concrete Wall Panels

- **Assembly Concepts for Precast Concrete Buildings**

- **Manufacture of Precast Concrete Structural Elements**
  Casting Beds
  Prestressing and Reinforcing Steel
  Carbon Fiber Reinforcing

Hollow-Core Slab Production
Column Production

- **Joining Precast Concrete Members**

  FASTENING TO CONCRETE

- **Composite Precast/Sitecast Concrete Construction**

- **The Construction Process**

  SUSTAINABILITY IN PRECAST CONCRETE CONSTRUCTION

- **Precast Concrete and the Building Codes**

- **Uniqueness of Precast Concrete**

Suction devices lift a precast concrete hollow-core slab from the casting bed where it was manufactured. (*Courtesy of The Flexicore Co., Inc.*)

Structural *precast concrete* elements—slabs, beams, girders, columns, and wall panels—are cast and cured in factories, transported to the construction job site, and erected as rigid components. Precasting offers many potential advantages over sitecasting of concrete: The production of precast elements is carried out conveniently at ground level. The mixing and pouring operations can be highly mechanized. Frequently, especially in difficult climates, these operations are carried out under shelter. Workmanship is generally better than on the construction job site. The concrete is cast in permanent forms made of steel, concrete, glass-fiber-reinforced plastic, or wood panels with smooth overlays, whose excellent surface properties are mirrored in the high-quality surfaces of the finished precast elements that they produce. The forms may be reused hundreds or thousands of times before they have to be renewed, so formwork costs per unit of finished concrete are low. The forms are equipped to pretension the steel in the precast elements for greater structural efficiency, which translates into longer spans, lesser depths, and lower weights than for comparable reinforced concrete elements. Concrete and steel of superior strength are used in precast elements: 270,000-psi (1860-MPa) prestressing steel and minimum 5000-psi (35-MPa) concrete.

For the fastest possible curing, the concrete for precast concrete elements is made with Type III portland cement, high early strength. In addition, the elements are usually *steam cured*. Steam furnishes heat to accelerate the hardening of the concrete and moisture for full hydration. Thus, a precasting plant is able to produce fully cured structural elements, from laying of the prestressing strands to removal of the finished elements from the casting bed, on a 24-hour cycle. Self-consolidating concrete, described in Chapter 13, is also finding increased use in precast concrete production, where it has the potential to increase productivity and produce architectural concrete elements with better finish quality.

When the elements produced by this expeditious technique are delivered to the construction job site, further advantages are realized: The erection process is similar to that used for structural steel, but it is often faster because most precast concrete systems include a deck as an integral part of the major spanning elements, without the need for placing additional joist or decking components (Figures 15.1 and 15.2). Erection is much faster than with sitecast concrete because there is no formwork to be erected and stripped, and little or no waiting for concrete to cure. Furthermore, erection of precast structures can take place under adverse weather conditions, such as high or low temperatures, which could complicate or delay the sitecasting of concrete.

When the designer is choosing between precast and sitecast concrete, the designer must weigh the potential advantages of precasting against some potential disadvantages. Precast structural elements, although lighter in weight than similar elements of sitecast concrete, are nevertheless heavy and bulky to transport over the roads and hoist into place. This restricts the size and proportions of most precast elements: They can be rather long, but only as wide as the maximum permissible vehicle width of 12 to 14 feet (3.66–4.27 m). This restricted width usually precludes utilization of the efficiencies of two-way structural action in precast slabs. Also, the three-dimensional sculptural possibilities of sitecast concrete are largely absent in precast concrete.

Although whole walls or rooms are sometimes precast as single units in concrete, this chapter will focus on the standard precast elements that are commonly mass-produced as structural components.

FIGURE 15.1
A fully precast building frame under construction. A poured concrete topping will cover the hollow-core slabs and the beams to create a smooth floor and tie the precast elements together. *(Courtesy of The Flexicore Co., Inc.)*

FIGURE 15.2
Workers guide two hollow-core slabs, lowered by a crane in wire rope slings, onto the precast concrete beams that will support them. *(Courtesy of The Flexicore Co., Inc.)*

PRECAST, PRESTRESSED CONCRETE STRUCTURAL ELEMENTS

Precast Concrete Slabs

The most fully standardized precast concrete elements are those used for making floor and roof slabs (Figure 15.3). These may be supported by bearing walls or frames made of precast concrete, sitecast concrete, masonry, or steel. Four kinds of precast slab elements are commonly produced: For short spans and minimum slab depths, *solid slabs* are appropriate. For longer spans, deeper elements must be used, and precast solid slabs, like their sitecast counterparts, become inefficient because they contain too much deadweight of nonworking concrete. In *hollow-core slabs*, precast elements suitable for intermediate spans, internal longitudinal voids replace much of the nonworking concrete. For the longest spans, still deeper elements are required, and *double tees* and *single tees* eliminate still more nonworking concrete.

For most applications, precast slab elements are manufactured with a rough top surface. After they have been erected, a concrete *topping* is poured over them and finished to a smooth surface. The topping, usually 2 inches (50 mm) thick, bonds during curing to the rough surface of the precast elements and becomes a working part of their structural action. The topping also helps the precast elements to act together as a structural unit, rather than as individual planks, in resisting concentrated loads and diaphragm loads, and it conceals the slight differences in camber that often occur in prestressed components. Structural continuity across a number of spans can be achieved by casting reinforcing bars into the topping over the supporting beams or walls. Underfloor electrical conduits may also be embedded in the topping. Smooth-top precast slabs are sometimes used without topping, as is discussed later in this chapter.

Either normal-density concrete or structural lightweight concrete may be selected for use in any of the precast slab elements. Lightweight concrete, approximately 20 percent less dense than normal concrete, reduces the load on the frame and foundations of a building but is more expensive than normal-density concrete.

There is considerable overlap in the economical span ranges of the different kinds of precast slab elements, allowing the designer some latitude in choosing which to use in a particular situation. Solid slabs and hollow-core slabs save on overall building height in multistory structures, and their smooth undersides can be painted and used as finish ceilings in many applications. For longer spans, double tees are preferred to the older single-tee design because they need not be supported against tipping during erection, allowing the construction to proceed more quickly and at less cost.

Precast Concrete Beams, Girders, and Columns

Precast concrete beams and girders are made in several standard shapes (Figure 15.4). The projecting *ledgers* on *L-shaped beams* and *inverted tees* provide direct support for precast slab elements. They conserve headroom in a building by supporting slabs near the bottoms of the beams, allowing the beam and slab to share some of the same structural depth. Precast concrete columns are usually square or rectangular in section and may be prestressed or simply reinforced.

SOLID FLAT SLAB
Widths vary

HOLLOW CORE SLAB
*2', 4', 8' wide
(610, 1220, 2440 mm)*

*(1'-4", 3'-4"
some manufacturers)*

DOUBLE TEE
*8', 10'' wide
(2440, 3050 mm)*

SINGLE TEE
*8', 10' wide
(2440, 3050 mm)*

FIGURE 15·3
The four major types of precast concrete slab elements. Hollow-core slabs are produced by different companies in a variety of cross-sectional patterns, using several different processes.

FOR PRELIMINARY DESIGN OF A PRECAST CONCRETE STRUCTURE

• Estimate the depth of a precast solid slab at ¹⁄₄₀ of its span. Depths typically range from 3½ to 8 inches (90–200 mm).

• An 8-inch (200-mm) precast hollow-core slab can span approximately 25 feet (7.6 m), a 10-inch (250-mm) slab 32 feet (9.8 m), and a 12-inch (300-mm) slab 40 feet (12 m).

• Estimate the depth of precast concrete double tees at ¹⁄₂₈ of their span. The most common depths of double tees are 12, 14, 16, 18, 20, 24, and 32 inches (300, 350, 400, 460, 510, 610, and 815 mm). Some manufacturers can provide double tees that are 48 inches (1220 mm) deep.

• A precast concrete single tee 36 inches (915 mm) deep spans approximately 85 feet (926 m) and a 48-inch (1220-mm) tee 105 feet (32 m).

• Estimate the depth of precast concrete beams and girders at ¹⁄₁₅ of their span for light loadings and ¹⁄₁₂ of their span for heavy loadings. These ratios apply to rectangular, inverted-tee, and L-shaped beams. The width of a beam or girder is usually about one-half of its depth. The projecting ledgers on inverted-tee and L-shaped beams are usually 6 inches (150 mm) wide and 12 inches (300 mm) deep.

• To estimate the size of a precast concrete column, add up the total roof and floor area supported by the column. A 10-inch (250-mm) column can support up to about 2300 square feet (215 m²) of area, a 12-inch (300-mm) column 3000 square feet (280 m²), a 16-inch (400-mm) column 5000 square feet (465 m²), and a 24-inch (600-mm) column 9000 square feet (835 m²).

These values may be interpolated to columns in 2-inch (50-mm) increments. Columns are usually square. These approximations are valid only for purposes of preliminary building layout and must not be used to select final member sizes. They apply to the normal range of building occupancies, such as residential, office, commercial, and institutional buildings, and parking garages.

For manufacturing and storage buildings, use somewhat larger members. For more comprehensive information on preliminary selection and layout of structural system and sizing of structural members, see Edward Allen and Joseph Iano, *The Architect's Studio Companion* (5th ed.), Hoboken, NJ, John Wiley & Sons, Inc., 2012.

RECTANGULAR BEAM   L-SHAPED BEAM   INVERTED TEE BEAM   AASHTO BEAM

FIGURE 15.4
Standard precast concrete beam and girder shapes. The heavier bars near the tops of the beams are mild steel reinforcing, and the smaller strands near the bottom are high-strength prestressing. Mild steel stirrups are also shown. Stirrups usually project above the top of the beam, as shown, to bond to a sitecast concrete topping, creating composite structural action. AASHTO (American Association of State Highway and Transportation Officials) beams were designed originally as efficient shapes for bridge structures. They have also found occasional use in parking garage structures or other utilitarian building types.

Precast Concrete Wall Panels

Precast concrete panels, either prestressed or conventionally reinforced, are used as loadbearing wall panels in many types of low-rise and high-rise buildings. Solid panels range from 3½ to 10 inches (90 to 250 mm) in thickness and can span one or two stories in height. When prestressed, strands are located in the vertical midplane of the wall panels to strengthen the panels against buckling and to eliminate camber. Ribbed or hollow-core panels, or sandwich panels with integral rigid foam insulation, may be as deep as 12 to 24 inches (305 to 610 mm) and can span up to four stories (Figure 15.12).

A number of manufacturers produce proprietary precast concrete wall panels for residential foundation construction. Vertical ribs, so-called concrete studs, spaced at 24 inches

(610 mm), provide faces with sheet metal or wood strips for the attachment of interior finishes and create cavity space that can accommodate insulation or the routing of services. Some manufacturers integrate rigid foam insulation or reinforced footings into the design of the panels.

Nonloadbearing precast concrete wall panels, called *architectural precast concrete*, are discussed in Chapter 20.

ASSEMBLY CONCEPTS FOR PRECAST CONCRETE BUILDINGS

Figure 15.5 shows a building whose precast slab elements (double tees in this example) are supported on a skeleton frame of L-shaped precast girders and precast columns. The slab elements in Figure 15.6 are supported on precast loadbearing

wall panels. Figure 15.7 illustrates a building whose slabs are supported on a combination of wall panels and girders. These three fundamental ways of supporting precast slabs—on a precast concrete skeleton, on precast loadbearing wall panels, and on a combination of the two—occur in endless variations in buildings. The skeleton may be one bay or many bays deep; the loadbearing walls are often constructed of reinforced

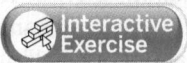

FIGURE 15.5
Double-tee slab elements supported on a frame of precast columns and L-shaped girders. *(Courtesy of Precast/Prestressed Concrete Institute.)*

FIGURE 15.6
Hollow-core slab elements supported on precast concrete loadbearing wall panels. *(Courtesy of Precast/Prestressed Concrete Institute.)*

FIGURE 15.7
Double-tee slab elements supported on a perimeter of precast concrete loadbearing wall panels and an interior structure of precast columns and inverted tee beams. *(Courtesy of Precast/Prestressed Concrete Institute.)*

masonry or of various configurations of precast concrete; the slab elements may be solid, hollow-core, or double tee, topped or untopped. One of the principal virtues of precast concrete as a structural material is that it is locally manufactured to order and is easily customized to an individual building design, usually at minimal additional cost.

MANUFACTURE OF PRECAST CONCRETE STRUCTURAL ELEMENTS

Casting Beds

Most precast concrete elements are produced in permanent forms called *casting beds*. Casting beds average 400 feet (125 m) in length but extend 800 feet (250 m) or more in some plants (Figure 15.8). A cycle of precasting usually begins in the morning, as soon as the elements that were cast the previous day have been lifted from the beds. High-strength steel reinforcing strands are strung between the abutments at the extreme ends of the bed. The strands are pretensioned with hydraulic jacks, during which process they stretch considerably. Once the strands are fully tensioned, transverse bulkhead separators may be placed along the bed at the required intervals to divide the individual elements from one another. (For solid slabs, cored slabs, and wall panels, the bulkheads are often omitted; the cured slab or panel is simply sawed into the required lengths before it is removed from the bed.)

(a)

(b)

FIGURE 15.8
Manufacturing double-tee slabs. (a) A worker inserts weld plates with their V-shaped anchors of reinforcing bar in the casting bed prior to pouring the concrete. The prestressing strands and wire fabric shear reinforcement have been installed in the stems of the double tees, and a portion of the welded wire fabric for the top slab can be seen in the foreground. Notice the great length of the casting bed; many elements can be cast end to end at the same time. (b) The top surface of the concrete is straightedged by machine. (c) The next morning, following overnight steam curing, a worker cuts the prestressing strands between bulkhead separators with an oxyacetylene cutting torch. The welded wire fabric is exposed at the ends of the elements because they will be used as untopped slabs (see Figure 15.20). (d) Using lifting loops cast into the ends, the slabs are lifted from the bed and stockpiled outdoors. The dapped stems will rest on inverted-tee and L-shaped girders; the notched corner will fit around a column. (e) Loading the double tees for trucking. (Photos by Alvin Ericson.)

(c)

(d)

(e)

When pretensioning and separator placement have been completed, mild steel reinforcing bars and welded wire fabric are placed as required. *Weld plates* and other embedments are installed. Then the concrete is placed in the bed, vibrated to eliminate voids, and struck off level. If the slabs are to be used without topping, the top surface is finished further with floats and trowels. Live steam or radiant heat is then applied to the concrete to accelerate its curing. Ten to twelve hours after pouring, the concrete has reached a compressive strength of 2500 to 4000 psi (24–28 MPa) and has bonded to the steel strands. The next morning, after the strength of the concrete has been verified by testing cylinders in the laboratory, the exposed strands are cut between the bulkhead separators, releasing the external force in the strands, which spring back to prestress the concrete. Asymmetrically prestressed slab and beam elements immediately arch up from the casting bed to take on a pronounced camber as the prestressing force is released into them. When the elements have been separated from one another, they are hoisted off the bed and stockpiled ready for shipment. Then a new cycle of casting begins.

Prestressing and Reinforcing Steel

Solid slabs, hollow-core slabs, and wall panels are cast around horizontal strands. Tees, double tees, beams, and girders are often cast around depressed or harped strands for greater efficiency of structural action (Figures 13.42 and 15.9).

Ordinary mild steel reinforcing is also cast into prestressed concrete elements for various purposes: Beams or slabs that will cantilever beyond their supports are given top reinforcing bars over the cantilever points. Welded wire reinforcing is used to reinforce the flanges of tees and double tees and for general reinforcing of wall panels. Where stirrups are required in the stems of beams and single or double tees, they are made of either mild steel reinforcing bars or welded wire reinforcing (Figure 15.10). Additional reinforcing may be installed to add strength around *dapped* (notched) *ends* and openings in panels or slabs for pipes, ducts, columns, and hatchways. Weld plates and other metal connecting devices are cast into the elements as required. Projecting steel loops are cast into many types of elements as crane attachments for lifting.

FIGURE 15.9
A precasting bed being readied for the pouring of a very long AASHTO girder. Side forms for the mold can be seen in the background to the right. The depressed strands are held down in the center of the beam by steel pulleys that will be left in the concrete after pouring. The bed is long enough that several girders are being cast end to end, with the depressed strands pulled up and down as required. Mild steel reinforcing bars are used for stirrups. The projecting tops of the stirrups will bond to the sitecast topping. Vertical twists of prestressing strand near the end of the girder will serve as lifting loops. *(Courtesy of Blakeslee Prestress, Inc.)*

Carbon Fiber Reinforcing

Carbon fiber reinforcing is gaining use as a substitute for mild steel reinforcing, such as for shear stirrups and temperature steel, in precast concrete products including wall panels, double tees, and deck panels for floors and roofs. Because carbon fiber, unlike steel, does not require protection from corrosion, less concrete cover is required than for steel reinforcing, significantly reducing the overall thickness and weight of carbon fiber–reinforced components. The low thermal conductivity of carbon fiber, combined with the thinner concrete sections possible with this type of reinforcing, permit

the casting of insulated panels with superior thermal performance. The much higher tensile strength and stiffness of carbon fiber in comparison to mild steel, and the innovative ways in which grids of carbon fiber reinforcing can be integrated into precast concrete components, yield improvements in structural efficiency as well.

Carbon fiber composite cable substitutes for high-strength steel posttensioning strands are finding use in experimental precast concrete highway structures. It is likely

that these materials will eventually enter the building construction market as well.

Hollow-Core Slab Production

The longitudinal voids in hollow-core slabs can be formed by a number of processes. In the *extruded process*, extrusion devices squeeze an extremely dry, stiff concrete mix through a moving extrusion die to produce the voided shape directly. This method has the disadvantage that vertical openings and weld

FIGURE 15.10
Workers install side forms for inverted-tee beams in an outdoor casting bed. Mild steel reinforcing bars are used extensively for stirrups. *(Courtesy of Blakeslee Prestress, Inc.)*

plates cannot easily be cast in; where openings are required in extruded slabs, they must be cut out of the stiff but still wet concrete just after extrusion or sawed after curing (Figure 15.11). Weld plates are added to the slabs by hand before the concrete has completely cured. In the *wet-cast process*, a bottom layer of wet concrete is deposited in the casting bed; then a second layer of concrete, with collapsible tubes, dry crushed stone, or lightweight aggregate carefully positioned to form the voids, is placed. Special forms may easily be placed in the bed to make openings as required, and weld plates may be cast in by this process. The tubes or aggregate are removed after the concrete has cured. In the *slip-form process*, a moving hopper deposits a zero-slump concrete mix in the casting bed. Tubes that form the slab cores move with the hopper and are pulled out of the slab as the casting process proceeds from one end of the slab to the other. This process falls somewhere between the wet-cast and extrusion processes in terms of the relative ease or difficulty of placing embeds and forming special openings.

After curing, the slabs are conveyed to an aggregate recovery area, where the dry stone is poured out of the voids and saved for reuse (Figure 15.11E). In another process, air-inflated tubes are used to form the voids. In some plants, hollow-core slabs are cast atop one another in stacks rather than in a single layer and are wet cured for 7 days rather than steam cured overnight.

Column Production

Precast concrete columns may be reinforced with ordinary mild steel or pretensioned strands. Pretensioned columns are often made and shipped in multistory lengths with *corbels* to support beams or slabs (Figures 15.16, 15.18). Columns with corbels on one side or on two opposing sides are easily cast in flat beds. If corbels are required on three sides or on two adjacent sides, box forms are set atop the upper side of the column form as it lies in the casting bed. For corbels on the fourth side, steel plate inserts are cast into the bottom face of the column in the bed, to which reinforcing bars are welded after the column is removed from the bed. The corbels on the fourth side are then cast around the reinforcing bars in a separate operation.

FIGURE 15.11

Steps in the manufacture of Span-Deck, a proprietary hollow-core slab. (*A*) A thin bottom slab of low-slump concrete is deposited into the casting bed with a traveling hopper. (*B*) A ridged roller compacts the bottom slab and makes indentations to key to the concrete webs. (*C*) Four prestressing strands are pulled onto the bed; these will be pretensioned with jacks. (*D*) After pretensioning, an extrusion device travels the length of the bed, forming the webs and the top of the slab and filling the cores with dry, lightweight aggregate that serves as temporary support for the fresh concrete. (*E*) The next day, after curing of the slab has been completed, a saw cuts it into the required lengths. Each length is then lifted from the bed and transported by an overhead crane to the aggregate recovery area, where the dry aggregate is poured out and saved for reuse. (*F*) Finished Span-Deck slabs stockpiled, ready for transportation to the construction site. The wads of paper in the cores prevent concrete from filling the cores during pouring of the topping. (*Courtesy of Blakeslee Prestress, Inc.*)

FIGURE 15.12

A tilting table being used for casting a foam-insulated concrete wall panel. To the left, a concrete panel face with welded wire fabric reinforcing is being cast. Sheets of rigid foam insulation with wire ties are seen in the center, and the welded wire fabric for the second panel face is seen at the right, with a pair of vertical bars and a pocket at the bottom being cast in as part of a system of connections. Notice the pipes at the left edge of the table for heating the mold to accelerate curing. (*Courtesy of Blakeslee Prestress, Inc.*)

JOINING PRECAST CONCRETE MEMBERS

Figures 15.13 through 15.22 and 15.27 illustrate examples of frequently used connection details in precast concrete construction. Bolting, welding, and grouting are all commonly employed. Connections can be posttensioned to produce continuous beam action at points of support (Figure 15.16). Exposed metal connectors that are not covered by topping are usually dry packed with portland cement grout (the grout is stiff but not actually dry at the time that it is installed) after being joined to protect them from fire and corrosion.

The simplest joints in precast concrete construction are those that rely on gravity by placing one element atop another, as is done where slab elements rest on a bearing wall or beam or where a beam rests on the corbel of a column. Column-to-footing or column-to-column joints are usually grouted, to allow full bearing between sections (Figures 15.13, 15.14, and 15.15). Bearing walls are joined with slabs in a similar manner (Figure 15.27). With spanning members, *bearing pads* are usually inserted at points of contact. These serve to avoid the grinding, concrete-to-concrete contact that might create points of high stress (Figures 15.16, 15.17, 15.18, and 15.19) while also allowing for movement caused by expansion and contraction or structural deflection of the members. For solid and hollow-core slabs, these pads are strips of high-density plastic. Under elements with higher point loadings, such as tees and beams, pads of synthetic rubber are used. For resistance to seismic and wind forces, the members in these simple joints must be tied together laterally. Slab elements are often joined over supports by reinforcing bars that are cast into the topping or, where no topping is used, into the grout keys between the slabs (Figures 15.18, 15.19, 15.20, 15.21, and 15.22). Beams, single tees, and double tees may additionally be connected by welding together steel plate inserts that have been embedded in the elements in the plant (Figures 15.18 and 15.19). The stems (lower portions) of single and double tees and the bottoms of beams are never welded to the supports, but are left free to move on their bearing pads as the slab elements deflect under load.

FASTENING TO CONCRETE

As a concrete frame is finished, many things must be fastened to it, including exterior wall panels and facings; interior partitions; hangers for pipes, ducts, and conduits; suspended ceilings; stair railings; cabinets; and machinery. Drawings *A–H* are examples of fastening systems that are cast into the concrete. *A* is a familiar *anchor bolt*. *B* is a steel plate welded to a bent rod or strap anchor; this weld plate or *embed plate* furnishes a surface to which steel components can be welded. The steel angle in *C* has a threaded stud welded to it so that another component can be attached by bolting. *D* is an adjustable insert of malleable iron that is nailed to the formwork through the slots in the ears on either side. A special nut twists and locks into the slot to accept a bolt or threaded rod from below. A slightly different form of this type of insert is shown in the detail of a masonry shelf angle attachment in Chapter 20. *E* and *F* are two different designs of threaded inserts that are cast into the

concrete. The sheet steel *dovetail slot* in *G* is used with special anchor straps as shown to tie masonry facings to a sitecast concrete frame or wall. *H* is simply a dovetailed wood nailer strip cast into the concrete, a detail that is risky because the wood may absorb moisture, swell, and crack the concrete, or dry out, shrink, and become loose. *A–F* are heavy-duty devices that are available in capacities sufficient to anchor heavy building components and machinery.

Details *I–P* depict fastening devices that are inserted in holes drilled into the cured concrete. *I* shows the steel post of a railing anchored into an oversized hole in a concrete slab using grout, poured lead, or epoxy; this detail is also used to fasten bolts to concrete and can carry heavy loads if properly designed and installed. *J* is a plastic sleeve, *K* a wood or fiber plug, and *L* a lead sleeve; all three are inserted into drilled holes and expand to grip the sides of the hole when a screw is driven into them. *M* is a similar type of metal sleeve but has a special nail that expands the sleeve as it is driven. *N* is a special bolt with a steel sleeve over a tapered shank at the inner end. The sleeve catches against the concrete as the bolt is driven into the hole and is expanded by the taper as the bolt is tightened. *O* is a special screw and *P* is a special nail, both designed to grip tightly when inserted into drilled holes of the correct diameter. *J–P* are light- to medium-duty fasteners, with the exception of *L* and *N*, which can carry rather heavy loads.

Q and *R* are driven anchors. *Q* is the familiar concrete nail or masonry nail, made of hardened steel. If driven through a strip of wood with a few blows of a very heavy hammer or inserted with a nail gun, it will penetrate concrete just enough to provide some shear resistance for furring strips and sleepers, but it has a tendency to loosen, particularly if driven with too many blows. Shown in *R* are three examples of *powder-driven* (often called

powder-actuated) *fasteners*, which are driven into steel or concrete by an exploding cartridge of gunpowder. The first fastener is a simple pin used for attaching wood or sheet metal components to a wall or slab. The middle one is threaded to accept a nut. The eye on the fastener to the right allows a wire, such as a hanger wire for a suspended ceiling, to be attached. Powder-driven fasteners are rapidly installed, economical, and have a moderately high load-carrying capacity.

Figure *S* typifies devices whose perforated metal plates adhere securely with a mastic-type adhesive to surfaces of concrete or masonry. The fastener shown here has a thin sheet metal spike, over which a panel of foam plastic insulation can be impaled. The tip of the spike is then bent across the face of the insulation panel to hold it in place. Roof edge systems employing perforated plates and adhesives to fasten them to the building are illustrated in Chapter 16.

Before Assembly   *Assembled*   *Grouted*

FIGURE 15.13

A simple base detail for precast concrete columns. Four anchor bolts project from the top of the sitecast foundation. Nuts and washers are placed on these bolts to support the column temporarily. The column, which was cast around steel dowels welded to a baseplate, is lowered by crane. Workers guide the column so that the anchor bolts come through the holes in the baseplate. Washers and nuts are added to the anchor bolts on top of the baseplate. The eight nuts are used to adjust the height of the column and to make it plumb. When this has been accomplished, the nuts are tightened and stiff grout is dry packed under the baseplate.

In some cases, precast slab elements, especially solid slabs, require temporary shoring at midspan to help support the weight of the topping until it has cured. After curing, the topping becomes part of the slabs and increases their strength and stiffness. For construction economy, smooth-topped precast slab elements are sometimes used without topping, especially for roofs, where any unevenness between elements is bridged by rigid thermal insulation placed on top. Untopped slabs may also be used for floors that will be finished with a pad and carpet and for parking garages. Untopped slabs require special connection details that do not rely on reinforcing bars placed in the topping (Figures 15.18 and 15.20).

Posttensioning can be used to combine large precast elements into even larger ones on the site. This is done to assemble precast concrete box segments into very long, deep girders for bridges (Figures 15.23 and 15.24) and to create tall shear walls from story-high precast panels in multistory buildings. In either case, ducts for the posttensioning tendons are placed accurately in the sections before casting so that they will mate perfectly end to end when the sections are assembled on the site. After assembly, tendons are inserted into the ducts, horizontally in the case of girders or vertically in the case of shear walls, tensioned with portable

Shims

Before Assembly

Assembled

Grouted

The entire joint is dry-packed with grout after alignment

FIGURE 15.14
Similar details serve both as an alternative way of placing a column on its foundation and for column-to-column connections. Metal shims support the upper column sections at the proper height until the grout cures. The open corners are dry packed with stiff grout after the column has been aligned and bolted; this protects the metal parts of the connection from fire and corrosion.

hydraulic jacks, then grouted if required.

Joint design is the area in which precast concrete technology is developing most rapidly. Historically, precast concrete building construction has been limited in regions of high seismic risk due to uncertainty regarding the performance of this structure type under seismic loadings. More recently, innovations in the joining of precast concrete members have led to connection systems that can more reliably absorb the energy imparted into the structure during a seismic event. Unbonded posttensioning strands that allow the structure to respond elastically to seismic displacements, hybrid joint connections that combine the ductility of mild steel reinforcing with the high strength of prestressed strands, ungrouted or open joint systems that permit controlled movements between members, joint friction dampers that limit frame movements while absorbing seismic energy, and other techniques have come into use or are under development.

Other development efforts focus on improving the ease and economy of joining precast members. New joining systems are patented each year, and as reinforcing techniques, grouts, and adhesives develop further, there will be additional simplifications and improvements of many kinds in precast concrete framing details.

Shims

Before Assembly

Assembled

Grouted

Splicing sleeve
before grouting

Passages for
injecting grout
into sleeves

Splice sleeve
after grouting

Grout

Shims

Section

FIGURE 15.15

This column-to-column connection uses proprietary sleeves that are cast into the lower end of the upper column section. Before the sections are assembled (*upper left*), the lower ends of the vertical bars from the upper column section, which reach down to the midheight of each sleeve, are the only contents of the sleeves. Assembly of the column sections starts with the placement of a stack of steel shims in the center of the top of the lower section. These shims serve to adjust the height of the column and to maintain a space for grouting between the two sections. In the next two drawings, the sections have been assembled by lowering the upper section onto the lower one. The sleeves mate with projecting reinforcing bars from the lower column section. After the upper column section has been shimmed to exactly the right height and plumbed up, a fluid grout is injected into each sleeve, where it cures and serves to connect the reinforcing bars. A stiff grout is dry packed between the ends of the columns. The grouted sleeves develop the full strength of the reinforcing bars that they connect.

Dry-pack grout

Posttensioned tendon

Anchorage plate

Pocket for tensioning jack

Bearing pad

Corbel

Pockets are grouted after tensioning is complete

FIGURE 15.16

A posttensioned, structurally continuous beam–column connection may be created by passing a tendon from a pocket in the top of one beam, through the column, to a pocket in the top of the other beam. The tendon is anchored to a plate in one pocket as it is tensioned by a jack in the other pocket.

Vertical hole
in beam

Precast hollow-
core slabs, grout,
and topping

Reinforcing bar
projects from
column

Bearing pad

Grout in the
upper half of the
hole anchors the
beam to the
reinforcing bar
that projects from
the column

Reinforcing loops
prevent vertical
bar from tearing
out of beam

Precast column

Mastic in the
lower half of the
hole allows for
structural
movement

FIGURE 15.17
Topped hollow-core roof slabs supported on beams are joined to a column with vertical rods. A similar connection can be used for floor beams resting on corbels.

Column bars

Column ties

Weld plates
cast into
columns

Weld plate
cast into beam

Reinforcing
for corbels

Prestressed
tendons

Stirrups

Bearing pads
on corbels

(*a*)

(*b*)

FIGURE 15.18

The beams in this system of framing rest on concrete corbels that are integrally cast with the column. The smooth-topped hollow-core slabs are detailed for use without topping. (*a*) Weld plates are cast into the column. (*b*) Beams are placed on bearing pads on the corbels. A weld plate is cast into the top of each beam at the end.

(*continued*)

Welded angle
connector

Untopped hollow-core slab

Grout

Reinforcing bar ties are
grouted into keys between
slab elements

Bearing pad

(c)

FIGURE 15.18 *continued*

(c) Short pieces of steel angle are welded to the plates to join the beams to the
columns. Smooth-top hollow-core precast concrete planks are placed on bearing pads
on top of each beam. Grout is poured into the gap between the ends of the planks
to unite loops of reinforcing that project from the tops of the beams, reinforcing
bars that are inserted through the loops, and lateral pieces of reinforcing bar that are
grouted into the keys between planks. The end result is a tightly connected assembly
that supports an untopped precast concrete floor or roof.

*Top bars through
holes in the
column connect
the beams*

Corbel

Column

(*a*)

*Weld plate
connections*

Topping

*Double tee slabs
bear on the
inverted tee beam*

Inverted tee beam

(*b*)

FIGURE 15.19

(*a* and *b*) **Topped double-tee floor slabs are supported by inverted-tee
beams in this detail. Reinforcing bars that pass through hollow tubes
cast into the column connect the beams and column.**

(*continued*)

(c)

Weld plate connections

Bearing pads

Field-placed reinforcing in grout

Inverted tee beam

Polyurethane sealant in control joints at lines of potential cracking

Double tee with dapped end and recessed top

FIGURE 15.19 *continued*
(c) The sitecast topping ties all the components together and gives a smooth, level surface.

FIGURE 15.20
A minimum-headroom, minimum-cost floor system for parking garages uses untopped double tees. Refer to Figure 15.8(c and d) for photographs of how the ends of the tees are detailed for use in this system. The stems of the tees are dapped so that the beam need be no deeper than the tees.

A sitecast concrete topping with welded wire reinforcing fabric bonds to the rough top of the double tees to form a composite structural unit

FIGURE 15.21
A cutaway view of a topped double-tee slab.

A sitecast concrete topping with welded wire reinforcing fabric bonds to the rough top of the precast slabs to form a composite structural unit

Adjacent precast slab elements are locked together by grout keys so they deflect equally and share concentrated loads

FIGURE 15.22
A cross section through a topped hollow-core slab.

Prestressing strands

FIGURE 15.23
The Linn Cove Viaduct in Linnville, North Carolina, was built of short precast segments that were posttensioned together as they were placed to form a continuous box girder deck. A section is being installed by the derrick at the extreme right. The maximum clear span is 180 feet (55 m). *(Engineer: Figg and Muller Engineers, Inc. Courtesy of Precast/Prestressed Concrete Institute.)*

FIGURE 15.24
The Linn Cove Viaduct was constructed with very little temporary shoring so as to disrupt the natural landscape below as little as possible. The precast hollow box sections formed a girder that could cantilever for long distances during construction. The box profile is highly resistant to the torsional forces that occur in a curving beam. *(Courtesy of Precast/Prestressed Concrete Institute.)*

COMPOSITE PRECAST/ SITECAST CONCRETE CONSTRUCTION

In the *filigree precast concrete system*, relatively thin precast elements that are either conventionally reinforced or prestressed are used as the formwork for site casting of beams and slabs. Once the process is complete, composite structural action between the sitecast concrete and precast units results in a unified, structurally efficient system. Because the precast units remain in place as part of the finished system, formwork costs are much less in comparison to conventional sitecast concrete construction methods (Figure 15.25).

THE CONSTRUCTION PROCESS

The construction process for precast concrete framing is analogous to that for steel framing. The structural drawings for the building are sent to the precasting plant, where engineers and drafters prepare shop drawings that show all the dimensions and details of the individual elements and how they are to be connected. These drawings are reviewed by the engineer and architect for conformance with their design intentions and corrected as necessary. Then the production of the precast components proceeds, beginning with construction of any special molds that are required and fabrication of reinforcing cages, then continuing through cycles of casting, curing, and stockpiling as previously described. The finished elements, marked to designate their positions in the building, are transported to the construction site as needed and placed by crane in accordance with erection drawings prepared by the precasting plant (Figures 15.26 and 15.27).

FIGURE 15.25

In the filigree precast concrete system, precast concrete elements serve as formwork for sitecast concrete. In this photograph, precast units for a filigree slab and beam system have been erected and the placement of steel reinforcing is in progress. Later, concrete will be sitecast on top of this formwork. Once the construction of this system is complete, the precast units and sitecast concrete work together as a unified structural system: Prestressed strands in the precast units provide bottom reinforcing. Top bars are added on site as conventional reinforcing. Lightweight bar joist reinforcing, cast into the precast units so as to remain partially exposed, will form an intimate structural bond between the precast and sitecast concrete portions of the system, creating composite structural action in a manner very similar to that achieved with shear studs in structural steel composite construction (Figures 11.62 and 11.63). Note also the shallow slab band form running from lower left to upper right in this photograph. (*Courtesy of Midstate Filigree Systems, www.filigreeinc.com.*)

FIGURE 15.26
A framing plan and elevation of a simple four-story building made of loadbearing precast concrete wall panels and hollow-core slab elements. *(Courtesy of Precast/ Prestressed Concrete Institute.)*

SUSTAINABILITY IN PRECAST CONCRETE CONSTRUCTION

• In addition to the issues of sustainability of concrete construction that were raised in Chapter 13, there are issues that pertain especially to precast concrete construction:

• Because of the higher-strength concrete mixes typically used in the production of precast concrete, its embodied energy is higher on a pound-for-pound basis than that of conventional concrete, generally falling in the range of 500 to 600 BTU per pound (1.1-1.4 MJ/kg).

• Precast concrete production encourages the reuse of formwork, reducing waste. Wood and fiberglass forms can be used up to 50 times without major maintenance. Concrete and steel forms can be reused hundreds or thousands of times.

• Because precast concrete is manufactured in a controlled, factory-like setting, raw materials are used more efficiently and less waste is produced. Gray water used in various production processes, sand used in finishing, and large aggregate used to create voids in hollow planks can all be readily reused.

• In many cases, the optimized design of precast concrete results in elements that use less material than comparable sitecast concrete systems.

• Precast concrete elements with high-quality architectural finishes reduce the need for volatile organic compound–emitting paints or other finish coatings. Concrete is not easily damaged by moisture and does not support the growth of mold.

• Precast concrete wall panels with properly sealed joints have low permeability to air leakage, reducing building heating and cooling costs and contributing to good indoor air quality.

• Sometimes, precast concrete wall panels can be reused when buildings are altered.

Grout

Anchor bolt

Steel connector plate in pocket fastens anchor to bolt

The pocket is grouted after the connection is made

Reinforcing bars in the grout keys between slab elements tie the floor structure together across the supporting wall

FIGURE 15.27

A typical detail for the slab–wall junctions in the structure shown in Figure 15.26. The reinforcing in the wall panels and the prestressing steel in the slabs have been omitted from these drawings for the sake of clarity.

PRECAST CONCRETE AND THE BUILDING CODES

The fire resistance of precast concrete building frames and bearing wall panels depends on whether they are made of structural lightweight concrete or normal concrete and on the amount of concrete cover that protects the prestressing strands and reinforcing bars. The building code table shown in Figure 1.4 indicates the fire resistance ratings in hours required for components of Type I and II construction. When the architect or engineer has determined the

construction type, the precaster can assist in determining how the necessary degrees of fire resistance can be achieved in each component of the building. Slab elements are readily available in 1- and 2-hour fire resistance ratings, and beams and columns in ratings ranging from 1 to 4 hours. The fire resistance ratings of precast concrete slab elements may be increased by adding a topping or by increasing the thickness of a required topping. Solid and hollow-core slabs may achieve ratings as high as 3 hours by this means. Single and double tees require the addition of applied fireproofing material or a membrane ceiling fire protection

beneath to achieve a fire resistance rating higher than 2 hours.

UNIQUENESS OF PRECAST CONCRETE

Precast, prestressed concrete structural elements are crisp, slender in relation to span, precise, repetitive, and highly finished. They combine the rapid all-weather erection of structural steel framing with the self-fireproofing of sitecast concrete framing to offer economical framing for many kinds of buildings (Figures 15.28–15.33). Solid and hollow-core slabs have become

FIGURE 15.28
A view of the construction of a building that uses precast concrete loadbearing wall panels. The rectangular pockets at the lower edges of the wall panels are for bolted connections of the type shown in Figure 15.27. Steel pipe braces support the panels until all the connections have been made and the structure becomes self-stable *(Courtesy of Blakeslee Prestress, Inc.)*

FIGURE 15.29
Exterior loadbearing wall panels are often made of specially colored concrete, with surface textures cast in. *(Courtesy of Blakeslee Prestress, Inc.)*

FIGURE 15.30
A crane hoists a single-piece precast stair for a high-rise building of loadbearing precast concrete wall panels. *(Courtesy of Blakeslee Prestress, Inc.)*

FIGURE 15.31
A close-up of the precast concrete building frame shown in Figure 15.32. Precast concrete columns with corbels support inverted L- and T-shaped beams. The floor structure is double tees. On the left-hand column, patches where lifting loops were burned off are can be seen. On the right-hand column, lifting loops are on one of the three faces of the column not visible in this photograph.

FIGURE 15.32
A precast concrete building frame with a long-span OWSJ roof structure. Details of the framing are shown in Figure 15.31.

FIGURE 15.33
**Placing a hollow-core slab deck
on a precast frame.** *(Courtesy of
Blakeslee Prestress, Inc.)*

a standard part of our structural
vocabulary in schools, hotels, apartment buildings, and hospitals, where
they are ideal both functionally and
economically. Engineers and architects have long been comfortable
with precast concrete in longer-span
building types, especially parking
structures, warehouses, and industrial plants, where its unique structural potential and efficient serial
production of identical elements
can be fully utilized and openly
expressed. Though less frequently,
precast concrete is also used to create public buildings of the highest
architectural quality, both inside and
out (Figures 15.34–15.39).

(a)

FIGURE 15.34
(*a*) **A crane lifts a column section from a flatbed truck to begin erection of a tennis
stadium in New Haven, Connecticut.**

(continued)

(b)

(d)

(c)

(e)

FIGURE 15.34 *continued*
(*b*) Grouted sleeve connectors of the type shown in Figure 15.15 were used to join the precast sections of the bents that support the grandstand seating. (*c*) Stepped sections of grandstand floor await lifting. (*d*) A precast stair section is placed. (*e*) Precast hollow-core planks will be used for miscellaneous areas of floor. (*f*) The relationship of the stairs, bents, and stepped floor sections is evident in this photograph of the finished stadium. (*g*) The stadium in use, after a construction process that took only 11 months. Seating capacity is **15,000.** (*Structural engineer: Spiegel Zamecnik & Shah, Inc. Photos by Clark Broadbent. Compliments of Blakeslee Prestress, Inc., P.O. Box 510, Branford, CT 06405; Tel: 203 481 5306.*)

(*f*)

(*g*)

FIGURE 15.35
A low-rise office building in Texas shows a carefully detailed frame of precast concrete. *(Architects: Omniplan Architects. Courtesy of Precast/Prestressed Concrete Institute.)*

FIGURE 15.36
Precast concrete girders span the roof of a paper mill in British Columbia.
(Engineer: Swan Wooster Engineering Co., Ltd. Courtesy of Precast/Prestressed Concrete Institute.)

FIGURE 15.37
The precast walls and slabs of these condominium apartments were erected during the winter in the New Mexico mountains. *(Architect: Antoine Predock. Courtesy of Precast/Prestressed Concrete Institute.)*

FIGURE 15.38
Highly customized precast concrete framing is used in this courthouse by architect Eduardo Catalano. *(Photo by Gordon H. Schenk, Jr. Courtesy of the architect.)*

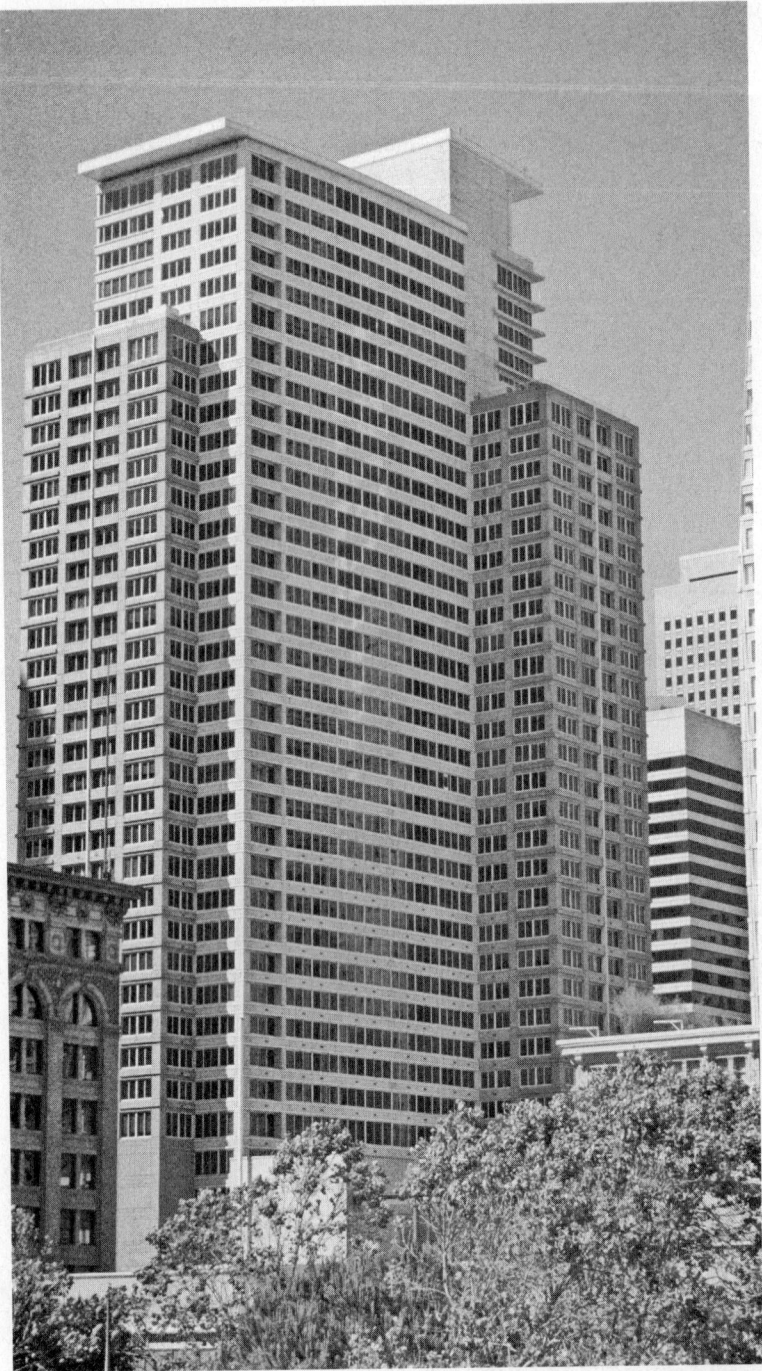

FIGURE 15.39
The 39 stories of the Paramount Building rise 420 feet (128 m) on a precast concrete frame. The building is located in San Francisco, California, in the zone of highest seismic risk in the United States. Beam–column connections are posttensioned to resist seismic forces. The precast concrete cladding panels act to resist lateral forces on the building. (*Architects: Elkus Manfredi and Kwan Henmi. Structural engineers: Robert Engelkirk Consulting Engineers, Inc. Photograph © Bernard André.*)

CSI/CSC MasterFormat Sections for Precast Concrete Framing Systems

03 41 00	**PRECAST STRUCTURAL CONCRETE**
03 41 13	Precast Concrete Hollow Core Planks
03 41 16	Precast Concrete Slabs
03 41 23	Precast Concrete Stairs
03 41 33	Precast Structural Pretensioned Concrete
03 41 36	Precast Structural Post-Tensioned Concrete
03 45 00	**PRECAST ARCHITECTURAL CONCRETE**
03 53 00	**CONCRETE TOPPING**

KEY TERMS

precast concrete
steam curing
solid slab
hollow-core slab
double tee
single tee
topping
ledger
L-shaped beam

inverted tee
architectural precast concrete
casting bed
weld plate
dapped end
carbon fiber reinforcing
(hollow-core slab) extruded
   process
(hollow-core slab) wet-cast process

(hollow-core slab) slip-form process
corbel
bearing pad
anchor bolt
embed plate
dovetail slot
powder-driven fastener,
   powder-actuated fastener
filigree precast concrete system

REVIEW QUESTIONS

1. Under what circumstances might a designer choose a precast concrete framing system over a sitecast system? Under what circumstances might a sitecast system be favored?

2. Why are precast concrete structural elements usually cured with steam?

3. Explain several methods of producing hollow-core slabs.

4. Diagram from memory several different ways of connecting precast concrete beams to columns.

5. Diagram from memory a method of connecting a pair of untopped double-tee slabs to an inverted-tee beam. Then work out a similar way of connecting a double tee to an L-shaped beam, as would

occur around the perimeter of the same building.

6. Explain the construction process for a filigree precast concrete system. What are the unique advantages of this system over other concrete systems?

EXERCISES

1. Design a simple two-story rectangular warehouse, 90 by 180 feet (27 by 54 m), using precast concrete for the floor and roof structure and for the walls as well. Use the preliminary structural design information given in this chapter to help determine the column spacing, the types of elements to use, and the depths of the elements. Draw a framing plan and typical connections for the building.

2. Locate a concrete precasting plant in your area and arrange a visit to view the production process. If possible, arrive at the plant early in the morning, when the strands are being cut and the elements are being lifted from the molds.

3. Learn from the management of the precasting plant where a precast concrete building is being erected; then

visit the building site. Trace a typical precast concrete structural element from raw materials through precasting, transporting, and erecting. Are there ways in which this process could be made more efficient? Sketch a few of the typical connections being used in the project.

SELECTED REFERENCES

Precast/Prestressed Concrete Institute. *Architectural Precast Concrete* (3rd ed.). Chicago, Author, 2007.

This design manual provides technical and design guidelines for achieving high-quality architectural finishes in precast concrete, and includes hundreds of images and line drawings.

Precast/Prestressed Concrete Institute. *Design and Typical Details of Connections for Precast and Prestressed Concrete* (2nd ed.). Chicago, Author, 1988.

This engineering design manual includes a large collection of drawings of standard connection details.

Precast/Prestressed Concrete Institute. *Erector's Manual: Standards and Guidelines for the Erection of Precast Concrete Products.* Chicago, Author, 1999.

The 96 pages of this manual describe and illustrate the best ways of hoisting and assembling the elements of a precast concrete building.

Precast/Prestressed Concrete Institute. *PCI Design Handbook.* Chicago, Author, updated regularly.

This is the major reference handbook for those engaged in designing precast, prestressed concrete buildings. It includes basic building assembly concepts, load tables for standard precast elements, engineering design methods, and a few suggested connection details.

WEB SITES

Precast Concrete Framing Systems

Author's supplementary web site: **www.ianosbackfill.com/15_precast_concrete_framing_systems**

Altus Group carbon reinforced precast concrete: **altusprecast.com**

National Precast Concrete Association: **precast.org**

Precast/Prestressed Concrete Institute (PCI): **www.pci.org**

Spancrete: **www.spancrete.com**

ROOFING

- **Low-Slope Roofs**
  Roof Decks
  Thermal Insulation

  BUILDING ENCLOSURE ESSENTIALS:
  THERMAL INSULATION AND VAPOR
  RETARDER

  Membranes for Low-Slope Roofs
  Ballasting and Traffic Decks
  Edge and Drainage Details
  Structural Panel Metal Roofing

- **Steep Roofs**
  Shingles
  Architectural Sheet Metal Roofing

  SUSTAINABILITY IN ROOFING

- **Cool Roofs**

- **Green Roofs**

- **Photovoltaic Systems**

- **Roofing and the Building Codes**

  BUILDING ENCLOSURE ESSENTIALS:
  DISSIMILAR METALS AND THE GALVANIC
  SERIES

Architect Tom Kundig, of Olson Sundberg Kundig Allen Architects, uses noncombustible, corrugated steel for the roof of a hilltop residence in a fire-prone area of southern California. The simple roof form is lifted above the main living space of the house, shading it from the sun and allowing naturally cool offshore breezes to circulate freely around and through it. *(Photograph by Tim Bies, courtesy of Olson Sundberg Kundig Allen Architects.)*

A building's roof is its first line of defense against the weather, protecting it from rain, snow, and sun. The roof helps to insulate the building from extremes of heat and cold and to control the accompanying problems of air leakage and water vapor condensation. And like any frontline defender, it must itself take the brunt of the attack: A roof is subject to the most intense solar radiation of any part of a building. At midday, the sun broils a roof with radiated heat and ultraviolet light. On clear nights, a roof radiates heat to the blackness of space and becomes colder than the surrounding air. From noon to midnight of the same day, it is possible for the surface temperature of a roof to vary from near boiling to below freezing. In cold climates, snow and ice cover a roof after winter storms, and cycles of freezing and thawing gnaw at the materials of the roof. A roof is vital to the sheltering function of a building, yet it is singularly vulnerable to the destructive forces of nature.

Roofs can be covered with many different materials. These can be organized conveniently into two groups: those that work on *steep roofs* and those that work on *low-slope roofs*, those that are nearly flat. The distinction is important: A steep roof drains itself quickly of water, giving wind and gravity little opportunity to push or pull water through the roofing material. Therefore, steep roofs can be covered with roofing materials that are fabricated and applied in small, overlapping units—shingles of wood, slate, or artificial composition; tiles of fired clay or concrete; or even tightly wrapped bundles of reeds, leaves, or grasses. There are several advantages to these materials: Many of them are inexpensive. The small, individual units are easy to handle and install. Repair of damage to the roof is easy. The effects of thermal expansion and contraction, and of movements in the structure that supports the roof, are minimized by the ability of the small roofing units to move with respect to one another. Water vapor vents itself easily from the interior of the building through the loose joints in the roofing material. In addition, a steep roof of well-chosen materials skillfully installed can be a delight to the eye.

Low-slope roofs have none of these advantages. Water drains relatively slowly from the surfaces, and small errors in design or construction can cause them to trap puddles of standing water. The membranes that cover low-slope roofs must be absolutely watertight. Even small punctures, tears, or gaps in seams, caused by defects in construction, physical wear and tear, or movements within the building structure, can allow significant quantities of water to enter the building structure and its interior, with potentially disastrous results. Or, water vapor pressure from within the building can blister and rupture the membrane. But low-slope roofs also have overriding advantages: A low-slope roof can cover a building of any horizontal dimension, whereas a steep roof becomes uneconomically tall when used on a very broad building. A building with a low-slope roof has a much simpler geometry that is often less expensive to construct. Low-slope roofs, when appropriately detailed, can also serve as balconies, decks, patios, and even landscaped gardens or parks.

FIGURE 16.1
A steep roof can be made waterproof with any of a variety of materials. This thatched roof is being constructed by fastening bundles of reeds to the roof structure in overlapping layers in such a way that only the butts of the reeds are left exposed to the weather.

FIGURE 16.2
The finished thatched roof has gently rounded contours and a pleasing surface texture. The decorative pattern of the ridge cap is the unique signature of the thatcher who made the roof. *(Photos of thatched roofs courtesy of Warwick Cottage Enterprises.)*

LOW-SLOPE ROOFS

A low-slope roof (often referred to, inaccurately, as a flat roof) is usually defined as one whose slope is less than 2:12, or 17 percent. It is a highly interactive assembly made up of multiple components. The *roof deck* is the structural surface that supports the roof. *Thermal insulation* is installed to slow the passage of heat into and out of the building. An *air barrier* restricts the leakage of air through the roof assembly, and a *vapor retarder*, used in colder climates or when enclosing humid spaces, prevents moisture vapor from condensing within it. The *roof membrane* is the impervious sheet of material that keeps water out of the building. Additional layers within the assembly may increase resistance to fire, protect soft insulation boards from damage by foot traffic, or separate replacement roof materials from older, incompatible existing layers. *Drainage* components, such as roof drains, gutters, and downspouts, remove the water that runs off the membrane. Around the membrane's edges and wherever it is penetrated by pipes, vents, expansion joints, electrical conduits, or roof hatches, special *flashings* and details must be designed and installed to prevent water penetration.

Roof Decks

Previous chapters of this book have presented the types of structural decks ordinarily used under low-slope roofs: wood panels over wood joists, solid wood decking over heavy timber framing, corrugated steel decking, panels of wood fiber bonded together with portland cement, site-cast concrete slab, and precast concrete slab. The structural deck and its supporting members must be strong and stiff enough to support rain and snow loads, as well as resistant to the forces of wind uplift. The deck itself, or the insulation materials on top of it, must slope steeply enough toward drainage points for water on top of the roof membrane to drain efficiently and reliably. A slope of at least ¼ inch per foot of run (1 in 48, or 2 percent) is normally required by the building code and most manufacturers of low-slope roof membranes. To achieve this, the beams that support the roof deck may be sloped by varying the height of their supporting columns, or the supporting structure may be constructed level and slope created by a layer of thermal insulation of varying thickness installed on the top of the deck. This layer may consist of lightweight insulating concrete or, more frequently, a system of tapered rigid insulation boards.

If a low-slope roof does not drain efficiently, *ponding* can occur, in which areas of water remain standing for extended periods of time. Except with roof membrane systems specially designed to tolerate standing water, this can cause deterioration of the membrane and increase the chance of leakage. Particularly with long-span roof structures, water accumulation in low spots caused by structural deflections is also a concern. As puddles deepen and increase in weight, the structure may deflect further, progressively attracting more and more water and becoming heavier and heavier. If water cannot drain from these areas and continues to accumulate, overloading and even structural collapse is a possibility.

The roof membrane must be laid over a smooth, continuous surface. A wood deck should have no large gaps, knotholes, or protruding fasteners, or it may be covered with a substrate board to provide a more suitable surface. A sitecast concrete deck should be troweled smooth, and a precast concrete plank deck, if not topped with a concrete fill, must be grouted at junctions between planks to fill the cracks. A corrugated steel deck is usually covered to create a continuous, flat surface. This is done with rigid insulation boards or some variety of *substrate board*, thin panels of wood fiber or gypsum that are attached to the top surface of the metal deck. Gypsum panel substrate boards are also installed over wood and metal roof decks to increase the fire resistance of the roof assembly.

It is important that the deck be dry at the time roofing operations commence, to avoid later problems with water trapped under the roof membrane. A deck should not be roofed when rain, snow, or frost is present in or on the deck material. Concrete decks and insulating fills must be cured and adequately dried.

If a deck is large in extent, the roofing system should be provided with enough movement joints to keep expansion and contraction or other movement within the deck from stressing the roof membrane. Where building separation joints occur within a building structure, these joints must carry through the roof membrane system as well (Figure 16.24). Where such joints do not occur or are too far apart to satisfy the requirements of the membrane, *area dividers*, which are much like building separation joints but do not extend below the surface of the roof deck, may be installed (Figure 16.25).

Thermal Insulation

Thermal insulation for a low-slope roof may be installed in any of three positions: below the structural deck, between the deck and the roof membrane, or above the membrane (Figure 16.3).

Insulation Below the Roof Deck

Below the roof deck, mineral or glass fiber batts are installed either between framing members or on top of a suspended ceiling assembly. With this type of insulation, the building code normally requires a ventilated airspace between the insulation and the underside of the deck to dissipate water vapor. Below-deck insulation is relatively economical and trouble-free, but it leaves the roof deck and the membrane exposed to the full range of outdoor temperature fluctuations. Where a vapor retarder is required, it is applied on the warm, conditioned side of the roof insulation to control the diffusion of water vapor into the insulated portions of the roof where condensation could occur.

Particularly in residential buildings, spray foam insulation may be applied directly to the underside of the roof deck. When air-impermeable foam insulation with sufficient vapor resistance and insulation value is used, the risk of moisture accumulation within the assembly is minimized, and ventilation between the deck and the insulation is not required. See Chapter 6 for more information about unventilated roof assemblies.

Insulation Between the Roof Deck and the Membrane

The traditional position for low-slope roof insulation is above the roof deck and underneath the membrane. Low-density rigid boards or

FIGURE 16.3

Low-slope roofs with thermal insulation in three different positions, shown here with a wood joist roof deck. At the left, insulation is located below the roof deck. In the center, insulation is placed between the deck and the membrane. At the right, a protected membrane roof is shown with insulation placed above the membrane. Vapor retarders are indicated in the appropriate location for roofs in cold regions. See the accompanying text for a further discussion of the use of vapor retarders in low-slope roof assemblies.

lightweight concrete that can support the membrane and withstand occasional foot traffic are used. The insulation protects the deck from outdoor temperatures and is itself protected from the weather by the membrane. The roof membrane remains exposed to the exterior environment.

A risk with this insulation strategy is the accumulation of moisture within the insulation. This moisture may come from roof membrane leaks, the leakage of air or diffusion of moisture vapor from within the building, or the roof structural components themselves if they were not sufficiently dry when the roofing system was installed. Moisture trapped within the insulation layers can damage the insulation and roof deck, as well as cause blistering and rupture of the roof membrane from high vapor pressure. Where vapor diffusion from the building interior is a concern, a vapor retarder may be installed below the insulation. Ventilation can also be provided under the roof membrane. One way to accomplish this is with the installation of *topside vents*, one per 1000 square feet (100 m²), that allow water vapor to pass to the exterior (Figures 16.4 and 16.5). Ventilated low-slope roof assemblies are most effective with roof membranes that are not fully adhered to the layers below, as this allows entrapped moisture to more easily find its way toward the vents from any location under the membrane.

Insulation Above the Roof Membrane: The Protected Membrane Roof

In a *protected membrane roof (PMR)* system, insulation is installed above the membrane. This offers two advantages: The membrane is protected from extremes of heat and cold, and the membrane is on the warm side of the insulation, where it is immune to vapor blistering problems. Because the insulation itself is exposed to water when placed above the membrane, it must be made of a material that retains its insulating value when wet and does not decay or disintegrate. Extruded polystyrene foam is the most commonly used insulation type (Figure 16.6). The panels of polystyrene are either embedded in a coat of hot asphalt to adhere them to the membrane below or are laid loose. They are held down and protected from sunlight (which disintegrates polystyrene) by a layer of *ballast*, which may consist of crushed stone or gravel, or concrete blocks (Figures 16.7 and 16.20). Because the membrane in a PMR system is shielded from sunlight and temperature by the insulation and ballast above, it can be expected to last roughly twice as long as in an assembly where it remains directly exposed to these elements. However, PMR systems do have potential disadvantages. When the roof insulation is exposed to precipitation, it may absorb moisture and lose some of its insulation value. When performing energy calculation for PMR systems, R-values for roof insulation may be slightly reduced from their standard values to account for such losses. PMR systems may not be appropriate for climates with extended periods of rainy, cold weather, as cool water continuously flowing around and under the insulation boards may negate much of their insulating value. Also, repairs to protected membrane roofs, if required, are more costly and time-consuming because accessing the roof membrane requires removal of the layers of material above it.

Because the roof membrane is located on the warm side of the roof insulation, a separate vapor retarder is not required in a protected membrane roof.

Vapor Retarders and Air Barriers in Low-Slope Roofs

Figure 16.3 shows typical locations for vapor retarders in low-slope roof assemblies. As discussed at greater length in the accompanying sidebar, the need for vapor retarders generally increases with colder climates and higher-humidity interior

FIGURE 16.4
Topside roof vents are being installed in this built-up roof membrane to release vapor pressure that may accumulate beneath it. *(Courtesy of Manville Corp. FP-10 One Way® Roof Vent.)*

6″

Two-Layer
Flashing of Felt

Rigid Insulation

Poured Fill

4½″

10″

Open

Closed

FIGURE 16.5
This topside roof vent, made of molded plastic with a synthetic rubber valve, allows moisture vapor to escape from beneath the membrane but closes to prevent water or air from entering. *(Courtesy of Manville Corp. FP-10 One Way® Roof Vent.)*

conditions. However, a consideration unique to low-slope roofing assemblies is the roofing membrane itself, which also has a low vapor permeance and can act as a vapor retarder. If an additional vapor retarder is included in the assembly, there is the possibility of entrapping moisture between these two impermeable layers. This complication makes the decision whether to include a separate vapor retarder in a low-slope roof more dependent on the case-by-case analysis of individual roofing installations rather than simple, generalized criteria.

The traditional material for a vapor retarder consists of two layers of asphalt-saturated roofing felt bonded together and adhered to the upper side of the roof deck with hot asphalt. Vapor retarder sheets made from factory-manufactured self-adhering bituminous membranes are also common, particularly with single-ply roofing systems or where roofing installation methods do not rely on hot asphalt.

A membrane or coating fully adhered to the structural roof deck is the simplest way to achieve an effective air barrier in a low-slope roofing system. The structural deck provides a continuous surface with a minimum of penetrations. By fully adhering the air barrier material to this surface, the chance for leaking air to travel under or around the membrane is minimized. When a vapor retarder is included in the roof assembly, the air barrier membrane and vapor retarder frequently are the same material. When an added vapor retarder is not desired, a vapor-permeable air barrier material may be used. The roof membrane itself can also act as the air barrier in the roof system, particularly when fully adhered to the materials below.

With any type of air barrier membrane, special attention must be given to the interface between the roof edges and the exterior walls. This is the most difficult location to maintain air barrier continuity and the most frequent location of air leakage. For a fuller discussion of air barrier systems, see Chapter 19.

FIGURE 16.6
Installing extruded polystyrene foam insulation over a roof membrane to create a protected membrane roof. *(Photograph provided by The Dow Chemical Company.)*

FIGURE 16.7
A cutaway detail of a protected membrane roof shows, from bottom to top, the roof deck, the membrane, polystyrene foam insulation, a polymeric fabric that separates the ballast from the insulation, and the ballast. *(Photograph provided by The Dow Chemical Company.)*

BUILDING ENCLOSURE ESSENTIALS: THERMAL INSULATION AND VAPOR RETARDER

Thermal Insulation

Thermal insulation is material added to a building assembly to slow the conduction of heat through the assembly. Insulation is installed in new roof and exterior wall assemblies, in floors over unheated spaces, around foundations and concrete slabs on grade, and wherever else heated or cooled interior space comes in contact with unconditioned space, the earth, or the outdoors. A well-insulated building enclosure increases occupant comfort and reduces the energy required to heat or cool the building.

A material's effectiveness in resisting the conduction of heat is called its *thermal resistance*, abbreviated as *R* and expressed as square foot-hour-degree Fahrenheit per BTU (ft²-hr-°F/BTU). In the metric system, thermal resistance is abbreviated as *RSI* (or sometimes also simply as R), expressed as square meter-degree Kelvin per watt (m²-°K/W). An insulating value of R-1 is equivalent to RSI 0.176. For lists of common insulating materials and their properties, see Figures 7.13 and 16.8. The higher a material's R-value, the higher its resistance to heat flow and the better its performance as a thermal insulator.

The thermal performance of a complete building assembly depends on the sum of the thermal resistances of the materials from which it is made. Every component of an assembly contributes in some measure to the assembly's overall thermal resistance, the amount of the contribution depending on the type of material and its thickness. Metals are poor insulators, and concrete and masonry are only slightly better. Wood has a higher thermal resistance, but not nearly as high as that of commonly used insulating materials. In conventional wall and roof assemblies, most of the thermal resistance comes from the insulation materials themselves. (When performing calculations to demonstrate compliance with energy code compliance, usually only the values of dedicated insulation materials are considered, and the minor contributions from other materials are ignored.)

In wintertime, it is warm inside a heated building and cold outside, and the inside surface of a wall or roof assembly is warm, while the outside surface is cold. Between the two surfaces, the temperature varies according to the thermal resistances of the various layers of the assembly. Because most of the insulating value of the assembly is in the insulation itself, most of the temperature change within the assembly occurs across the thickness of the insulating materials (Figure A).

Water Vapor and Condensation

Water exists in three physical states, depending on its temperature and pressure: solid (ice), liquid, and vapor. Air always contains some water in the form of *water vapor*, an invisible gas. The higher the temperature of the air, the more water vapor it can contain. At a given temperature, the amount of water vapor the air actually contains, in proportion to the maximum amount it can contain, is the *relative humidity* of the air. For example, air at 50 percent relative humidity contains half as much water vapor as it is capable of holding at that particular temperature.

If a mass of air is cooled, its relative humidity rises. The amount of water vapor in the air mass has not changed, but the ability of the air mass to hold water vapor has diminished because the air has become cooler. If the cooling of the air mass continues, a temperature will be reached at which the humidity is 100 percent. This is the temperature at which the air is fully saturated with water vapor, also

FIGURE A

known as the *dew point temperature*. The dew point is different for every air mass. A roomful of very humid air has a high dew point, which is another way of saying that the air in the room would not have to be cooled very much before it reached 100 percent humidity. In comparison, a roomful of dry air has a lower dew point and can be cooled to a lower temperature before it reaches saturation.

When a mass of air is cooled below its dew point, it can no longer contain all its water vapor and some of the vapor turns to liquid. The further the air mass is cooled below this point, the less water vapor it can hold and the more vapor turns to liquid. This process of converting water vapor to liquid by cooling is called *condensation*.

Condensation takes place in buildings in many different ways. In winter, room air circulating against a cold pane of glass may be cooled to below its dew point, and water droplets will form on the glass. If the air is very humid, the droplets will grow in size, then run down the glass to accumulate in puddles on the window sill. If the glass is very cold, the condensate may freeze into patterns of ice crystals on the glass. In a similar fashion, on a hot, humid summer day, water vapor in the air may condense on the surface of a cold water pipe, a cool basement wall, or an ice-cold glass of lemonade. Though hidden from view, condensation can also occur within the walls, roofs, and other assemblies that enclose a building.

Water Vapor in Building Assemblies

Water vapor is a gas and exerts pressure, called *vapor pressure*, on the surfaces that contain it. The more water vapor an air mass contains, the greater the vapor pressure. Under wintertime heating conditions, the air inside a building is at a higher temperature and contains more water vapor than the air outside. This is especially true in areas of the building with many occupants, or where cooking, bathing, or washing take place, or wet industrial processes occur. The result is a net difference in vapor pressure acting from inside to outside, causing water vapor to diffuse outward directly through the various material layers of the enclosing assembly. If the rate of diffusion is high enough and the drop in temperature within the assembly is great enough, water vapor will reach its dew point and condense within the assembly. Moisture condensation within a building exterior is generally undesirable. If it accumulates, insulation may lose its effectiveness and materials may be damaged by rusting, mildew, decay, freeze-thaw, or other harmful processes. If water runs out of the assembly, finishes or building contents may be harmed. (Water vapor can also be carried through an assembly by air leaking through gaps in materials. This is a different mechanism than the diffusion of water vapor through materials and is discussed under the topic of air barriers in Chapter 19.)

Under summertime cooling conditions in hot, humid weather, the diffusion of water vapor through building assemblies is reversed. Water vapor is driven from the warm, damp outside air toward the cooler, drier air within the building. In most of North America, this summer condition is not as severe as that in winter: Differences in summer temperature and humidity between indoors and outdoors are not as great as those in winter, and the cooling season is short compared to the heating season. Where heating conditions predominate, control of water vapor in building assemblies focuses primarily on the flow of vapor from interior to exterior, though conditions of reverse, inward flow may also warrant some consideration. In areas of the American South and in the Hawaiian Islands, however, the summer condition is the more severe, and the flow of water vapor from exterior to interior is the predominant problem to be solved.

Vapor Retarder

A *vapor retarder* (traditionally also referred to as a *vapor barrier*) is a material used to slow the diffusion of water vapor through a building assembly. Vapor retarders are continuous sheets or coatings made of plastic, metal foil, coated paper, or other material resistant to the passage of water vapor.

Vapor retarders are located toward the warmer side of the insulation in a building assembly. In this position, they restrict the diffusion of water vapor into the assembly from the side of higher vapor pressure, limiting chances for dew point conditions and condensation to occur within the assembly's cooler portions. For buildings in most parts of North America, where winter heating conditions predominate, vapor retarders are placed toward the interior, heated side of insulation in the assembly. In humid regions where warm-weather cooling conditions predominate within buildings, the vapor retarder should be located toward the exterior side of insulation. In relatively mild or balanced climates, or where assemblies are designed to minimize condensation conditions, a vapor retarder may not be necessary at all.

The better a material's resistance to water vapor diffusion—that is, the lower its *vapor permeance*—the more effective it is as a vapor retarder. Permeance is measured in *perms*, defined by ASTM E96 as the passage of one grain of water vapor per hour through 1 square foot of material at a pressure differential of 1 inch of mercury between the two sides of the material (grains/hr-ft^2-in. Hg). In metric units, a perm is measured in grams per second per square meter per Pascal of pressure difference (g/s-m^2-Pa). One U.S. perm equals 5.72×10^{-8} metric perms. To arrive at more workable units, metric perms may also be calculated with nanograms (billionths of a gram) in place of grams. In this case, one U.S. perm equals 57.2 ng/s-m²-Pa.

In the International Building Code, vapor retarders are classified according to their relative permeance. Class I vapor retarders are the most resistant to vapor diffusion, and have a vapor permeance of 0.1 perms (6 ng/s-m²-Pa) or less. Class II vapor retarders may have a permeance as great as 1 perm (60 ng/s-m²-Pa) and Class III vapor retarders as great as 10 perms (600 ng/s-m²-Pa).

When evaluating vapor retarders, vapor permeance should not be confused with *vapor permeability*. Vapor permeability is defined as a material's vapor permeance for a unit of thickness. For example, the vapor permeability of a particular rigid insulation may be given as 0.75 perm-inch. The vapor permeance of an insulation board made of this material is then found by dividing the material's permeability by the actual thickness of the board used. For example, if the insulation board is ½ inch thick, its permeance is 1.5 perms (0.75 perm-in./0.5 inch), or if 2 inches thick, 0.375 perms (0.75 perm-in./2 inches). In metric units, vapor permeability is calculated as ng-m/Pa-s-m², which reduces to ng/Pa-s-m.

The vapor permeance of some materials varies. Materials that absorb moisture, such as plywood or the kraft paper facing found on much glass fiber batt insulation, tend to increase in permeance when damp. At least one manufacturer's proprietary vapor retarder membrane is engineered purposely to increase in permeance under conditions of high humidity. Such materials can provide benefits, inhibiting vapor diffusion into the assembly under normal conditions but becoming more permeable and enhancing the ability of an assembly to dry outward if moisture accumulates within it.

Commonly used vapor retarder materials are polyethylene plastic sheet, kraft paper facing on glass fiber batt insulation, aluminum foil facing on various types of insulation, and specially formulated paint primers with low water vapor permeance. Some foam insulation materials, depending on their formulation and thickness, can also act as vapor retarders. In low-slope roof construction, vapor retarders may be made from roofing felts layered in hot asphalt or of adhered rubberized asphalt sheets (Figure B).

Vapor Retarder Usage

Vapor retarders are placed on the warm side of insulated building assemblies to prevent condensation within the assembly. This is most likely to occur where the wall or roof is exposed to large temperature differences from one side to the other and the relative humidity is high on the warmer side. North American model building codes generally require Type I or II vapor retarders in insulated assemblies for roughly the upper half to two-thirds of the continental United States and all of Canada. Alternatively, assemblies that are designed to tolerate higher vapor diffusion rates without harmful condensation or damage to the assembly may be constructed with Type III vapor retarders or possibly none at all.

Vapor retarders are also especially important when enclosing high-humidity spaces, such as indoor swimming pools or museum galleries in which relative humidity is kept high to protect artwork. With higher interior humidity levels, the risk of condensation within the wall also increases.

Where vapor retarders are used, the cooler side of the assembly, opposite the vapor retarder, should remain breathable. Thus, if moisture does find its way into the assembly, it can be dispersed by means of ventilation or diffusion. Examples of this strategy include ventilated roof systems, ventilated rainscreen cladding, and the use of highly vapor-permeable house wraps. Assemblies with multiple vapor-impermeable layers should be avoided: They can trap moisture with no means for escape. A rule of thumb sometimes used for the selection of materials on the opposite side of an assembly from a vapor retarder is that they should have a vapor permeance at least 10 times that of the vapor retarder itself.

The overuse or misapplication of vapor retarders should also be avoided. For example, in mixed and hot climates, the use of vapor retarders where none is needed or of vapor retarders with a much lower permeance than required can lead to the unintended entrapment of moisture. In air-conditioned buildings in hot, humid climates, care should be taken not to inadvertently create a vapor retarder on the interior (cold) side of insulated walls and roofs. For example, interior vinyl wallpaper (a vapor retarder) used in these circumstances can lead to mold and mildew growth underneath the vinyl. Insulated below-grade walls may also benefit from more permeable vapor retarders or none at all.

Additional factors to consider in the use of vapor retarders include the following:

• *The temperature and humidity conditions to which a building assembly is exposed over time are not static.* A wall system designed exclusively for winter heating conditions may perform poorly when vapor migrates from the exterior inward during summer months with air conditioning in operation. Solar heating of a rain-soaked exterior cladding system can cause strong inward water vapor movement at any time of the year.

• *Building assemblies must be able to dry out when wet.* Water introduced into a building assembly, such as from penetrating rain or from construction with wet building materials, must have a means to escape so that the assembly can dry.

• *Air barriers are also important to the control of water vapor.* Under many circumstances, air leakage can transport many times greater amounts of water vapor into an assembly than can vapor diffusion. Air barriers are discussed in Chapter 19.

• *Design for water vapor control should consider all components of an assembly.* The permeability of exterior sheathing, cladding or roofing, provisions for ventilation, and the location and types of insulating materials within an assembly can all influence the choice and placement of vapor retarders.

For more information on the design of vapor retarders in insulated building assemblies, see the references listed at the end of this chapter.

	Vapor Permeance		
	U.S. Perms **grains/hr-ft^2-in. Hg**	**Metric Perms** **ng/s-m^2-Pa**	**Class**
Aluminum foil, 1 mil (0.025 mm)	~0	~0	I
Double-ply roof felt with hot asphalt	0.005	0.29	I
Self-adhered bituminous flashing and waterproofing membranes	0.05	2.8	I
Polyethylene sheet, 4 mil (0.08 mm)	0.08	4.6	I
Vapor retarder under concrete slab on grade (ASTM E1745)	0.3	17	II
Paint, exterior alkyd (oil), three coats over wood	0.3–1.0	17–57	II
Kraft paper facing for glass fiber batt insulation			
Low humidity	0.4–1.0	23–57	II
High humidity	8	460	III
Paint, latex vapor barrier primer	0.4	23	II
EPDM single-ply roofing, 45 mil (1.1 mm)	0.4	23	II
High-density, closed cell spray polyurethane foam, 2 in. (50 mm)	0.6	34	II
Plywood, exterior glue, ½ in. (12 mm) thick			
Low humidity	0.5	29	II
High humidity	3–10	170–570	III
Brick masonry, 4 in. (100 mm) thick	0.8–1.1	46–63	II
Proprietary variable-permeance membrane			
Low humidity	< 1	< 57	II
High humidity	> 10	> 570	n/a
Paint, interior alkyd, interior primer plus one flat finish coat over plaster	1.6–3.0	92–172	III
Gypsum wallboard, ½ in. (12 mm) thick			
Painted, interior latex	2–3	110–172	III
Unpainted	25	1400	n/a
Building paper, asphalt-impregnated	3.3	190	III
Housewraps			
Low permeance	5–10	290–570	III
High permeance	> 10	> 570	n/a
Plaster on metal lath, ¾ in. (19 mm)	15	860	n/a
Low-density, open cell spray polyurethane foam insulation, 2 in. (50 mm)	17	970	n/a

FIGURE B
Vapor permeance of common building materials. Values are typical, and may vary between products from different manufacturers. Class designations are according to IBC criteria, as explained in the accompanying text.

	R-Value[a]	Composition	Advantages	Disadvantages
Cellulose fiber board	2.8 *19*	A rigid, low-density board of wood or vegetable fibers and a binder, which may be coated or impregnated with asphalt	Economical; compatible with hot bitumens	Lower insulating efficiency than plastic foams; susceptible to absorption of moisture and decay
Perlite board	2.8 *19*	Granules of expanded volcanic glass and a binder pressed into a rigid board	Inert; fire resistant; compatible with hot bitumens; dimensionally stable	Lower insulating efficiency than plastic foams
Cellular glass board	2.9 *20*	A rigid, low-density board of heat-fused, closed-glass cells, with kraft paper facers	Inert; fire resistant; compatible with hot bitumens; dimensionally stable	Lower insulating efficiency than plastic foams
Mineral fiber board	4 *28*	A rigid, low-density board of rock, slag, or glass fibers and binder, usually with glass fiber mat facers	Inert; fire resistant; dimensionally stable; compatible with hot bitumens; ventilates moisture freely	
Expanded polystyrene foam board (EPS)	3.6–4.2 *25–29*	A closed-cell rigid foam of polystyrene plastic manufactured through a molding process	Less expensive than extruded polystyrene foam; compatible with asphalt at low temperatures	Combustible; high coefficient of thermal expansion; greater moisture absorption in comparison to extruded polystyrene foam; not compatible with hot bitumens
Extruded polystyrene foam board (XPS)	5 *35*	A closed-cell rigid foam of polystyrene plastic manufactured through an extrusion process	Resistant to moisture; suitable for protected membrane roof systems; compatible with asphalt at low temperatures; available in high densities suitable for roofs subject to traffic loads or heavy overburdens	Combustible; high coefficient of thermal expansion; not compatible with hot bitumens
Polyisocyanurate foam board	5–6 *35–42*	A closed-cell rigid foam of polyisocyanurate, with facings of aluminum foil, glass fiber mats, or cellulosic mats	High insulating efficiency; compatible with hot bitumens; moderate fire resistance	Expensive; rated R-value may decrease over time
Composite insulating boards	varies	Sandwich layers of foam plastic and other materials such as perlite board, mineral fiber board, and saturated felt	Combine the high insulating efficiency and moisture resistance of foam plastics with the fire resistance, structural rigidity, and/or bitumen compatibility of other materials	
Lightweight insulating concrete	1–1.5 *6.9–10*	Concretes made from lightweight mineral aggregates (perlite or vermiculite) or foaming, air-entraining agents	Inert; can easily produce a tapered insulation layer for positive roof drainage; fire resistant; dimensionally stable; often combined with embedded foam plastic insulation to achieve higher insulating efficiency	Residual moisture from mixing water can cause blistering of membrane

[a] R-values are per unit of thickness, expressed first in U.S. customary units of ft²-hr-°F/Btu-in. and second in metric units (italicized) of m²-°K/W-m. For example, for an insulation material with R-values of 5 ft²-hr-°F/Btu-in. (35 m²-°K/W-m), 3 inches (76 mm) of the material would have a total insulation value of 5 ft²-hr-°F/Btu-in. × 3 in. = R-15 or 35 m²-°K/W-m × 0.076 m = RSI-2.7.

FIGURE 16.8

A comparative summary of some rigid insulating materials for low-slope roofs.

Rigid Insulating Materials for Low-Slope Roofs

An insulating material for low-slope roofs should have high thermal resistance; adequate resistance to compression, denting, gouging, moisture decay, and fire; and, if part of a hot-applied system, resistance to melting or dissolving when hot asphalt is mopped onto it. No single material has all these virtues. Some rigid insulating materials commonly used on low-slope roofs in North America are listed in Figure 16.8, along with a summary of their advantages and disadvantages. The best choice is often a combination of materials, or a composite board that combines two or more materials into one product, to exploit the best qualities of each. A composite insulating board for installation under a built-up bituminous roof membrane might include, for example, a bottom layer of polyisocyanurate foam with high insulating value and a top layer of perlitic board resistant to hot bitumens.

If rigid insulating boards are located below the roof membrane, they may be adhered to the deck with hot asphalt or other adhesives, or fastened to the deck mechanically with screws. Insurance companies favor the mechanical fasteners because they are more secure against wind uplift (Figures 16.9 and 16.10).

Lightweight insulating concrete is an economical insulating material that also creates a nailable roof deck. Formulated with lightweight aggregates or foaming, air-entraining agents, this material has densities ranging from 20 to 40 lb/ft^3 (320 to 640 kg/m^3) compared to 145 lb/ft^3 (2320 kg/m^3) for conventional concrete. Lightweight concrete may be applied directly to corrugated steel decking or over rough concrete decks and can easily be tapered during installation to slope toward points of roof drainage. Thermal resistance per inch is not as high for this material as for most other types of roof insulation. However, plastic foam boards may be embedded in the insulating concrete to achieve higher

FIGURE 16.9
Workers bed rigid insulation boards in strips of hot asphalt over a corrugated metal roof deck. *(Courtesy of GAF Corporation.)*

FIGURE 16.10
Screws and large sheet metal washers attach insulation more securely to a metal deck than can hot asphalt. *(Courtesy of GAF Corporation.)*

FIGURE 16.11
Two built-up roof constructions, as seen from above. The top diagram is a cutaway view of a protected membrane roof over a sitecast concrete roof deck. The membrane is made from plies of felt overlapped in such a way that at no location is it less than four plies thick. Rigid foam insulation boards are bedded in hot asphalt over the membrane and ballasted with stone aggregate to keep them in place and protect them from sunlight. The bottom diagram shows rigid insulation boards attached to a corrugated steel roof deck in two staggered layers to provide a firm, smooth base for application of the membrane. A three-ply membrane is shown. In cold climates, a vapor retarder may be installed between the layers of insulation or under the insulation over a substrate board fastened to the steel deck.

insulating values within reasonable thicknesses. Lightweight concrete fill insulation contains large amounts of free water at the time it is placed. It must be cured and dried as thoroughly as possible before being covered by the roof membrane, and some form of venting to allow the escape of moisture vapor from the insulation during the life of the roof, via either topside vents or bottomside slotted *vented metal decking*, is usually advisable. Poured-in-place gypsum, another decking material popular in the past for forming lightweight, nailable sloping roof decks, is no longer used in new construction.

An *insulation coverboard* made of materials similar to those described for substrate boards may be placed over the rigid insulation prior to installation of the roof membrane. The coverboard protects the insulation from damage by foot traffic on the finished roof, increases the fire resistance of the roof assembly, and separates the membrane from incompatible insulation materials. Thin, synthetic fabrics may also be used for similar purposes.

Membranes for Low-Slope Roofs

Membranes for low-slope roofing can have life expectancies ranging from 10 to 30 years, depending on the membrane material and thickness, its exposure to extremes of temperature and UV radiation, and the quality of the roof's installation and maintenance. These membranes fall into three categories: bituminous roof, single-ply, and fluid-applied.

Bituminous Roof Membranes

Bituminous roof membranes are of two types, built-up or modified bitumen. A *built-up roof (BUR) membrane* is assembled in place from multiple layers of asphalt-impregnated roofing felt bedded in layers of hot bitumen (Figures 16.11, 16.12, and 16.13). The felt, made from cellulose, glass, or synthetic fibers, is saturated with asphalt at the factory and delivered to the site in rolls. The hot bitumen applied in the field is usually asphalt derived from the distillation of petroleum. But for dead-level or very low-slope roofs, coal tar pitch is used instead because of its greater resistance to standing water. Polymer-modified asphalts, as described later for modified bitumen roofs, may also be used. The felts are laminated in overlapping layers (plies), with the hot bitumen welding them into a unified membrane, two to four plies thick. The more plies used, the more durable the roof.

To protect the membrane from sunlight and physical wear, a layer of crushed stone or other mineral granules is embedded in the top surface. Less commonly, a built-up roof may be made from felt plies bedded in *cold-applied mastics* (*solvent-based asphalts*). These are asphalt compounds applied by spray or brush at ambient temperatures and then cured through solvent evaporation.

A *modified bitumen roof membrane* is made from factory-manufactured sheets of polymer-modified bitumens. Modified bitumens are asphalt materials to which compounds such as *atactic polypropylene (APP)* or *styrene-butadiene-styrene (SBS)* have been added to increase the material's flexibility, cohesion, toughness, and resistance to flow. Modified bitumen roof membrane sheets are also reinforced with plastic or glass fibers or fibrous mats. Sheet thickness typically ranges from 0.040 to 0.160 inch (1.0–4.0 mm).

Like a built-up roof, modified bitumen sheets are assembled in place in overlapping layers to form a multi-ply system, usually two or three plies thick. The sheets are bonded to one another in a number of possible ways: With a *torch-applied* membrane, as a roofing sheet is unrolled, an open-flame apparatus is used to thermally fuse the underside of the sheet to the top surface of the substrate or underlying sheet. A *hot-mopped* membrane relies on the application of hot asphalt to bond the sheets, a *cold process* or *cold-applied adhesive* membrane uses liquid adhesives, and a *self-adhered* membrane relies on factory-applied adhesives (Figure 16.14).

The top or *cap sheet* in a modified bitumen roof is surfaced with mineral granules, thin metallic laminates, or asphaltic or elastomeric coatings for greater resistance to ultraviolet deterioration, wear, and fire (Figure 16.15). Cap sheets with reflective white coatings that comply with cool roof

FIGURE 16.12

A base sheet of asphalt-saturated felt is installed over rigid insulation, using a machine that unrolls the felt and presses it into a layer of hot asphalt. (*Courtesy of The Celotex Corporation.*)

FIGURE 16.13

Overlapping layers of roofing felt are hot-mopped with asphalt to create a four-ply membrane. (*Courtesy of Manville Corporation.*)

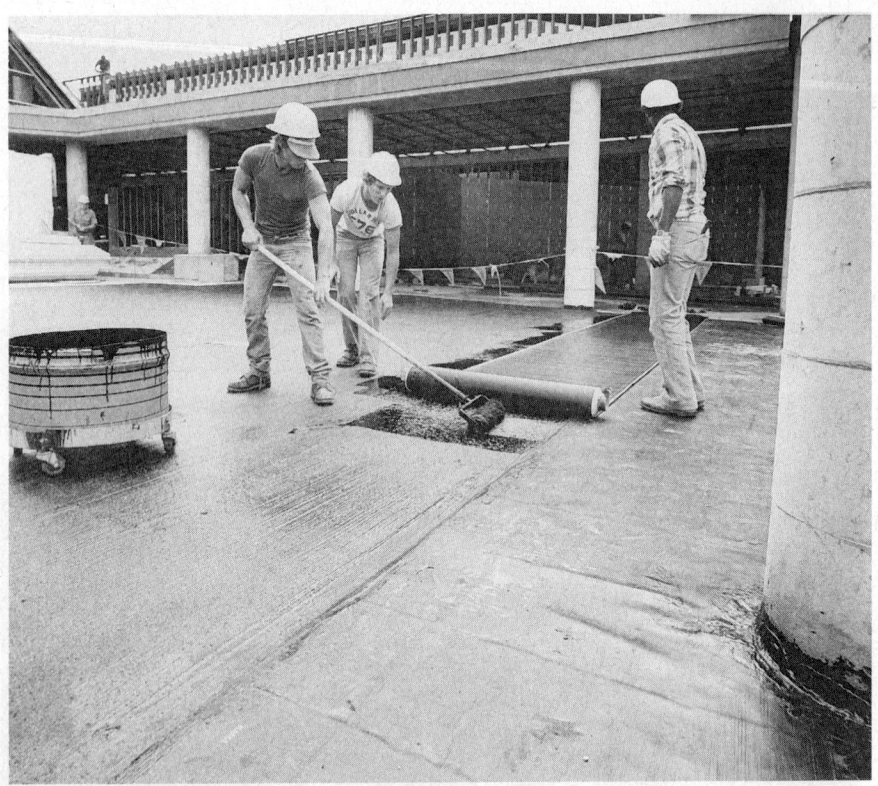

FIGURE 16.14
Roofers bond a polymer-modified bitumen membrane to a concrete deck with a cold-applied adhesive. The seams will be heat-fused together. *(Courtesy of Koppers Company, Inc.)*

FIGURE 16.15
A roofer heat-fuses a seam between two sheets of an aluminum-faced modified bitumen membrane. The aluminum foil protects the membrane from the sun and reflects solar heat from the roof. *(Courtesy of Koppers Company, Inc.)*

standards are also available. In comparison to built-up roofs, modified bitumen roofs combine the toughness and redundancy of multi-ply field application with the improved material qualities of factory-manufactured sheets. Built-up and modified bitumen systems may also be combined, with a modified bitumen cap sheet applied over several plies of built-up roofing to create a *hybrid membrane bituminous roof*. Bituminous roofing systems account for approximately 40 percent of the market for low-slope roofing membranes, with the larger portion of this share belonging to modified bitumen systems.

Single-Ply Roof Membranes

Single-ply roof membranes are a diverse and seemingly ever-evolving group of sheet materials that are applied to the roof in a single layer (Figure 16.16). Compared to bituminous roof membranes, they require less on-site labor, and especially in comparison to built-up roof membranes, they are more elastic and therefore less prone to cracking or tearing as they age. However, single-ply membranes lack the redundancy of bituminous membranes, making them potentially more vulnerable to leakage through small holes or seaming defects. Common membrane thicknesses vary from 0.035 to 0.120 inch (0.9 to 3.0 mm), depending on the membrane material type and the requirements of the roofing application. The membrane sheets come in rolls in standard widths ranging from 3 to 20 feet (0.9 to 6 m). They are affixed to the roof deck by adhesives, the weight of ballast, or fasteners concealed in the seams between sheets (Figure 16.17).

The materials used for single-ply membranes fall into two groups: thermoplastic and thermosetting. *Thermoplastic materials* can be softened by the application of heat and may be joined at the seams by heat (or solvent) welding. This welding process, which fully fuses one sheet to another, results in seams between sheets that are as strong

FIGURE 16.16

Workers unfold a wide single-ply, EPDM roof membrane sheet. Using wider sheets when roofing larger areas reduces the amount of field seaming work required. *(Photo courtesy of Carlisle SynTec Systems.)*

and permanent as the sheets themselves (Figures 16.17, 16.18, and 16.19). *Thermosetting materials*, also sometimes referred to as *elastomerics*, have a more tightly linked molecular structure and cannot be softened by heat. Thermosetting sheets are joined at the seams by liquid adhesives or pressure-sensitive tapes. Single-ply membranes, unlike multi-ply bituminous membranes, have no redundancy.

The most commonly used thermoplastic roof membrane materials are *polyvinyl chloride (PVC)* and *thermoplastic polyolefin (TPO)*. PVC roof membranes, made of PVC resins, plasticizers, stabilizers, and reinforcing fibers or fabrics, have a track record of successful performance established over many decades. However, concerns over toxic chemicals associated with the manufacture and formulation of PVC have led to questions regarding the appropriateness of PVC for use in buildings. Although PVC manufacturers continue to improve

FIGURE 16.17

A thermoplastic membrane is attached with closely spaced fasteners along its edge. The screws are long enough to reach the wooden deck beneath the insulation layers. Oversized washers keep the membrane from tearing when subjected to wind uplift forces. *(Photo by Joseph Iano.)*

FIGURE 16.18
After the first sheet of membrane is fastened in place, the next sheet is unrolled alongside so that it overlaps the edge fasteners in the first sheet. After the seam is welded, water running on the surface of the membrane cannot reach the fastener penetrations. The surface under the roofing membrane is a gypsum insulation coverboard. (*Photo by Joseph Iano.*)

FIGURE 16.19
A hot-air welder used to weld the seams in a thermoplastic membrane. The machine delivers air at the proper temperature and travels along the seam at the prescribed rate to maintain the quality of the weld. (*Photo by Joseph Iano.*)

their manufacturing processes and institute materials recycling programs in response to public concerns, segments of the construction industry have moved toward avoidance of this material in building products. TPO roof membranes are somewhat newer to the North American roofing market. They are made from blends of polyethylene, polypropylene, and ethylene-propylene rubber polymers reinforced with fibers or fabrics. TPO membranes exhibit good resistance to heat and ultraviolet (UV) radiation, characteristics more commonly associated with thermosetting membranes, but as thermoplastics, their seams can be heat welded. PVC and TPO together account for approximately 20 percent of the market for low-slope roofing materials. Other, less widely used thermoplastic roof membrane materials include *ketone ethylene ester (KEE)* and a class of materials referred to as PVC alloys or PVC compounded thermoplastics, made from various blends of PVC and other polymers. Thermoplastic roof membranes are generally available in a broad range of colors, including reflective white for cool roofs.

The most commonly installed thermosetting roof membrane material is *ethylene propylene diene monomer (EPDM)*, a synthetic rubber that may or may not include fiber or fabric reinforcing. EPDM has a stable chemical structure with excellent resistance to ozone, heat, UV radiation, and weathering. It is most commonly black, but is also available in cool white from some manufacturers. Because EPDM cannot be heat welded, seaming is performed with tapes or adhesives. Like PVC, EPDM has a many-decades-long track record of successful performance. It is the most widely used material in North America for low-slope roofing of any type, accounting on its own for more than one-third of the market for such applications. Other thermosetting roof membrane materials include chlorosulfonated polyethylene (CSPE) and polyisobutylene (PIB).

Fluid-Applied Roof Membranes

Fluid-applied roof membranes are frequently used for domes, shells, and other complex shapes that are difficult to roof by other means. Such shapes are often too flat on top for shingles but too steep on the sides for built-up roof membranes; if doubly curved, they are difficult to cover with preformed sheets. Fluid-applied membranes are installed with a roller or spray gun, usually in several coats, and cure to form a solid membrane. Materials applied by this method include neoprene (with a protective layer of CSPE), silicone, polyurethane, butyl rubber, asphalt emulsion, and polymethyl methacrylate (PMMA).

Fluid-applied membranes are also used as a waterproofing layer over sprayed-on polyurethane foam insulation in proprietary roofing systems designed for surfaces that are hard to fit with flat sheets of insulation and membrane. These systems are also a convenient means for adding thermal insulation and a new roof membrane over existing but deteriorated built-up roofs on any shape of building.

Ballasting and Traffic Decks

Roof membranes may be covered after installation with a ballast of loose, rounded stone aggregate ranging in size from 1½ to 2½ inches

(38 to 64 mm) in diameter or with precast concrete paver blocks. The ballast holds the membrane down against wind uplift and protects the membrane from ultraviolet light and physical wear. It may also contribute to the fire resistance of the roof covering.

Traffic decks are installed over flat roof membranes for walks, roof terraces, and sometimes driveways or parking surfaces. Two different systems are used: In one, plastic pedestals are placed on top of the roof membrane to support the corners of square paving stones or slabs with open joints (Figure 16.20). In the other, a drainage layer of gravel or pervious concrete is leveled over the membrane, and open-jointed paving blocks are installed on top. In

either detail, water falls through the joints in the paving and is caught and drained away by the membrane below. Notice that the membrane is not pierced in either system. To protect the membrane from accidental damage, a protective layer made of roof membrane or other material is frequently placed over the membrane before the traffic deck components are installed.

Edge and Drainage Details

Some typical details of low-slope roofs are presented in Figures 16.21 through 16.30. All are shown with built-up roof membranes. Details for single-ply membranes are similar, though the membranes themselves are only one-ply in thickness.

FIGURE 16.20

Plastic pedestals support stone or precast concrete paving blocks over a low-slope roof membrane, permitting the use of a low-slope roof as an outdoor terrace. Each high-density polyethylene pedestal supports the adjacent corners of four paving blocks. The vertical fins on the pedestal provide a uniform drainage space between the blocks. Matching polyethylene leveling plates (not shown) can be placed under the pedestals to compensate for irregularities in the roof surface. Adjustable-height pedestals that rely on threaded, telescoping sections are also available. (*Photo courtesy of Envirospec, Inc., Buffalo, NY.*)

Metal roof edge in 10'(3 m) maximum lengths

Cover plate at joints in the roof edge

Base flashing

Wood curb

Sealant

Insulation

Roof membrane

Deck

FIGURE 16.21

A roof edge for a conventional built-up roof. The membrane consists of four plies of felt bedded in asphalt with aggregate surfacing. The *base flashing* is composed of two additional plies of felt that seal the edge of the membrane and reinforce it where it bends over the curb. The curb directs water toward interior drains or scuppers rather than allowing it to spill over the edge. The exposed vertical face of the metal roof edge is called a *fascia*.

FIGURE 16.22

A roof edge system for low-slope roofs. The perforated metal strip is fastened to the roof with a mastic adhesive that oozes through the perforations to create a tighter bond. When the adhesive has hardened, a galvanized steel curb is fastened in place with the tabs of perforated metal and an aluminum roof edge is hooked on, with a backup piece at the end joints as shown to prevent leakage. Lastly, the roof edge and the membrane are locked in place simultaneously by installing a clamping strip that engages the hook on the top of the aluminum roof edge. The clamping strip is held in place by screws that pass through the edge of the membrane into the galvanized curb, as seen at the top of the photograph. (*Product of W. P. Hickman Company, Asheville, NC, USA.*)

A notched metal angle keeps the aggregate on the roof

A sheet metal gutter and downspout catch and drain the water

2'-0" (600 mm) maximum

Metal roof edge

Stripping

Roof membrane

Deck

Insulation

FIGURE 16.23

Detail of a *scupper*. The curb is discontinued to allow water to spill off the roof into a gutter and downspout. Additional layers of felt, called *stripping*, seal around the sheet metal components. Most roofs use interior drains (Figure 16.29) as their primary means of drainage, with scuppers more frequently used as secondary drainage to limit ponding in the case of a clog in the primary drain.

Flexible
waterproof
expansion joint
cover

Vapor retarder

Compressible
insulation

Base
flashing

Wood curb
and cant

Separation in
building
Structure

FIGURE 16.24
A building separation joint in a low-slope
roof. Large differential movements
between the adjoining parts of the
structure can be tolerated with this
type of joint because of the ability of
the flexible joint cover to adjust to
movement without tearing. High curbs
keep standing water away from the edge
of the membrane, which is sealed with a
two-ply base flashing.

Sheet metal
flashing

Wood curb and
cant strip

Base flashing

FIGURE 16.25
An area divider is designed to allow for
some movement only in the membrane
itself, not in the entire structure. It is
used to subdivide a very large membrane
to allow for thermal movement.

The coping is held in place by steel dowels set in the wall by the mason and capped by the sheet metal contractor

Cut stone or precast concrete coping

A drip grove prevents water from running back under the coping

A continuous through-wall flashing prevents leakage through joints or cracks in the coping

A continuous metal counter flashing is interlocked with the through-wall flashing

The turned-up edge of the membrane is sealed to the wall with fabric and mastic

A wood or fiber cant strip eases the bending of the membrane

Base flashing

Roof membrane

Insulation

Deck

FIGURE 16.26

A *parapet*—a low wall that projects above the roof edge—with conventional *counterflashing* and *coping. Cant strips*, as seen in this and many other illustrated details in this chapter where the roof membrane turns up onto a vertical surface, are commonly used with bituminous roofing, in which the less flexible membranes and multiple plies cannot easily make sharp right-angle bends. Single-ply membranes usually do not require cant strips at similar conditions.

Interactive Exercise

FIGURE 16.27
A parapet coping system. The perforated metal channel is fastened to the masonry with a mastic adhesive or mechanical fasteners. Sections of metal coping are snapped over the channel, with a special pan element beneath the joints to drain leakage. *(Product of W. P. Hickman Company, Asheville, NC, USA.)*

FIGURE 16.28
A conventional cast iron interior *roof drain* for a low-slope roof. Two plies of felt stripping seal around the sheet metal flashing.

Strainer

Stripping

Copper or lead flashing, at least 30" (750 mm) square

The drain unit clamps securely to the deck and to the roof membrane

Drain pipe

The insulation tapers toward the drain

Roof membrane

Deck

FIGURE 16.29
A proprietary single-piece roof drain made of molded plastic. *(Product of W. P. Hickman Company, Asheville, NC, USA.)*

Sealant

A metal draw band clamps the flashing to the pipe

Sheet metal flashing

The top of the curb is sealed with fabric and mastic

Base flashing

Wood curb and cant strip

FIGURE 16.30

A *roof penetration* for a plumbing vent stack. Notice how this and all the previous edge and penetration details for a flat roof use the curb, cant strip, and stripping to keep standing water away from the edges of the membrane.

Structural Panel Metal Roofing

Manufacturers of prefabricated metal building systems have developed proprietary systems of metal roofing panels that can be used as low-slope roofs at pitches as shallow as ¼:12 (1 in 48, or 2 percent). These systems can be applied not only to prefabricated metal buildings but to buildings constructed of other materials as well (Figures 16.31, 16.32, and 16.33). They are called *structural panel metal roofing* because the folded shape of the metal roofing gives it sufficient stiffness to support itself and normal rain and snow loads between purlins without the need for a supporting deck beneath. This name also distinguishes it from *architectural sheet metal roofing*, the traditional forms of metal roofing that are not self-supporting. Architectural sheet metal roofing is utilized largely on steep roofs and is described later in this chapter.

FIGURE 16.31
As a first step in reroofing a building with a structural metal roof, steel Z-purlins are erected over the old roof on tubular metal posts. *(Courtesy of Metal Building Manufacturers Association.)*

FIGURE 16.32
A proprietary metal clip is used to fasten the metal roofing sheets to the Z-purlins while allowing for thermal movement in the sheets. The plastic foam collar avoids thermal bridging. *(Courtesy of Metal Building Manufacturers Association.)*

FIGURE 16.33
The completed structural metal roof has a slope of only ¼:12 (1:48). The numerous penetrations for plumbing vents, air vents, and ductwork are typical of low-slope roofs. *(Courtesy of Metal Building Manufacturers Association.)*

STEEP ROOFS

Roofs with a pitch of 2:12 (17 percent) or greater are referred to as steep roofs. Traditional *thatch*, occasionally still used for the restoration of historic structures, is an attractive and effective, but highly labor-intensive, roofing material consisting of bundles of reeds, grasses, or leaves (Figures 16.1 and 16.2). Modern steep-slope shingle and sheet metal roofs of many types are common to every type of building and range from the most economical roof coverings to the most durable and costly.

Insulation and vapor retarders for steep roofs are most often installed below the roof sheathing. Alternatively, the underside of the roof deck may be left exposed in the finished construction, with a vapor retarder and rigid insulation boards installed above. In this second scenario, a nailable surface for the finish roofing is provided by fastening a layer of plywood or OSB over the insulation boards as a *nail base* for the roofing, or a composite insulation board with a nailable panel factory-bonded to the rigid insulation may be used. Examples of both of these approaches to steep-slope roofing are illustrated in Chapters 6 and 7.

In cold climates, steep roofs may have a tendency to form ice dams at the eaves under wintertime conditions. Where the risk of ice damming is high, building codes require rubberized underlayment or other ice barrier material along the eaves to prevent trapped water from entering the building (see Chapter 6).

Shingles

The word *shingle* is used here in a generic sense to include wood shingles and shakes, asphalt shingles, slates, clay tiles, and concrete tiles. What these materials have in common is that they are applied to the roof in small, overlapping units with staggered vertical joints.

Wood shingles are thin, tapered slabs of wood sawn from short pieces of tree trunk with the grain of the wood running approximately parallel to the face of the shingle (Figure 16.58). *Wood shakes* are thicker than shingles, exhibit a rougher texture, and are frequently split from the wood rather than sawn. Most wood shingles and shakes in North America are made from red cedar, white cedar, or redwood to take advantage of these woods' natural decay resistance.

Traditionally, wood shingles and shakes were installed over *skip sheathing* (*spaced sheathing*), spaced boards running perpendicularly across the sloped roof rafters. Skip sheathing allows air flow underneath the shingles that prevents condensation and deterioration of the shingles (Figure 16.34). When applied over continuous, solid sheathing, airflow

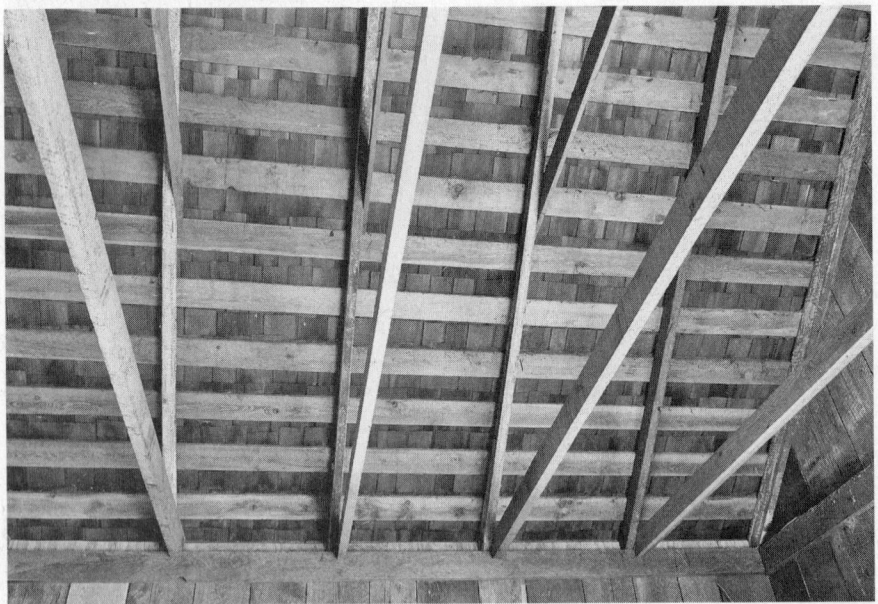

FIGURE 16.34
Wood shingles applied over skip sheathing, as seen from below. Even without a continuous roof sheathing or roofing underlayment, this shingle installation provides reliable protection from the weather. *(Photo by Joseph Iano.)*

FIGURE 16.35
Wood shingles and shakes are installed
so that any part of the roof is always
covered by at least three shingles.
Coverage recommendations also vary
with the slope of the roof and the quality
of the wood used in the shingles or
shakes. *(Photo by Joseph Iano.)*

underneath the wood shingles and shakes can be provided by adding spaced wood strips over the sheathing or with a *breather mat*, a wiry plastic mat placed under the shingles that creates a thin airspace. Wood roof coverings are moderately expensive and are not highly resistant to fire unless the shakes or shingles have been pressure treated with fire-retardant chemicals. They eventually fail from erosion of the wood fibers and may be expected to last 15 to 25 years under average conditions (Figures 16.35 and 16.36).

Asphalt or *composition shingles* are die-cut from heavy sheets of asphalt-impregnated felt faced with mineral granules to act as a wearing layer and decorative finish. Most felts are based on glass fibers, but some still retain the older cellulose composition. The most common type of asphalt shingle, which covers probably 90 percent of the single-family houses in North America, is 12 inches by 36 inches (305 mm by 914 mm) in size. (A metric-size shingle 337 mm by 1000 mm is also widely marketed.) In the most popular pattern, each shingle is slotted twice to produce a roof that looks as though it were made of smaller shingles (Figures 16.37–16.41). Other sizes and many other styles are available, including thicker

shingles that are laminated from several layers of material. Asphalt shingles are inexpensive to buy, quick to install, moderately fire resistant, and have a life expectancy of 15 to 25 years, depending on their composition.

The same sheet material from which asphalt shingles are cut is also manufactured in rolls 3 feet (914 mm) wide as asphalt *roll roofing*.

Roll roofing is very inexpensive and is used primarily on utility and agricultural buildings. Its chief drawbacks are that thermal expansion of the roofing or shrinkage of the wood deck can cause unsightly ridges to form in the roofing and that thermal contraction can tear it.

Slate shingles for roofing are delivered to the site split, trimmed to size,

FIGURE 16.36
Shake installation over a new roof deck. Strips of asphalt-saturated felt have been placed in advance with their lower edges unfastened. Each course of shakes is laid out, slipped up under its felt strip, then quickly fastened with pneumatic powered staplers as the roofers walk across the roof. *(Photograph courtesy of Senco Products, Inc.)*

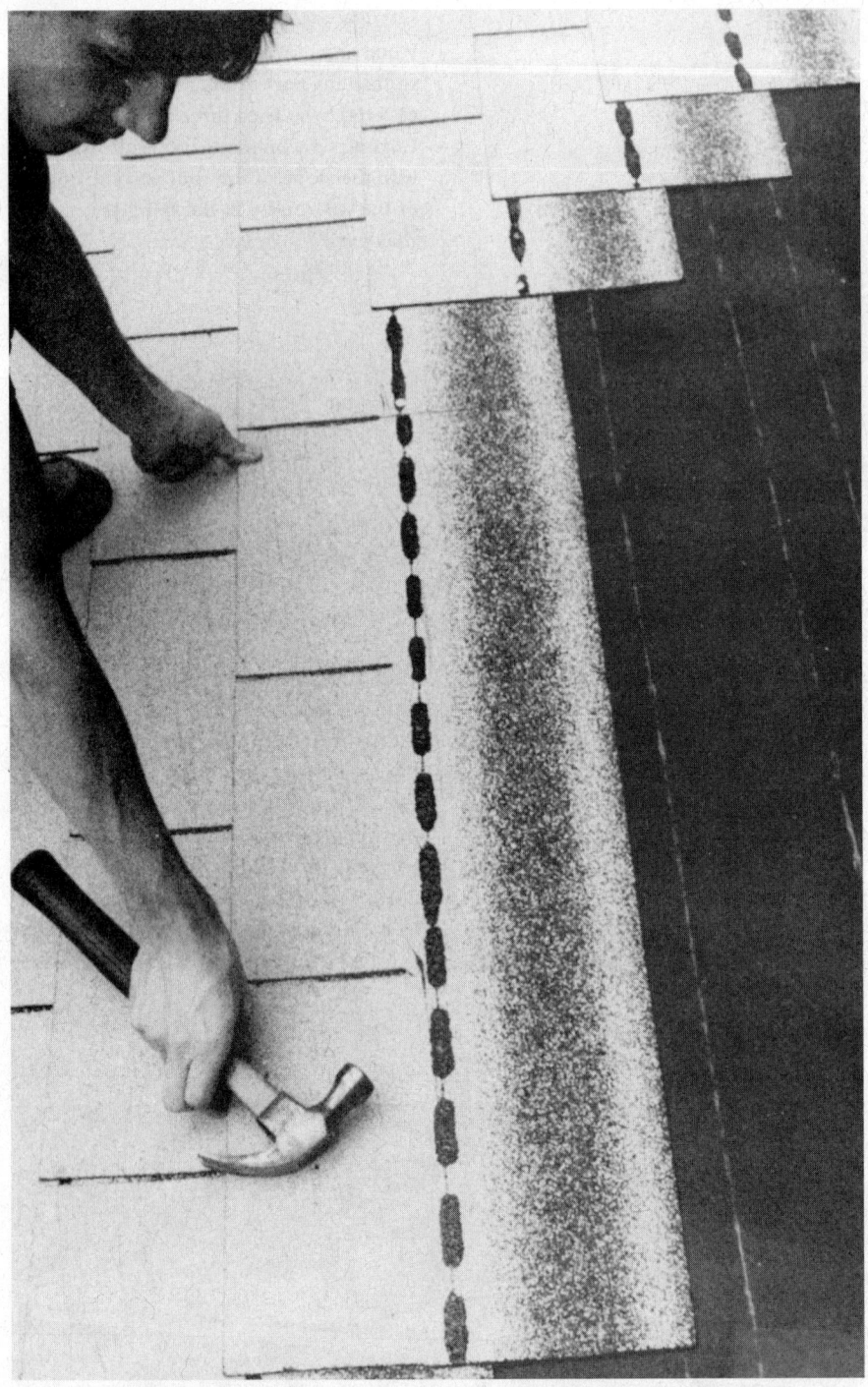

FIGURE 16.37
Installing asphalt shingles. To give a finer visual scale to the roof, the slots make each shingle appear to be three smaller shingles when the roof is finished. Many different patterns of asphalt shingles are available, including ones that do not have slots. (*Photo by Edward Allen.*)

and punched or drilled for nailing (Figures 16.42 and 16.43). They form a fire-resistant, long-lasting roof that is suitable for buildings of the highest quality. Its first cost is high, but a slate roof typically lasts 60 to 80 years.

Clay tiles have been used on roofs for thousands of years. It is said that the tapered barrel tiles traditional to the Mediterranean region (similar to the mission tiles in Figure 16.44) were originally formed on the thighs of the tilemakers. Many other patterns of clay tiles are now available, both glazed and unglazed. *Concrete tiles* are generally less expensive than those of clay and are available in some of the same patterns. Tile roofs are heavy, durable, highly resistant to fire, and relatively expensive in first cost. Expected lifetimes range from 30 to 75 years, depending on climate and the resistance of the tiles to water absorption.

Other materials used for roof shingles include sheet metal, rubber, fiber-reinforced cement, and plastic. Each type of shingle must be laid on a roof deck that slopes sufficiently to ensure leakproof performance. Minimum slopes for each material are specified by the manufacturer and the building codes. The minimum slope for a standard asphalt shingle roof is usually 4:12 (33 percent); with specially protective underlayments, slopes as low as 2:12 (17 percent) may be acceptable in some circumstances. For any shingle type, slopes greater than the minimum can better resist leakage from water driven up the roof surface by heavy winds.

FIGURE 16.38
Laminated asphalt shingles create a pleasing, textured roof surface. The cupola roof is copper sheet metal with standing seams.

Asphalt-saturated felt paper

Nailable deck (plywood or boards)

Self-sealing strips soften in the sun's heat and bonds the shingles together against wind uplift

Each successive course starts with a shingle cut 1/2 tab shorter than the first shingle in the course below, to stagger the slots from one course to the next

The starter course has its tabs cut off to maintain a uniform thickness of roofing. It provides a waterproof layer under the slots in the first course of shingles

Metal drip edge supports the projecting edge of the shingles and forces water to drop clear of the fascia

Profile of metal drip edge

FIGURE 16.39
Starting an asphalt shingle roof. As explained in Chapter 6, building codes require the installation of an ice barrier beneath the shingles along the eave in regions with cold winters prone to ice damming. Where required, the ice barrier would replace the lowest course of asphalt-saturated felt paper shown in this illustration.

Single shingle tabs with tapered butts close the ridge

Aluminum step flashing prevents leakage at wall intersections

FIGURE 16.40
Completing an asphalt shingle roof. A metal or plastic ridge vent strip (see Chapter 6) is often substituted for the shingle tabs on the ridge to provide an outlet for ventilation under the roof sheathing.

A. OPEN VALLEY

(a)

B. WOVEN VALLEY

(b)

C. CLOSED CUT VALLEY

(c)

FIGURE 16.41

A *valley* is formed in a roof where two sloping roof planes meet above an inside corner of the building. Three alternative methods of making a valley in an asphalt shingle roof are shown here. (*a*) The *open valley* uses a sheet metal flashing; the ridge in the middle of the flashing helps prevent water that is coming off one slope from washing up under the shingles on the opposite slope. The *woven valley* (*b*) and *cut valley* (*c*) are favorites of roofing contractors because they require no sheet metal. The solid black areas on shingles in the open and cut valleys indicate areas to which asphaltic roofing cement is applied to adhere shingles to each other.

Architectural Sheet Metal Roofing

Thin sheets of metal have been used for roofing since ancient times and remain a popular roofing material to this day. They are installed using ingenious systems of joining and fastening to maintain watertightness (Figures 16.45–16.50). Seams between sheets must be spaced closely enough to hold the sheets securely against wind uplift and minimize *oil canning* (unsightly waviness). These seams also create strong

FIGURE 16.42
Splitting slate for roofing. The thin slates in the background will next be trimmed square and to dimension, after which nail holes will be punched in them. *(Photo by Flournoy. Courtesy of Buckingham-Virginia Slate Corporation.)*

FIGURE 16.43
A slate roof during installation. *(Courtesy of Buckingham-Virginia Slate Corporation.)*

visual patterns that can be manipulated by the designer to emphasize the qualities of the roof shape. Architectural sheet metal roofing is relatively high in first cost but, when properly installed, can be expected to last for many decades.

Various types of metal may be used in the production of architectural sheet metal roofs:

• *Lead* is a soft, easily formed, very long-lasting metal that oxidizes over time to a dull white color. It is also toxic to humans and its use in building materials is discouraged by leading green building programs. Where it is used or encountered, special health precautions are required during its handling.

• *Copper* is a relatively soft metal that turns a beautiful blue-green in clean air and a dignified black in an industrial atmosphere; various chemical treatments can be used to obtain and preserve the desired color.

• *Lead-coated copper* sheet has historically been used to combine the greater strength of copper with the gray-white color of lead and to avoid the staining of wall materials by oxides of copper that can occur with rainwater runoff from uncoated copper roofs. As noted earlier, the presence of lead in building materials is discouraged by green building programs.

• *Zinc* metal roofing is a long-lasting roofing metal made of zinc alloyed with small amounts of copper and titanium to improve its workability. It normally ages to a dark gray color and can also be treated in various ways to alter or preserve its appearance.

MISSION TILES

SHINGLE TILES

FIGURE 16.44
Two styles of clay tile roofs. The mission tile has ancient origins.

FIGURE 16.45
A roofer forms standing-seam pans from a coil of copper sheet metal and a portable pan-forming machine mounted in the back of a pickup truck. *(Photo by Joseph Iano)*

FIGURE 16.46
Standing-seam copper roofs. *(Designer: Emil Hanslin. Courtesy of Copper Development Association, Inc.)*

FIGURE 16.47
An automatic roll seamer, moving under
its own power, locks standing seams in
a copper roof. A cleat is just visible at
the lower right. *(Photo courtesy of DBI/
SALA, Red Wing, MN; Courtesy of Copper
Development Association, Inc.)*

*Asphalt saturated
felt paper*

Metal pans

Cleats

Rosin paper

FIGURE 16.48
Installing a *flat-seam metal roof.* The
three diagrams at the bottom of the
illustration show the steps in forming
the seam, viewed in cross section. The
cleats, which fasten the roofing to the
deck, are completely concealed when the
roof is finished. To ensure that seams
remain watertight, they are sealed with
elastomeric sealant. For low slopes,
seams may be soldered.

Step 1:   Each pan is formed in
the sheet metal shop
with folded edges.

Step 2:   Sheet metal cleats
interlock with the
folded edges and are
nailed to the deck.
The cleat is folded
back over the nail
head to protect the
pan.

Step 3:   The next pan is
interlocked with the
first. When all pans
are in place, the
edges are beaten flat
and soldered or
sealed.

Metal pans

Cleats

Rosin paper

Step 1: The nailed cleats and pans are aligned.

Step 2: The edges of the pans and the cleats are rolled together into a standing seam.

Step 3

Step 4

FIGURE 16.49
Installing an architectural *standing-seam metal roof*. For low slopes, seams may be made taller or sealed with elastomeric sealant.

Metal pans

Wood battens

Cleats

Cleat

Wood batten

Nails

Step 1

Step 2

Step 3

Step 4

FIGURE 16.50
Installing a *batten-seam metal roof*. The battens are tapered in cross section to allow for expansion of the roofing metal. For low slopes, seams may be sealed with elastomeric sealant.

• *Stainless steel* and *titanium* are strong and long-lasting but less easily worked than other roofing metals. Both are silvery-white.

• *Zinc-tin alloy coated stainless steel* has a darker, duller appearance than uncoated stainless steel. This material is very similar in appearance to—and is sometimes equated with—a lead-tin alloy coated metal called *terne-coated stainless steel* that is no longer manufactured but may still be found on older buildings.

The metals in the preceding list all form self-protecting oxide coatings (patinas) that provide long-lasting resistance to corrosion. They are usually installed uncoated and allowed to patinate naturally. Other, less expensive metals, which are not as long-lasting in an uncoated condition, are commonly coated at the factory with high-performance organic (paint-like) coatings that extend their life expectancy and provide a wide range of color choice. These include *aluminum*, *metallic-coated steel* (steel coated with alloys of zinc or zinc-aluminum), or *zinc-tin alloy coated steel*, and even, occasionally, plain or *ferrous steel* without any protective metallic coating. For more about architectural metals, see Chapter 12.

The thickness of steel sheet traditionally has been specified by *gauge* (also spelled *gage*), a system of whole numbers in which lower numbers correspond to greater metal thickness. However, due to the absence of uniformity in the translation between gauge and actual metal thickness, sheet metal standards now specify metal thickness in decimal or fractional inches. The exception to this rule is copper, which is specified by weight, expressed in ounces per square foot (0.092 m^2). In general, thicker metal sheets are longer-lasting, less prone to distortion, often more difficult to form into shapes, and more expensive than thinner sheets. Figure 16.51 lists typical thicknesses for some common sheet metal roofing materials.

When metal roofing is custom-fabricated by the sheet metal contractor from unformed metal stock, it is specified as sheet metal roofing (Figure 16.45). Alternatively, roofing panels may be factory-formed into families of shapes that can be selected from a catalog. In this case, the roofing is specified as metal roof panels. Factory-made sheet metal roof panels are usually made from aluminum or metallic-coated steel and finished with organic coatings. They may rely on interlocking seams with concealed fastener systems similar in appearance to traditional standing-seam or batten-seam sheet metal roofing (Figure 16.52), or they may consist

Steel Sheet	Gauge	Nominal Thickness of Sheet Metal	
		Metallic-coated steel	Stainless steel
	18	0.052 in.	0.050 in.
		1.32 mm	*1.27 mm*
	20	0.040 in.	0.038 in.
		1.02 mm	*0.95 mm*
	22	0.034 in.	0.031 in.
		0.86 mm	*0.79 mm*
	24	0.028 in.	0.025 in.
		0.71 mm	*0.64 mm*
	25	0.025 in.	0.022 in.
		0.64 mm	*0.56 mm*
	26	0.022 in.	0.019 in.
		0.56 mm	*0.48 mm*
Copper Sheet	Weight per square foot (0.092 m^2)	Nominal Thickness of Sheet Metal	
	24 oz.	0.032 in.	
		0.82 mm	
	20 oz.	0.027 in.	
		0.68 mm	
	16 oz.	0.022 in.	
		0.55 mm	
Aluminum Sheet		Nominal Thickness of Sheet Metal	
		0.060 in.	
		1.52 mm	
		0.050 in.	
		1.27 mm	
		0.040 in.	
		1.02 mm	
		0.032 in.	
		0.81 mm	

FIGURE 16.51

Thicknesses of sheet metals commonly used in architectural sheet metal roofing.

FIGURE 16.52
The raised seams in the architectural panel metal roof accentuate the complex roof geometries of the International Center in Brattleboro, Vermont. *(Architect: William A. Hall Partnership. Photo by Stanley Jesudowich.)*

of simpler corrugated or folded profiles fastened with exposed screws and rubber washers (Figure 16.53). Manufactured metal roof panels are generally less expensive than custom-formed sheet metal roofing.

The minimum recommended slope for architectural sheet metal roofing is 1:12 (8 percent) for standing-seam roofing and 3:12 (25 percent) for flat-seam or batten-seam roofing. In some configurations, lower slopes can be achieved with special seam treatments, such as making seams taller, soldering seams, or treating seams with elastomeric sealant. As noted earlier in this chapter, some structural configurations of structural sheet metal roofing are designed to remain watertight at slopes otherwise reserved for low-slope membrane roof systems.

FIGURE 16.53
This lightweight steel Quonset hut structure is clad with curved roof panels on its roof and flat panels on its walls. The panels are attached with exposed mechanical fasteners made watertight with resilient washers.

Protection from Corrosion Between Dissimilar Metals

When different metals come in contact in the presence of moisture, corrosion can occur. This phenomenon is discussed in greater detail later in this chapter. The ideal way to avoid such corrosion is to use the same metal for every component of a sheet metal roof system, including its fasteners, anchor clips, roofing metal, flashings, gutters, and downspouts. However, sheet metal roofs made with relatively soft copper, lead, or zinc are frequently anchored with harder, stronger stainless steel clips and fasteners. Because the stainless steel components are relatively small in area and electrochemically noble (passive) in relation to these other metals, the risk of corrosion between metals is minimized.

SUSTAINABILITY IN ROOFING

The roof can contribute in many ways to the sustainability of a building:

• A roof can capture rainwater and snowmelt and conduct them to a cistern, tank, or pond for use as domestic water, industrial water, or irrigation.

• A properly proportioned overhang can shade south-facing windows from the high summer sun but admit warming light from the low winter sun.

• A light-colored roof covering, if kept clean, can reflect half or more of the solar radiation striking its surface, improving occupant comfort and reducing the heating load on the occupied space below. Even darker-hued roof materials, when coated with specially formulated cool color pigments, can reflect 25 percent or more of solar radiation. Such cool roofs can reduce cooling energy costs for buildings by 10 to 25 percent and extend the life of the roofing materials.

• Reflective roofs that reduce the absorption of solar heat can decrease the elevation of air temperatures in densely built areas, thereby reducing a building's indirect contributions to smog, degraded air quality, environmental discomfort, and other heat island effects.

• In a hot climate, a shading layer above a roof, with a freely ventilated space between, can eliminate most solar heat gain through the roof surfaces. The shading layer might consist of latticework, fabric, or corrugated metal; the exact material is less important than providing both shading and ventilation.

• A roof surface can support flat-plate solar heat collectors used to reduce building heating costs or arrays of photovoltaic solar modules to provide electrical power. Electrical power for building use can also be produced from thin-film photovoltaic materials laminated directly onto a variety of conventional roof coverings.

• Roof membranes coated with photocatalytic materials can convert various air pollutants into harmless organic materials in the presence of sunlight. Rain then washes the resulting organic material off the roof.

Roof Insulation

Thermal insulating materials in roofs and walls are probably the most cost-effective, planet-saving materials used in buildings. They increase occupant comfort by moderating the radiant temperatures of ceilings and walls. They reduce heating and cooling energy requirements to a fraction of what they would otherwise be. They pay for themselves through energy savings in a very short period of time. Additional considerations related to the sustainability of insulation materials used in roofing are discussed in Chapter 7.

Low-Slope Membranes

Low-slope roof membranes have varying impacts on the environment:

• Bituminous roofing is largely based on asphaltic compounds derived from coal and petroleum.

COOL ROOFS

Roofs are exposed to solar radiation daily, and as that radiation is absorbed and converted to heat, the temperature of the roof covering rises. Depending on the intensity of the radiation and the portion of it retained by the covering, roof surfaces may routinely reach temperatures of 150 degrees Fahrenheit (65°C) or higher. High roof temperatures can lead to overheating of interior spaces; reduced comfort for building occupants; increased building energy consumption; the need for larger, more expensive cooling equipment; shortened life span of roofing

materials; and an increased contribution to urban heat island effects through elevation of the surrounding air temperature. Selecting a *cool roof* covering that minimizes such heating can reduce these effects.

Solar heating of roofs is principally affected by two properties of the roofing material. A material's *solar reflectance*, or *albedo*, is a measure of its tendency to reflect solar radiation rather than absorb it. Solar reflectance is measured on a unitless scale from 0 to 1, where 1 represents a material that reflects all solar radiation and 0 represents one that absorbs all solar radiation. A higher solar reflectance corresponds to a cooler roof. *Thermal emittance* is a

measure of a material's capacity to radiate infrared heat energy and cool itself as its temperature rises. Like solar reflectance, thermal emittance is measured on a scale of 0 to 1, and a higher thermal emittance implies a cooler roof.

Cool roof criteria differ among energy conservation standards and green building programs. Requirements for the U.S. Environmental Protection Agency's (EPA) Energy Star program are based solely on a roof covering's solar reflectance, measured both when the covering is new and after it has weathered. Requirements for the U.S. Green Building Council's LEED for New Buildings are based on a roof covering's *solar reflective index*

• Roofing operations with hot asphalt and pitch give off plentiful quantities of fumes that are decidedly unpleasant and potentially unhealthful, but once the roof has cooled, these emissions subside.

• Most roofing felts today are made with cellulose or glass fibers, but very old felts in buildings being demolished or reroofed may contain asbestos, a proven carcinogen.

• The various rubber and plastic formulations used in single-ply membranes generally utilize petroleum as the primary raw material. Each has its own characteristics with regard to embodied energy and outgassing.

• Chemicals associated with the manufacture, use, and disposal of polyvinyl chloride are associated with a variety of severe health effects and serious environmental problems. PVC is on the Living Building Challenge Red List, meaning it should not be used at all in buildings meeting this standard. The use of materials containing vinyl chloride, an ingredient in the manufacture of PVC, is also discouraged in a variety of the LEED standards.

• Adhesive bonding, solvent welding, and heat welding of seams may give off volatile organic compounds (VOCs). Some single-ply membrane materials also emit VOCs.

• Materials from demolition of built-up roof membranes are generally incinerated or taken to landfills. Thermoplastic single-ply membranes can be recycled, although at present most are not. Thermosetting membranes cannot be recycled.

• Low-slope roofs may be developed as green roofs, in which the membrane is covered with soil and plants. These living constituents of the roof consume carbon dioxide, generate oxygen, shield the membrane and the inhabited space below from temperature extremes, cool the roof surface by evaporation and transpiration, delay the passage of stormwater into sewers and reduce its volume, and create a pleasant roofscape.

Steep Roof Materials

Steep roof materials—shingles, slates, tiles, sheet metal—come from a variety of sources with different environmental impacts:

• Asphalt shingles consist mostly of petroleum.

• The environmental impact of wood shingles is much the same as that of other wood products; see Chapter 3.

• Slates are quarried stone, and concrete tiles are, of course, concrete, the sustainability of which are discussed in Chapter 9 and Chapter 13, respectively.

• Copper, stainless steel, aluminum, terne, and lead roofing materials originate as ores that must be mined, refined, and manufactured into finished form, all at various environmental costs.

• Metal roofing materials can be recycled. There is little recycling of the other steep roof materials. There are available, however, proprietary shingles made almost entirely of recycled tires or other recycled materials.

(SRI). SRI is a measure of solar heating potential, derived according to ASTM E1980, that accounts for a material's reflective and emittive properties, as well as for its ability to lose heat through thermal conductance to the surrounding air. Two roofing materials with the same SRI are expected to achieve the same surface temperature under comparable exposures. Higher SRI values correspond to cooler roof coverings, with an SRI value of 0 corresponding to a standard reference black surface and a value of 100 corresponding to a standard reference white surface (Figure 16.54).

Comparative solar heating properties for common roofing materials are listed in Figure 16.55. Product-specific data can be obtained from the manufacturer's product literature or the Cool Roof Rating Council, an independent organization that maintains a roof material rating program and publishes the properties of tested products (see the list of web sites at the end of this chapter). In comparison to traditional dark-colored EPDM or bituminous membranes, reflective cool membranes on low-slope roofs can reduce roof surface temperatures by as much as 50 to 75 degrees Fahrenheit (25–40°C) and cut building cooling costs by an estimated 15 to 25 percent. Cool roofing materials on steep roofs have the potential to save an estimated 5 to 10 percent of building cooling costs.

Cool color roofing materials are nonwhite in color but nevertheless reflect a significant portion of the sun's radiation. Cool colors are formulated with pigments that are selectively reflective to different portions of the solar spectrum. They are highly reflective of *near-infrared (NIR) radiation*, an invisible component of solar radiation that accounts for more than half of the total heat energy radiated by the sun, while they remain selectively absorptive in the visible light spectrum, which accounts for their apparent color. Cool color pigments can be applied to aggregate granules used to coat asphalt shingles, as well as to sheet metal, clay or concrete tile, fiber-cement shingles, and other

	EPA Energy Star	USGBC LEED
Low-Slope Roofs	Minimum solar reflectance: 0.65 unaged 0.50 aged 3 years	Minimum SRI: 78
Steep Roofs	Minimum solar reflectance: 0.25 unaged 0.15 aged 3 years	Minimum SRI: 29

FIGURE 16.54
Cool roof requirements for low-slope and steep roofs. For EPA Energy Star programs, aged values are based on testing of material samples taken from installed, weathered roofs or from samples that have undergone accelerated aging to simulate natural weathering.

	Solar Reflectance	Thermal Emissivity	Solar Reflective Index (SRI)	Roof Surface Temperature Rise
Low-Slope Roofing				
White thermoplastic	0.85	0.90	105	10°F (5°C)
Bituminous membrane with reflective elastomeric coating or reflective film cap sheet	0.80	0.90	100	15°F (8°C)
White EPDM	0.75	0.90	95	20°F (11°C)
Bituminous membrane with white-coated granules	0.65	0.90	80	30°F (16°C)
Bituminous membrane with light granules	0.35	0.90	35	60°F (30°C)
Bituminous membrane with dark granules	0.25	0.90	25	70°F (39°C)
Black EPDM	0.10	0.85	5	87°F (48°C)
Steep Roofing				
Metal panel with cool white coating	0.70	0.85	85	25°F (14°C)
Aluminum, uncoated	0.60	0.05	40	56°F (31°C)
Red clay tile	0.35	0.90	35	60°F (33°C)
Metal panel with cool color coating	0.30	0.85	30	66°F (37°C)
Asphalt shingle with cool color granulated surface	0.25	0.90	25	70°F (39°C)
Unpainted concrete tile	0.25	0.90	25	70°F (39°C)
Grey asphalt shingle	0.22	0.90	20	73°F (41°C)
Black asphalt shingle	0.05	0.90	0	90°F (50°C)

FIGURE 16.55
Solar heating properties for common roof materials. Materials in shaded rows do not meet the criteria for cool roofs listed in the previous figure. Figures are approximate and should not be used to document compliance with cool roof rating programs.

roofing materials to produce products meeting cool roof standards for steep roofs. As the formulation of cool color pigments continues to evolve, smooth-surfaced roofing materials with reflectance values as high as 0.45 and granule-surfaced materials with values as high as 0.30 are anticipated.

GREEN ROOFS

Green roofs, also called *vegetated roofs*, are roofing systems covered with vegetation, soil, and additional materials needed to support plant growth. Like protected membrane roofs, green roofs extend the life of the roof membrane by shielding it from UV radiation and extremes of temperature. Green roofs may also reduce heating and cooling costs by moderating temperature swings in the roof assembly. They reduce the transmission of noise through the roof system and decrease the reflection of exterior noise. They reduce stormwater runoff and provide habitat for birds and insects. By supporting plant growth and reducing heat island effects, green roofs improve air quality. They provide aesthetic value and, in some cases, create pleasant, usable space.

Extensive green roofs are relatively shallow, with soil depths of 2 to 6 inches (25–150 mm). They are planted with herbs, grasses, mosses, sedums, or other drought-tolerant plants that do not require irrigation or frequent maintenance (Figures 16.56 and 16.59). *Intensive green roofs* may have soils as deep as 30 inches (750 mm) and are designed to support a broader variety of plant types and shrubs. Intensive roofs require irrigation and regular maintenance such as weeding, trimming, pest management, and fertilization.

Planning for the structural loads of soil and plant materials is an important part of green roof design. Because of their lesser depth, extensive roofs are relatively light in weight, imposing loads, when saturated with water, ranging from 12 to 35 psf (0.57 to 1.7 kPa) on the supporting roof structure. Intensive roofs impose loads of 50 psf (2.4 kPa) or more. Although most green roofs are essentially flat, extensive roofs with slopes as great as 12:12 (100 percent) are technically feasible with special soil retention measures.

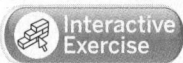

FIGURE 16.56

A green roof over a sloping roof deck. This extensive planted roof has a 6-inch (150-mm)-deep soil layer and supports native and ornamental grasses. *(Design by John Senhauser Architects. Photo courtesy of Steve Grim.)*

From the top down, typical components of a green roof system include the following:

- **Plant materials** may be selected on the basis of hardiness, climate, depth of soil, maintenance expectations, and appearance.

- The **growth medium** (soil) must provide long-lasting, optimal growing conditions. Particulars of its formulation vary with the depth of soil and the types of vegetation supported. Soil stability, drainage properties, and, in extensive roofs, drought resistance are important considerations.

- A geotextile **filter fabric** prevents soil particles from being washed out of the growth medium and clogging the drainage layer below.

- **Soil restraints** made from perforated plastic or metal allow the free flow of water while confining the growth medium at the roof perimeter and around drains or scuppers.

- **Drainage layer** materials, such as a molded plastic panel or an entangled plastic filament mat, provide efficient drainage and aeration beneath the soil. Some products also provide **water retention**, benefiting the plants' subsoil environment. Crushed rock or other aggregate material may also be used, although at the cost of added weight in comparison to the use of manufactured drainage components.

- Rigid foam **insulation** boards may be positioned above or below the roof membrane. Where insulation is placed above the membrane, extruded polystyrene is used because it retains its insulating value when wet. Insulation boards must be sufficiently dense and strong in compression to resist the weight of the soil and plant materials above.

- Depending on the membrane system, one or more **protection layers** to prevent root invasion and to relieve physical stress may be laid over the membrane.

- Because access to the **waterproofing membrane** becomes difficult once it is covered with the green roof components, it should be chosen after careful consideration of its long-term performance in a buried, continuously damp environment. Especially for intensive green roofs, robust waterproofing systems are recommended in place of lighter-duty, conventional roof membranes. Though many membrane materials may be used, hot rubberized asphalt, PVC, and multi-ply modified bitumen have the longest track records of successful installations. Where modular green roof systems are used (see discussion following) and the membrane will remain more easily accessible, less expensive conventional roof membrane types may also be suitable. Prior to being covered, green roof membranes may be *flood tested*, that is, tested for watertightness by placing the membrane under continuously submerged conditions for a period of hours or days, to check for leaks prior to placing the overlying components.

- Requirements for the **vapor retarder** and **air barrier** are no different for green roofs than for conventional low-slope roof assemblies.

- The **roof deck** and **supporting structure** must be engineered to carry the additional loads imposed by green roof components.

With *modular green roof systems*, all the components of the green roof system above the membrane are preassembled in easily transported trays or modules. These trays, typically 2 to 4 feet (600–1200 mm) in plan dimension and 2 to 8 inches (50–200 mm) in depth, are preplanted and arrive on the construction site ready to be placed directly over the roof membrane. Modular green roof systems are relatively lightweight, easy to specify, easy to assemble on site, and easy to remove or adjust at a later date.

PHOTOVOLTAIC SYSTEMS

Photovoltaic (PV) materials directly convert sunlight to electricity. *PV modules*, also called *solar panels*, are made of collections of individual photovoltaic semiconductor wafers interconnected to produce direct current

FIGURE 16.57

A rooftop-mounted array of photovoltaic modules. Each module is composed of a series of round tubes coated with thin-film PV material. The PV tubes convert light coming from any direction, such as directly from the sun, from the diffuse sky on cloudy days, or reflected back from the cool white roof membrane. This array, situated in the relatively cloudy Pacific Northwest, covers 2.5 acres of roof area and is capable of generating more than 830,000 kWh of electricity annually. *(Photo by Joseph Iano.)*

electricity at a specified voltage. Collections of such modules, called *PV arrays,* are mounted to the building roof or exterior walls (Figure 16.57). *Building-integrated photovoltaics (BIPVs)* are conventional building materials laminated with thin-film PV semiconductor materials, also interconnected into arrays. BIPVs serve both the traditional functions of cladding, roofing, or glazing, as well as the production of electricity from sunlight.

Electricity produced by any type of PV array is fed through equipment that regulates and converts the direct current output of the PVs to alternating current at typical building power voltages. Where permitted by the public utility, this power may be sold back into the utility grid and used to offset power consumption charges. Or, in a *stand-alone* or *off-grid PV system,* PV power is fed into rechargeable batteries or some other type of storage system, and then drawn for use on demand.

Estimating the potential power contribution of a photovoltaic system and evaluating its potential economic payback for a particular project require an analysis of the project's solar exposure, projected power needs, and utility costs. See the list of web sites at the end of this chapter for more information.

ROOFING AND THE BUILDING CODES

For most types of roofing, manufacturing standards, minimum slopes, underlayment materials, and installation requirements are specified by building codes. Codes also regulate a roof's required level of resistance to flame spread and fire penetration, tested according to standards ASTM E108 or UL 790 and rated as *Class A, B, or C roof coverings* (listed here in decreasing order of resistance). The International Building Code requires roofs to meet Class B requirements on most buildings of Construction Types I, II-A, III-A, IV, or V-A, or Class C requirements on other types (see Figure 1.4 for an explanation of construction types). Roofs for single-family homes and other small residential or utility buildings generally may be nonclassified, except that where portions of such roofs are located close to property lines, a minimum Class C rating is required. Property insurance policies or local building regulations, such as may apply in dense urban areas or in areas prone to wildfire, may also impose roof class rating requirements.

Roof class rating tests apply to whole roof assemblies, including the membrane, shingles or other covering, underlayments, insulation, decking, and ballast, if any. Broadly speaking, most low-slope roof membranes and noncombustible tiles (such as concrete or clay) can meet Class A requirements, as can some metal roof coverings and asphalt shingles made of glass felts. Asphalt shingles made of organic felts can meet Class B requirements. Fire-retardant wood shingles and shakes can meet Class C requirements.

FIGURE 16.58
A house is both roofed and sided with red cedar shingles to feature its sculptural qualities. *(Architect: William Isley. Photo by Paul Harper. Courtesy of Red Cedar Shingle and Handsplit Shake Bureau.)*

BUILDING ENCLOSURE ESSENTIALS: DISSIMILAR METALS AND THE GALVANIC SERIES

When different metals come into contact in the presence of moisture or some other conducting medium, an electric current flows between them. Such metal pairs, or *galvanic couples*, form the basis, for example, for the storage of electrical energy in batteries. In building construction, galvanic couples are important for the manner in which the exchange of electrons affects the corrosion rates of the metals involved—a phenomenon that, if applied properly, can be used to positive effect or, if ignored, can lead to the premature deterioration of critical building components.

The Galvanic Series

In any galvanic couple, the metal donating electrons, called the *anode*, experiences an accelerated rate of corrosion; conversely, the metal receiving electrons, called the *cathode*, experiences corrosion at a reduced rate. This effect is relatively great for pairs of metals with large differences between their electrochemical potentials and proportionally smaller for those with lesser potential difference.

To facilitate the prediction of the corrosion potential of metal pairs, metals may be listed in a *galvanic series*, that is, in order of their relative electrochemical potential, with the most anodic, or *active*, metals at one end of the list and the most cathodic, *noble* or *passive*, ones at the other. Metal pairs positioned far apart in the series have greater corrosion potential than pairs closer to each other. Any number of galvanic series can be compiled for a group of metals in different conducting mediums and under different environmental conditions. For building construction, a galvanic series based on metals immersed in flowing sea water is used (Figure A).

Applying the Galvanic Series to Problems in Building Construction

Corrosion in the Anode and the Cathode

In Figure A, any metal listed higher in the series behaves as the anode, and experiences accelerated corrosion, when paired with any other metal listed lower in the series, which behaves as the cathode and experiences a reduced rate of corrosion.

A common application of galvanic couples is in the use of anodic metals as protective coatings for other more cathodic metals. For example, consider *galvanized* (zinc-coated) *steel* fasteners. Zinc appears higher in the galvanic series and is therefore the anode in this pair of metals. Even if the zinc coating is damaged and the steel itself becomes exposed to moisture and air, the steel can remain protected. As long as there is sufficient surrounding zinc to sustain galvanic action between the metals, only the zinc will corrode, in effect sacrificing itself to preserve the steel. For the same reason, zinc and zinc-aluminum alloys are frequently applied as protective metallic coatings to steel sheet used for roofing and flashing. These alloys form sacrificial, protective coatings for the underlying steel.

Environmental Conditions

Corrosion rates of metals are affected by the environment. Exterior metals exposed to marine, salt-laden air or to atmospheres laden with industrial pollutants will corrode at a faster rate than others in less aggressive or more protected environments. The more severe the environment, the more important it is to avoid contact between dissimilar metals. Where dissimilar metals must be combined in such environments, metal pairs with the least electrochemical potential difference (that is, as close to each other as possible in the galvanic series) are preferable.

Relative Surface Area of Anode to Cathode

The corrosive effects within a galvanic couple are proportional to the relative surface areas of the metals involved. For example, consider copper roofing fastened with stainless steel fasteners. In this metal pair, copper is the anode and at risk of corrosion. However, the relative surface area of anode to cathode, that is, the surface area of the copper roofing, is very large in relation to the surface area of the stainless steel fasteners. This means that the electrochemical effects are diluted over a very large area of the copper and are therefore relatively weak. In fact, the use of stainless steel fasteners for the attachment of most types of metal roofing, including copper, is quite common. In contrast, consider the use of galvanized steel fasteners for the attachment of the same copper roofing. In this case, the fasteners are the anode in the galvanic couple, and the ratio of surface areas of anode (fastener) to cathode (roof metal) is very low. This condition will lead to accelerated corrosion of the fasteners and is not recommended. As a general rule, when dissimilar metals are paired, the ratio of surface areas of anode to cathode should be kept as large as possible.

Cathodic Fasteners

As a corollary to the previous rule, where fasteners of one metal are used to attach metals of another type, the fasteners should be the cathodic metal in the pair. Because

Most Anodic or Active
Magnesium
Zinc
Aluminum
Cadmium
Steel, iron
Aluminum bronze
Naval brass, yellow brass, red brass
Tin
Copper
Lead-tin solder
Admiralty brass, aluminum brass
Manganese bronze, silicon bronze, tin bronze
Types 410 and 416 stainless steel (passive)
Nickel silver
High copper content nickel-copper alloys
Lead
Nickel-aluminum-bronze
Nickel
Silver
Monel, high nickel content nickel-copper alloys
Types 304 and 316 stainless steel (passive)
Titanium
Platinum
Gold
Graphite
Most Cathodic or Noble

FIGURE A

The galvanic series for selected metals, immersed in flowing sea water, adapted from ASTM G82. When paired as galvanic couples, metals positioned higher on the list will sacrificially corrode while protecting metals positioned lower on the list. The horizontal lines in the list separate metals into groups, with roughly equal ranges of potential difference. For example, although aluminum bronze and lead appear relatively far apart on the list, their electrochemical potential difference is roughly as great as that between nickel-aluminum-bronze and titanium or cadmium and steel. Some metals, such as the stainless steel alloys, can appear in different locations within the series, depending on their chemical state and the environment in which they are placed. In this series, stainless steels are listed in their so-called passive states, their expected condition under normal usage.

BUILDING ENCLOSURE ESSENTIALS: DISSIMILAR METALS AND THE GALVANIC SERIES (CONTINUED)

fasteners are typically small in surface area in comparison to the materials they attach, this approach ensures a relatively large anode-to-cathode surface area ratio. In practice, stainless steel fasteners are normally considered suitable for fastening construction metals of virtually all other types. This is a safe practice because stainless steel is the most cathodic metal normally found on the building site.

Water Flowing from One Metal to Another

Where rainwater flows from one metal type to another, the first metal should be anodic in relation to the second. Consider again a copper roof, in this case draining into a stainless steel gutter. As rainwater washes off the roof metal, it will carry dissolved copper ions into the gutter. Because the stainless steel is cathodic in relation to the copper, the gutter metal itself will remain unharmed. Conversely, consider the same roof with a galvanized steel gutter. In this case, the galvanized steel gutter metal is anodic in relation to the dissolved copper washing through it and may experience accelerated corrosion.

Insulating Dissimilar Metals

For galvanic action to occur, electric current must flow between the metals. Thus, by insulating dissimilar metals from each other, galvanic corrosion can be prevented. For example, a layer of felt paper or rubberized asphalt membrane may be used to separate dissimilar sheet metal materials used in roofing and flashing work.

Paints, aluminum anodizing, and other coatings can also frequently act as insulators between one metal and another. However, when using coatings to separate dissimilar metals, the metals to be coated should be chosen carefully. Where an insulating coating is applied only to the anodic metal in a metal pair, that metal is at risk of accelerated corrosion if the coating becomes damaged. This is because if the coating fails in only limited areas, the exposed area of anodic metal is relatively small, and the galvanic action becomes concentrated in these limited areas (in other words, the anode-to-cathode surface area ratio is very small). For this reason, where coatings are used to separate dissimilar metals, it is safer to coat only the cathodic metal in the pair or to coat both metals.

FIGURE 16.59
Built on a steep slope, the roofs of the building below have a strong presence in the landscape and view beyond. Plantings on the green roof integrate with the terraced landscaping in the foreground while the standing-seam sheet metal contrasts with the body of water beyond.
(Photo by Joseph Iano.)

CSI/CSC MasterFormat Sections for Roofing

07 22 00	**ROOF AND DECK INSULATION**
07 31 00	**SHINGLES AND SHAKES**
07 31 13	Asphalt Shingles
07 31 16	Metal Shingles
07 31 19	Mineral-Fiber Cement Shingles
07 31 26	Slate Shingles
07 31 29	Wood Shingles and Shakes
07 31 33	Composite Rubber Shingles
07 31 53	Plastic Shingles
07 32 00	**ROOF TILES**
07 32 13	Clay Roof Tiles
07 32 16	Concrete Roof Tiles
07 33 00	**NATURAL ROOF COVERINGS**
07 33 16	Thatched Roofing
07 33 63	Vegetated Roofing
07 41 00	**ROOF PANELS**
07 41 13	Metal Roof Panels
07 51 00	**BUILT-UP BITUMINOUS ROOFING**
07 52 00	**MODIFIED BITUMINOUS MEMBRANE ROOFING**
07 53 00	**ELASTOMERIC MEMBRANE ROOFING**
07 53 29	Ethylene-Propylene-Diene-Monomer Roofing
07 54 00	**THERMOPLASTIC MEMBRANE ROOFING**
07 54 19	Polyvinyl-Chloride Roofing
07 54 23	Thermoplastic-Polyolefin Roofing
07 55 00	**PROTECTED MEMBRANE ROOFING**
07 56 00	**FLUID-APPLIED ROOFING**
07 58 00	**ROLL ROOFING**
07 61 00	**SHEET METAL ROOFING**
07 61	Standing Seam Sheet Metal Roofing
07 61 16	Batten Seam Sheet Metal Roofing
07 61 18	Flat Seam Sheet Metal Roofing
07 62 00	**SHEET METAL FLASHING AND TRIM**
07 71 00	**ROOF SPECIALTIES**
07 71 13	Manufactured Copings
07 72 00	**ROOF ACCESSORIES**
07 72 23	Relief Vents
07 72 26	Ridge Vents
07 76 00	**ROOF PAVERS**

KEY TERMS

steep roof
low-slope roof
roof deck
thermal insulation
air barrier
vapor retarder
roof membrane
drainage
flashing
ponding
substrate board
area divider
topside vent
protected membrane roof (PMR)
ballast
thermal resistance (R, RSI)
water vapor
relative humidity
dew point temperature
condensation
vapor pressure
vapor retarder, vapor barrier
vapor permeance
perm
vapor permeability
lightweight insulating concrete
vented metal roof decking
insulation coverboard
bituminous roof membrane
built-up roof (BUR) membrane
cold-applied mastic, solvent-based
    asphalt
modified bitumen roof membrane
atactic polypropylene (APP)
styrene-butadiene-styrene (SBS)
torch-applied modified bitumen
    membrane
hot-mopped modified bitumen
    membrane
cold process modified bitumen
    membrane, cold-applied adhesive
    modified bitumen membrane

self-adhered modified bitumen
    membrane
cap sheet
hybrid membrane bituminous roof
single-ply roof membrane
thermoplastic material
thermosetting material, elastomeric
polyvinyl chloride (PVC)
thermoplastic polyolefin (TPO)
ketone ethylene ester (KEE)
ethylene propylene diene monomer
    (EPDM)
fluid-applied roof membrane
traffic deck
base flashing
fascia
scupper
stripping
parapet
counterflashing
coping
cant strip
roof drain
roof penetration
structural panel metal roofing
architectural sheet metal roofing
thatch
nail base
shingle
wood shingle
wood shake
skip sheathing, spaced sheathing
breather mat
asphalt shingle, composition shingle
valley
open valley
woven valley
cut valley
roll roofing
slate shingle
clay tile
concrete tile

oil canning
flat-seam metal roof
standing-seam metal roof
batten-seam metal roof
lead
copper
lead-coated copper
zinc
stainless steel
titanium
zinc-tin alloy coated stainless steel
terne-coated stainless steel
aluminum
metallic-coated steel
zinc-tin alloy coated steel
ferrous steel
gauge, gage
cool roof
solar reflectance, albedo
thermal emittance
solar reflective index (SRI)
cool color
near-infrared (NIR) radiation
green roof, vegetated roof
extensive green roof
intensive green roof
flood test
modular green roof system
photovoltaic (PV) material
PV module, solar panel
PV array
building-integrated photovoltaic (BIPV)
stand-alone PV system, off-grid PV system
Class A, B, C roof coverings
galvanic couple
anode
cathode
galvanic series
active metal
noble metal, passive metal
galvanized steel

REVIEW QUESTIONS

1. What are the major differences between a low-slope roof and a steep roof? What are the advantages and disadvantages of each type?

2. Discuss the three positions in which thermal insulation may be installed in a low-slope roof, and the advantages and disadvantages of each.

3. Explain in precise terms the function of a vapor retarder in an exterior wall or roof assembly.

4. Compare a bituminous roof membrane to a single-ply roof membrane.

5. What is the difference between cedar shingles and cedar shakes?

6. What metals are used for architectural sheet metal roofing? What are the strengths and drawbacks of each?

7. What are the benefits of a cool roof? What properties of a roofing material affect its solar heating and how?

8. List the major components of a green roof system and describe their functions.

9. Which of the following metal combinations are generally safe, and why?

a. Copper sheet metal roofing with stainless steel fasteners

b. Copper sheet metal roofing with copper fasteners

c. Copper sheet metal roofing with galvanized steel fasteners

d. Galvanized steel sheet metal roofing with stainless steel fasteners

e. Galvanized steel sheet metal roofing with galvanized steel fasteners

EXERCISES

1. For a low-slope-roofed university classroom building with a masonry bearing wall, steel interior frame, corrugated steel roof deck, and parapet:

a. Show two ways of achieving a 1:50 roof slope on structural bays 36 feet (11 m) square.

b. Sketch a set of details of the parapet edge, building separation joint, area divider, and roof drain for a low-slope roof system of your choice. Show insulation, vapor retarder (if any), roof membrane, and flashings.

2. Sketch a fascia detail for a low-slope roof system of your choice, assuming that the wall below is made of precast concrete panels and the roof deck of precast concrete slab elements.

3. Find a low-slope roof system being installed and take notes on the process until the roof is completed. Ask questions of the roofers, the architect, or your instructor about anything you don't understand.

4. Examine a number of low-slope roofs around your campus or neighborhood, looking for problems such as cracking, blistering, tearing, and leaking. Attempt to explain the reasons for each problem that you discover.

SELECTED REFERENCES

ASHRAE. *ASHRAE Handbook—Fundamentals*. Atlanta, GA, American Society of Heating, Refrigerating, and Air-Conditioning Engineers, updated regularly.

This handbook provides a definitive treatment of the physics of heat and moisture transfer through building assemblies and the application of these principles to building construction methods.

National Roofing Contractors Association. *The NRCA Roofing Manual: Steep-Slope Roof Systems; The NRCA Roofing Manual: Architectural Metal Flashing, Condensation Control and Reroofing; The NRCA Roofing Manual: Membrane Roof Systems; The NRCA Roofing Manual: Metal Panel and SPF Roof Systems*. Rosemont, IL, Author, updated regularly.

These manuals, all updated regularly, are the most comprehensive guides to current U.S. practice for low-slope and steep-slope roofing systems. The treatment is exhaustive, and both diagrams and text are excellent. These products may also be available in combined editions.

Sheet Metal and Air Conditioning Contractors National Association. *Architectural Sheet Metal*. Chantilly, VA, Author, updated regularly.

Architectural sheet metal roofs are copiously detailed in this excellent reference, along with every conceivable flashing, fascia, gravel stop, and gutter for flat and shingled roofs.

Zahner, L. William. *Architectural Metals: A Guide to Selection, Specification, and Performance* (3rd ed.). Hoboken, NJ, John Wiley & Sons, 1995.

This is one of the most comprehensive treatments available regarding architectural metal types and their uses in construction.

WEB SITES

Roofing

Author's supplementary web site: **www.ianosbackfill.com/16_roofing**

Low-Slope Roofs

Asphalt Roofing Manufacturers Association: **www.asphaltroofing.org**
Carlisle SynTec: **www.carlisle-syntec.com**
Firestone Building Products: **www.firestonebpco.com**
National Roofing Contractors Association: **www.nrca.net**
Polyisocyanurate Insulation Manufacturers Association: **www.polyiso.org**
Whole Building Design Guide, Roofing Systems: **www.wbdg.org/design/env_roofing.php**

Steep Roofs

Asphalt Roofing Manufacturers Association: **www.asphaltroofing.org**
Cedar Shingle and Shake Bureau: **www.cedarbureau.org**
Certainteed Roofing Products: **www.certainteed.com**
Copper Development Association: **www.copper.org**
GAF Roofing Products: **www.gaf.com**
Metal Construction Association: **www.metalconstruction.org**
Sheet Metal and Air Conditioning Contractors National Association: **www.smacna.org**
Umicore Building Products (zinc roofing): **www.vmzinc-us.com**

Cool Roofs

Cool Metal Roofing Coalition: **www.coolmetalroofing.org**
Cool Roof Rating Council: **www.coolroofs.org**

Photovoltaic Roofs

Renewable Resource Data Center (RReDC): **www.nrel.gov/rredc/**

17

GLASS AND GLAZING

- **History**

- **The Material Glass**

  SUSTAINABILITY OF GLASS

  Thicknesses of Glass
  Heat-Treated Glass
  Laminated Glass
  Chemically Strengthened Glass
  Safety Glazing
  Fire-Rated Glass
  Fritted Glass
  Spandrel Glass
  Tinted and Reflective Coated Glass
  Insulating Glass
  Low-Emissivity Coated Glass
  Self-Cleaning Glass

  Glass That Changes Its Properties
  Other Types of Glass
  Plastic Glazing
  Aerogel-Filled Glazing

- **Glazing**
  Glazing Small Lites
  Glazing Large Lites
  Advanced Glazing Systems

- **Glass and Energy**

- **Glass and the Building Codes**

  FROM CONCEPT TO REALITY
  Skating Rink at Yerba Buena Gardens

Glass staircase, designed by Eckersley O'Callaghan Structural Design and Bohlin Cywinski Jackson Architects Planning Interior Design. The stair is made of structural laminated glass and titanium fittings. *(Photo by Joseph Iano.)*

Glass plays many roles and takes many forms in buildings: Gothic church windows made of thousands of jewel-like pieces of colored glass; breathtaking expanses of smooth, uninterrupted glass that fill whole walls of today's buildings; Elizabethan casement windows with tiny diamond panes set in lead; skyscrapers that shimmer in facets of reflective glass mirroring the sky; cozy windows; comfortable windows; windows that bring soft, natural light; windows that frame spectacular views; windows that welcome winter sunlight to warm a room; windows that dissolve the boundaries between inside and outside. But glass can also form windows that make privacy impossible; windows that admit a harsh, glaring light; winter-cold surfaces that chill the body and tax the heating system; windows that broil a room in summer afternoon sunlight. Glass skillfully used in building contributes strongly to our enjoyment of architecture, but glass thoughtlessly used can make a building unattractive, uneconomical, and uncomfortable to inhabit.

HISTORY

The origins of glass are lost in prehistory. Initially a material for colored beads and small bottles, glass was first used in windows in Roman times. The largest known piece of Roman glass, a crudely cast sheet used for a window in a public bath at Pompeii, was nearly 3 by 4 feet (800 by 1100 mm) in size.

By the 10th century AD, the Venetian island of Murano had become the major center of glassmaking, producing *crown glass* and *cylinder glass* for windows. Both the crown and cylinder processes began by blowing a large glass sphere. In the crown process, the heated glass sphere was adhered to an iron rod called a *punty* opposite the blowpipe. The blowpipe was then removed, leaving a hole opposite the punty. Next, the sphere was reheated, whereupon the glassworker spun the punty rapidly, causing centrifugal force to open the sphere into a large disk, or crown, 30 inches (750 mm) or more in diameter (Figure 17.2). When the crown was cut into panes, one pane always contained the "bullseye" where the punty had been attached before being cracked off. In the cylinder process, the sphere, heated to a molten state, was swung back and forth, pendulum fashion, on the end of the blowpipe to elongate it into a cylinder. The hemispherical ends were cut off and the remaining cylinder was slit lengthwise, reheated, opened, and flattened into a rectangular sheet of glass that was later cut into panes of any desired size (Figure 17.3). Before the introduction of modern

FIGURE 17.1
An entire wall of the Baltimore Convention Center is made of low-emissivity double glazing supported by a tubular steel substructure. Adjustable stainless steel fittings with rubber washers connect the sheets of glass to the substructure. *(Architects: Cochran, Stevenson, & Donkervoet. Photograph of Pilkington Planar System by W & W Glass Systems, Inc.)*

FIGURE 17.2
The glassworker in this old engraving wore a face shield (*a*) and hand shield (*b*) to protect against the heat of the large glass crown (*c*) that he had just spun on the end of a punty. After cooling, the crown was cut into small lites of window glass. (*Courtesy of the Corning Museum of Glass, Corning, NY.*)

FIGURE 17.3
Making cylinder glass in the 19th century, Pittsburgh, Pennsylvania. Elongated glass bottles were blown by swinging the blowpipe back and forth in the pit in front of the furnace (*center*). As each bottle solidified (*left*), it was brought to another area where the ends were cut off to produce cylinders (*right*). The cylinders were reheated and flattened into sheets from which window glass was cut. (*Courtesy of the Corning Museum of Glass, Corning, NY.*)

glassmaking techniques, crown glass was favored over cylinder glass for its surface finish, which was smooth and brilliant because it was formed without contacting another material. Cylinder glass, though more economical to produce, was limited in surface quality by the texture and cleanliness of the surface on which it was flattened.

Neither crown glass nor cylinder glass was of sufficient optical quality for the fine mirrors desired by the 17th-century nobility. For this reason, *plate glass* was first produced, in France, in the late 17th century. Molten glass was cast into frames, spread into sheets by rollers, cooled, then ground flat and polished with abrasives, first on one side and then the other. The result was a costly glass of near-perfect optical quality in sheets of unprecedentedly large size. Mechanization of the grinding and polishing operations in the 19th century reduced the price of plate glass to a level that allowed it to be used for storefronts in both Europe and America.

In the 19th century, the cylinder process evolved into a method of drawing cylinders of molten glass vertically from a crucible. This made possible the routine, economical production of cylinders 40 to 50 feet (12–15 m) long. In 1851, the Crystal Palace in London (Figure 11.1) was glazed with 900,000 square feet (84,000 m²) of cylinder glass supported on a cast iron frame.

In the early years of the 20th century, cylinder glass production was gradually replaced by processes that pulled flat sheets of *drawn glass* directly from a container of molten glass. Highly mechanized production lines for the grinding and polishing of plate glass were established, with rough glass sheets entering the line continuously at one end and finished sheets emerging at the other.

In 1959, the English firm of Pilkington Brothers Ltd. started production of *float glass*, which has since been licensed to other glassmakers and has become the worldwide standard, replacing both drawn glass and plate glass. In this process, a ribbon of molten glass is floated across a bath of molten tin, where it hardens before touching a solid surface (Figures 17.4, 17.5, and 17.6). The resulting sheets of glass have parallel surfaces, high optical quality (virtually indistinguishable from that of plate glass), and a brilliant surface finish. Nearly all glass used in buildings is now produced as float glass. Float glass has been produced in North America since 1963 and now accounts for nearly all domestic flat glass production.

The terminology associated with glass developed early in this long history. The term *glazing* as it applies to building refers to the installation of glass in an opening or to the material itself (usually glass, but also plastic or other transparent or translucent material) in a glazed opening. The installer of glass is known as a *glazier*. Individual pieces of glass are known as *lights*, or often, to avoid confusion with visible light, *lites*.

THE MATERIAL GLASS

The major ingredient of glass is sand (silicon dioxide). Sand is mixed with soda ash (sodium hydroxide or sodium carbonate), lime, and small amounts of alumina, potassium oxide, and various elements to control color, then heated to form glass. The finished material, while seemingly crystalline and rigid, is actually an amorphous solid, with no fixed melting point and an open, noncrystalline microstructure.

When drawn into small fibers, glass is stronger than steel, though not nearly as stiff. In larger pieces, the microscopic imperfections that are an inherent characteristic of glass reduce its useful strength to significantly lower levels, particularly in tension. When a surface of a sheet of glass is placed in sufficient tension, as happens when an object strikes the glass, cracks propagate from an imperfection near the point of maximum tension and the glass shatters.

FIGURE 17.4

In the float glass process, molten glass from the furnace is floated on a bath of liquid tin to form a continuous sheet of glass. The annealing *lehr* cools the glass at a controlled rate to avoid internal stresses, after which it is cut into smaller sheets. (*Courtesy of PPG Industries.*)

FIGURE 17.5
The superior flatness and bright surface finish of float glass are readily seen in the reflections on the glass ribbon emerging from the annealing lehr. *(Photo courtesy of LOF Glass, a Libby-Owens-Ford Company.)*

FIGURE 17.6
Track-mounted cutting devices score the ribbon of cooled float glass as part of a computer-controlled cutting operation that automatically produces the glass sizes ordered by customers. *(Courtesy of PPG Industries.)*

SUSTAINABILITY OF GLASS

Glass Production

- The major raw materials for glass—sand, limestone, and sodium carbonate—are finite but abundant minerals.

- The high embodied energy of glass manufactured using traditional methods, roughly 7000 BTU per pound (16 MJ/kg), can be reduced by as much as 30 to 65 percent as new, more energy-efficient manufacturing technologies are introduced.

- Some glass production involves the generation of potentially unhealthful or pollution-causing waste materials. Traditional mirror glass manufacturing, for example, generates an acidic waste effluent with high concentrations of copper or lead. However, more recently, mirror glass manufactured with more environmentally friendly production techniques has become available.

- Although glass bottles and containers are recycled into new containers at a high rate, there is little recycling of flat glass at the present time. Most old glass goes to landfills.

- Efforts are underway to find new uses for waste glass. For example, vitrified glass aggregate (glass that has been melted and rapidly quenched to trap heavy metals and other contaminants) can be reused in asphalt, concrete, construction backfill, roofing shingles, and ceramic tiles.

Uses of Glass

- If it is not broken by accident or improper installation, glass lasts for a very long time with little degradation of quality, often much longer than most other building components.

- Glass is inert and does not affect indoor air quality. It is easily kept clean and free of molds and bacteria.

- The impact of glass on energy consumption can be very detrimental, very beneficial, or anything in between, depending on how intelligently it is used.

- If badly used, glass can contribute to summertime overheating from unwanted solar gain, excessive wintertime heat losses due to inherently low R-values, visual glare, wintertime discomfort caused by radiant heat loss from the body to cold glass surfaces, and condensation of moisture that can damage other building components.

- If well used, glass can bring solar heat into a building in winter and exclude it in summer, with attendant savings in heating and cooling energy. It can bring daylight into a building without glare, reducing both the use of electricity for lighting and the cooling load produced by that lighting.

- These benefits accrue over the entire life of the building, and the payoffs can be huge. Thus, glass is a key component of every energy-efficient building and a chief accomplice of the ill-informed designer in most energy-wasting buildings.

Thicknesses of Glass

Glass is manufactured in a series of thicknesses typically ranging from ³⁄₃₂ inch (2.5 mm), called *single-strength*, through ⅛ inch (3 mm), called *double-strength*, up to as much as 1 inch (25.4 mm). Glass thickness for a particular window is determined by the size of the lite and the expected maximum wind loads on the glass. For low buildings with relatively small windows, glass ⅛ inch thick is usually sufficient. For larger windows and for windows in tall buildings, thicker glass is generally required, along with increased attention to how the glass is supported in its frame.

Because of unavoidable manufacturing defects in the glass, as well as the probability of damage to the glass during installation and while it is in service, a certain amount of breakage must always be anticipated in a large building. ASTM E1300 establishes standard procedures for evaluating the probability of breakage in glass. These are used to determine the glass thickness that will result in an acceptably low occurrence of breakage for a window of given dimensions, support conditions, and wind pressure.

During its manufacture, ordinary window glass is *annealed*, meaning that it is cooled slowly under controlled conditions to avoid locked-in thermal stresses that might cause it to behave unpredictably in use.

Heat-Treated Glass

Heat-treated glass is produced by reheating annealed glass in an oven to approximately 1150 degrees Fahrenheit (620°C) and then cooling (*quenching*) both of its surfaces rapidly with blasts of air while its core cools much more slowly. This process induces permanent compressive stresses in the edges and faces of the glass and tensile stresses in the core. The resulting glass is stronger in bending than annealed glass and more resistant to thermal stress and impact. These properties make heat-treated glass useful for windows exposed to heavy wind pressures, impact, or intense heat or cold. By adjusting the quenching process, greater or lesser degrees of residual stress may be introduced into the glass, producing products referred to as either tempered glass or heat-strengthened glass.

Tempered Glass

Tempered glass has higher residual stresses than heat-strengthened glass and is about four times as strong in bending as annealed glass. If it does break, the sudden release of its internal stresses reduces tempered glass almost instantaneously to small, square-edged granules rather than long, sharp-edged shards. This characteristic, combined with its high strength, qualifies it for use as safety glazing (discussed later) in locations of possible occupant impact. Tempered glass is also used for all-glass doors that have no frame at all (Figure 17.7), for whole walls of squash and handball courts, for hockey rink enclosures, and for basketball backboards.

Tempered glass is more costly than annealed glass. It often has noticeable optical distortions created by the tempering process. In addition, all cutting to size, drilling, and edging must be done before the heat treatment of the glass, because any such operations after tempering will release the stresses in the glass and cause it to disintegrate. Tempered glass is also sometimes referred to as *fully tempered glass* to distinguish it more clearly from heat-strengthened glass.

Heat-Strengthened Glass

For many applications, lower-cost *heat-strengthened glass* may be used instead of tempered glass. The induced compressive stresses in the surface and edges of heat-strengthened glass are about one-third as high as those in fully tempered glass (typically 5000 psi compared to 15,000 psi for tempered glass, or 34 MPa versus 104 MPa). Heat-strengthened glass is about twice as strong in bending as annealed glass and is more resistant to thermal stress. It usually has fewer distortions than tempered glass and cutting and other fabrication operations can be performed after it has been heat treated without shattering the glass. Its breakage behavior is more like that of annealed glass than tempered glass. For that reason, it cannot be used as safety glazing except in laminated form (laminated glass is discussed in the next section).

Laminated Glass

Laminated glass is made by sandwiching a transparent *polyvinyl butyral (PVB) interlayer* between sheets of glass and bonding the three layers together under heat and pressure.

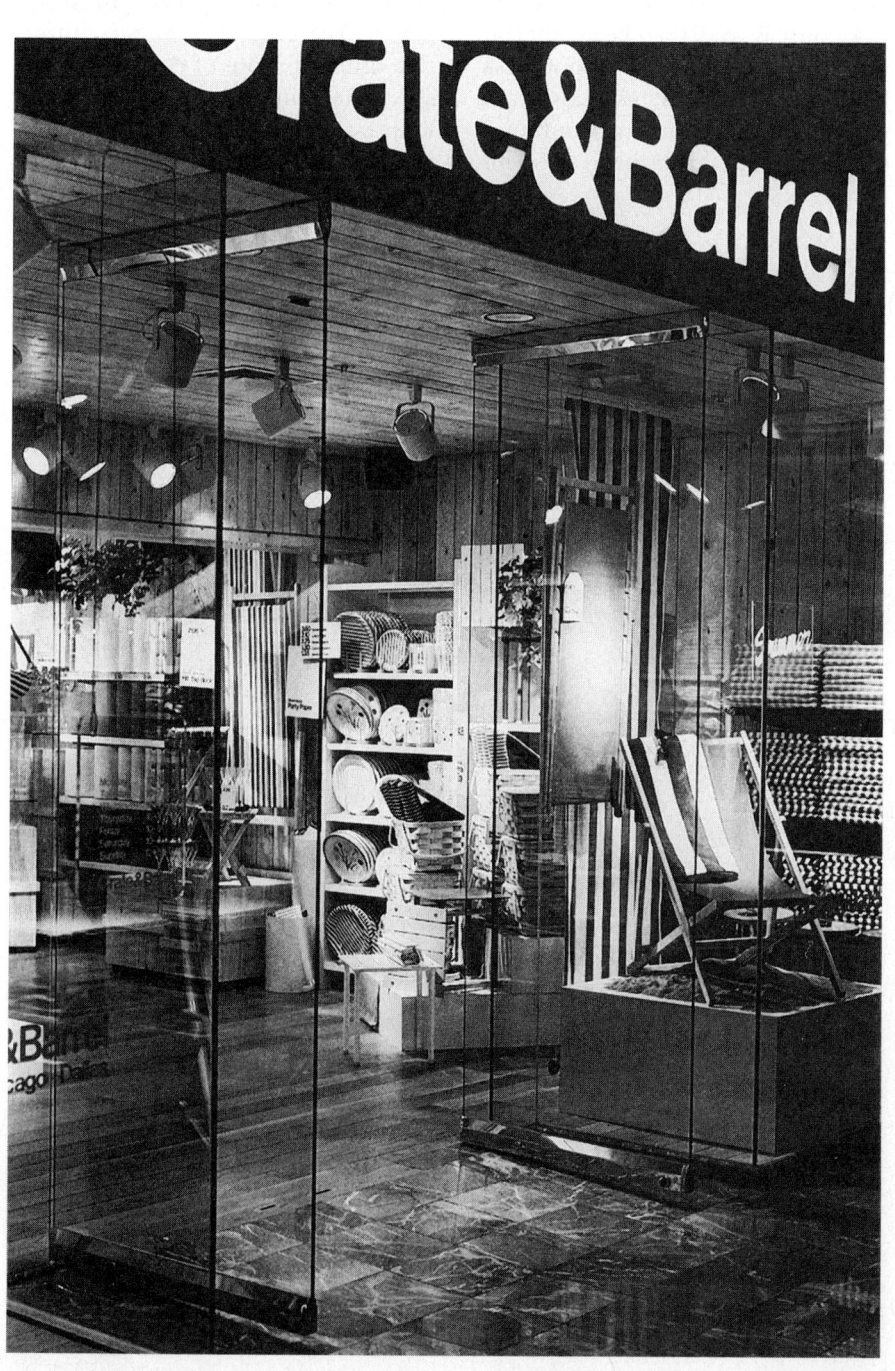

FIGURE 17.7

Tempered glass is used for strength and breakage safety in both the doors and windows of this store in a downtown shopping mall. *(Photo by Edward Allen.)*

Laminated glass is not as strong as annealed glass of the same thickness, but when laminated glass breaks, the soft interlayer holds the shards of glass in place rather than allowing them to fall out of the frame of the window. This makes laminated glass useful for skylights and overhead glazing, because it reduces the risk of injury to people below in case of breakage (Figures 17.8 and 17.9). The PVB interlayer may be colored or patterned to produce a wide range of visual effects in laminated glass. Because laminated glass does not create dangerous, loose shards of glass when it breaks, it also can qualify as safety glazing (discussed later).

Laminated glass is a better barrier to the transmission of sound than solid glass. It is used to glaze windows of residences, classrooms, hospital rooms, and other spaces that must be kept quiet in the midst of noisy environments. It is especially effective when installed in two or more layers with airspaces between.

Security glass, used for drive-in banking windows and other facilities that must be resistant to burglary, is made of multiple layers of glass and PVB, and is available in a range of thicknesses to stop any desired caliber of bullet. Laminated glass is also used in blast-resistant and windborne debris-resistant glazing systems, which are described in more detail in Chapter 18.

Protection from Fading Damage

Laminated glass also blocks the transmission of most ultraviolet (UV) radiation, an invisible component of sunlight that contributes to the fading and degradation of interior finishes, fabrics, and some types of artwork. In the past, where protection from fading was desired, laminated glass was often selected for this property. However, it is now recognized that other portions of the visible light spectrum contribute significantly to fading as well, and where such damage is a concern, the full spectrum of

FIGURE 17.8

The entrance canopy of the Newport, Rhode Island, Hospital is made of laminated glass supported by stainless steel spider fittings that transmit the weight of the roof to cantilevered steel beams. The heads of the bolts that fasten the glass to the fittings are recessed unobtrusively within the thickness of the glass. *(Architects: Taylor and Partners. Photograph of Pilkington Planar System by W & W Glass Systems, Inc.)*

light transmitted by the glazing material must be taken into account. This full-spectrum behavior is expressed as a glazing material's *Damage Weighted Transmittance* (*Tdw*), a value that ranges from 1 to 0. For example, a wall opening without any glazing inserted has a Tdw of 1, and an opaque material that blocks the passage of all light, a Tdw of 0. Ordinary clear glass has a Tdw of approximately 0.85 and laminated glass, of 0.65. Tinted, reflective, and low-emissivity coated glass products, described later in this chapter, are also capable of Tdw values as low as approximately 0.35. Where protection against damage from exposure to sunlight is a concern, glass products with low Tdw values should be selected.

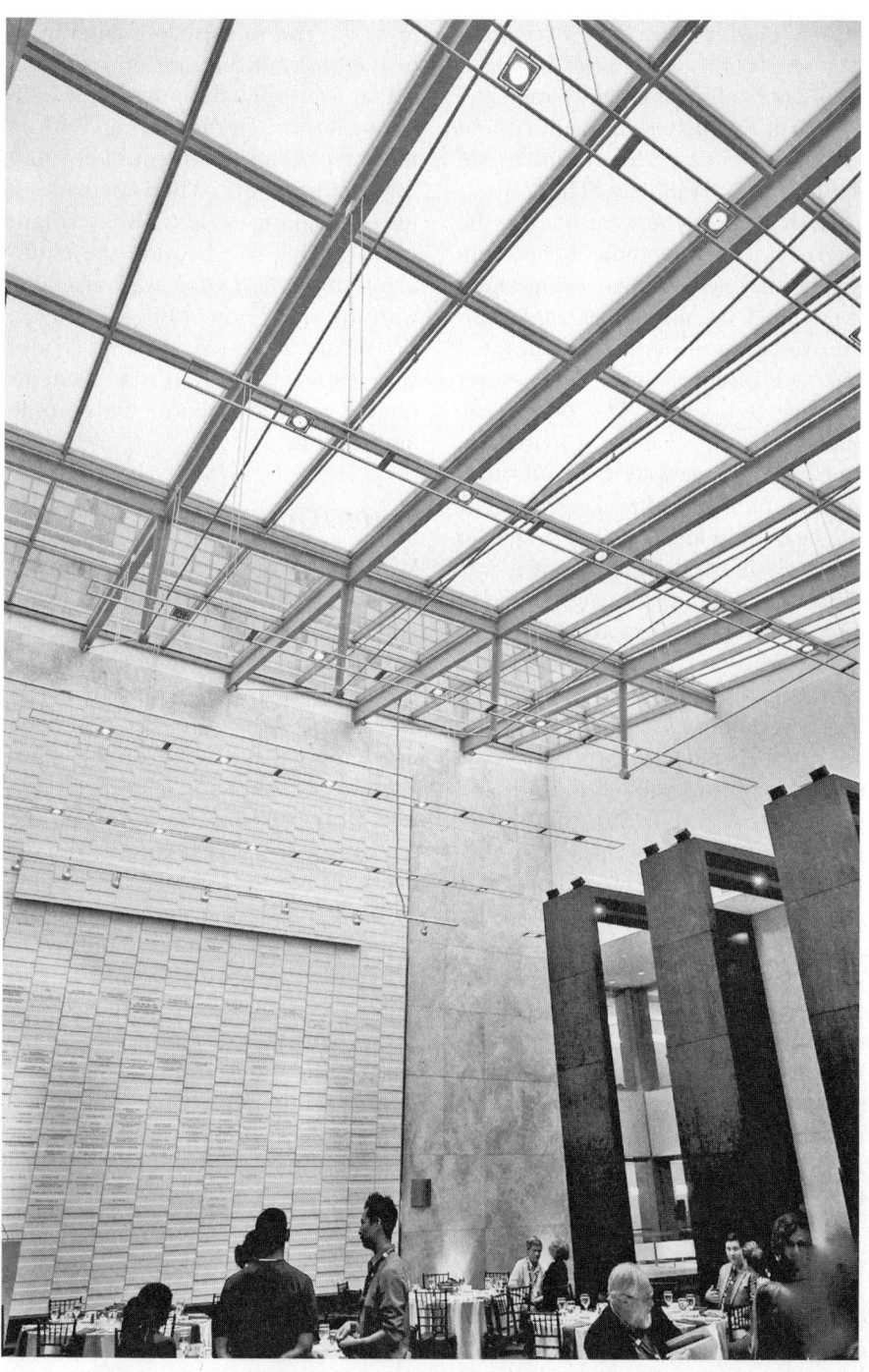

Chemically Strengthened Glass

Chemically strengthened glass is produced by an ion exchange process that takes place when annealed glass is immersed in a molten salt bath. As smaller sodium ions in the glass are replaced with larger potassium ions from the salt solution, the faces of the glass are put into compression relative to the core, and the glass is prestressed in a manner similar to the one that occurs with heat treating. However, because the temperatures involved in chemical strengthening are lower, chemically strengthened glass does not experience the optical distortions or warping that are common with heat-treated glass. Depending on the particulars of the treatment process, the strength and toughness of chemically strengthened glass can exceed those of tempered glass.

Unlike tempered glass, chemically strengthened glass can be cut after strengthening, although its strength is diminished along the cut edges. When chemically strengthened glass breaks, it produces large, hazardous shards. So, like heat-strengthened glass, it cannot be used as safety glazing unless it is laminated. Chemical strengthening is used for pieces of glass that are not easily heat treated, such as those that are small, thin, or oddly shaped. It is also used in some fire-rated glass products and in laminated form for security glass, blast-resistant glass, and windborne debris-resistant glass.

Safety Glazing

Glass and other transparent glazing materials installed where they could be subject to human impact must meet requirements for *safety glazing*. This means that the material must be resistant to impact and must be

FIGURE 17.9
Laminated glass provides safety against falling shards in an overhead glazing installation. *(Photo by Joseph Iano.)*

constituted such that, if it does fail, it does not break into knife-like, potentially life-threatening shards. Examples of *hazardous locations* subject to safety glazing requirements include glazing in and adjacent to doors, floor-to-ceiling glazing, shower and other wet area enclosures, glazing in guards and railings, glazing adjacent to stairs and ramps, and windows having individual lites larger than 9 square feet (0.9 m²) in area and with the lowest edge less than 18 inches (450 mm) above the floor.

Most safety glazing must comply with Consumer Products and Safety Commission requirements identified as *CPSC 16 CFR Part 1201*. This standard establishes two categories of performance. *Category II glazing* has a higher impact resistance than *Category I glazing* and is required for larger pieces of glazing and glazing in what are considered more hazardous locations. In other locations, Category I glazing may be used. In some instances the building code also recognizes a separate but comparable standard, *ANSI Z97.1*, for compliance. Safety glazing must be permanently etched or marked in the factory with an indication of its testing compliance so that, once installed, it is possible to verify that appropriate products have been used where required. Tempered glass, heat-strengthened laminated glass, and some plastics can all be used in safety glazing applications.

Fire-Rated Glass

Fire-rated glass may be used in buildings in two ways: in its traditional role as glazing within fire doors and fire windows, or, with more recently available higher-performing glass products, as a complete substitute for rated wall assemblies.

Fire-protection rated glazing is used within fire-rated door and window assemblies. Depending on the specific application, the glazing must pass one of several tests that demonstrate resistance to the passage of smoke

and flame for the required duration, and if in a door or other location considered hazardous for human impact, also meet safety glazing criteria. Fire-protection rated glazing used in fire doors must achieve ratings of 20 to 90 minutes, depending on the requirements of the wall in which the door is located. Glazing in fire windows must achieve ratings of 20 to 45 minutes, and is limited to windows in walls with a 1-hour fire-resistance rating or less, with a glazing area not exceeding 25 percent of the wall overall.

Historically, *wired glass*—annealed glass with a wire mesh embedded in it while hot—was used for glazing in fire doors and fire windows. Though this glass shatters in the heat of fire, the internal wire reinforcement continues to hold the glass in the opening where it can block the passage of smoke and fire. Today, most fire-protection rated glazing is wire-free, made from either specially tempered glass or optical-quality ceramic, both capable of meeting the necessary fire and safety glazing criteria. Traditional wired glass can be used in locations not subject to safety glazing requirements; if a protective film is added, it can be used in safety glazing locations as well.

Fire-resistance rated glazing can substitute, in full, for solid, fire-rated wall assemblies. To qualify for this application, the glazing must pass the same fire testing applied to walls and partitions, ASTM E119, rather than the less stringent tests applied to fire-rated doors and windows. (This test is explained further in Chapters 1 and 19.) In addition to preventing the passage of flame and smoke for the prescribed time period, the glazing must also protect combustible materials on the side of the glazing opposite the fire from ignition by blocking radiant heat from the fire and limiting the rise in surface temperature of the glass itself on its non-fire side.

Two types of glazing can pass ASTM E119 testing. In one, a heat-absorbing polymer gel is contained between two sheets of tempered glass. In the other, thin layers

of transparent intumescent material are sandwiched between multiple layers of annealed glass. The glazing units range in thickness from ¾ in. (19 mm) to a little over 2 in. (50 mm), depending on the fire rating and type of unit. When either glass type is exposed to the heat of fire, the gel or intumescent interlayers react to form opaque, insulating layers. As a result, these products not only resist the passage of flame and smoke, they also limit the rise in surface temperature of the glass on the side opposite the fire and prevent the transfer of radiant heat through the glazing. Both of these types can also meet safety glazing requirements. When used with a tested framing system, fire-resistant rated glazing can provide the equivalent protection of a wall assembly with up to a 2-hour rating. However, fire-resistance rated glazing is also very expensive, so its use remains limited to specialized or high-profile applications.

Fritted Glass

With *fritted* or *silkscreened glass*, the surface of the glass is imprinted with silkscreened patterns of ceramic-based paints. The paints consist primarily of pigmented glass particles called *frit*. After the frit has been applied to the surface of the glass, the glass is dried and then fired in a tempering furnace, transforming the frit into a hard, permanent ceramic coating. Many colors are possible in both translucent and opaque finishes. The most common patterns are dot and stripe motifs (Figure 17.10), but custom-designed patterns and even text are easily reproduced. By controlling the transparency and density of the fritted pattern, fritted glass can also be used to control the penetration of solar light and heat into a space.

Spandrel Glass

Frits are also used to create special opaque glass sheets for covering

FIGURE 17.10
Fritted patterns modulate the sunlight that enters a theater lobby. *(Photograph of Pilkington Planar System by W & W Glass Systems, Inc.)*

spandrel areas (the bands of wall around the edges of floors) in glass curtain wall construction (Figure 17.11). A uniform coating of frit is applied to what will be the interior surface of the glass. Some *spandrel glasses* are made as similar as possible in exterior appearance to the window glass that will be used above and below the spandrels. But, even with reflective coated glass it is difficult to make the spandrels indistinguishable from the windows under all lighting conditions, so most spandrel glasses are made to contrast with the adjacent window areas. In some cases, spandrel glass is supplied with thermal insulation and vapor retarder already applied to the interior side of the glass. Spandrel glass is usually tempered or heat strengthened to resist the thermal stresses caused by solar heating behind the spandrel.

Tinted and Reflective Coated Glass

Tinted and reflective glass products are one method used to reduce unwanted sunlight from entering a building, to reduce visual glare and unwanted solar heating within the building.

Tinted Glass

The transparency of glass to visible light is called its *visible light transmittance (VT)*. It is measured as the ratio of visible light that passes through the glass relative to the amount of light striking the glass. Clear glass has a visible light transmittance in the range of 0.80 to 0.90, meaning that 80 to 90 percent of the visible light striking the glass passes through. The

FIGURE 17.11
The 1301 Second Avenue tower, Seattle, by architects NBBJ, combines blue spandrel glass with lighter, blue-tinted insulated glass units for the windows. *(Photo by Joseph Iano.)*

heat admitted through a particular glass to the total heat energy striking the glass. SHGC accounts for the solar radiation that passes through glass, as well as for heat that is conducted or radiated into the space due to heating of the glass itself. Clear glass has a solar heat gain coefficient ranging from about 0.90 to 0.70, depending on its clarity and thickness. Solar heat gain coefficients for tinted glasses range from about 0.70 to 0.35, meaning that these glasses allow 70 to 35 percent of the solar energy striking the glass to pass through. Generally speaking, for buildings dominated by a heating load, glass with a high SHGC is desirable to take advantage of passive solar heat gains. In buildings dominated by cooling, glass with a low SHGC is preferable to minimize unwanted solar heating. (*Shading coefficient*, a measure similar to SHGC, is an older measure of reduction in solar heat gain that has been mostly replaced by SHGC.)

Visible transmittance and solar heat gain coefficient can be combined to determine the *light to solar gain (LSG) ratio*, a useful measure of the overall energy-conserving potential of glass. The LSG ratio is defined as the visible light transmittance divided by the solar heat gain coefficient. A glass with a high LSG ratio admits a relatively large portion of visible light in comparison to the amount of solar heat admitted, combining the greatest daylighting potential with the least solar heating potential. Green- and blue-tinted glasses tend to have high LSG ratios, whereas those of bronze, gold, and gray tints tend to be lower.

Reflective Coated Glass

Thin, durable films of metal or metal oxide can be deposited on the surface of either clear or tinted glass

remaining 10 to 20 percent is either reflected or absorbed by the glass and converted to heat.

Tinting glass reduces its visible light transmittance. *Tinted glass* is made by adding small amounts of selected chemical elements to the molten glass mixture to produce the desired hue and intensity of grays, bronzes, blues, greens, and golds. The visible light transmittance of commercially available tinted glasses ranges from about 0.75 in the lightest

tints to 0.10 for dark gray. The overall reduction in solar heat gain is often significantly less, however, because the solar radiation absorbed by the glass and converted to heat must go somewhere, and a substantial portion of it is conducted or reradiated to the interior of the building (Figure 17.12).

The effectiveness of glass in reducing heat gain from solar radiation is described as its *solar heat gain coefficient (SHGC)*; it is the ratio of solar

CLEAR FLOAT GLASS   REFLECTIVE GLASS   DOUBLE GLAZING WITH REFLECTIVE COATING INSIDE THE OUTER LIGHT

FIGURE 17.12

A schematic representation of the effect of three different glazing assemblies on incoming sunlight. Outdoors is to the left. The relative widths of the arrows indicate the relative percentages of the incoming light transmitted, reflected, and absorbed. In clear float glass, to the left, most of the light is transmitted, with small quantities reflected, or absorbed and reradiated as heat. Reflective coated glass, at the center, bounces a large proportion of the light back to the outdoors, and also absorbs and reradiates a significant amount. (Because the reflective coating is on the inner glass surface, the larger portion of reflected light is reflected from this surface.) In double glazing, many different combinations of types of glass are possible; the one shown on the right of this diagram utilizes glass with a reflective coating on the inner side of the outer lite.

sheets under closely controlled conditions to make *reflective coated glass,* also called *solar control glass.* Depending on its composition, the film may be applied to either the inside of the glass or the outside. In double glazing, it may also be applied to either of the surfaces that face the space between the layers of glass. While remaining thin enough to see through, the film reflects a substantial portion of the incident visible light. Visible light transmittance and solar heat gain coefficients for reflective coated glasses vary significantly, depending on the density of the

metallic coating and the tinting of the glass to which it is applied. Reflective coated glasses appear as mirrors from the outside on a bright day and are often chosen by architects for this property alone. At night, with lights on inside the building, they appear as dark but transparent glass.

The sunlight reflected by a building glazed with reflective coated glass can be helpful in some circumstances by lighting an otherwise dark urban street space. It can also create problems in other situations by bouncing solar heat and glare into neighboring buildings and onto the street.

> Who when he first saw the sand and ashes . . . would have imagined that in this shapeless lump lay concealed so many conveniences in life. . . . [B]y some such fortuitous liquefaction was mankind taught to procure a body at once in a high degree solid and transparent; which might admit the light of the sun, and exclude the violence of the wind; which might extend the sight of the philosopher to new ranges of existence . . . .
>
> —Dr. Samuel Johnson, writer and lexicographer, *The Rambler,* April 17, 1750

Insulating Glass

Glass, on its own, is a poor thermal insulator. A single sheet of glass (*single glazing*) conducts heat about

5 times as fast as 1 inch (25 mm) of polystyrene foam insulation and 20 times as fast as a well-insulated wall. A second sheet of glass applied to a window with an airspace between the two sheets (*double glazing*) cuts this rate of heat loss in half, and a third sheet with its additional airspace (*triple glazing*) reduces the rate of heat loss to about a third of the rate through a single sheet. A triple-glazed window, however, still loses heat many times faster than the wall in which it is placed. Continuing to add more sheets of glass and airspaces to further improve the thermal performance of a glazing unit becomes increasingly impractical, as it continues to add weight, bulk, and expense to both the glass unit and the frame that holds it. As discussed later in this chapter, triple and quadruple glazing can also be made using thin, specially coated plastic films for the inner layers of the unit, resulting in lighter, thinner units.

To prevent moisture condensation within such *insulating glass units*, or *IGUs*, the units are hermetically sealed at the time of manufacture with dry air pumped into the space between the lites. A hollow metal *edge spacer* or *spline* is inserted between the edges of the sheets of glass, and the edges are closed with a sealant compound. A small amount of a chemical drying agent, or *desiccant*, inside the spacer remains in the finished unit to remove residual moisture from the trapped air or from air that seeps into the unit after manufacture. When an insulating glass unit does exhibit internal condensation, it is usually a sign of failure of the edge seal, at which point the unit must be replaced. To increase the longevity of the insulating glass unit, a *dual seal spacer*, combining a primary butyl seal with a secondary seal of silicone, urethane, or other organic sealant may be used (Figure 17.13).

Because the edge seals are more conductive of heat than the glass-only portions of an insulating glass unit, they tend to lower the insulating

efficiency of the unit as a whole. Making the spacer from a less conductive metal, such as stainless steel instead of aluminum, is one way to improve thermal performance. To achieve more significant efficiency gains, *warm edge spacers*, made of thermally broken aluminum and plastic or of extruded rubber or dense foam, are used. Warm edge spacers also result in a more uniform surface temperature across the entire inner surface of the glazing, increasing occupant comfort and reducing the chance of condensation around the inner edges of the unit. Very high-efficiency warm edge spacers may have thermal conductivities as little as one-tenth that of traditional aluminum edge seals (Figure 17.13).

The thickness of the airspace in insulating glass units is less critical to the units' insulating value than the mere presence of the airspace: From ⅜ inch (9 mm) up to about 1 inch (25 mm) of thickness, the insulating value of the airspace increases somewhat, but above that thickness little additional benefit is gained. The most common overall thickness for large, double-glazed lites is 1 inch (25.4 mm), which results in an airspace ½ inch (13 mm) thick if ¼-inch (6-mm) glass is used.

The thermal performance of insulated glazing units can also be improved by introducing gases with greater density and lower thermal conductivity than that of ordinary air into the space between the sheets of glass. Depending on the gas used and the thickness of the space, improvements in thermal performance of 12 to 18 percent are possible. Argon and krypton are the gases most commonly used.

The performance of glazing as a thermal insulator is quantified as its *U-Factor*. U-Factor is expressed as BTUs per square foot-hour-degree Fahrenheit (BTU/ft²-hr-°F) or, in metric units, as watts per square meter-degree Kelvin (W/m²-°K). U-Factor is the mathematical reciprocal of R-value, and as such, lower values represent improved thermal performance. Some examples of glazing configurations and their U-Factors and R-values are listed in Figure 17.14. For more detailed information, consult glass manufacturers' product literature.

The performance of insulated glazing units can also be improved by evacuating most of the air from the space between the glass sheets. In such *vacuum-insulated glazing*, glass sheets are spaced relatively

FIGURE 17.13

Left, a dual-sealed, metal spacer for a double-glazed unit. The spacer is filled with desiccant. Small holes in the inward-facing side of the spacer allow the desiccant to absorb moisture from the airspace. Right, one type of warm edge spacer, also dual-sealed. The thin foil vapor barrier between the foam spacer and seals prevents moisture diffusion into the dry air space. Desiccant is embedded within the foam.

	VT	SHGC	LSG	U[1]	R[2]
Single glazing, clear	0.90	0.85	1.3	1.1	0.91
				6.3	*0.16*
Double glazing, clear	0.79	0.70	1.1	0.47	2.1
				2.7	*0.37*
Double glazing, medium gray tinted	0.40	0.45	0.9	0.47	2.1
				2.7	*0.37*
Triple glazing, clear	0.53	0.52	1.0	0.34	2.9
				1.9	*0.52*
Double glazing, clear, low-e, low SHGC	0.64	0.27	2.4	0.28	3.6
				1.6	*0.63*
Double glazing, clear, low-e, high SHGC, argon filled	0.78	0.63	1.2	0.27	3.7
				1.5	*0.65*
Triple glazing, clear, low-e, medium SHGC, krypton filled	0.65	0.50	1.3	0.15	6.7
				0.85	*1.2*

[1] U-Factor: Btu/ft^2-hr-$^\circ F$ followed by *W/m^2-$^\circ K$*
[2] R-Value: ft^2-hr-$^\circ F/Btu$ followed by *m^2-$^\circ K/W$*

FIGURE 17.14
Comparative properties of sample glass units. Note the high light-to-solar-gain ratio possible with low-e double glazing. Higher LSG values indicate better overall energy efficiency of the glazing unit (in buildings where cooling loads dominate). U-Factors and R-values are center-of-glass values and, for the insulating glass units listed, do not account for reduced thermal performance around the edges due to the greater conductivity of the spacers. The lower the U-Factor (higher the R-value), the better the thermal performance of the unit.

close together, separated by a barely perceptible array of small-diameter spacers that prevent the sheets from bowing inward due to the difference in air pressure on their two sides. Units as little as ¼ inch (6 mm) in thickness can replace single glass in existing windows, with minimal visual impact but providing thermal performance comparable to conventional double glazing. They can also be used as one lite in an otherwise conventionally constructed double- or triple-glazed insulating unit. Higher-performing units that incorporate low-emissivity coatings (see the next section) are expected to arrive on the construction market soon. These are predicted to achieve U-Factors as low as 0.08 BTU/ft²-hr-°F (0.5 W/m²-°K) and in units only roughly ½ inch (12 mm) thick.

Low-Emissivity Coated Glass

The thermal performance of glazing can be improved substantially by the use of glass with a *low-emissivity* (*low-e*) *coating*. Low-e coatings are ultrathin, virtually transparent, and almost colorless metallic coatings that selectively reflect solar radiation at different wavelengths. They have a high visible light transmittance and, depending on the particular coating, a low transmittance for some or all types of infrared radiation (heat).

Low-e coated glass is most commonly used as one of the two lites in double glazing, where it offers several benefits: By reducing the radiant transfer of heat between individual lites, the overall thermal transmittance of the glazing unit is reduced. This allows low-e double glazing to meet or exceed the thermal performance of ordinary triple glazing. By reflecting the majority of the infrared component of solar radiation, low-e double glazing can also simultaneously provide high visible light transmittance with low solar heat gain, allowing such units to achieve the highest LSG ratios of any insulated glass type.

By varying the properties of the low-e coating and by combining it with different types of tinted glass, the performance characteristics of the glazing unit can be tailored to meet different needs. For buildings dominated by wintertime heating loads, low-e units with high U-Factors (to minimize heat loss) and high solar heat gain coefficients (to promote wintertime solar heat gains) may be selected. For buildings dominated by cooling loads, units with low solar heat gain coefficients (to minimize solar heating) and lower visible light transmittance are used (Figures 17.14 and 17.15). Like laminated glass, low-e coated glass can reduce the transmittance of solar wavelengths responsible for fading of materials, thereby reducing damage to interior finishes and furnishings. Low-e coated glass may also be used in single or triple glazing to improve the thermal performance of these glass types.

When specifying glass with any type of low-e glass or reflective coating, it is necessary to specify on which glass surface the coating is to be located. By convention, glass surfaces are numbered starting from the exterior side of a glazing unit and working inward. In single glazing, the outward face is *surface number* 1 and the inward face is surface number 2. In double glazing, the outward face of the outer glass lite is surface number 1 and the inward face of this lite is surface number 2; the outward face of the inner glass lite is surface number 3, and its inward face is surface number 4. In low-e double glazing, the low-e coating is most commonly located on the number 2 surface, although, where a high solar heat gain coefficient is desired, it may be located on the number 3 surface instead.

Low-e coatings can also be applied to very thin membranes of transparent plastic. One or two of these plastic films can be installed within the center of the airspace or gas space of a double-glazed unit, stretched tight, parallel to the sheets of glass, where they act as virtually weightless additional glazing elements. Combined with selective properties of the low-e coating, thermal performance values ranging from U-0.17 to U-0.05 BTU/ft²-hr-°F (0.95 to 0.28 W/m²-°K) are reported by manufacturers of glazing units incorporating these films (see Figure 18.11).

Self-Cleaning Glass

Glass tends to attract dirt, and must be washed periodically both inside and out to maintain its transparency. *Self-cleaning glass* is coated with titanium oxide on its exterior surface. This coating acts as a catalyst that enables sunlight to convert organic particles to carbon dioxide and water. It also causes rainwater to run down the surface in sheets rather than to bead up. Nonorganic material, such as sand, is unaffected by the catalyst, but the sheets of water are more effective at removing such matter from self-cleaning glass than on conventional glass. The coating is applied only to the outside of the glass; therefore, the interior surface of the glass must be washed manually.

Glass That Changes Its Properties

Glass that can change its optical properties is called *chromogenic glass*. *Thermochromic glass* becomes darker when it is warmed by the sun. *Photochromic glass* becomes darker when exposed to bright light. Both types are potentially useful as passive devices to reduce cooling loads in buildings.

Electrochromic glass changes its transparency in response to the passage of electric current. Also called *switchable glass*, it can be actively controlled by building occupants or

FIGURE 17.15
Low-e, low-SHGC glazing on the southwest facade of a university lab sciences building.
(Architect: ZGF. Photo by Joseph Iano.)

automated systems, allowing, in comparison to passive technologies, more precise response to requirements for control of solar heat gain, daylighting, or occupant privacy. Electrochromic glass products typically consist of two or more laminated sheets of glass with one of several varieties of solid-state or ion-conducting interlayers that change their optical properties in response to electrical stimulation. Products for both interior and exterior applications are available.

Gasochromic glass is another developing switchable technology for altering the light transmittance of glass, in which the transparency of a reactive coating on the number 2 surface of an insulated glass unit is altered by the pumping of gas into or out of the interstitial space of the unit.

Other Types of Glass

Glass can be produced with an amazing range of physical properties and variations in appearance, and new products with unique characteristics continue to be developed. Fins of *structural glass* function as beams to resist wind loads in very tall or wide curtain walls, and one glass manufacturer has produced hollow glass cylinders prestressed with steel wire running through their center axes that it claims can take the place of concrete or steel elements in resisting structural compressive loads.

Antireflective glass minimizes residual reflections that normally occur when light levels differ significantly on the opposite sides of glass. It is used for glazing in showrooms, display areas, sports stadiums, artwork framing, and other applications where the highest possible optical quality is desired. Mirrors are made of *mirror glass*, which has a thin silver-based coating on its back side. A thin layer of copper applied over the silver prevents corrosion, and a second layer of backing paint provides additional protection. *Patterned glass*, which is hot glass rolled into sheets with many different surface patterns

and textures, is used where light transmission is desired but vision must be obscured for privacy. Glass manufactured with a high percentage of lead oxide can be used as *radiation-shielding glass*. *Photovoltaic glass* is coated with a thin film of amorphous silicon that generates electricity from sunlight, and can thus help reduce a building's net electricity consumption. (See Chapter 16 for more information about photovoltaic systems in buildings.) Traditional *stained glass* and contemporary *colored glass*, formulated with ingredients that alter the color of the glass, can be used in a wide range of artistic and architectural applications. Glass may be blown, molded, fused, and colored to produce an infinite variety of types of *art glass* used for decorative and sculptural purposes.

Plastic Glazing

The two most common transparent plastic glazing materials are *acrylic* and *polycarbonate*. Both are more expensive than ordinary float glass. Both have very high coefficients of thermal expansion, which cause them not merely to expand and contract with temperature changes, but also to bow visibly toward the warm side when subjected to high indoor–outdoor temperature differentials. This, in turn, requires that plastic sheet materials be installed in their frames with relatively expensive glazing details that allow plenty of linear movement and rotation. Both polycarbonate and acrylic are soft and easily scratched, although more scratch-resistant formulations are available.

Plastic glazing is most commonly used where glass is inappropriate: Plastics can be cut to shapes with inside corners (L-shapes and T-shapes, for example) that are likely to crack if cut from glass. They can be bent easily to fit in curved frames. They can be heat-formed into domed glazing for skylights. Plastics such as polycarbonate, which are literally impossible

to break under ordinary conditions, are also widely used for windows in buildings where vandalism is a problem or high impact resistance is required. Polycarbonate plastics can be manufactured in a variety of colors and varying degrees of transparency. They can also be manufactured in a double-walled configuration, called *cellular polycarbonate glazing*, creating hollow panels, roughly ¼ inch to 1½ inches (6–40 mm) thick, with greater stiffness and better thermal performance than solid sheets. Plastic glazing sheets can also meet safety requirements for glazing used in areas subject to human impact.

Translucent but nontransparent plastic sheets reinforced with glass fibers (*fiberglass-reinforced polyester glazing*) are also used in buildings. Corrugated sheets are used for industrial skylights and utilitarian roofs. Thin, flat sheets of plastic specially formulated with a high translucency to solar energy are used for skylights and low-cost solar collector glazing.

Aerogel-Filled Glazing

Aerogel, a silicon-based foam that is more than 99 percent air, can be used to fill the cavity in double-glazed glass or plastic products. Though aerogel was invented many decades ago, its commercialization was delayed by its fragility and high cost of manufacture—problems more recently solved. Aerogel is milky in color, not fully transparent, and has a visible transmittance that varies with its thickness. Aerogel-filled glazing has a good light-to-solar-gain ratio, making it an efficient source of diffuse, low-contrast, natural daylight. Currently available aerogel products can achieve insulating values of about R-8 per inch (RSI-1.4 per 25 mm), more than twice that of glass fiber insulation. Products under development that rely on nanotechnology to improve thermal performance are predicted to have insulating values as high as R-40 per inch (RSI-7 per 25 mm).

GLAZING

Glazing Small Lites

Small lites of glass are subjected neither to large wind force stresses nor to large amounts of thermal expansion and contraction. They may be glazed by very simple means (Figures 17.16 and 17.17). In a traditional wood sash, the glass is first held in place by small metal *glazier's points* and then sealed on the outside with *glazing putty*, a simple compound of linseed oil and pigment that gradually hardens by oxidation of the oil. Putty must be protected from the weather by subsequent painting, and tends to harden and become brittle with age.

As an alternative to glazing putty, newer latex and silicone caulking compounds that can be applied more quickly and need not be painted may be used for field glazing. Improved, more adhesive, and more elastic putties or *glazing compounds* are employed for factory-glazed sash as well.

Glazing Large Lites

Large lites of glass, those over 6 square feet (0.6 m²) in area, require more care in glazing. Wind load stresses on each lite of glass are higher, and the glass must span farther between its supporting edges. Any irregularities in the frame of the window may result in distortion of the glass, highly concentrated pressures on small areas of the glass, or glass-to-frame contact, any of which can lead to abrasion or fracture of the glass. In large lites, thermal expansion and contraction can also cause stresses to build up in the glass.

The design objectives for a large-lite glazing system are:

1. To support the weight of the glass in such a way that the glass is not subjected to intense or abnormal stress patterns

2. To support the glass against wind pressure and suction

3. To isolate the glass from the effects of structural deflections in the frame of the building and in the smaller framework of *mullions* that supports the glass

4. To allow for expansion and contraction of both glass and frame without damage to either

5. To avoid contact of the glass with the frame of the window or with any other material that could abrade or stress the glass

The weight of the glass is supported in the frame by synthetic rubber *setting blocks*, normally two per lite,

FIGURE 17.16
Alternative methods of single glazing lites of glass. Glass is traditionally mounted in wood sash using glazier's points and glazing compound (A) or a wood stop nailed to the sash (B). Metal sash was once glazed in the manner shown in (C), but most metal sash is now glazed with snap-in beads and synthetic rubber gaskets, as in (D).

FIGURE 17.17
A steel sash system, with single glass glazed as shown in Figure 17.16C, creates a thin, delicate pattern of glass and steel. However, single glass and solid steel mullion window systems cannot meet contemporary energy code requirements for thermal efficiency. (*Photo by Joseph Iano.*)

located at the quarter points of the bottom edge of the lite. For support against wind loads, a specified amount of *bite* (depth of grip on the edge of the glass) is provided by the supporting mullions. If the bite is too little, the glass may pop out under wind loading; if too much, the glass may not be able to deflect enough under heavy wind loads without being stressed at the edge. The mullions, of course, must be stiff enough to transmit the wind

loads from the glass to the frame of the building without deflecting so far as to overstress the glass. The resilient glazing material used to seal the glass-to-mullion joint must be of sufficient dimension and elasticity to allow for thermal movement and irregularities in the mullions.

Glazing materials used between the mullions and the glass may be *wet glazing components* and *dry glazing components*. The wet components are mas-

tic sealants of various formulations. The dry components are rubber or elastomeric gaskets. Wet glazing, with good workmanship, is more effective in sealing against the penetration of water and air. Dry glazing is faster, easier, and less dependent on workmanship than wet glazing. The two types are often used in combination to utilize the best properties of each.

In the large-lite glazing systems shown in Figure 17.18, rubber

Outdoor side at left in all cases. System numbers have no significance as to order of preference.

KEY TO MATERIALS
1 Pre-shimmed butyl or polyisobutylene tape
2 Acrylic or butyl sealant
3 Butyl or polyisobutylene tape
4 Cellular neoprene
5 Dense neoprene roll-in gasket
6 Dense neoprene gasket
7 High range sealants (polysulfide, silicone or acrylic sealant (see Note)
8 Dense neoprene roll-in rod
9 Cellular tape

Note: Choice of cap sealant depends on opening size and other conditions; consult sealant mfr.

FIGURE 17.18

An assortment of typical large-lite glazing details, with the outdoor side to the left. (*Courtesy of The Standard Products Co. and American Architectural Manufacturers Association Windows and Selection Guide—1988.*)

setting blocks are indicated by the rectangles with Xs in them under the lower edge of the glass. Systems 1, 2, 3, 6, 9, and 10 use *preformed solid tape sealant*, a thick ribbon of very sticky synthetic that is adhered by pressure to the glass and the mullions. Several different formulations (butyl, polyisobutylene) are usually lumped together under the name *polybutene glazing tape*. The tape sealant material exerts an extremely strong hold on the glass and stays plastic indefinitely to allow for movement in the glazing system.

Systems 1, 2, 3, 8, and 10 utilize *wedge* or *roll-in gaskets*, strips of elastomeric material that are simply pushed into the gap between the glass and the mullion on the interior side to wedge the assembly tightly together and seal against air leakage. Systems 5–10 seal the outside gap with a wet glazing material. System 4 comprises a two-piece *lockstrip gasket* that is a completely self-contained glazing system.

The properties that solid tape sealants, mastic sealants, and compression gaskets have in common are that they possess the required degree of resiliency, they can be installed in the thickness required to cushion the glass against expected movements, and they form a watertight seal to both glass and frame. To guard against leakage and moisture condensation, however, *weep holes* are also provided to drain water from the horizontal mullions to the exterior of the window frame, as seen directly below the bottom edge of the glass in Systems 9 and 10 in Figure 17.18.

Advanced Glazing Systems

In their quest to design ever more minimal details for buildings, architects have encouraged the development of systems of glazing that seem, in varying degrees, to defy gravity. In *butt-joint glazing*, the head and sill of the glass sheets are supported conventionally in metal frames, but vertical mullions are eliminated, the vertical joints between sheets of glass being made by the injection of a colorless silicone sealant. This gives a strong effect of unbroken horizontal bands of glass wrapping continuously around the building (Figures 17.20, 17.21, and 17.22).

In *structural glazing*, the metal mullions lie entirely inside the glass, with the glass adhered to the mullions with *structural silicone sealant* or, more recently, *acrylic foam structural glazing tape*. Structural glazing allows the outside skin of the building to be completely flush, unbroken by protruding mullions (Figures 17.23 –17.25). Notice in Figure 17.23 that the critical silicone sealant work is done in a factory, not on a

FIGURE 17.19
Insulated glass units weighing close to 200 pounds (90 kg) each are stacked on a wheeled dolly. Vacuum suction cups with handles are used to grasp or lift the units. *(Photo by Joseph Iano.)*

FIGURE 17.20
A floor-to-ceiling mullionless butt-joint glazing system uses only a bead of silicone sealant at the vertical joints in the glass. *(Photo by Joseph Iano.)*

FIGURE 17.21
A mullionless butt-joint glazing installation, seen from the outside.
(Architects: Neuhaus & Taylor. Courtesy of PPG Industries.)

Spacers

Structural silicone sealant

Extruded aluminum mullion

Glass

Snap-on aluminum sill cover

Synthetic rubber glazing gaskets

Setting block (two per light)

Weep hole and porous baffle

FIGURE 17.22
Horizontal strip windows that must appear mullionless only from the exterior can be created by adhering the glass to interior mullions with structural silicone sealant. The sill and head are conventionally glazed, using snap-on aluminum covers to hold the interior glazing gaskets. Either single glazing, as shown, or double glazing can be used with this type of system. *(Courtesy of PPG Industries.)*

Machine screw

Extruded
aluminum
structural
mullion

Extruded
aluminum
attachment
strip

Neoprene
gasket

Glazing

Glazing strip

Backup

Structural
silicone sealant

*1. Glazing is
factory
fabricated with
metal glazing
strips adhered
to the glass
using structural
silicone sealant.*

*2. The machine
screw and
attachment
strip clamp the
glazing unit
into place.*

Snap-on
aluminum
mullion cover

Polyethylene
foam backup
rod

Structural
silicone sealant

*3. A snap-on
mullion cover
and an exterior
sealant joint
complete the
assembly.*

Figure 17.23

**Steps in the assembly of a mullion for a four-sided structural
silicone exterior flush glazing system. This system is used to
construct multistory glass walls with no metal exposed on the
exterior of the building. The adhesive action of structural silicone
sealant is the sole means by which the glass is held in place. This
system is applicable either to double glazing, as shown, or to single
glazing. Notice that the internal complexities of the aluminum
components and sealant are completely concealed when the
installation is finished. From the inside, one sees only a simple
rectangular aluminum mullion, and from the outside, only glass and
a thin bead of silicone sealant.** *(Copied by permission from PPG EFG
System 712 details. Courtesy of PPG Industries.)*

FIGURE 17.24
Structural silicone glazing shows no metal framing and only thin lines of sealant on the exterior of this marine port public arrival building. Solid wall areas are clad with vertical Douglas fir 4 × 4 (89 × 89 mm) screening, prefinished wood veneer panels, and composite aluminum panels. The structure is framed with glue-laminated wood and structural steel. *(Courtesy of Checkwitch Poiron Architects—www.checkwitchpoiron.com)*

Extruded aluminum structural mullion

Gasket

Butyl-dessicant seal

Glazing

Aluminum spacer

Extruded aluminum pressure plate

Backup rod

Silicone sealant

FIGURE 17.25
Structural spacer glazing is a patented system of flush glazing that provides a more positive attachment of the double-glazed units to the building. The glass is fastened to the mullion with an aluminum pressure plate that engages a slot in the spacer strip between the sheets of glass. The desiccant required to remove residual moisture from the airspace is mixed with the butyl sealant material in this system.

FIGURE 17.26
The low-iron glass in the Magic Johnson Theater in New York City is highly transparent, an effect that is accentuated by suspending the glass from above, using only vertical glass stiffeners to resist wind loads. Where a single sheet of glass spans from the top of the window to the bottom, the glass and stiffeners may be joined with sealant alone. To create walls that are taller than a single sheet, stainless steel fittings, as seen here, are used to join sheets at the corners and edges. *(Photograph of Pilkington Planar System by W & W Glass Systems, Inc.)*

construction site. The lites of glass are transported to the site already adhered to the aluminum channels that will connect them to the mullions. In comparison to structural silicone glazing, acrylic foam structural glazing tape exhibits superior elastic properties, it can be applied more quickly and with less waste, it develops its adhesive bond more quickly, and it produces a cleaner visual appearance.

Most striking of all are the *suspended glazing systems*, used primarily for high walls of glass around building lobbies and enclosed grandstands. In the *glass mullion system*, the tempered glass sheets are suspended from above on special clamps and are stabilized against wind pressure by perpendicular stiffeners, also made of tempered glass (Figure 17.26). Alternatively, structural columns (Figures 17.1, 17.10, and 17.27) or

systems of stainless tension cables and glass fittings (Figure 17.28) may be used to brace the glass.

Stainless steel cable suspension systems may also be used to support overhead glass (Figures 17.29, 17.30, and 17.34). In any tension cable system, the building structure around the perimeter of the opening must be strong and stiff enough to resist the pull of the cables that sustain the glass.

FIGURE 17.27
This suspended glazing system relies on anticlastic curvature (compound concave and convex curvature) and a system of tension cables to give the glass wall resistance to both positive and negative wind forces. *(Photo by Joseph Iano.)*

FIGURE 17.28
One of the spider fittings used in the suspended glass wall seen in Figure 17.27. The combination of parts allows rotation in any plane and also prevents the fitting from introducing bending forces into the glass when the wall deflects under varying wind loads.
(Photo by Joseph Iano.)

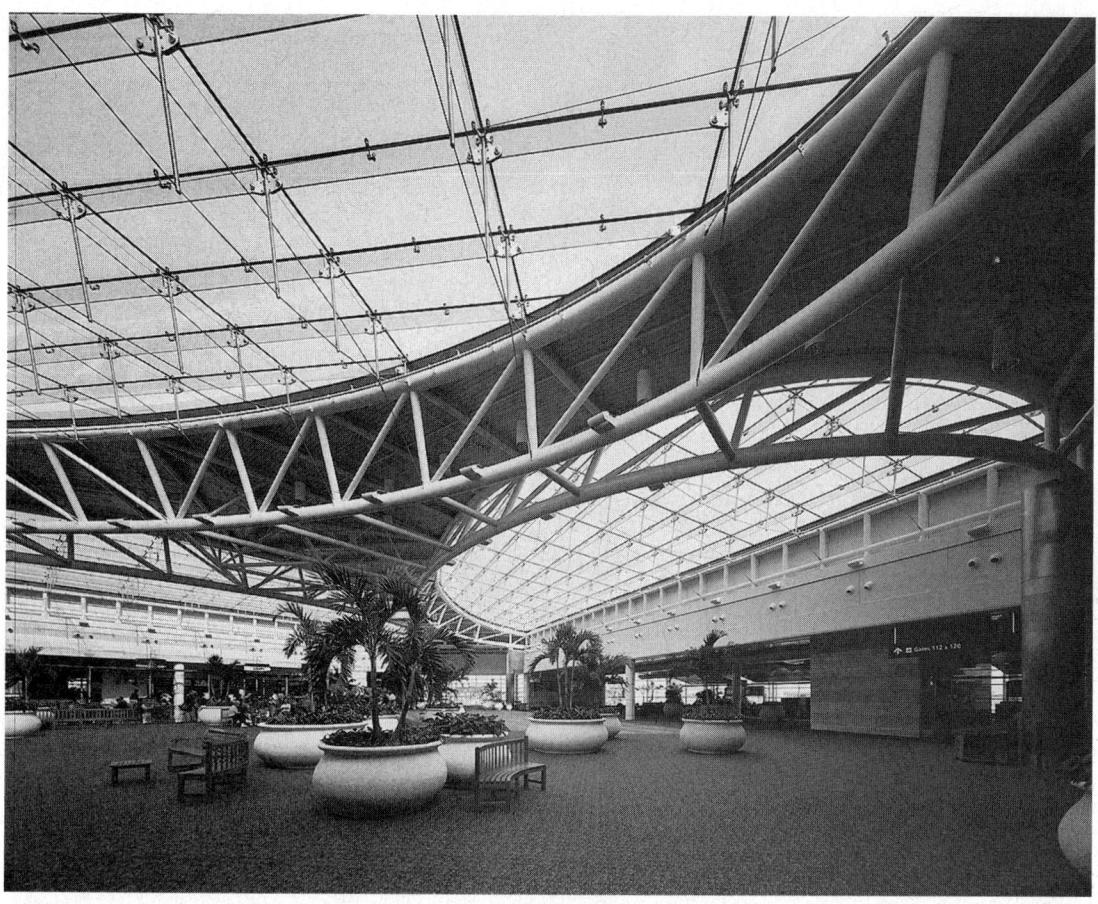

FIGURE 17.29
Airside 2 Terminal at Orlando International Airport, designed by HOK Architects, features large areas of insulating laminated glass units. Laminated glass fins serve as beams to conduct the weight of the roof to vertical stainless steel rods that are supported by a stainless steel cable tensile structure. The downward-hanging cables transmit the weight to the stiff steel pipe trusses around the perimeter, which also resist the inward pull of the cables. The upward-arching cables hold the glass surface down against possible suction forces exerted by wind. Frit was used to diminish reflected glare on glass facing the control tower, and low-e coatings were applied to the insulating glass. *(Photograph of Pilkington Planar System by W & W Glass Systems, Inc.)*

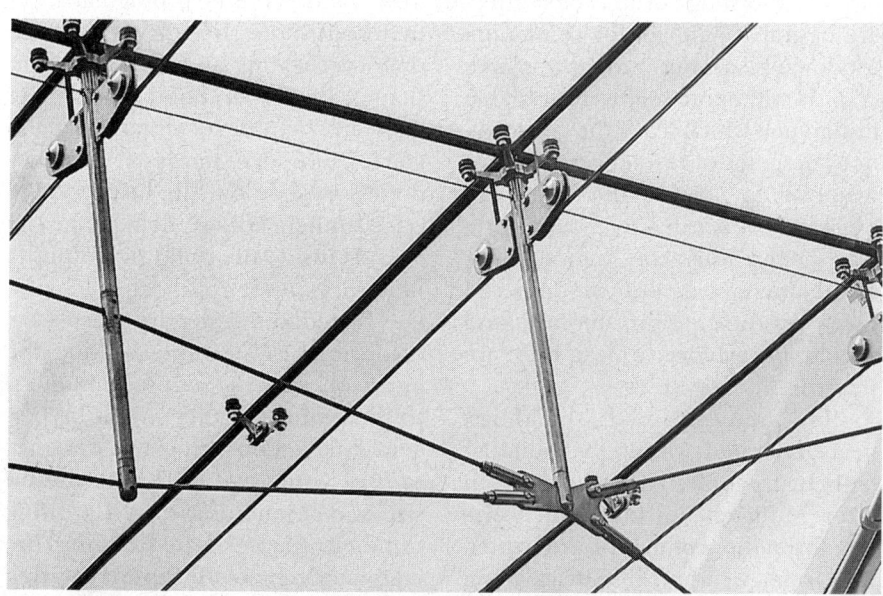

FIGURE 17.30
A detail of the glass roof supports. Notice the four-point and two-point stainless steel spider fittings that attach the glass components to each other and to the metal supporting structure. *(Photograph of Pilkington Planar System by W & W Glass Systems, Inc.)*

FIGURE 17.31
Leaded diamond-pane windows of handmade rolled glass in an Elizabethan English bay window. *(Photo by Edward Allen.)*

Imagine a city iridescent by day, luminous by night, imperishable! Buildings, shimmering fabrics, woven of rich glass; glass all clear or part opaque and part clear, patterned in color or stamped to harmonize with the metal tracery that is to hold it all together, the metal tracery to be, in itself, a thing of delicate beauty consistent with slender steel construction . . . .

—Frank Lloyd Wright, in *Architectural Record*, April 1928

GLASS AND ENERGY

Glass is a two-way pipeline for the flow of both conducted and radiated heat. As noted previously, glass, even when doubled or tripled, conducts heat rapidly into or out of a building. It can also collect and trap large amounts of solar heat inside a building.

In residential buildings, the conduction of heat through glass should be minimized in the extremely hot or cold seasons of the year. Double glazing, low-e coatings, low-conductivity gas fills, and snug curtains or shutters are desirable features for residential windows. Warming sunlight is welcome in winter but highly undesirable in summer, which leads the conscientious designer of residences to orient major windows toward the south, with overhangs or sunshades above to protect them against the high summer sun. Large east or west windows can cause overheating in summer and should be avoided unless they are shaded by nearby trees.

In nonresidential buildings, heat generated within the building by lights, people, and machinery is often sufficient to maintain comfort throughout much of the winter. In warmer weather, this heat, along with any solar heat that has entered the windows, must be removed from the building by a cooling system. In this situation, north-facing windows contribute least to the problem of cooling the building, and south-facing windows with horizontal sunshades above allow the entry of solar heat only in the winter. East and west windows are problematic, as they contribute strongly to summertime overheating and are difficult to shade. Shades or blinds inside the glass are helpful in eliminating the glare from such windows, but they do less to keep out the heat because once sunlight strikes them, its heat is already inside the building and little of it will escape.

Tinted and reflective glasses are of obvious value in controlling the entry of solar heat into buildings, to the point that they might be perceived as encouraging the designer to pay little attention to window size and orientation. But a growing number of larger-scale buildings are characterized by different glazing

section 'f'

sec. 'e'

sec. 'g'

Metal ring

Radius

Brick Relieving Arch

section at center line

section 'h'

FIGURE 17.32
Limestone and metal mullions for a Gothic church window. (*Courtesy of Indiana Limestone Institute.*)

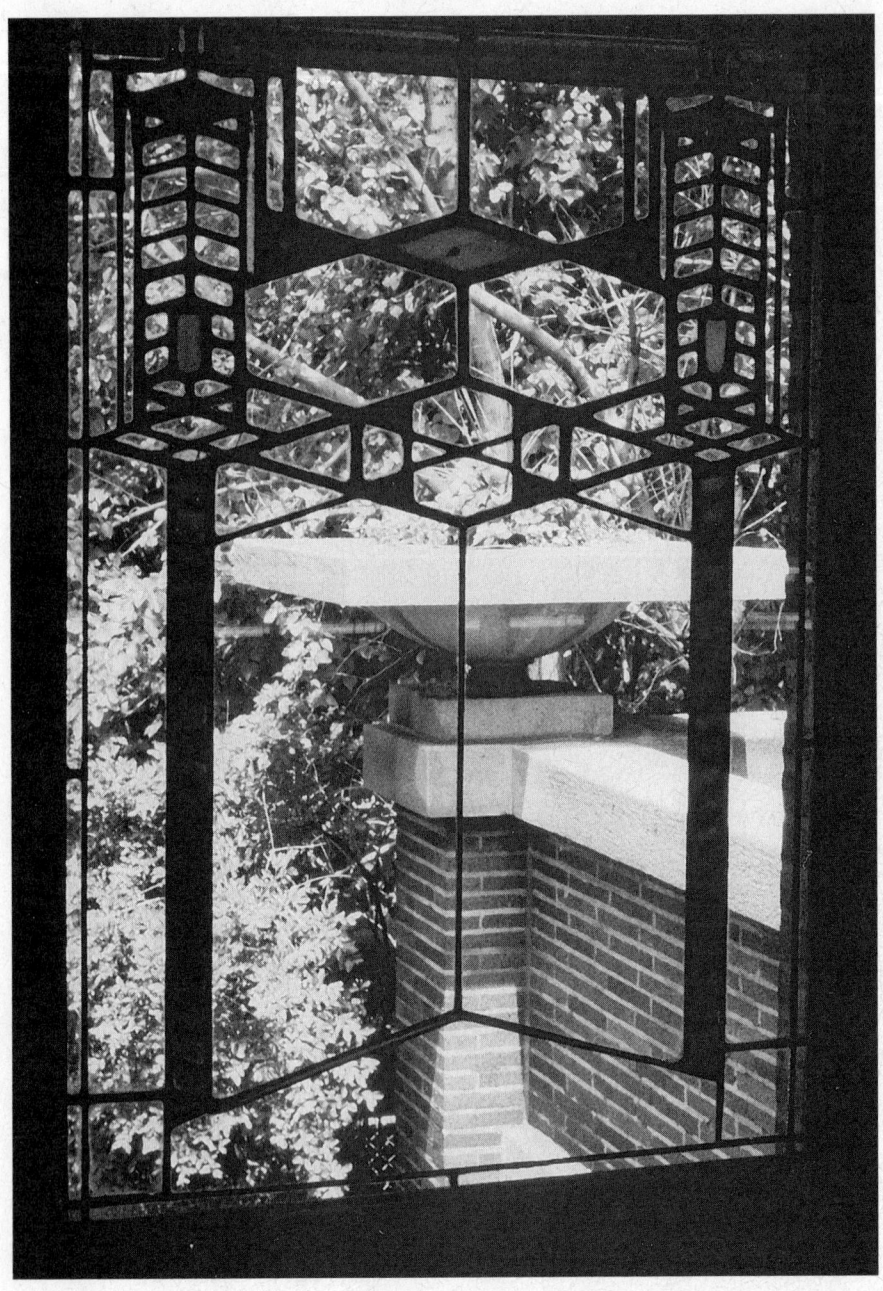

FIGURE 17.33
Leaded stained glass in the Robie House, Chicago, 1906. *(Architect: Frank Lloyd Wright.)*

schemes for the different sides of the building, each designed to create an optimal flow of heat into and out of the building for that orientation, and each making creative use of the available types of glass for this purpose. The results, as measured in occupant comfort and energy savings, are generally impressive, and the aesthetic possibilities are intriguing.

This last statement could apply equally well to the role of glass in admitting light to a building. Electric lighting is often the major consumer of energy in a commercial building, especially when the heat generated by the lights must be removed from the building by a cooling system. Daylight shining through windows and skylights, distributed throughout a space by reflecting and diffus-

ing surfaces, can reduce or eliminate the need for electric lights under many circumstances and is often more pleasant than artificial illumination. Low-cost computer models make it easy to predict the levels of daylighting that can be achieved with alternative designs, enabling more and more architects and engineers to readily apply these design strategies.

FIGURE 17.34
Airside 2 Terminal at Orlando International Airport. *(Photograph of Pilkington Planar System by W & W Glass Systems, Inc.)*

GLASS AND THE BUILDING CODES

Building codes are concerned with several functional aspects of glass: its structural adequacy against wind and impact loads; its role in providing natural light in habitable rooms; its breakage safety; its ability to prevent the spread of fire through a building; and its role in determining the energy consumption of a building.

The International Building Code provides structural criteria for determining the necessary thicknesses of glass to resist wind and other structural loads. In coastal regions where hurricanes are common, the code also requires windows or window coverings to meet requirements for resistance to the impact of objects blown against glazed areas by high winds.

The International Residential Code requires all habitable rooms to have a net exterior glazed area equal to at least 8 percent of their floor area. The use of natural light to provide interior illumination and the provision of views to the exterior for building occupants are recognized components of healthy, energy-efficient buildings and are encouraged by LEED for New Construction and other sustainable design programs.

Other regulatory concerns related to the use of glass in buildings have already been discussed: Vertical glass that is subject to human impact must be of a type that does not break into dangerous shards. Overhead glass, if it breaks, must not fall onto building occupants. Glass in fire-rated walls, doors, and windows must meet various standards to ensure that it remains sufficiently intact and performs as needed during a building fire.

Lastly, energy conservation codes limit the loss of heat (or cooling) through windows and other glazed openings, either by limiting the area of glazing or by mandating minimum levels of resistance to heat flow by the glass and its framing system.

CSI/CSC MasterFormat Sections for Glass and Glazing	
08 44 00	**CURTAIN WALL AND GLAZED ASSEMBLIES**
08 44 23	Structural Sealant Glazed Curtain Walls
08 44 26	Structural Glass Curtain Walls
08 80 00	**GLAZING**
08 81 00	**GLASS GLAZING**
08 83 00	**MIRRORS**
08 84 00	**PLASTIC GLAZING**
08 85 00	**GLAZING ACCESSORIES**
08 87 00	**GLAZING SURFACE FILMS**
08 88 00	**SPECIAL FUNCTION GLAZING**
08 88 13	Fire-Resistant Glazing
08 88 19	Hurricane-Resistant Glazing
08 88 23	Cable Suspended Glazing
08 88 36	Switchable Glass
08 88 53	Security Glazing

KEY TERMS

crown glass
cylinder glass
punty
plate glass
drawn glass
float glass
lehr
glazing
glazier
light, lite
single-strength glass
double-strength glass
annealed glass
heat-treated glass
quenching
tempered glass, fully tempered glass
heat-strengthened glass
laminated glass
polyvinyl butyral (PVB) interlayer
security glass
damage weighted transmission (Tdw)
chemically strengthened glass
safety glazing
hazardous locations
CPSC 16 CFR Part 1201
Category II glazing
Category I glazing
ANSI Z97.1
fire-rated glass
fire-protection rated glazing
wired glass

fire-resistance rated glazing
fritted glass, silkscreened glass
frit
spandrel glass
visible light transmittance (VT)
tinted glass
solar heat gain coefficient (SHGC)
shading coefficient
light to solar gain (LSG) ratio
reflective coated glass, solar control glass
single glazing
double glazing
triple glazing
insulating glass unit (IGU)
edge spacer, spline
desiccant
dual seal spacer
warm edge spacer
U-Factor
vacuum-insulated glazing units
low-emissivity coating, low-e coating
surface number
self-cleaning glass
chromogenic glass
thermochromic glass
photochromic glass
electrochromic glass, switchable glass
gasochromic glass
structural glass
antireflective glass
mirror glass

patterned glass
radiation-shielding glass
photovoltaic glass
stained glass
colored glass
art glass
acrylic glazing
polycarbonate glazing
cellular polycarbonate glazing
fiberglass-reinforced polyester glazing
aerogel
glazier's point
glazing putty
glazing compound
mullion
setting block
bite
wet glazing component
dry glazing component
preformed solid tape sealant
polybutene glazing tape
wedge gasket, roll-in gasket
lockstrip gasket
weep hole
butt-joint glazing
structural glazing
structural silicone sealant
acrylic foam structural glazing tape
suspended glazing system
glass mullion system

REVIEW QUESTIONS

1. What are the advantages of float glass over drawn glass? Over plate glass?

2. Name two situations in which you might use each of the following types of glass: (a) tempered glass, (b) laminated glass, (c) wired glass, (d) patterned glass, (e) reflective glass, (f) polycarbonate plastic glazing sheet.

3. What are the design objectives for a large-light glazing system?

4. Discuss the role of glass that faces each of the principal directions of the compass in adding solar heat to an air-conditioned office building in summer. How should windows be treated on each of these façades to minimize summertime solar heat gain?

5. In what ways does a typical building code regulate the use of glass and why?

6. What is a low-e coating, and what are its benefits?

EXERCISES

1. Examine the ways in which glass is mounted in several actual buildings and sketch a detail of each. Explain why each detail was used in its situation, and why you agree or disagree with the choice of detail used.

2. Find a book on passive solar heating of houses, and in it, a table of solar heat gains for your area. For windows facing each of the four directions of the compass, plot a curve that represents the solar heat coming through a square foot of clear glass on an average day in each month of the year. Which window orientation maximizes wintertime heat and minimizes summertime heat? Is there an undesirable orientation that maximizes summertime heat and minimizes wintertime heat?

SELECTED REFERENCES

The most current information on glass will be found in manufacturers' product literature, available either in hard copy or on the Internet.
Glass Association of North America. *Glazing Manual.* Topeka, KS, Author, updated regularly.

This handbook summarizes current practice in the production and use of glass in buildings.
Schittich, Christian, et al. *Glass Construction Manual.* Munich, Birkhauser Verlag, 2007.

This is a beautifully produced and comprehensive treatment of modern uses of glass in architecture, including the material properties of glass, glazing and building energy consumption, and the details of glazing systems.

WEB SITES

Glass and Glazing

Author's supplementary web site: **www.ianosbackfill.com/17_glass_and_glazing**
Cardinal Glass: **www.cardinalcorp.com**
Corning Museum of Glass: **www.cmog.org**
Glass Association of North America: **www.glasswebsite.com**
Nathan Allan Glass Studios: **www.nathanallan.com**
National Glass Association: **www.glass.org**
Pilkington: **www.pilkington.com**
Pilkington Planar: **www.pilkington.com/international+products/planar/**
PPG Industries: **www.ppg.com/corporate/ideascapes**
TGP Fire Rated Glass and Framing: **www.fireglass.com**
Whole Building Design Guide, Glazing: **www.wbdg.org/design/env_fenestration_glz.php**

FROM CONCEPT TO REALITY

PROJECT: Skating Rink at Yerba Buena Gardens

DESIGN ARCHITECT: Santos Prescott and Associates

EXECUTIVE ARCHITECT: LDA Architects

STRUCTURAL ENGINEER: Johnson and Neilsen/SOH&A

The project concept was to make the skating rink at Yerba Buena Gardens, San Francisco, a great urban room, intimately connected with its civic surroundings. Early in design, as part of their strategy to achieve this goal, architects Santos Prescott and Associates envisioned the north wall of the rink constructed entirely of glass. This would maximize the rink's connection to the neighboring outside garden and capitalize on views of the city skyline beyond. Finding an economical and elegant solution to the design and construction of this curtain wall became one of the essential ingredients in realizing the project (Figure A).

As the design developed, translucent glazing panels were proposed for the remaining sides of the rink enclosure to provide balanced natural illumination. As a consequence, Santos Prescott was now looking for a glazing system that could accommodate both 2¾-inch-thick translucent panels as well as standard 1-inch clear glazing for the south wall. As the building continued to take shape, additional criteria were established:

- At 30 feet in height, the south wall was too tall for the aluminum curtain wall to span on its own. A structural steel frame would be required behind the curtain wall to provide support against wind loads.

- Once the decision was made to fully support the curtain wall, structural requirements for the mullion system itself became minimal. This freed the architect to choose a relatively simple mullion profile based primarily on its ability to accommodate the glazing materials of varying thicknesses.

- To maximize transparency and fit the curtain wall within the spacing of the roof supports, a mullion grid 11 feet 3 inches wide by 5 feet high was established. These dimensions were

approaching the upper size limits for sealed insulating glass units. However, pushing for larger glass units also permitted the least number of curtain wall mullions and steel backup elements to be used.

Once a curtain wall system had been selected, work proceeded in close coordination with the structural engineer to develop the steel backup structure in greater detail (Figure B). Still striving for transparency, the goal was to find the slenderest support elements possible. Because deflection under wind load was the limiting factor in the design, a tube section was selected for its efficient sectional properties, and final sizes were determined: $3 \times 8 \times \frac{1}{2}$ inches for the vertical tubes and $3 \times 3 \times \frac{3}{8}$ inches for the horizontal tubes.

In the construction documents, the aluminum curtain wall and steel backup assembly were detailed (see Figure C) and its various components were specified. In the case of the aluminum mullion system, the specification (see excerpt) delegated the final engineering of this part of the assembly to its supplier. Specified performance criteria included resistance to structural loads, accommodation of movement, allowance for dimensional tolerances within the supporting elements, allowance for thermal expansion and contraction, and more.

Also included in the specification were requirements for the supplier to submit shop drawings and product information describing in detail the proposed curtain wall components and their attachment to the structural backup. Later in the project, these documents were reviewed by the general contractor, the architects, and the structural engineer. This review ensured that the proposed system indeed met the specified requirements and that the various parts of the curtain wall/glazing/structural system would come together as intended.

FIGURE A
Concept diagram.

1″ GLASS

GASKET

11″ +/-

2″

T.S. 2 × 8

ALUM SHIM

ALUMINUM
MULLION BY DRUSA

FIGURE B
**Schematic study of the curtain wall
and steel support assembly.**

SECTION 089 10

GLAZED ALUMINUM CURTAIN WALLS

PART 1—GENERAL

1.03 PERFORMANCE REQUIREMENTS

A. General: Provide glazed aluminum curtain wall systems, including anchorage, capable of withstanding, without failure, the effects of the following:
1. Structural loads
2. Movements of supporting structure including, but not limited to, story drift, twist, column shortening, long-term creep, and deflection from uniformly distributed and concentrated live loads
3. Dimensional tolerances of building frame and other adjacent construction
4. Thermal expansion and contraction

During the construction phase, a change was required. While preparing shop drawings, the curtain wall installer uncovered potential difficulties in attaching the curtain wall to the steel structure. When erection tolerances for the steel tubes were taken into consideration, it became apparent that it might not be possible to position the tubes with sufficient precision to meet the fastening requirements of the curtain wall. This problem was exacerbated by the ½-inch-thick side walls of the vertical steel tubes, which reduced the allowable area for fasteners on the face of the tubes to a narrow central strip. At this relatively late stage in the project, the decision was made to switch to a 4-inch-wide tube with a thinner side wall to provide a wider target area for fasteners (Figure D). In the end, the change had little if any noticeable visual impact.

This example illustrates how a successful project outcome depends on good communication among the various team members throughout the design and construction of the project, and how even seemingly minor technical considerations can affect essential aspects of a project's design.

METAL ROOFING

SM GUTTER

SM CAP

1
830

WEATHER CLOSURE

SUSPENDED CEMENT
PLASTER SOFFIT

12
850

11
850

SUSPENDED CEMENT PLASTER
CEILING ASSEMBLY

TS VERT
CURTAIN WALL SUPPORT
BEYOND © EA MULLION

4
850

PFS COLUMN

TS HOR
CURTAIN WALL SUPPORT
@ EA MULLION

6'–6"

STOREFRONT
ENCLOSURE

10
820

10
850

SUNSCREEN

4
DA890

GYP BD

15
850

COLS BEYOND

STOREFRONT

CONC CURB

3
850

CER TILE BASE

PLANTING (SEE
LANDSCAPE)

1'–0"

GARDEN LEVEL
EL 36'–6"

FIGURE C

Wall section from the construction drawing set.

Special thanks to Santos Prescott and Associates and Bruce Prescott, Principal, for assistance with the preparation of this case study.

FIGURE D

Detail study of the curtain wall attachment to tube steel.

FLAT AREA FOR ATTACHMENT

TS 3 × 8 × ½

TS 4 × 8 × ⅜

R = ¾"

FASTENERS

4" TUBE

3" TUBE

ALUMINUM MULLION

FIGURE E

Interior view of the completed building.

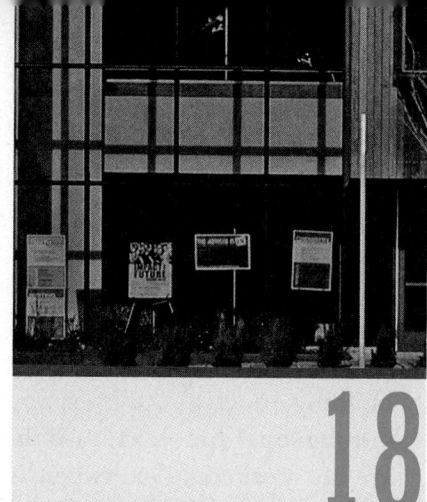

18

WINDOWS AND DOORS

- **Windows**
  Types of Windows
  Window Frames

  ───────────────────────

  PLASTICS IN BUILDING CONSTRUCTION

  ───────────────────────

  Muntins
  Glazing
  Installing Windows

  ───────────────────────

  SUSTAINABILITY OF WINDOWS AND DOORS

  ───────────────────────

- **Doors**
  Wood Doors
  Steel Flush Doors

  Fire Doors
  Egress Doors and Accessible Doors
  Door Hardware

- **Other Window and Door Requirements**
  Safety Considerations in Windows and Doors
  Structural Performance and Resistance to Wind and Rain
  Thermal Performance
  Impact Resistance
  Blast Resistance

High-performance windows and glazing contribute to a LEED Gold rating for the University of Washington's Poplar Hall. *(Architect: Mahlum. Photo by Joseph Iano.)*

Doors and windows are very special components of walls. Doors permit people, goods, and, in some cases, vehicles to pass through walls. Windows allow for simultaneous control of the passage of light, air, and sight through walls. Windows and doors, in addition to performing these important functions, play a large role in establishing the character and personality of a building, much as our eyes, nose, and mouth play a large role in our personal appearance. At the same time, windows and doors are the most complex, expensive, and potentially troublesome parts of a wall. To achieve satisfactory results, the experienced designer exercises great care and wisdom in the selection of doors and windows and in the specification of their installation.

WINDOWS

The word *window* is thought to have originated in an old English expression that means "wind eye." The earliest windows in buildings were open holes through which smoke could escape and fresh air could enter. Devices were soon added to the holes to give greater control: hanging skins, mats, or fabric to regulate airflow; shutters for shading and to keep out burglars; translucent membranes of oiled paper or cloth, and eventually of glass, to admit light while preventing the passage of air, water, and snow. When a translucent membrane was eventually mounted in a moving sash, light and air could be controlled independently of each other. With the addition of woven insect screens, windows permitted air movement while keeping out mosquitoes and flies. Further improvements followed over the centuries. A typical window today is an intricate, sophisticated mechanism with many layers of control: curtains, shade or blind, sash, glazings, insulating airspace, low-emissivity and other coatings, insect screen, weatherstripping, and perhaps a storm sash or shutters.

Windows were formerly made on the construction site by highly skilled carpenters, but today nearly all of them are produced in factories. The primary reasons for factory production are higher production efficiency, lower cost, and, most important, better quality. Windows must be made to a very high standard of precision if they are to operate easily and maintain a high degree of weathertightness for many years. In cold climates especially, a loosely fitted window with single glass and a frame that is highly conductive of heat will significantly increase heating fuel consumption for a building, cause noticeable discomfort to the occupants, and condense large quantities of water that will stain and decay materials in and around the window.

Types of Windows

Figure 18.1 illustrates in diagrammatic form the window types used most commonly in residential buildings, and Figure 18.2 shows additional types that are found largely in commercial and institutional buildings. *Fixed windows* are the least expensive and the least likely to leak air or water, because they have no operable components. *Single-hung* and *double-hung windows* have one or two moving *sashes*, which are the frames in which the glass is mounted (Figure 18.3). The sashes slide up and down in tracks that are part of the window frame. In older windows, the sashes were held in position by cords and counterweights, but today's double-hung windows usually rely on a system of springs to counterbalance the weight of the sashes. A *sliding window* is essentially a single-hung window on its side, and shares with single-hung and double-hung windows the advantage that tracks in the frame hold the sashes securely along two opposite sides. This inherently stable construction allows single-hung, double-hung, and sliding windows to be designed in an almost unlimited range of sizes and proportions. It also allows the sashes to be more lightly built than those in *projected windows*, a category that includes principally *casement, awning, hopper, inswinging,* and *pivot windows*. All projected windows have sashes that rotate outward or inward from their frames and therefore must be stiff enough to resist wind loads while being supported at only two points.

With the exception of the rare *triple-hung window*, no window with sashes that slide can be opened to more than half of its total area. By contrast, many projected windows can be opened to almost their full area. Projecting casement windows assist in catching passing breezes and inducing ventilation through the building. They are generally narrow in width, but can be joined to one another and to sashes of fixed glass to fill wider openings. Awning windows can be broad but are not usually very tall. They have the advantages of providing protection from water during a rainstorm, even when open, and of lending themselves to a building-block approach to the design of window walls (Figure 18.4). Hopper windows are more common in commercial buildings than in residential ones. Like awning windows, they will admit little or no rainwater if left open during a rainstorm. *Tilt/turn windows* (not illustrated) are a type

Fixed

Single-Hung

Double-Hung

Sliding

Casement

Awning

Hopper

Skylight

Roof Window

Terrace Door

French Door

Sliding Door

FIGURE 18.1
Basic window types.

Top-Hinged Inswinging

Side-Hinged Inswinging

Pivoting

FIGURE 18.2

Additional window types that are used mainly in larger buildings.

Interior casing

Head jamb

Exterior casing

Sash

Stool

Sill

Apron

FIGURE 18.3

Basic window nomenclature follows a tradition that has developed over many centuries. The *jamb* consists of the head *jamb* across the top of the window and the *side jamb* to either side. In practice, the head jamb is usually referred to simply as the *head* and the side jambs as jambs. The sill frames the bottom of the opening on the exterior side, and the stool does the same on the interior. *Interior casings* and *exterior casings* cover the gaps between the jambs and the rough opening, and *aprons* do the same below the sill and stool.

of projected window with clever but concealed hardware that allows each sash to operate both as a side-hinged inswinging window and as a hopper. Window types that are used almost exclusively in commercial and institutional work include horizontally and vertically *pivoting windows* and *top-hinged inswinging windows*.

Most tall buildings do not provide operable windows, due to the complexity of balancing varying air pressures on the outside of the building with those of the internal air-handling system. These make it difficult to maintain steady, comfortable air temperatures and speeds within

the building. However, with proper design (and in appropriate climates), operable units can be part of a natural ventilation strategy that provides comfortable airflows in place of a mechanical air-conditioning system. On tall buildings, inswinging window types may be preferred, as they are less vulnerable to damage by high winds. For safety reasons, such windows may be fitted with devices that limit the extent to which they can be opened.

A projected window is usually provided with synthetic rubber weatherstripping that compresses snugly around the edges of the sash when the

FIGURE 18.4
Awning and fixed windows in coordinated sizes offer the architect the possibility of creating patterned walls of glass. *(Photo courtesy of Marvin Windows & Doors.)*

window is closed. Single-hung, double-hung, and sliding windows must rely on brush-type or very low-modulus (highly flexible) weatherstripping that does not exert so much friction as to make operation of the sash difficult. However, these weatherstripping types also may not seal as tightly and are subject to greater wear over the life of the windows. As a result, projected windows are frequently more resistant to air leakage than windows that slide in their frames.

Glazed units installed in roofs are specially constructed and flashed to remain watertight in their sloping or horizontal orientation. *Skylights* may be either fixed or operable (*venting skylight*). The term *roof window* is also sometimes applied to venting skylights; at other times, it is applied more narrowly only to windowlike units that include some kind of inward rotation capability, making glass cleaning easier.

Large glass doors (which are most often supplied by window manufacturers) may slide in tracks or swing open on hinges (Figure 18.1). The hinged *French door* opens fully and, with its arms flung wide, is a more welcoming type of door than the *sliding door*, but it cannot be used to regulate airflow through the room unless it is fitted with a doorstop that can hold it securely in an open position. The French door is prone to air leakage along its seven separate edges, which must be carefully fitted and weatherstripped. The *terrace door*, with only one operating door, minimizes this problem but, like the sliding door, can open to only half its area.

Insect screens may be mounted only inside the sash in casement and awning windows (because the sash swings outward). Screens are usually positioned to the exterior side of other window types. Sliding patio doors and terrace doors have exterior sliding screens, and French doors require a pair of hinged screen doors on the exterior. Pivoting windows cannot be fitted with insect screens.

Glass must be washed at intervals if it is to remain transparent and attractive. Inside surfaces of glass are relatively easy to reach. Outside surfaces are often harder to reach, requiring ladders, scaffolding, or window-washing platforms that hang from the top of the building on cables. Accordingly, most operable windows are designed to allow washing of the outside glass surface from inside the building. Casement and awning windows are usually hinged in such a way that there is sufficient space between the hinged edge of the sash and the frame when the window is open to allow one's arm to reach the outer surface of glass. Double-hung and sliding windows are often designed to allow sash to be rotated or tilted out of their tracks to allow easy access to exterior glass (Figure 18.15). Inward-swinging window types naturally expose their outer glass surfaces to the interior when opened.

As seen in many of the following figures, windows and glass doors may also be combined side by side or stacked vertically to create larger glazed areas with any of a great variety of fixed and operable component configurations.

Window Frames

Wood

Wood is the traditional frame material for windows. It is a fairly good thermal insulator, changes size relatively little with changes in temperature, and, if free of knots, is easily worked and consistently strong. In service, though, wood shrinks and swells with changing moisture content and requires repainting every few years. When wetted by weather, leakage, or condensate, *wood windows* are subject to decay, though their resistance to decay can be improved with preservative treatments. Knot-free wood is becoming increasingly rare and expensive, so composite wood products are increasingly used. These include lumber made of short lengths of defect-free wood finger-jointed and glued together, oriented strand lumber, and laminated veneer lumber. These materials, although functionally satisfactory, are not attractive, so they are covered with wood veneer on the interior and clad with plastic or aluminum on the exterior (Figures 18.5, 18.6, and 18.7). *Clad wood windows* account for the largest share of the market for wood-framed windows.

FIGURE 18.5
Cutaway sample of an aluminum-clad wood-framed window. (*Photo courtesy of Marvin Windows & Doors.*)

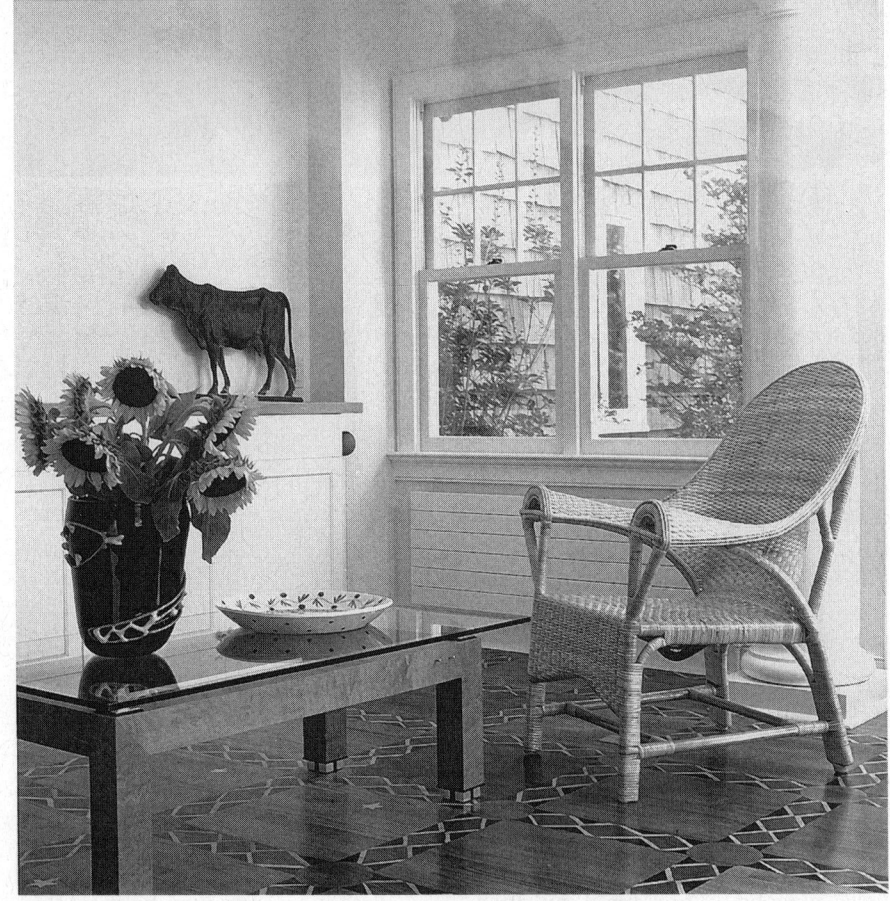

FIGURE 18.6
A pair of double-hung wood-framed windows in a dwelling. (*Photo courtesy of Marvin Windows & Doors.*)

Aluminum

Aluminum, when used in window construction, is strong, easy to form and join, and, in comparison to wood, much less vulnerable to moisture damage. The extrusion process by which aluminum sections are formed results in shapes with crisp, attractive profiles, and durable factory finishes eliminate the need for periodic repainting after installation.

Aluminum conducts heat so rapidly, however, that unless an aluminum frame is constructed with a *thermal break* made of plastic or synthetic rubber components to interrupt the flow of heat through the metal, condensate and sometimes even frost will form on interior frame surfaces during cold winter weather. *Aluminum windows* are also more costly than wood or plastic windows. The majority of commercial and institutional windows, as well as many residential windows, are framed with aluminum (Figures 18.8–18.11). Aluminum frames are usually anodized or permanently coated, as described in Chapter 21.

Plastics

Plastic window frames, though relatively new, now account for more than half of all windows sold in the U.S. residential market. Plastic windows never need painting, and they are fairly good thermal insulators. Many also cost less than wood or clad wood windows. The disadvantages of plastics are that they are not as stiff or strong as other window materials and they have very high coefficients of thermal expansion (Figure 18.12). The most common material for plastic window

FIGURE 18.7
Large double-hung wood windows and a triangular fixed window bring sunlight and views. *(Photo courtesy of Marvin Windows & Doors.)*

OPTIONAL AUTOMATIC LOCK

3-3/8" (85.7)

HEAD - DOUBLE HUNG
4

SWEEP LOCK

2-7/8" (73)

MEETING RAIL - DOUBLE HUNG
9

4
16
9
16
3
DOUBLE HUNG

4" (TYP) (101.6)

3-11/32" (84.9)

JAMB - DOUBLE HUNG
16

OPTIONAL AUTOMATIC LOCK

3-3/4" (95.3)

OPTIONAL 4-3/16" (106.4)

SILL
SINGLE / DOUBLE HUNG
3

©KAWNEER COMPANY, INC., 1997

FIGURE 18.8

The details of this commercial-grade double-hung aluminum window are keyed to the numbers on the small elevation view at the upper left. Cast and debridged thermal breaks, which are shown on the drawings as small white areas gripped by a "claw" configuration of aluminum on either side, separate the outdoor and indoor portions of all the sash and frame extrusions. Pile weatherstripping seals against air leaks at all the interfaces between sashes and frame. For help in understanding the complexities of aluminum extrusions, see Chapter 21. *(Courtesy of Kawneer Company, Inc.)*

FIGURE 18.9
Two aluminum double-hung window units in back, with an aluminum sliding window in front. *(Courtesy of Kawneer Company, Inc.)*

FIGURE 18.11
A sample of quadruple glazing in a high-thermal-performance aluminum frame. The glazing consists of two outer lites of ¼-inch (6-mm) glass and two inner plastic films, to create three airspaces within the unit. The exterior lite and the innermost film both have low-emissivity coatings. The metal foil vapor barrier in the warm edge spacer is also visible. The aluminum frame thermal break is made from two strips of polyamide plastic filled with glass fiber insulation between the strips. This window can achieve an overall thermal performance of approximately U-0.13 (U-0.75 metric). For comparison with other common window types, see Figure 18.21. *(Photo by Joseph Iano.)*

FIGURE 18.10
A cutaway sample of a standard cast and debridged plastic thermal break in an aluminum window frame. *(Photo by Joseph Iano.)*

FIGURE 18.12
A comparison of the coefficients of thermal expansion of wood, glass-fiber-reinforced plastic (GFRP), aluminum, and vinyl. Vinyl expands 15 times as much as wood, 8 times as much as GFRP, and 3 times as much as aluminum. Units on the graph are in./in./°F × 10⁻⁶ on the left of the vertical axis and mm/mm/°C × 10⁻⁶ on the right.

FIGURE 18.13
Comparative details of a single-hung residential window with an aluminum frame (*left*) and a double-hung residential window with a PVC plastic frame (*right*). The small inset drawing at the top center of the illustration shows an elevation view of the window with numbers that are keyed to the detail sections below. The comparatively thick sections of plastic are indicative of its lesser stiffness in comparison to aluminum. The diagonally hatched areas of the aluminum details are plastic thermal breaks. The inherently low thermal conductivity of the PVC and the multichambered construction eliminate the need for thermal breaks in the plastic window. (*Reprinted with permission from AAMA Aluminum Curtain Wall Design Guide.*)

FIGURE 18.14
Cutaway sample of a plastic double-hung window with double glazing and an external half-screen. (*Courtesy of Vinyl Building Products, Inc.*)

FIGURE 18.15
For ease of washing the exterior surfaces of the glass, the sashes of this plastic window can be unlocked from the frame and tilted inward. (*Courtesy of Vinyl Building Products, Inc.*)

frames is polyvinyl chloride (PVC, vinyl), which is formulated with a high proportion of inert filler material to minimize thermal expansion and contraction. Some typical PVC window details are shown in Figures 18.13, 18.14, and 18.15. For a discussion of concerns regarding the environmental and health effects of PVC, see the sidebar on plastics later in this chapter.

Glass-fiber-reinforced plastic (GFRP) windows, frequently referred to as *fiberglass windows*, are the newest product in the window market. GFRP frame sections are produced by a process of *pultrusion*: Continuous lengths of glass fiber are pulled through a bath of plastic resin, usually polyester, and then through a shaped, heated die in which the resin hardens. The resulting sash pieces are strong, stiff, and relatively low in thermal expansion. Like PVC, they are fairly good thermal insulators. However, GFRP windows are more expensive than those made of wood or plastic.

The thermal performance of both vinyl and GFRP window frames can be enhanced with foam insulation injected into the hollow spaces within the frame sections.

PLASTICS IN BUILDING CONSTRUCTION

Plastics may be loosely defined as giant molecules (polymers or copolymers) made up of large numbers of small, repetitive chemical units. Such molecular chains occur naturally, for example in the cellulose from which tree cells are structured. The first synthetic plastics were formulated in the 19th century and the development of modern plastic materials began in earnest in the 1930s. In the United States today, the construction industry consumes more than 10 billion pounds (5 billion kg) of plastics annually.

Synthetic plastics are manufactured largely from organic molecules obtained from oil, natural gas, and coal. Most are based on carbon chains, except for the silicones, which are structured around silicon (Figures A-F). *Synthetic rubber* compounds are usually classified separately from plastics, although chemically they are similar; they are often referred to as *elastomers*.

A *polymer* is composed of many identical chemical units or *monomers*. Polyvinyl chloride (PVC), for example, is a polymer produced by polymerizing vinyl chloride monomers into a long chain. A *copolymer* consists of repeating patterns of two or more monomers. High-impact polystyrene is a copolymer made up of both polystyrene and polybutadiene.

The molecular structure of a polymer or copolymer bears an important relationship to its physical properties. A high-density polyethylene molecule, for instance, is a long, compact, single chain containing up to 200,000 carbon atoms. Low-density polyethylene has a branching molecular structure that does not pack together as tightly as the single chain.

There are two broad classes of plastics. *Thermoplastic* plastics may be softened by reheating at any time after their manufacture. Upon cooling, they regain their original properties. *Thermosetting* plastics have a molecular structure that is strongly *crosslinked* in three dimensions. They cannot be remelted after manufacture. Thermosetting plastics are generally harder, stronger, and more chemically stable than thermoplastics.

Many modifiers are added to plastics to change their properties or reduce their cost. *Plasticizers* are organic compounds that impart flexibility and softness to otherwise brittle plastics. *Stabilizers* are added to resist deterioration of polymers from the effects of sunlight, heat, oxygen, and electromagnetic radiation. *Fillers*—inexpensive, nonreacting materials such as talc or marble dust—are added to reduce the cost or to improve toughness or resistance to high temperatures. *Extenders* are waxes or oils that add bulk to the plastic at low cost. *Reinforcing* fibers of glass, metal, carbon, or minerals can increase strength, impact resistance, stiffness, abrasion resistance, hardness, and other mechanical properties. *Flame retardants* are often introduced into plastics that are destined for interior use in buildings. Color can be added to plastics

FIGURE A
Polystyrene.

FIGURE B
A silicone.

FIGURE C
Polyvinyl chloride (PVC).

FIGURE D
High-impact polystyrene.

FIGURE E
High-density polyethylene.

FIGURE F
Low-density polyethylene.

with dyes or pigments. Some examples of useful modifiers are the carbon black that is often added to polyethylene as a stabilizer to improve its resistance to sunlight, the lead carbonate that is used to stabilize PVC products for use outdoors, the di-isooctyl plasticizer that converts brittle PVC to a stretchable, rubbery compound, and the chopped glass fibers that reinforce polyester to make it suitable for use in boat hulls and building components.

The plastics used in buildings range from dense solids such as those used in floor tiles to the lightweight cellular foams used for thermal insulation. They include the soft, pliable sheets used for roofing membranes and flashings and the hard, rigid plastics used for plumbing pipes. Glazing sheets are made from highly transparent plastics, whereas most plastics manufactured for other purposes are opaque.

People readily recognize the solid plastic materials used in construction, but most tend not to realize that liquid plastics are major ingredients of many protective paints and coatings. Plastics show up also in the form of *composites*, in which they are teamed with nonplastic materials: *laminates* consisting of paper and melamine formaldehyde, used for countertops and facings; sandwiches such as the foam-core plywood panels used as cladding for heavy timber frame buildings; and mixes of plastics with particulate materials, such as polyester concrete (stone aggregates cemented with a polyester binder rather than portland cement) and particleboard (wood chips and phenolic resins).

Plastics are given form through any of an almost endless list of processes. *Extrusion* manufactures long shaped sections by forcing the plastic through a shaped die.

Pultrusion, used for certain fiber-reinforced products, is much the same as extrusion, except that the section is pulled through the die rather than pushed. A host of *molding* processes cast plastic into shaped cavities to give it form. Many foam plastics can be foamed directly into the mold, which they expand to fill. Polyethylene film is produced by *film blowing*, in which air is pumped into a small extruded tube of plastic to expand and stretch it into a very thin-walled tube many feet in diameter. The large tube is then slit lengthwise, folded, and rolled for distribution. Films and sheets can be made by casting plastic onto a chilled roller or a chilled moving belt. Some plastic sheet products and many plastic laminates are produced by *calendaring*, a process in which a material or a sandwich of materials is pressed first through hot rollers, then cold rollers.

Thermoplastic sheets may be further formed by heating them and pressing them against shaped dies. The pressing force may be furnished mechanically by a matching die or by air pressure. If compressed air pushes the plastic into the die, the process is called *blow forming*. In *vacuum forming*, a pump draws the air from between the heated plastic sheet and the die, and atmospheric pressure does the rest.

Many plastics are amenable to machining processes—sawing, drilling, milling, planing, turning, sanding—like those used to shape wood or metal. Nylon and acrylics are often given final form in this way. Thermoplastic plastics can be joined with heat or solvents that soften the surfaces to be joined and weld them together as a single piece. The crosslinked chemistry of thermoset plastics does not permit welding, so these materials must be joined with

Some Synthetic Rubbers Used in Construction

Chlorinated polyethylene, chlorosulfonated polyethylene (Hypalon)	Roof membranes
Ethylene-propylene-diene monomer (EPDM)	Roof membranes, flashings
Isobutylene-isoprene copolymer (butyl rubber)	Flashings, waterproofing
Polychloroprene (Neoprene)	Gaskets, waterproofing
Polyisobutylene (PIB)	Roof membranes
Polysiloxane (silicone rubber)	Sealants, adhesives, coatings, roof membranes
Polyurethanes	Sealants, insulation, coatings
Sodium polysulfide (polysulfide, thiokol)	Sealants

PLASTICS IN BUILDING CONSTRUCTION (*CONTINUED*)

Some Plastics Used in Construction

Chemical Name (common names and trade names in parentheses)	Thermoplastic or Thermosetting	Characteristics	Some Uses in Buildings
Acrylonitrile-butadienestyrene terpolymer (ABS)	tp	Hard, dimensionally stable	Plumbing pipes, shower stalls
Alkyd	ts	Tough, transparent, weather resistant	Coatings
Epoxy	ts	Tough, hard, high adhesion	Adhesives, coatings, binders
Melamine formaldehyde	ts	Hard, very durable	Plastic laminates
Phenol formaldehyde (phenolic, Bakelite)	ts	Hard, very durable	Coatings, laminates, insulating foams, electrical boxes
Polyamides (Nylon)	tp	Tough, strong, elastic, machinable	Membranes for air-supported and tensile roofs, hardware for doors and cabinets
Polybutene	tp	Sticky, soft, weather resistant	Sealant tapes, sealants
Polycarbonate (Lexan)	tp	Hard, transparent, unbreakable	Glazing sheet, lighting fixtures, door sills
Polyesters (Mylar, Dacron)	ts	Durable, combustible	Fabrics for roofing membranes and geotextiles, matrix for glass-fiber-reinforced plastics
Polyethylene (polythene)	tp	Tough, flexible, impermeable	Vapor retarders, moisture barriers, piping, tarpaulins
Polyisocyanurate	ts	Heat resistant, fire retardant	Insulating foams
Polymethyl methacrylate (acrylic, Plexiglas, Lucite)	tp	Tough, transparent, machinable	Glazing sheet, dome skylights, lighting fixtures, illuminated signs, coatings, adhesives
Polypropylene	tp	Tough, hard, does not fatigue	Crack-control fibers in concrete, one-piece molded hinges
Polystyrene (Styrofoam, bead foam)	tp	Foam is excellent thermal insulator, solid is transparent	Insulating foams, glazing sheet
Polytetrafluorethylene (PTFE, Teflon)	tp	Heat resistant, low friction	Sliding joint bearings, pipe thread tape
Polyurethanes	ts or tp	Vary widely with form of plastic, from low-density foams to hard, dense coatings to synthetic rubbers	Coatings, sealants, insulating foams, adhesives
Polyvinyl chloride (PVC, vinyl), polyvinyl acetate (PVA, vinyl)	tp	Vary widely with form of plastic, from flexible rubberlike sheet to hard, rigid solid	Pipes, conduits, ductwork, siding, gutters, window frames, flooring sheets and tiles, coatings, wallcoverings, geotextiles, roof membranes

adhesives. Plastics can also be joined mechanically with screws or bolts. Many plastic products are designed with ingenious snap-together features so that they can be joined without fasteners.

As a group, plastics exhibit some common advantages and some common problems when used as building materials. Among the advantages, plastics are low in density; they are often cheaper than other materials that will do the same job; they have good surface and appearance qualities; and, because their molecules are made to order for each end use, they can often offer a more optimal solution to a building problem than other materials. Plastics generally are little affected by water or biological decay, and they do not corrode. They tend to have low thermal and electrical conductivity. Many have high strength-to-weight ratios. Most plastics are essentially impermeable to water and water vapor. Many are very tough and resistant to abrasion.

Their disadvantages are also numerous. Synthetic plastics are derived chiefly from oil, coal, and natural gas, all nonrenewable resources. There are also environmental and health concerns with some plastics and their manufacturing processes. For example, formaldehyde (CH_2O), a common ingredient in the manufacture of many polymers, is a human carcinogen and associated with other adverse health effects.

Polyvinyl chloride, in particular, has been singled out by some as a plastic with potentially severe hazards: Its manufacture utilizes hazardous chemicals and generates dioxin, PCBs, and other byproducts associated with cancer, neurological damage, and other serious health effects. Heavy metals, phthalates, and other harmful ingredients, used as stabilizers, plasticizers, and coloring agents, can leach out of the material and find their way into human or animal biosystems. When PVC is burned in municipal or medical waste incinerators, dioxins and other harmful chemicals are released directly into the atmosphere. As a consequence, PVC is included on the Living Building Challenge Red List, meaning it should not be used at all in buildings meeting this standard. Materials containing vinyl chloride, an ingredient in the manufacturer of PVC, are also discouraged in LEED standards. In an effort to address such concerns, the PVC manufacturing industry has responded by improving manufacturing processes, searching for substitutes for phthalates and other harmful ingredients, and instituting recycling programs to divert PVC from disposal in landfills or incineration.

All plastics can be destroyed by fire, and many give off toxic combustion products. Some burn very rapidly, but others are slow-burning, self-extinguishing, or do not ignite at all, so careful selection of polymers and modifiers can be crucial in building applications. Flame-spread ratings, smoke developed ratings, and toxicity of combustion products should be checked for each use of plastics within a building.

Plastics have much higher coefficients of thermal expansion than other building materials and require close attention to the detailing of control joints, expansion joints, and other devices for accommodating volume change. Plastics tend not to be very stiff. They deflect considerably more under load than most conventional materials. Taken together with their combustibility, this severely limits their application as primary structural materials. Many also creep under prolonged loading, especially at elevated temperatures. In strength, plastics vary from fiber-reinforced composites that are as strong as many metals (but not nearly as stiff) to cellular foams that can be crushed easily between two fingers.

Plastics tend to degrade in the outdoor environment. They are especially susceptible to attack by the ultraviolet component of sunlight, oxygen, and ozone. In many plastic products (acrylic glazing sheets and some synthetic rubbers, for example), these problems have largely been solved by adjustments to the chemistry of the material that have resulted from a prolonged program of testing and research. However, it is usually a mistake for a designer to use an untried plastic material in an exposed location without consulting its manufacturer.

Selected References

Dietz, Albert G. H. *Plastics for Architects and Builders*. Cambridge, MA, MIT Press, 1969.

Despite its age, this deservedly famous little book is still the best introduction to the subject for building professionals.

Hornbostel, Caleb. *Construction Materials: Types, Uses and Applications* (2nd ed.). Hoboken, NJ, John Wiley & Sons, Inc., 1991.

The section on plastics of this monumental book gives excellent summaries of the plastics most used in buildings.

Steel and Bronze

The chief advantage of steel (and to some extent bronze) as a frame material for windows is its strength, which permits steel sash sections to be slenderer than those of wood and aluminum (Figures 18.16 through 18.20). *Steel windows* may be made of steel coated in the factory with a long-lasting paint coating, galvanized steel, or stainless steel, depending on requirements for durability and appearance. Bronze windows are made in configurations similar to those for steel, usually with a natural patina finish. Steel and bronze are both less conductive of heat than aluminum, so windows made of these metals are less prone to forming condensation in cold weather. Where improved thermal performance is required, thermal break systems are also available.

Muntins

In earlier times, because of the difficulty of manufacturing large sheets of glass that were free of significant defects, windows were necessarily divided into small lites by *muntins*, thin wooden bars in which the glass was mounted within each sash. (The upper sashes in Figure 18.6 have muntins.) A typical double-hung window had its upper sash and lower sash each divided into six lites and was referred to as a *six over six*. Muntin arrangements changed with changing architectural styles and improvements in glass manufacture. Today's windows, glazed with large, virtually flawless lites of glass, need no muntins at all, but many building owners and designers prefer the look of traditional muntined windows. This desire for muntins is greatly complicated by the necessity of using insulated glazing to meet energy conservation code requirements. Some manufacturers offer the option of individual small lites of double glazing held in deep muntins. This is relatively expensive, and the muntins tend to look thick and heavy. The least expensive option utilizes grids of imitation muntin

FIGURE 18.16
Samples of hot-rolled steel window frame sections, non-thermally broken. The nearest sample includes a snap-in aluminum bead for holding the glass in place. *(Steel windows by Hope's; Hope's photography by David Moog.)*

FIGURE 18.17
Cutaway sample of a non-thermally broken steel-framed window with aluminum glazing beads and factory finish. *(Steel windows by Hope's; Hope's photography by David Moog.)*

bars, made of wood or plastic, that are clipped into each sash against the interior surface of the glass. These are designed to be removed easily for washing the glass. Other alternatives are imitation muntin grids between the sheets of glass, which are not very convincing replicas of the real thing; and grids, either removable or permanently bonded to the glass, on both the outside and inside faces of the window. Another option is to use a primary window with authentic divided lights of single glazing and to increase its thermal performance with an outer storm sash or other secondary window with improved thermal performance. Of all the options, this one looks best from the inside, but reflections in the outer sash or window largely obscure the muntins from the outside.

Glazing

Single glazing is acceptable only in the mildest of climates, because of its low resistance to heat flow and the likelihood that moisture will condense on its interior surface in cool weather. Double glazing is the minimum permitted by energy codes in most climate zones. Where low-e coatings, dense gas fills, and thermally improved spacers were once considered exotic, high-performance options, these features are now becoming increasingly common in standard window units, especially as energy codes continue to raise thermal efficiency standards. And even higher-performing systems, with triple and quadruple glazing, are becoming increasingly available.

FIGURE 18.18
Cutaway section through a thermally broken, stainless steel window frame, with double glazing. Note how exterior steel components (*left-hand side in photo*) are separated by plastic thermal breaks or other low-density materials from steel parts on the inside (*right-hand side*). *(Photo by Joseph Iano.)*

FIGURE 18.19
Exterior detail view of the thermally broken stainless steel window frame seen in Figure 18.18. At bottom right is a fixed window unit, above that is an operable unit, and to the left is a door. *(Photo by Joseph Iano.)*

Figure 18.21 lists thermal transmittance properties for some example combinations of window frame and glazing options. The listed U-Factors are overall values for complete window assemblies, accounting for differences in the thermal properties of the center of glass, edge of glass, and frames. When selecting actual windows, whole-window U-Factors for the particular window are provided by the window manufacturer as determined by laboratory testing or computer simulation. In addition to thermal transmittance, solar heat gain coefficient (SHGC) and visible light transmittance (VT) are other important measures of a window system's performance. See Chapter 17 for a discussion of these properties.

Installing Windows

Some catalog pages for windows are reproduced in Figures 18.22, 18.23, and 18.24 to give an idea of the information on window configurations that is available to the designer. Important dimensions given in catalogs are those of the rough opening and masonry opening. The *rough opening* height and

FIGURE 18.20
The narrow sight lines of steel windows and doors are evident in this photograph.
(Steel windows by Hope's; Hope's photography by David Moog.)

width are the dimensions of the hole that must be left in a framed wall for installation of the window. They are slightly larger than the corresponding outside dimensions of the window unit itself, to allow the installer to locate and level the unit accurately and to ensure that the window unit is isolated from structural stresses within the wall system. The *masonry opening* dimensions indicate the size of the hole that must be provided if the window is mounted in a masonry wall.

A rough opening or masonry opening should be flashed carefully before the window is installed to avoid later problems with leakage of water or air (for example, Figures 6.13 through 6.15). Adhesive-backed window flashing materials or sheet metal are the commonly used materials. Adhesive-backed flashings may be made of rubberized asphalt, similar in composition to the rubberized roof underlayment frequently used along the eaves of the roof (described in Chapter 7), reinforced plastic, or synthetic fibers designed for compatibility with proprietary building wrap products. Metal used for flashings must be corrosion resistant.

Most factory-made windows are easy to install, often requiring only a few minutes per window. Windows that are framed or clad in aluminum or plastic are usually provided with a continuous flange around the perimeter of the window unit. When the unit is pushed into the rough opening from the outside, the flange bears against the sheathing along all four edges. After the unit has been located and plumbed (made level and square) in the opening, it is attached to the frame by means of nails driven through the flanges. Then all the edges of the flanges should be made airtight, as shown in Figure 6.13. The flanges are eventually concealed by the exterior cladding or trim.

Methods for anchoring window units into masonry walls vary widely, from nailing the unit to wood strips that have been fastened inside the masonry with bolts or powder-driven fasteners to attaching the unit to steel clips that have been laid into the mortar joints of the masonry. The window manufacturer's recommendations should be followed in each case.

Window Frame	Single-Glazed	Overall U-Factor	
		Double-Glazed, Clear	Double-Glazed, Low-e, Argon Gas
Aluminum, without thermal break	1.2	0.65	0.52
	6.8	*3.7*	*3.0*
Aluminum, thermal break	1.0	0.60	0.42
	5.7	*3.4*	*2.4*
Steel, without thermal break	0.92	0.55	0.45
	5.2	*3.1*	*2.6*
Steel, with thermal break		0.40	0.30
		2.3	*1.7*
Wood, vinyl	0.84	0.32	0.28
	4.8	*1.8*	*1.6*
GFRP	0.65	0.30	0.28
	3.7	*1.7*	*1.6*

[a] U-factor: Btu/ft^2-hr-°F followed by W/m^2-°K (in *italicized* type)

FIGURE 18.21

Comparative U-Factors for various window frame and glazing combinations. Lower values correspond to better thermal performance. Note that frame material has a significant impact on the performance of the overall window assembly. Very high performance windows, with even lower overall U-factors (approximately 0.1 Btu/ft^2-hr-°F or 0.6 W/m^2-°K) are available from some manufacturers. See Chapter 17 for a more in-depth discussion of glazing types.

ROUGH OPENING

FRAME SIZE

MASONRY OPENING

1/2"[13]

1/4"[6]

1/2"[13]

1/4"[6]

OPERATOR JAMB DETAIL

NOTE: INTERIOR TRIM
SUPPLIED BY MARVIN
APPLIED BY OTHERS.

VERTICAL MULLION

HORIZONTAL MULLION

ROUGH OPENING

FRAME SIZE

1/2"[13]

1/4"[6]

HEAD JAMB & SILL DETAIL

MASONRY OPENING

FIGURE 18.22
Manufacturer's catalog details for an aluminum-clad, wood-framed casement window with double glazing and an interior insect screen.
(Courtesy of Marvin Windows & Doors.)

FIGURE 18.23

One of several pages in a manufacturer's catalog that show stock sizes and configurations of aluminum-clad wood casement windows. (*Courtesy of Marvin Windows & Doors.*)

*THIS UNIT MAY BE ORDERED AS EITHER A CASEMASTER OR AN AWNING PICTURE UNIT. THE CASEMASTER CODE NUMBER IS SHOWN HERE. TO ORDER AN AWNING PICTURE UNIT, SUBSTITUTE THE CODE CAWNP.

NOTE: Contact your distributor for Masonry Opening on Clad Casemaster/Awning Picture Units with factory installed brick mould casing.

Not to scale.

CCMP3224	CCMP3624	CCMP4024	CCMP4824*	CCMP6024*	CCMP6424*	CCMP7224*			
		CAWNP5628	CAWNP6028	CAWNP6428	CAWNP7228				
		CCMP4032	CCMP4832	CCMP5632*	CCMP6032*	CCMP6432*	CCMP7232*		
		CCMP4036	CCMP4836	CCMP5636*	CCMP6036*	CCMP6436*	CCMP7236*		
		CCMP4040	CCMP4840	CCMP5640*	CCMP6040*	CCMP6440*	CCMP7240*		
		CCMP4048	CCMP4848	CCMP5648*	CCMP6048*	CCMP6448*	CCMP7248*		
CAWNP2456	CAWNP2856 CCMP3656*	CCMP4056*	CCMP4856*	CCMP5656*	CCMP6056*	CCMP6456*	CCMP7256*		
CAWNP2460	CAWNP2860 CCMP3660*	CCMP4060*	CCMP4860*	CCMP5660*	CCMP6060*	CCMP6460*	CCMP7260*		
CAWNP2464	CAWNP2864 CCMP3664*	CCMP4064*	CCMP4864*	CCMP5664*	CCMP6064*	CCMP6464*	CCMP7264*		
CAWNP2472	CAWNP2872 CCMP3672*	CCMP4072*	CCMP4872*	CCMP5672*	CCMP6072*	CCMP6472*	CCMP7272*		

Mas. Opg.	2-0½ (622)	2-4½ (724)	2-8½ (826)	3-8½ (927)	3-4½ (1029)	4-0½ (1233)	4-8½ (1435)	5-0½ (1537)	5-4½ (1638)	6-0½ (1842)
Rgh. Opg.	2-1 (635)	2-5 (737)	2-9 (838)	3-1 (940)	3-5 (1041)	4-1 (1245)	4-9 (1448)	5-1 (1549)	5-5 (1651)	6-1 (1854)
Frame Size	2-0 (610)	2-4 (711)	2-8 (813)	3-0 (914)	3-4 (1016)	4-0 (1219)	4-8 (1442)	5-0 (1524)	5-4 (1626)	6-0 (1829)

CCMP2224

CCMP3224

Mas. Opg. 1-11⅜(587) 1-11⅜(600) 1-11⅜(594) 1-11⅜(595)
Rgh. Opg. 2-3⅞(703) 2-3⅞(695)
Frame Size 2-3⅞(689)

2-11⅜(892) 2-7⅞(791)
2-11⅜(898) 2-7⅞(797)
2-11⅜(905)

3-3⅜(1006)
3-3⅜(1000)
3-3⅜(994)

3-11⅜(1210)
3-11⅜(1203)
3-11⅜(1197)

4-7⅞(1413)
4-7⅞(1406)
4-7⅞(1400)

4-11⅜(1514)
4-11⅜(1508)
4-11⅜(1502)

5-3⅜(1616)
5-3⅜(1610)
5-3⅜(1603)

5-11⅜(1819)
5-11⅜(1813)
5-11⅜(1807)

☀ = Available in 3/4" IG Picture Units

☐ = Available in 3/4" or 1" Picture Units

☀☀ = Available in 1" Picture Units

NOTE: A PICTURE UNIT CANNOT BE CONVERTED TO AN OPERATING UNIT.

FIGURE 18.24

These fixed windows are sized to match stock sizes of casement windows by the same manufacturer, allowing the designer to mix and match. (*Courtesy of Marvin Windows & Doors.*)

SUSTAINABILITY OF WINDOWS AND DOORS

Chapter 17 covers issues of sustainability that pertain to glass and glazing. With regard to windows and doors:

- According to the National Renewable Energy Laboratory, solar heat gain and wintertime heat losses through windows account for roughly 30 percent of U.S. building heating and cooling electrical loads.

- In addition to the thermal properties of the glass in a window, the thermal conductivity of the frame and the air leakage of the window or door unit have very significant effects on the amount of energy that will be required to heat and cool the building.

- Doors can leak significant quantities of heat by conduction through the material of the door. Foam-core doors have better thermal performance than other types. The performance of any exterior residential door can be improved substantially by adding a storm door during the cold season of the year. Airlock vestibules can limit the amount of outdoor air that enters a building when the exterior door is open, as well as improve the comfort of building occupants in the vicinity of the vestibule. Revolving doors, which maintain an air seal regardless of their position, are suitable alternatives to vestibules. All doors should be tightly weatherstripped to limit loss of conditioned air.

- With respect to frame materials:

  - Issues of sustainability of wood production are covered in Chapter 3. When a building with wood windows is demolished, the windows are generally sent to landfills or incinerators and are not recycled.

  - Aluminum frames must be thermally broken for the sake of energy efficiency. They are often recycled during demolition and should be recycled in every case. Chapter 21 discusses sustainability issues relating to the material aluminum itself.

  - PVC window frames are thermally efficient. They can be recycled during demolition, and a significant percentage is being recycled at present. However, sustainability programs, such as LEED and the Living Building Challenge, discourage the use of PVC in buildings due to concerns over this material's environmental and health effects.

  - Steel window and door frames are made from recycled steel and can be recycled again when a building is demolished. Their thermal performance is moderate and can be improved greatly by the insertion of thermal breaks.

DOORS

Doors fall into two general categories, *exterior* and *interior*. Weather resistance is usually the most important functional factor in choosing exterior doors, whereas resistance to the passage of sound or fire and smoke are frequently important criteria in the selection of interior doors. Many different modes of door operation are possible (Figure 18.25).

There are numerous types of exterior doors: solid entrance doors, entrance doors that contain glass, storefront doors that are mostly or entirely made of glass, storm doors, screen doors, vehicular doors for residential garages and industrial use, revolving doors, and cellar doors, to name just a few. Interior doors come in dozens of additional types. To simplify our discussion, we will focus on swinging doors for both residential and commercial use.

Wood Doors

At one time, nearly all doors were made of wood. In simple buildings, primitive doors made of planks and *Z-bracing* were once common. In more finished buildings, *stile-and-rail doors* gave a more sophisticated appearance while avoiding the worst problems of moisture expansion and contraction to which plank doors are subject (Figures 7.23 and 7.24; Figure 18.26). The panels are not glued to the stiles and rails, but instead "float" in unglued grooves that allow them to move. The doors may be made of solid wood or of wood composite materials with veneered faces and edges. In either case, they are available in many different species of woods.

In recent decades, stile-and-rail doors have continued to be popular in higher-quality buildings. However, *flush doors* have captured the majority of the market, chiefly because they

are easier to manufacture and therefore less costly. For exterior use in small buildings, and for both exterior and interior use in institutional and commercial buildings, flush doors are constructed with a *solid core* of wood blocks or wood composite material (Figure 7.24). Interior doors in residences often have a *hollow core*. These consist of two veneered wood faces that are bonded to a concealed grid of interior spacers made of paperboard or wood. The perimeters of the faces are glued to wood edge strips. Flush doors with wood faces are also available with a solid *mineral core* that allows them to perform as fire doors.

Most flush wood doors are manufactured and specified according to the Window and Door Manufacturers *ANSI/WDMA I.S.1-A-04 Architectural Wood Flush Doors*. This standard addresses door appearance and durability, and includes three performance grades—Standard Duty, Heavy Duty, and Extra Heavy Duty—intended for

FIGURE 18.25
Some modes of door operation.

Swinging

Bifold

Accordion

Pocket Sliding

Bypass Sliding

Surface Sliding

Overhead

Coiling

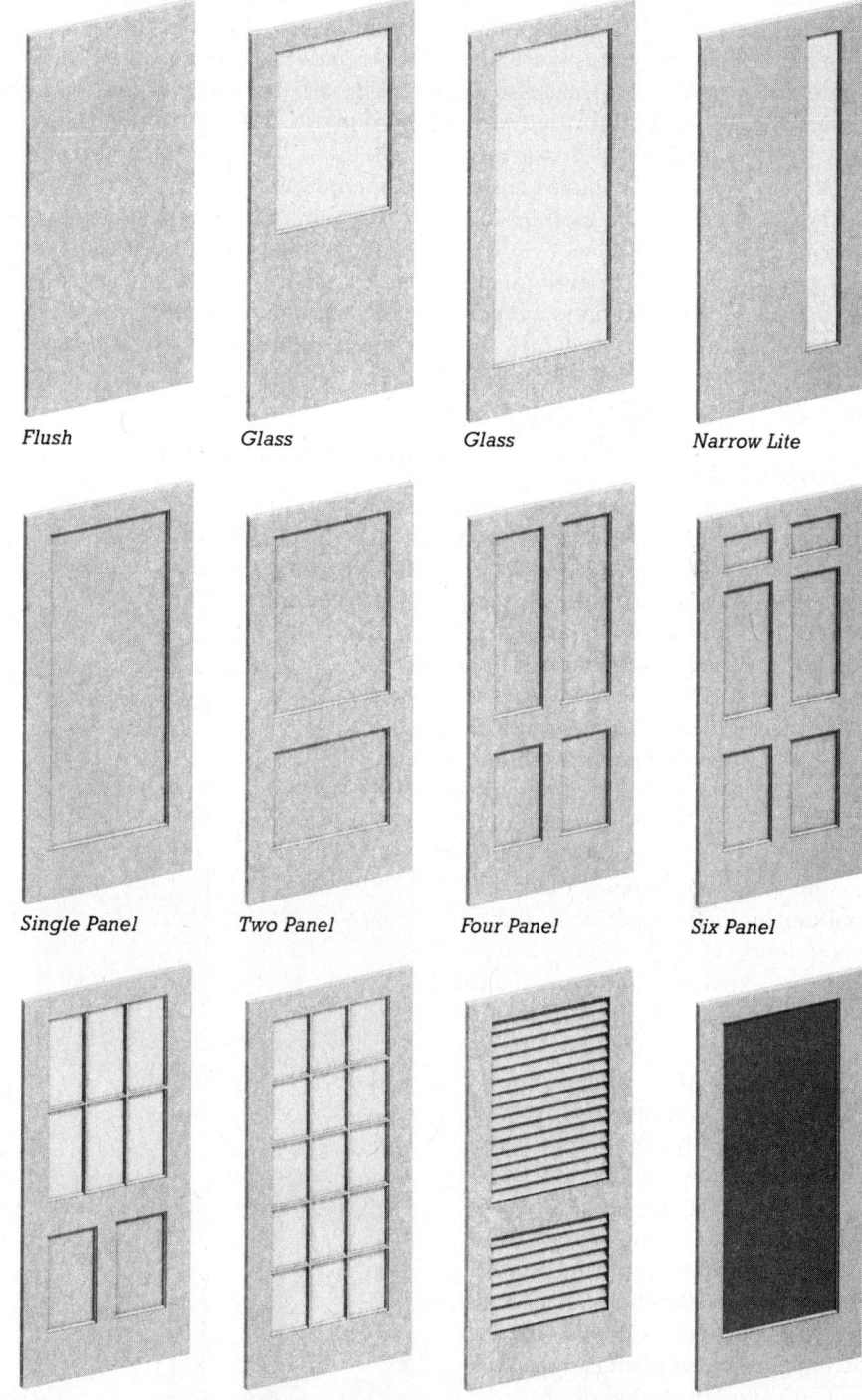

FIGURE 18.26
Some typical configurations for wood doors. The top row consists of flush doors. The middle row is made up of stile-and-rail doors.

Flush   Glass   Glass   Narrow Lite

Single Panel   Two Panel   Four Panel   Six Panel

Glass   Glass   Louver   Storm/Screen

doors used in applications of increasingly heavy usage. High-quality wood doors, especially those custom made in the woodworker's shop, can also be made to the Architectural Woodworking Institute standard *Architectural Woodwork Standards*, a comprehensive manual of materials and methods for custom woodwork.

A relatively recent development is a door made of wood fiber composite material that is pressed into the shape of a stile-and-rail door. Usually, the faces of the door may be given an artificial wood grain texture or faced with real wood veneer.

Entrance doors must be well constructed and tightly weatherstripped if they are not to leak air and water. Properly installed and finished wood panel or solid-core doors are excellent for exterior residential use (Figures 6.16, 6.17). Pressed *sheet metal doors* and molded *GFRP doors*, usually embossed to resemble wood stile-and-rail doors, are popular alternatives to wood exterior residential doors. Their cores are filled with insulating plastic foam, making their thermal performance superior to that of wood doors. They do not suffer from moisture expansion and contraction, as wood doors do. They are often furnished *prehung*, meaning that they are already mounted on hinges in a surrounding frame, complete with weatherstripping, ready to install by merely nailing the frame into the wall. Wood doors can also be purchased prehung, although many are still hung and weatherstripped on the building site. The major disadvantage of metal and plastic exterior doors is that they do not have the satisfying appearance, feel, or sound of a wood door.

Residential entrance doors almost always swing inward and are mounted on the interior side of the door frame. This makes them less vulnerable to thieves who would remove hinge pins or use a thin blade to push back the latch to gain entrance. In cold climates, it also prevents snow that may accumulate against the door from preventing the door from

opening. For improved wintertime thermal performance of the entrance, a *storm door* may be mounted on the outside of the same frame, swinging outward. The storm door usually includes at least one large panel of tempered glass.

In summer, a *screen door* may be substituted for the storm door. A *combination door*, which has easily interchangeable screen and storm panels, is more convenient than separate screen and storm doors.

Steel Flush Doors

Flush doors with faces of painted sheet steel, called *hollow metal doors*, are the most common type of door in nonresidential buildings (Figure 18.27). For

economy, interior steel doors in many situations have hollow cores. Solid-core doors are required for exterior use and in situations that demand increased fire resistance, more rugged construction, or better acoustical privacy between rooms.

Metal doors and most non-residential wood doors are usually hinged to steel frames called *hollow metal frames* (Figure 18.28). *Knocked down (KD) steel frames* arrive on the construction site in three separate sections for the two jambs and head, and are assembled in place into the previously prepared wall opening. *Welded steel frames* arrive on the construction site preassembled and welded at the corners. When used with gypsum-faced partitions, these

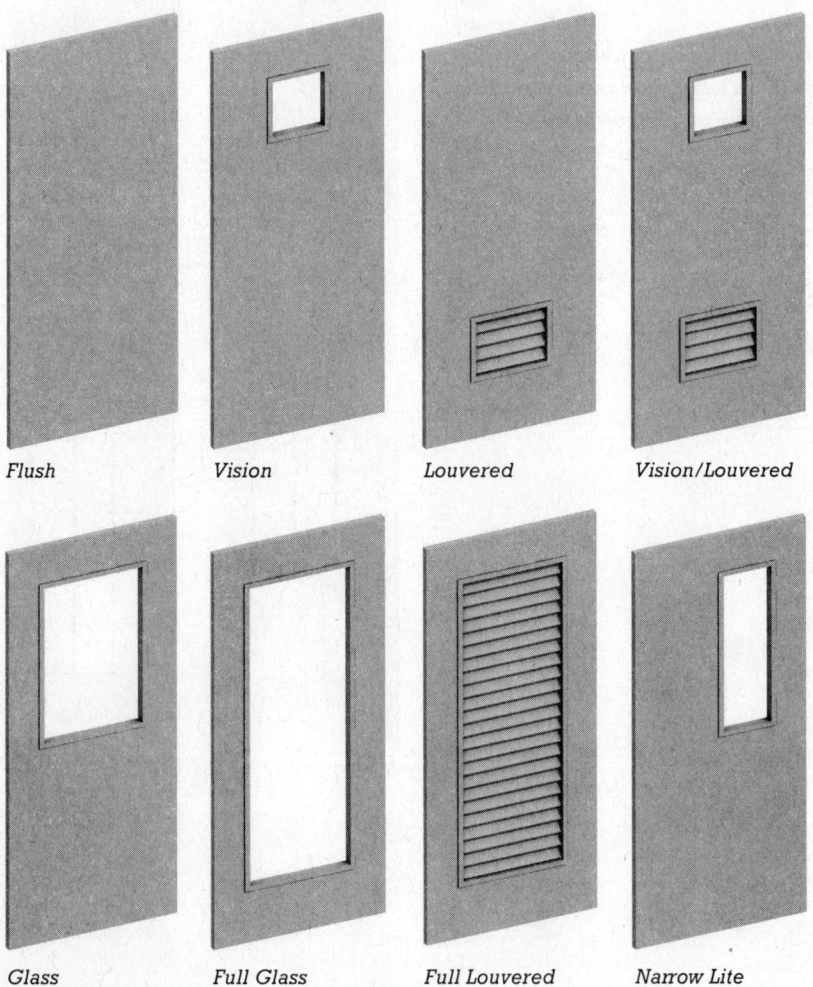

Flush *Vision* *Louvered* *Vision/Louvered*

Glass *Full Glass* *Full Louvered* *Narrow Lite*

FIGURE 18.27

Some typical configurations for steel doors.

Anchor clips

Elevation of door frame

A

C

B

B. Reinforcment of jamb at hinge attachments

C. For jamb anchorage to masonry walls, loose sheet metal tees are inserted into the frame and built into mortar joints

C. For jamb anchorage to steel studs, sheet metal zees are factory-welded to the jambs to receive screws driven through the studs

C. For jamb anchorage to wire truss studs, notches key to the vertical members of the studs, and holes provide for tie wires

C. Jams are anchored to wood studs by nailing through holes in the jamb inserts

FIGURE 18.28

Details of hollow steel door frames. The lettered circles on the elevation at the upper left correspond to the details on the rest of the page.

Solid-core door

A. Jambs can be attached to the floor with powder-driven fasteners

Hollow-core door

A. Jambs can be attached to the floor by pouring floor topping concrete around the door frame frames must be erected before the wall, because of the way in which the frame overlaps with the adjacent wall surfaces. Welded steel frames are also stronger than KD frames and the absence of any visual seam at the frame corners is frequently considered more attractive. Where hollow metal door frames are installed within masonry walls, they may be filled with cementitious grout to improve sound deadening and to make the door frame more resistant to tampering or forced entry. As an alternative to hollow metal, frames of wood or aluminum can also be used.

The sheet metal used in the manufacture of hollow metal doors and frames can be varied in thickness, to achieve varying levels of durability. Where corrosion resistance is a concern, galvanized steel or stainless steel may be used in place of ordinary steel.

Standard steel doors and frames are manufactured and specified according to the Steel Door Institute's *ANSI/SDI A250.8 Recommended Specifications for Standard Steel Doors and Frames.* Custom, high-quality hollow metal doors and frames follow the Hollow Metal Manufacturers Association's *ANSI/NAAMM-HMMA 861 Guide Specifications for Commercial Hollow Metal Doors and Frames.*

Fire Doors

Fire doors have a noncombustible mineral core and are rated according to the period of time for which they are able to resist specified time and temperature conditions, as defined by *NFPA 252 Standard Methods of Fire Tests of Door Assemblies,* or several similar tests defined by Underwriters Laboratories. In general, doors within fire resistance-rated walls must themselves also be fire rated. However, because doors constitute only a limited area of most walls, and because combustible furnishings or materials are not normally located directly in front of door openings, the required fire resistance rating for fire doors is often less than that required for the walls in which they are located. Figure 18.29 gives fire resistance ratings for fire doors as required by the International Building Code (IBC). For example, a door in a 2-hour rated exit stairway enclosure must be 1½-hour rated, a door in a 1-hour rated exit stairway enclosure must be 1-hour rated, and a door in a 1-hour rated exit corridor must be 20-minute (⅓-hour) rated. Doors in walls that are 2-, 3-, and 4-hour fire resistance rated, such as those separating uses within a building or separating buildings from one another, must be 1½- or 3-hour rated. A standardized label is permanently affixed to the edge of each fire door at the time of manufacture to designate its degree of fire resistance. (The building code requires that these labels not be painted over during construction so that the fire rating of the door can always be verified during subsequent building inspections.)

Glass used in fire doors must itself be fire rated (see Chapter 17) so that it will not break and fall out of the opening for a specified length

Type of Assembly	Required Assembly Rating (hours)	Minimum Fire Door and Fire Shutter Rating (hours)
Shaft, exit enclosure, and exit passageway walls	2	1½
	1	1
Fire walls and fire barriers	4	3
	3	3
	2	1½
	1½	1½
	1	¾
Fire partitions at corridor walls	1	20 minutes
	½	20 minutes
Other fire partitions	1	¾
	½	20 minutes
Exterior walls	3	1½
	2	1½
	1	½
Smoke barriers	1	20 minutes

FIGURE 18.29
Required fire resistance ratings for doors, according to the International Building Code. *(Part of Table 716.5; Excerpted from the 2012 International Building Code, Copyright 2011. Washington, D.C.: International Code Council. Reproduced with permission. All rights reserved.* **www.ICCSAFE.org***)*

of time when exposed to the heat of fire. The maximum size of glass may also be restricted, depending on the fire classification of the door and the type of rated assembly into which it is installed. Like glass in any door, glass in fire doors must meet the requirements of safety glazing so that if broken, it does not create dangerous shards.

Egress Doors and Accessible Doors

Many doorways act as components of a building's egress system, the path that occupants take when exiting a building during a fire or other emergency. Building codes require that such doorways be sufficiently wide to allow occupants to exit in a timely manner, with the width of any particular door dependent on the number of occupants served. For ease of operation, most egress doors must be side-hinged; they must not be too large; and when equipped with closers, they must not require too much force to swing open. The International Building Code requires egress doors serving 50 or more building occupants, as well as doors serving Hazardous Occupancy spaces, to swing in the direction of egress travel so that they do not become impediments to occupants attempting to exit quickly. Even when locked, egress doors must remain readily openable from the side from which occupants may approach the door when exiting. To ensure the simplest possible operation under emergency conditions, some egress doors are required to be fitted with *panic hardware*, horizontal bars or similar devices installed across the face of the door that unlock and unlatch the door whenever the bar is depressed.

Doorways along buildings routes that must be accessible to persons with disabilities must meet requirements for minimum width, ease of operation, maximum height of sill, and adequate clearance for approaching and opening the door.

There are many types of special-purpose doors. Among the most common are X-ray shielding doors, which contain a layer of lead foil; electric field shielding doors, with an internal layer of metal mesh that is electrically grounded through the hinges; heavily insulated cold storage doors; and bank vault doors.

Door Hardware

Unlike windows, in which operating hardware is selected by the manufacturer and installed in the factory, most door hardware is selected by the designer independently from the door itself, and installed on the construction site. Every door requires at least some hardware to control its motion and the means by which occupants manipulate it. Swinging doors are usually hung with two, three, or four *butt hinges*, depending on their weight. Each hinge consists of two leafs, one attached to the edge of the door and the other to the frame, and joined by a pin around which the leafs rotate. For door operation, either a round door *knob* or flatter *lever* is provided, located a little below mid-height of the door and close to the swinging edge of the door. Knobs are frequently used in residential occupancies, but wherever accessibility is a concern, levers that are more universally operable must be used. The knob or lever controls a *latch* in the door edge, disengaging it from the mating *strike* in the door frame and allowing the door to be swung open.

Depending on the functional requirements of the door, a bewildering array of additional hardware choices may be made, including, for example, locking and keying mechanisms to secure a door when closed, closing and opening devices to automate or provide assistance in door operation, stops mounted to walls or floors to limit door swing, silencers (small cushions) attached to the door frame that reduce door closing noise, viewers to see through a door, knockers, thresholds to manage transitions

between materials on either side of the door, weatherstripping to control the flow of air and water around the door edges, soundstripping to control the passage of sound, kick plates to protect the door from damage, electronic interfaces with building communications and controls, and more. All of these come in various styles and operational configurations. Hardware can also be specified to differing levels of durability and with many different finishes. On larger projects, hardware specification is frequently performed by a consultant with specialized knowledge in these areas.

OTHER WINDOW AND DOOR REQUIREMENTS

Safety Considerations in Windows and Doors

To prevent accidental breakage and injuries, building codes require glass within doors, and large lites within windows that are near enough to the floor or to doors to be mistaken for open doorways, to be made of breakage-resistant material. Tempered glass is most often used for this purpose, but some laminated glass and plastic glazing sheets can also meet the necessary requirements. For more information, see the discussion of safety glazing in Chapter 17.

In residences, buildings codes require at least one *emergency escape and rescue opening* in each bedroom, consisting of either a door to the exterior or a window that can be opened to an aperture large enough to permit occupants of the bedroom to escape through it and firefighters to enter through it. In the International Building Code, where a window is used for this purpose, it must have a net clear opening area of at least 5.7 square feet ($0.53 \, m^2$), a clear width of at least 20 inches (510 mm), a clear height of at least 24 inches (610 mm), and no sill higher than 44 inches (1.12 m) above the floor.

Where operable windows in apartments, residential dwelling units, and similar residential occupancies are more than 6 feet (1829 mm) above the exterior finished grade, the International Building Code requires that they be designed to minimize the risk of a child accidentally falling through them. Such windows must have sills not less than 36 inches (610 mm) above the interior finish floor. Where glazing is closer to the floor, the window unit must fixed, it must have openings sufficiently limited in size that a 4-inch (102-mm)-diameter sphere cannot pass through, or it must be protected with guards or other fall prevention devices.

Casement and awning windows should not be used adjacent to porches or walkways, where someone might be injured by running into the projecting sash. Similarly, inswinging windows should not be used in corridors unless they are above head level.

Structural Performance and Resistance to Wind and Rain

The performance of windows and exterior doors is defined by the *Standard/Specification for Windows, Doors, and Unit Skylights*, jointly published by the American Architectural Manufacturers Association (AAMA), the Window and Door Manufacturers Association (WDMA), and the Canadian Standards Association (CSA), officially designated as AAMA/WDMA/CSA 101/I.S.2/A440. This specification establishes minimum requirements for air leakage, water penetration, structural strength, operating force, and forced-entry resistance of aluminum, plastic, and wood-framed windows, doors, and unit skylights.

The AAMA/WDMA/CSA Standard/Specification uses a letter designation called *Performance Class* and a numeric designation called *Performance Grade* to indicate the minimum capabilities of fenestration products. Performance Classes, in order of increasing capability, are R, LC, C,

HC, and AW. In previous editions of the standard, these letter designations were associated with the terms "residential," "light commercial," "commercial," "heavy commercial," and "architectural," respectively. Although these plain word descriptions have been removed from newer versions of the standard, knowledge of them is still helpful in recalling the intended ranking of the letter designations themselves. Each Performance Class sets criteria for resistance to wind loads, resistance to water penetration, and maximum air leakage.

Numeric Performance Grades correspond to maximum design wind pressures. For example, Grade 30 indicates a unit suitable for design wind pressures up to 30 psf (1440 Pa). Grades are specified starting at 15 psf (720 Pa) and increasing in increments of 5 psf (240 Pa). Each Class has a minimum acceptable Grade, and higher than minimum Grades can be specified where resistance to higher wind forces is required.

An example of a manufacturer's complete tested product designation is *Class R—PG30: Size tested: 760 × 1520 mm (~30 × 60 in)—Casement*, where R is the Performance Class; 30 is the Performance Grade; the pairs of numbers indicate the maximum size of the tested unit that meets these criteria, expressed as width by height, first in millimeters and then, in parentheses, in inches; and finally, the type of window as a casement. In practice, the designer may choose a Performance Class based on the building type and general expectations for durability of the system. For example, a Class LC window may be specified for a low-rise multifamily building, a Class HC window for a hospital or school, or a Class AW window for a large institutional or high-rise building. The required Performance Grade should be determined based on the design wind pressures acting on the building, information that is usually provided by the structural engineer.

Thermal Performance

With regard to energy efficiency, the National Fenestration Rating Council (NFRC) defines a program of testing and labeling based on two standards: NFRC 100 *Procedure for Determining Fenestration Product U-Factors*, and NFRC 200 *Procedure for Determining Fenestration Product Solar Heat Gain Coefficient and Visible Transmittance at Normal Incidence*. The two most important properties included in these standards are thermal transmittance (U-Factor) and solar heat gain coefficient, both of which directly affect building energy consumption and are regulated by energy codes. Importantly, U-Factors represent the *overall thermal transmittance*, or *whole product heat loss*, of complete window, door, and skylight products. That is, they account for the combined contributions to thermal transmittance of the center of glass, edge of glass, framing, and other components. Visible light transmittance, air leakage, and condensation resistance ratings may also be included in NFRC ratings. An example of a standard label that is affixed to each NFRC-rated window is shown in Figure 18.30.

Windows sold in the Canadian market are also given an Energy Rating (ER), a relative indicator of heating season thermal performance. The ER combines U-value, air leakage, and solar heat gain measurements into a single number indicating net heat loss (negative ER) or heat gain (positive ER) over a defined set of conditions. ERs may be used to compare the relative thermal efficiency of windows, with a higher (or more positive) ER indicating better thermal performance than a window with a lower (or more negative) ER.

The AAMA/WDMA/CSA and NFRC standards are referenced by the International Building Code and National Building Code of Canada, making them the de facto standards for the selection of most North American building fenestration products.

	World's Best Window Co.
National Fenestration Rating Council **CERTIFIED**	Millennium 2000⁺ Vinyl-Clad Wood Frame Double Glazing • Argon Fill • Low E Product Type: **Vertical Slider**

ENERGY PERFORMANCE RATINGS

U-Factor (U.S./I-P)	Solar Heat Gain Coefficient
0.34	**0.25**

ADDITIONAL PERFORMANCE RATINGS

Visible Transmittance	Air Leakage (U.S./I-P)
0.41	**0.2**

Manufacturer stipulates that these ratings conform to applicable NFRC procedures for determining whole product performance. NFRC ratings are determined for a fixed set of environmental conditions and a specific product size. Consult manufacturer's literature for other product performance information.
www.nfrc.org

FIGURE 18.30
A sample NFRC certification label that is affixed to a window unit so that buyers may compare energy efficiencies.

Impact Resistance

Buildings in hurricane-prone regions can be subject to extremely powerful and destructive winds, and glazed openings in such buildings are especially vulnerable. The pressure of high-speed winds can cause glass to break, or it can suck whole lites out of their sashes, whole sashes out of their surrounding frames, or whole frames out of their rough openings. Glass can also be broken by rocks, severed tree limbs, and other debris launched by the wind with missile-like force. Once openings in a building are breached, the force of the wind can, in extreme cases, literally blow the roof off a structure. Even where a building structure remains intact, failed openings can admit large amounts of rainwater that can severely damage the building and its contents.

In the International Building Code, glazed openings in *wind-borne debris regions* must meet special standards for resistance to high wind forces and debris impact. These regions include portions of the U.S. Atlantic Ocean and Gulf of Mexico coasts, the islands of Hawaii, and certain other U.S. territorial islands that are frequently subjected to hurricane-force winds. In these regions, most glazed openings must meet the requirements of two tests, ASTM E1996 and E1886, which subject assemblies to airborne "missiles" and cyclical air pressures to determine their ability to remain in place under hurricane-like conditions. The testing can be quite dramatic. For windows destined for installed locations not more than 30 feet (9.1 m) above grade, a 9-pound, roughly 8-foot-long (4.1 kg, roughly 2.4 m long) 2 × 4 is fired endwise toward the window from a special cannon at a speed of 34 mph (55 kph). Although the glass is permitted to crack, it must survive in place, without being penetrated by the wood member.

Such *impact-resistant openings* (also sometimes referred to as *hurricane-rated openings*) are fitted with laminated glass with a heavy interlayer of PVB or other similarly tough, viscous plastic. They also have stronger glazing (gasketing) systems to better hold the glass units in place, their frames are structurally reinforced, and they are fastened into their rough openings with extra-strong attachment hardware. As an alternative to providing impact-resistant openings in one- and two-story buildings, the code permits the use of precut plywood or OSB panels that can be fastened into place over the outside of such openings when needed to act as temporary storm shutters.

Blast Resistance

In buildings subject to special security requirements, windows, curtain walls, and other glazing may be designed for resistance to the force of explosive blasts. Two federal standards for such *blast-resistant glazing systems* are in place, one published by the GSA/Interagency Committee Security Design Criteria and the other by the U.S. Department of Defense.

Design for blast resistance involves defining the size of the blast and its distance from the glazing system, as well as the glazing system's response to the blast. Of particular concern is the extent to which glass remains intact within the assembly or to which it shatters and disperses as hazardous fragments that could injure building occupants. Like impact-resistant glazing, blast-resistant glazing typically relies on laminated glass and reinforced framing and attachment systems.

FIGURE 18.31
Custom Alaskan yellow cedar wood doors
on architect Steven Holl's The Chapel of
St. Ignatius. *(Photo by Joseph Iano.)*

FIGURE 18.32
The Blanchard Road Alliance Church in
Wheaton, Illinois, silhouettes laminated
wood trusses against a wall of vinyl-clad
wood frame fixed windows. *(Architect:
Walter C. Carlson Associates. Photo courtesy
Andersen Windows, Inc. Andersen is a
registered trademark of Andersen Corporation,
copyright 1997. All rights reserved.)*

CSI/CSC MasterFormat Sections for Windows and Doors	
08 11 00	**METAL DOORS AND FRAMES**
08 11 13	Hollow Metal Doors and Frames
08 14 00	**WOOD DOORS**
08 14 16	Flush Wood Doors
08 14 33	Stile and Rail Wood Doors
08 15 00	**PLASTIC DOORS**
08 16 00	**COMPOSITE DOORS**
08 16 13	Fiberglass Doors
08 51 00	**METAL WINDOWS**
08 51 13	Aluminum Windows
08 51 16	Bronze Windows
08 51 19	Stainless-Steel Windows
08 51 23	Steel Windows
08 52 00	**WOOD WINDOWS**
08 53 00	**PLASTIC WINDOWS**
08 53 13	Vinyl Windows
08 54 00	**COMPOSITE WINDOWS**
08 54 13	Fiberglass Windows
08 61 00	**ROOF WINDOWS**
08 62 00	**UNIT SKYLIGHTS**
08 71 00	**DOOR HARDWARE**

KEY TERMS

window
fixed window
single-hung window
double-hung window
sash
sliding window
projected window
casement window
awning window
hopper window
inswinging window
pivot window
jamb
head jamb, head
side jamb
sill
stool
interior casing
exterior casing
apron
triple-hung window
tilt/turn window
pivoting window
top-hinged inswinging window

skylight
venting skylight
roof window
French door
sliding door
terrace door
wood window
clad wood window
thermal break
aluminum window
plastic window
glass-fiber-reinforced plastic (GFRP)
    window, fiberglass window
pultrusion
plastic
synthetic rubber, elastomer
polymer
monomer
copolymer
thermoplastic
thermosetting
crosslinked
plasticizer
stabilizer

filler
extender
reinforcing (fibers)
flame retardant
(plastic) composite
(plastic) laminate
extrusion
molding
film blowing
calendaring
blow forming
vacuum forming
steel window
muntin
six over six
rough opening
masonry opening
door
exterior door
interior door
Z-bracing
stile-and-rail door
flush door
solid core door

hollow-core door
mineral core door
sheet metal door
GFRP door
prehung door
storm door
screen door
combination door
hollow metal door
hollow metal frame

knocked down steel frame, KD steel frame
welded steel frame
fire door
panic hardware
butt hinge
door knob
door lever
latch
strike
emergency escape and rescue opening

Performance Class (AAMA/WDMA/CSA
  101/I.S.2/A440)
Performance Grade (AAMA/WDMA/
  CSA 101/I.S.2/A440)
overall thermal transmittance, whole
  product heat loss
wind-borne debris region
impact-resistant opening,
  hurricane-rated opening
blast-resistant glazing system

REVIEW QUESTIONS

1. List in detail the primary functional requirements for windows in each of the following situations:

 a. A residential bathroom in an urban apartment building

 b. A jail cell

 c. A display window in a department store

 d. A teller's window in a drive-in banking facility

 e. A bedroom in Nome, Alaska

 f. A living room in Hilo, Hawaii

2. Select a type of window operation for each of the following situations:

 a. A window that can be left open in the rain

 b. A window that must induce the maximum possible ventilation from passing breezes

 c. A window in a high-rise office building

 d. A window to frame an expansive view of distant mountains

 e. A window that can be operated as either a casement or a hopper

3. Compare the advantages and disadvantages of wood, plastic-clad wood, PVC, aluminum, and steel as window frame materials.

4. A well-insulated residential wall has a U-value of 0.05 and an R-value of 20 (in U.S. units). Compare the heat loss per square foot of the worst- and best-performing window and glazing combinations listed in Figure 18.21 with that of this wall.

5. Select a type of door for each of the following situations:

 a. Your bedroom closet

 b. A front door of a house

 c. A front door of a department store

 d. A door between the industrial arts shops and the cafeteria in a high school

 e. A door on a warehouse loading dock

EXERCISES

1. Obtain a copy of the building code that applies to the area in which you currently live. What fire resistance ratings are required for doors in the following situations?

 a. A door between a hotel room and a public corridor

 b. A door in an egress stair enclosure

 c. A door between an iron foundry and an office building

 d. A door between a single-family residence and its attached garage

2. Obtain catalogs from several window manufacturers. From them, select a set of windows for a one-room wilderness cabin of your own design.

3. Examine closely the windows in the room in which you are now sitting. What type of glazing do they have? What type of frame? How do they operate? How are they weatherstripped? Do these windows make sense to you in terms of today's energy efficiency requirements and your own feelings about the room? How would you change them?

SELECTED REFERENCES

Carmody, John, Stephen Selkowitz, Dariush Arasteh, and Lisa Heschong. *Residential Windows: A Guide to New Technologies and Energy Performance* (3rd ed.). New York, W. W. Norton, 2007.

This book is a clearly written, well-illustrated introduction to considerations of energy efficiency in residential windows.

Hollow Metal Manufacturers Association. *Hollow Metal Manual*. Chicago, Author, various dates.

This binder includes standards for custom hollow metal doors and others useful to the designer and specifier of steel doors and frames.

Selkowitz, Stephen, Eleanor S. Lee, Dariush Arasteh, Todd Willmert, John Carmody, and Eleanor Lee. *Window Systems for High-Performance Buildings*. New York, W. W. Norton, 2003.

This book addresses the myriad performance requirements and selection criteria for commercial glazing and window systems.

Window & Door Manufacturers Association. *Specifiers Guide to Windows and Doors*. Des Plaines, IL, Author, various dates.

This compilation of documents, useful to the designer and specifier of window and door systems, includes two important standards discussed in this chapter: AAMA/WDMA/CSA 101/I.S.2/A440 and ANSI/WDMA I.S.1-A.

WEB SITES

Windows and Doors

Author's supplementary web site: **www.ianosbackfill.com/18_windows_and_doors**

American Architectural Manufacturers Association: **www.aamanet.org**

Andersen Windows: **www.andersenwindows.com**

Ceco Steel Doors: **www.cecodoor.com**

Efficient Windows Collaborative: **www.efficientwindows.org**

Hollow Metal Manufacturers Association: **www.naamm.org/hmma**

Hope's Steel Windows & Doors: **www.hopeswindows.com**

Impact Grade Windows (EFCO Corporation): **www.impactgrade.com**

Marvin Windows: **www.marvin.com**

National Fenestration Rating Council: **www.nfrc.org**

Steel Door Institute: **www.steeldoor.org**

Steel Window Institute: **www.steelwindows.com**

Whole Building Design Guide, Windows: **www.wbdg.org/design/env_fenestration_win.php**

Window & Door Manufacturers Association: **www.wdma.com**

Windows and Daylighting (Lawrence Berkeley National Laboratory): **windows.lbl.gov**

Window Systems for High-Performance Buildings (Center for Sustainable Building Research): **www.commercialwindows.org**

## 19

DESIGNING EXTERIOR WALL SYSTEMS

- **Design Requirements for the Exterior Wall**
  Keeping Water Out
  Controlling Light
  Controlling Air Leakage
  Controlling Water Vapor
  Controlling the Radiation of Heat
  Controlling the Conduction of Heat
  Controlling Sound
  Secondary Functions of the Exterior Wall

  SUSTAINABILITY OF EXTERIOR WALL SYSTEMS

- **Conceptual Approaches to Watertightness in the Exterior Wall**
  Rainscreen Cladding and Pressure-Equalized Wall Design

- **Sealing Joints in the Exterior Wall**
  Sealant Materials
  Sealant Joint Design

- **Loadbearing Walls and Curtain Walls**
  The Loadbearing Wall
  The Curtain Wall

  BUILDING ENCLOSURE ESSENTIALS:
  AIR BARRIER

  Curtain Wall Testing and Standards

- **The Exterior Wall and the Building Codes**

The de Young Museum, San Francisco, by architect Herzog & de Meuron. The building outer skin consists of perforated copper panels allowed to patina naturally. *(Photo by Joseph Iano.)*

The *exterior wall enclosure*, or *building envelope*, is the part of a building that must defend the interior spaces against invasion by water, wind, sunlight, heat and cold, and all the other forces of nature. Its design is an intricate process that merges art, science, and craft to solve a long list of difficult problems. The exterior wall also typifies a paradox of building: Those parts of a building that are exposed to our view are also those that are exposed to wear and weathering. The outermost layer of the exterior wall is the most visible part of a building—one to which architects devote a great deal of time to achieve the desired visual effect. It is also the part of the building that is most subject to attack by natural forces that can spoil its appearance.

DESIGN REQUIREMENTS FOR THE EXTERIOR WALL

The major purpose of the exterior wall is to separate the indoor environment of a building from the outdoors in such a way that indoor temperature, humidity, and other conditions are maintained at levels suitable for the building's intended use. This translates into a number of separate and diverse functional requirements.

Keeping Water Out

The exterior wall must prevent the entry of rain, snow, and ice into a building. This requirement is complicated by the fact that water on the face of a building is often driven by wind at high velocities and air pressures, not just in a downward direction but in every direction, even upward. Water problems are especially acute on tall buildings, which present a large profile to the wind at altitudes where wind velocities are much higher than at ground level. Enormous amounts of water must be drained from the windward face of a tall building during a heavy rainstorm, and the water, pushed by wind, tends to accumulate in crevices and against projecting mullions, where it will readily penetrate the smallest crack or hole and enter the building. A considerable portion of this chapter is devoted to methods for keeping water out.

FIGURE 19.1
A steel-framed Chicago office building during the installation of its aluminum, stainless steel, and glass curtain wall cladding. Notice the diagonal wind braces in the steel frame. *(Architects: Kohn Pederson Fox/Perkins and Will. Photo by Architectural Camera. Permission of American Institute of Steel Construction.)*

Controlling Light

The exterior wall of a building must control the passage of light, especially sunlight. Sunlight is heat that may be welcome or unwelcome. Sunlight is visible light, useful for illumination but bothersome if it causes glare. Sunlight includes destructive ultraviolet (and other) wavelengths that must be kept off human skin and away from interior materials that will fade or deteriorate. Windows should be placed and proportioned with these considerations in mind. Exterior wall systems sometimes include external shading devices to keep light and solar heat away from windows. The glass in windows is often selected to control light and heat, as discussed in Chapter 17. Interior shades, blinds, and curtains may be added for further control.

Controlling Air Leakage

The exterior wall of a building must limit the unintended passage of air between indoors and outdoors. At a gross scale, this is necessary to regulate air velocities and temperatures within the building. Air leaks waste conditioned (heated or cooled) air and increase the chance of water vapor condensation within the wall. Air leakage can also result in reduced occupant comfort and higher concentrations of indoor air pollutants. Sealants, gaskets, weatherstrips, and air barrier membranes of various types are all used to prevent air leakage through the exterior wall. The role of air barriers in controlling air leakage through the building enclosure is discussed in more detail later in this chapter.

Controlling Water Vapor

The exterior wall of a building must retard the passage of water vapor by diffusion through the wall materials. In the heat of summer or the cold of winter, vapor moving through a wall assembly may condense inside the assembly and cause staining, loss of insulating value, corrosion of metals, and decay of wood.

Controlling the Radiation of Heat

Beyond its role in regulating the flow of radiant heat from the sun, the exterior wall of a building should also present interior surfaces at temperatures that will not cause radiant discomfort for building occupants. A cold interior surface will make people feel chilly when they are near the wall even if the air in the building is warmed to a comfortable level. A hot interior surface or direct sunlight in summer can cause overheating of the body despite the coolness of the interior air. External sun shading

FIGURE 19.2

The curtain wall of Chicago's Reliance Building, built in 1894–1895, has spandrels constructed of white terracotta tiles. (*Architect: Charles Atwood, of Daniel H. Burnham and Company. Photo by Wm. T. Barnum. Courtesy of Chicago Historical Society ICHi-18294.*)

devices, adequate thermal insulation and thermal breaks, and appropriate selection of glass are potential strategies for controlling heat radiation.

Controlling the Conduction of Heat

The exterior wall of a building must resist the conduction of heat into and out of the building. This requires not merely satisfactory overall resistance of the wall to the passage of heat, but also avoidance of *thermal bridges,* wall components such as metal framing members that are highly conductive of heat and therefore likely to cause localized condensation on interior surfaces. Thermal insulation, appropriate glazing, and thermal breaks are used to control heat conduction through the exterior wall, as we will observe in Chapters 20 and 21. Building codes specify minimum values of thermal resistance of wall components as a way of limiting the conduction of heat and also as a way of controlling the condensation of moisture on cold interior surfaces.

Controlling Sound

The exterior wall further serves to isolate the inside of a building from noises outside and vice versa. Noise isolation is best achieved by walls that are airtight and massive, and possibly with the inclusion of interior layers that are heavy yet resilient. The required degree of noise isolation varies from one building to another. For example, the exterior wall for a hospital near a major airport requires more noise isolation than that for a commercial building in a suburban office park.

Secondary Functions of the Exterior Wall

Fulfillment of the primary functional requirements of the exterior wall leads unavoidably to a secondary but equally important set of requirements.

SAME WALL UNDER DIFFERENT WIND DIRECTIONS

GROUND LEVEL

MAXIMUM POSITIVE PRESSURES

MAXIMUM NEGATIVE PRESSURES

(UNITON ISOBARS ARE POUNDS PER SQUARE FOOT)

Resisting Wind Forces

The exterior wall of a building must be strong and stiff enough to sustain the pressures and suctions imposed upon it by the wind. For low buildings, which are exposed to relatively predictable winds, this requirement is fairly easily met. The upper reaches of taller buildings are beset by much faster winds, whose directions and velocities are often made more complicated by aerodynamic effects from surrounding buildings. Both high pressure and suction forces can occur on some portions of the exterior wall, especially near corners of the building (Figure 19.3).

FIGURE 19.3

An example of positive and negative wind pressures acting on the cladding of a tall building, shown here in elevation, as predicted by wind tunnel testing. The building in this case is 64 stories tall and triangular in plan. Notice the high negative pressures (suctions) that can occur on the upper regions of the facade. The wind pressures on a building are dependent on many factors, including the shape of the building, its orientation, topography, wind direction, and the effects of surrounding buildings. Each building must be modeled and tested individually to determine the pressures it should be expected to encounter. *(Reprinted with permission from AAMA Aluminum Curtain Wall Design Guide.)*

Adjusting to Movement

Various forces are always at work throughout a building, tugging and pushing both the frame and the exterior wall: thermal expansion and contraction, moisture expansion and contraction, and structural deflections, to name a few. These forces must be accommodated in the design of the building enclosure.

The exterior wall must tolerate movements caused by changes in temperature: Indoor/outdoor temperature differences can cause warping of cladding panels due to differences in differential expansion and contraction on opposite sides of the panels (Figure 19.4A). Different materials within the exterior wall, such as aluminum, steel, glass, and concrete, each expand and contract at different rates in response to changes in temperature. Portions of the exterior wall as a whole, exposed to outdoor temperature variations, expand and contract constantly with respect to the structural frame of the building. Where parts of a building frame are exposed to the exterior, they exhibit their own expansion and contraction behavior. And even where the building frame is within the insulated envelope of the building, it expands and contracts to some extent, especially between the time the exterior wall is installed and when the building is first occupied and its indoor temperature is controlled.

The exterior wall must accommodate expansion and contraction caused by variations in moisture content: Fired clay bricks, which leave the kiln with a very low moisture content, expand after they are installed as they absorb moisture from the atmosphere. Many building stones also expand slightly

FIGURE 19.4
Distortions of curtain wall panels, illustrated in cross section. (*A*) Bowing, caused in this case by greater thermal expansion of the outside skin of the panels than of the inside skin under hot summertime conditions. (*B*) Twisting of spandrel beams because of the weight of the curtain wall.

HOT   COLD

A

B

COLUMN SHORTENING

SPANDREL BEAM DEFLECTION

WIND AND EARTHQUAKE DEFORMATIONS

SPANDREL BEAM DEFLECTION

DIFFERENTIAL FOUNDATION SETTLEMENT

DIFFERENTIAL SPANDREL BEAM DEFLECTION

FIGURE 19.5

Forces on curtain wall panels caused by movements in the frame of the building, illustrated in elevation. In each of the six examples, the drawing to the left shows the movement in the overall frame of the building, and the larger-scale drawing to the right shows its consequences on the curtain wall panels (shaded in gray) covering one bay of the building. Points of attachment between the panels and the frame are shown as crosses. The black arrows indicate forces on the wall panels caused by the movement in the structure. The magnitude of the structural movements is exaggerated for clarity, and some inadvisable attachment schemes are shown to demonstrate their consequences. Forces such as these, if not taken into account in the design of the frame and cladding, can result in glass breakage, panel failures, and failure of the attachments between the panels and the frame.

after they are installed. Concrete masonry units and precast concrete shrink by small amounts after installation as their curing is completed. Such movements, though small, can nevertheless accumulate to the extent that materials become distressed or distorted, or even fail completely. In smaller buildings, wood cladding components are most susceptible to moisture movement, as discussed in Chapters 3 and 6.

The exterior wall must adjust to structural movements in the frame of the building: Building foundations may settle unevenly, causing distortions of the frame. Gravity forces shorten columns and cause beams and girders to which the exterior wall system is attached to sag slightly. Wind and earthquake forces push laterally on building frames and wrack panels attached to the faces. Long-term creep causes significant shortening of concrete columns and sagging of concrete beams and slabs during the first year or two of a building's life.

Posttensioning of concrete components causes significant dimensional contraction in the early life of the structure.

If the exterior wall design does not adequately account for movement between it and the building frame, wall materials can be damaged or the performance of the wall compromised, with results varying from minor aesthetic effects to the need for wholesale replacement of the wall system. Where joints in the wall system become overstressed, leakage of air or water can cause exterior staining, damage to interior finishes, or mold and corrosion within the wall. Where glass panels are impinged by their surrounding frames, they may break. Where cladding materials become overstressed, they may buckle, crack, or fail completely and fall from the building. Strategies for dealing with movement from all these causes are evident in the details of exterior wall systems presented in Chapters 20 and 21 (Figure 19.5).

SUSTAINABILITY OF EXTERIOR WALL SYSTEMS

For many if not most buildings, the design of the exterior wall has a greater effect on lifetime energy consumption than any other building system. A poorly designed all-glass box loses excessive amounts of heat in winter and gains excessive solar heat in summer. Its undifferentiated faces show no awareness on the designer's part of the effects of orientation on energy transactions through the walls of a building.

• Glass should be used where it can supply daylighting and provide views. If it cannot be effectively shaded, it should be avoided where summertime overheating could otherwise occur or where occupants could be subject to excessive glare at times of the day when the sun is low in the sky.

• In many buildings, windows that can be opened and closed by the occupants can help reduce energy costs.

• Opaque areas of the exterior wall should be well insulated.

• Thermal bridges should be eliminated from the exterior wall.

• The entire building envelope should be detailed for airtightness. Fresh air should be provided by the building's ventilation system, not by air leakage through the exterior wall.

• Where appropriate, south-facing glass can be used to provide solar heat to the building in winter, but care must be taken to avoid glare, local overheating, and ultraviolet deterioration of interior surfaces and furnishings that are exposed to sunlight.

• As photovoltaic cells become more economical, consideration should be given to using south-facing surfaces of the exterior wall to generate electrical energy.

Resisting Fire

Building codes regulate the combustibility and fire resistance of the exterior wall. These issues are addressed in greater length at the end of this chapter.

Weathering Gracefully

To maintain the visual quality of a building, its cladding must weather gracefully. The inevitable dirt and grime should accumulate evenly, without streaking or splotching. Functional provisions must be made for maintenance operations such as glass and sealant replacement and for periodic cleaning, including scaffolding supports and safety equipment attachment points for window washers. The cladding must resist oxidation, ultraviolet degradation, breakdown of organic materials, corrosion of metallic components, chemical attack from air pollutants, and freeze-thaw damage of stone, brick, concrete, concrete block, and tile.

Constructability Requirements for the Exterior Wall

The exterior wall system should be easy to install. There should be secure places for installers to stand, preferably on the floors of the building rather than on scaffolding outside. There must be built-in adjustment mechanisms in the fastenings between components of the wall system and the frame to allow for the inaccuracies that are normally present between these systems. Sufficient clearance must be provided so that components can be installed easily, without interference by other parts. Most importantly, the system must be sufficiently forgiving in its design to allow for a lifetime of trouble-free performance despite the lapses in workmanship or material quality that inevitably occur.

CONCEPTUAL APPROACHES TO WATERTIGHTNESS IN THE EXTERIOR WALL

In detailing the exterior wall for water resistance, we work from a secure theoretical base. In order for water to penetrate a wall, three conditions must be satisfied simultaneously:

1. There must be water present at the outer face of the wall.

2. There must be an opening through which the water can move.

3. There must be a force to move the water through the opening.

If any one of these conditions is not satisfied, the wall will not leak. This suggests three conceptual approaches to making a wall watertight:

1. We can try to keep water completely away from the wall. A very broad roof overhang can keep a one- or two-story wall dry under most conditions. When designing the exterior wall of a taller building, however, we must assume that major portions of the wall will get wet.

2. We can try to eliminate the openings in a wall. We can build very carefully, sealing every seam in the wall with membranes, sealants, or gaskets, attempting to eliminate every hole and crack.

This latter approach, called a *barrier wall* or *face-sealed wall*, is simple in concept but often difficult in practical application. In a wall made up of many jointed components, all these joints are unlikely to be perfect. If a surface is a bit damp, dirty, icy, or oily, joint sealants may not reliably stick to it. If a worker applying a sealant or gasket is insufficiently skilled, has to reach a bit too far to complete the joint, or is working under cold, wet, or windy conditions that interfere

with careful installation, the joint may be left compromised. Even if joints are well made, building movements can tear sealant or pull gaskets loose. Because, in a barrier wall, sealed joints are on the outside of the building, they are exposed to the full effects of sun, wind, water, and ice and may fail over time from weathering and age. Further, whatever the cause of joint failure, joints on the outside face of the wall are difficult to reach for inspection and repair. Thus, in practice, long-term watertightness is often difficult to achieve with barrier wall construction.

3. We can try to eliminate or neutralize all of the five forces that can move water through the wall: gravity, momentum, surface tension, capillary action, and air currents (Figure 19.6).

Gravity is a factor in pulling water through a wall only if the wall contains an inclined plane that slopes into, rather than out of, the building. It is usually a simple matter to detail the exterior wall system so that no such inclined planes exist, though sometimes a loose gasket or an errant bead of sealant can create one despite the best efforts of the designer.

The *momentum* of falling raindrops can drive water through a wall only if there is a suitably oriented slot or hole that goes completely through the wall. Momentum is neutralized by applying a cover to each joint in the wall or by designing each joint as a *labyrinth*.

The *surface tension* of water, which causes it to adhere to the underside of a cladding component, allows water to be drawn into the building. A *drip* on any underside surface to which water might adhere eliminates the problem.

Capillary action is the surface tension effect that pulls water through any opening that can be bridged by

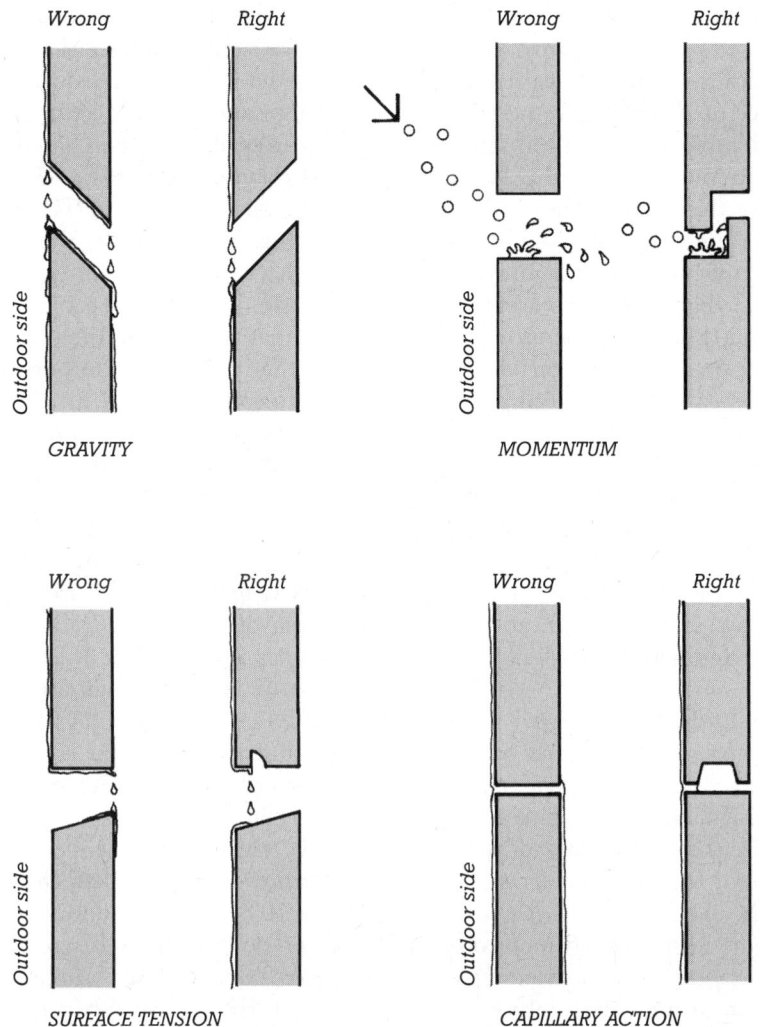

GRAVITY

MOMENTUM

SURFACE TENSION

CAPILLARY ACTION

a water drop. It is, for example, the primary force that transports water through the pores of a masonry wall. It can be eliminated as a factor in the entry of water through a wall by making each of the openings in a wall wider than a drop of water can bridge or, if this is not feasible or desirable, by providing a concealed *capillary break* somewhere inside the opening. In porous materials such as brick, capillary action can be counteracted by applying an invisible water-repellent coating that destroys the adhesive force between water and the walls of the pores in the brick.

The solutions described in the four preceding paragraphs are easy to implement. With relatively straightforward geometric manipulations of the joint, the possibility of leakage caused by four of the five forces that can move water through an opening in a wall can be eliminated. The fifth force, *wind pressure*, is the most difficult to deal with in designing a wall for watertightness. We can neutralize it by employing pressure-equalized wall design.

AIR PRESSURE
DIFFERENCES
(WIND CURRENTS)

Pressure
equalization
chamber

FIGURE 19.6

The five forces that can move water through an opening in a wall, illustrated in cross section with the outdoors to the left. Each pair of drawings shows first a horizontal joint between wall panels in which a force moves water through the joint, then an alternative design for that joint that neutralizes this force. Leakage caused by gravity is avoided by sloping internal surfaces of joints toward the outside; the slope is called a *wash*. Momentum can be counteracted with a labyrinth. A drip and a capillary break are shown here as means for stopping leakage from surface tension and capillary action, which are closely related forces. Air pressure differences between the outside and inside of the joint will result in air currents that can transport water through the joint. This is prevented by closing the area behind the joint with a pressure equalization chamber (PEC). When wind strikes the face of the building, pressure in the PEC quickly rises until it is equal to the pressure outside the wall, after which all air movement through the joint ceases. Each joint in an exterior wall, window, or door must be designed to neutralize all five of these forces.

Rainscreen Cladding and Pressure-Equalized Wall Design

The solution to the wind current problem is to allow wind pressure differences between the outside and inside of the exterior wall to neutralize themselves through a concept known as *pressure-equalized wall design*. This involves the creation of an airtight plane, the *air barrier*, behind the outer face of the wall. The air barrier is protected from direct exposure to the outdoors by an unsealed, labyrinth-jointed layer known as the *rainscreen*. Between the rainscreen and the air barrier is a space known as the *pressure equalization chamber (PEC)*.

As the wind fluctuates and exerts varying pressures on the exterior wall, small currents of air pass back and forth through each unsealed joint in the rainscreen, just enough to equalize the pressure inside the PEC with the pressure immediately outside it (Figure 19.6). These currents are far too weak to carry water with them. A small flaw in the air barrier, such as a sealant bead that has pulled away from one side of the joint, is unlikely to cause a water leak because the volume of air that can pass through the flaw is insufficient to carry water. By contrast, any flaw, no matter how small, in an external sealant joint in a barrier wall design will allow water

to leak through, because the sealant joint itself is wetted (Figure 19.7).

Because wind pressures across the face of a building frequently vary considerably at any given moment between one area of the face and another, the PEC must be divided into compartments small enough that volumes of air cannot rush through the joints in higher-pressure areas of the face and flow across the air chamber to lower-pressure areas, carrying water with them as they go. The appropriate size of these chambers may vary considerably, depending on the design of the wall system and the wind forces to which it is exposed. Broadly speaking, PECs are normally no taller than one story or wider than one or two column bays. In some applications they may be significantly smaller.

Simple Rainscreen Design

The term *rainscreen principal* originated with the concept of pressure-equalized wall design, and at one time was used exclusively in reference to pressure-equalized cladding systems. More recently, the term *rainscreen cladding* has come to be applied more broadly to any cladding with a system of internal drainage, regardless of the extent of compartmentalization of the drainage space and the degree of pressure equalization that can be achieved. In practice, varying degrees of pressure equalization or moderation occur, and the lines

between cladding systems best characterized as simple rainscreens and those considered fully pressure-equalized are often indistinct. To avoid confusion with pressure-equalized systems, simple rainscreen cladding may also be referred to as *drained and back-ventilated cladding*.

Figure 19.9 illustrates the components of a commercial rainscreen cladding system, and Figure 6.19, a typical residential application. In either case, water that penetrates the outer cladding is captured in the furred space and readily drained out the bottom of the wall. Air circulation behind the cladding further helps to extract moisture from within the furred space. Masonry cavity walls, used on buildings of virtually any size, are another example of rainscreen cladding that frequently incorporates at least some degree of pressure-equalization (Figure 20.3). As we will see in Chapter 21, internal drainage systems are an integral component of every metal-and-glass curtain wall system.

A Pressure-Equalized Wall Design

Figure 19.8 depicts a cladding design that embodies the rainscreen and pressure-equalization principles in simple form. No surface joint sealants or gaskets are used. The metal rainscreen panels do not touch one another, but are separated by generous gaps that preclude capillary movement of water, provide installation

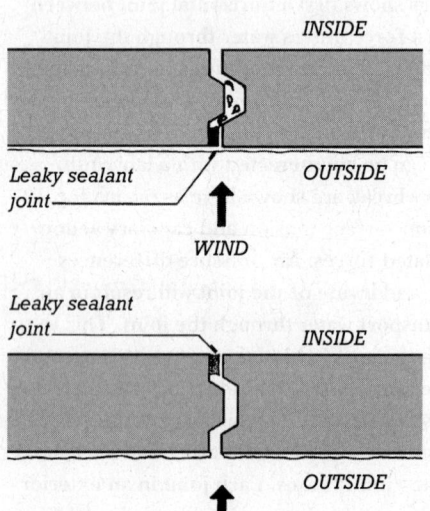

FIGURE 19.7

Leakage through a defective vertical sealant joint between curtain wall panels, shown in plan view. In the upper example, the sealant joint is at the outside face of the panels, where it is wetted during a storm. Even a small current of air passing through the defective joint carries water with it. In the lower example, with the defective sealant located on the inside of the panels where it remains dry, air leakage through the joint is insufficient to transport water through the joint, and no water penetrates.

FIGURE 19.8

The rainscreen in this exemplary cladding system is made up of metal panels, each formed from sheet metal. The design team included Wallace, Floyd Associates, Inc., Bechtel/ Parsons Brinckerhoff, Stull and Lee, Inc., Gannett Fleming/URS/TAMS Consultants, and the Massachusetts Highway Department.

Metal cladding

Vertical metal channel

Air barrier

Plastic foam thermal
insulation

SECTION THROUGH HORIZONTAL JOINT

PLAN OF VERTICAL JOINT

clearances, and allow for expansion and contraction. All four edges of each panel are shaped so as to create labyrinth joints. The forces of surface tension and gravity are counteracted by sloping surfaces and drips.

Installation is simple and forgiving of minor lapses in workmanship: Metal U-shaped clips are bolted to the backup wall, which is coated with an airtight mastic to create an air barrier. Rigid insulation panels are adhered to the wall, allowing the clips to project through. Vertical metal channels are bolted to the clips. Finally, the metal panels that make up the rainscreen are simply hung on horizontal rods that are supported by the channels, much as pictures are hung on hooks on a wall. The space between the metal rainscreen panels and the insulation acts as an internal drainage space.

To achieve a pressure-equalized design, horizontal metal angles, not shown, are installed between the channels at one- or two-story intervals. The vertical channels further divide the PEC into narrow compartments. The spaces between the edges of the panels and the channels are narrow enough to restrict airflow sufficiently to achieve a pressurized design. If more complete compartmentalization is needed, compressible foam rods or gaskets can be installed alongside the channels to create more airtight boundaries at these locations. When wind drives rain against this wall, small quantities of air flow through the open joints in the rainscreen until the pressure in the PEC equals the pressure outside. These airflows are insufficient in volume or velocity to carry water with them.

Pressure Equalization at Smaller Scale

The principles of rainscreen design and pressure equalization may also be applied on a small scale to guide us in many aspects of exterior detailing of buildings. Figure 19.10 demonstrates how this practice is embodied in the placement of weatherstrip in a window sill detail.

In the correct detail, the weatherstrip, whose function is to act as an air barrier, is placed to the inside of the lower rail of the sash. The open joint under the sash rail, which is provided with a capillary break, acts as the PEC. Unless the weatherstrip is grossly defective, water cannot be blown through the joint by air pressure differentials. Notice how the other forces that could transport water through the joint are counteracted: A slope on the sill prevents gravity from pulling water in. The groove in the lower edge of the sash that acts as a capillary break also acts as a drip to counteract surface tension. The L-shaped joint between the sash and the sill acts as a labyrinth to prevent entry by momentum. In the incorrect detail, the weatherstrip can be wetted by rain. Any minor flaw in the weatherstrip will allow water to be blown through the joint.

Relatively few buildings rely completely on the rainscreen principle

Figure 19.9

A cutaway mockup of a rainscreen wall system. Beginning from the top right and moving down and to the left: The light-colored, diagonal band is glass-mat-faced gypsum sheathing. The narrow dark strip is liquid primer applied to the sheathing. The primer improves the adhesion of the self-adhered membrane (the shinier, reflective dark surface starting below the primer) which acts as an air barrier, water barrier, and vapor retarder in this assembly. Semirigid mineral wool batt insulation is applied continuously over the self-adhered membrane. This material is sufficiently dense and water repellent to maintain its thermal insulation value even in the relatively damp rainscreen drainage space. Vertical galvanized steel furring strips are held away from the sheathing by fiberglass standoffs that have low thermal conductivity and act as thermal breaks between the furring and the remainder of the wall. Finish cladding may be a cementitious or metal panel with gapped joints, directly fastened to the furring strips.

FIGURE 19.10
Applying the rainscreen principle to the detailing of the sill of a double-hung window. *(From Edward Allen,* Architectural Detailing: Function, Constructability, Aesthetics, *Hoboken, NJ, John Wiley & Sons, Inc.; reproduced with permission of the publisher.)*

and pressure-equalized design for watertightness. However, there are very few contemporary cladding systems that do not employ these principles as an important part of their defense against water penetration. Consider again the familiar example of the masonry cavity wall: These principles can be seen in the brick facing wythe acting as a rainscreen, the backup wall as an air barrier, and the cavity, partially pressurized through weep and vent holes, as the PEC. Nevertheless, the surface of the outer masonry veneer is also frequently sealed with compounds to reduce its absorbency (an application of the barrier wall approach), and the cavity is flashed and provided with weep holes so that water that does penetrate the veneer can be safely channeled back to the exterior (an application of internal drainage).

SEALING JOINTS IN THE EXTERIOR WALL

Joint seals are an important part of all exterior wall systems, whether located at the face of the wall or deeper within its construction. The role of the seal is to prevent the passage of air and water while still allowing for necessary dimensional tolerances during assembly and reasonable amounts of movement once construction is complete. Seals are often used between panels of stone or precast concrete in a curtain wall (Figures 20.8 and 20.13), beneath shelf angle in brick curtain walls (Figure 20.3), and between dissimilar materials, such as where a metal-and-glass cladding system ends against a masonry wall (Figure 21.12, details 6, 9, and 9A). Specially formulated sealants are used to seal between lites of glass and the frames that support them (Figure 17.18) and even to prevent the passage of sound around the edges of interior partitions (Figures 23.21, 23.22, 23.34, and 23.37). In general, sealed joints are usually ⅜ to ¾ inch (9–19 mm) in width. But they can be small as ¼ inch (6 mm), and, where anticipated movements are great, 1 inch (25 mm) or wider.

Sealant Materials

Gunnable Joint Sealants
Gunnable joint sealant materials are viscous, sticky liquids that are injected into the joints of a building with a *sealant gun* (Figure 19.11). They cure within the joint to become rubberlike materials that adhere to the surrounding surfaces and seal the joint against the passage of air and water. Gunnable joint sealants can be grouped conveniently in three categories according to the amount of change in joint size that each can safely withstand after curing:

- *Low-range sealants,* also called *caulks,* have very limited *elongation* (stretching and squeezing) capabilities, up to plus or minus 5 percent of the width of the joint. They are used for filling minor cracks or nonmoving joints, especially in preparation for painting. Most caulks cure by evaporation of water or an organic solvent and shrink substantially as they do so. None are used for sealing joints in building exterior wall systems. (Although the term "caulk" is properly applied only to low-range sealants, in common usage it is frequently applied more broadly to mean any type of sealant, regardless of elongation capability.)

- *Medium-range sealants,* such as butyl rubber or acrylic, have elongation capacity in the range of plus or minus 5 to 10 percent. They are used in the building exterior wall for sealing nonworking joints (joints that are fastened together mechanically as well as being filled with sealant, such as the preformed solid tape sealant shown in Figure 19.12*b*). Because these sealants cure by the evaporation of water or an organic solvent, they undergo some shrinkage during curing.

- *High-range sealants* can safely sustain elongations up to plus or minus

FIGURE 19.11
Applying polysulfide, a high-range gunnable sealant, to a joint between exposed-aggregate precast concrete curtain wall panels, using a sealant gun. The operator moves the gun slowly so that a bulge of sealant is maintained just ahead of the nozzle. This exerts enough pressure on the sealant so that it fully penetrates the joint. Following application, the operator will return to smooth and compress the wet sealant into the joint with a convex tool, much as a mason tools the mortar joints between masonry units. (*Courtesy of Morton Thiokol, Inc., Morton Chemical Division.*)

50 to 100 percent. They include various *polysulfides*, which are usually site mixed from two components to effect a chemical cure; *polyurethanes*, which may also cure by a two-component reaction or by reacting with moisture vapor from the air, depending on the formulation; and *silicones*, which cure by reacting with moisture vapor from the air. None of these sealants shrink upon curing because none relies on the evaporation of water or a solvent. All adhere tenaciously to the sides of properly prepared joints. All are highly resilient materials that return to their original size and shape after being stretched or compressed, and

all are durable for many years if properly formulated and installed. Sealants for the working joints in exterior wall systems are selected from among this group. Polysulfide sealants have the longest history of use in such applications. However, improved formulations of polyurethanes and silicones now account for 90 percent or more of the high-range construction sealant market, with silicones generally considered the longest lasting and highest performing of the three.

Gunnable joint sealants are specified according to ASTM standard C920, which defines designations for sealant *Type, Grade, Class,* and *Use.*

Type S sealants are *single-component* and require no job site mixing. Type M sealants are *multicomponent* and must be mixed on the job site before installation. Multicomponent sealants generally cure faster than single-component sealants. They also allow a greater variety in color choice, as dye packs can be added during mixing. Grade P sealants, also called *self-leveling*, are *pourable.* They are easily installed in horizontal paving joints. For vertical wall joints, Type NS, *nonsag*, sealants must be used. Class defines the elongation capability of a sealant. A Class 25 sealant can tolerate 25 percent expansion and contraction under normal usage. A Class 100/50 sealant (the highest Class designation in the current standard) can tolerate 100 percent expansion and 50 percent contraction. A Use T, *traffic*, sealant can tolerate wear and physical abuse of pedestrian or vehicular traffic (most pourable sealants are also Use T); a Use NT, *nontraffic*, sealant is not suitable for traffic exposure and is normally intended for use in vertical wall joints; a Use I, *immersible*, sealant is suitable for sealing applications that will be submerged once the sealant has cured. Sealants may also be classified as Use M, G, A, or O, meaning that they have passed a series of tests demonstrating satisfactory adherence to mortar, glass, aluminum, or other materials, respectively. As an example, a multicomponent sealant intended for expansion joints between sections of aluminum curtain wall, which must be capable of 50 percent elongation, can be specified as Type M, Grade NS, Class 50, Uses NT and A.

Solid Joint Seal Materials

As an alternative to gunnable sealants, preformed solid materials may be used for various joint sealing needs in exterior wall systems (Figure 19.12):

- *Gaskets* are strips of fully cured elastomeric (rubberlike) materials manufactured in different configurations and sizes for different purposes, but most commonly dry glazing. They are compressed into a joint to seal tightly against the surfaces on either side.

- *Preformed cellular tape sealant* is a strip of polyurethane sponge material that has been impregnated with a mastic sealant. It is delivered to the construction site in an airtight wrapper, compressed to one-fifth or one-sixth of its original volume. When a strip is unwrapped and inserted, it expands to fill the joint, and its sealant material cures with moisture from the air to form a watertight seal. Preformed cellular tapes can be used to seal very

FIGURE 19.12

(*a*) Preformed solid tape is adhered to the vertical leg of the aluminum frame. Next, the glazing is inserted and an aluminum glazing bead with a solid gasket is snapped into place, pressing the glazing tightly between the gasket and the solid tape. Lastly, gunnable sealant is installed over the solid tape for a better weather seal. (*b*) Preformed solid tape seals the nonmoving joint between two adjacent metal panels. (*c*) Preformed solid tape holds the aluminum panels at a controlled distance from the backing plate. Structural sealant adheres the panels to the backing. (Though not shown here, the backing plate is also fastened to the wall.) A backer rod and gunnable sealant form a weather seal at the face of the panels. Backer rods are discussed in the next section of the text. (*d*) Preformed solid tape holds sheets of single glass between an aluminum mullion and its coverplate.

wide joints, such as occur at seismic separations between large portions of a building.

- *Preformed solid tape sealants* are used in nonmoving lap joints, as in mounting glass in a metal frame or overlapping two sheets of metal cladding. They are thick, sticky ribbons of polybutene or polyisobutylene that both seal and cushion the joint. They are installed as the joint is assembled, as they are too sticky to be inserted later into the finished joint. They are also stiff enough to maintain a predictable spacing between the elements being joined.

Sealant Joint Design

Figure 19.13 illustrates the principles of joint design for gunnable sealants. Before the sealant is installed, the joint surfaces must be cleaned of oil, dirt, oxide, moisture, concrete form release compound, or other contaminants. Where necessary, the edges of the joint are *primed* with a suitable coating to improve adhesion between the sealant and the joint surfaces.

Next, *backer rod* or *backup rod* is inserted into the joint. This is a cylindrical strip of compressible, flexible plastic foam that is just a bit larger in diameter than the width of the joint. It is pushed into the joint, where it is held by friction. The backer rod limits the depth to which the sealant will penetrate. This maintains the optimum proportions of the sealant bead and avoids waste of sealant material. It also ensures that only two sides of the sealant adhere to the joint itself. As illustrated in Figure 19.14, sealant adhered on three sides has limited movement capability. Backer rod is available in a large range of diameters to fit every joint.

Sealant is extruded into the joint from the nozzle of a sealant gun and then mechanically tooled, much as a masonry mortar joint is tooled, to compress the sealant material firmly against the sides of the joint and the backer rod. Tooling also produces a smoother surface and creates the desired surface profile to the sealant. The backer rod's role is now finished but, being inaccessible, it remains in the joint.

The proper width of the sealant joint depends on the expected range of movement in the joint and the movement capacity of the sealant. For example, a Class 50 sealant (capable of 50 percent elongation or compression) installed into a 1-inch (25-mm) joint is capable of accommodating no more than ½ inch (13 mm) of movement. In practice, this calculation is made more complicated by consideration

A. UNTOOLED   TOOLED

B. JOINT TOO NARROW

C. SEALANT BEAD TOO DEEP

D. SEALANT BEAD CORRECTLY PROPORTIONED

E. LAP JOINT CORRECTLY PROPORTIONED

FIGURE 9.13
Good and bad examples of sealant joint design. (*A*) This properly proportioned joint is shown both untooled and tooled. The untooled sealant fails to penetrate completely around the backer rod and does not adhere fully to the sides of the joint. (*B*) A narrow joint may cause the sealant to elongate beyond its capacity when the panels on either side contract, as shown to the right. (*C*) If the sealant bead is too deep, sealant is wasted, and the four edges of the sealant bead are stressed excessively when the joint enlarges. (*D*) A correctly proportioned sealant bead. The backer rod, made of a spongy material that does not stick to the sealant, is inserted into the joint to maintain the desired depth. The width is calculated so that the expected elongation will not exceed the safe range of the sealant, and the depth is between ⅛ and ⅜ inch (3 and 9.5 mm). (*E*) A correctly proportioned lap joint. The width of the joint (the distance between the panels) should be twice the depth of the sealant bead and twice the expected movement in the joint.

A. NO BOND BREAKER

B. BOND BREAKER

FIGURE 19.14
In a three-sided joint, the sealant is likely to tear unless a nonadhering plastic bond breaker strip is placed in the bottom of the joint before the sealant is installed.

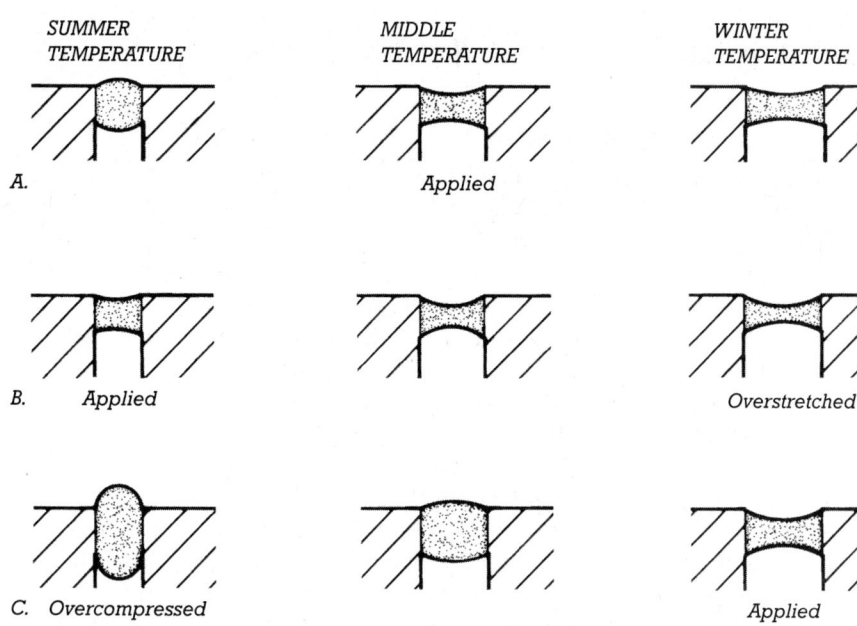

SUMMER TEMPERATURE

A.

MIDDLE TEMPERATURE

Applied

WINTER TEMPERATURE

B.　*Applied*

Overstretched

C.　*Overcompressed*

Applied

FIGURE 19.15
Sealants are best applied at moderate temperatures. If cold- or hot-weather applications of sealants are anticipated, the joints should be proportioned to minimize overstretching or overcompression. Row *A* shows the behavior of a sealant applied at a medium temperature. Rows *B* and *C* show sealant applied at summer and winter temperatures, respectively.

of variations in joint width due to construction tolerances and other factors. For example, for a joint between materials with high coefficients of thermal expansion, the time of year when the sealant is installed must be taken into account. Sealant installed in cold weather will have to stretch less during its lifetime but will have to compress more in summer than the same sealant installed in hot weather, which will have to stretch more and compress less (Figure 19.15).

LOADBEARING WALLS AND CURTAIN WALLS

The Loadbearing Wall

Until the late 19th century, nearly all large buildings were built with loadbearing exterior walls. These walls supported a substantial portion of the floor and roof loads of the building, as well as separating the indoor environment from the outdoors. In noncombustible buildings, these walls were built of brick or stone masonry. Functionally, such walls had several

inherent limitations. They were poor thermal insulators, and they were heavy, requiring large foundations and limiting their height to a few stories.

The *loadbearing wall* has been brought up to date with higher-strength masonry and concrete; components such as thermal insulating materials, cavities, flashings, air barriers, and vapor retarders have been added to make the wall more resistant to the passage of water, air, and heat; and the addition of steel reinforcing has allowed the wall to become thinner, lighter, and better able to resist wind and seismic loads. Loadbearing masonry and concrete exterior walls are often attractive and economical for low- and medium-rise buildings. High-rise residential towers with exterior loadbearing masonry walls also continue to be built, especially in Asia. These types of construction are illustrated and discussed in more detail in Chapters 8, 10, and 14.

The Curtain Wall

The first steel-framed skyscrapers, built in the late 19th century,

introduced the concept of the *curtain wall*, an exterior wall supported at each story by the frame. The name "curtain wall" derives from the idea that the wall is thin and "hangs" like a curtain on the structural frame. (Most curtain wall panels do not actually hang in tension from the frame; rather, they are supported from the bottom at each floor level.) The earliest curtain walls were constructed of masonry (Figure 19.2). The principal advantage of the curtain wall is that, because it bears no vertical load, it can be thin and lightweight regardless of the height of the building, compared to a masonry loadbearing wall, which may become prohibitively thick and weighty at the base of a very tall building.

Curtain walls may be constructed of any noncombustible material that is suitable for exposure to the weather. They may be either constructed in place or prefabricated. In Chapter 20, we will examine curtain walls that are made of masonry and concrete. In Chapter 21, we will look at curtain walls that are made of metal and glass.

BUILDING ENCLOSURE ESSENTIALS: AIR BARRIER

Air Leakage

Air can move through a building wall, floor/ceiling assembly, or roof wherever air pressure differences exist between one side of the assembly and the other and one or more openings exist for the air to pass through. Air pressure differences can be created by wind acting on the external surfaces of a building, by *stack effect* (the tendency of tall buildings to act like chimneys, drawing air in at either the top or bottom and expelling it at the other end), and by building mechanical equipment such as air handling systems and exhaust fans (Figure A).

When outside air infiltrates a building through exterior walls and roofs, it increases building energy consumption. A 2005 study by the U.S. National Institute of Standards and Technology estimated that air infiltration accounts for as much as 40 percent of a building's heating and cooling costs. Outside air infiltration also introduces unfiltered air pollutants and unconditioned air into the building's interior, where it can degrade indoor air quality and reduce occupant comfort. Air leakage transports water vapor into insulated walls and roofs, increasing the risk of condensation and moisture damage to building components. When air leaks between spaces within a building, it can disrupt pressure differentials maintained by HVAC systems for the purpose of controlling the spread of odors or contaminants between separately zoned parts of a building. For example, unpleasant cooking odors can be drawn from one apartment building living unit to another, car exhaust or gas fumes from a parking garage can infiltrate adjacent occupied areas, dust particles can be carried into a laboratory clean room, or bacteria can be introduced into a hospital operating suite.

Air Barriers

Air barrier materials control air leakage through building assemblies. Examples of air barrier materials include building wraps, gypsum wallboard, polyethylene sheet plastic, rigid foam insulation, liquid-applied membranes of various formulations, flashings, sealants, gaskets, tapes, and more. To function as an air barrier, a material must be resistant to the passage of air; it must have sufficient strength and rigidity to withstand the air pressure differentials that act upon it; where it spans joints, it must be sufficiently resilient to accommodate movement without failure; and it must be durable enough to perform its function throughout the life of the building.

The greater a material's resistance to the passage of air, the lower is its *air permeance* and the better is its performance as an air barrier. Air permeance is measured according to ASTM E2178 and is expressed as a volume of air passing through a standard area of material in a 1-minute period, at a prescribed air pressure (cubic feet of air per minute per square foot of area at 1.57 pounds per square foot, or 0.3 inch of water, of air pressure, or, in metric units, as liters per second per square meter of area at 75 Pascals of air pressure). An air permeance of 1 cfm/sf@1.57 psf is equal to approximately 5 L/s-m^2@75 Pa. The most commonly cited standard for air barrier materials is an air permeance not greater than 0.004 cfm/sf@1.57 psf or 0.02 L/s-m^2@75 Pa (Figure B).

An air barrier material must be able to resist air pressure, acting either inward or outward across the building assembly, without damage or excessive deflection. Flexible sheet materials such as building wraps, plastic sheets, roofing membranes, and flexible flashings are especially

FIGURE A

Air pressure differences in a building can be caused by (*a*) wind, (*b*) the stack effect, such as within an elevator shaft, and (*c*) building mechanical systems.

Material	Air Permeance	
	cfm/sf@1.57 psf	L/s-m²@75 Pa
Polyethylene sheet, 6 mil (0.15 mm)	~0	~0
Aluminum foil, 1 mil (0.025 mm)	~0	~0
Self-adhered modified asphalt membranes	~0	~0
Plywood, ⅜ in. (9.5 mm)	~0	~0
Extruded polystyrene rigid foam insulation, 1½ in. (38 mm)	~0	~0
Most low-slope roofing membranes	~0–0.002	~0–0.01
Proprietary high-density spray polyurethane foam insulation, 1½ in. (38 mm)	0.0002	0.001
Fluid-applied, vapor permeable air barrier membrane	<0.0004	<0.002
Nonperforated polyolefin building wrap		
Commercial grade	0.001	0.005
Residential grade	0.007	0.036
Low-density spray polyurethane foam insulation, 3 in. (75 mm)	0.002–0.32	0.01–1.6
Gypsum wallboard, ½ in. (12 mm)		
Exterior sheathing	0.002	0.0091
Interior wallboard, unpainted	0.004	0.0196
Oriented strand board, ⅜ in. (11 mm)	<0.004	<0.02
Roofing felt, #30	0.037	0.1873
Nonperforated asphalt felt, #15	0.078	0.3962
Asphalt-impregnated fiberboard	0.163	0.8285
Tongue & groove planks	3.7	19
Glass fiber batt insulation	7.3	37
Cellulose insulation, sprayed	17	87

FIGURE B

Air permeance of common building materials. Shaded rows indicate materials exceeding recommended air permeance for air barrier materials.

vulnerable. If not properly supported or adequately fastened, these materials can tear, stretch, or pull loose and become ineffective as air barriers. Damage caused by air pressures can also lead to the failure of materials to perform other important functions, such as keeping liquid water out of the building or resisting the diffusion of water vapor. If an air barrier material remains intact but deflects excessively under alternating cycles of positive and negative air pressure, it can pump air in and out of a building assembly, reducing the effectiveness of insulation and increasing the risk of water vapor transport. Air barrier materials must also be able to accommodate the normal thermal and structural movements that occur within building systems without undue wear or failure.

Air Barrier Assemblies and Systems

Air permeance standards for the performance of air barrier materials used in combination are less stringent than those for individual materials themselves, reflecting the reality that flawless, continuous sealing between components is never possible. Recommendations for the maximum air permeance of *air barrier assemblies*, that is, collections of materials responsible for the performance of a wall, roof, or floor, are in the range of 0.01 to 0.04 cfm/sf@1.57 psf (0.05 to 0.2 L/s-m²@75 Pa).

To effectively control air leakage into and out of a whole building, its conditioned space must be completely surrounded by combinations of air barrier assemblies and materials, creating an uninterrupted *air barrier system*

BUILDING ENCLOSURE ESSENTIALS: AIR BARRIER (CONTINUED)

capable of resisting air pressure differentials acting anywhere across the heated and cooled boundary of the building. Careful attention to design details and quality of construction is required to create such complete systems of enclosure. All potential discontinuities—gaps between panels, laps in sheet materials, transitions between dissimilar substrates, fastener penetrations, movement joints, penetrations for structure or services, installation space around window and doors frames, junctions between foundation, wall, and roof assemblies, gaps between operable doors and windows and their frames, and so on—must be made airtight by the use of tapes, sealants, caulks, flashings, gaskets, and other materials that can themselves meet air leakage performance requirements. The International Energy Conservation Code and the Canadian National Building Code both include standards for air leakage through such whole building air barrier systems.

Air Barrier Location

Air barrier materials can be located anywhere in an assembly as long as they form an interconnected, continuous system. At the inside surface of the building enclosure, the airtight drywall approach and simple caulk and seal are air barrier systems consisting of gypsum wallboard combined with caulks, sealants, and gaskets to seal leakage paths around wallboard penetrations and between underlying framing members (see Chapter 7). These systems are relatively easy and inexpensive to install, making them especially popular for residential construction. They are less favored for commercial building types where frequent changes to interior partitions, finishes, and wiring make it unlikely that the continuity of a system depending on the careful detailing of these elements will be maintained over the life of the building.

Plastic sheeting, which is frequently used as a vapor retarder behind gypsum wallboard, also has low air permeance and can act as an air barrier. However, difficulty in sealing plastic sheet seams and penetrations, as well as a tendency for the plastic to stretch and deflect between supporting framing, limit this material's suitability in air barrier systems, especially for taller buildings or wherever else high air pressure differentials are expected.

Toward the middle of a building enclosure assembly, foam insulation can be sprayed into the space between studs, joists, and rafters, acting as part of an air barrier in combination with caulks or sealants to seal leakage paths around framing members.

Toward the outside of the building enclosure, air barrier materials are installed over sheathing in framed

Curtain Wall Testing and Standards

Structural Performance and Resistance to Wind and Rain

For any new curtain wall design, it is advisable to build and test a full-scale section of wall to determine its resistance to infiltration of air and water and its structural performance under heavy wind loads. There are several outdoor laboratories in North America that are equipped to conduct these tests. A full-scale specimen of the wall system, often two stories high and a bay wide, is constructed as the exterior wall of a chamber that can be pressurized or evacuated by a calibrated blower system.

Curtain wall testing is conducted according to the American Architectural Manufacturers Association standard *AAMA 501 Methods of Tests for Exterior Walls*, which itself references numerous other standards for specific aspects of the testing. The specimen is tested first for air infiltration, using ASTM E283, in which it is subjected to a static air pressure that corresponds to the pressure that will be created by the anticipated maximum wind velocity in the vicinity of the building. Air that leaks through the wall is carefully measured, and the rate of leakage is compared to specified standards.

A static test for water penetration is next, using ASTM E331: The wall is subjected to a static air pressure while being wetted uniformly across its surface at a rate of 5 gallons per hour per square foot (3.4 L/m²-min). Points of water leakage are noted, and leaking water is carefully collected and measured. A dynamic water penetration test may also be performed in accordance with AAMA 501.1, using an aircraft engine and propeller to drive water against the wall.

The structural performance of the wall is tested according to ASTM E330, in which a calibrated blower subjects the wall specimen to air pressures and suctions as high as 50 percent over the specified wind load, and the deflections of the structural members in the wall are measured. Optionally, tests for thermal performance, sound transmission, and the effects of thermal cycling, seismic loads, and movement of the structure to which the curtain wall is attached may also be performed.

construction or on the exterior face of masonry or concrete backup walls. Building wraps, plywood or gypsum board sheathing panels, and fluid-applied or fully adhered sheet membranes may all be used in combination with various sealing and taping materials. In this location, air barrier materials are easy to install, with a minimum of complex intersections. Where penetrations occur for the anchoring of cladding or sheathing, they are sealed to ensure airtightness.

Where air barriers are located to the exterior of the building insulation, they can protect the insulation against *wind washing*, in which air currents passing through the porous insulation materials reduce their effectiveness. However, materials close to the exterior of the building assembly must also be sufficiently durable and able to perform satisfactorily while exposed to the effects of exterior moisture and large temperature fluctuations. In taller buildings, systems consisting of liquid-applied or sheet membranes that are fully adhered to rigid substrates are superior to flexible sheets such as building wrap. Exterior-side air barrier materials also frequently act as the primary moisture barrier in the wall, creating a water-resistant drainage surface behind the outer cladding.

Unlike vapor retarders, there is no harm in installing multiple air barriers within one assembly. Multiple air barriers can provide the particular advantages of each type and provide redundancy, lessening the chance of a flaw in one material compromising a building's overall air leakage performance.

Air Barriers and Water Vapor Control

When air passes through a building assembly, water vapor in the air is transported with it. In fact, uncontrolled air leakage through the exterior wall is capable of transporting a volume of water vapor that is orders of magnitude greater than can occur by water vapor diffusion, as discussed in relation to vapor retarders in walls. By preventing the flow of warm, moist air to the cooler side of a building assembly, air barriers can play a major role in protecting against condensation within the assembly.

Depending on an air barrier's location within the exterior wall, it may have to be either vapor permeable or a vapor retarder. When the air barrier is located on the warm side of the building insulation, it can also perform as the vapor retarder in the wall, and an air barrier material with a low vapor permeance can be used. However, when the air barrier is located closer to the cold side of the wall, it must be vapor permeable, so as to avoid trapping moisture within the wall.

Although all these tests yield numerical results, it is also important that the behavior of the specimen be observed closely during each test so that specific problems with the design, materials, detailing, and installation can be identified and corrected. Most wall system specimens fail one or more of the tests for air and water leakage on the first attempt. By observing the sources of leakage during the test, it is usually possible to modify the flashings, sealants, weep holes, or other components of the design so that the modified specimen will pass the subsequent test. These modifications are then incorporated into the final details for the actual building.

After testing has been completed and final design adjustments have been made, production of the wall components begins, and deliveries to the site can commence as soon as the frame is ready to receive the system.

Curtain wall systems require careful inspection during installation to be sure that there are no defects in workmanship. Even seemingly small imperfections in assembly can lead to large, expensive problems later. As work progresses, installed portions of the curtain wall can be checked for water leakage according to AAMA 501.2. This involves directing water at the joints in the wall with a hose that has a specified nozzle and following specified procedures to isolate the causes of any leaks. Where deemed necessary, more elaborate instrumented field tests for water and air leakage can also be performed.

Thermal Performance and Other Properties

The thermal properties of curtain wall systems are most commonly tested according to AAMA 1503, for thermal transmittance, and AAMA 507, for solar heat gain, visible light transmittance, and condensation resistance, although comparable NFRC standards may also sometimes be used. Curtain wall systems are adaptable to a great variety of glass types, frame sizes, and configurations. For this reason, determining precise U-Factors and other properties for a particular system design usually requires more detailed analysis than, for example, when standard window configurations are specified.

Where impact or blast resistance is required, curtain wall systems can be tested to the same standards described in Chapter 18 for doors and windows.

THE EXTERIOR WALL AND THE BUILDING CODES

The major impact of building codes on the design of the exterior wall is in the areas of structural strength, fire resistance, and energy efficiency. Strength requirements relate to the strength and stiffness of the wall system itself and to the adequacy of its attachments to the building frame, with special reference to wind and seismic loadings.

Fire requirements are concerned with the combustibility of the wall materials, the fire resistance ratings and vertical dimensions of parapets and spandrels, the fire resistance ratings of exterior walls facing other buildings that are near enough to raise questions of fire spread from one building to the other, and the closing-off (*firestopping*) of any vertical passages in the wall that are more than one story in height. At each floor, the space inside column covers and the space between the exterior wall system and the edges of floors must be firestopped, using mineral wool *safing* or other material that can restrict the passage of smoke and fire through these gaps (Figures 19.16

and 20.16). See Chapter 1 for more about fire resistance rating requirements for building exterior walls.

As energy conservation codes, sustainability programs, and high-performance building standards continue to become increasingly stringent in their limits on building energy consumption, exterior wall designs continue to evolve and improve to keep step with these demands. Improvements in air barrier and insulation materials, more attention to the proper detailing of these components, improved glazing materials, and enhanced window and curtain wall framing all contribute to buildings that are more energy efficient, comfortable, and durable.

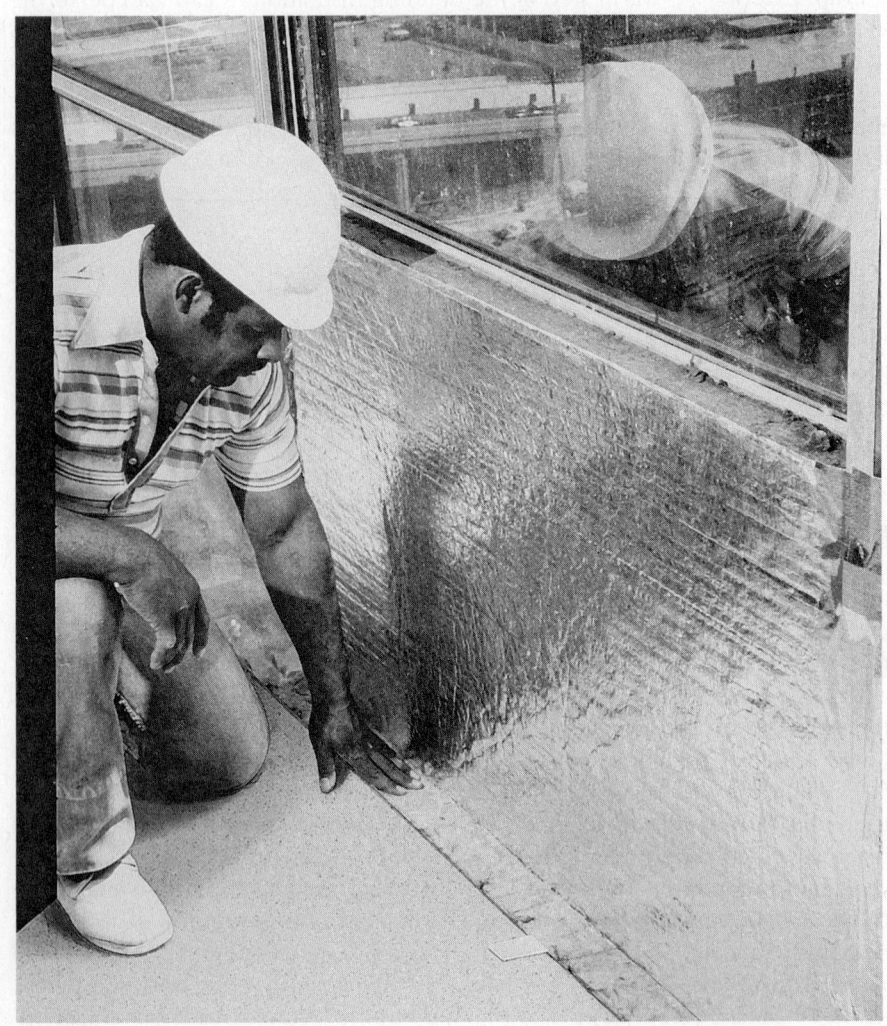

FIGURE 19.16

Safing is a high-temperature, fire-resistant mineral batt material that is inserted between a curtain wall panel and the edge of the floor slab to block the passage of fire from one floor to the next. It is seen here behind a metal-and-glass curtain wall with insulated spandrel panels. The safing is held in place by metal clips such as the one seen in the foreground. *(Photo courtesy of United States Gypsum Company.)*

CSI/CSC MasterFormat Sections for Designing for Exterior Wall Systems

07 27 00	**AIR BARRIERS**
07 84 00	**FIRESTOPPING**
07 84 53	Building Perimeter Firestopping
07 91 00	**PREFORMED JOINT SEALS**
07 91 13	Compression Seals
07 91 16	Joint Gaskets
07 91 23	Backer Rods
07 92 00	**JOINT SEALANTS**
07 92 13	Elastomeric Joint Sealants

KEY TERMS

exterior wall enclosure, building envelope
thermal bridge
barrier wall, face sealed wall
wash
gravity
momentum
labyrinth
surface tension
drip
capillary action
capillary break
wind pressure
pressure-equalized wall design
air barrier
rainscreen
pressure equalization chamber (PEC)
rainscreen principal

rainscreen cladding, drained and back-ventilated cladding
gunnable joint sealant
sealant gun
low-range sealant, caulk
elongation
medium-range sealant
high-range sealant
polysulfide sealant
polyurethane sealant
silicone sealant
sealant Type
sealant Grade
sealant Class
sealant Use
single-component sealant
multicomponent sealant
self-leveling sealant, pourable sealant

nonsag sealant
traffic sealant
nontraffic sealant
immersible sealant
gasket
preformed cellular tape sealant
preformed solid tape sealant
priming (of sealant joints)
backer rod, backup rod
loadbearing wall
curtain wall
stack effect
air barrier
air permeance
air barrier assembly
air barrier systemwind washing
firestopping
safing

REVIEW QUESTIONS

1. Why is it so difficult to make cladding watertight?

2. List the functions that cladding performs and list one or two ways in which each of these functions is typically satisfied in a cladding design.

3. Using a series of simple sketches, explain the principles of sealant joint design. List several sealant materials suitable for use in the joints that you have shown.

4. What are the forces that can move water through a joint in an exterior wall? How can each of these forces be neutralized?

EXERCISES

1. Examine the cladding of a building with which you are familiar. Look especially for features that have to do with insulation, condensation, drainage, and movement. Sketch a detail of how this cladding is installed and how it works. You will probably have to guess at some of the hidden features, but try to produce a complete, plausible detail. Add explanatory notes to make everything clear.

2. Work out a way to add a rainscreen window with fixed double glazing to the cladding system shown in Figure 19.8.

3. Prepare a sample sealant joint, using a backer rod and silicone sealant obtained from a hardware store or building materials supplier. Apply the sealant to two parallel pieces of quarry tile or glass that are taped together with a spacer between them. After the sealant has had time to cure (a week or so), remove the tape and spacer and test the joint by pulling and twisting it to find out how elastic it is and how well the sealant has adhered to the substrate.

SELECTED REFERENCES

The Institute for Research in Construction of the National Research Council Canada has done pioneering work in theorizing about exterior wall design and performing tests and field observations to back up the theory. This work is summarized in a large library of reports on specific topics that may be viewed online at www.nrc.ca/irc/ircpubs. See, for example, the document entitled *Evolution of Wall Design for Controlling Rain Penetration.*

Amstock, Joseph S. *Handbook of Adhesives and Sealants in Construction.* New York, McGraw-Hill, 2000.

This book offers detailed information on all types of sealants, construction adhesives, joint and crack control in concrete, and firestopping, as well as the design, specification, and testing of sealant joints.

Brock, Linda. *Designing the Exterior Wall: An Architectural Guide to the Vertical Envelope.* Hoboken, NJ, John Wiley & Sons, Inc., 2005.

This book covers the building science underlying the performance of the exterior wall and provides examples of its application to the design of wall types ranging from light wood construction to metal curtain wall.

Brookes, Alan, and Maarten Meijs. *Cladding of Buildings* (4th ed.). New York, Taylor & Francis, 2008.

This book provides a clear general introduction to cladding principles and material types.

WEB SITES

Designing Exterior Wall Systems

Author's supplementary web site: **www.ianosbackfill.com/19_designing_cladding_systems**

Dow-Corning sealants: **www.dowcorning.com**

GE sealants: **www.siliconeforbuilding.com**

Institute for Research in Construction: **www.nrc.ca/irc/ircpubs**

Whole Building Design Guide—Wall Systems: **www.wbdg.org/design/env_wall.php**

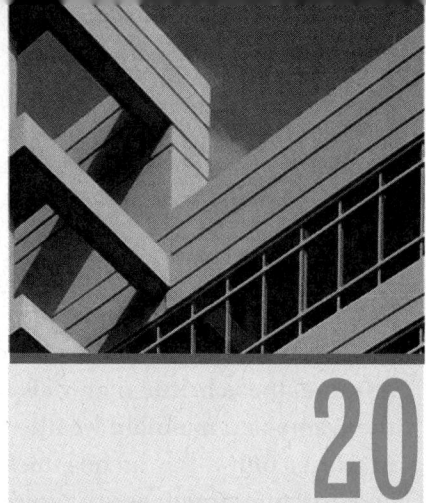

## 20

CLADDING WITH MASONRY AND CONCRETE

- **Masonry Veneer Curtain Walls**

  Prefabricated Brick Panel Curtain Walls

- **Stone Curtain Walls**

  Stone Panels Mounted on a Steel Subframe

  Monolithic Stone Cladding Panels

  Stone Cladding on Steel Trusses

  Posttensioned Limestone Spandrel Panels

  Very Thin Stone Facings

- **Precast Concrete Curtain Walls**

  Glass-Fiber-Reinforced Concrete Curtain Walls

- **Exterior Insulation and Finish System**

- **Keeping Water Out with Masonry and Concrete Cladding**

  FROM CONCEPT TO REALITY
  Seattle University School of Law

Architects Thompson Ventulett Stainback & Associates created a bold pattern of precast concrete sunshades and spandrel panels for the facades of the United Parcel Service Headquarters Building in Atlanta, Georgia. *(Photo by: Brian Gassel/TVS & Assoc.)*

Buildings framed with structural steel or concrete are often clad with brick, stone, or precast concrete. These substantial materials, though hung from the loadbearing frame of the building, impart a sense of mass, solidity, and permanence. These thin, suspended facings, however, do not behave the same way as the solid loadbearing walls they sometimes mimic. When mounted on a frame, these brittle materials must adjust to the movements of the frame and maintain weathertightness despite being applied in a layer only a few inches thick. Careful detailing and good construction practices are required to make this possible.

MASONRY VENEER CURTAIN WALLS

Figure 20.1 illustrates steps in the construction of brick *masonry veneer* cladding (a single wythe of brick masonry separated by a cavity from a backup wall) applied to a reinforced concrete frame. The veneer may also be made of stone. The veneer is erected brick by brick or stone by stone with conventional mortar, starting from

(*a*)

Interactive Exercise

FIGURE 20.1

Construction sequence for a brick veneer curtain wall supported by a reinforced concrete frame. (*a*) Before the concrete frame of the building was cast, inserts were put into the formwork to form attachments for the brick veneer, including wedge anchors along the line of each shelf angle, two vertical dovetail slots in each column, short vertical dovetail slots in the spandrel beams, and horizontal reglets in the centers of the spandrel beams to accept the inner edge of a flashing over each window head (see the Chapter 15 sidebar "Fastening to Concrete" for pictures of inserts such as these). To begin installation of the brick veneer, a steel shelf angle is bolted to the spandrel beam, using, for example, malleable iron wedge inserts as shown in Figure 20.2. A slab of polystyrene foam or mineral wool board thermal insulation (gray) is placed over the upright leg of the shelf angle, and a continuous flashing (white) is installed over the shelf angle, the insulation, and the edge of the floor slab. This flashing also wraps around the front of the column. All the seams in the flashing are overlapped and made watertight with sealant. The first course of brickwork is laid directly on the shelf angle and flashing, without a bed joint of mortar. Every third head joint is left open in this first course to form a weep hole.

Three courses of brickwork bring the veneer up to the level of the floor slab. (*b*) The first course of the concrete masonry backup wall is laid. Vertical reinforcing bars are grouted into the hollow cores of the backup wall at intervals specified by the structural engineer. A liquid coating or adhered sheet is applied to the backup wall to act as an air and moisture barrier. Three more courses of brick veneer bring the top of the veneer up to the level of the top of the first course of concrete masonry. Additional insulation is placed against the concrete masonry. A combination joint reinforcing and masonry tie made of heavy stainless steel wires is laid on top of the masonry, tying the brick veneer to the backup wall. Plastic clips are snapped onto the tie rods of the joint reinforcing to hold the insulation in position. A vertical expansion joint in the brick veneer is provided at the centerline of each column. A heavy wire masonry tie in a dovetail slot anchors the brick veneer to the column on each side of the joint; another such anchor is lying loose on top of the bricks, ready to be installed, in this view.

(b)

(c)

(c) The wall rises in vertical increments
of 16 inches (400 mm), which equals six
courses of brick or two of concrete masonry. This is
also the vertical distance between ties and the height of the
insulating panels. A-blocks (see Figure 9.23) are utilized as needed
in the backup wall to avoid having to thread blocks over the tops of the
vertical reinforcing bars. The vertical expansion joint is sealed with backer rod
and sealant. As an alternative to the sequence of operations illustrated here, the
backup wall and air barrier may be installed to their full height first, followed
later by the installation of insulation and veneer.

A malleable iron wedge insert is cast into the concrete spandrel beam

An askew-head bolt permits vertical adjustment in the wedge insert

Steel shelf angle

(a)

Spandrel beam

Clip angle

Wedge-shaped nut in wedge-shaped slot

Shelf angle

Steel shim plates as required to level the shelf angle

Detail of shim plate

(b)

FIGURE 20.2

(a) An example of a cast-in-place anchoring system for attaching a steel shelf angle to a concrete spandrel beam. Steel shims are added as necessary between the shelf angle and the spandrel beam to place the angle exactly in the plane of the veneer wythe. (b) The traditional method for attaching shelf angles to steel spandrel beams uses steel clip angles with shim plates as needed to make up for dimensional inaccuracies in the components. In practice, providing anchoring systems with adequate adjustability to account for deviations in the structural frame is often a difficult challenge.

a steel *shelf angle* that is attached to the structural frame at each floor (Figure 20.2).

Many steps in the process and details of the assembly are essentially the same as for a masonry cavity wall of a single-story building, but there are some crucial differences: To prevent normal movements in the frame of the building or expansion and contraction in the masonry veneer from causing stress in the veneer, the veneer is divided into rectangular panels of reasonable size surrounded by expansion joints that can absorb these movements. At each floor level, a *horizontal expansion joint* (*soft joint*) is provided beneath the shelf angle (Figure 20.3). This joint is made wide enough to absorb the sum of column creep, brick expansion, spandrel beam deflection, and tolerances for

construction inaccuracies while not exceeding the safe movement capacity of the sealant. Similarly, *vertical expansion joints* protect the veneer from stresses caused by movements in the horizontal plane (Figure 20.4).

Stone and brick veneer cladding may also be constructed with a *backup*

wall constructed of light gauge steel studs covered with water-resistant sheathing panels of gypsum or cementitious materials (Figure 20.5). In comparison to concrete masonry, stud-framed backup walls are lighter in weight, faster to erect, easier to insulate, more accommodating of

FIGURE 20.3

A complete detail section of the brick veneer wall that was begun in Figure 20.1 shows how the top of the backup wall is fastened to the underside of the spandrel beam with a series of steel restraint clips that brace the top of the wall against wind loads but allow the spandrel beam to deflect under load. This joint is sealed with two lines of backer rod and sealant. A soft joint of sealant beneath the shelf angle permits the spandrel beam to deflect without applying force to the brick veneer. The brick ties nearest the underside of the shelf angle are anchored to dovetail slots in the spandrel beam. An additional plastic clip on each wire tie in the center of the cavity acts as a drip to prevent water from clinging to the tie and running toward the backup wall. The interior of the building is finished with gypsum veneer plaster mounted on steel furring channels, similar to the assembly shown in Figure 23.4.

MASTIC AIR BARRIER

1″ POLYSTYRENE INSULATION

2″ CAVITY

BRICK FACING
WITH CONCAVE
MORTAR JOINTS

PLASTIC CLIPS

CONTINUOUS FLASHING

WEDGE ANCHOR INSERT

SHELF ANGLE

WEEP HOLES 16″ O.C.

BOND BREAKER AND SEALANT

COMPRESSIBLE FILLER STRIP

TIES IN DOVETAIL
SLOTS 16″ O.C.

BACKER ROD AND SEALANT

COMPRESSIBLE FILLER STRIP

BACKER ROD AND SEALANT

JOINT REINFORCING/TIE

VERTICAL REINFORCING

8″ CONCRETE MASONRY

METAL FURRING 16″ O.C.

GAP IN FURRING FOR CONDUIT

½″ VENEER BASE AND PLASTER

HORIZONTAL FURRING CHANNEL

WOOD BASEBOARD

CARPET AND PAD

RESILIENT CHANNEL 16″ O.C.

⅝″ VENEER BASE
AND PLASTER

METAL FURRING 16″ O.C.

RESTRAINT CLIP

RESILIENT TRIM BEAD

METAL FURRING 16″ O.C.

½″ VENEER BASE
AND PLASTER

TYPICAL SPANDREL SECTION

(a)

(b)

FIGURE 20.4
(a) A carefully detailed brick curtain wall by architects Kallman, McKinnell and Wood covers the steel frame of Hauser Hall at Harvard University. Notice the vertical expansion joint near the far-right corner. (b) At the base of Hauser Hall, the facing wythe is made of limestone blocks. The backup wall consists of steel studs and gypsum sheathing panels. A vertical expansion joint is visible in the far-left corner in this view. *(Photos © Steve Rosenthal.)*

electrical wiring, and readily covered with a great variety of interior finish materials. However, a steel stud framed wall is more flexible than one made of concrete masonry and prone to greater deflections under the pressures of the wind. If these deflections are too great, mortar joints in the masonry veneer begin to crack, and water leakage through the veneer increases. This in turn can subject the gypsum sheathing panels to moisture deterioration and the steel studs and fasteners to corrosion. In contrast, a concrete masonry backup wall is easily made stiffer than the veneer that it supports, greatly reducing the risk of wind load stresses on the veneer. A concrete masonry backup wall can

also, if necessary, sustain prolonged periods of wetting with less risk of deterioration.

For these reasons, where steel stud backup systems are selected, the metal framing, masonry ties, and fasteners should be sized conservatively to ensure that the completed system is stiff enough to protect the veneer from wind load stresses. The sheathing material, anchors, and fasteners should be selected for their durability under damp conditions. Each metal tie that connects the masonry veneer to the studs should be attached to a stud with at least two corrosion-resistant screws. The wall must be detailed carefully to keep leaked water away from the backup components.

Inspection should be required during construction to ensure that all these details are faithfully executed and the cavity is kept clean so that it will drain freely.

The structural frame of a building is never absolutely flat or plumb. Thus, the system for attaching shelf angles must allow adjustment, so that the veneer may be constructed in a more precisely vertical plane with level courses. Figure 20.2 illustrates examples of how this is accomplished with both concrete and steel frames. The attachment system shown in Figures 20.5 and 20.6, which is designed to suspend a masonry veneer spandrel wall over a continuous band of windows, also provides

Sill of aluminum window

Rowlock brick sill

Flashing and weep holes

Brick veneer

Cavity

Masonry anchor screwed to steel stud

Steel studs welded to slab edge

Rigid plastic foam or semirigid mineral wool cavity insulation

Air- and water-barrier membrane

Gypsum sheathing panels

Steel studs and glass fiber batt insulation

Flashing and weep holes

Steel shelf angle

Head of aluminum window

Wood stool

Wood subsill

Gypsum board

Vapor retarder

Resilient base and flooring

Bent plate slab edge with rod anchors, cast into concrete topping

Steel angle supporting frame for shelf angle

Suspended gypsum board ceiling

FIGURE 20.5

A detail section of a brick curtain wall. The veneer is supported on a shelf angle suspended below the level of the spandrel beam on a frame also made of steel angles. The supporting frame becomes necessary when continuous horizontal bands of windows are to be installed between brick spandrels. All its connections are made with bolts in slotted holes to allow alignment of the shelf angle. After the frame has been aligned and before the masonry work begins, the connections are welded to prevent slippage. Continuous insulation applied to the outside of the metal stud framed backup wall minimizes energy losses from thermal bridging through the metal framing components.

 Watch the Animation

for free adjustment of the shelf angle location.

The flashing above the shelf angle plays an important role in capturing water that reaches the cavity and conducting it back out of the wall. This flashing runs continuously over the shelf angle and should project beyond the face of the veneer sufficiently to form an effective drip. If a flexible flashing material is used, it should be cemented to a strip of sheet metal, with the sheet metal projecting from the wall. Efforts to hide this important flashing should be approached with caution. Terminating the flashing at the edge of, or even slightly behind, the face of the veneer invites water to

FIGURE 20.6

The section shown in Figure 20.5 allows the construction of brick spandrels between continuous horizontal bands of glass. *(Photo by Edward Allen.)*

be drawn under the flashing by capillary action. Water that gets behind the flashing can corrode the shelf angle and its fastenings; in cold climates, it also invites freeze-thaw of the masonry veneer. Over time, damage to the shelf angle or its supports can cause unsightliness, and if more severe, imperil the stability of the veneer itself.

Prefabricated Brick Panel Curtain Walls

Figure 20.7 shows the use of *prefabricated reinforced brick panels* for cladding.

FIGURE 20.7

Fabrication and installation of a brick panel curtain wall. (*a*) Masons construct the panels in a factory, using conventional bricks and mortar. Both horizontal and vertical reinforcing are used, the vertical bars being grouted into the hollow cores of the bricks. (*b*) Brick panels are stored to await shipment, complete with thermal insulation attached to the back of the panels. The welded metal brackets are for attachment to the building; the structural strength of the panel comes primarily from the reinforced masonry, not the brackets. (*c*) At the construction site, panels are lifted into position by crane. (*d*) Corners can be constructed as single panels. (*Panelized masonry by Vet-O-Vitz Masonry Systems, Inc., Brunswick, Ohio.*)

(*a*)

(*b*)

(*c*)

(*d*)

Masons construct the panels while working comfortably at ground level in a dry, climate-controlled factory or shop. Horizontal reinforcing may be laid into the mortar joints or grouted into channel-shaped bricks. Vertical reinforcing bars are placed in grouted cavities of hollow-core bricks. These panels are self-rigid; they need no structural backup and can be fastened to the building in much the same way as precast concrete panels. A steel stud backup wall is required to carry thermal insulation, air barrier, electrical wiring, and an interior finish layer, but it has no structural role.

STONE CURTAIN WALLS

Chapter 9 discusses types of stone and illustrates conventionally set stone facing systems that tie relatively small blocks of cut stone to a concrete masonry backup wall. Slabs of stone that are larger in surface area may be fastened to framed buildings in several different ways.

Stone Panels Mounted on a Steel Subframe

Figure 20.8 shows a system for mounting stone panels on a steel subframe, called *grid-system-supported stone cladding*. The vertical members of the subframe are erected first. They are designed to transmit gravity and wind loads from the stone slabs to the frame of the building. The horizontal members are aluminum shapes that engage slots in the upper and lower edges of each panel to attach

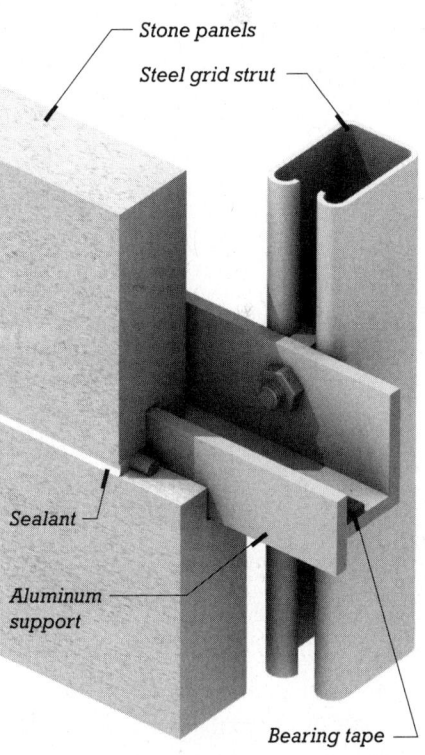

FIGURE 20.8

(*Left*) **A subframe of vertical steel struts supports a facing of stone panels by means of horizontal metal clips that engage slots in the upper and lower edges of the panels. To avoid corrosion and staining problems, the steel struts should be galvanized, and the clips should be made of a nonferrous metal (aluminum or stainless steel) that is chemically compatible with the type of stone being used. (*Right*) A detail of the support-to-panel connection.**

FIGURE 20.9

(a) Parapet and (b) spandrel details for a stone panel curtain wall made of limestone, marble, or granite. The broken lines indicate the outline of the interior finish and thermal insulation components, which are not shown. Each support plate holds edges of two adjacent wall panels, which are pocketed as shown to rest on the plate. The plate should be made of a noncorroding metal. The vertical joints between panels are closed with a backer rod and sealant.

(a)

Sealant

Flashing

Cant strip

Roofing membrane

Rigid insulation

Expansion anchor

Slotted hole in angle for expansion bolt

Welded stud in slotted hole

Cut stone wall panel

FIGURE 20.10

A granite panel curtain wall of the type illustrated in Figure 20.9 wraps around the corner of a Boston office building. The upper-floor windows have not yet been installed, but the window frames have been mounted in two of the middle floors, and the lower floors have been glazed.

(b)

Aluminum window sill

Cut stone wall panel

The support plate is shimmed and bolted, then welded to an anchor plate cast into the edge of the floor topping

Mortar setting bed

Epoxy

Support plate with rod welded on

Pocket milled into edges of panel (see detail)

Safing

Steel angle struts brace the panels against tilting

Bolt in expansion anchor

Drip slot

Aluminum window head

DETAIL OF EDGE POCKET IN STONE PANEL

them firmly to the building. They are added as the installation of the stone panels progresses. Backer rods and sealant fill the spaces between the panels, allowing for necessary movement. A nonstructural backup wall, usually made of steel studs and gypsum sheathing panels, is constructed within the frame of the building but is not attached to the subframe. Its functions are to provide an air barrier, to house thermal insulation batts and electrical wiring, and to support the interior wall finish layer, which is usually plaster or gypsum board.

A weakness of this system is its dependence on the integrity of the sealant joints. If leakage occurs, water may accumulate in the slots in the tops of the stone panels, and freeze-thaw deterioration may risk damage to the relatively fragile, narrow tongues of stone on either side of the groove.

Monolithic Stone Cladding Panels

Figures 20.9 and 20.10 illustrate the use of *monolithic stone cladding panels* that are fastened directly to the frame of the building. The weight of each panel is transferred to two steel support plates by means of edge pockets that are cut into both sides of each panel at the stone mill. Each panel is stabilized by a pair of steel angle struts that are bolted to the stone with expansion anchors in drilled holes. Joints are closed with backer rod and sealant, and a nonstructural backup wall is required.

FIGURE 20.11
A steel truss system for stone cladding.
(*a*) Masons working in a fabrication yard
attach thin sheets of stone to welded
steel trusses. The vertical joints are
closed with backer rods and sealant.
(*b*) The fabricated spandrel panel is
lifted onto a truck using a crane. The
metal clips that are just visible along
the top and bottom edges of the panel
engage slots in the edges of the sheets
of stone to hold the stone securely to
the truss. The steel angle clips at the two
upper corners of the truss will support
the panel on brackets welded to the steel
columns of the building frame. (*c*) The
panel is installed. *(Courtesy of International
Masonry Institute, Washington, D.C.)*

(*a*)

(*b*)

Stone Cladding on Steel Trusses

In *truss-supported stone cladding*, sheets
of stone are combined into large prefabricated panels by mounting them
on steel trusses (Figure 20.11). Each
truss is designed to carry both wind
loads and the dead load of the stone
to connection brackets that transfer
these loads to the frame of the building. Sealant joints and a nonstructural
backup wall finish the installation.

(*c*)

Posttensioned Limestone Spandrel Panels

Thick blocks of limestone may be joined with adhesives into long spandrel panels and posttensioned with high-strength steel tendons so that the assembly is self-supporting between columns (Figure 20.12). Such *posttensioned limestone spandrel panels* are relatively costly, because of their use of comparatively large quantities of stone per unit area of cladding.

Very Thin Stone Facings

Extremely thin sheets of stone (as thin as ¼ inch, or 6.5 mm, for granite) may be stiffened with a structural backing such as a metal honeycomb and mounted as spandrel panels in an aluminum mullion system such as those described in Chapter 21.

Very thin sheets of stone may also be used as facings for precast concrete curtain wall panels. The stone sheets are laid face down in the forms. Stainless steel clips are inserted into holes drilled in the backs of the stone. A grid of steel reinforcing bars is added, and then the concrete is poured and cured to complete the panel. The clips anchor the stone to the concrete.

When specifying the thickness of stone for any exterior cladding application, the designer should work closely with the stone supplier and also consult the relevant standards of the building stone industry. Stone that has been sliced thinner than industry standards has caused a number of failures of cladding systems.

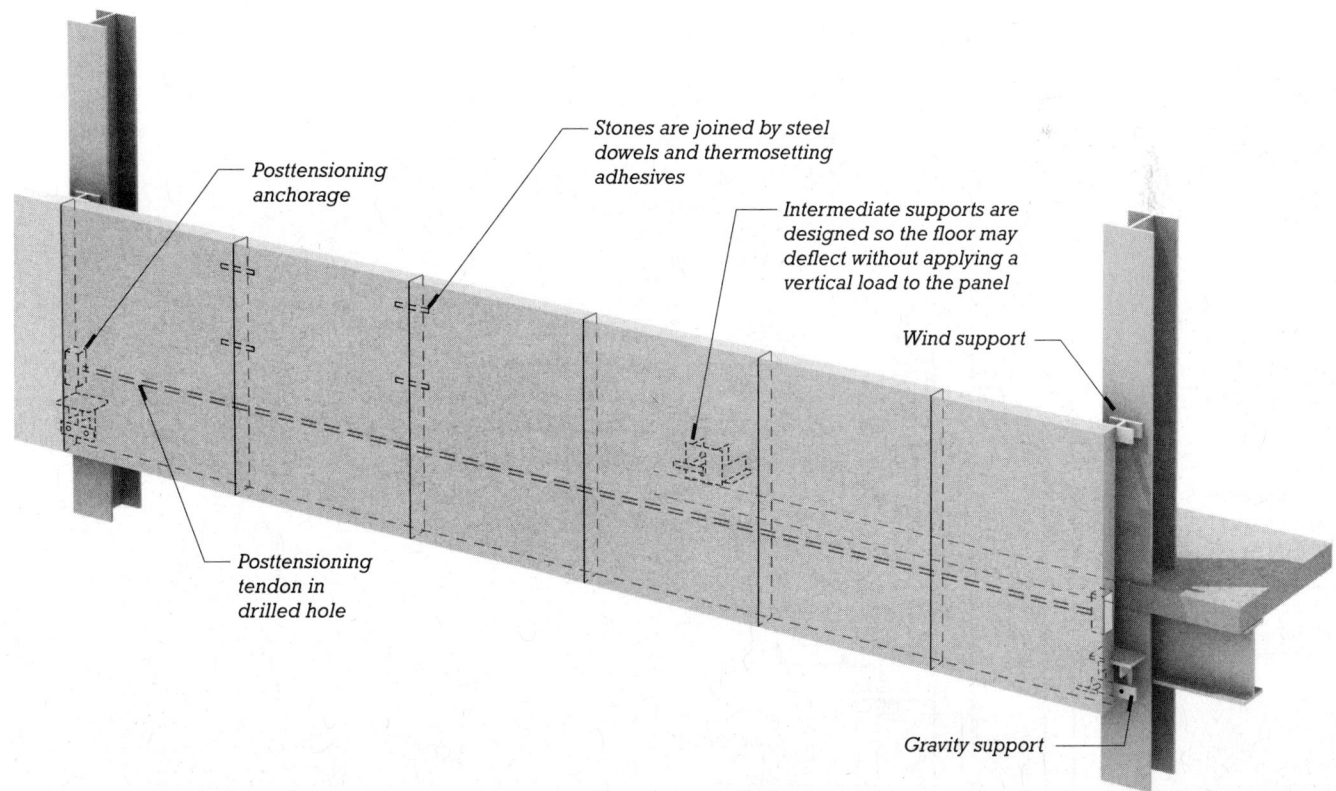

FIGURE 20.12

Thicker blocks of Indiana limestone may be posttensioned together to make spandrel panels that span from column to column but require little steel. The posttensioning tendon is threaded through matching holes that are drilled in the individual stones prior to assembly.

PRECAST CONCRETE CURTAIN WALLS

Precast concrete cladding panels, both conventionally reinforced and pre-stressed (this chapter's opening photograph and Figures 20.13–20.16), are simple in concept but require close attention to matters of surface finish, mold design, thermal insulation,

attachment to the building frame, and sufficient strength and rigidity in the building frame to support the weight of the panels.

The factory production of concrete cladding panels makes it possible to utilize high-quality molds and a variety of surface finishes, from glassy smooth to rough, exposed aggregates. Ceramic tiles, thin bricks, or thin stone facings may be attached

to precast concrete panels. In *precast concrete sandwich panels*, thermal insulation is incorporated as an inner layer of the panel (Figures 20.17 and 20.18). Alternatively, insulation may be affixed to the back of the panel or may be provided in a nonstructural backup wall that is constructed in place. Reinforcing or prestressing of the panel must be designed to resist wind, gravity, seismic, and lifting

Glazing in lockstrip gasket

Drip

Precast concrete curtain wall panel

Cast-in anchor

Angle support clip with side stiffener plates welded on

Bolts

Shims

Drip

Sealant

Compressible tape interior sealant

Safing

Bolt in cast-in anchor slot

Bolts

Angles with slotted holes allow for adjustment

Drip

Glazing in lockstrip gasket

Cast-in anchor

FIGURE 20.13

A typical detail of a precast concrete curtain wall on a sitecast concrete frame. Panels in this example are a full story high, each containing a fixed window. The reinforcing has been omitted from the panel for the sake of clarity, and the outline of the thermal insulation and interior finishes, which are not shown, is indicated by the broken lines.

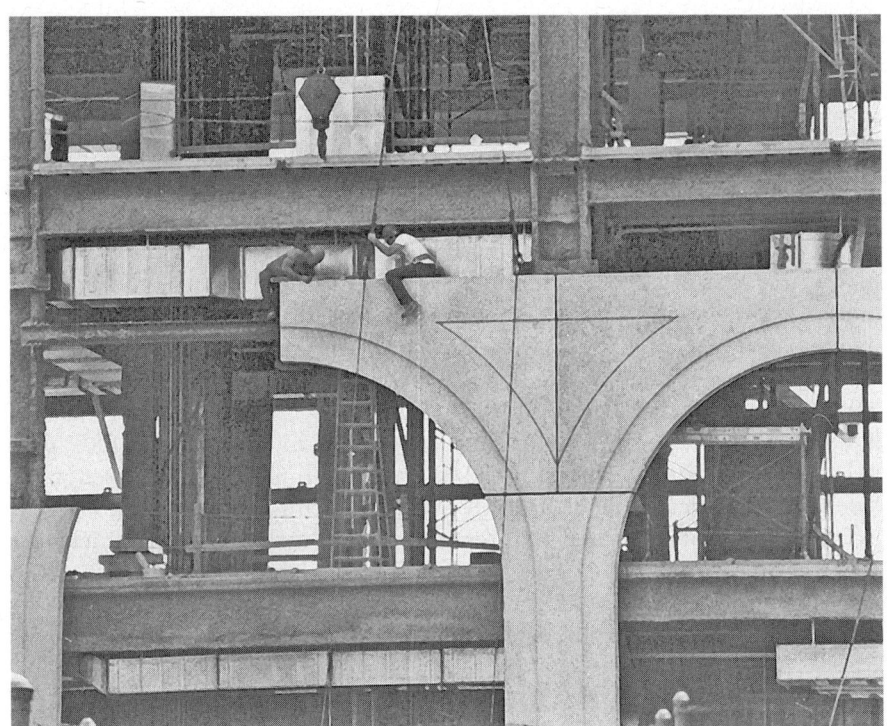

FIGURE 20.14
Workers install a precast concrete curtain wall panel. *(Architects and engineers: Ander-Nichols Company, Inc. Photo by Edward Allen.)*

FIGURE 20.15
A Chicago hotel is clad in precast concrete panels. *(Photo by Hedrich Blessing, Ltd.)*

forces and control cracking of the concrete. Attachments must transfer all these forces to the building frame while allowing for installation adjustment and for relative movements of the frame and the cladding.

More recently developed materials, such as carbon fiber reinforcing (Chapter 15) or ultra-high-performance concrete (Chapter 13), allow the manufacture of panels that are thinner and lighter than those made of conventional materials.

Glass-Fiber-Reinforced Concrete Curtain Walls

Glass-fiber-reinforced concrete (GFRC) is a relatively new cladding material that has several advantages over conventional precast concrete panels. Its admixture of short glass fibers furnishes enough tensile strength that no steel reinforcing is required. Panel thicknesses and weights are about one-quarter of those for conventional precast concrete panels, which saves money on shipping, makes the panels easier to handle, and allows the use of lighter attachment hardware. The light weight of the cladding also allows the load-bearing frame of the building to be lighter and less expensive. GFRC can be molded into three-dimensional forms with intricate detail and an extensive range of colors and textures (Figures 20.19 and 20.20).

The fibers in GFRC must be manufactured from a special alkali-resistant type of glass to prevent their disintegration in the concrete. The panels may be self-stiffened with GFRC ribs, but the usual practice is to attach a welded frame made of light gauge steel studs to the back of each GFRC facing in the factory. The attachment is made by means of thin steel rod anchors that flex slightly as needed to permit small amounts of relative movement between the facing and the frame. Figure 20.21 shows typical ways of attaching metal-framed GFRC panels to the building. The edges of the GFRC facing, which is usually only about ½ inch (13 mm) thick, are flanged as shown in Figure 20.22 so that backer rods and sealant may be inserted between panels.

 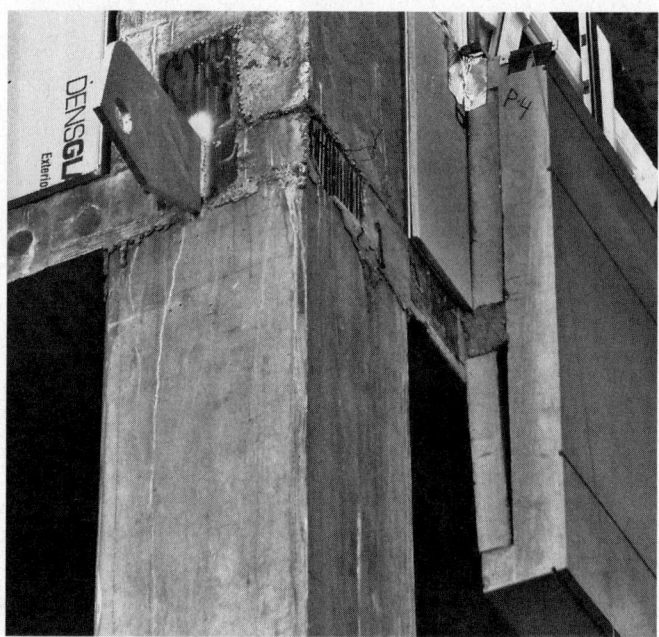

FIGURE 20.16

(*Left*) Precast concrete panels and aluminum windows being installed on a concrete frame. (*Right*) A close-up view of the panel installation. The back of the panel is insulated with polyisocyanurate plastic foam insulation. At the edge of the concrete floor slab, safing (to protect against fire easily passing from one floor level to the next) seals the gap between the slab edge and the back of the panel. The metal plate welded to the face of the column will support a future metal canopy. (*Photo by Joseph Iano.*)

FIGURE 20.17
Manufacturing a Corewall® insulated
panel, a proprietary foam-core precast
concrete sandwich panel. Rollers apply a
ribbed texture to the outside of a panel
that includes a layer of foam plastic
insulation sandwiched between layers of
concrete. *(Photos reproduced with permission
of COREWALL LIMITED.)*

FIGURE 20.18
A completed panel is lifted from the
casting bed. *(Photos reproduced with
permission of COREWALL LIMITED.)*

FIGURE 20.19
Fabrication of a GFRC wall panel.
(*a*) Concrete and chopped glass fibers
are sprayed into a mold and compacted
with a hand roller to create a panel
facing. Only the top half of the facing
has been applied to the mold in this
illustration. (*b*) A welded frame of steel
studs with L-shaped steel rod anchors
is lowered onto the back of the facing
and held just above it by spacers. Pads
of GFRC are placed over the anchors
by hand to join the facing to the
frame. (*c*) After overnight curing, the
completed panel is removed from
the mold and stored for further curing
before installation.

(*a*)

FIGURE 20.20
Fabrication of a GFRC curtain wall
panel. (*a*) A special gun deposits a layer
of sand–cement slurry simultaneously
with 1.5-inch (38-mm) lengths of alkali-
resistant glass fiber reinforcing. Three
layers are usually required to make up
the full thickness of the panel facing;
each is compacted with a small hand
roller before the next layer is applied.
The overall thickness is usually ½ inch
(13 mm). (*b*) After the GFRC facing layer
has been completed, the steel frame is
lowered over it and the operator hand-
applies pads of wet GFRC over the rod
anchors to bond the frame to the GFRC
facing. (*Courtesy of Precast/Prestressed
Concrete Institute.*)

(*b*)

EXTERIOR INSULATION AND FINISH SYSTEM

An *exterior insulation and finish system (EIFS)* consists of a layer of plastic foam insulation that is adhered or mechanically fastened to a backup substrate, a reinforcing mesh that is applied to the outer surface of the foam by embedment in a base coat of a stuccolike material, and an exterior finish coat of a similar material that is troweled over the reinforced base coat. In most cases the substrate is concrete masonry or site-erected steel studs covered with water-resistant sheathing (Figures 20.23 and 20.24). The system also adapts readily to prefabrication (Figure 20.25).

EIFS is an unusually versatile type of cladding, used for building types as diverse as single-family residences of wood or masonry construction as well as the largest buildings of noncombustible construction. It is used both for new construction and for refacing and insulating existing buildings. The insulating foam layer may be up to 4 inches (100 mm) thick, and there is little or no thermal bridging. The finish layer may be applied in a range of colors and textures. In appearance, at least from a distance, EIFS is virtually indistinguishable from conventional stucco.

Two types of EIFS can be distinguished by differences in insulation and coating materials. *Polymer-based* or *Class PB EIFS* most frequently uses a low-density expanded polystyrene bead foam insulation, a glass fiber reinforcing mesh embedded in a base coat that is formulated primarily from either portland cement or acrylic polymer, and a finish coat that consists of texture granules in an acrylic polymer vehicle. The foam insulation is usually adhered to the backup wall, although in some cases it may be mechanically attached. *Polymer-modified* or *Class PM EIFS* uses a slightly higher density, extruded polystyrene foam insulation rather than expanded bead foam. The foam panels are mechanically attached to the backup wall with metal or plastic screws (plastic screws minimize thermal bridging through the insulation). A metal reinforcing mesh is embedded in a relatively thick portland cement base coat, and the finish coat is formulated of portland cement with acrylic modifiers. Class PM EIFS is also sometimes referred to as *hard coat EIFS*.

Note: Attach window to frame free of skin

Blockout in Slab for Panel Bearing Angle (filled in after panel installed)

Bolt Adj. Elev.

Shear Plate

Flex Anchors
Trussed
Gravity Anchor

Stiffener Plate or Brace by Steel Supplier if Required

Angle With Oversized Hole Provided by Structural Steel Supplier

Diagonal Brace & Downrigger Provided by Structural Steel Supplier

FIGURE 20.21

Typical connections of GFRC panels to a steel building frame. The lower connection in each case is a threaded rod that can flex as necessary as the height of the upper connection is adjusted with shims. *(Courtesy of Precast/Prestressed Concrete Institute.)*

In North America, Class PB is the most commonly specified type of EIFS. The relatively thin coating is elastic and resistant to cracking. However, this type is also the most vulnerable to dents or puncture when applied to areas of a building that may come in contact with passersby or vehicles. Class PM EIFS is more durable (and more expensive) than polymer-based systems. It is more susceptible to shrinkage cracking during curing, but also more resistant to physical abuse.

In the past, EIFS has at times proven vulnerable to water intrusion, leading, in some cases, to costly failures of the building enclosure. In response, materials and installation methods have evolved. For example, where paper-faced gypsum panels were once used as the system substrate, today more durable, moisture-resistant panel types, such as glass mat faced gypsum sheathing, are used. To prevent sealant joints around the edges of the panels from peeling away the EIFS finish coat, sealant is now installed against the base coat of the system before the finish coat is applied. Careful attention is given

FIGURE 20.22

Typical edge details for GFRC cladding panels. *(Courtesy of Precast/Prestressed Concrete Institute.)*

to sealant material selection and joint design, to limit the forces exerted on the base coat by the sealant.

Additionally, EIFS installations have evolved two approaches to the control of moisture penetration. *Barrier-wall EIFS* relies solely on the coating layers and joint sealants between panels to protect against water penetration. *Drainage-wall* or *water-managed EIFS* creates a drainage plane between the insulation and substrate to which it is attached, allowing water that does penetrate to flow downward and back out through weeps provided at the base of each panel. The drainage plane consists of a water-resistant membrane applied over the substrate and a thin space between this layer and the insulation. The most common way to create the drainage space is by applying the insulation adhesive in vertical strips, with the gaps remaining in between acting as drainage channels. Alternatively, insulation boards with integral vertical channels may be used, or a thin, pervious drain mat may be

FIGURE 20.23
Four steps in installing an EIFS over a building with walls of masonry or solid sheathing. (*A*) A panel of foam is daubed with polymer-modified portland cement mortar. The foam may be as thick as required to achieve the desired thermal performance. (*B*) The foam panel is pressed into place, where it is held permanently by the adhesive mortar. (*C*) A thin base coat of polymer-modified stucco is applied to the surface of the foam panels, with an embedded mesh of glass fiber to act as reinforcing. (*D*) After the base coat has hardened, a finish coat in any desired color is troweled on. (*Used by permission of Dryvit® System, Inc.*)

FIGURE 20.24
An early 20th-century warehouse, reclad with EIFS to improve its energy efficiency and renew its appearance. *(Photo by Joseph Iano.)*

FIGURE 20.25
EIFS cladding can be shop-fabricated and erected in panel form. (A) Steel studs are welded together to make panel frames. (B) Rigid sheathing is screwed to the panel frames and finished with EIFS as shown in Figure 20.23. (C) The finished panels are bolted to the frame of the building. *(Photo: John Bare. Used by permission of Dryvit® System, Inc.)*

installed between the insulation and substrate (Figure 20.26).

Class PB EIFS is installed to meet the performance requirements of ASTM E 2568. The International Building Code requires water-managed EIFS where applied over light wood frame buildings of any residential occupancy type. Careful supervision of EIFS during its installation is also required, to ensure that proper installation methods are followed and the necessary workmanship standards are maintained.

KEEPING WATER OUT WITH MASONRY AND CONCRETE CLADDING

The choice of barrier wall or rainscreen design for watertightness (strategies described in Chapter 19) varies with the characteristics of the cladding system. For example, brick masonry veneer is best suited to rainscreen design. The large area of mortar joints within the wall surface tends to make the veneer layer leaky, especially as it ages. For this reason, brick veneer cladding is designed with an internal drainage plane that intercepts water that penetrates the masonry and redirects the water back to the exterior (Figure 20.2). Stone cladding systems made up of relatively small stone units, individually installed in the field, also typically rely on rainscreen design.

On the other hand, precast concrete cladding often relies, in large part, on a barrier wall strategy. The panels themselves are manufactured in factory-controlled conditions, using high-quality ingredients and methods of fabrication. The result is a panel that is durable, is highly resistant to water penetration, and does not necessarily benefit from the addition of a secondary drainage plane behind the panel (Figures 20.9 and 20.16). However, where panels meet, and joints must be sealed in the field, a two-stage joint that does incorporate rainscreen principles is more

FIGURE 20.26
A mockup demonstrates the features of a water-managed EIFS system. From interior to exterior, the layers are a water-resistant substrate board, an asphalt-saturated felt air and moisture barrier, a drainage mat composed of plastic fibers, plastic foam insulation, reinforcing mesh, a base coat, and a finish coat. A continuous plastic flashing under a gap at the bottom of the wall drains any leakage out to the face of the wall. Fluid-applied or adhered sheet membranes that can act as air and moisture barriers are also frequently used in place of the building felt shown here. *(Photograph of Senergy CD System, courtesy of Senergy, Cranston, RI.)*

likely to remain reliably watertight over the life of the cladding than a simple surface-sealed joint.

Considerations of constructability also play a role in the choice of design for watertightness with the cladding systems discussed in this chapter. When large, prefabricated panels are installed from the exterior side of the building, the relatively intricate flashing and drainage systems that are part of rainscreen systems may become difficult or impractical to achieve, and barrier wall design may be the more economical and practical solution.

CSI/CSC MasterFormat Sections for Cladding with Masonry and Concrete	
03 45 00	**PRECAST ARCHITECTURAL CONCRETE**
03 45 13	Faced Architectural Precast Concrete
03 49 00	**GLASS-FIBER-REINFORCED CONCRETE**
03 49 16	Glass-Fiber-Reinforced Concrete Spandrels
04 25 00	**UNIT MASONRY PANELS**
04 25 13	Metal-Supported Unit Masonry Panels
04 26 00	**SINGLE-WYTHE UNIT MASONRY**
04 26 13	Masonry Veneer
04 42 00	**EXTERIOR STONE CLADDING**
04 42 23	Truss-Supported Stone Cladding
04 42 26	Grid-System-Supported Stone Cladding
04 42 43	Stone Panels for Curtain Walls
07 24 00	**EXTERIOR INSULATION AND FINISH SYSTEMS**
07 24 13	Polymer-Based Exterior Insulation and Finish System
07 24 16	Polymer-Modified Exterior Insulation and Finish System
07 24 19	Water-Drainage Exterior Insulation and Finish System
07 24 23	Direct-Applied Finish Systems

KEY TERMS

masonry veneer
shelf angle
horizontal expansion joint, soft joint
vertical expansion joint
backup wall
prefabricated reinforced brick panels
grid-system-supported stone cladding

monolithic stone cladding panel
truss-supported stone cladding
posttensioned limestone spandrel panel
precast concrete cladding panel
precast concrete sandwich panel
glass-fiber-reinforced concrete (GFRC)
   cladding panel

exterior insulation and finish system
   (EIFS)
polymer-based EIFS, Class PB EIFS
polymer-modified EIFS, Class PM EIFS,
   hard coat EIFS
barrier-wall EIFS
drainage-wall EIFS, water-managed EIFS

REVIEW QUESTIONS

1. List all the common ways of attaching stone cladding to a building. Make a simple sketch to explain each system.

2. Working from memory, sketch all the details of a brick veneer wall over a concrete frame.

3. What are some options of surface finishes for precast concrete cladding panels?

4. Describe the process of producing GFRC panels, illustrating your account with simple sketches.

5. Name two types of EIFS. Describe two ways of applying EIFS to a building. Why should barrier wall EIFS be avoided?

EXERCISES

1. Design and detail a brick veneer cladding for a multistory building that you are designing. Rather than trying to conceal the flashings and soft joints, work out a way of expressing them boldly as part of the architecture of the building.

2. Visit one or more buildings under construction that are being clad with masonry, concrete, GFRC, or EIFS. Make sketches of how the materials are detailed, especially how they are anchored to the

building. What will happen to any water that leaks through the cladding?

3. Adapt the brick veneer details in this chapter to installation on a building framed with structural steel.

SELECTED REFERENCES

All the references on stone and concrete masonry listed at the end of Chapter 9 are also relevant to this chapter.

Brick Industry Association. *Technical Notes on Brick Construction*, Nos. 18, 18A, 21, 21A, 21B, 21C, 27, 28B. Reston, VA, Author, various dates.

These detailed pamphlets cover every aspect of brick veneer cladding systems.

Precast/Prestressed Concrete Institute. *Architectural Precast Concrete* (3rd ed.). Chicago, Author, 2007.

This is a well-illustrated hardbound book that covers all aspects of the design, manufacture, and installation of precast concrete curtain walls. Also available from the same source is *Architectural Precast Concrete—Color and Texture Selection Guide* (2003), an extensive set of full-color

plates of finishes for precast concrete panels.

Precast/Prestressed Concrete Institute. *GFRC: Recommended Practice for Glass Fiber Reinforced Concrete Panels* (4th ed.). Chicago, Author, 2001.

This 104-page booklet is a clear, complete guide to the design and manufacture of GFRC cladding systems.

WEB SITES

Cladding with Masonry and Concrete

Author's supplementary web site: **www.ianosbackfill.com/20_cladding_with_masonry_and_concrete**

Brick Industry Association: **www.gobrick.com/**

Dry-Vit Systems: **www.dryvit.com**

EIFS Industry Members Association: **www.eima.com**

Precast/Prestressed Concrete Institute: **www.pci.org**

Whole Building Design Guide, Wall Systems: **www.wbdg.org/design/env_wall.php**

FROM CONCEPT TO REALITY

PROJECT: Seattle University School of Law

ARCHITECT: Olson Sundberg Kundig Allen Architects

ASSOCIATED ARCHITECT: Yost Grube Hall Architecture

STRUCTURAL ENGINEER: Putnam Collins Scott Associates

When Seattle University founded its new school of law, it commissioned Olson Sundberg Kundig Allen Architects to design the program's new home. The university wanted a building that would give the school prominence on the campus, one that would both fit within the context of existing masonry structures and reflect the modern roots of this newly instituted program.

In response, the architects chose a building cladding system combining brick veneer and aluminum curtain wall—a balance of old and new materials. As the design took shape, they also strove to design and detail the veneer in such a way as to express the modern skinlike qualities of this cladding system. Brick panels were to be arranged in horizontal bands, visibly supported on structural steel channels. Integrated exterior steel sunshades would reinforce the ribbonlike character of the system.

An important decision came when the architects had to design the veneer's steel channel supports. One option was

to support the brick with steel shelf angles concealed behind the veneer and then add a visible channel shape that would express the loadbearing function while not actually acting in a structural role. The second option was to design the steel channel as the true means of veneer support, allowing the channel to perform in both the architectural and structural roles.

There were pros and cons to both options. The first option relied on a more conventional method of construction, one that would be easier to communicate to the builder. In addition, because the channel would be nonstructural, it could be fabricated from materials that would be lighter and would accept a wider range of finish options, such as sheet metal formed to imitate the shape of a structural channel. However, a structural channel would have a heavier and more genuine appearance. It would not be at risk of being seen as an unnecessary adornment that could be removed from the project

FIGURE A

Seattle University School of Law.

SHOP OR FIELD WELD

THREADED ROD-WELD
TO BACK OF I

CROSS-SECTION DETAIL

5 × 5 TUBE
STEEL

SLAB EDGE

SHOP WELDED
TAB SYSTEM

WIDE SLOTS FOR
SIDE/SIDE; UP/DOWN
ADJUSTMENT
(SHIMS FOR IN/OUT)

A/B

FLASHING

CUT SLOTS IN
SHEATHING
TO ACCEPT
TABS, SEAL
ALL AROUND

THREADED
ROAD, BACK
SIDE OF
CHANNEL

BOLT
FROM
BACK SIDE

① _ERECT STRUCT. STEEL_   ② _ERECT WALL FRAMING/SHEATHING_   ③ _INSTALL STEEL I CHANNEL_

BRICK LEDGER CHANNEL CONSTRUCTION "A"

5 × 5 T.S.

TAB

FIELD WELD
IN BLDG. ENVELOPE
(OR BOLT)

CROSS-SECTION DETAIL

5 × 5 TUBE STEEL

SLAB EDGE

FIELD WELD/BOLT
TAB TO 5 × 5 T.S.

[CHANNEL
BRICK SUPPORT

WALL FRAMING

A/B

FLASHING

SEAL JOINT

① _ERECT STRUCT. STEEL_   ② _ERECT WALL FRAMING_   ③ _SHEATHING_

BRICK LEDGER CHANNEL CONSTRUCTION "B"

as a budget-cutting measure. To the architects, this second option seemed a more authentic way to express the weight of the veneer and its means of support.

If the second option was to be considered, the architect needed to validate this unconventional method of veneer support. Working with the structural engineer, design of the channel and options for its attachment to the building frame were studied (Figure B). Sequences of assembly were explored to verify that the various building systems, including the structural frame, steel stud backup, sheathing, exterior enclosure materials, and veneer, could be erected in a logical way. These studies convinced the architect that the structural channel was practical, and the decision was made to proceed with this option. The final construction drawings showed a structural steel channel blind bolted to connection plates that were, in turn, welded to the structural steel frame (Figure C).

FIGURE C

Construction drawing set detail.

W1

(SEE SHEET A0.1 FOR
WALL ASSEMBLIES)

LAP MOISTURE BARRIER
OVER SAF AND LEDGER FLASHING

MASONRY CAVITY
FLASHING (TYPE 2)
W/ SELF-ADHESIVE MEMBRANE
FLASHING OVERLAY

PEA GRAVEL

WEEP JOINTS @24″ O.C.

STEEL CHANNEL LEDGER
AND CONN. PLATE PER STRUCT.

SUNSCREEN SUPPORT BEYOND
WHERE OCCURS-
SEE SHEET A8.10

COMPENSATING CHANNEL W/
CUSTOM EXTENDED SNAP-ON
PIECE
UNIT WINDOW

3/16″

9″

2″

LINE

F.O. MASONRY NOMINAL

6 1/2″

1′-6″ TYP.
7′-0″ @ WEST WALL

6″
STUD WALL

5 1/2″

(COLUMN)

SHIM

1′-2″ TYP.

6′-8″ @ WEST WALL

BLINDS WHERE OCCUR
SEE FINISH SCHEDULE

As expected, extra effort was required on the part of the architect to address questions that arose during construction. One issue concerned the sequence of assembly. After studying various options during the design phase, the architect had completed the design and detailing based on the assumption that the supporting channels would be erected along with the brickwork, after the backup steel studs and sheathing had been put in place. The contractors based their construction schedule on attachment of the structural channels much earlier in the process, at the same time as the erection of the structural steel frame. Though both methods were feasible, adopting this alternative sequence required the architect to revise various other parts of the construction documents to coordinate the effects of this change with other aspects of the exterior wall assembly.

A second difficulty was related to maintaining erection tolerances for the structural channels. The channel supports had to be located with sufficient precision to achieve the dimensional accuracy required by the brickmasons for their work. Recognizing that tolerances for the placement of the structural steel columns were significantly greater than could be allowed for the channels themselves, attachments between the columns and

channels had been designed to allow for adjustability between these elements. Despite these preparations, column misalignments in the field in some cases exceeded the limits of adjustment that had been designed into the connections.

This problem was compounded by unanticipated irregularities in the channels themselves. The channels had been specified to be hot-dip galvanized, a process by which a heavy zinc coating is applied to the steel to protect against corrosion. Heating of the channels during the hot-dip process caused them to warp subtly. Although not sufficient to make the channels unusable, this added distortion made it even more difficult to mount the channels on the columns with the required precision. To solve these problems, the connection design was revised to allow welding as an alternative to the original bolted design, thereby permitting greater adjustability. In the end, constructional difficulties were resolved, the brick veneer was successfully completed, and the final wall system achieved the functional and architectural goals of both the client and the architect. This project illustrates the important role that planning for constructability can play in the smooth and successful outcome of a project and the realization of the project's design goals.

FIGURE D
Photo detail of veneer and channel support.

## 21

CLADDING WITH METAL AND GLASS

- **Aluminum Extrusions**
  Thermal Breaks

  SUSTAINABILITY OF ALUMINUM CLADDING
  COMPONENTS

  Surface Finishes for Aluminum
  Other Curtain Wall Frame Materials

- **Aluminum and Glass Framing Systems**
  Modes of Assembly
  Outside Glazing and Inside Glazing
  An Outside Glazed Curtain Wall
  System

An Inside Glazed Curtain Wall System
The Rainscreen Principle in Metal-
and-Glass Curtain Walls
Expansion Joints in Metal-and-Glass
Walls

- **Double-Skin Facades**
- **Sloped Glazing**
- **The Curtain Wall Design and Construction Process**
- **Metal Panel Cladding**

Faceted planes and diamond-patterned mullions of the Seattle Public Library create a striking geometrical composition. The makeup of the double glazing varies depending on exposure. Where the highest thermal efficiency is required, units include krypton gas fill and low-e coating. Where the potential for solar heating is greatest, an aluminum expanded metal mesh interlayer reduces solar transmission. Structural steel, framed to mirror the diamond pattern, carries the weight of the curtain wall and resists wind loads. *(Architect: OMA/LMN. Curtain wall design build: Seele GmbH & Co. Photo by Joseph Iano.)*

The contemporary metal-and-glass curtain wall is a descendant of the cast-iron-and-glass walls that were common features of commercial buildings in the 19th century. Today's walls, however, are vastly more sophisticated in every respect. They are carefully isolated from the frame of the building so that they support only their own weight and the forces of the wind. They are insulated and thermally broken to maximize comfort and minimize heating and cooling costs and moisture condensation. They utilize advanced glazing and spandrel materials that offer precise control of luminous and thermal properties. They are carefully gasketed and drained to discourage water leaks. Their intricate inner features are concealed behind smooth snap-on covers. They are designed for easy, forgiving installation and maintenance. And they are made of aluminum, light and strong, in the form of sleek extrusions that glisten with long-lasting anodized or organic finishes.

FIGURE 21.1
The concept of extrusion.

ALUMINUM EXTRUSIONS

Aluminum is the metal of choice for metal cladding systems for three primary reasons: It protects itself against corrosion. It accepts and holds a variety of attractive surface finishes. It can be fabricated economically into elaborately detailed shapes by means of the process of extrusion.

The principle of *extrusion* is easily visualized: It is like squeezing toothpaste from a tube. In response to the squeezing pressure, the tube extrudes a column of toothpaste that is cylindrical in shape because the orifice of the tube is round. If the shape of the orifice were changed, one could produce many other shapes of toothpaste as well—square, triangular, flat, and so on.

To manufacture an aluminum extrusion, a large cylindrical *billet* of aluminum is heated to a temperature at which the metal flows under pressure but still retains its shape when the pressure is released. The heated billet is placed in the cylinder of a large press where a piston squeezes it under enormous pressure through a *die*, a steel plate with a shaped metal orifice. The orifice imparts its shape to a long extruded column of aluminum that is supported on rollers, cooled, straightened as necessary, and cut into convenient lengths (Figures 21.1, 21.2).

Very intricate aluminum sections can be extruded for a variety of purposes, including not only curtain wall components but also door frames, window frames, entrances and storefronts, handrails, grillwork, and structural shapes such as wide flanges, channels, and angles. The high precision of the extrusion process permits it to be used for close-tolerance details such as snap-in glazing beads, snap-on mullion covers, *screw slots*, and *screw ports* (Figures 21.3–21.6). Hollow shapes may be extruded by mounting the portion of the die that forms the interior of the shape on a steel "spider" that is attached to the inside of the die. The metal

FIGURE 21.2
Making aluminum extrusions for curtain wall components. (*a*) Hundreds of extrusion dies for the many components of curtain walls are organized in racks. (*b*) A heated billet of aluminum is inserted into the cylinder of the extrusion press. (*c*) An extrusion emerges from the die. (*d*) Long aluminum extrusions cool on rollers, ready for straightening and cutting to length. (*Photographs courtesy of Kawneer Company, Inc.*)

Extrusion die · Cylinder · Piston · Heated billet of aluminum · Pressure · Extrusion

(a)

(b)

(c)

(d)

FIGURE 21.3
Screw ports are extruded cylindrical features that allow a screw to be driven parallel to the long axis of an extrusion. In the upper example, the screws pass through slightly oversized holes drilled through one box-shaped extrusion into screw ports in another box-shaped extrusion that are slightly smaller in diameter than the screws, thus engaging the screw threads tightly in the screw ports and making a snug joint. In the lower example, the screw ports are slightly larger than the outside diameter of the screws. The screws pass through the ports in a short piece of aluminum known as a "shear block" and engage slightly undersized holes in the vertical extrusion. Then the horizontal member is slipped over the shear block and screwed to it through drilled holes.

FIGURE 21.4
Self-tapping screws are driven through screw ports to fasten an extruded aluminum shear block to a vertical mullion. *(Courtesy of Kawneer Company, Inc.)*

FIGURE 21.5
Screw slots allow screws to be driven perpendicular to the long axis of an extrusion. In this example, screws pass through oversized holes in an extruded aluminum pressure plate and pull the plate down toward the screw slot. Extruded gaskets of synthetic rubber are pressed into channels in both extrusions to seal tightly against the glass.

SCREW SLOT

BEFORE

AFTER

RATCHET
ACTION

FIGURE 21.6

Snap-on and snap-together features are commonly used in extruded aluminum curtain wall components. Assembly is accomplished simply by aligning the components and tapping firmly with a rubber mallet or squeezing with a rubber-cushioned clamp.

flows around the legs of the spider before it passes through the orifice. The accompanying illustrations show some of the ways in which extruded aluminum details are utilized in building cladding. Extrusion dies are easily produced for custom-designed sections if the production run will be long enough to amortize their expense.

Thermal Breaks

Aluminum conducts heat rapidly. In very cold weather, the indoor surfaces of an aluminum member that passes from the outside of the building to the inside, such as a window frame, would be so cold that moisture and possibly frost would condense on them. In very hot, humid weather, in an air-conditioned building, the outdoor surfaces of the same

window frame might be cool enough to condense moisture from the air. This is why all but the simplest aluminum framing systems are manufactured with *thermal breaks*, which are internal components of insulating material that isolate the aluminum on the interior side of the component from aluminum on the exterior side. These dramatically reduce the flow of heat through the member (Figures 21.7 and 21.8).

There are several ways of creating thermal breaks. Figure 21.7 shows a *cast and debridged thermal break*, in which molten plastic is poured into a deep channel in the center of an aluminum member, where it hardens. Then the aluminum that forms the bottom of the channel is cut away in a *debridging* process that leaves only the plastic to connect the two halves of the member. In Figure 21.8, thermal breaks are created with

preformed polyamide thermal strips that are mechanically inserted into the aluminum extrusions during extrusion manufacture. The hollow space between the strips is filled with insulation material. *Polyamide strip thermal breaks* are more thermally efficient than cast and debridged beaks. Other thermal break configurations from rubber or plastic gaskets, plastic strips, and plastic clips are illustrated by examples elsewhere in this chapter.

FIGURE 21.7
This thermally broken aluminum mullion extrusion is sitting on a block of dry ice in a humid room. Frost has formed on the portion of the mullion nearest the dry ice, but the thermal break keeps the rest of the mullion warm enough that moisture does not condense on it. *(Photo courtesy of the H.B. Fuller Company, St. Paul, Minnesota.)*

FIGURE 21.8
Thermal breaks in this aluminum window frame system are made from extruded polyamide plastic strips locked into preformed slots in the aluminum extrusions. The space between the thermal break strips is filled with plastic foam or mineral fiber insulation. *(Photo by Joseph Iano.)*

SUSTAINABILITY OF ALUMINUM CLADDING COMPONENTS

Manufacture

• The ore from which aluminum is refined, bauxite, is finite but relatively plentiful. The richest deposits are generally found in tropical areas, often where rain forests must be clearcut to facilitate mining operations.

• Aluminum is refined from bauxite by an electrolytic process that uses huge quantities of electricity. For this reason, aluminum smelters are often located near plentiful supplies of inexpensive hydroelectric power.

• The embodied energy in aluminum is roughly 100,000 BTU per pound (230 MJ/kg), seven times that of steel, making it one of the most energy-intensive materials used in construction.

• Large volumes of water are required for smelting. Wastewater from aluminum manufacture contains cyanide, antimony, nickel, fluorides, and other pollutants.

• Aluminum is recycled at a very high rate, due largely to industry efforts. Recycled aluminum is produced using only a fraction of the energy, approximately 5000 BTU per pound (12 MJ/kg), required to convert ore to aluminum.

• Aluminum extrusions are easy to produce and to form into cladding components. Their light weight saves transportation energy.

• Powder coatings for aluminum, which release no solvents into the atmosphere, are preferable environmentally to solvent-based coatings.

Construction

• Aluminum cladding is easy to erect because of its light weight and simple connections. Little waste or pollution is associated with the process. Scrap is readily recycled.

In Service

• Aluminum cladding seldom requires maintenance, lasts for a very long time, and can be recycled when a building is demolished.

• Because aluminum is highly conductive of heat, cladding components must be thermally broken.

• Aluminum foils used as vapor retarders, components of insulation systems, and radiant heat barriers save large amounts of heating and cooling energy. They are so thin that they consume little metal relative to the energy they can save over the lifetime of the building.

Surface Finishes for Aluminum

Aluminum, though it is a very active metal chemically, does not corrode away in service because it protects itself with a thin, tenacious oxide film that seals the surface of the metal and discourages further oxidation. Although this film does an adequate job of protecting the aluminum, in the outdoor environment it develops a chalky or spotty appearance that can come to look rather shabby over time.

Anodizing is a manufacturing process that produces an integral oxide coating on aluminum that is thousands of times thicker and more durable than the natural oxide film that would otherwise form. The component to be anodized is immersed in an acid bath and becomes the anode in an electrolytic process that takes oxygen from the acid and combines it with the aluminum. Color can be added to the coating by means of dyes, pigments, special electrolytes, or special aluminum alloys. The colors most frequently used in buildings are the natural aluminum color, golds, bronzes, grays, and black, but other colors are possible. The advantages of anodized finishes are their extreme hardness and, in most colors, their extreme resistance to weather and fading.

Aluminum cladding components can also be finished with a variety of organic coatings. *Fluoropolymer coatings* are based on highly inert synthetic resins, such as *polyvinylidene fluoride (PVDF)*, which are exceptionally colorfast and resistant to all forms of weathering, including ultraviolet deterioration. In a typical finishing process, the aluminum is first chemically cleaned and then primed. Next, one or two fluoropolymer finish coats, first a color coat and, optionally, a second clear coat, are spray applied. After the application of each of these finish coats, the aluminum piece is passed through an oven and baked at 450 degrees Fahrenheit (230°C), a process that causes the resin molecules to intertwine and fuse into a tightly bonded matrix. Fluoropolymer coatings are available in a broad spectrum of colors, including bright metallic finishes. They are the most expensive of the organic coatings and the longest lasting; the best of them can be expected to last 20 years or more under normal service conditions.

Powder coatings are manufactured with thermosetting powders that are composed of plastic resins, such as polyester, and pigments. The powder is electrically charged and then sprayed onto the aluminum component, which is grounded so that the powder adheres to it electrostatically. The component is then passed through an oven, where the powder fuses to produce a hard, resistant coating, usually in a single application. Among the advantages of powder coatings are their lower cost in comparison to fluoropolymers, their durability, the wide range of colors and finishes in which they are available, and their freedom from organic solvents that cause air pollution.

Baked enamel coatings, consisting of spray-applied acrylic or polyester polymers, sometimes modified with silicone, are also used as aluminum coatings. They are inexpensive and provide finishes with very high glosses in a wide selection of colors.

Aluminum finishes are specified according to standards published by the American Architectural Manufacturers Association (AAMA) that establish minimum requirements for coating thickness, colorfastness, gloss retention, corrosion resistance, resistance to chalking, abrasion resistance, and other characteristics related to the durability and performance of the finish. Anodized finishes are specified according to AAMA 611. Class I anodized finishes are thicker and longer lasting than Class II anodized finishes. Class I finishes are suitable, for example, for curtain wall components on a tall building, whereas Class II anodized finishes are more commonly specified for residential aluminum products or low-rise aluminum storefront systems. Organic coatings are specified according to AAMA standards 2605, 2604, and 2603. Fluoropolymers with a high PVDF content and some powder coatings can meet the requirements of AAMA 2605, the most stringent of the three. Coatings meeting the requirements of either AAMA 2604 or 2603 are progressively less expensive and less longlasting.

A wide range of surface effects may be applied to aluminum by mechanical and chemical processes. Mechanical finishes are produced by such means as wire brushing, wheel or belt polishing, buffing, grinding, burnishing, barrel tumbling, sandblasting, blasting with steel shot or glass beads, and abrasive blasting. Each produces a different surface texture. Chemical finishes include bright dipping, which produces mirrorlike surfaces; etching; and chemical conversion coatings such as oxides, phosphates, or chromates. Mechanical and chemical finishes may be done in preparation for the application of other types of finishes or, in some cases, may act as final finishes.

Other Curtain Wall Frame Metals

Although the vast majority of contemporary curtain wall framing is made from aluminum extrusions, other metals, such as bronze, galvanized steel, or stainless steel, may also be used. In comparison to aluminum curtain wall framing, stainless steel allows thinner sight lines or greater spans between supports. The lower thermal conductivity of stainless steel improves thermal performance. Also, stainless steel curtain wall systems are capable of achieving up to 2-hour fire resistance ratings for applications where a rated exterior enclosure is required.

ALUMINUM AND GLASS FRAMING SYSTEMS

Aluminum framing and glass are used to construct enclosure systems of various types, including windows, entrances, storefronts, window walls, and curtain walls. Aluminum-framed windows are discussed in Chapter 18. *Entrances* are systems of aluminum-framed doors, hardware, aluminum framing, and glass typically used for commercial buildings. They may also include framing for vestibules, sidelights, transoms, and the like.

Aluminum *storefronts*, though similar in appearance to curtain wall systems, are based on simplified, lighter framing elements that are less expensive and quicker to assemble. Most storefront systems span vertically no more than 10 to 12 feet (3.0–3.7 m), although some, with heavier framing members or internal reinforcing, may span farther. Rather than being hung off the face of the building, as is common with curtain wall systems, storefront framing is installed between floor slabs or within wall openings (Figure 21.9). Most importantly, the use of lighter framing elements and simplified internal construction make storefront systems, in comparison to curtain wall systems, more limited in their ability to resist wind loads and water penetration. For these reasons, even improved storefront framing systems are not used more than three to four stories high on a building.

Aluminum *window walls*, like storefront, span between the top of one slab to the underside of the next. But more like curtain wall, they have internal water management features that make them suitable for use on tall buildings with exposure to higher wind pressures. In comparison to curtain wall, window walls more readily accommodate a variety of operable window types. They also allow more flexibility in installation, as they need not necessarily be erected in sequence from bottom to top of the building structure.

Aluminum *curtain wall* systems are the most high-performance, sophisticated, and costly of the aluminum-and-glass framing systems. Curtain wall systems can be readily applied to any number of stories. They use stronger, stiffer aluminum sections that can resist the higher wind pressures that act on taller buildings. Their methods of joining and attachment to the building structure can tolerate the larger movements that arise

FIGURE 21.9

As the name implies, storefront systems are intended primarily for ground-level commercial glazing applications. These simple, field-assembled systems lack the greater capacity to resist wind loads and the more sophisticated water management features of curtain wall systems. *(Photo by Joseph Iano.)*

from structural deflections as well as from thermal expansion and contraction of the aluminum itself. They also include sophisticated internal construction and systems of sealing and water drainage that enable them to resist water leakage even under the more severe conditions of wind-driven rain to which they are exposed. It is to these types of systems that the majority of this chapter is devoted.

Modes of Assembly

Metal curtain wall systems can be classified according to their mode of assembly at the time of installation on the building (Figure 21.10). Many metal-and-glass curtain walls are furnished as *stick systems* whose principal components are metal mullions, and rectangular panels of glass and spandrel material that are assembled in place on the building (Figure 21.10*a*). Stick systems have the advantages of low shipping bulk and high degree of adjustability to unforeseen site conditions. But they must be assembled on site, under highly variable conditions, rather than in a factory with its ideal tooling, controlled environmental conditions, and lower wage rates. Aluminum entrances and storefronts are also typically installed as stick systems.

The *unit* (or *unitized*) *system* of curtain wall installation takes full advantage of factory assembly and minimizes on-site labor, but the units require more space during shipping and more protection from damage than stick system components (Figure 21.10*b*). The *unit-and-mullion system* (Figure 21.10*c*), which is seldom used today, offers a middle ground between the stick and unit systems.

The *panel system* (Figure 21.10*d*) is made up of homogeneous units that are formed from metal sheet. Its advantages and disadvantages are similar to those of the unit system, but its production involves the higher tooling costs of a custom-made die or mold, which makes it advantageous only for a building that requires a large number of identical panels.

STICK SYSTEM– Schematic of typical version
1: Anchors.   2: Mullion.   3: Horizontal rail (gutter section at window head).
4: Spandrel panel (may be installed from inside building).   5: Horizontal rail (window sill section).   6: Vision glass (installed from inside building).
7: Interior mullion trim.

Other variations: Mullion and rail sections may be longer or shorter than shown. Vision glass may be set directly in recesses in framing members, may be set with applied stops, may be set in sub-frame, or may include operable sash.

(*a*)

(*continued*)

FIGURE 21.10
Modes of assembly for curtain walls. (*a*) Stick system.

The *column-cover-and-spandrel system* (Figure 21.10*e*) emphasizes the structural module of the building rather than creating its own grid on the facade, as the previously described systems do. A custom design must be created for each project because there is no standard column or floor spacing for buildings. Special care is required in detailing the spandrel panel support to ensure that the panels do not deflect when loads are applied to the spandrel beams of the building frame; otherwise, the window strips could be subjected to loadings that would deform the mullions and crack the glass.

UNIT SYSTEM– Schematic of typical version

1: Anchors.   2: Pre-assembled framed unit.

Other variations: Mullion sections may be interlocking "split" type or may be channel shapes with applied inside and outside joint covers. Units may be unglazed when installed or may be per-glazed. Spandrel panel may be either at top or bottom of unit.

(b)

UNIT-AND-MULLION SYSTEM– Schematic of typical version

1: Anchors.   2: Mullion (either one- or two-story lengths).   3: Per-assembled unit –lowered into place behind mullion from floor above.   4: Interior mullion trim.

Other variations: Framed units may be full-story height (as shown), either unglazed or per-glazed, or may be separate spandrel cover units and vision glass units. Horizontal rail sections are sometimes used between units.

(c)

FIGURE 21.10

(b) Unit system. (c) Unit-and-mullion system. (d) Panel system. (e) Column-cover-and-spandrel system. *(Reprinted with permission from AAMA Aluminum Curtain Wall Design Guide.)*

Outside Glazing and Inside Glazing

An *outside glazed* curtain wall system requires that glass be installed or replaced by installers working from scaffolding or staging on the outside the building. An *inside glazed* cladding system allows glass installation to be performed from the inside. Inside glazing is more convenient and more economical for a tall building, but it requires a somewhat more elaborate set of extrusions. Outside glazing systems utilize simpler, less expensive shapes but are usually reserved for buildings only up to approximately three stories in height. Some curtain wall systems are designed so that they can be glazed from either side.

PANEL SYSTEM– Schematic of typical version

1: Anchors.   2: Panel.

Other variations: Panels may be formed sheet or castings, may be full story height (as shown) or smaller units, and may be either pre-glazed or glazed after installation.

(*d*)

COLUMN COVER AND SPANDREL SYSTEM– Schematic of typical version

1: Column cover.   2: Spandrel panel.   3: Glazing infill.

Other variations: Column covers may be one piece or an assembly, may be of any corss-sectional profile, and either one or two stories in height. Spendrel panel may be plain, textured or patterned. Glazing infill may be a pre-assembly, either glazed or unglazed , or be assembled in place.

(*e*)

An Outside Glazed Curtain Wall System

An off-the-shelf, externally glazed stick system for aluminum-and-glass curtain walls is illustrated in Figures 21.11–21.16. This system is suitable for a low building whose walls workers can easily reach from external scaffolding.

An Inside Glazed Curtain Wall System

An off-the-shelf, internally glazed stick system for aluminum-and-glass curtain walls is illustrated in Figures 21.17–21.23. This system is suitable for use on tall buildings, because it is installed entirely by workers who stand inside the building. Replacement of glass may also be done from within the building.

©KAWNEER COMPANY, INC., 1994

FIGURE 21.11

The Kawneer 1600 System1® is an outside glazed stick system. Vertical mullions that run continuously from floor to floor support discontinuous horizontal mullions that are connected to them by means of shear blocks and screws. Each lite of glass sits on two rubber glazing blocks in the gutter of the horizontal mullion (not shown in this drawing). The inner surface of the glass rests against extruded rubber glazing gaskets pressed into small channels in both the vertical and horizontal mullions. Extruded aluminum pressure plates with rubber glazing gaskets are applied to both horizontal and vertical mullions to clamp the glass into place and create a weathertight seal. Each pressure plate is attached by means of screws that pass through drilled holes into an extruded screw slot. A thick rubber gasket in the screw slot acts as a thermal break. Snap-on covers conceal the screw heads and give a neat exterior appearance. A molded rubber plug at either end of each horizontal mullion contains any leakage or condensate within the horizontal mullion, from which it escapes via 5/16-inch-diameter (8-mm) weep holes (not shown here) drilled through the pressure plate and the bottom edge of the snap-on cover. Figures 21.12–21.16 further illustrate this system. *(Courtesy of Kawneer Company, Inc.)*

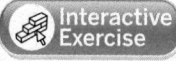

FIGURE 21.12

The manufacturer's details for the Kawneer 1600 System1 curtain wall are all keyed to the small elevation view at the upper left corner of this page. The details are reproduced here at one-quarter of their actual size. *(Courtesy of Kawneer Company, Inc.)*

SCALE 3″ = 1′–0″

ELEVATION IS NUMBER KEYED TO DETAILS

©KAWNEER COMPANY, INC., 1994

162-004

162-002

5¼"
(133.4)

3¾"
(95.3)

7½"
(190.5)

6"
(152.4)

¾"
(19.0)
TYP

2½"
(63.5)
TYP

©KAWNEER COMPANY, INC., 1994

FIGURE 21.13
In this full-size detail of the vertical mullion for the Kawneer 1600 System1, we see that the basic extrusion is a rectangular box shape that is structurally stiff and presents a neat appearance inside the building. The broken lines indicate a smaller box mullion (162-002) that may be used for buildings with shorter floor-to-floor heights or smaller wind loads. The extruded plastic thermal break attaches to the mullion with a projecting "pine tree" spline that is pushed into the screw slot. Holes are drilled through the thermal break for the screws (not shown) that attach the pressure plate. The four rubber glazing gaskets attach to the aluminum pieces with projecting splines. *(Courtesy of Kawneer Company, Inc.)*

FIGURE 21.14
Two different horizontal mullion extrusions are available for the Kawneer 1600 System1. The closed extrusion *(top)* is used in most situations. The open-back extrusion *(bottom)* allows access to the shear block on the vertical mullion for purposes of assembly; it is used for portions of the wall that butt against other materials, such as heads, sills, and end bays. A snap-on cover closes the open back of the mullion, leaving only a virtually invisible seam.

Rubber setting blocks are shown beneath the edges of the glass in both mullions. In the upper-left corner of the upper detail, a small aluminum extrusion has been added to allow the mullion to hold a single layer of spandrel glass or vision glass rather than the standard 1-inch (25-mm) double-glazing assembly. Weep holes are not shown; they are drilled horizontally through the pressure plate just above the thermal break and vertically through the bottom edge of the outside snap-on cover.

(Courtesy of Kawneer Company, Inc.)

162-001

162-003

6″
(152.4)

7½″
(192.5)

162-008

162-009

©KAWNEER COMPANY, INC., 1994

27-477
Separator
between anchor
and mullion

Weld
plate

Size and spacing
of weld per approved shop drawings

FIGURE 21.15
Vertical mullions of the Kawneer 1600 System1 are attached to the edges of the building's floors with the angle anchor shown in the top drawing. Sections of vertical mullion are spliced with the aluminum internal spline shown in the lower drawing. The spline is screwed to the lower section of mullion but the upper section is free to slide, which allows for thermal expansion and contraction. *(Courtesy of Kawneer Company, Inc.)*

Field-applied
fixing screw
128-348
#10 × 1-1/4″
flat head
self-drilling

Shop-applied stop
screw 128-267
#12 × 1″ pan
head type AB

FIGURE 21.16
This building for the Harley-Davidson Motor Company in Wauwatosa, Wisconsin, features the Kawneer 1600 System1, whose details are shown in the previous series of figures. *(Architect: Flad & Associates, Madison, Wisconsin. Courtesy of Kawneer Company, Inc.)*

©KAWNEER COMPANY, INC., 1997

FIGURE 21.17
The Kawneer 1600 System4® may be glazed from either inside
or outside the building. Inside glazing is usually preferred in
taller buildings for its speed, safety, and convenience. *(Courtesy of
Kawneer Company, Inc.)*

SCALE 3″ = 1′ 0″

ELEVATION IS NUMBER KEYED TO DETAILS

OPTIONAL
HORIZONTAL

ALTERNATE
JAMB DETAIL

©KAWNEER COMPANY, INC., 1997

FIGURE 21.19
The vertical mullion for the Kawneer 1600 System4 is rigid, with no removable components except the glazing gaskets. The thermal break is an H-shaped extrusion of rigid plastic that is crimped securely into small grooves in the inner and outer portions of the mullion. Notice that the glazing pockets are of two different depths. This facilitates inside glazing, as explained in Figure 21.20. *(Courtesy of Kawneer Company, Inc.)*

FIGURE 21.18
A complete set of details for the Kawneer 1600 System4, reproduced in the manufacturer's detail book at one-quarter of the full size. *(Courtesy of Kawneer Company, Inc.)*

Vision

FIGURE 21.20
To glaze the Kawneer 1600 System4 from inside the building, rubber setting blocks are applied inside the shallow glazing pocket on the vertical mullion to cushion the edge of the glass. Rubber setting blocks are also installed in the horizontal mullion, and all the exterior-side glazing gaskets are inserted. Then the glass unit is pushed obliquely into the deep glazing pocket, squared away, and backed into the shallow pocket. *(Courtesy of Kawneer Company, Inc.)*

Anchors at spandrels

FIGURE 21.21
To make this mode of glass installation possible, the bottom half of the interior of each horizontal mullion is left off until the glass has been pushed into place. Then the bottom-half extrusion is added, and the interior glazing gaskets are installed. The interior gaskets are called "glazing wedges" because of their blunt wedge shape; this allows them to be installed simply by pressing them into place, wedging them between the glass and the mullion, after the glass has been installed. Notice how this shape is different from that of the exterior gaskets, which are installed before the glass. *(Courtesy of Kawneer Company, Inc.)*

7½″
(190.5)

6″
(152.5)

164-208   164-209

©KAWNEER COMPANY, INC., 1997

27-908
"W"
side
block

Installed
position
vision

Flatten "W" block
and insert into
glazing pocket

Installed
position
vision

Locate at center
of glass height

FIGURE 21.22
Before the glazing wedges are installed in the deep-pocket side of the vertical mullion, a W-shaped rubber block is flattened and slipped between the glass and the interior side of the mullion until it passes into the empty pocket. There it springs back into shape and prevents the glass from moving too deeply into the pocket. *(Courtesy of Kawneer Company, Inc.)*

FIGURE 21.23
This Las Vegas casino hotel features the Kawneer 1600 System4, whose details are shown in Figures 21.17–21.22.
(Courtesy of Kawneer Company, Inc.)

The Rainscreen Principle in Metal-and-Glass Curtain Walls

At first glance, neither of the curtain wall systems presented in this chapter might seem to be a rainscreen design, because both use rubber gaskets to seal around the glass on the outside of the wall as well as on the inside. However, consider what happens in either of these systems if both the inner and outer gaskets around a lite of glass are defective: During a wind-driven rain, gravity and capillary action would be likely to draw some water past the outer gasket into the spaces between the edges of the glass and the aluminum. However, as we discovered in Chapter 19, an imperfect inner gasket is unlikely to allow air currents strong enough to carry water farther toward the interior of the assembly. If water has leaked in along a vertical edge of the glass, it is contained within the vertical mullion and will fall by gravity to the bottom, where it drains out through weep holes. If water accumulates in the horizontal mullion, it is prevented from running out of the ends of the mullion by sealed rubber end plugs (Figure 21.24). Its only recourse is to drain to the outdoors through weep holes that are drilled horizontally through the pressure plate and vertically through the bottom edge of the external snap-on cover. Thus, the external gasket need serve only as a *deterrent seal*—essentially a rainscreen—to discourage water from entering without necessarily barring its entry altogether. The internal gasket serves as an air barrier, and the hollow spaces between the edges of the glass and the mullions act as pressure equalization chambers. The entire system functions as a rainscreen assembly. In practice, of course, the manufacturer and installer take every precaution to ensure that the external gaskets are properly installed and will act as positive barriers to the passage of water, but curtain wall systems do not depend on a perfect seal to retain their watertightness.

FIGURE 21.24
A curtain wall installer uses a putty knife to smooth silicone sealant into the joint between horizontal and vertical mullions. A rubber plug, as shown in Figure 21.11, previously inserted into the end of the horizontal mullion, is concealed by the sealant.

Expansion Joints in Metal-and-Glass Walls

Aluminum has a relatively high coefficient of thermal expansion. The coefficient for glass is less than half as much. Because the cladding of a building is exposed to air temperature fluctuations as well as direct heating by the sun, it must be provided with expansion joints to allow thermal movement to occur without damaging the cladding or the frame of the building.

The differences in thermal movement between the glass and the aluminum are generally accommodated by very small sliding and flexing motions that occur between the glass and the gaskets in which it is mounted. Rubber blocks placed between the edge of the glass and the mullion on either side of each lite prevent the glass from "walking" too far in either direction during repeated cycles of heating and cooling.

In the curtain wall systems illustrated in this chapter, vertical thermal movement in the aluminum is absorbed by telescoping joints that are provided at regular intervals in the vertical mullions (Figure 21.15). Horizontal thermal movement is accommodated by intentionally cutting horizontal aluminum components slightly short by a calculated fraction of an inch at each vertical connection. Because the horizontal mullions are interrupted at each vertical mullion, there are many of these joints to work together in absorbing horizontal expansion and contraction.

DOUBLE-SKIN FACADES

In a *double-skin facade*, the wall system is constructed of two separate glass cladding systems separated by an airspace that is wide enough in most cases to allow service personnel to pass between them. Double-skin facades have long been popular in European building construction and are now appearing in some North American buildings as well. Various configurations of glazing and treatment of the space between the two skins are possible. Most often, one of the two skins is double-glazed, while the other relies on single glass. The interstitial airspace may be ventilated to the interior of the building (when the outer skin is double-glazed) or to the exterior (when the inner skin is double-glazed). Natural ventilation of the building interior through operable openings in one or both skins is also possible. Ventilation of the interstitial space may be passive, that is, powered by natural convection; or active, accomplished with the aid of fans, and in some cases may be coupled to the building's HVAC system. Shading devices, such as louvered blinds or roller shades, manually controlled or driven by automatic timers or solar tracking devices, are also frequently integrated into the interstitial space.

Double-skin facades can reduce unwanted solar heat gain, increase daylighting potential, minimize thermal conduction and radiation through the wall system, provide a space for shading elements that is protected from the weather, permit natural ventilation designs in tall buildings, and create a quieter building interior than is possible with conventional glazing systems. Against these potential benefits must be weighed the disadvantages: The airspace may subtract as much as 5 to 10 percent of the usable floor area of the building. Though the second skin of the facade may be constructed of less expensive framing and glass than the first skin, this still constitutes a significant increase in construction cost over a traditional curtain wall. Also, during the life of the building, twice as much glass surface must be washed on a regular basis.

Given these trade-offs, double-skin facade systems are most appropriate for glass walls that must meet stringent energy conservation targets, where the building owner's financial planning allows the higher cost of construction to be amortized over a relatively long time period in which it can be offset by increased energy savings, and where the creation of a high-performance building is an implicit goal of the building program.

SLOPED GLAZING

Many buildings feature glass roofs over such areas as lobbies, restaurants, cafes, swimming pools, garden courtyards, and other amenities. A glass roof presents particular problems with respect to potential water leakage, because it is impossible to neutralize the force of gravity on a surface that is not vertical. Furthermore, moisture that condenses on the interior surfaces of the glass can accumulate and drip onto occupants below. Therefore, *sloped glazing* systems are designed to include an internal drainage system. This drainage system collects water that results from leakage

or condensation and channels it to the outdoors. The glass surface is sloped rather than absolutely flat because the slope enables gravity to assist in keeping water from ponding on the roof, in causing condensate to run to the lower edge of each lite of glass before dripping off, and in moving water through the drainage channels to the weep holes through which it is conducted back to the outdoors.

The sloped glazing illustrated in Figures 21.25 and 21.26 is designed to adapt to a range of slopes from 15 to 60 degrees. Water leakage is discouraged by a well-designed system of glazing gaskets. But if leakage should occur because of gasket deterioration or faulty installation, the internal drainage system will catch it along any purlin or rafter and drain it away. Moisture condensation is minimized by double glazing and thermal breaks. Any condensate that may form under extreme conditions is also caught and drained by the same system of channels in the aluminum members.

For glass systems sloped more than 15 degrees from vertical, building codes restrict the types of glass used, to protect occupants below from the risk of injury from falling broken glass. The only glazing materials permitted without any limitations are laminated glass and plastic. Other glass types are permitted, subject to restrictions relating to the height of the glass above the floor, size of the lites, type of occupancy, and other factors. Alternatively, any glass type is permitted if a metal screen is installed below the glass that can catch shards should the glass break.

THE CURTAIN WALL DESIGN AND CONSTRUCTION PROCESS

Metal curtain wall design is not undertaken by an architect alone; it is too complex and specialized a process. For many buildings, the architect simply adopts a proprietary system

that meets design requirements. This puts the primary responsibility on the manufacturer and installer of the system to ensure that the proper components are chosen and the wall is correctly installed.

When an architect sets out to design a new system of metal cladding, as is often done for large, important buildings, other professionals are brought into the process. A cladding consultant brings unique experience and specialized expertise to the effort. The structural engineer must understand the weight and attachment requirements of the system and how wind loads acting on the curtain wall will be transferred to the building frame. This engineer must also define the deflection criteria of the building structure and communicate this information to those responsible for the design of the curtain wall. A curtain wall manufacturer is also brought into the design team early in the process.

The architect and building owner may elect either of two methods for selecting a manufacturer: One curtain wall manufacturer may be chosen on the basis of reputation and past experience. Alternatively, the architect and cladding consultant may prepare a rough design and performance specifications and submit these to several manufacturers for proposals. Each manufacturer will then submit a more detailed design and a financial proposal, and one will be selected on the basis of these proposals to proceed with the project.

The manufacturer understands better than any other member of the team the manufacturing, assembly, installation, and cost implications of a new curtain wall design. The manufacturer often does the installation as well as the manufacturing of the components, or it subcontracts the installation work to companies that it certifies to be well qualified and familiar with the manufacturer's products and standards. This installation experience is also invaluable during the design process.

©KAWNEER COMPANY, INC., 1991

FIGURE 21.25
The Kawneer 1600 S.G.® sloped glazing system is framed with aluminum extrusions that incorporate internal gutters to drain away condensate and incidental leakage.
(Courtesy of Kawneer Company, Inc.)

From conceptual drawings prepared by the architect and the cladding consultant, the manufacturer prepares a more detailed set of design drawings as a basis for reaching preliminary agreement on the design. The manufacturer then prepares a very detailed set of shop drawings and installation drawings. These are checked carefully by the architect, engineer, and cladding consultant to ensure compliance with the design intentions and the structural capabilities of the building frame. Full-scale testing of the curtain wall, as described in Chapter 19, is usually carried out before the manufacture of a custom-designed curtain wall system is authorized to begin. The curtain wall manufacturer may also visit the construction site during the erection of the building frame to become familiar with the level of dimensional accuracy of the structural surfaces to which the curtain wall will be fastened.

FIGURE 21.26

This hotel in Orlando, Florida, includes massive areas of Kawneer 1600 S.G. sloped glazing. Reflective glass is used to minimize solar overheating of the interior space.

(Courtesy of Kawneer Company, Inc.)

METAL PANEL CLADDING

Although the primary focus of this chapter is aluminum-framed glass curtain walls, metal is also used to form cladding panels of a variety of types. These panels can be glazed into curtain wall and other aluminum framing systems, much as glass is, or they can be used as independent cladding systems in their own right.

Architectural sheet metal cladding, made of thin, flexible metal sheets, is applied to walls in very much the same way as this type of metal is used for roofing, illustrated in Chapter 16. Because the metal is thin and requires closely spaced seams and clips to hold the panels in place, the choices of seam type and arrangement tend to figure prominently in the finished expression of the cladding (Figure 21.27). Sheet metal can also be factory-formed into various ribbed or corrugated shapes called *formed metal wall panels.* These are usually attached with regularly spaced exposed fasteners and resilient washers that deter water leakage at the penetrations through the panels. The profile of corrugation or ribs is the dominant

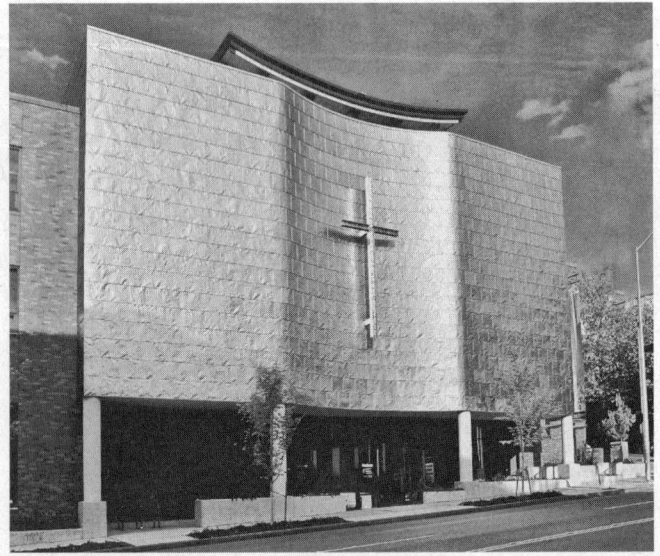

FIGURE 21.27
Titanium sheet metal cladding, purposefully "crinkled" to create a surface with many specular highlights. See Chapter 16 for a more in-depth discussion of architectural sheet metal materials and methods of installation. *(First Methodist Church, Seattle, WA; Bassetti Architects. Photo by Joseph Iano.)*

FIGURE 21.28
Diamond-shaped stainless steel metal plate panels are used to clad a conical dome. *(Museum of Glass, Tacoma, WA; Arthur Erickson, architect. Photo by Joseph Iano.)*

architectural feature of this panel type (Figure 16.53).

Metal plate wall panels are made from heavier metal sheet, usually about ⅛-inch (3 mm) thick. When properly detailed for thermal expansion and contraction, these stiffer panels are capable of producing consistently flat or smoothly curved exterior surfaces with precise, uniform edges. Seams can be spaced at greater distances than sheet metal cladding. Rather than overlapping the edges of adjacent sheets as is common in sheet metal cladding, panel edges can be kept flush, with a uniform gap in between (Figure 21.28). *Composite metal wall panels*, similar in thickness to solid metal panels, are made from sheet metal laminated to both sides of a thermoset plastic core. Though lower in cost than solid metal panels, the detailing and architectural expression are quite similar. Composite metal panels made with other core materials, such as metal foil honeycomb, are made in a range of thicknesses up to 1 to 2 inches (25–50 mm). Both of these panel types can be attached to metal furring or framing components behind the panels that, in turn, are fastened to a backup wall that also supports insulation and other components of the exterior enclosure. Both exposed and hidden fastener systems are possible, and installations can be designed to function as either a barrier wall or rainscreen (see, for example, Figures 19.8 and 19.9).

Insulated metal panels consist of sheet metal laminated to both sides of a rigid foam plastic insulating core. When fabricated with a complete system of interlocking edge seams, these lightweight, easily erected panels can provide many of the separate enclosure functions of the exterior wall in one factory-made component, including control of the passage of heat, exterior moisture, air, and interior water vapor. Panels can be designed as a barrier wall system, or, with the addition of an airspace and airtight drainage plane behind the panels, as a rainscreen system (Figures 21.29, 21.30).

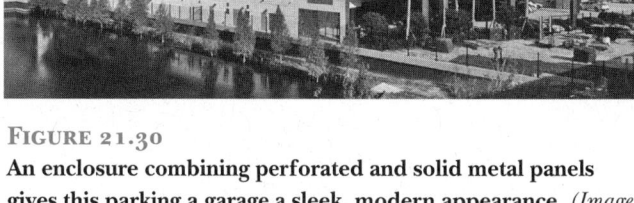

FIGURE 21.30
An enclosure combining perforated and solid metal panels gives this parking a garage a sleek, modern appearance. *(Image courtesy of CENTRIA.)*

FIGURE 21.29
CENTRIA Formawall® aluminum-skinned, insulating metal panels are supported by vertical framing members aligned behind the panel joints. Each panel is mechanically fastened to the frame at the top corners, seen here in the middle of the picture. The bottom edge of the panel is held in place by its interlocking joint with the panel below. Solid gaskets between panels and sealant between panels and backing complete the air and watertightness of the assembly. *(Images courtesy of CENTRIA. Formawall® is registered trademark of Centria.)*

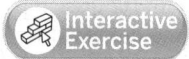

CSI/CSC MasterFormat Sections for Cladding with Metal and Glass	
07 42 00	**WALL PANELS**
07 42 13.13	Formed Metal Wall Panels
07 42 13.16	Metal Plate Wall Panels
07 42 13.19	Insulated Metal Wall Panels
07 42 13.23	Metal Composite Material Wall Panels
07 61 00	**SHEET METAL WALL CLADDING**
08 41 00	**ENTRANCES AND STOREFRONTS**
08 44 00	**CURTAIN WALL AND GLAZED ASSEMBLIES**
08 44 13	Glazed Aluminum Curtain Walls
08 44 16	Glazed Bronze Curtain Walls
08 44 19	Glazed Stainless-Steel Curtain Walls
08 44 33	Sloped Glazing Assemblies

KEY TERMS

extrusion
billet
die
screw slot
screw port
thermal break
cast and debridged thermal break
debridging
polyamide strip thermal break
anodizing
fluoropolymer coating

polyvinylidene fluoride (PVDF)
powder coating
baked enamel coating
(aluminum-framed) entrance
(aluminum-framed) storefront
window wall
curtain wall
stick system
unit system, unitized system
unit-and-mullion system
panel system

column-cover-and-spandrel system
outside glazed
inside glazed
deterrent seal
double-skin facade
sloped glazing
architectural sheet metal cladding
formed metal wall panel
metal plate wall panel
composite metal wall panel
insulated metal wall panel

REVIEW QUESTIONS

1. For what reasons is aluminum the metal most frequently used in metal-and-glass cladding systems?

2. What are the relative advantages and disadvantages of a stick system and a panel system?

3. What are some characteristic features of aluminum extrusions, and what is the function of each?

4. What are the most common finishes applied to aluminum curtain wall components? What are the advantages of each?

5. How would you choose between an inside glazed system and an outside glazed one?

6. Explain the differences between aluminum-framed entrances, storefronts, and curtain walls. What is an appropriate application for each?

7. In what way is a slope glazing system more difficult to make watertight than a wall cladding system? How is this reflected in the details of a slope glazing system?

EXERCISES

1. Make photocopies of full-scale details of the vertical and horizontal mullions of a metal curtain wall system from a manufacturer's catalog. Paste the details on a larger sheet of paper and add notes and arrows to explain every aspect of them: the features of the extrusions, the gaskets and sealants, the glazing materials and methods, drainage, insulation, thermal breaks, rainscreen features, and so on. From your examination of the details, list the order in which the components would be assembled. Is the system glazed from inside the building or outside? How can you tell?

2. Design a coffee table that is made up of aluminum extrusions and a glass top. Design and draw complete details of the extrusions and connections. Select a surface finish for the aluminum. What type of glass will you use?

Selected References

American Architectural Manufacturers Association. *Aluminum Curtain Wall Design Guide Manual.* Schaumburg, IL, Author, 2005.

This publication covers its topic in exemplary fashion with clear text and beautifully prepared illustrations. The same organization also publishes a series of more specialized publications on various aspects of metal-and-glass curtain walls: rainscreen design, care and handling of aluminum components, design of wind loads and wind tunnel testing, fire safety, test methods, installation procedures, finishes, and so on.

Web Sites

Cladding with Metal and Glass

Author's supplementary web site: **www.ianosbackfill.com/21_cladding_with_metal_and_glass**

Aluminum Extruders Council: **www.aec.org**

American Architectural Manufacturers Association: **www.aamanet.org**

Centria curtain wall: **www.centriaperformance.com**

EFCO curtain wall: **www.efcocorp.com**

Formacore Curtain Walls: **www.portafab.com/curtain_walls.shtml**

Kawneer curtain wall and windows: **www.kawneer.com**

Whole Building Design Guide, Curtain Walls: **www.wbdg.org/design/env_fenestration_cw.php**

## 22

SELECTING INTERIOR FINISHES

- **Installation of Mechanical and Electrical Services**

- **The Sequence of Interior Finishing Operations**

  SUSTAINABILITY OF INTERIOR FINISHES

- **Selecting Interior Finishes**
  Appearance
  Durability and Maintenance

Acoustic Criteria
Fire Criteria
Relationship to Mechanical and
  Electrical Services
Changeability
Cost
Avoiding Unhealthful Materials

- **Trends in Interior Finish Systems**

Workers complete installation of an elaborate ceiling made of gypsum board. The corners of the steplike construction have been reinforced with metal corner bead, and the joints and nail heads have been filled and sanded, ready for painting.
(Courtesy of United States Gypsum Company.)

The installation of interior finish materials—ceilings, walls, partitions, flooring, casework, finish carpentry—cannot proceed at full speed until the roof and exterior walls of a building are complete and mechanical and electrical services have been installed. The roof and walls are needed to shelter the moisture-sensitive finish materials. The mechanical and electrical services will be (in most cases) covered by the interior finish materials, and thus must precede them. The finish materials themselves must be selected to meet a bewildering range of functional parameters: durability, acoustical performance, fire safety, relationship to mechanical and electrical services, and changeability over time. They must also look good, presenting a neat appearance and meeting the architectural goals of the building.

INSTALLATION OF MECHANICAL AND ELECTRICAL SERVICES

When a building has been roofed and most of its exterior cladding has been installed, its interior is sufficiently protected from the weather that work can begin on the mechanical and electrical systems. The waste lines and water supply lines of the plumbing system are installed and, if required, the pipes for an automatic sprinkler fire suppression system. The major part of the work for the heating, ventilating, and air conditioning system is carried out, including the installation of boilers, chillers, cooling towers, pumps, fans, piping, and ductwork. Electrical, communications, and control wiring are routed through the building. Elevators and escalators are installed in the structural openings provided for them.

The vertical runs of pipes, ducts, wires, and elevators through a multistory building are made through vertical *shafts* whose sizes and locations were determined at the time the building was designed. Before the building is finished, each shaft will be enclosed with fire-resistive walls to prevent the vertical spread of fire (Figure 22.1).

FIGURE 22.1

A worker constructs a fire-resistant wall around an elevator shaft, using gypsum panels and steel C–H studs. Chapter 23 contains more detailed information on shaft walls.

(Courtesy of United States Gypsum Company.)

FIGURE 22.2

Three diagrammatic plans for an actual three-story suburban office building show the principal arrangements for plumbing, communications, electricity, heating, and cooling. Heating and cooling are accomplished by means of air ducted downward through two shafts from equipment mounted on the roof. The conditioned air from the vertical ducts is distributed around each floor by a system of horizontal ducts that run above a suspended ceiling, as shown on the plan of the intermediate floor. A row of doubled columns divides the building into two independent structures at the building separation joint to allow for differential foundation settlement and thermal expansion and contraction.

(Courtesy of ADD Inc. Architects.)

Air conditioning equipment

Elevator Overtravel

Air conditioning equipment

Mechanical penthouse

Elevator Overtravel

Skylights

Plumbing vents

Air conditioning equipment

Building separation joint

ROOF

Air conditioning shaft

Elevators

Air conditioning shaft

Electric room with electrical riser

Electric room with electrical riser

Toilets

Freight elevator

Plumbing space

Typical branch ducts at ceiling

INTERMEDIATE FLOOR

Main air conditioning supply duct at ceiling

Elevators

Elevators machinery

Water meter room

Stair

Duct risers at ceiling

Duct risers at ceiling

Emergency electric generator

Toilets

Stair

Freight elevator

Electrical and telephone equipment

Stair

Plumbing space

Electric room

GROUND FLOOR

Horizontal runs of pipes, ducts, and wires are usually located just below each floor slab, above the ceiling of the floor below, to keep them up out of the way. These may be left exposed in the finished building or, as is more common, hidden above suspended ceilings. Sometimes these services, especially wiring, are concealed within a hollow floor structure such as cellular metal decking or cellular raceways. Sometimes services are run between the structural floor deck and a raised access flooring system. (For a more complete explanation of these systems, see Chapter 24.) Where space is needed for supply and waste piping behind wall-mounted plumbing fixtures, double walls are constructed that readily accommodate these items.

Specific floor areas are reserved for mechanical and electrical functions in larger buildings (Figure 22.2). Distribution equipment for electrical and communications wiring and fiberoptic networks is housed in special rooms or closets.

Fan rooms are often provided on each floor for air-handling machinery. In a large multistory building, space is set aside, usually at a basement or subbasement level, for pumps, boilers, chillers, electrical transformers, and other heavy equipment. At the roof are penthouses for elevator machinery and mechanical system cooling towers and ventilating fans. In very tall buildings, one or more whole floors may be set aside for mechanical equipment, and the building is zoned vertically into groups of floors that can be reached by ducts and pipes that run up and down from each of these dedicated mechanical floors.

THE SEQUENCE OF INTERIOR FINISHING OPERATIONS

Interior finishing operations follow a carefully ordered sequence that varies somewhat from one building to another, depending on the specific requirements of each project. The first finish items to be installed are usually hanger wires for suspended ceilings, and full-height partitions and enclosures, especially those around mechanical and electrical shafts, elevator shafts, mechanical equipment rooms, and stairways. *Firestopping* is inserted around pipes, conduits, and ducts where they penetrate floors and fire-rated walls (Figure 22.3). The full-height partitions and enclosures, firestopping, joint covers (Figure 22.4), and safing around the perimeters of the floors constitute a very important system for keeping fire from spreading through the building.

After the major horizontal electrical conduits and air ducts have been installed, the grid for the suspended ceiling is attached to the hanger wires so that the lights and ventilating louvers can be mounted. Then, typically, the ceilings are finished, and framing for the partitions that do not penetrate the finish ceiling is installed.

(a) (b) (c)

FIGURE 22.3
Applying firestopping materials to floor penetrations. (*a*) Within a plumbing wall, a layer of safing insulation is cut to fit and inserted by hand into a large slab opening around a cast iron waste pipe. Then a mastic firestopping compound is applied over the safing to make the opening airtight (*b*). (*c*) Applying firestopping compound around an electrical conduit at the base of a partition. (*Courtesy of United States Gypsum Company.*)

FIGURE 22.4
Building separation joints must be covered on the interior of the building to make them safe, attractive, and airtight. The covers must be able to adjust readily to expected movements between the separate parts of the building. Shown here are joint covers for a floor and a wall. Both are ingeniously designed to remain tightly in place while accommodating differential movements of any type. Firestopping, to prevent the easy passage of smoke and fire from one floor to the next, not shown here, is installed behind the joint cover. For a general discussion of movement joints in buildings, see the Chapter 10 sidebar, "Movement Joints in Buildings." *(Courtesy of Architectural Art Manufacturing, Inc., Wichita, KS.)*

Resilient strips can compress as movement occurs

A spring-loaded aluminum cover plate bridges the gap tightly despite differential movement

The metal frame of the joint cover is cast into the concrete floor

SUSTAINABILITY OF INTERIOR FINISHES

In addition to sustainability considerations related to specific materials and systems discussed in Chapters 23 and 24, selection of interior finishes for sustainable buildings should be guided by the following general principles.

Materials

• Finish materials that have a high recycled content reduce the demand for virgin materials and make productive use of materials that would otherwise be treated as waste. The availability of finish materials with recycled content continues to increase in many finish product categories.

• Finish materials that can be reused or recycled when they reach the end of their useful life also reduce waste. Some manufacturers, such as those of ceiling tile, carpet, and gypsum products, have established recycling or reclamation programs to divert these products from the waste stream.

• Interior finishes derived from rapidly renewable sources, such as bamboo flooring, or from certified woods reduce the depletion of raw materials of limited supply and protect forest ecosystems.

• Finish materials that are extracted, processed, and manufactured locally require less energy to transport, and their use helps to support local economies.

Indoor Air Quality

• Indoor finish materials and coatings present large surface areas to the interior environment of a building, making them potentially significant sources of emissions and indoor air quality problems. Potential emitters include glues and binders used in wood panels and other manufactured wood products, leveling compounds applied to subflooring, carpet fabrics and backings, carpet cushions, carpet adhesives, antimicrobial and mothproofing carpet treatments, wall covering adhesives, resilient flooring adhesives, vinyl in all its forms, gypsum board joint compounds, curtain and upholstery fabrics, paints, varnishes, stains, and more.

• Formaldehyde gas is an irritant to building occupants, causes nausea and headaches, and can exacerbate asthma. Potential sources include processed wood products, glues, adhesives, carpets, and permanent press fabrics.

• Volatile organic compounds are air pollutants; they can act as irritants, and some are significant greenhouse gases. Common emitters include processed wood products, glues, adhesives, paints and other coatings, carpets, and plastic welding processes.

• Increasingly, manufacturers are publishing emissions data for their products, offering products with reduced emissions, and participating in rating systems meeting the low-emission standards of LEED and other green building programs, making it easier for designers and specifiers to select green products.

• Mold and mildew growth in carpets, wall coverings, gypsum board assemblies, and fabrics can cause acute respiratory distress in many people. Generally, this problem occurs only when these materials are repeatedly wetted by leakage or condensation. In response to this concern, many manufacturers now offer finish materials with improved resistance to moisture and mold growth.

• Construction dust, if not fully removed before the building is occupied, can be a source of irritating particulates after occupancy.

Design

• Floor plans that are flexible and easily adapted to new uses and partition systems that are easy to modify encourage building reuse.

• The strategic use of high ceilings, low partitions, transparency, reflective surfaces, and light colors can maximize daylighting potential and views to the exterior.

• Spaces designed with exposed structure and without suspended ceilings save materials.

Electrical and communications wiring is brought down from the conduits above the ceilings to serve outlets in the partitions. The walls are finished and painted. The last major finishing operation is the installation of the flooring materials. This is delayed as long as possible to let the other trades complete their work and get out of the building; otherwise, the floor materials could be damaged by dropped tools, spilled paint, heavy construction equipment, weld spatter, coffee stains, and construction debris ground underfoot.

SELECTING INTERIOR FINISHES

Appearance

A major function of interior finish components is to make the interior of the building look neat and clean by covering the rougher and less organized portions of the framing, insulation, vapor retarder, electrical wiring, ductwork, and piping. Beyond this, the architect designs the finishes to carry out a particular concept of interior space, light, color, pattern, and texture. The form and height of the ceiling, changes in floor level, interpenetrations of space from one floor to another, and the configurations of the partitions are primary factors in determining the character of the interior space. Light originates from windows and electric lighting fixtures and is propagated by successive reflections off the interior surfaces of the building. Lighter-colored materials raise interior levels of illumination; darker colors and heavier textures result in a darker interior. Patterns and textures of interior finish materials are important in bringing the building down to a scale of interest that can be appreciated readily by the human eye and hand. No two buildings have the same requirements: Deep carpets and rich, polished marbles in muted tones may be chosen to give an air of affluence to a corporate lobby, brightly colored surfaces to create a happy atmosphere in a day care center, or slick plastic and highly reflective surfaces to provide a trendy ambience for the sale of designer clothing.

Durability and Maintenance

Expected levels of wear and tear must be considered carefully in selecting finishes for a building. Highly durable finishes generally cost more than shorter-lived ones and are not always required. In a courthouse, a transportation terminal, a recreation building, or a retail store, traffic is intense, and long-wearing materials are essential. In a private office or an apartment, more economical finishes are usually adequate. Water resistance is an important attribute of finish materials in kitchens, bathrooms, locker and shower rooms, entrance lobbies, and some industrial buildings. In hospitals, medical offices, kitchens, and laboratories, finish surfaces must not trap dirt and must be easily cleaned and disinfected. Maintenance procedures and costs should be considered in selecting finishes for any building: How often will each surface be cleaned, with what type of equipment, and how much will this procedure add to the cost of owning the building? How long will each surface last, and what will it cost to replace it?

Acoustic Criteria

Interior finish materials strongly affect noise levels, the quality of listening conditions, and levels of acoustic privacy inside a building. In noisy environments, interior surfaces that are highly absorptive of sound can decrease the noise intensity to a tolerable level. In lecture rooms, classrooms, meeting rooms, theaters, and concert halls, acoustically reflective and absorptive surfaces must be proportioned and placed to create optimum hearing conditions.

Between rooms, acoustic privacy is created by partitions that are both heavy and airtight. The acoustic isolation properties of lighter-weight partitions can be enhanced by additional layers of wallboard, partition details that damp the transmission of sound vibrations by means of resilient mountings on one of the partition surfaces, and sound-absorbing batts of mineral wool in the interior cavity of the partition. Manufacturers test full-scale sample partitions of every type of material for their ability to reduce the passage of sound between rooms in a procedure outlined in ASTM E90. The results of this test are converted to *Sound Transmission Class* (STC) numbers that can be related to accepted standards of acoustic privacy. In an actual building, however, if the cracks around the edges of a partition are not completely sealed, or if a loosely fitted door or even an unsealed electric outlet is inserted into the partition, its airtightness is compromised and its acoustical performance severely compromised. Similarly, a partition with a high STC is of little benefit if the rooms on both sides are served by a common air duct that acts incidentally as a conduit for sound, or if the partition reaches only to a lightweight, porous suspended ceiling that allows sound to pass over the top of the partition.

Transmission of impact noise from footsteps and machinery through floor–ceiling assemblies is measured according to ASTM E492, in which a standard machine taps on a floor above while instruments in a chamber below record sound levels. The results are reported as *Impact Isolation Class (IIC)* ratings. Impact noise transmission can be reduced by floor details that rely on soft materials that do not transmit vibration readily, such as carpeting, soft underlayment boards, or resilient underlayment matting.

Fire Criteria

Building codes devote many pages to provisions that control the materials and details for interior finishes in buildings. These requirements are aimed at several important characteristics of interior finishes with respect to fire.

Flammability and Smoke Generation

The surface burning characteristics of interior wall and ceiling finish materials are tested in accordance with ASTM E84, also called the *Steiner Tunnel Test*. In this test, a sample of material 20 inches wide by 24 feet long (500 × 7300 mm) forms the ceiling of a rectangular furnace into which a controlled flame is introduced at one end. The time the flame takes to spread across the face of the material from one end of the furnace to the other is recorded, along with the density of smoke developed. The results of this test are given as a *flame-spread rating*, indicating the rapidity with which fire will spread across a surface of a given material, and a *smoke-developed rating*, which classifies a material according to the amount of smoke it gives off when it burns.

Figure 22.5 defines allowable flame-spread ratings for interior finish materials for various building occupancies according to the International Building Code. It assigns each material to one of three classes: A, B, or C. Class A materials are those with flame-spread ratings between 0 and 25, Class B between 26 and 75, and Class C between 76 and 200. (The scale of flame-spread numbers is established arbitrarily by assigning a value of 0 to cement–asbestos board and 100 to a red oak board.) For all three classes, the smoke-developed rating may not exceed 450. Materials exceeding this rating are not permitted to be used as interior finishes, because smoke, not heat or flame, is the primary killer in building fires. Interior trim materials, if their surface area does not exceed 10 percent

TABLE 803.9
INTERIOR WALL AND CEILING FINISH REQUIREMENTS BY OCCUPANCY[k]

GROUP	SPRINKLERED[l]			NONSPRINKLERED		
	Interior exit stairways, interior exit ramps and exit passageways[a, b]	Corridors and enclosure for exit access stairways and exit access ramps	Rooms and enclosed spaces[c]	Interior exit stairways, Interior exit ramps and exit passageways[a, b]	Corridors and enclosure for exit access stairways and exit access ramps	Rooms and enclosed spaces[c]
A-1 & A-2	B	B	C	A	A[d]	B[e]
A-3[f], A-4, A-5	B	B	C	A	A[d]	C
B, E, M, R-1	B	C	C	A	B	C
R-4	B	C	C	A	B	B
F	C	C	C	B	C	C
H	B	B	C[g]	A	A	B
I-1	B	C	C	A	B	B
I-2	B	B	B[h,i]	A	A	B
I-3	A	A[j]	C	A	A	B
I-4	B	B	B[h,j]	A	A	B
R-2	C	C	C	B	B	C
R-3	C	C	C	C	C	C
S	C	C	C	B	B	C
U	No restrictions			No restrictions		

For SI: 1 inch = 25.4 mm, 1 square foot = 0.0929m².

a. Class C interior finish materials shall be permitted for wainscotting or paneling of not more than 1,000 square feet of applied surface area in the grade lobby where applied directly to a noncombustible base or over furring strips applied to a noncombustible base and fireblocked as required by Section 803.11.1.

b. In other than Group I-2 occupanicies in buildings less than three stories above grade plane of other than Group I-3, Class B interior finish for nonsprinklered buildings and Class C interior finish for sprinklered buildings shall be permitted in interior exit stairways and ramps.

c. Requirements for rooms and enclosed spaces shall be based upon spaces enclosed by partitions. Where a fire-resistance rating is required for structural elements, the enclosing partitions shall extend from the floor to the ceiling. Partitions that do not comply with this shall be considered enclosing spaces and the rooms or spaces on both sides shall be considered one. In determining the applicable requirements for rooms and enclosed spaces, the specific occupancy thereof shall be the governing factor regardless of the group classification of the building or structure.

d. Lobby areas in Group A-1, A-2, and A-3 occupancies shall not be less than Class B materials.

e. Class C interior finish materials shall be permitted in places of assembly with an occupant load of 300 persons or less.

f. For places of religious worship, wood used for ornamental purposes, trusses, paneling, or chancel furnishing shall be permitted.

g. Class B material is required where the building exceeds two stories.

h. Class C interior finish materials shall be permitted in administrative spaces.

i. Class C interior finish materials shall be permitted in rooms with a capacity of four persons or less.

j. Class B materials shall be permitted as wainscotting extending not more than 48 inches above the finished floor in corridors and exit access stairways and ramps.

k. Finish materials as provided for in other sections of this code.

l. Applies when protected by an automatic sprinkler system installed in accordance with Section 903.3.1.1 or 903.3.1.2.

FIGURE 22.5

Flame-spread rating requirements for interior wall and ceiling finish materials, taken from the IBC. The Class ratings A, B, and C are explained in the accompanying text. (*Table 803.9 excerpted from the 2012 International Building Code, Copyright 2011. Washington, D.C.: International Code Council. Reproduced with permission. All rights reserved. www.ICCSAFE.org*)

of the total wall and ceiling area of the room, may be of Class A, B, or C in any type of building.

Alternatively, interior wall and ceiling finish materials can be tested for *room fire-growth contribution* according to NFPA 265. In this test, the finish material is mounted to several adjacent wall and ceiling surfaces, and then (as in the Steiner Tunnel Test) subjected to a flame of controlled intensities and durations. For the material to pass, flame spread and growth must be controlled, and the quantities of heat and smoke emitted must not exceed stated limits. Materials passing this test are considered comparable to ASTM E84 Class A materials. Some materials, such as textile or expanded vinyl wall coverings, foam plastics, and combustible draperies, wall hangings, and other decorative materials, are subject to other special tests or limitations.

The combustibility of many flooring materials used in exits, corridors, and areas connected to these spaces must be tested according to NFPA 253 for *minimum critical radiant flux exposure*. The purpose of this test is to ensure that flooring in essential parts of the egress system cannot be easily ignited by the radiant heat of fire and hot gases in adjacent spaces. Materials must meet either Class I (most resistant to radiant heat) or Class II (moderately resistant) ratings, depending on the Occupancy and whether or not the area is protected with an automatic sprinkler system. Some traditional flooring materials, such as solid wood, resilient materials, and terrazzo, which have historically demonstrated satisfactory performance, are not required to meet this standard. In other areas of the building, flooring materials are subject to the *pill test* (Consumer Product

Safety Commission DOC FF-1), which evaluates a material's propensity for flame spread when exposed to a burning tablet intended to simulate a dropped lit cigarette, match, or similar hazard.

Fire Resistance

Fire resistance of a wall, ceiling, or floor assembly refers not to the assembly's own combustibility, but rather to its ability to resist the passage of fire from one side of the assembly to the other. The building code regulates the fire resistance of assemblies used to protect the structure of the building, to separate various parts of a building from one another, and to separate one building from another.

The table in Figure 22.6 specifies the required *fire resistance rating*, in hours of separation, between different Occupancies within the same building according to the

TABLE 508.4
REQUIRED SEPARATION OF OCCUPANCIES (HOURS)

OCCUPANCY	A, E		I-1, I-3, I-4		I-2		Rᵃ		F-2, S-2ᵇ, U		B, F-1, M, S-1		H-1		H-2		H-3, H-4		H-5	
	S	NS	S	NS	S	NS	S	NS	S	NS	S	NS	S	NS	S	NS	S	NS	S	NS
A, E	N	N	1	2	2	NP	1	2	N	1	1	2	NP	NP	3	4	2	3	2	NP
I-1, I-3, I-4	—	—	N	N	2	NP	1	NP	1	2	1	2	NP	NP	3	NP	2	NP	2	NP
I-2	—	—	—	—	N	N	2	NP	2	NP	2	NP	NP	NP	3	NP	2	NP	2	NP
Rᵃ	—	—	—	—	—	—	N	N	1ᶜ	2ᶜ	1	2	NP	NP	3	NP	2	NP	2	NP
F-2, S-2ᵇ, U	—	—	—	—	—	—	—	—	N	N	1	2	NP	NP	3	4	2	3	2	NP
B, F-1, M, S-1	—	—	—	—	—	—	—	—	—	—	N	N	NP	NP	2	3	1	2	1	NP
H-1	—	—	—	—	—	—	—	—	—	—	—	—	N	NP	NP	NP	NP	NP	NP	NP
H-2	—	—	—	—	—	—	—	—	—	—	—	—	—	—	N	NP	1	NP	1	NP
H-3, H-4	—	—	—	—	—	—	—	—	—	—	—	—	—	—	—	—	1ᵈ	NP	1	NP
H-5	—	—	—	—	—	—	—	—	—	—	—	—	—	—	—	—	—	—	N	NP

S = Buildings equipped throughout with an automatic sprinkler system installed in accordance with Section 903.3.1.1.
NS = Buildings not equipped throughout with an automatic sprinkler system installed in accordance with Section 903.3.1.1.
N = No separation requirement.
NP = Not permitted.
a. See Section 420.
b. The required separation from areas used only for private or pleasure vehicles shall be reduced by 1 hour but to not less than 1 hour.
c. See Section 406.3.4.
d. Separation is not required between occupancies of the same classification.

FIGURE 22.6

IBC requirements for fire resistance ratings, in hours, for fire separation assemblies between differing Occupancies. The requirements of this table apply when occupancies are treated as *separated*. An alternative approach, particularly suitable to smaller buildings, allows occupancies to be *nonseparated*, in which case rated separations are not required. *(Table 508.4 excerpted from the 2012 International Building Code, Copyright 2011. Washington, D.C.: International Code Council. Reproduced with permission. All rights reserved. www.ICCSAFE.org)*

FIGURE 22.7

FIGURE 22.7
This table summarizes IBC fire resistance rating requirements for various types of assemblies not included in other code tables reproduced in this book. Fire areas, listed in the last row of the table, are portions of a building that are limited in area or occupant number for the purpose of determining sprinkler requirements. Some of these wall types are also discussed in greater detail in Chapter 23.

Assembly Type	Required Fire Resistance
Shaft enclosures, elevator hoistway enclosures	2 hours where connecting four or more floors 1 hour where connecting 3 or few floors
Exit stairway and exit passageway enclosures	Same as shaft enclosures
Corridor and elevator lobby separations	1 hour in unsprinklered buildings; 0–1 hour in sprinklered buildings, depending on the Occupancy and number of occupants served
Dwelling unit or sleeping unit separations in multiunit buildings	½–1 hour, depending on the Construction Type and presence of automatic sprinklers
Mall tenant separations	1 hour
Fire area separations	1–4 hours, depending on the Occupancy

requirements of the International Building Code. (See the figure caption for comments regarding when these requirements apply.) Fire resistance rating requirements found elsewhere in the code for separations such as shaft walls, corridors, exit enclosures, dwelling unit separations, and other nonbearing partitions, are summarized in Figure 22.7. Requirements for the protection of structure, for the fire resistance of exterior walls, and for fire walls that separate buildings are shown in Figures 1.4 and 1.7. Such requirements can be related to fire resistance information provided in manufacturers' literature similar to the examples shown in Figures 1.5 and 1.6.

Fire resistance ratings are determined by full-scale endurance tests conducted in accordance with ASTM E119, which applies not only to partitions and walls, but also to beams, girders, columns, and floor–ceiling assemblies. In this test, the assembly is constructed in a large laboratory furnace and subjected to the structural load (if any) for which it is designed. The furnace is then heated according to a standard time–temperature curve, reaching 1700 degrees Fahrenheit (925°C) at 1 hour and 2000 degrees Fahrenheit (1093°C) after 4 hours. To achieve a given fire resistance rating in hours, an assembly must safely carry its design structural load for the designated period, must not develop any openings that permit the passage of flame or hot gases, and must insulate sufficiently against the heat of the fire to maintain surface temperatures on the side away from the fire within specified maximum levels. Wall and partition assemblies must also pass a test, called the *hose stream test*, intended to assess their durability while exposed to fire conditions. A duplicate sample of the assembly is subjected to half of its rated fire exposure, then sprayed with water from a calibrated fire nozzle for a specified period at a specified pressure. To pass this test, the assembly must not allow passage of the water stream.

Openings in floors, ceilings, and partitions with required fire resistance ratings are restricted in size by most codes and must be protected against the passage of fire in various ways. Doors must be rated for fire resistance in accordance with a table such as that shown in Figure 18.29. Ducts that pass through rated assemblies must be equipped with sheet metal dampers (*fire dampers*) that close automatically if hot gases from a fire enter the duct. Penetrations for pipes and conduits must be sealed tightly with fire-resistive firestopping material.

As an example of the use of these tables, consider a multistory vocational high school building of Type IIA construction that includes both a number of classrooms and a woodworking shop and that is fully sprinklered. For purposes of the table reproduced in Figure 22.6, the International Building Code places classrooms in Occupancy Group E (Educational) and the shop in Occupancy Group F-1 (Industrial, Moderate Hazard). Assuming that the building is large enough that the requirements of this table apply, the two uses must be separated from each other by walls (called "fire barriers") and, if applicable, floor–ceiling assemblies, of 1-hour construction. Doors through such walls must be rated at ¾ of an hour (Figure 18.29). Figure 22.5 indicates that, in the Occupancy E portion of the building, finish materials in exit stairway enclosures must have at least a Class B rating, while Class C finish materials are permitted throughout the remainder of the building. According to Figure 22.7, walls and floor–ceiling assemblies separating corridors from adjacent spaces must have a fire resistance rating between 0 and 1 hours, and exit stairways must be enclosed in construction rated between 1 and 2 hours (the final determination of these requirements depending also on other provisions of the code). Referring to Figure 1.4, we can see that the building's structural system must be protected with 1-hour rated assemblies or, as explained in the footnotes to this table and depending on other requirements of the code, it may be permissible to leave the structure unprotected due to the presence of a sprinkler system.

Fireblocking and Draftstopping of Combustible Concealed Spaces

Building codes require that concealed hollow spaces within combustible assemblies (that is, within wood-framed walls, floors, ceilings, and roofs) be internally partitioned so that fire burning concealed within these spaces cannot rapidly spread undetected over large areas or from one floor to the next. Within combustible vertical spaces, the International Building Code requires *fireblocking* at every floor level; at vertical intervals of 10 feet (3 m) in tall walls; at the intersection of wall framing with floor, ceiling, and roof framing; where stair tops and bottoms meet floors; and other similar conditions. Although some of these requirements are satisfied by the top and bottom plates that are a normal part of structural framing, in other cases, additional materials are needed. Materials permitted for this use include solid lumber, plywood, OSB, particleboard, gypsum board, cement fiberboard, and even some mineral fiber and cellulose insulation products.

In a similar fashion, large, concealed, combustible horizontal spaces must be partitioned with *draftstopping* materials. For example, draftstopping is required in line with dwelling unit separations in multifamily structures, and in flooring, attic, and roof framing that exceed certain area limits. Specific requirements vary with building Occupancy and the presence of sprinklers. The list of materials permitted for draftstopping is similar to that for firestopping. Where the depth of a framed area is too great to easily draftstop, it may also be protected with fire sprinklers within the concealed space.

Relationship to Mechanical and Electrical Services

Interior finish materials join the mechanical and electrical services of a building at the points of delivery of the services—the electrical outlets, the lighting fixtures, the ventilating diffusers and grills, the convectors, the lavatories and water closets. Leading up to these points, the services may or may not be concealed by the finish materials. If the service lines are to be concealed, the finish systems must provide space for them, as well as for maintenance access points in the form of access doors, panels, hatches, cover plates, or ceiling or floor components that can be lifted out to expose the lines. If service lines are to be left exposed, the architect should organize them visually and specify a sufficiently high standard of workmanship in their installation so that their appearance will be satisfactory.

Changeability

How often are the use patterns of a building likely to change? In a concert hall, a chapel, or a hotel, major changes will be infrequent, so fixed, unchangeable interior partitions are appropriate. Appropriate finishes include many of the heavier, more expensive, more luxurious materials (such as tile, marble, masonry, and plaster) that are considered desirable by many building owners. In a rental office building or a retail shopping mall, changes will be frequent; lighting and partitions should be easily and economically adjustable to new use patterns without long delays or unnecessary disruption. The likelihood of frequent change may lead the designer to select either relatively inexpensive, easily removed construction such as gypsum wallboard partitions or relatively expensive but durable and reusable construction such as proprietary systems of modular, relocatable partitions. The functional and financial choices must be weighed for each building.

Cost

The cost of interior finish systems may be measured in two different ways. *First cost* is the installed cost. First cost is often of paramount importance when the construction budget is tight or the expected life of the investment in a building is short. *Life-cycle cost* takes into account not only first cost, but also the expected lifetime of the finish system, maintenance costs and fuel costs (if any) over that lifetime, replacement cost, an assumed rate of economic inflation, and the time value of money. Life-cycle cost is important to building owners who expect to retain ownership for an extended period of time. Because of its higher maintenance and replacement costs, a material that is inexpensive to buy and install may be more costly over the lifetime of a building than a material that is initially more expensive.

Avoiding Unhealthful Materials

A building's interior finish materials reside within the same environment as the building's occupants. Thus, unhealthful ingredients in these materials have the potential to readily interact with and adversely affect users of the building. For example, formaldehyde gas emitted by resins and adhesives used in some manufactured wood and other building products is a known carcinogen and irritant. A host of solvents from paints, varnishes, flooring adhesives, and other such materials, capable of a range of potential adverse health impacts, permeate the air of a building, especially when it is new. Plasticizers and stabilizers that leach from some plastics have been identified as carcinogenic, toxic, or asthma-causing. Airborne mineral and glass fibers can irritate skin, eyes, and mucous membranes. Airborne spores from molds and mildews that grow on moisture-sensitive materials can cause reactions from mild to severe in some portions of the population. On occasion, natural stone and masonry materials have been known to emit radioactive radon gas. Even ordinary construction dust can inflame respiratory passages.

Some long-recognized health hazards have been eliminated from construction materials by state or federal health regulations, and only remain a concern when encountered in older buildings. Examples include lead-containing paint; asbestos fibers in plaster, flooring, or fireproofing materials; and arsenic-containing wood preservatives. Other materials have only more recently garnered recognition as significant health concerns. In these instances, it remains incumbent upon building designers and materials specifiers to be knowledgeable of these hazards and endeavor to keep them out of buildings.

Fortunately, as awareness of these concerns has grown, the task of recognizing and avoiding unhealthful substances has become easier. For example, the Living Building Challenge Health petal includes the imperatives of Civilized Environment, ensuring access to fresh air and daylight, and Healthy Air, requiring management of indoor air quality. The Materials petal includes the Red List imperative, which identifies materials and chemicals, deemed damaging to human health or the environment, that must be excluded from the building. Similarly, the LEED rating system requires the protection of interior air quality, selection of low-emitting materials, and avoidance of known unhealthful or toxic materials. In response to initiatives such as these, there is today more data readily available to the building designer to assist in identifying and avoiding undesirable materials. And, as the construction industry responds to these market pressures, it becomes increasingly easy to find safer, appropriate alternatives.

TRENDS IN INTERIOR FINISH SYSTEMS

Interior finish systems have undergone a transformation over the past roughly 75 years. Formerly, the installation of finishes for a commercial office began with the construction of partitions of heavy clay tiles or gypsum blocks set in mortar. These were covered with two or three coats of plaster and joined to a three-coat plaster ceiling. The floor was commonly made of hardwood strips with a wood baseboard, or perhaps of poured terrazzo with an integral terrazzo base. Today, the same office might be framed in light metal studs and walled with gypsum board. The ceiling might be a separate assembly of lightweight, acoustically absorbent tiles, and the floor might be a thin layer of vinyl composition tile glued to a smooth concrete slab.

Several trends can be discerned in these changes. One is away from an integral, single-piece system of finishes toward a system made up of discrete components. In the old office, the walls, ceiling, and floor were all joined, and none could be changed without disrupting the others. In the new office, ceiling and floor finishes often extend uninterrupted from one side of the building to the other, so that partitions can be changed at will without affecting either the ceiling or the floor. The trend toward discrete components is epitomized by partitions made of modular, demountable, relocatable panels.

Another discernible trend is away from heavy finish materials to lighter ones. A partition of metal studs and gypsum board has a fraction of the weight of one of clay tiles and plaster, and a vinyl composition tile installation is many times lighter than a traditional terrazzo one of equal area. Lighter finishes reduce the dead load the structure of the building must carry. This enables the structure itself to be lighter and less expensive. Lighter finish materials reduce shipping, handling, and installation costs, and are easier to move or remove when changes are required.

"Wet" systems of interior finish, made of materials mixed with water on the building site, have been largely replaced by "dry" ones. Plaster has been replaced by gypsum board and ceiling tiles in most areas of new buildings, and tile and terrazzo floors by resilient materials or carpet. The installation of dry systems is fast and less dependent on weather conditions. Dry systems require less skill on the part of the installer than wet systems, because the skilled work is transferred from the job site to the factory, where it is done by machines. All these differences tend to result in a lower installed cost.

Traditional finishes, however, are far from obsolete. Gypsum board cannot rival three-coat plaster over metal lath for surface quality, durability, or design flexibility. Tile and terrazzo floorings are unsurpassed for wearing quality and appearance. In many situations, the life-cycle costs of traditional finishes compare favorably with those of lighter-weight alternatives whose first cost is considerably less. In addition, the aesthetic qualities of, for example, marble floors, wood wainscoting, and sculpted plaster ceilings, where such qualities are called for, cannot be imitated by any other material.

CSI/CSC MasterFormat Sections for Selecting Interior Finishes	
07 84 00	**FIRESTOPPING**
07 84 13	Penetration Firestopping
07 84 43	Joint Firestopping

KEY TERMS

shaft	room fire-growth contribution	hose stream test
firestopping	minimum critical radiant flux exposure	fire damper
Sound Transmission Class (STC)	pill test	fireblocking
Impact Isolation Class (IIC)	fire resistance	draftstopping
Steiner Tunnel Test	fire resistance rating	first cost
flame-spread rating	separated occupancy	life-cycle cost
smoke-developed rating	nonseparated occupancy	

REVIEW QUESTIONS

1. Draw a flow diagram of the approximate sequence in which finishing operations are carried out on a large building of Type IIA construction.

2. List the major considerations that an architect should keep in mind while selecting interior finish materials and systems.

3. What are the two major types of fire tests conducted on interior finish systems? What measures of performance are derived from each?

4. What is the difference between first cost and life-cycle cost?

EXERCISES

1. You are designing a 31-story apartment building (Occupancy R-2) in a large city. Assume that it will be fully sprinklered. What types of construction are you permitted to use under the International Building Code? What fire resistance rating will be required for the separation between the apartment floors and the retail stores on the ground floor, assuming that the different occupancy areas must be separated? What classes of finish materials can you use in the exit stairway? In the corridors to those stairways? Within the individual apartments? If a red oak board has a flame-spread rating of 100, can you panel an apartment in red oak? What fire resistance ratings are required for partitions between apartments? Between an egress corridor and an apartment? What type of fire door is required between the apartment and the exit corridor? What is the required fire resistance rating for the walls around the elevator shafts? For the purposes of this exercise, when referring to Figure 22.7, assume the highest rating requirement where a range of requirements is provided.

2. Select a compound from the Living Building Challenge Red List. Research examples of building products that contain this compound. Explain where these products are commonly used in buildings. Find products, free of Red List compounds, that may be used as alternatives.

SELECTED REFERENCES

Allen, Edward, and Joseph Iano. *The Architect's Studio Companion* (4th ed.). Hoboken, NJ, John Wiley & Sons, Inc., 2006.

The fourth section of this book gives extensive information on providing space for mechanical and electrical equipment in buildings.

International Code Council, Inc. *International Building Code.* Falls Church, VA, Author, updated regularly.

The reader is referred to Chapters 7 and 8 of this model code, which deal with fire-resistive construction requirements for interior finishing systems.

Juracek, Judy A. *Surfaces: Visual Research for Artists, Architects, and Designers.* New York, W. W. Norton, 1996.

This book is a wonderful catalog of the vast expressive potential of the architectural surface, containing more than 1200 color photographs of finish surfaces of differing material types, patterns, textures, colors, and forms.

WEB SITES

Selecting Interior Finishes

Author's supplementary web site: www.ianosbackfill.com/22_selecting_interior_finishes
Greenguard Environmental Institute: www.greenguard.org
Health Product Declaration Forum: www.hpdworkinggroup.org
Healthy Building Network: www.healthybuilding.net

## 23

INTERIOR WALLS AND PARTITIONS

- **Types of Interior Walls**
  Fire Walls
  Shaft Walls
  Fire Barriers and Fire Partitions
  Smoke Barriers and Smoke Partitions
  Other Nonbearing Partitions

- **Framed Partition Systems**
  Partition Framing
  Plaster

  SUSTAINABILITY OF GYPSUM PRODUCTS

Plaster Partition Systems

PLASTER ORNAMENT

Gypsum Board
Gypsum Board Partition Systems

- **Masonry Partition Systems**

- **Wall and Partition Facings**

A plasterer applies a scratch coat of gypsum plaster to expanded metal lath. The extrusions of plaster forced through the small openings in the lath will, once the plaster dries, form a strong mechanical bond with the lath. The partition is framed with open-truss wire studs, a type of stud seen in older buildings but no longer used in new construction. Soft-annealed galvanized steel wire was used to make all the connections in this type of partition, minimizing the need for screw fasteners. *(Courtesy of United States Gypsum Company.)*

There is more to interior walls and partitions than meets the eye. Behind their simple surfaces lie assemblies of materials carefully chosen and combined to meet specific performance requirements relating to structural strength, fire resistance, durability, and acoustical isolation. A partition may be framed with steel or wood studs and faced with plaster or gypsum board. Alternatively, masons may construct it of concrete blocks or structural clay tiles. For improved appearance or durability, a partition may be faced with ceramic tiles, masonry veneer, wood paneling, or any of a long list of other finish materials to tailor it to the requirements of a specific application.

TYPES OF INTERIOR WALLS

Fire Walls

A *fire wall* is a wall that forms a required separation to restrict the spread of fire through a building, and extends continuously from the foundation to or through the roof. A fire wall is used to divide a single building into smaller units, each of which may be considered as a separate building when calculating allowable heights and areas under the building code. A fire wall must either meet a noncombustible roof structure at the top or extend through and above the roof by a specified minimum distance, 30 inches (760 mm) in the case of the International Building Code (IBC). In this code, a fire wall must also extend horizontally at least 18 inches (450 mm) beyond the exterior walls of the building unless these exterior walls meet certain fire resistance and combustibility requirements. Except in buildings of Type V construction, a fire wall must be framed with noncombustible materials such as steel studs, concrete, or masonry. A fire wall must also have sufficient structural stability during a building fire to allow collapse of the construction on either side without itself collapsing.

Openings in fire walls are restricted in size and aggregate area and must be closed with fire doors or fire-rated glass. The required fire resistance ratings for fire walls under the International Building Code are defined in Figure 1.7.

Shaft Walls

A *shaft wall* is used to enclose vertical openings that extend through multiple floors of a building, such as for elevators, exit stairs, or vertical runs of ductwork, conduits, or pipes. In the International Building Code, a shaft wall connecting four or more floors must have a fire resistance rating of 2 hours or, if connecting fewer floors, a rating of 1 hour. Walls for elevator shafts must be able to withstand the air pressure and suction loads placed on them by the movements of the elevator cars within the shaft, and should be designed to prevent the noise of the elevator machinery from reaching other areas of the building.

Fire Barriers and Fire Partitions

Fire-rated walls are also used to restrict the spread of fire and smoke within a single building. Depending on the type of separation, the International Building Code requires such walls to be constructed as either fire barriers or fire partitions. Unlike fire walls, these wall types do not necessarily extend from foundation to roof. A *fire barrier* must extend vertically from the top of one floor slab to the underside of the next. Fire barriers are used to separate different occupancies, and to limit the extent of *fire areas* (areas bounded by fire-resistant construction, the size and location within the building of which are related to automatic sprinkler requirements). Requirements for a *fire partition* are less stringent than for fire barriers. For example, in some cases, a fire partition may terminate at the underside of a suspended ceiling. Fire partitions are used to enclose corridors and to separate tenant spaces in mall buildings or dwelling units in hotels, dormitories, and other multidwelling unit buildings.

Openings in fire barriers and partitions are restricted in size and must be closed with fire doors or fire-rated glass. Required fire resistance ratings for the various types of fire barriers and partitions according to the International Building Code are listed in Figures 22.6 and 22.7.

The structural elements that support fire barriers and fire partitions must have a fire resistance rating at least as great as that of the wall being supported. For example, in a building of Type IIB construction, the structure as a whole is permitted to be unprotected. However, columns, bearing walls, and portions of the structure that support a 1-hour rated corridor wall must themselves also be protected with at least a 1-hour fire resistance rating.

Smoke Barriers and Smoke Partitions

In certain institutional occupancies such as hospitals and prisons, where occupants are unable to leave the building in case of fire, special partitions called *smoke barriers* are required. This type of wall divides floors of buildings in such a way that occupants may take refuge in case of fire by moving to the side of the smoke barrier that is away from the fire, without having to exit the building. A smoke barrier is a 1-hour rated partition that is continuous from one side of the building to the other and from the top of a floor slab to the bottom of the slab on the floor above. It must be sealed at all edges. Penetrations for air ducts must be protected with dampers that close automatically if smoke is detected in the air. Other penetrations, such as those for pipes and conduits, must be sealed as well. Doors through smoke barriers are necessary to allow movement of people in case of fire. They must be close fitting, without grilles or louvers, and must close automatically.

A *smoke partition* is a wall constructed—like a smoke barrier—to resist the passage of smoke, but without any fire resistance rating. For example, when walls for corridors and elevator lobbies need not be fire-rated, they are constructed as smoke partitions.

Other Nonbearing Partitions

Many of the partitions in a building neither bear a structural load nor are required as fire separation walls. These may be made of any material that meets the combustibility provisions of the building code for the selected type of construction, as explained in the next section.

FRAMED PARTITION SYSTEMS

Partition Framing

Partitions that will be finished in plaster or gypsum board are usually framed with wood or metal studs. Framing with wood studs is permitted by the building code only in buildings of certain combustible construction types, including Types III and V. Partitions in buildings of Type I or Type II (noncombustible)

construction must be framed with light gauge metal studs. Partitions in Type IV, Heavy Timber construction, must either be framed with metal studs or constructed with wood members assembled into solid, laminated partitions as specified in the code. With certain limitations, fire-retardant-treated wood is also permitted for partition framing in Types I, II, and IV construction.

Metal partition framing is directly analogous to wood light framing, but constructed of *light gauge steel studs* and *tracks* made of galvanized steel sheet metal, usually from 0.018 to 0.030 inches (0.45–0.75 mm) thick (Figures 23.1 and 23.2). Light gauge steel members and framing methods are detailed in Chapter 12. Because of its noncombustibility, metal partition framing is permitted in all building code construction types.

Nonloadbearing partitions must be isolated from movements in the surrounding building structure that could otherwise cause distortion of the partition framing or stresses in partition finishes. Resilient sealant joints may be required where partition ends meet structural walls, columns, or curtain wall framing. Where partition tops abut structural framing or slabs above, deflection tracks isolate the partition from movement in those elements (Figure 23.2*b*).

If plaster or gypsum board surfaces are to be applied over a masonry wall, they may be spaced away from the wall with either wood or metal *furring strips* (Figures 23.3, 23.4, and 23.5). Furring allows the installation of a flat wall finish over an irregular masonry surface, and provides a concealed space between the finish and the masonry for installing plumbing, wiring, and thermal insulation.

Plaster

Plaster is a generic term that refers to any of a number of cement-like substances that are applied to a surface in paste form and then harden into a solid material. Plaster may be

applied directly to a masonry surface or, more generally, to any of a group of plaster bases known collectively as *lath* (rhymes with "math"). Plastering began in prehistoric times with the smearing of mud over masonry walls or over a mesh of woven sticks and vines to create a construction known as *wattle and daub*, the wattle being the mesh and the daub the mud. The early Egyptians and Mesopotamians developed finer, more durable plasters based on gypsum and lime. Portland cement plasters evolved in the 19th century. It is from these latter three materials—gypsum, lime, and portland cement—that the plasters used in buildings today are prepared.

Gypsum Plaster

Gypsum is an abundant mineral in nature, a crystalline hydrous calcium sulfate. It is quarried, crushed, dried, ground to a fine powder, and heated to 350 degrees Fahrenheit (175°C) in a process known as *calcining* to drive off about three-quarters of its water of hydration. The *calcined gypsum*, ground to a fine white powder, is known as *plaster of Paris*. When plaster of Paris is mixed with water, it rehydrates and recrystallizes rapidly to return to its original solid state. As it hardens, it gives off heat and expands slightly.

Gypsum is a major component of interior finish materials in most buildings. It has but one major disadvantage—its solubility in water. Among its advantages are that it is durable and light in weight compared to many other materials. It resists the passage of sound better than most materials. It has a very fine grain, is easily worked in either its wet or dry state, and can be fashioned into surfaces that range from smooth to heavily textured. Above all, it is inexpensive, and it is highly resistant to the passage of fire.

When a gypsum building component is subjected to the intense heat of a fire, a thin surface layer is calcined and gradually disintegrates. In

FIGURE 23.1
Noncombustible partitions are assembled with light gauge steel studs attached to top and bottom track sections (runners), usually with self-drilling screws. Preformed holes in the studs provide openings for wiring, small-diameter piping, or metal bracing.

FIGURE 23.2
(*a*) Attaching a runner to a concrete floor using powder-driven fasteners. The gun explodes a small charge of gunpowder to drive a steel pin through the metal and into the concrete to make a secure connection. (*b*) Where a nonloadbearing partition abuts the underside of a structural member, a slotted *deflection track* isolates the partition from deflections in that member, preventing structural loads from distorting the partition or causing damage to finishes. Note how the screws are fastened at the bottom of the slots, to leave the track free to move downward without impinging on the studs. Deflection tracks can also be made by nesting one track inside another or by leaving studs unattached to the top track and adding metal bracing near the top of the studs to hold them laterally. A sheet metal angle connects the deflection track to a tab welded to the HSS structural steel member above. (*c*) A twisting motion is used to ease studs into place between the top and bottom runners. The snug fit between the vertical legs of the runners and the flanges of the studs temporarily holds the studs in place until they are fastened. On the right side of the photograph, a stack of gypsum board awaits installation. *(Photos (a) and (c) courtesy of United States Gypsum Company; Photo (b) by Joseph Iano.)*

Vinyl base

Gypsum board

Rigid foam Insulation

Z-furring channels hold the insulation boards and provide a flange for attachment of the gypsum board

FIGURE 23.3
A furred gypsum board finish over a concrete block wall. The *Z-furring channels* are attached to the masonry with powder-driven fasteners. The plastic foam insulation is tucked in behind the flange of the channel, and the gypsum board is screwed to the face of the flange. Long slots punched from the web of the channel (not visible in this drawing) help to reduce the thermal bridging effect of the Z-furring channel.

Gypsum board is screwed to furring channels

Wood base

Metal furring channels are fastened to the masonry with powder driven fasteners

FIGURE 23.4
A furred gypsum board finish using a standard hat-shaped metal furring *hat channel.*

the process, it absorbs considerable heat and gives off steam, both of which cool the fire (Figure 23.6). Layer by layer, the fire works its way through the gypsum, but the process is slow. The uncalcined gypsum never reaches a temperature more than a few degrees above the boiling point of water, so areas behind the gypsum component are well protected from the fire's heat. Any required degree of fire resistance can be created by increasing the thickness of the gypsum as necessary. The fire resistance of gypsum can also be increased by adding lightweight aggregates to reduce its thermal conductivity and by adding reinforcing fibers to retain the calcined gypsum in place as a fire barrier.

For use in construction, calcined gypsum is carefully formulated with various admixtures to control its setting time and other properties. Gypsum plaster is made by mixing the appropriate dry plaster formulation with water and an aggregate, either fine sand or a lightweight aggregate such as perlite or vermiculite. Because of its expansion during setting, gypsum plaster, unlike portland cement plaster, is not prone to shrinkage cracking as it dries.

An adjustable masonry furring bracket is fastened to the masonry with powder driven fasteners

Metal lath and plaster

A special metal channel serves as a plaster ground as well as a permanent baseboard

$^3/_4$" (19 mm) metal channels

A sheet metal base clip fastenes the channel stud and the metal base to the floor

6" (152 mm)

1040°C — 1900°F
510°C — 950°F
104°C — 220°F
82°C — 180°F
54°C — 130°F

FIGURE 23.5
A furred plaster finish using adjustable furring brackets. Each bracket has a series of teeth along its upper edge so that a metal channel can be wired securely to it in any of a number of positions, allowing the lather to produce a flat wall regardless of the surface quality of the masonry.

FIGURE 23.6
The effect of fire on gypsum, based on data from Underwriters Laboratories, Inc. After a 2-hour exposure to heat following the ASTM E119 time–temperature curve, less than half of the gypsum on the side toward the fire, shown here by shading, has calcined. The portions of the gypsum to the right of the line of calcination remain at temperatures below the boiling point of water.

SUSTAINABILITY OF GYPSUM PRODUCTS

Sources of Gypsum

• Naturally occurring gypsum is not renewable, but it is plentiful and widely distributed geographically.

• The majority of newly extracted gypsum is quarried in surface mines, with attendant loss of wildlife habitat, surface erosion, and water pollution, as well as the problem of disposing of overburden and mine tailings.

• There is increasing use of *synthetic gypsum*, material recovered from power plant flue gases that would otherwise be sent to landfills, in the manufacture of gypsum construction materials. According to the Gypsum Association, approximately 1.5 million tons (1.4 million metric tons) of synthetic gypsum is used annually to produce about 7 percent of the U.S. construction industry's calcined gypsum. Some synthetic gypsums, however, contain toxic byproducts from the manufacturing processes in which they are produced and cannot be safely recycled into new construction materials.

Gypsum Products Manufacturing

• The calcining of gypsum involves temperatures that are not much higher than the boiling point of water, which means that the embodied energy of gypsum is relatively low, about 1200 BTU per pound (2.8 MJ/kg) for plaster and 2600 BTU per pound (6.0 MJ/kg) for gypsum board.

• The calcining process emits particulates of calcium sulfate, an inert, benign chemical, as dust.

• The paper faces of gypsum board are composed primarily of recycled newspapers.

• Some manufacturers produce gypsum board products made with as much as 95 percent recycled materials, including synthetic gypsum and recycled postconsumer waste paper.

Gypsum Products on the Building Site

• Approximately 15 million tons (14 million metric tons) of gypsum board are manufactured annually in the United States. On a typical construction site, about 10 to 12 percent of this material becomes waste.

• Gypsum board waste generated during construction can be minimized by sizing walls and ceilings to make efficient use of whole boards or by ordering custom-sized boards for nonstandard-size surfaces.

• Gypsum board scrap can be permanently stored in the hollow cavities of finished walls, eliminating disposal and transportation costs and reducing the amount of material destined for landfills (though care must be taken not to create interference with the pulling of electrical wires at a later date).

• Some dust is generated by the cutting and sanding of gypsum board and plaster. This dust has not been tied to any specific illnesses, but it is a nuisance and a source of discomfort until the work is done and all the dust has been swept up and removed from the building. Remodeling and demolition also create large quantities of gypsum dust.

• Most installed gypsum products have extremely low emissions. Some joint compounds, however, may emit volatile organic compounds (VOCs).

• Additives used in the manufacture of moisture-resistant and fire-resistant gypsum board are potential sources of VOC emissions.

• Paints, wallcovering adhesives, and other products used to finish gypsum surfaces can be significant emitters of VOCs, and thus require care in selection.

Gypsum Disposal and Recycling

• Gypsum board waste can be recycled back into the manufacture of new gypsum board products. Current efforts limit recycled content to no more than 15 or 20 percent, due to the amount of paper waste that can be safely introduced into the new gypsum without impairing its fire resistance.

• Gypsum board waste from the demolition of older buildings may be contaminated with nails, drywall tape, joint compound, and paint. Gypsum board demolished from buildings constructed prior to 1978 may be coated with lead-based paint. These foreign materials must be removed from the waste; their presence may limit the material's recycling potential.

• Gypsum board waste can be used as a soil amendment and plant nutrient. With the recent advent of mobile grinders, construction site recycling of gypsum board waste for use as a soil amendment on the same building site is feasible.

• Gypsum is an ingredient in many manufacturing and industrial processes. Studies and small-scale tests currently underway to identify potential uses of gypsum board waste in such processes are likely to lead to additional recycling opportunities in the future.

Gypsum plasters are manufactured in accordance with ASTM C28 and fall into two general categories: *base-coat plasters*, used for the underlying preparatory coats of a plaster application (as explained later), and *finish-coat plasters*. Base-coat plasters are provided either *mill-mixed*, also called *ready-mixed*, with aggregate added at the manufacturing plant; or *neat*, for use with aggregate added at the job site. The more common base-coat gypsum plasters are:

• Ordinary *gypsum plaster*, in various formulations suitable for either hand or machine application

• *Wood-fibered gypsum plaster*, gypsum plaster blended with chopped wood fibers for lighter weight, and greater strength and fire resistance

• *Lightweight gypsum plaster* with perlite or vermiculite aggregate, for lighter weight and greater fire resistance

• *High-strength base-coat plaster*, for use under high-strength finish coats

Finish-coat plasters are typically a blend of gypsum plaster and lime. The *lime* provides superior workability and finishing qualities, while the gypsum provides greater hardness and strength and prevents shrinkage cracking. (For more information about the manufacture of lime and its use in plasters, see Chapter 8.) Common finish-coat plasters include:

• *Ready-mixed finish plaster*, with factory-blended lime and other ingredients

• *Gauging plaster*, gypsum plaster for job site mixing with *hydrated finishing lime* (also called *lime putty*)

• *High-strength gauging plaster*, formulated to produce a finish plaster with higher compressive strength

• *Keenes cement*, a proprietary gauging plaster that produces an exceptionally dense, crack-resistant, low-absorbency finish

• *Molding plaster*, a fast-setting, fine-textured material for molding plaster ornament and running cornices (see the sidebar on "Plaster Ornament" later in this chapter)

Retarders and accelerators can also be added to plaster mixes on the job site to adjust the setting time to job site temperature and humidity conditions.

Portland Cement Plaster

Portland cement–lime plaster, also known as *stucco*, is similar to masonry mortar. It is used where the plaster is likely to be subjected to moisture, as on exterior wall surfaces or in commercial kitchens, industrial plants, and shower rooms. Because freshly mixed stucco is not as buttery and smooth as gypsum and lime plasters, it is not as easy to apply and finish. It shrinks slightly during curing, so it should be installed with frequent control joints to regulate cracking.

Plastering

Plaster can be applied either by machine or by hand. Machine application is essentially a spraying process (Figure 23.7). Hand application is done with two very simple tools: a *hawk* in one hand to hold a small

FIGURE 23.7
Spray-applying a scratch coat of plaster onto gypsum lath. (*Courtesy of United States Gypsum Company.*)

quantity of plaster ready for use and a *trowel* in the other hand to lift the plaster from the hawk, apply it to the surface, and smooth it into place (Figures 23.8, 23.16, and 23.20). Plaster is transferred from the hawk to the trowel with a quick, practiced motion of both hands, and the trowel is moved up the wall or across the ceiling to spread the plaster, much as one uses a table knife to spread soft butter. After a surface is covered with plaster, it is leveled by drawing a straightedge called a *darby* across it, after which the trowel is used again to smooth the surface.

Lathing

Historically, the most common form of lath consisted of thin strips of wood nailed to wood framing with small spaces left between the strips to allow *keying* of the plaster. Most lath today is made of either expanded metal or preformed gypsum boards. The skilled tradesperson who applies lath and trim accessories is known as a *lather*.

Expanded metal lath is made from thin sheets of steel that are slit and stretched in such a way as to produce a mesh of diamond-shaped openings (Figure 23.9). It is applied to light gauge steel studs with *self-drilling, self-tapping screws* or to wood studs with large-headed *lathing nails*. Lath used with portland cement stucco, and likely to be used in a damp exposure, is galvanized to deter corrosion.

Gypsum lath is made in gypsum board sheets (see discussion later in this section), usually 16 by 48 inches (406 × 1220 mm) and ⅜ inch (9.5 mm) thick. It consists of sheets of hardened gypsum plaster faced with outer layers of a special absorbent paper to which fresh plaster readily adheres and inner layers of water-resistant paper to protect the gypsum core. Gypsum lath is attached to steel or wood studs with screws (Figure 23.10). Gypsum lath cannot be used as a base for gauged gypsum-lime finish plasters or portland cement stucco, as these materials will not bond adequately to the paper facing.

Veneer plaster base (*gypsum veneer base*) is a paper-faced gypsum board that comes in sheets 4 feet (1220 mm) wide, 8 to 14 feet (2440–4270 mm)

FIGURE 23.8

Applying a scratch coat over gypsum lath with a hawk (a corner of which is visible behind the plasterer's stomach) and trowel. Notice the wire clips that hold the lath to the open-truss studs and the sheet metal clips that strengthen the end joints between panels. The end joints do not occur over studs. *(Courtesy of United States Gypsum Company.)*

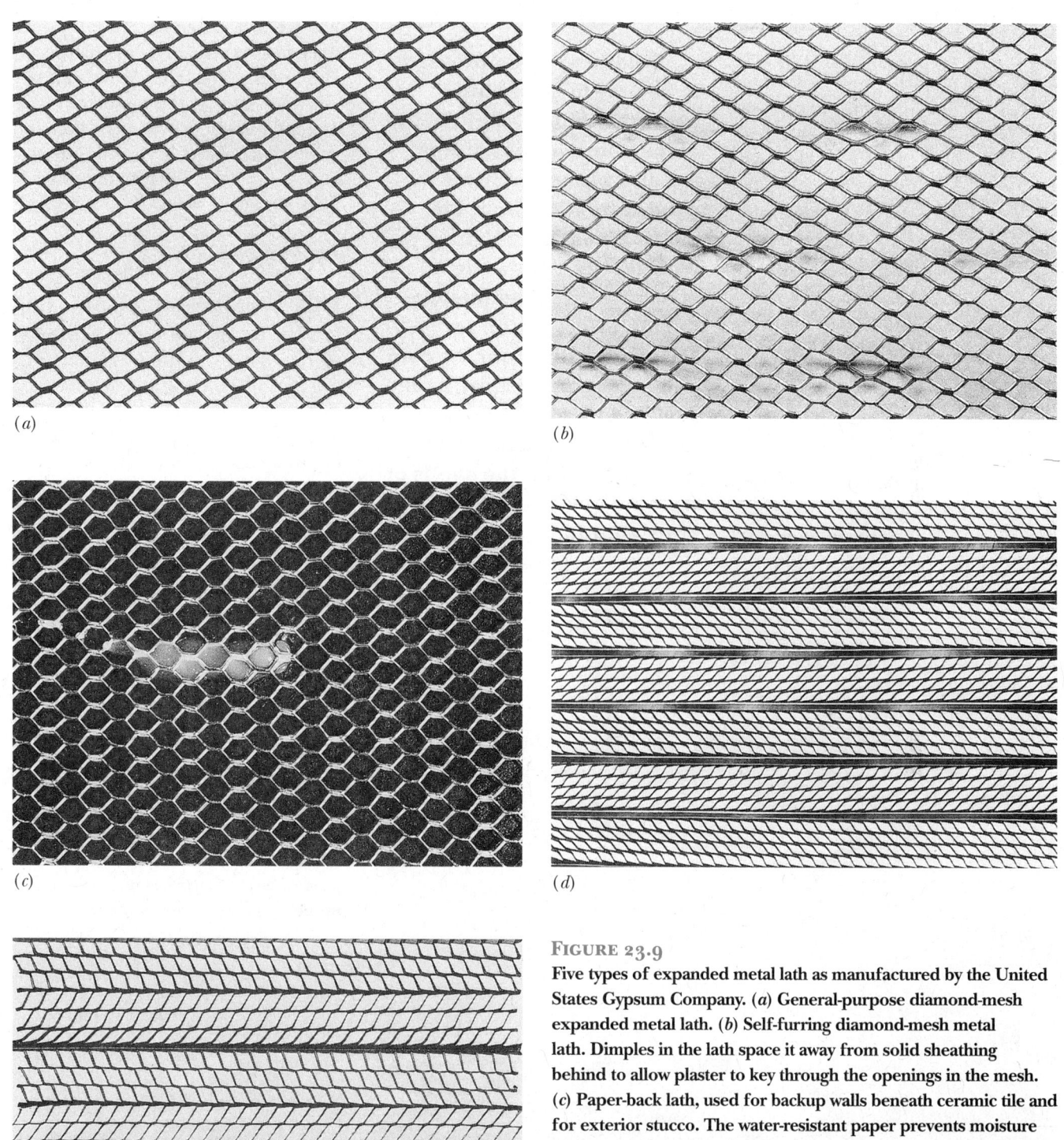

(a)

(b)

(c)

(d)

(e)

FIGURE 23.9

Five types of expanded metal lath as manufactured by the United States Gypsum Company. (*a*) General-purpose diamond-mesh expanded metal lath. (*b*) Self-furring diamond-mesh metal lath. Dimples in the lath space it away from solid sheathing behind to allow plaster to key through the openings in the mesh. (*c*) Paper-back lath, used for backup walls beneath ceramic tile and for exterior stucco. The water-resistant paper prevents moisture that seeps through the plaster from penetrating further into the wall. (*d*) Four-mesh Z-riblath is stiffer than ordinary diamond-mesh lath, making it suitable for ceilings. (*e*) Three-eighths-inch (10-mm) riblath has V-shaped ribs for exceptional rigidity; it is used for ceilings or concrete formwork where supports are widely spaced. (*Courtesy of United States Gypsum Company.*)

long, and ½ to ⅝ inch (13–16 mm) thick. It is screwed to wood or steel studs, or nailed to wood studs, and used as a base specifically for the application of gypsum veneer plaster (discussed later in this section).

Various *lathing trim accessories*, most frequently made of galvanized steel, are used at the edges of a plaster surface to make a neat, durable edge or corner (Figure 23.11). These are installed by the lather at the same time as the lath. In very long or tall plaster surfaces, metal *control joint* accessories are mounted over seams in the lath at predetermined intervals to control cracking. Trim accessories are also designed to act as lines that gauge the proper thickness and plane of plaster surfaces. A straightedge may be run across them to level the wet plaster. In this role, the trim accessories are known collectively as *grounds*. Trim accessories are made in several different thicknesses to match the required plaster thicknesses over the different types of lath.

In lathing I was pleased to be able to send home each nail with a single blow of the hammer, and it was my ambition to transfer the plaster from the board to the wall neatly and rapidly. . . . I admired anew the economy and convenience of plastering, which so effectually shuts out the cold and takes a handsome finish. . . . I had the previous winter made a small quantity of lime by burning the shells of the *Unio fluviatilis*, which our river affords.

—Henry David Thoreau,
***Walden*, 1854**

Trim accessories are also produced as extrusions of plastic or aluminum. The aluminum accessories and some of the plastic ones are designed for improved precision and appearance when used in innovative details for bases, edges, and shadow lines in plaster walls.

Plaster Systems

Plaster over Expanded Metal Lath

Plaster is applied over expanded metal lath in three coats (Figure 23.12). The first, called the *scratch coat*, is troweled on rather roughly and cannot be made completely flat because the uncoated lath moves in and out considerably under the pressure of the trowel (see the chapter opening photograph). This first coat is scratched while still wet, using a notched darby, a broom, or a special rake, to create a rough surface to which the second coat can bond mechanically (Figure 23.13).

FIGURE 23.10
Installing gypsum lath over light gauge steel studs with self-drilling, self-tapping screws. The electric screw gun disengages automatically from the screw head when the screw has reached the proper depth. *(Courtesy of United States Gypsum Company.)*

FIGURE 23.11
Trim accessories for lath and plaster construction, as manufactured by the United States Gypsum Company. *(Courtesy of United States Gypsum Company.)*

USG Corner Beads, Trim, Control Joints

description

USG Corner Beads and Trim, made from top-quality galvanized steel, enjoy the industry's top acceptance because of their dependability and continual improvement in design. Corner beads are available in 8 and 10-ft. lengths, metal trim in 7 and 10-ft. lengths, casing beads in 7, 8 and 10-ft. lengths.

1-A Expanded Corner Bead has $2\frac{7}{8}''$ wide expanded flanges that are easily flexed. Preferred for irregular corners. Provides increased reinforcement close to nose of bead. Made from galvanized steel or zinc alloy for exterior applications.

X-2 Corner Bead has full $3\frac{1}{4}''$ flanges easily adjusted for plaster depth on columns. Ideal for finishing corners of structural tile and rough masonry. Has perforated stiffening ribs along expanded flange.

4-A Flexible Corner Bead is an economical general purpose bead. By snipping flanges, this bead may be bent to any curved design (for archways, telephone niches, etc.).

800 Corner Bead gives $\frac{1}{16}''$ grounds needed for one-coat veneer finishes. Approx. 90 keys per lin. in. provide superior bonding and strong, secure corners. The $1\frac{1}{4}''$ fine-mesh flange eliminates shadowing, is easily nailed or stapled.

900 Corner Bead is used with two-coat veneer systems, gives $\frac{3}{32}''$ grounds. Its $1\frac{1}{4}''$ fine-mesh flange can be either stapled or nailed. Provides superior plaster key and eliminates shadowing.

Cornerite and Striplath are strips of painted Diamond Mesh Lath used as reinforcement. **Cornerite** is bent lengthwise in the center to form a 100° angle. It should be used in all interior angles where metal lath is not lapped or carried around, over non-ferrous lath anchored to the lath, and over internal angles of masonry constructions to reduce plaster cracking. Sizes: $2'' \times 2'' \times 96''$ and $3'' \times 3'' \times 96''$. **Striplath** is a similar flat strip, used as a plaster reinforcement over joints of non-metallic lathing bases and where dissimilar bases join; also to span pipe chases. **Size:** $4'' \times 96''$.

USG Metal Trim comes in two styles and two grounds to provide neat edge protection for veneer finishing at cased openings and ceiling or wall intersections. All have fine-mesh expanded flanges to strengthen plaster bond and eliminate shadowing. **No. 701-A,** channel-type, and **No. 701-B,** angle edge trim, provide $\frac{3}{32}''$ grounds for two-coat systems. **No. 801-A,** channel-type, and **No. 801-B,** angle edge trim, provide $\frac{1}{16}''$ grounds for one-coat systems. Sizes: for $\frac{1}{2}''$ and $\frac{5}{8}''$ IMPERIAL Gypsum Base.

USG P-1 Vinyl Trim is a channel-shaped rigid trim with flexible vinyl fins which compress on installation to provide a positive acoustical seal comparable in performance to one bead of acoustical sealant. For veneer finish partition perimeters. **Lengths:** 8, 9 and 10 ft. Sizes: for $\frac{1}{2}''$ and $\frac{5}{8}''$ gypsum base.

USG P-2 Vinyl Trim is a channel-shaped vinyl trim with a pressure-sensitive adhesive backing for attachment to the wall at wall-ceiling intersections. Provides positive perimeter relief in radiant heat and veneer finish systems. Allow $\frac{1}{8}''$ and $\frac{1}{4}''$ clear space for insertion.

Length: 10 ft.

USG Control Joint relieves stresses of expansion and contraction in large plastered areas. Made from roll-formed zinc, it is resistant to corrosion in both interior and exterior uses with gypsum or portland cement plaster. An open slot, $\frac{1}{4}''$ wide and $\frac{1}{2}''$ deep, is protected with plastic tape which is removed after plastering is completed. The perforated short flanges are wire-tied to metal lath, screwed or stapled to gypsum lath. Thus the plaster is key-locked to the control joint, which not only provides plastering grounds but can also be used to create decorative panel designs. **Limitations:** Where sound and/or fire ratings are prime considerations, adequate protection must be

provided behind the control joint. USG Control Joints should not be used with magnesium oxychloride cement stuccos or stuccos containing calcium chloride additives. **Sizes and grounds: No. 50, $\frac{1}{2}''$; No. 75, $\frac{3}{4}''$; No. 100, 1″ (for exterior stucco curtain walls)—10-ft. lengths.**

1-A

X-2

4-A

800
900

Cornerite

701-A
801-A

701-B
801-B

USG P-1
Vinyl Trim

USG P-2
Vinyl Trim

USG Control
Joints #50,
75, & 100

USG Control
Joints #093

USG Expansion
Joints #15

#66 Square Edge

#66 Square Edge

After the scratch coat has hardened, it works together with the lath as a rigid base for the second application of plaster, which is called the *brown coat*. The purpose of the brown coat is to build strength and thickness and to present a level surface for the application of the third or finish coat. The level surface is produced by drawing a long straightedge across the surfaces of the grounds (the *edge beads, corner beads,* and control joints) to strike off the wet plaster. On large, uninterrupted plaster surfaces, *plaster screeds,* intermittent spots or strips of plaster, are leveled up to the grounds in advance of brown coat plastering to serve as intermediate reference points for setting the thickness of the plaster during the striking-off operation. Base-coat plasters are used for scratch coats and brown coats.

FIGURE 23.12
Sections through the three common lath-and-plaster systems, reproduced at full scale. Metal lath (*left*) requires three coats of plaster; the surface of the first coat is scratched for a better bond to the second coat. Gypsum lath (*middle*) may be finished with three coats or with two, as shown. Veneer plaster (*right*) usually consists of only a single thin finish coat, although two thin coats may be applied over rougher or more uneven substrates, such as concrete masonry or sitecast concrete.

FIGURE 23.13
Scratching the scratch coat while it is still soft to create a better bond to the brown coat. (*Courtesy of United States Gypsum Company.*)

The *finish coat* is a very thin application of finish-coat plaster, about ¹⁄₁₆ inch (1.5 mm) thick. It may be troweled smooth or worked into any desired texture (Figures 23.14 and 23.15). The total thickness of the plaster that results from this three-coat process, as measured from the face of the lath, is about ⅝ inch (16 mm). Three-coat work over metal lath is the premium-quality plaster system, extremely strong and resis-tant to fire. The only disadvantage of three-coat plaster work is its cost, which can be attributed largely to the labor involved in applying the lath and the three separate coats of plaster.

Plaster over Gypsum Lath

The best plaster work over gypsum lath is applied in three coats, but gypsum lath is sufficiently rigid that if it is firmly mounted to the studs, only a brown coat and a finish coat need be applied. The elimination of the scratch coat has obvious economic advantages. Even with three coats of plaster, gypsum lath is often less expensive than metal lath because the gypsum in the lath replaces much of the plaster that would otherwise have to be mixed and applied by hand in the scratch coat. The total thickness of plaster applied over gypsum lath is ½ inch (13 mm).

FIGURE 23.14
A sponge-faced float can be used to create various rough surface textures on plaster.
(Reprinted with permission of the Portland Cement Association, from Design and Control of Concrete Mixtures, *12th edition; PHOTO: Portland Cement Association, Skokie, IL.)*

(*a*) (*b*) (*c*)

FIGURE 23.15
Three different plaster surface textures from among many: (*a*) **Float finish.** (*b*) **Spray finish.** (*c*) **Texture finish.** *(Courtesy of United States Gypsum Company.)*

Plaster Direct-Applied to Masonry

Where plaster is applied directly over brick, concrete masonry, or poured concrete walls, the walls should be dampened thoroughly in advance of plastering to prevent premature dehydration of the plaster. A bonding agent may have to be applied to some smooth masonry surfaces to ensure good adhesion of the plaster. The number of coats of plaster required to cover a wall is determined by the degree of unevenness of the masonry surface. For the best work, three coats totaling ⅝ inch (16 mm) should be applied, but for many walls two coats will suffice (Figure 23.16).

Veneer Plaster

Veneer plaster is the least expensive of the gypsum plaster systems and is competitive in price with gypsum board finishes in many regions. The veneer base and accessories create a very flat surface that can be finished with a layer of a specially formulated dense gypsum plaster (manufactured to a separate standard, ASTM C587) that is applied in one or, occasionally, two coats usually no more than ⅙ to ⅛ inch (2–3 mm) in total thickness (Figures 23.17–23.20). A typical single-coat application is applied in a "double-back" process in which a thin coat is followed immediately by a second "skim" coat that is finish-troweled to the desired texture. The plaster veneer hardens and dries so rapidly that it may be painted the following day. A two-coat application of veneer plaster can also be directly applied to surface of sitecast concrete or concrete masonry walls.

Stucco

Stucco (portland cement-lime plaster) is applied over galvanized metal lath, using accessories of galvanized steel, or in wet areas or exterior applications, of solid zinc or plastic, which are less prone to corrosion. Whereas gypsum plaster expands during hardening, portland cement stucco shrinks and is prone to cracking as it dries. Stucco walls are therefore provided with control joints at regular intervals to channel the shrinkage into predetermined lines rather than allowing it to occur randomly. The curing reaction in stucco is the same as that of concrete and is very slow relative to that of gypsum plaster. Stucco must be kept moist for a week before it is allowed to dry in order to attain maximum hardness and

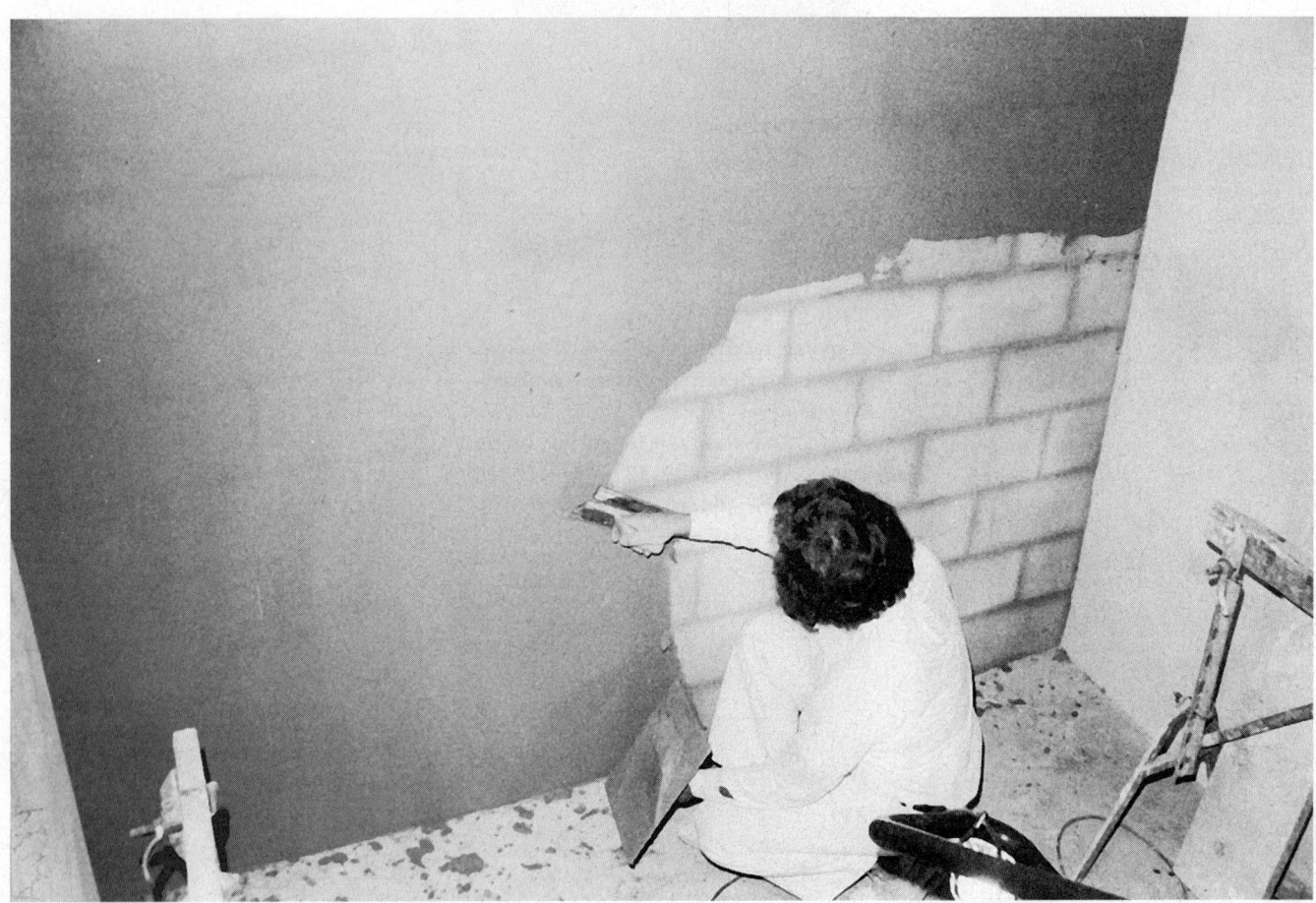

FIGURE 23.16

Applying a finish coat of portland cement plaster over a concrete masonry partition. The block joints are visible in the base coat of plaster because of a difference in the rate of water absorption between the blocks and the mortar joints (*Reprinted with permission of the Portland Cement Association, from* Design and Control of Concrete Mixtures, *12th edition; PHOTO: Portland Cement Association, Skokie, IL.*)

strength through full hydration of its portland cement binder.

In exterior applications over metal or wood studs, stucco may be applied over sheathing or without sheathing. Over sheathing, two layers of asphalt-saturated building felt or building paper are first applied as an air and moisture barrier. Then a *self-furring metal lath*, which is formed with "dimples" that hold the lath away from the surface of the wall a fraction of an inch to allow the stucco to key to the lath, is attached with nails or screws (Figure 23.9b). If no sheathing

FIGURE 23.17
Installing veneer plaster base with a screw gun. *(Courtesy of United States Gypsum Company.)*

FIGURE 23.18
Stapling a corner bead to veneer plaster base to create a straight, durable corner. *(Courtesy of United States Gypsum Company.)*

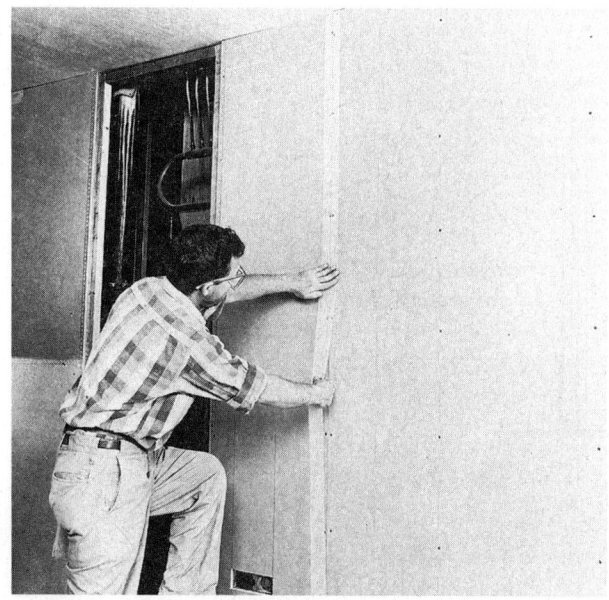

FIGURE 23.19
Reinforcing the panel joints of veneer plaster base with a self-adhesive glass fiber mesh tape. A panel opening for access to mechanical equipment is visible behind the installer. *(Courtesy of United States Gypsum Company.)*

FIGURE 23.20
Applying veneer plaster with a hawk and trowel. *(Courtesy of United States Gypsum Company.)*

TOP OF PARTITION

Clearance for structural movement

Acoustic sealant

Casing bead

Steel stud

Stud opening for wiring or other services

DOOR HEAD

Expanded metal lath and plaster

Gypsum lath and plaster

Veneer plaster base and veneer plaster

DOOR JAMB

For heavy or oversize doors, the hollow metal frame may be fully grouted

Studs are screw-attached to anchors welded into the jambs of the door frame at the factory

Lighter-duty frames may be spot grouted only at the anchors or left fully ungrouted

Door openings may be framed with a single or double stud, depending on stiffness requirements

BASE

Wood base screwed to flange of metal runner

Casing bead

Wood base, screwed to wood plate under metal track

Resilient base, adhered to plaster

Acoustic sealant

A   B   C

FIGURE 23.21

Three plaster partition systems. (*A*) Three coats of plaster on metal lath, rated at 1 hour of fire resistance and STC 39. (*B*) Two coats of plaster over gypsum lath, rated at 1 hour and STC 41. (*C*) Veneer plaster on light gauge steel studs, rated at 1 hour and STC 40. Notice especially the provisions for airtightness and structural movement at the top and bottom of each partition. Door frames can also be strengthened by grouting, as shown in (*A*) and (*B*). However, care must be taken to avoid introducing excessive moisture that could cause corrosion of the door frame.

is used, the wall is laced tightly with strands of *line wire* a few inches apart and paperbacked metal lath is attached to the line wire, after which stucco is applied to encase the building in a thin layer of what amounts to reinforced concrete.

In exterior applications, where stucco is exposed to wind-driven rain, a thin, tangled filament matting may be inserted directly behind the plaster and lath, to create a drainage space and lessen the risk of water penetration further into the wall assembly.

Stucco is usually applied in three coats over metal lath, with a total thickness of ⅞ inch (22 mm), or in two coats when applied directly to the surface of concrete or concrete masonry, with a thickness of ⅜ to ½ inch (10 mm). It is applied either with a hawk and trowel or by spraying (Figure 6.33). In exterior work, pigments or dyes may be added to give an integral color, and rough textures are frequently used.

Plaster Partition Systems

Types of plaster partitions are detailed in Figures 23.21 and 23.22. These diagrams show some of the ways in which the various trim accessories are used and the precautions taken to isolate the partitions from structural or thermal movements in the frame of the building. For larger, heavier doors, or for doors that are heavily used, the door opening is made stronger and stiffer by framing the jambs with double studs and/or using studs made from thicker sheet metal. As illustrated in Figure 23.21, door frames may also be grouted, either fully or only spot-grouted at the jamb anchors. Although this can strengthen the door-partition assembly, it adds cost and care must be taken to use proper grouting materials and methods to avoid trapping moisture in the door frame and causing corrosion.

Gypsum base and plaster

Mineral fiber batt acoustic insulation absorbs sound within the partition cavity

Resilient channels spaced at 24″ (610 mm) reduce sound transmission through the partition

SCALE

The lath is suspended from a special ceiling angle with a perforated vertical leg

After the scratch coat, plastering is done on both sides of the lath

Vertical 3/4″ (19 mm) steel channels strengthen the partition around opening, and are joined by a horizontal steel bar thin enough to fit within the layer of plaster

A special metal base anchors the lath to the floor

A

B

FIGURE 23.22

Two plaster partition systems. (*A*) This partition achieves an STC 54 rating by mounting the gypsum lath on one side with resilient metal clips and filling the hollow space in the partition with a sound attenuation blanket. (*B*) A traditional, solid plaster partition rated only STC 37. Because this assembly requires skilled plasterers to construct and is relatively costly, it is not likely to be seen in contemporary new construction. However, it illustrates the potential of traditional techniques to achieve an unusually thin partition assembly. Both of these partitions are rated at 1 hour.

PLASTER ORNAMENT

Gypsum plaster, with its fine grain and even texture, has more sculptural potential than any other material used in architecture. While wet, it is easily formed with trowels and spatulas, molds, or templates. When dry, it is readily worked by sawing, sanding, machining, and carving. Plaster building ornament has been created for many centuries by two economical but powerful techniques, *casting* and *running*, and continues to be used in buildings of every size and every historic style.

Cast plaster ornament is made by pouring soupy plaster into molds. The plaster hardens in a few minutes, allowing the mold to be stripped and reused (Figure A). Both rigid molds and soft rubber molds are used. The rubber molds are very flexible, so even undercut shapes can be cast without encountering difficulties in mold removal. Traditional rubber molds are created by first carving a plaster original, then brushing layers of latex over the original to build up the required wall thickness. More recently, two-component synthetic rubber compounds have replaced latex in most applications; their advantage is that they can be spread over the original in a single application rather than in layers.

Cast ornament is adhered to the brown coat of plaster in walls and ceilings with gobs of wet plaster or a mixture of plaster and glue. Once the ornament has been securely fastened in place, the finish coat of plaster is applied around it, and the plaster surfaces and adjacent pieces of ornament are merged by skillful trowel work and sanding to create a single-piece finish.

Running is used to make linear ornaments (*run plaster ornament*) such as classic cornice moldings (Figure B). A rigid *blade* made of sheet metal or sheet plastic is cut to the profile of the molding. The blade is attached to a sliding wooden frame to create a *template*. The template is pushed back and forth along a guide strip mounted temporarily on the wall or ceiling while a mixture of lime putty and gauging plaster is inserted in front of the blade, which strikes it off to the desired profile. Repeated passes of the template are required to finish the molding smoothly and perfectly. These passes must be completed before the plaster begins to harden, or the setting expansion of the gypsum will cause the template to bind and spoil the plaster surface. The template may also be attached to a radius guide to produce circular moldings (Figures C and D).

Casting and running are often used to reproduce plaster ornament during restoration of historic buildings. Rubber molds for casting can be made directly from existing ornaments, and templates for running duplicates of existing profiles are easily and cheaply produced.

New designs for plaster ornament are readily translated from the architect's paper sketches into carved plaster originals, from which rubber molds are made and duplicates cast. New profiles for moldings are quickly

FIGURE A
Removing the flexible rubber mold from a cast plaster ornament. *(Courtesy of Dovetail, Inc.)*

FIGURE B
Running a plaster cornice molding in place. *(Courtesy of Dovetail, Inc.)*

converted into template blades. The possibilities are almost limitless, yet few contemporary architects have chosen to explore them. This is surprising because plaster ornament is inexpensive compared to ornament carved from wood or stone, and there are few technical constraints on what can be accomplished.

Cast plaster ornament can be reinforced with short fibers of alkali-resistant glass. These greatly increase its strength and toughness and allow it to be produced in thinner sections and larger pieces than unreinforced plaster. This recent development has dramatically changed the economics and methods of ornamentation in plaster that is based on stock designs. A number of manufacturers issue catalogs of stock designs for ornaments made by this process. Much of the on-site assembly work for elaborate ornaments can be eliminated by combining what were formerly a number of small, thick, brittle castings into a single, larger, thinner casting that is light in weight and highly resistant to breakage. On the construction site, the lightweight castings are glued in place over gypsum board or veneer plaster base, using an ordinary mastic adhesive. The edges of the ornaments are feathered into the wall or ceiling surfaces with joint compound or veneer plaster, and the joints between pieces of ornament are smoothed over and sanded.

FIGURE C
Sculptor David Flaharty runs a circular plaster medallion on a bench in his shop. (© Brian McNeil.)

FIGURE D
A close-up of Flaharty's blade and template. The template rides on the sledlike runner of the portion under his hand, which is called the "slipper." The long portion of the template, called the "stock," is a radius guide that is fastened to a pin at the end to create the circular form of the medallion. (© Brian McNeil.)

FIGURE E
The medallion having been removed from the bench and glued in place on the ceiling, Flaharty adds cast components to complete the ornament. (© Brian McNeil.)

Gypsum Board

Gypsum board is a prefabricated plaster sheet material that is manufactured in widths of 4 feet (1220 mm) and lengths of 8 to 14 feet (2440–4270 mm). It is also known as *gypsum wallboard*, *plasterboard*, and *drywall*. (The term "sheetrock" is a registered trademark of one manufacturer of gypsum board and should not be used in a generic sense.)

Gypsum board is the least expensive of all interior finishing materials for walls and ceilings. For this reason alone, it has found wide acceptance throughout North America as a sub-stitute for plaster in buildings of every type. It retains the fire-resistive characteristics of gypsum plaster, but it is installed with less labor by less skilled workers than lathers and plasterers. And because it is installed largely in the form of dry materials, it eliminates some of the construction delay that may be associated with the curing and drying of plaster.

The core of gypsum board is formulated as a slurry of calcined gypsum, starch, water, pregenerated foam to reduce the density of the mixture, and various additives. This slurry is sandwiched between special paper faces and passed between sets of rollers to reduce it to the desired thickness. Within 2 or 3 minutes, the core material has hardened and bonded to the paper faces. The board is cut to length and heated to drive off residual moisture, then bundled for shipping (Figure 23.23).

Types of Gypsum Board

Gypsum board is produced in a variety of types suitable to a wide range of requirements:

- Regular gypsum board is used wherever the unique properties of other special board types are not required.

FIGURE 23.23
Sheets of gypsum board roll off the manufacturing line, trimmed and ready for packaging. *(Courtesy of United States Gypsum Company.)*

- For most fire-rated assemblies, *Type X gypsum board* is required. The core material of Type X board is reinforced with short glass fibers. When exposed to fire, the fibers hold the gypsum in place to continue to act as a barrier to fire rather than permitting it to erode or fall out.

- *Type C gypsum board* is a proprietary, fire-resistant alternative to Type X board. Often, a thinner board of Type C may be substituted for a thicker one of Type X.

- In locations exposed to moderate amounts of moisture, *water-resistant gypsum backing board*, with facings of water-repellant paper or glass matt and a moisture-resistant core formulation, is used.

- *Abuse-resistant* or *impact-resistant gypsum board* provides greater resistance to indentation and penetration and is installed where it is likely to be subjected to rough usage. It may be manufactured with heavier facing paper and backing sheets, have its core reinforced with cellulose fibers, be faced with glass fiber mesh, or be backed with polycarbonate film.

- The common paper facing on gypsum board can, when wetted, provide conditions suitable to the growth of mold spores. *Mold-resistant gypsum board* combines moisture-resistant cores with chemically treated paper or glass-matt facings that are not conducive to mold growth. Alternatively, panels may be manufactured with strengthened cores that eliminate the need for facing materials of any kind.

- *Coreboard* is a 1-inch (25.4-mm)-thick panel that is used for shaft walls and solid gypsum board partitions. To facilitate handling, it is fabricated in sheets 24 inches (610 mm) wide.

- Sag-resistant *ceiling gypsum board*, ½ inch (13 mm) thick, is substantially lighter than, but just as resistant to sagging when used in ceiling assemblies, as ⅝-inch (15-mm) panels.

- *Foil-backed gypsum board* can be used to eliminate the need for a separate vapor retarder in exterior wall assemblies. If the back of the board faces a dead airspace at least ¾ inch (19 mm) thick, the bright foil can also act as a thermal radiant barrier.

- *Predecorated gypsum board* is covered with paint, printed paper, or decorative plastic film during its manufacture. If handled carefully and installed with small nails, it requires no further finishing. This product is used in a number of demountable office partitioning systems.

- Weather-resistant *exterior gypsum soffit board* is used for exterior soffits, carport ceilings, undersides of exterior canopies, and other sheltered exterior applications.

- Gypsum lath and gypsum veneer base are used in conjunction with plastering systems, as previously discussed.

Most gypsum board product types are manufactured in conformance with ASTM standard C1396. Some related products, such as fibercement backing board or glass matt–faced gypsum board, are manufactured to separate standards.

Gypsum board is manufactured with a variety of edge profiles along its longer sides. The most common by far is the *tapered edge*, which permits sheets to be joined with a flush, invisible seam by means of subsequent joint finishing operations. Rounded and beveled edges are useful in predecorated panels, and tongue-and-groove edges serve to join coreboard panels in concealed locations. The edges of the shorter ends of gypsum board panels are not tapered or shaped, as the individual panels are cut from longer continuous sheets during manufacturing.

FIGURE 23.24

Attaching gypsum board to studs with a screw gun. (*Courtesy of United States Gypsum Company.*)

*Stud at 19%
moisture content*   *Stud at 10%
moisture content*

FIGURE 23.25
**When wood studs dry and shrink during
a building's first heating season, nail
heads may pop through the surface of
gypsum board walls.** *(Courtesy of United
States Gypsum Company.)*

Many different thicknesses of
board are produced:

• ¼-inch (6.4-mm) board is used as a
gypsum backing board in certain sound
control applications. Boards of this
thickness may also be used for tight-
radius bends (Figure 23.28).

• ⁵⁄₁₆-inch (8-mm) board is made for
manufactured housing, where weight
reduction to facilitate shipping is an
important consideration.

• ⅜-inch (9.5-mm) board is largely
used in double-layer wall finishes.
Though less durable than thicker

panels, it may also sometimes be applied as a single-layer finish over ceiling joists or wall studs spaced no more
than 16 inches (400 mm) apart.

• ½-inch (12.7-mm) board is the most
commonly used thickness. It may be
applied to studs and joists spaced up
to 24 inches (610 mm).

• ⅝-inch (16-mm) board is also
limited to stud spacings of not more
than 24 inches (610 mm). It is used
where additional fire resistance, stiffness, durability, or sound deadening is
required.

(a)

(b)

FIGURE 23.26
**Cutting gypsum board. (*a*) A sharp knife and metal T-square are used to score a straight line
through one paper face of the panel. (*b*) The scored board is easily "snapped," and the knife is
used a second time to slit the second paper face.** *(Courtesy of United States Gypsum Company.)*

- ¾-inch (19-mm) Type X board is produced by some manufacturers. It is used to create 2-hour partitions with a single layer of gypsum board on each face and 4-hour partitions that include two layers on each face.

- 1-inch (25.4-mm)-thick board is made as coreboard, for use in shaft wall construction (Figure 23.37).

Installing Gypsum Board

Hanging the Board

Gypsum board may be installed over either wood or light gauge steel studs, using self-drilling, self-tapping screws to fasten to steel and either screws or nails to fasten to wood (Figure 23.24). Wood studs can be troublesome with gypsum, as they usually shrink somewhat after the board is installed, causing nails to loosen slightly and "pop" through the finished surface of the board (Figure 23.25). *Nail popping* can be minimized by using fully dried framing lumber, ring-shank nails that have extra gripping power in the wood, and the shortest nail that will do the job. Screws have less of a tendency to pop than nails. When screws or nails are driven into gypsum board, their heads are driven to a level slightly below the surface of the board but not enough to tear the paper surface.

To minimize the length of joints that must be finished and to create the stiffest wall possible, gypsum board is usually installed with the long dimension of the boards horizontal. The longest possible boards are used to minimize end joints between boards, which are difficult to finish because the ends are not tapered. Gypsum board is cut rapidly and easily by scoring one paper face with a utility knife, snapping the brittle core along the score line with a blow from the heel of the hand, and cutting the other paper face along the fold created by the snapped core (Figure 23.26). A large metal T-square is used to make straight cuts that are perpendicular to the edges of the board. Notches, irregular cuts, and holes for electric boxes are made with a small hand saw or small electric router.

When two or more layers of gypsum board are installed on a surface, the joints between layers are staggered to create a stiffer wall, and a mastic adhesive is often used to join the layers to one another. Adhesive is also sometimes used between the studs and the gypsum board in single-layer installations to make a stronger joint.

Gypsum board can be curved when a design requires it. For gentle curves, the board can be bent into place dry (Figure 23.27). For somewhat sharper curves, the paper faces are moistened to decrease the stiffness of the board before it is installed.

FIGURE 23.27

Gypsum board can be curved to a large radius by simply bending it around a curving line of studs (*Courtesy of United States Gypsum Company.*)

When the paper dries, the board is as stiff as before. Special high-flex ¼-inch (6.5-mm) board is available that can be bent dry to even smaller radii (Figure 23.28).

Metal trim accessories are required at exposed edges and external corners to protect the brittle board and present a neat edge (Figure 23.29). These are similar to lathing accessories for plaster.

Finishing the Joints and Fastener Holes

Joints and holes in gypsum board are finished to create the appearance of a monolithic surface almost indistinguishable from plaster. The finishing process is based on the use of a *joint compound* that resembles a smooth, sticky plaster. For most purposes, a *drying-type joint compound* is used; this is a mixture of marble dust, binder, and admixtures, furnished either as a dry powder to be mixed with water or as a premixed paste. In some high-production commercial work, *setting compounds* that cure rapidly by chemical reaction are used to minimize the waiting time between applications. Joint compounds of different weights and strengths may be used for different stages of the joint finishing process, or a single all-purpose compound may be used for all steps.

The finishing of a joint between panels of gypsum board begins with the troweling of a layer of joint compound into the tapered edge joint and the bedding of a paper reinforcing tape in the compound (Figures 23.30–23.33). Compound is also troweled over the nail or screw holes. After drying (usually overnight), a second layer of compound is applied to the joint to bring it level with the face of the board and to fill the space left by the slight drying shrinkage of the joint compound. When

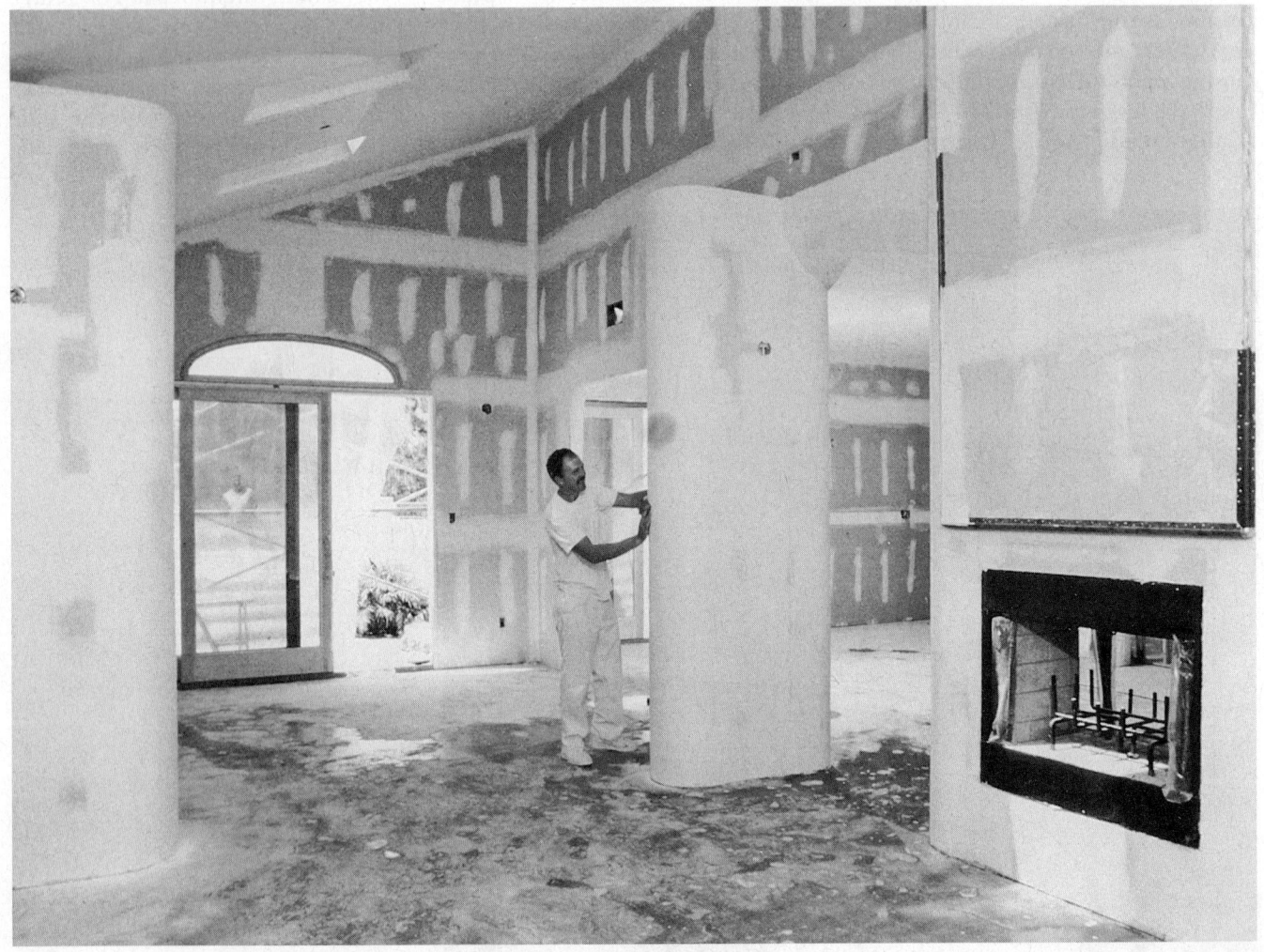

FIGURE 23.28
Tighter radii of curvature can be achieved by using ¼-inch (6.5-mm) high-flex gypsum board. *(Courtesy of United States Gypsum Company.)*

this second coat is dry, the joints are lightly sanded before a very thin final coat is applied to fill any remaining voids. The final coat is feathered out (tapered down to zero thickness) to create an invisible edge. Before painting, the wall is again sanded lightly to remove any roughness or ridges. If the finishing is done properly, the painted or papered wall will show no signs at all that it is made of discrete panels of material.

Gypsum board has a smooth surface finish, but where desired, spray-on textures or textured paints can be applied to give a rougher surface. In residential work, many gypsum board contractors prefer to texture ceilings; the texture conceals the minor irregularities in workmanship that are likely to occur because of the

FIGURE 23.29
Accessories for gypsum board construction, as manufactured by the United States Gypsum Company.
(Courtesy of United States Gypsum Company.)

DUR-A-BEAD
Reinforcement

#800 Corner Bead

USG Control
Joint #093

200-A
Metal Trim

200-B
Metal Trim

#400 Series
Metal Trim

P-1 Vinyl
Trim

P-2 Vinyl
Trim

801-A
Metal Trim

801-B
Metal Trim

USG Trim Accessories

difficulty of working in the overhead position.

Standardized gypsum board *finish levels* have been developed by the Gypsum Association and are also specified in ASTM C840. These enable the designer to specify quickly and simply the minimum level of finish that is acceptable for any project or portion of a project.

• Level 0, the minimum, consists of just the boards, without taping, finishing, or accessories. It is usually used only for temporary construction or where finishing is postponed until a later date.

• Level 1 requires only that joints be covered with tape set in joint compound. Its primary use is in areas of the building that are not open to view, such as above ceilings, in attics, and in service corridors. Level 1 is also the minimum finish level for fire-resistance-rated gypsum board assemblies, in which applications it may also be referred to as *fire-taping*.

• Level 2 adds to a Level 1 finish a coat of joint compound over the accessories and fasteners. After joint tape is set in compound, these joints are also immediately wiped with a joint knife a second time to add a thin coat of compound over the tape. A Level 2 finish is appropriate in garages, warehouses, and storage areas, and for boards used as a backer for ceramic tile.

• Level 3 adds a full second coat of compound over tape, accessories, and fasteners after the first coat has dried. It is intended for surfaces that will be textured or covered with heavy wallcoverings.

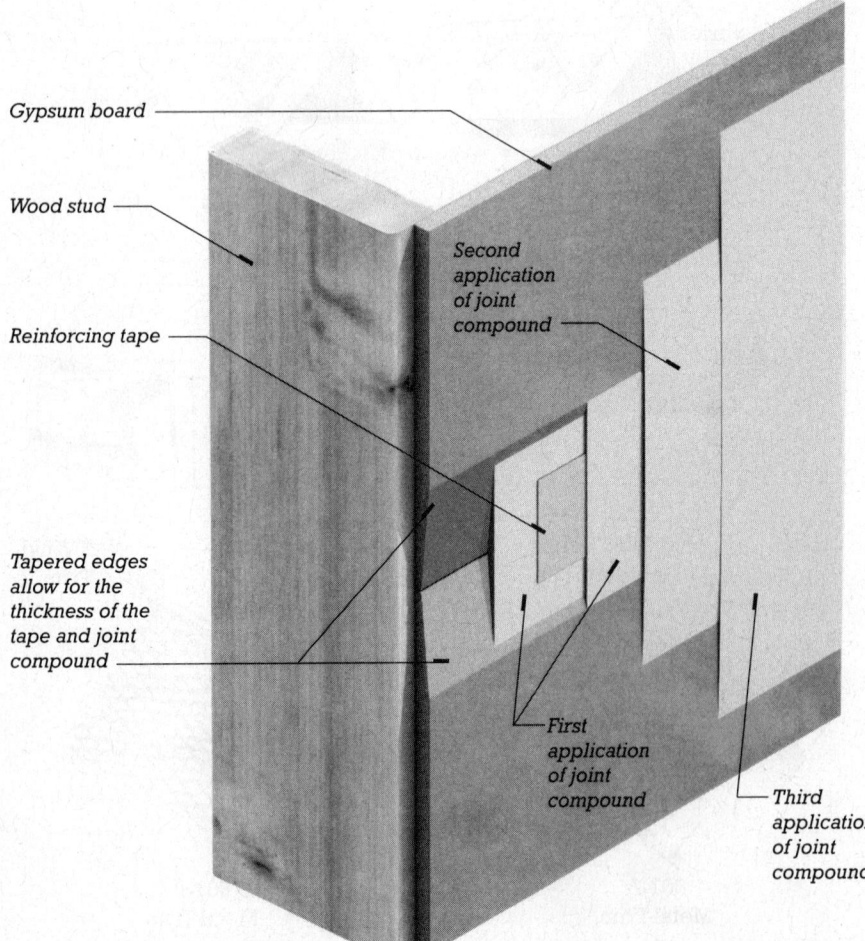

FIGURE 23.30

Finishing a joint between panels of gypsum board. *(Courtesy of United States Gypsum Company.)*

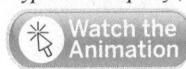

Gypsum board

Wood stud

Reinforcing tape

Tapered edges allow for the thickness of the tape and joint compound

Second application of joint compound

First application of joint compound

Third application of joint compound

FIGURE 23.31
Applying paper joint reinforcing tape to gypsum board. *(Courtesy of United States Gypsum Company.)*

FIGURE 23.32
An automatic taper simultaneously applies tape and joint compound to gypsum board joints. *(Courtesy of United States Gypsum Company.)*

• Level 4 is designed for surfaces to be finished with flat paints, light textures, or thin wallcoverings. It adds a third distinct coat of joint compound over taped seams, fasteners, and accessories.

• Level 5, the highest, adds a very thin *skim coat* of joint compound over the entire surface of the board. The skim coat has no measurable thickness, because its purpose is only to fill pores and low spots in the wall to produce a very smooth surface. It is recommended for surfaces that will receive gloss or semigloss paints and for surfaces that will be lit in such a way as to cast shadows that can highlight even slight imperfections.

Gypsum Board Partition Systems

Gypsum board partition assemblies have been designed and tested with fire resistance ratings of up to 4 hours and sound transmission performance of up to at least STC 69. Notice in these details how provisions are made to prevent the gypsum panels from being subjected to structural loadings caused by movements in the loadbearing frame of the building—structural deflections, concrete creep, moisture expansion, and temperature expansion and contraction. Notice also the use of sealant to eliminate sound transmission around the edges of the partitions and *sound attenuation blankets* to reduce the passage of sound through the partition itself (Figures 23.34 and 23.35).

Demountable partition systems of gypsum board use concealed mechanical fasteners that can be disassembled and reassembled easily without damage to the panels (Figure 23.36). These systems are used in buildings where partitions must be rearranged at frequent intervals.

FIGURE 23.33
Plasterers and drywall finishers often work on stilts to avoid having to erect and move scaffolding. The stilts, which are strapped to the legs of the workers, are sophisticated, stable devices that keep the worker's foot fully supported and parallel to the floor. *(Photo by Rob Thallon.)*

FIGURE 23.34
Four gypsum board partition systems. (A) A 1-hour partition, STC 40, using Type X gypsum board over light gauge steel studs. (B) This 1-hour partition on wood studs achieves an STC of 60 to 64 through heavy laminations of gypsum board, a sound attenuation blanket, and resilient channel mounting for one face of the partition. (C) A 2-hour partition with an STC of 48. (D) A 4-hour partition, STC 58. These are representative of a large number of gypsum board partition systems in a range of fire ratings and sound transmission classes.

A

B

SCALE

0   1″   3″   6″   12″

0   50 mm   100 mm   200 mm   300 mm

C

CEILING

DOOR HEAD

Acoustic sealant

Metal edge trim

Resilient channel

Sound attenuation
blanket

Wood studs

⅝″ (16 mm)
Type X gypsum
board

2 ⅝″ (67 mm)
steel studs

DOOR JAMB

The door frame is
spot grouted at
the anchors only,
at the time the
gypsum board is
installed

⅝″ (16 mm)
Type X gypsum
board

Conduit and
electrical wires

⅝″ (16 mm)
Type X gypsum
board

½″ (12.7 mm)
Type X gypsum
board

⅜″ (9.5 mm)
gypsum board

1 ⅝″ (41 mm)
steel studs and
sound attenuation
blanket

D

BASE

Vinyl base

Acoustic sealant

Wood base

FIGURE 23.35
Installing a dense, mineral fiber sound attenuation blanket. Conventional glass fiber insulation batts are also sometimes used for the same purpose. (*Courtesy of United States Gypsum Company.*)

FIGURE 23.36
Two photographs of relocatable (demountable) partition systems. (*Courtesy of United States Gypsum Company.*)

Gypsum Shaft Wall Systems

Walls around elevator shafts, stairways, and mechanical chases can be made of any masonry, lath and plaster, or gypsum panel assembly that meets fire resistance and structural requirements.

However, gypsum shaft wall systems are lighter in weight, installed dry, and built entirely from the floor outside the shaft, with no need to erect scaffolding inside. Depending on the requirements for fire resistance rating, air pressure resistance, acoustic performance, and floor-to-ceiling height, a variety of designs may be used. Figure 23.37 shows representative shaft wall details, and Figure 22.1 shows a shaft wall being installed.

FIGURE 23.37

Three gypsum shaft wall systems, all framed with steel C–H studs. The H portion of the stud holds a 1-inch (25-mm) shaft wall coreboard panel, while the C portion accepts the screws used to attach finish layers of gypsum board. (A) A 1-hour system. (B) A 2-hour system, STC 47. (C) A 3-hour system.

MASONRY PARTITION SYSTEMS

A century or more ago, interior partitions were often made of common brick masonry plastered on both sides. These had excellent acoustic properties and fire resistance ratings but were labor intensive and heavy. Partition systems of hollow clay tile and hollow gypsum tile (Figure 23.38) were developed to meet these objections and continued to be used extensively until the 1950s. Both have now become obsolete in North America, replaced by plaster, gypsum board, and concrete masonry, although they are still frequently encountered in the restoration of older buildings.

Concrete masonry partitions may be plastered or faced with gypsum board but are more often left exposed, either painted or unpainted. Several types of lightweight aggregate may be used to reduce the dead weight of the masonry units. Decorative concrete masonry units, as described in Chapter 9, may also be used. Electrical wiring is relatively difficult to conceal in concrete masonry partitions; the electrician and the mason must coordinate their work, or the wiring must be mounted on the surface of the wall after the mason has finished.

Glazed structural clay tiles make very durable partitions, especially in areas with heavy wear, moisture problems, or strict sanitation requirements (Figure 23.39). The ceramic glazes are nonfading and virtually indestructible.

WALL AND PARTITION FACINGS

The vast majority of gypsum board partitions are finished with several coats of paint. For more information about paints and coatings, see the Chapter 6 sidebar, "Painting."

Ceramic tile facings are often added to walls for reasons of appearance, durability, sanitation, or moisture resistance. In a *thickset* or *mortar bed* application, tile is applied to a base of portland cement mortar (Figure 23.40).

Lower-cost tile wall facings eliminate the mortar base and are *thin-set* onto *tile backing boards*, also called *backer boards*, most frequently made of fiber-reinforced lightweight cement or glass-matt-faced water-resistant gypsum board, similar to the floor tile assembly illustrated in Figure 24.28. Cement backer board is the more water resistant, but is more difficult to cut and handle than lighter-weight gypsum backer board.

Tiles are bonded to the backer board with a variety of compounds, the most common of which are dryset mortar, latex/polymer modified

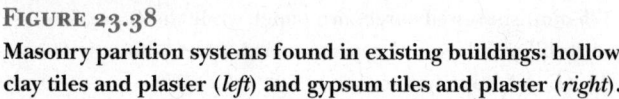

FIGURE 23.38
Masonry partition systems found in existing buildings: hollow clay tiles and plaster (*left*) and gypsum tiles and plaster (*right*).

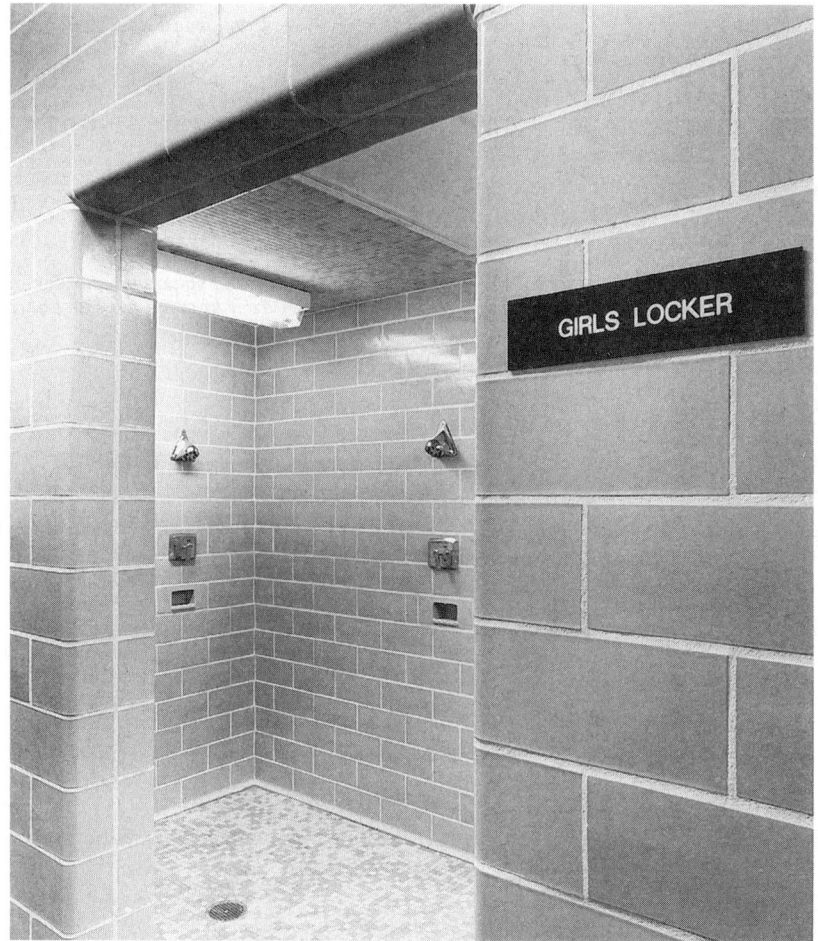

FIGURE 23.39
A glazed structural clay tile partition installation. The floor is finished with glazed ceramic tiles. *(Courtesy of Stark Ceramics, Inc.)*

Concrete block partition

Bullnose tile cap

Ceramic tiles

Grout between tiles

Bond coat of portland cement paste

Mortar bed

Metal lath reinforcing

The mortar bed is reinforced, and a cleavage membrane of roofing felt provided beneath, over wood subflooring

Cove tiles

Mortar bed

FIGURE 23.40
Example details of a thickset ceramic tile installation. Thickset applications are used where the face of the partition or the surface of the floor is cracked, coated, rough, unstable, or so uneven as to make it unsuitable for direct bonding of tile using thin-set methods. Depending on the quality of the substrate, metal reinforcing of the portland cement mortar base may or may not be required, and the mortar base may or may not be isolated from the substrate with a layer of felt paper acting as a slip sheet. Thickset mortar beds for wall tiling are typically ¾ to 1 inch (19–25 mm) thick, while those for floors are typically 1¼ to 2 inches (32–50 mm) thick.

Light gauge steel stud

TYPICAL WALL ANCHOR

Two layers of gypsum board

Plaster spots shim the slab away from the gypsum board

#8 (3.25 mm) copper or stainless wire

Gypsum plaster is used to embed one end of the wire in a hole drilled in the edge of the slab, and to embed the other end in a hole drilled in the gypsum board

FIGURE 23.41
Attaching a stone facing over a backup of gypsum board and steel studs.

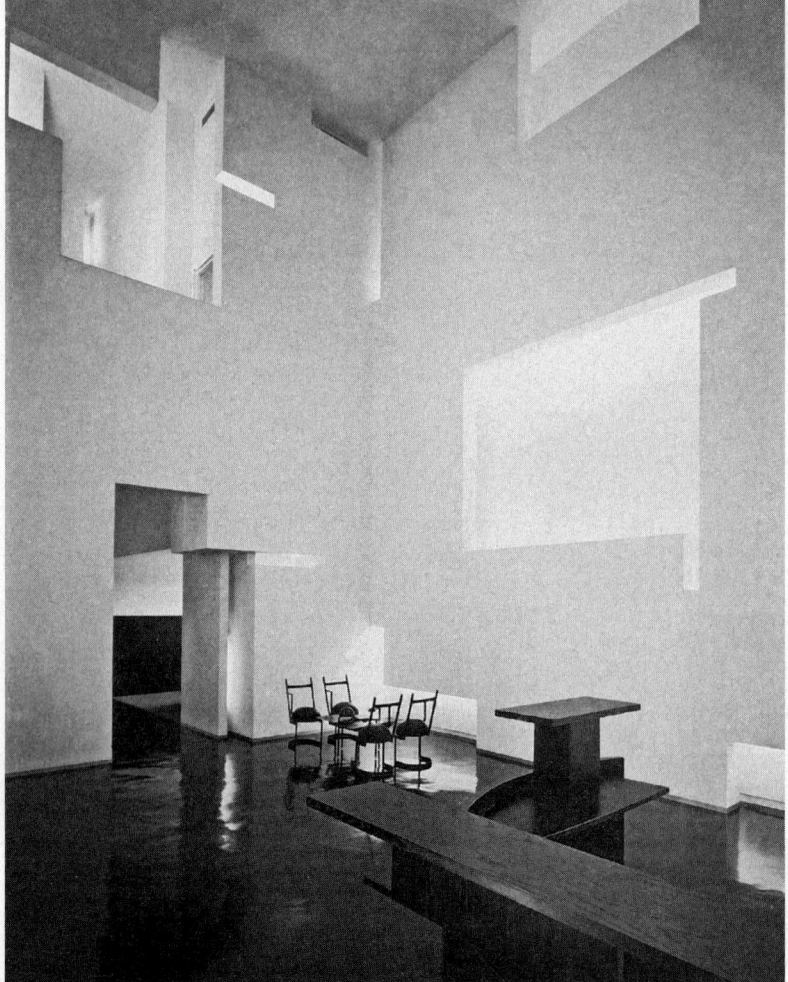

FIGURE 23.42
Imaginative use of gypsum products is coordinated with concealed sources of light to create this dramatic office space.
(Stephen Holl Architects. Photo by Paul Warchol.)

portland cement mortar, and organic adhesive. *Dry-set mortar* is a mix of cement, fine sand, and water retention compounds that allow the thin mortar layer to cure properly. *Latex/ polymer modified portland cement mortar* is similar to dry-set mortar, but with additives that improve the cured mortar's freeze-thaw resistance, flexibility, and adhesion. *Organic adhesives* are various proprietary synthetic polymer adhesives used for light-duty applications. After the tiles have become fully adhered, a cementitious *grout* of any desired color is wiped into the tile joints with a rubber-faced trowel. Thin-set compounds and grouts formulated with epoxies or furan resins (colorless, highly volatile solvents distilled from wood) may be used for tiling applications where greater strength, impact resistance, or chemical resistance is required.

In showers, steam rooms, and other wet locations, a *tile waterproofing membrane* is added to the assembly to prevent water from seeping through the tile and into the wall behind. Either liquid-applied or flexible sheet membranes may be used, usually positioned beneath the mortar bed in a thickset application or over the backer board when tile is thin-set.

Facings of granite, limestone, marble, or slate are sometimes used in public areas of major buildings. One method of mounting is shown in Figure 23.41.

Wood wainscoting and paneling is applied over gypsum wallboard, to gain the acoustic and fire-resistance benefits of the gypsum panels. In noncombustible buildings, wood finishes such as these are subject to the fire and smoke criteria discussed in Chapter 22.

CSI/CSC MasterFormat Sections for Interior Walls and Partitions

04 21 00	**CLAY UNIT MASONRY**
04 21 26	Glazed Structural Clay Tile Masonry
04 22 00	**CONCRETE UNIT MASONRY**
04 22 23	Architectural Concrete Unit Masonry
06 42 00	**WOOD PANELING**
09 21 00	**PLASTER AND GYPSUM BOARD ASSEMBLIES**
09 21 16.23	Gypsum Board Shaft Wall Assemblies
09 22 00	**SUPPORTS FOR PLASTER AND GYPSUM BOARD**
09 22 13	Metal Furring
09 22 16	Non-Structural Metal Framing
09 22 36.13	Gypsum Lath
09 22 36.33	Metal Lath
09 22 39	Veneer Plaster Base
09 23 00	**GYPSUM PLASTERING**
09 24 00	**CEMENT PLASTERING**
09 26 00	**VENEER PLASTERING**
09 28 00	**BACKING BOARDS AND UNDERLAYMENTS**
09 28 13	Cementitious Backing Boards
09 28 16	Glass-Mat Faced Gypsum Backing Boards
09 29 00	**GYPSUM BOARD**
09 30 00	**TILING**
09 31 00	**THIN-SET TILING**
09 32 00	**MORTAR-BED TILING**
09 75 00	**STONE FACING**
09 81 00	**ACOUSTIC INSULATION**
09 91 00	**PAINTING**
09 91 23	Interior Painting
09 94 00	**DECORATIVE FINISHING**
09 94 13	Textured Finishing
10 22 00	**PARTITIONS**
10 22 19	Demountable Partitions

KEY TERMS

fire wall
shaft wall
fire barrier
fire area
fire partition
smoke barrier
smoke partition
light gauge steel stud
light gauge steel track
deflection track
furring strip
Z-furring channel
hat channel
plaster
lath
wattle and daub
gypsum
calcining
calcined gypsum
plaster of Paris
synthetic gypsum
base-coat plaster
finish-coat plaster
mill-mixed base-coat plaster, ready-mixed
  base-coat plaster
neat base-coat plaster
gypsum plaster
wood-fibered gypsum plaster
lightweight gypsum plaster
high-strength base-coat plaster
lime
ready-mixed finish plaster
gauging plaster
hydrated finishing lime, lime putty

high-strength gauging plaster
Keenes cement
molding plaster
portland cement–lime plaster, stucco
hawk
trowel
darby
keying
lather
expanded metal lath
self-drilling, self-tapping screw
lathing nail
gypsum lath
veneer plaster base, gypsum veneer base
lathing trim accessories
control joint
ground
scratch coat
brown coat
edge bead
corner bead
plaster screed
finish coat
veneer plaster
self-furring metal lath
line wire
casting
running
cast plaster ornament
run plaster ornament
blade
template
gypsum board, gypsum wallboard,
  plasterboard, drywall

Type X gypsum board
Type C gypsum board
water-resistant gypsum backing board
abuse-resistant gypsum board, impact-
  resistant gypsum board
mold-resistant gypsum board
coreboard
ceiling gypsum board
foil-backed gypsum board
predecorated gypsum board
exterior gypsum soffit board
tapered edge
gypsum backing board
nail popping
joint compound
drying-type joint compound
setting compound
finish level
fire-taping
skim coat
sound attenuation blanket
demountable partition system
glazed structural clay tile
ceramic tile
thickset tile, mortar bed tile
thin-set tile
tile backing board, backer board
dry-set mortar
latex/polymer modified portland
  cement mortar
organic adhesive
grout
tile waterproofing membrane

REVIEW QUESTIONS

1. What are the major types of interior walls and partitions in a larger building, such as a hospital, classroom building, apartment building, or office building? How do these types differ from one another?

2. Why is gypsum used so much in interior finishes?

3. Name the coats of plaster used over expanded metal lath and explain the role of each.

4. Under what circumstances would you specify the use of portland cement plaster? Keenes cement plaster? Veneer plaster?

5. Describe step by step how the joints between sheets of gypsum board are made invisible.

EXERCISES

1. Determine the construction of a number of partitions in the places where you live and work. What materials are used? What accessories? Why were these chosen for their particular situations? Sketch a detail of each partition.

2. Sketch typical details showing how the various metal lath trim accessories are used in a plaster wall.

3. Repeat Exercise 2 for gypsum board trim accessories.

4. What type of gypsum wall finish system would you specify for a major art museum? For a low-cost rental office building? Outline a complete specification of wall and partition construction for a building you are presently working on.

SELECTED REFERENCES

Gypsum Association. *GA-600, Fire Resistance Design Manual*. Washington, DC, Author, updated regularly.

Fire resistance ratings and STCs are given in this booklet for a large number of wall and ceiling assemblies that use either gypsum plaster or gypsum board. This and many other useful resources are also available for free download from the Gypsum Association's web site.

Marble Institute of America. *Dimension Stone Design Manual*. Cleveland, OH, Author, updated regularly.

This binder provides comprehensive guidance on the design, specification, and installation of all types of stone wall facing.

Pegg, Brian, and W. D. Stagg. *Plastering: An Encyclopedia* (4th ed.). Oxford, Blackwell Publishing/New York, Crown Publishers, 2007.

Techniques of ornamental plastering are covered in complete detail in this 276-page reference work.

Portland Cement Association. *Portland Cement Plaster (Stucco) Manual*. Skokie, IL, Author, updated regularly.

A complete, illustrated guide to stucco.

Tile Council of North America, Inc. *TCNA Handbook for Ceramic, Glass, and Stone Tile Installation*. Anderson, SC, Author, updated regularly.

This is the definitive standard for ceramic tile installation materials and methods. More than 100 methods of installation for floors and walls, both interior and exterior, are illustrated and specified. Guidelines for selecting appropriate installation methods based on project requirements are also included.

USG Company. *Gypsum Construction Handbook*. Chicago, Author, revised regularly.

This manual represents manufacturers' literature at its best—close to 600 well-illustrated pages crammed with every important fact about gypsum wallboard, gypsum plaster, and associated products. Available for purchase or as a free download from the USG web site.

USG Company. *SA100—Fire-Resistant Assemblies*. Chicago, Author, revised regularly.

Though limited to this company's products, this brochure provides an extensive and easy-to-use listing of fire-rated and acoustically rated wall and ceiling assemblies. Along with many other useful technical documents, it is available for free download from the USG web site.

WEB SITES

Interior Walls and Partitions

Author's supplementary web site: **www.ianosbackfill.com/23_interior_walls_and_partitions**

Dietrich Metal Framing: **www.dietrichindustries.com**

Drywallrecycling.org: **www.drywallrecycling.org**

Georgia-Pacific: **www.gp.com/build/**

Gypsum Association: **www.gypsum.org**

Marble Institute of America: **www.marble-institute.com**

National Gypsum: **www.nationalgypsum.com**

Portland Cement Association: **www.cement.org**

Tile Council of North America (TCNA): **www.tcnatile.com**

USG: **www.usg.com**

24

FINISH CEILINGS AND FLOORS

- **Finish Ceilings**
  Functions of Finish Ceilings

- **Types of Ceilings**
  Exposed Structural and Mechanical
      Components
  Tightly Attached Ceilings
  Suspended Ceilings
  Interstitial Ceilings

  SUSTAINABILITY OF FINISH CEILINGS AND
  FLOORS

- **Finish Flooring**
  Functions of Finish Flooring
  Underfloor Services

- **Types of Finish Flooring
  Materials**
  Hard Flooring Materials
  Wood and Bamboo
  Resilient Flooring
  Carpet
  Fluid-Applied Flooring

- **Flooring Thickness**

A worker lays acoustical panels in a recessed grid to create a suspended ceiling. *(Photo courtesy of United States Gypsum Company.)*

As the ceilings and finish floors are being installed, the construction of a building is drawing to a close. The components of the mechanical and electrical systems that remain exposed are either finished or concealed, intersections of interior surfaces are neatly trimmed, and painters work their magic to reveal for the first time the interior character of the building. The architect, engineers, and municipal building officials make their last inspections, and, following last-minute corrections of minor defects, the contractor turns the building over to its owner.

FINISH CEILINGS

Functions of Finish Ceilings

The ceiling surface is an important functional component of a room. It helps control the diffusion of light and sound about the room. It may play a role in preventing the passage of sound vertically between the rooms above and below, and horizontally between rooms on either side of a partition. It is often designed to resist the passage of fire and must itself be appropriately noncombustible. Frequently, it is called upon to assist in the distribution of conditioned air, artificial light, and electrical energy. In many buildings, it must accommodate sprinkler heads for fire suppression and loudspeakers for intercommunication systems. And its color, texture, pattern, and shape are prominent in the overall visual impression of the room. A ceiling can be a simple, level plane, a series of sloping planes that give a sense of the roof above, a luminous surface, a richly coffered ornamental ceiling, or even a frescoed plaster vault such as Michelangelo's famous ceiling in the Sistine Chapel in Rome; the possibilities are endless.

TYPES OF CEILINGS

Exposed Structural and Mechanical Components

In many buildings, it makes sense to omit finished ceiling surfaces altogether and simply expose the structural and mechanical components of the floor or roof above (Figures 24.1, 4.20). In industrial and agricultural buildings, where appearance is not of prime importance, this approach offers the advantages of economy and ease of access for maintenance. Many types of floor and roof structures are inherently attractive if left exposed, such as heavy timber beams and decking, concrete waffle slabs, and steel trusses. Other types of structures, such as concrete flat plates and precast concrete planks, have little visual interest, but their undersurfaces can be painted and left exposed as finished ceilings in apartment buildings and hotels, which have little need for mechanical services at the ceiling. This saves money and reduces the overall height of the building. In some buildings, the structural and mechanical elements at the ceiling, if carefully designed, installed, and painted, can create a powerful aesthetic of their own.

Exposing structural and mechanical components rather than covering them with a finished ceiling does not always save money. Mechanical and structural work is not normally done in a precise, attractive fashion because it is not usually expected to be visible, and it is less expensive for workers to take only as much care in installation as is required for satisfactory functional performance. To achieve perfectly straight, neatly sealed ductwork

that is free of dents, steel decks without rust and weld spatter, and square, well-organized runs of electrical conduit and plumbing, the drawings and specifications for the project must tell exactly the results that are expected, and a higher labor cost must be anticipated.

Tightly Attached Ceilings

Ceilings of almost any material can be attached directly to the underside of the structure above. Commonplace examples include gypsum board or plaster mounted directly to the underside of wood or steel joists and rafters, acoustical tile adhered to the underside of a concrete deck, or textured, acoustic materials spray-applied to the underside of concrete slabs or corrugated metal decking (Figure 24.2). When finish ceilings are in direct contact with the structure above, special finishing arrangements must be worked out for any beams and girders that protrude

through the plane of the ceiling, and for ducts, conduits, pipes, and sprinkler heads if these elements are located below the ceiling.

Suspended Ceilings

A ceiling that is suspended on wires some distance below the floor or roof structure can hang level and flat despite varying sizes of girders, beams, joists, and slabs above, and even under a roof structure that slopes toward roof drains. Ducts, pipes, and conduits can run freely in the *plenum* space between the ceiling and the structure above. Lighting fixtures, sprinkler heads, loudspeakers, and fire detection devices may be recessed into the ceiling. Such a ceiling can also—at additional cost—serve as a *membrane ceiling* or *membrane fire protection* for the floor or roof structure above, eliminating the need for fussy individual fireproofing of steel joists, or imparting a higher fire resistance rating to wood or

FIGURE 24.1
Air conditioning ductwork, electrical conduits, fluorescent lighting fixtures, skylights, curtain gymnasium area dividers, and athletic equipment are exposed in a ceiling structure of painted open-web steel joists and corrugated steel decking. *(Weinstein A|U, Architects. Photo by Joseph Iano.)*

FIGURE 24.2
Spraying a textured finish onto the underside of a concrete slab in a residential building, where there are no pipes, ducts, or wires to be concealed below the plane of the floor structure. *(Courtesy of United States Gypsum Company.)*

precast concrete structures. For these reasons, *suspended ceilings* are a popular and economical feature in many types of buildings, especially office and retail structures.

Suspended ceilings can be made of almost any material; the ones most widely used are gypsum board, plaster, and various proprietary panels and tiles composed of incombustible fibers or other material. Each of these is supported on its own system of small steel framing members that hang from the structure on heavy steel wires.

Suspended Gypsum Board and Plaster Ceilings

Suspended plaster ceilings, assembled from steel *hanger wire* and cold-rolled channels, have been in use for many decades; typical details are shown in Figure 24.3. Although most suspended plaster ceilings are flat, lathers are capable of constructing ceilings that are richly sculpted, ranging from configurations resembling highly ornamented Greek or Roman coffered ceilings to nearly any form that the contemporary designer can draw. This capability is especially useful in auditoriums, theaters, lobbies of public buildings, and other uniquely shaped rooms.

Gypsum board suspended ceilings may be screwed to the same type of suspension system as shown in Figure 24.3 for plaster, or to light gauge steel framing made up of stud or joist members. Special components have been developed that make it easy to suspend more complex shapes, such as cylindrical vaults, undulating surfaces, and deep coffers.

Perforated gypsum panels, manufactured with attractively arranged patterns of small holes, are used in gypsum board ceilings where enhanced acoustic performance is required. Panels are mounted to special suspension systems that eliminate the need for normal taping and finishing. With the addition of a layer of acoustic insulation over the top of the panels, acoustic ratings comparable

to those of suspended acoustical ceilings, discussed below, are possible.

Suspended Acoustical Ceilings

Ceilings made from fibrous materials in the form of lightweight tiles or panels are customarily referred to as *acoustical ceilings* because most of them are absorptive of sound energy, unlike plaster and gypsum board, which are highly reflective of sound. They are also often less costly than either plaster or gypsum board ceilings. The sound absorption performance of a ceiling material is measured as its *Noise Reduction Coefficient (NRC)*. NRC is a number between 0 and 1, with higher values indicating better sound absorption at four specific frequencies ranging from 250 Hz to 2000 Hz. An NRC of 0.85 indicates that the material absorbs 85 percent of the sound that reaches it and reflects only 15 percent back into the room. NRCs for most acoustical ceiling materials range from 0.50 to 0.90, compared to values below 0.10 for plaster and gypsum board ceilings. This makes acoustical ceilings valuable for reducing noise levels in lobbies, office spaces, restaurants, retail stores, recreational spaces, and noisy industrial environments.

The lightweight, porous materials that produce high NRC ratings allow most sound energy to pass through. In other words, a ceiling made of porous materials will not furnish very good acoustic privacy between adjacent rooms unless a suitable full-height wall separates the rooms and blocks the ceiling plenum. The ability of a ceiling system to reduce sound transmission from one room to another through a shared plenum is measured by its *Ceiling Attenuation Class (CAC)*. CAC is measured in decibels, with higher values representing greater reductions in sound transmission. For closed offices with shared ceiling plenums, a ceiling system with a CAC of not less than 35 to 40 is recommended. Dense, nonporous ceiling materials tend to have higher CACs than lighter, more porous materials.

FIGURE 24.3

A suspended metal lath-and-plaster ceiling. At the top of the page is a cutaway isometric drawing, as viewed from below, of the essential ceiling components. Across the center of the page are details of six ways of supporting the hanger wires: (A) A pin is powder-driven into a concrete structure. (B) A corrugated sheet metal tab with a hole punched in it is nailed to the formwork before the concrete is poured. When the formwork is stripped, the tab bends down and the hanger wire is threaded through the hole. (C) A sharp, daggerlike tab of sheet metal is driven through the corrugated metal decking before the concrete topping is poured. (D) A sheet metal hook is hung onto the lap joints in the metal decking. (E) The hanger wire is wrapped around the lower chord of an open-web steel joist. (F) The hanger wire is passed through a hole drilled near the bottom of a wood joist. At the bottom of the page is a section through a furred, insulated plaster wall and a suspended plaster ceiling.

Steel hanger wires support 1½″ (38 mm) steel runner channels 36″ to 54″ (915 mm to 1370 mm) apart

¾″ (19 mm) steel furring channels are wired to the runner channels for support and are spaced 16″ to 24″ (400 mm to 600 mm) apart, depending on the type of lath used

A casing bead terminates the plaster neatly at the walls

Three coats of plaster are applied to the metal lath

Metal lath is wired to the furring channels

A   B   C   D   E   F

Adjustable wall furring bracket

Sealant or a resilient gasket allows for expansion and contraction

Acoustical tile or panel

Wall angle

EXPOSED GRID

RECESSED GRID

CONCEALED GRID

FIGURE 24.4

Acoustical ceilings are supported on suspended grids of tees formed from sheet metal. At the top of the figure is a cutaway view looking up at an acoustical ceiling of lay-in panels. Below are sections illustrating how the grid may be exposed, recessed, or concealed for different appearances.

Hanger wire

Main runner

Cross tee

FIGURE 24.5

The grid for an acoustical ceiling is assembled with a simple interlocking joint.

A third measure of ceiling acoustical performance is *Articulation Class* (*AC*). Like NRC, AC is a measure of sound reflection and absorption. However, AC is intended specifically to measure a ceiling system's contribution to speech clarity and privacy in a typical open office environment. It measures a ceiling's absorption and reflection of sound over a 60-inch (1500-mm)-high partition at frequencies ranging from 500 to 4000 Hz, those particularly critical to normal speech. Higher AC values represent greater acoustical clarity and privacy, with minimum recommended values falling in the range of 170 to 200.

Where both noise reduction within a space and sound attenuation between spaces are required simultaneously, composite ceiling panels with an absorbent material laminated to a dense substrate may be used; these have high values for both noise reduction (NRC or AC) and sound attenuation (CAC). The same result can be achieved by mounting acoustically absorbent tiles on a suspended ceiling of plaster or gypsum board.

The most economical acoustical ceiling systems consist of *lay-in panels* that are supported by an *exposed grid* (Figures 24.4 and 24.5). Any panel in the ceiling can be lifted and removed for access to services in the plenum space. For a neater appearance, a *concealed grid* system may be used. Concealed grid systems require special panels for plenum access. For curved ceilings, curvilinear grid systems and flexible tiles are used to create ceilings that warp along one or both axes of the ceiling plane. Suspended acoustical ceilings are available in hundreds of different designs. Figures 24.6–24.10 show some typical examples.

FIGURE 24.6
Many acoustical ceilings are manufactured as *integrated ceiling systems* that incorporate the lighting fixtures and air conditioning outlets into the module of the grid. In this integrated ceiling, viewed from above, the hanger wires, grid, acoustical panels, fluorescent lighting fixture, and distribution boot for conditioned air have been installed. The boot will be connected to the main ductwork with a flexible oval duct.
(Photo courtesy Armstrong World Industries.)

FIGURE 24.7
Air distribution in this integrated acoustical ceiling is through slots that occur between panels. Lighting fixtures, loudspeakers, and smoke detectors are incorporated simply and unobtrusively.
(Courtesy of United States Gypsum Company.)

FIGURE 24.8
Locations of HVAC diffusers and suspended lighting fixtures are carefully coordinated with the grid layout of this suspended acoustical tile ceiling. Fire sprinklers, a smoke detector, and a fire annunciator are mounted flush to the underside of the ceiling. *(Photo by Joseph Iano.)*

> **Economy has worked so great a change in our dwellings, that their ceilings are, of late years, little more than miserable naked surfaces of plaster. [A discussion of ceiling design will] possess little interest in the eye of speculating builders of the wretched houses erected about the suburbs of the metropolis, and let to unsuspecting tenants at rents usually about three times their actual value. To the student it is more important, inasmuch as a well-designed ceiling is one of the most pleasing features of a room.**
>
> **—Joseph Gwilt, *The Encyclopedia of Architecture*, London, 1842**

Other Suspended Ceiling Materials

Almost any material can be incorporated into a suspended ceiling system. *Metal panel ceilings* are made from sheet metal suspended in a manner similar to that for acoustic panels (Figure 24.11). *Linear metal ceilings* are made from longer metal strips suspended on a concealed grid system (Figure 24.12). The metal sheets may be solid or perforated, and formed flat, stamped with intricate patterns, or curved, depending on aesthetic and acoustic requirements. Due to its ease of formability and light weight, aluminum is the metal most commonly used.

Suspended wood ceilings, made from solid wood or wood veneered over a variety of substrates, are produced in both panel and linear forms, similar to those of metal (Figure 24.13). Textile ceilings may be made from *fabric-wrapped ceiling panels,* consisting of natural or synthetic fabrics mounted over acoustic panels. Larger, seamless textile ceilings are constructed as *stretched-fabric ceiling systems* in which a synthetic fabric is stretched across a special mounting system with acoustic insulation behind the fabric. Wherever flammable materials are used in ceilings, these materials must meet the fire criteria requirements discussed in Chapter 22.

With the advent of three-dimensional modeling and computer-driven fabrication techniques, the ability to produce economical yet unique ceiling panel forms continues to grow, further expanding the range of possibilities in suspended ceiling types and designs.

Suspended Fire-Resistance Rated Ceilings

Membrane ceilings, designed to contribute to the fire resistance of the roof or floor structure above, may be made of gypsum board, plaster,

FIGURES 24.9 AND 24.10
A variety of patterns and textures are available in acoustical ceilings. *(Photos courtesy Armstrong World Industries.)*

or lay-in panel and grid systems. These ceilings must be carefully detailed and installed to maintain the required fire resistance throughout the extents of the ceiling: Lighting fixtures must be backed up with fire-resistive material, air conditioning grills must be isolated from the ducts that feed them by automatic fire dampers, access panels provided for maintenance of above-ceiling services must themselves meet the same fire resistance rating as the ceiling system, and any other penetrations must be appropriately treated.

FIGURE 24.11
Trapezoidal panels in a curved, stepped, metal panel ceiling. Air distribution, recessed lighting, and sprinklers are neatly integrated into the ceiling system. *(Photo by Joseph Iano.)*

Interstitial Ceilings

Many hospital and laboratory buildings have extremely elaborate mechanical and electrical systems, including not just the usual air conditioning ducts, water and waste piping, and electrical and communications wiring, but also such services as fume hood ducting, fuel gas lines, compressed air lines, oxygen piping, chilled water piping, vacuum piping, and chemical waste piping. These ducts and tubes occupy a considerable volume of space in the building, often in an amount that virtually equals the inhabited volume. Furthermore, all these systems require continual maintenance and are subject to frequent change. As a consequence, many laboratory and hospital buildings are designed with *interstitial ceilings*. An interstitial ceiling is suspended at a level that allows workers to travel freely in the plenum space, usually while walking erect, and is structured strongly enough to safely support the weight of the workers and their tools. In effect, the plenum space becomes another floor of the building, slipped in between the other floors, and the overall height of the building must be increased accordingly. Its advantage is that maintenance and updating work on the mechanical and electrical systems of the building can be carried on without interrupting the activities below. Interstitial ceilings are made of gypsum or lightweight concrete and combine the construction details of poured gypsum roof decks and suspended plaster ceilings. Figure 24.14 shows the installation of an interstitial ceiling.

FIGURE 24.12
A mirror-finish linear metal suspended
ceiling, with an exposed steel space truss
structure beyond. *(Architect: DeWinter and
Associates. Photo courtesy of Alcan Building
Products, division of Alcan Aluminum
Corporation.)*

FIGURE 24.14

The final steps in constructing an interstitial ceiling: The ceiling plane consists of gypsum reinforced with hexagonal steel mesh. It is framed with steel truss tee subpurlins, the tops of which are visible at the right of the picture, supported by steel wide-flange beams suspended on rods from the sitecast concrete framing above. The final layer of gypsum is being pumped onto the ceiling from the hose near the center of the picture. The wet gypsum is struck off level with the wooden straightedge seen here hanging on the beams, and is then troweled to a smooth walking surface. When the gypsum has hardened, installation of the ductwork, piping, and wiring in the interstitial plenum space can begin, with workers using the gypsum ceiling as a walking surface. *(Photo compliments Keystone Steel & Wire Co.)*

SUSTAINABILITY OF FINISH CEILINGS AND FLOORS

Acoustical Ceilings

• Acoustical ceiling tile can be a source of volatile organic compound (VOC) emissions as well as a reservoir for emissions from other sources.

• Low-emitting acoustic tiles are available, as are tiles with high recycled materials content.

• When the space above suspended ceilings is used as a return air plenum for the building's HVAC system, contaminants and emissions from the acoustical tile can be introduced into the system's airstream and redistributed to other parts of the building.

Hard Flooring Materials

• Concrete, stone, masonry, ceramic tile, and cementitious mortars and grouts are chemically inert and generally free of emissions.

• Organic adhesives used in tiling and resins used in thin-set terrazzo may be sources of emissions.

• Sealers applied to hard flooring materials to provide water repellency and protection from staining are potential sources of emissions. Solvent-based sealers generally have higher emissions than water-based products.

Wood and Bamboo Flooring

• Considerations of sustainability in wood products are discussed in Chapter 3 of this text.

• Bamboo is harvested in a 4- to 6-year cycle and is considered a rapidly renewable material. Some bamboo flooring products are produced with urea-formaldehyde glues, a potential source of emissions. Others are manufactured with alternative low-emitting adhesives.

• Finishes for wood and bamboo flooring are potential sources of emissions. Water-based finishes and waxes are generally lower emitting than solvent-based finishes, but they may not be as durable or as easy to apply.

Resilient Flooring

• Self-leveling cements used to prepare subfloors for resilient flooring coverings are potential sources of emissions.

• Vinyl (polyvinyl chloride) is a component of many resilient floor coverings and other interior finish products. It has an embodied energy of approximately 30,000 BTU per pound (70 MJ/kg). Issues related to the environmental sustainability of this material are discussed at greater length in Chapter 18.

• The embodied energy of synthetic rubber is about 50,000 BTU per pound (120 MJ/kg), and that of natural latex rubber is about 30,000 BTU per pound (70 MJ/kg).

• Some rubber flooring is manufactured from recycled car tires. Natural rubber is a renewable resource extracted from tropical rubber plants without harming the plant.

• Linoleum is made from linseed oil, pine rosin, clay, cork, limestone, and jute. Many of its ingredients are renewable, although the extraction of clay and limestone can cause ecological disruption.

• Cork is a renewable material harvested in a 9-year cycle.

• VOC emissions from the stripping and rewaxing of resilient flooring over its lifetime can be many times greater than those originating from the flooring material itself.

• Resilient flooring meeting the Resilient Floor Institute's *Floor Score* requirements for low VOC emissions satisfy current LEED requirements for low-emitting floor materials.

Carpet

• The embodied energy of nylon carpet is roughly 65,000 BTU per pound (150 MJ/kg), that of polypropylene is 40,000 BTU per pound (93 MJ/kg), and that of wool is 45,000 BTU per pound (105 MJ/kg). In general, natural fibers tend to be lower in embodied energy than synthetic fibers but also less durable.

• Many carpets and cushions are made with at least some recycled material.

• Carpet, cushions, and adhesives meeting the requirements of the Carpet and Rug Institute's *Green Label Plus* program satisfy current LEED requirements for low-emitting carpet systems.

• Factory-applied adhesives tend to have lower VOC emissions than adhesives applied on the construction site.

• Stretch-in installation of sheet carpet and free-lay installation of carpet tile (methods of installation explained later in this chapter) eliminate the need for carpet adhesive.

• Carpets can become reservoirs for VOCs emitted by other materials, as well as for bacteria, microbes, dust mites, and other contaminants.

• Carpet tile, which allows easy spot replacement, lessens the need for full carpet replacement when a small area becomes worn or damaged, thus extending the life of the carpet installation and reducing waste.

• Recycling of used carpet is complicated by the fact that many different fibers, which must be recycled separately, are used. Some manufacturers are introducing "close-the-loop" reclamation programs in which they will recycle or repurpose their own products to prevent them from being disposed of in landfills.

FINISH FLOORING

Functions of Finish Flooring

Floors have a lot to do with our visual and tactile appreciation of a building. We sense their colors, patterns, and textures, their "feel" underfoot, and the noises they make in response to footsteps. Floors affect the acoustics of a room, contributing to a noisy or hushed quality, depending on whether a hard or soft flooring material is used. Floors also interact in various ways with light: Some floor materials give mirror-like reflections; others give diffuse reflections or none at all. Dark flooring materials absorb more of the light incident upon them and contribute to the creation of a darker room, whereas light materials help create a brighter space and in some cases, can even contribute to reduced energy consumption by allowing artificial lighting levels to be lowered.

Floors are also a major functional component of a building. They are its primary wearing surfaces, subject to water, grit, dust, and the abrasive and penetrating actions of feet and furniture. They require more cleaning and maintenance effort than any other component of a building. They must be designed to deal with problems of skid resistance, sanitation, noise reduction between floors of a building, and even electrical conductivity in occupancies such as computer rooms and hospital operating rooms, where static electricity would pose a threat. Also, like other interior finish components, floors must be selected with an eye to combustibility, fire resistance ratings, and the structural loads that they will place on the frame of the building.

Reducing Noise Transmission Through Floors

In multistory buildings, it is sometimes necessary to take precautions to reduce the amount of impact noise transmitted through a floor to the room below. This is particularly true of hotels, condominiums, and apartment buildings, where people are sleeping in rooms below the rooms of others who may be awake and moving about. Impact noise is generated by footsteps or machinery and is transmitted as *structureborne vibration* through the material of the floor to become airborne noise in the room below.

There are several strategies for dealing with impact noise; these may be employed individually or in various combinations. One is to use padded carpeting or cushioned resilient flooring to reduce the amount of impact noise that is generated. A second is to add an *acoustic underlayment,* a layer of resilient material that is not highly conductive of impact noise, beneath the finish flooring. Cellulose fiber panels and nonwoven plastic filament matting are two materials used for this purpose. A third mechanism is to make an airtight ceiling below of a heavy, dense material such as plaster or gypsum board, and to mount this ceiling on resilient clips or on hanger wires with springs. The springs or clips absorb much of the sound energy that would otherwise travel through the structure.

Many floor–ceiling assemblies have been tested for sound transmission and rated for both Sound Transmission Class (STC), which is concerned with transmission of airborne sound, and Impact Insulation Class (IIC). (These ratings are explained in more detail in Chapter 22.) Product ratings will be found in trade literature concerned with various types of floor construction, and offer a ready comparison of acoustic performance.

Resistance to Fire

Floor finish materials must also meet building code requirements for resistance to ignition by radiant heat (NFPA 253) and flame spread (DOC FF-1), as explained in Chapter 22.

Slip Resistance

The slip resistance of a floor surface may be expressed as the *coefficient of friction (COF)* between the floor and some reference material, such as rubber or leather. On a level surface, COF is measured as the force required to pull an object across the floor, divided by the weight of the object. For example, if 10 lb of force is needed to drag a 50-lb rubber block across a floor, the COF between the block and the floor is 0.20 (10 lb/50 lb = 0.20). A lower COF implies a more slippery surface. A higher coefficient implies a less slippery one. COF may be measured starting with an object at rest, the *static coefficient of friction (SCOF),* or once the object has begun to slide, the *dynamic coefficient of friction (DCOF).* Usually, more force is required to initiate sliding of a stationary object than is required to keep an object sliding once it starts. So, for any combination of materials and conditions, SCOF is typically higher than DCOF.

Historically, in North America, there has been no defining regulatory requirement for minimum floor surface slip resistance. For example, the Americans for Disabilities Act standards require that a floor surface be stable, firm, and slip resistant but do not provide further guidance as to the determination of slip resistance. Some standards reference minimum SCOFs of 0.50 or 0.60 for level floor surfaces or 0.80 for sloped ramps. However, a lack of uniformity in how these measurements are made or agreement on appropriate values have prevented a more widespread application of COFs to floor safety requirements.

More recently, the National Floor Safety Institute has developed a revised standard for measuring floor slip resistance, ANSI/NFSI Standard B101. This test measures slip resistance of a wet floor surface, using different methods than previous tests. Surfaces are rated as High-Traction (SCOF greater than 0.6), Moderate Traction (SCOF from 0.4 to 0.6), or Low Traction (SCOF less than 0.4). The Tile Council of North America has also issued a slip resistance standard for tiled floor surfaces, called the *DCOF AcuTest method.* For interior tiled floors that are expected to be walked on when wet (e.g., tiled surfaces surrounding an indoor swimming pool), a minimum DCOF of 0.42 is recommended. This figure is measured using a special rubber block, with the floor lubricated with slightly soapy water, and using standardized equipment and methods so as to produce consistent and repeatable results. It is likely that more standardized slip resistance guidelines for other types of floor surfaces will be released in the foreseeable future.

Electrical Resistivity

In areas housing equipment that could be damaged by static electrical discharge, *static control flooring* provides a moderately conductive pathway for the dissipation of electrical charge between occupants and equipment. In this way, the flooring prevents sudden, larger, damaging discharges.

Underfloor Services

Floor structures are frequently used for the distribution of electrical and communications wiring, especially in floor areas that are broad and have few fixed partitions. If the needs for electrical and telephone service are minimal and predictable, the most economical horizontal distribution system for wiring in a floor consists of conventional conduits of metal tubing that are embedded in the floor slab or concrete topping. In most commercial buildings, however, greater flexibility is required to accommodate wiring changes that will have to occur from time to time during the life of the building. There are several alternative systems for creating this flexibility. In buildings with concrete structural systems, *cellular raceways* may be cast into the floor slabs (Figure 24.15). These are sheet metal ducts that can carry many wires. Working through access boxes that reach from the top of the raceway to the surface of the slab, electricians can add or remove wiring at any time. Electrical and communications outlets can be installed in any of the access boxes.

In steel-framed buildings, *cellular steel decking* provides the same functional advantages as cellular raceways (Figure 24.16). *Poke-through fittings*

(Figure 24.17) allow wiring flexibility over time without the need for raceways or cellular decking. Poke-through systems, however, require the electrician to work from the floor below the one on which changes are being made, which can be an inconvenience for the tenants of the lower floor.

Raised access flooring is advantageous in buildings where wiring changes are frequent and unpredictable, such as equipment-intensive offices and dedicated server rooms (Figures 24.18 and 24.19). Raised access flooring has a virtually unlimited capability to meet future wiring needs, and changes in wiring are extremely easy to make. If the access floor is raised high enough, ductwork for air distribution can be run beneath it, possibly eliminating the need for a suspended ceiling. This can be useful in buildings where the structural system is meant to be left exposed as a finish ceiling, or in older buildings with beautiful plaster or wood ceilings. Raised access flooring also works well in older buildings because its pedestal heights can be adjusted to compensate for uneven floor surfaces. Several systems for providing individualized control of air conditioning in large office buildings utilize the space below a raised access floor as a large distribution chamber for conditioned air; each workstation is provided with a small outlet diffuser in the floor surface.

FIGURE 24.15

A worker levels a cellular raceway over steel dome formwork for a concrete waffle slab floor structure. The concrete will be poured so that its surface is level with the tops of the access boxes, completely embedding the raceways and allowing each box to be opened merely by removing its metal cover. Electrical and communications outlets can be installed in any access box. The covers of unused access boxes will be concealed beneath vinyl composition floor tiles or carpeting. (*Courtesy of American Electric— Construction Materials Group.*)

FIGURE 24.16
Cellular steel decking is often used for underfloor electrical and communications wiring. A transverse feeder trench, near the top of the picture, brings the wiring across the floor from the electrical risers to the cells in the deck. Boxes cast into the topping give access to the cells for the installation of electrical outlets. Notice the shear studs on the steel beam at the top of the picture. *(Photo courtesy of H. H. Robertson Company.)*

FIGURE 24.17
(*a*) A poke-through fitting is designed to be installed in a hole drilled through a concrete floor slab. The junction box at the bottom connects to a wiring conduit above the suspended ceiling of the floor below. Wires pass from the junction box through the vertical tube to the outlet above. (*b*) Gaskets act as firestopping to seal the hole through the slab so that fire cannot pass, thus maintaining the fire resistance rating of the floor structure. If a poke-through fitting is removed in a later renovation, a special fire-resistant abandonment plug is installed to seal the hole. *(Courtesy of American Electric— Construction Materials Group.)*

(*a*)

(*b*)

FIGURE 24.18
Raised access flooring provides unlimited capacity for wiring, piping, and ductwork. The space below the flooring can serve as a plenum for air distribution. Changes in any of the underfloor systems are easily made, and wiring outlets can be installed at any point in the floor. *(Courtesy of Tate Architectural Products, Inc.)*

With *undercarpet wiring systems*, electrical power and communications wiring can be delivered within the floor system using flat conductors rather than conventional round wires installed between the finish flooring and the slab below. Undercarpet wiring is especially useful for retrofit projects where the previously discussed systems for delivering in-floor service are less practical (Figures 24.20 and 24.21).

FIGURE 24.19
Conditioned air is supplied to this server room through the space below the raised access flooring and is fed upward through perforated floor panels. Air is returned through slots in the suspended ceiling. *(Photo courtesy Armstrong World Industries.)*

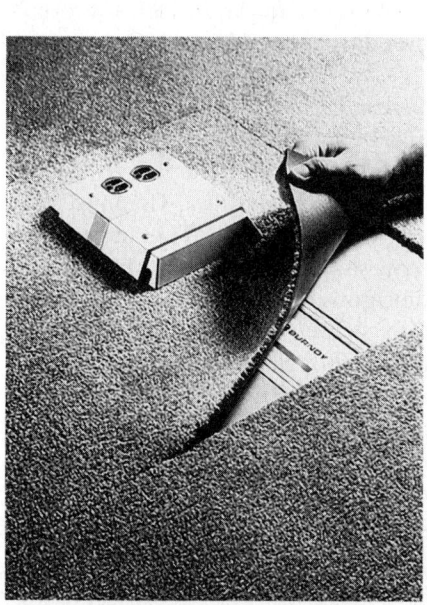

FIGURE 24.20
The flat conductors for an undercarpet wiring system lie unseen beneath the carpet and are accessed through projecting boxes. *(Photo: Courtesy of Burndy Corporation.)*

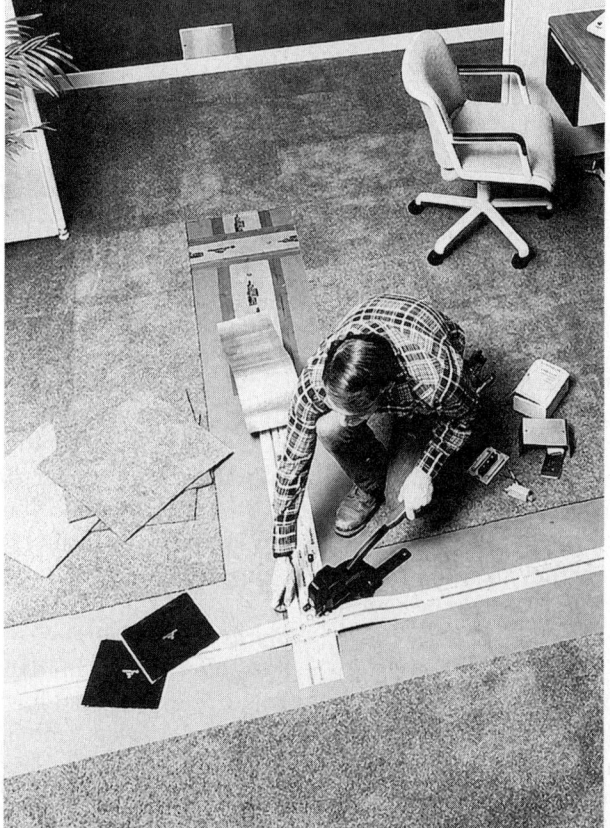

FIGURE 24.21
The flat conductors are ribbons of copper laminated between insulating layers of plastic sheet. These conductors are connected with the splicing tool shown in this photograph and covered with a grounded metallic shield before being taped to the floor. *(Photo: Courtesy of Burndy Corporation.)*

> **The brick floors, because the bricks may be made of diverse forms and of diverse colors by reason of the diversity of the chalks, will be very agreeable and beautiful to the eye. . . . The ceilings are also diversely made, because many take delight to have them of beautiful and well-wrought beams . . . [T]hese beams ought to be distant one from another one thickness and a half of the beam, because the ceilings appear thus very beautiful to the eye.**
>
> —Andrea Palladio, *The Four Books of Architecture*, 1570

TYPES OF FINISH FLOORING MATERIALS

Hard Flooring Materials

Hard finish flooring materials (concrete, stone, brick, tile, and terrazzo) are often chosen for their resistance to wear and moisture. Being rigid and unyielding, they are not comfortable to stand on for extended periods of time, and they contribute to a live, noisy acoustic environment. Many of these materials, however, are so beautiful in their colors and patterns, and so durable, that they are often considered highly desirable types by designers and building owners alike.

Concrete

With a lightly textured wood float finish for traction, concrete makes an excellent finish floor for parking garages and many types of agricultural and industrial buildings. With a smooth, hard, steel trowel finish, concrete finds its way into a vast assortment of commercial and institutional buildings, and even into homes and offices. Machine-polished concrete floors can achieve an almost glass-like level of smoothness and sheen. Color can be added with a colorant admixture, a concrete stain, or a couple of coats of floor paint. For industrial floors where hardness and durability are a special concern, *concrete hardeners* that densify the wearing surface of the concrete are added during slab finishing operations.

The chief advantages of concrete as a finish flooring material are its low initial cost and its durability. On the minus side, extremely good workmanship is required to make an acceptable floor finish, and, unless applied as a finish topping very close to the end of construction, even the best concrete surface is likely to sustain some damage and staining during construction.

Stone

Many types of building stone are used as flooring materials, in surface textures ranging from mirror-polished marble and granite to split-face slate and sandstone (Figures 24.22 and 24.23). Installation is a relatively simple but highly skilled procedure of bedding the stone pieces in mortar and filling the joints with grout (Figure 24.24). Most stone floorings are coated with applications of a clear sealer, and are waxed periodically throughout their life building to bring out the color and figure of the stone while protecting it from stains.

Bricks and Brick Pavers

Both bricks and half-thickness bricks called *pavers* are used for finish flooring, with pavers often preferred because they add less thickness and dead weight to the floor (Figures 10.23, 10.24, and 24.25). Bricks may be laid with their largest surface horizontal or on edge. As with stone and tile flooring, decorative joint patterns can be designed especially for each installation.

Quarry Tiles

Quarry tiles are not, as might be guessed from their name, made of stone. They are simply large, fired clay tiles, usually square but sometimes rectangular, hexagonal, octagonal, or other shapes (Figures 24.24, 24.26, and 24.27). Sizes range from about 4 inches (100 mm) to 12 inches (300 mm) square, with thicknesses ranging from ⅜ inch (9 mm) to a full inch (25 mm) for some handmade tiles. Quarry tiles are available in myriad earth colors, as well as in certain kiln-applied colorations. They are usually set in a reinforced mortar bed, although in residential work they may be thin-set directly to a subfloor of wood panels or tile backer boards (Figure 24.28). (Tile-setting methods are discussed in more detail in Chapter 23.) It is important that any subfloor to which tiles are glued be sufficiently stiff; otherwise, flexing of the subfloor under changing loads will pop the tiles loose. Additional subfloor thickness and/or a stiff underlayment are advisable.

Ceramic Tiles

Fired clay tiles that are smaller than quarry tiles are referred to collectively as *ceramic tiles*. Ceramic tiles are usually glazed. The most common shape is square, but rectangles, hexagons, circles, and more elaborate shapes are also available (Figure 24.29). Sizes range from ½ inch (13 mm) to 4 inches (100 mm) and more. Smaller-sized tiles are shipped from the factory with their faces adhered to larger backing sheets of plastic mesh or perforated paper. The tilesetter lays whole sheets of 100 or more tiles together in a single step rather than as individual units (Figure 7.39). The backing sheet is easily removed by wetting it after the tile adhesive has cured.

FIGURE 24.22
A slate floor in an automobile showroom. *(Photo by Bill Engdahl, Hedrich Blessing. Courtesy of Buckingham-Virginia Slate Corporation.)*

FIGURE 24.23
Flooring of matched, polished triangles of white-veined red marble gives a kaleidoscopic effect. *(Architect: The Architects Collaborative. Photo by Edward Allen.)*

Grout color has a strong influence on the appearance of tile surfaces, just as the mortar used to lay brick and stone does. Many different premixed colors are available, or the tilesetter may custom-color a grout with pigments.

Methods of installing ceramic tile on interior wall surfaces, discussed in Chapter 23, apply to floor tiling as well, and an example of ceramic tile installation on floors and walls is shown in Figure 23.40. As in wall tiling applications, waterproofing membranes are integrated into floor tile assemblies in wet-use locations. Where tile is thickset over substrates that are cracked or prone to excessive deflection, a *slip sheet* or *cleavage membrane*, usually consisting of ordinary building felt, may be inserted between the mortar base and the substrate to isolate the tile assembly and reduce the chance of tile cracking. Where tile is thin-set over difficult substrates, *crack isolation membranes* or *uncoupling membranes* that preserve the necessary bond between the thin-set compound and the substrate, but limit the transfer of stresses into the tile assembly, may be used.

Terrazzo

Terrazzo is an exceptionally durable flooring. It is made by grinding and polishing a concrete that consists of marble or granite chips selected for size and color in a matrix of colored portland cement or another binding agent. The polishing brings out the pattern and color of the stone chips. A sealer is usually applied to further enhance the appearance of the floor (Figure 24.30). Terrazzo may be formed in place or installed as factory-made tiles. For stair treads, window sills, and other large components, terrazzo is often precast. Because of its endless variety of colors and textures, terrazzo is often used in decorative flooring patterns. The colors are separated from one another by *divider strips* of metal, plastic, or marble. The divider strips are installed in the underbed prior to placing the terrazzo, and are ground and polished flush in the same operation as the terrazzo itself.

Traditionally, terrazzo was installed over a thin bed of sand that isolates it from the structural floor slab, thus protecting it to some extent from movements in the building frame.

QUARRY TILE WOOD
SUBFLOOR

- Quarry tiles
- Bond coat of portland cement paste
- ¾″ to 1¼″ (19 to 32 mm) reinforced mortar bed over paper cleavage membrane

BRICKS ON CONCRETE SLAB

- Bricks
- ¾″ (19 mm) mortar bed

BONDED TERRAZZO

- ½″ (13 mm) terrazzo topping
- ¾″ to 1½″ (19 to 38 mm) underbed

MONOLITHIC TERRAZZO

- Zinc, brass, or plastic divider strip
- ½″ (13 mm) terrazzo topping
- Divider strip in saw-cut control joint

FIGURE 24.24

Typical details of tile, brick, stone, and terrazzo flooring. Traditional sand cushion terrazzo flooring is not illustrated. The terrazzo systems shown here are thinner and lighter than sand cushion terrazzo, but they perform as well on a properly engineered floor structure. Thin-set terrazzo, the lightest of the systems, uses very small stone chips in a mortar that is strengthened with epoxy, polyester, or polyacrylate.

STONE ON CONCRETE SLAB

THIN-SET TERRAZZO

FIGURE 24.25
A floor of glazed brick pavers meets a planting bed constructed of brick masonry.
(Courtesy of Stark Ceramics, Inc.)

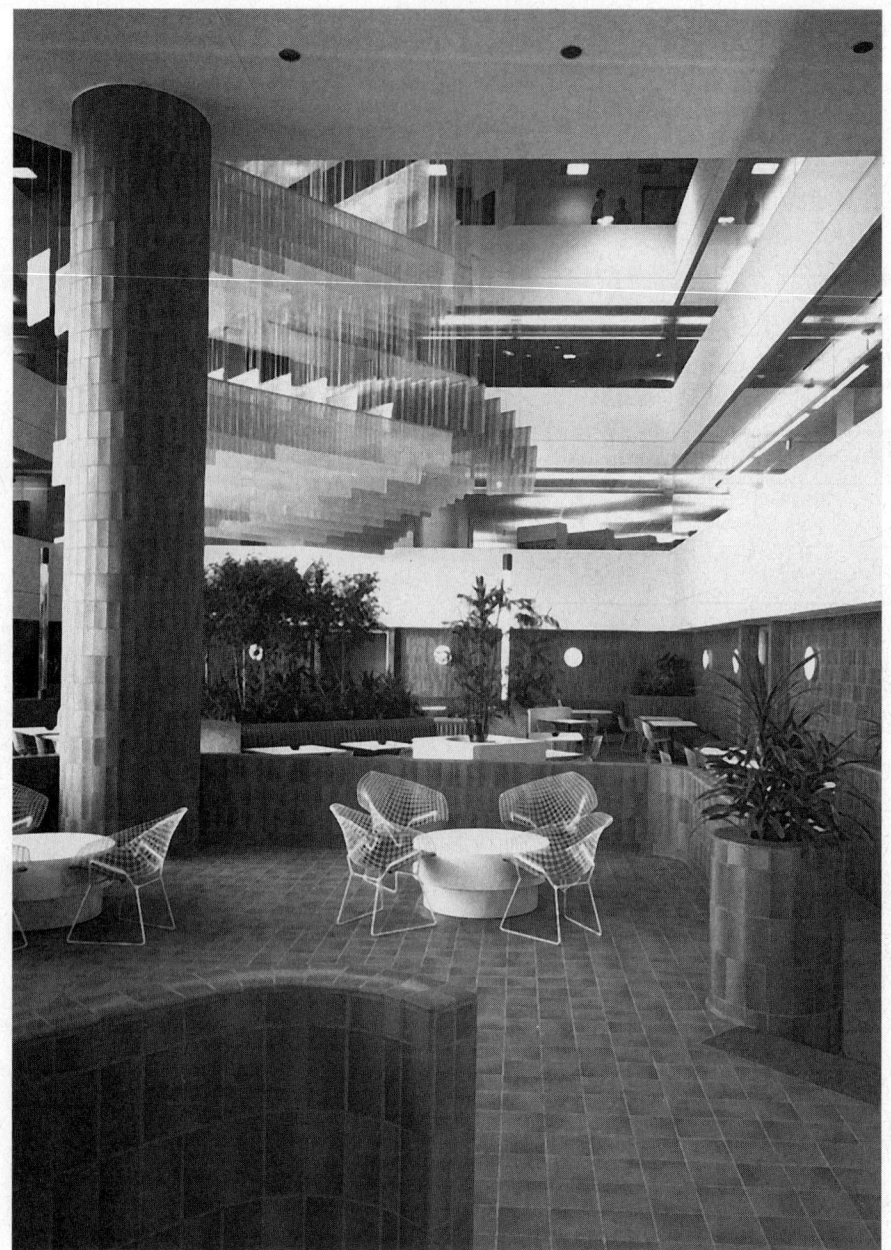

FIGURE 24.26
**Unglazed quarry tiles used as flooring
and as a facing for columns and railings.**
*(Architect: Skidmore, Owings & Merrill.
Interior designer: Duffy, Inc. Courtesy of
American Olean Tile.)*

FIGURE 24.27
Square and rectangular quarry tiles
in contrasting colors create a pattern
on the floor of a retail mall. *(Architect:
Edward J. DeBartolo Corp. Courtesy of
American Olean Tile.)*

Grout

Tile

Adhesive or
thin-set mortar

Backer board is
glued and/or
screwed to
subfloor

Adhesive

Subfloor

FIGURE 24.28
Quarry or ceramic tiles may be adhered
to a cementitious backer board that is
applied as an underlayment to a wood
panel subfloor.

FIGURE 24.29
Ceramic tile wainscoting and flooring in a bar. *(Architect: Daughn/Salisbury, Inc. Designer: Morris Nathanson Design. Courtesy of American Olean Tile.)*

FIGURE 24.30
A terrazzo floor in a residential entry uses divider strips and contrasting colors to create a custom floor pattern. *(Courtesy of National Terrazzo and Mosaic Association, Inc.)*

However, this *sand cushion terrazzo* is thick, usually 2½ inches (64 mm), and heavy. For greater economy and a thinner application, the sand bed is eliminated to produce *bonded terrazzo*, or both the sand bed and underbed may be eliminated with *monolithic terrazzo*. *Thin-set terrazzo*, made from epoxy resins, polyester resins, or polymer-modified cements, is the thinnest of all terrazzo installation methods (Figure 24.24). In any of these systems, a terrazzo baseboard can be formed and finished as an integral part of the floor, thus eliminating a dirt-catching seam where the floor meets the wall.

Wood and Bamboo

Wood Flooring

Wood is used in several different forms as a finish flooring material, the most common of which is solid wood tongue-and-groove *strip flooring*, typically ¾ inch (19 mm) thick and 1½ to 2¼ inches (38–57 mm) wide. Strip flooring can be made from many hardwood and softwood species, some of the more commonly used being white oak, red oak, pecan, and maple (Figures 24.31, 24.32, and 24.33). The wood strips are held tightly together and *blind nailed* by driving nails diagonally through the

upper interior corners of the tongues, where they are concealed from view once the next strip is installed. The entire floor is then sanded smooth, stained if desired, and finished with a varnish or other clear coating. When its surface becomes worn, the flooring can be restored to a new appearance by sanding and refinishing.

Solid wood flooring is also available in widths ranging from 3 to 8 inches (75–200 mm), called *plank flooring*. Because the wider planks are more prone to distortion with changes in moisture content, they are usually fastened through the face with countersunk and plugged screws

The base molding is thin and flexible, to conform tightly to irregularities in the surface of the wall

The baseboard protects the wall against damage by feet, furniture, or cleaning equipment. It is too large and stiff to conform to irregularities in the flooring or wall. The depression in the back allows it to lie flat even if it is cupped or the plaster is irregular

The shoe is thin and flexible, to conform tightly to irregularities in the floor surface

Building paper beneath the finish flooring seals off air leakage through the floor

The profile of each strip of flooring is designed so only the wood near the top surface makes full contact; this assures that a tightly driven floor will have no visible gaps

A one-piece baseboard is economical, but cannot fit as closely as a three-piece baseboard

Wood sleepers are fastened to the concrete block wall and concrete slab with concrete nails or powder-driven fasteners. Wood paneling and wood flooring are then nailed to the sleepers

If the concrete slab lies directly on grade, a sheet of polyethylene is laid beneath the sleepers to prevent moisture from entering the building

Figure 24.31

Details of hardwood strip flooring installation. At the left, the flooring is applied to a wood joist floor, and at the right, to wood sleepers over a concrete slab. The blind nailing of the flooring is shown only for the first several strips of flooring at the left. The baseboard makes a neat junction between the floor and the wall, covering up the rough edges of the wall material and the flooring. The three-piece baseboard shown at the left does this job somewhat better than the one-piece baseboard, but it is more expensive and elaborate. For a more stable, higher-quality installation, an additional sheet of plywood underlayment or sheathing may be added under the wood flooring in each of these examples.

FIGURE 24.32
An installer of hardwood strip flooring uses a pneumatic nail gun to drive the diagonal blind nails that fasten the flooring to the subfloor. The nail gun is a special type that is activated by a blow from a mallet that also serves to drive the flooring pieces tightly together. Asphalt-saturated felt paper cushions the flooring and helps to prevent squeaking. *(Photo by Rob Thallon.)*

FIGURE 24.33
Oak strip flooring in a hair salon. Notice the use of exposed ducts and a lighting track at the ceiling. *(Architects: Michael Rubin and Henry Smith-Miller in association with Kenneth Cohen. Courtesy of the Oak Flooring Institute.)*

in addition to, or instead of, being blind nailed along their edges.

For greater economy, factory-made wood flooring consisting of finish wood veneers laminated to a plywood-like core, called *engineered wood flooring*, is used. Typically ⅜ or ½ inch (10 or 13 mm) thick, it is glued to the subfloor with a mastic adhesive rather than nailed. The laminated construction of engineered wood flooring makes it less sensitive to changes in moisture content and more dimensionally stable than solid wood flooring. For this reason, it is considered better suited for use on basement slabs on grade or in other locations exposed to high humidity levels. Because the finish surface is a thin veneer, most engineered wood floorings are not able to withstand subsequent sanding and refinishing.

Parquet is wood flooring of varying hues arranged in patterns. It may be made of solid wood strips assembled in the field, of factory-preassembled blocks made from solid wood strips, or of engineered wood.

Some wood flooring systems are not nailed or glued to the subfloor, but instead "float" above it on a thin pad of resilient foam. These so-called *floating floors* are made by connecting the individual pieces of flooring together at the edges to make one continuous piece as large as the room in which the flooring is laid. Edge gluing is the most common way of making this connection, but systems are available that use metal clips or interlocking edge details. A gap is left at the edges to permit expansion and contraction of the floor. This is later covered with baseboard trim. Most floating floors are made of engineered wood flooring, which is more dimensionally stable than solid wood flooring.

Many types of wood flooring are available with factory-applied finishes. When the installation of conventional unfinished flooring begins on a project, it may require the complete cessation of other construction activities for a number of days while the flooring is installed and sanded, and several coats of finish are applied and allowed to dry. With factory-finished flooring, the time required for a complete installation and the impact on the other activities are minimized. Prefinished flooring, however, cannot be sanded after installation, so these products are supplied with eased edges (a slight beveling of the corners along the finish face edges) that hide minor differences in level between adjacent pieces of flooring after installation. Factory finishing is especially common with engineered wood flooring. Because the finish veneer is thin, and can never be sanded and refinished, an especially hard, wear-resistant acrylic resin finish is applied in the factory that helps to prolong the life of the floor.

Plastic laminate flooring is composed of planks or large tiles that have a wood composite core and a wearing layer of high-density plastic laminate much like that used on countertops. It is almost always laid as a floating floor. The laminate is usually patterned to resemble wood, but other patterns are also available. Most laminate floorings may be used in kitchens or baths if a sealant is applied around the perimeter of the floor to keep water from getting beneath it.

Exceptionally long-wearing industrial *wood block flooring* is made of small blocks of wood set in adhesive with their grain oriented vertically. Although this type of floor is relatively high in first cost, it is economical for heavily used floors and is sometimes chosen for use in public spaces because of the beauty of its pattern and grain.

Bamboo Flooring

Bamboo, a rapidly growing grass, can be used to produce flooring products very much like those of wood. The manufacturing process entails slicing the hollow bamboo shoots into strips, processing the strips to remove starch, laminating and gluing the strips under pressure, and then machining the laminated stock into the final flooring profile. Laminations may be oriented either vertically or horizontally within the strip, creating a surface appearance analogous to either flat grain or edge grain in solid woods.

Bamboo flooring is harder and more dimensionally stable than conventional wood flooring. Its natural color is light, akin to that of maple. A darker amber hue can be achieved by pressure steaming, a process called "carbonization." Like wood, bamboo flooring can be provided either as solid strip, made entirely of bamboo laminations, or as an engineered product consisting of a roughly ⅛-inch (3-mm)-thick bamboo finish layer adhered to a laminated base of conventional wood. It may be provided unfinished or factory finished.

Resilient Flooring

The oldest *resilient flooring* material is *linoleum*, a sheet material made of ground cork in a linseed oil binder over a burlap backing. *Asphalt tiles* were later developed as an alternative to linoleum, but the majority of today's resilient sheet floorings and tiles are made of compounds of vinyl or rubber. The primary advantages of resilient floorings are the wide range of available colors and patterns, moderately high durability, and low initial cost.

Vinyl composition tile (VCT), made of one or more vinyl resins in combination with binders, pigments, and fillers (VCT may consist of as much as 85 percent limestone filler), has the lowest installed cost of any flooring material except concrete and is used in vast quantities on the floors of residences, offices, classrooms, and retail spaces. Other common resilient tile flooring materials include *solid vinyl tile (SVT)*, with higher vinyl content and greater durability than VCT; and *rubber floor tile*, made from vulcanized natural or synthetic rubber compounds and various additives. Floor tile thickness is typically ⅛ inch (3 mm) or slightly less. The most common tile size is 12 inches (305 mm) square, although other sizes, up to 36 inches (914 mm) square, are also available.

The most common *resilient sheet flooring* materials are solid vinyl and rubber. Resilient floorings of linoleum, cork, and other materials are also available. Each offers particular characteristics of durability and appearance. Thicknesses of sheet flooring are also on the order of ⅛ inch (3 mm), slightly thinner for lighter-duty floorings, and slightly thicker if a cushioned back is added to the product. Sheet floorings are furnished in rolls 4.9 to 12 feet (1.5–3.6 m) wide. If they are skillfully installed, the seams between sheet strips are virtually invisible.

Most resilient flooring materials are glued to the concrete or wood surface of the structural floor (Figures 24.34, 24.35, and 24.36). The resilient materials themselves are so thin and deformable that they show even the slightest irregularities in the floor deck beneath. Concrete surfaces to which resilient materials will be applied must first be scraped clean of construction debris and spatters. Wood panel decks are covered with a layer of smooth *underlayment panels*, usually of hardboard, particleboard, or sanded plywood, to provide a smoother substrate for the resilient flooring materials. Joints between underlayment panels are offset from joints in the subfloor to eliminate soft spots. The thickness of the underlayment is chosen to make the surface of the resilient flooring level with the surfaces of flooring materials

such as hardwood and ceramic tile that are used in surrounding areas of the building. Leveling compounds, gypsum or cement-based plaster-like materials that can be feathered to a very thin edge, are also used to fill in minor low spots in the floor surface before the resilient materials are installed.

Various flooring accessories, such as bases, stair treads, and stair nosings, are also made of vinyl and rubber compounds. *Cove base* is the base most commonly used with resilient flooring (Figure 24.34). *Straight base*, also called *flat* or *toeless base*, has no cove or toe and is most commonly used with carpet flooring. *Fit-to-floor base*, or *butt-to base*, has a square-edged toe the same thickness as the floor covering; it butts tightly to the finish flooring, creating a flush transition between the two.

FIGURE 24.34
Typical installation details for resilient flooring. At the left, resilient tiles applied directly to a steel-trowel-finish concrete slab, with a cove base adhered to a concrete masonry partition. To the right, resilient sheet flooring on underlayment and a wood joist floor structure.

FIGURE 24.35
Vinyl composition tile and a vinyl cove base. Notice how the base is simply folded around the outside corner of the partition. *(Photo courtesy Armstrong World Industries.)*

FIGURE 24.36
Sheet vinyl flooring can be flash coved to create an integral base that is easily cleaned, for use in health care facilities, kitchens, and bathrooms. The seams are welded to eliminate dirt-catching cracks *(Photo courtesy Armstrong World Industries.)*

Carpet

Carpet is manufactured in fibers, styles, and patterns to meet almost any flooring requirement, indoors or out, except for rooms that require thorough sanitation, such as hospital rooms, food-processing facilities, and toilet rooms. Some carpets are tough enough to wear for years in public corridors (Figure 24.37); others are soft enough for intimate residential interiors. The costs of carpeting are often competitive with those of other flooring materials of similar quality, whether they are measured on an installed-cost or life-cycle-cost basis. Slightly more than one-half of all carpet sold in North America is made with nylon fibers; more than an additional one-third is made with polypropylene; the remainder is made with other synthetics and natural fibers.

There are four ways to install carpet: The two most common are gluing directly to the floor deck (*direct glue-down installation*) or stretching over a *carpet pad* or *cushion* and attaching the perimeter by means of a *tackless strip* (*stretch-in installation*). The tackless strip (also called a *tackstrip*) is a continuous length of wood, fastened to the floor, that has protruding spikes along the top to catch the backing of the carpet and hold it taut as it is stretched into place by the installers. Less frequently, carpet pad is glued to the floor deck and the carpet is then glued to the pad (*double glue-down installation*), or carpet with a factory-attached pad is glued to the floor deck (*attached cushion installation*).

If carpet or carpet tile is laid directly over a wood panel subfloor such as plywood, the panel joints perpendicular to the floor joists should be blocked beneath to prevent movement between sheets. Tongue-and-groove plywood subflooring accomplishes the same result without blocking. Or, a layer of underlayment panels may be nailed over the subfloor with its joints offset from those in the subfloor.

Carpets, carpet pads, and adhesives are all potentially significant sources of VOC emissions. Products meeting the requirements of the Carpet & Rug Institute's *Green Label Plus* program are certified as meeting low-emission standards complying with current requirements of LEED NC and other green building programs. A newer standard is the NSF International's *NSF/ANSI Standard 140 Sustainable Carpet Assessment Standard.* NSF 140 provides a more comprehensive life-cycle rating system for carpet that addresses

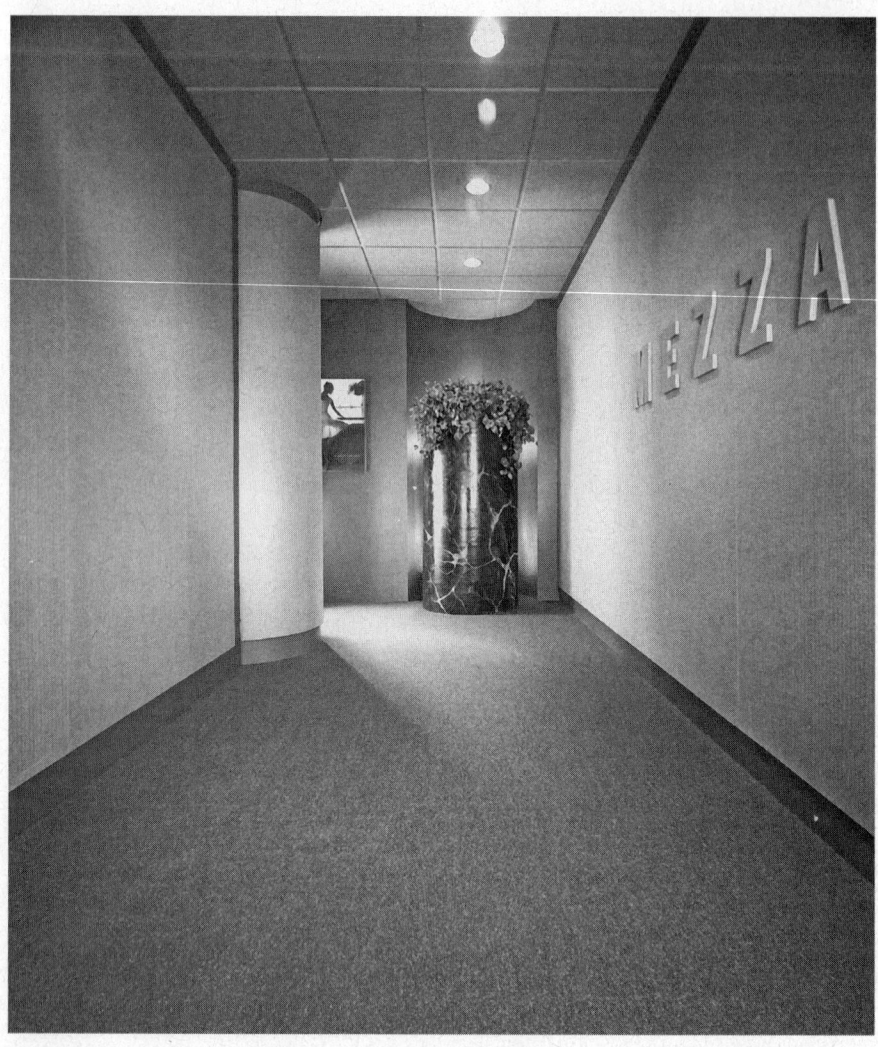

FIGURE 24.37
Wall-to-wall carpeting in a commercial installation (*Photo courtesy Armstrong World Industries.*)

not only indoor air quality, but also other public health and environmental concerns, energy efficiency, recycled or biobased materials content, end-of-life reclamation, and other considerations.

Carpet Tile

Carpet materials are also manufactured in tile form, with typical sizes ranging from 18 to 36 inches (457–914 mm) square. In comparison to sheet carpet, tiles are easier to deliver, store, handle, and install; they allow easier spot replacement; they permit ready underfloor access; and they are compatible with raised-access flooring systems. Carpet tile installation

methods include *glue-down*, in which every tile is adhered to the subfloor; *partial glue-down*, in which only periodically spaced tiles are adhered to the subfloor; and *free-lay*, in which interlocking tiles are laid without any adhesive.

Fluid-Applied Flooring

Fluid-applied flooring is applied as a thin liquid coating over concrete or wood decking and then cures to a seamless, easy-to-clean finish floor surface. Depending on formulation, coverings satisfying a wide range of requirements for durability, chemical resistance, ease of installation, and

appearance are possible. Though more expensive to install than other resilient flooring options, their greater durability makes these resinous flooring systems often less expensive when costs are considered over the full life of the flooring.

Urethane and *epoxy floor coatings* are the most common. A typical application over concrete consists of preparation of the concrete surface by grinding or blasting, application of primer to ensure good adhesion between the substrate and the following layers, one or two coats of the primary flooring material, possibly with the addition of sand, quartz aggregate, rubber chips, or other

Figure 24.38
A seamless, epoxy traffic coating provides a durable, easy-to-clean, and chemical-resistant finish flooring for this automobile repair garage.

fillers, and a top coat and/or sealer. Thickness of the finished flooring typically ranges from ³⁄₁₆ to ³⁄₈ inch (5–10 mm), although both thinner and thicker application are also possible.

Acrylic (methyl methacrylate monomer or *MMA*) *floor coatings*, though high in VOC emissions, cure in as little as one hour. They are popular where rapid turn-around time in the floor coating process is especially important. Thin, *UV-curable coatings* emit no VOCs and cure virtually instantaneously as a machine with a controlled-intensity ultraviolet lamp is rolled across the wet material.

Cementitious overlays are formulations of polymers, portland cement, fine sand, and other ingredients that are applied as self-leveling fluids. Once cured, they form durable, polishable, integrally colored, finished floor surfaces.

Traffic coatings are fluid-applied products, most commonly urethanes, formulated for exterior exposure, and able to tolerate both foot and vehicular traffic. They find frequent use as coatings for parking garage decks, exterior residential balconies, and elsewhere a low-cost, waterproof, and durable coating is required (Figure 24.38).

FLOORING THICKNESS

Thicknesses of floor finishes vary from ⅛ inch (3 mm) or less for resilient flooring to 3 inches (76 mm) or more for brick or stone flooring. Frequently, several different types of flooring are used on different areas of the same floor level of a building. If the differences in thickness of the flooring materials are not great, they can be resolved by using tapered edgings or thresholds at changes of material. Greater differences may be resolved with variations in the thickness of underlayment panels. Alternatively, *self-leveling* gypsum or cementitious *underlayments* can be poured up to several inches thick over portions of a subfloor to raise the level before application of the finish flooring.

If none of these solutions can satisfactorily resolve the differences, the level of the top of the floor deck must be adjusted from one part of the building to the next to bring the finish floor surfaces to the same elevation. The architect should work out the necessary level changes in advance and indicate them clearly on the construction drawings. In many cases, structural details must be drawn to indicate how the level changes should be made. In wood framing, they can usually be made either by notching the ends of the floor joists to lower the subfloor in parts of the building with thicker floor materials or by adding sheets of underlayment material of the proper thickness to areas of thinner flooring. In steel and concrete buildings, slab or topping thicknesses can change, or whole areas of the structure can be raised or lowered by the necessary amount.

CSI/CSC MasterFormat Sections for Interior Walls and Partitions

03 54 00	**CAST UNDERLAYMENT**
07 18 00	**TRAFFIC COATINGS**
09 22 00	**SUPPORTS FOR PLASTER AND GYPSUM BOARD**
09 22 27	Suspension Systems
09 30 00	**TILING**
09 30 13	Ceramic Tiling
09 30 16	Quarry Tiling
09 30 33	Stone Tiling
09 30 39	Brick Tiling
09 31 00	**THIN-SET TILING**
09 32 00	**MORTAR-BED TILING**
09 51 00	**ACOUSTICAL CEILINGS**
09 51 23	Acoustical Tile Ceilings
09 51 53	Direct-Applied Acoustical Ceilings
09 53 00	**ACOUSTICAL CEILING SUSPENSION ASSEMBLIES**
09 54 00	**SPECIALTY CEILINGS**
09 54 23	Linear Metal Ceilings
09 54 26	Suspended Wood Ceilings
09 54 33	Decorative Panel Ceilings
09 54 43	Stretched-Fabric Ceiling Systems
09 54 46	Fabric-Wrapped Ceiling Panels
09 56 00	**TEXTURED CEILINGS**
09 58 00	**INTEGRATED CEILING SYSTEMS**
09 60 00	**FLOORING**
09 60 13	Acoustic Underlayment
09 62 00	**SPECIALTY FLOORING**
09 62 23	Bamboo Flooring
09 62 29	Cork Flooring
09 63 00	**MASONRY FLOORING**
09 63 13	Brick Flooring
09 63 40	Stone Flooring
09 64 00	**WOOD FLOORING**
09 65 00	**RESILIENT FLOORING**
09 65 13	Resilient Base and Accessories
09 65 16	Resilient Sheet Flooring
09 65 19	Resilient Tile Flooring
09 65 36	Static-Control Resilient Flooring
09 65 43	Linoleum Flooring
09 66 00	**TERRAZZO FLOORING**
09 66 13	Portland Cement Terrazzo Flooring
09 66 16	Terrazzo Floor Tile
09 66 23	Resinous Matrix Terrazzo Flooring
09 67 00	**FLUID-APPLIED FLOORING**
09 69 00	**ACCESS FLOORING**

KEY TERMS

plenum
membrane ceiling
membrane fire protection
suspended ceiling
hanger wire
perforated gypsum panel
acoustical ceiling
Noise Reduction Coefficient (NRC)
Ceiling Attenuation Class (CAC)
Articulation Class (AC)
lay-in panel
exposed grid
concealed grid
integrated ceiling system
metal panel ceiling
linear metal ceiling
suspended wood ceiling
fabric-wrapped ceiling panel
stretched-fabric ceiling system
interstitial ceiling
structureborne vibration
acoustic underlayment
static coefficient of friction
  (SCOF)
static control flooring
cellular raceway
cellular steel decking

poke-through fitting
raised access flooring
undercarpet wiring system
concrete hardener
paver
quarry tile
ceramic tile
slip sheet, cleavage membrane
crack isolation membrane,
  uncoupling membrane
terrazzo
divider strip
sand cushion terrazzo
bonded terrazzo
monolithic terrazzo
thin-set terrazzo
strip flooring
blind nail
plank flooring
engineered wood flooring
parquet
floating floor
plastic laminate flooring
wood block flooring
resilient flooring
linoleum
asphalt tile

vinyl composition tile (VCT)
solid vinyl tile (SVT)
rubber floor tile
resilient sheet flooring
underlayment panel
cove base
straight base, flat base,
  toeless base
fit-to-floor base, butt-to base
carpet
direct glue-down installation
carpet pad, cushion
tackless strip, tackstrip
stretch-in installation
double glue-down installation
attached cushion installation
glue-down tile
partial glue-down tile
free-lay tile
fluid-applied flooring
urethane floor coating
epoxy floor coating
acrylic (MMA) floor coating
cementitious overlays
UV-curable floor coating
traffic coating
self-leveling underlayment

REVIEW QUESTIONS

1. List the potential range of functions of a finish ceiling and of a finish floor.

2. What are the advantages and disadvantages of a suspended ceiling compared to those of a tightly attached ceiling?

3. When designing a building with its structure and mechanical equipment left exposed at the ceiling, what precautions should you take to ensure a satisfactory appearance?

4. What does underlayment do? How should it be laid? What does acoustical underlayment do?

5. List several different approaches to the problem of running electrical and communications wiring beneath a floor of a building framed with steel or concrete.

Exercises

1. Visit some rooms that you happen to like in nearby buildings: classrooms, auditoriums, theaters, restaurants, bars, museums, shopping centers—both new buildings and old. For each room, list the ceiling and floor finishes used. Why was each material chosen? Sketch details of critical junctions between materials.

What does each room sound like—noisy, hushed, "live"? How does this acoustical quality relate to the floor and ceiling materials? What is the quality of the illumination in the room, and what role do ceiling and floor materials play in creating this quality?

2. Consult current architectural magazines for interior photographs of buildings that appeal to you. What ceiling and floor materials are used in each? Why?

Selected References

Because so many ceiling and flooring materials and systems are proprietary, much of the best information on ceiling and floor finishes is to be found in manufacturers' literature. Certain generic products are, however, well documented in trade association literature:

Carpet and Rug Institute. *Carpet Installation Standard.* Dalton, GA, Author, updated regularly.

Industry-standard recommendations for the installation of carpet of all types, available from this association's web site as free downloads.

Ceilings and Interior Systems Construction Association. *Ceiling Systems Handbook.* St. Charles, IL, Author, updated frequently.

Aimed at construction workers, this manual covers everything pertaining to the installation of ceilings.

Marble Institute of America. *Dimension Stone Design Manual.* Cleveland, OH, Author, updated regularly.

This binder provides comprehensive guidance on the design, specification, and installation of all types of stone flooring.

National Wood Flooring Association. *Installation Guidelines.* Chesterfield, MO, Author, updated regularly.

This manual provides industry standard recommendations for the installation of all kinds of wood flooring.

Tile Council of North America, Inc. *TCNA Handbook for Ceramic Tile, Glass, and Stone Tile Installation.* Anderson, SC, Author, updated regularly.

This is the definitive standard for ceramic tile installation materials and methods. More than 100 methods of installation for floors and walls, both interior and exterior, are illustrated and specified. Guidelines for selecting appropriate installation methods based on project requirements are also included.

WEB SITES

Author's supplementary web site: **www.ianosbackfill.com/24_finish_ceilings_and_floors**

Armstrong Floors and Ceilings: **www.armstrong.com**

Carpet and Rug Institute (CRI): **www.carpet-rug.org**

Ceilings and Interior Systems Construction Association: **www.cisca.org**

Forbo Floor Coverings: **www.forbo-flooring.us**

H. H. Robertson: **www.hhrobertson.com**

Mannington Floors: **www.mannington.com**

Maple Flooring Manufacturers Association (MFMA): **www.maplefloor.org**

Marble Institute of America (MIA): **www.marble-institute.com**

National Terrazzo & Mosaic Association (NTMA): **www.ntma.com**

National Wood Flooring Association (NWFA): **www.nwfa.org**

Resilient Floor Covering Institute (RFI): **www.rfci.com**

Tile Council of North America (TCNA): **www.tcnatile.com**

Densities and Coefficients of Thermal Expansion of Common Building Materials

Material	Density		Coefficient of Thermal Expansion		
	lb/ft³	kg/m³	in./in.-°F × 10⁶	mm/mm-°C × 10⁶	
Wood (seasoned)					
Douglas fir	32	510	2.1	3.8	parallel to grain
			32	58	perpendicular to grain
Pine	26	415	3.0	5.4	parallel to grain
			19	34	perpendicular to grain
Oak, red or white	41–46	655–735	2.7	4.9	parallel to grain
			30	54	perpendicular to grain
Masonry					
Limestone	130–170	2080–2720	4.4	7.9	
Granite	165–170	2640–2720	4.7	8.5	
Marble	165–170	2640–2720	7.3	13.1	
Brick	100–140	1600–2240	3.6	6.5	
Concrete masonry units	75–145	1200–2320	5.2	9.4	
Concrete					
Normal-weight concrete	145	2320	5.5	9.9	
Metals					
Steel	490	7850	6.5	11.7	
Stainless steel, Type 304	490	7850	9.9	17.8	
Aluminum	165	2640	12.8	23.1	
Copper	556	8900	9.3	16.8	
Finish Materials					
Gypsum board	43–50	690–800	9.0	16.2	
Gypsum plaster, sand	105	1680	7.0	12.6	
Glass	156	2500	5.0	9.0	
Acrylic glazing sheet	72	1150	41.0	74.2	
Polycarbonate glazing sheet	75	1200	44.0	79.6	
Polyethylene	57–61	910–970	85.0	153	
Polyvinyl chloride (PVC)	75–106	1200–1700	40.0	72.0	

APPENDIX

Unit Conversions

U.S. Customary	Metric		Metric	U.S. Customary
1 in.	25.4 mm		1 mm	0.394 in.
1 ft	304.8 mm		1 m	39.37 in.
1 ft	0.3048 m		1 m	3.2808 ft
1 lb	0.4536 kg		1 kg	2.205 lb
1 lb	4.448 N		1 N	0.2248 lb
1 ton	0.9072 metric ton (tonne)		1 metric ton (tonne)	2205 lb
1 ft^2	0.0920 m^2		1 m^2	10.76 ft^2
1 ft^3	0.02832 m^3		1 m^3	35.31 ft^3
1 ft^3	28.32 L		1 L	0.03531 ft^3
1 psi	6.894 kPa		1 kPa	0.1451 psi
1 psi	0.006894 MPa		1 MPa	145.1 psi
1 lb/ft^2	4.882 kg/m^2		1 kg/m^2	0.2048 lb/ft^2
1 lb/ft^2	0.04788 kPa		1 kPa	20.89 lb/ft^2
1 lb/ft^3	16.02 kg/m^3		1 kg/m^3	0.06243 lb/ft^3
1 lb/ft^3	0.01602 g/cm^3		1 g/cm^3	62.43 lb/ft^3
1 ft/min	0.00508 m/s		1 m/s	197 ft/min
1 cfm	0.0004719 m^3/s		1 m^3/s	2119 cfm
1 cfm	0.4739 L/s		1 L/s	2.119 cfm
1 BTU	0.2522 kcal		1 kcal	3.966 BTU
1 BTU	1.055 kJ		1 kJ	0.9478 BTU
1 BTUH	0.2931 W		1 W	3.412 BTUH

Thermal resistance and conductance				
R 1 (ft^2-hr-°F/BTU)	RSI 0.1761 (m^2-°K/W)[a]		RSI 1 (m^2-°K/W)[a]	R5.678 (ft^2-hr-°F/BTU)
U 1 (BTU/ft^2-hr-°F)	U5.678 (W/m^2-°K)[a]		U 1 (W/m^2-°K)[a]	U 0.1761 (BTU/ft^2-hr-°F)

Vapor permeance				
1 perm (grains/hr-ft^2-in. Hg)	57.21 ng/s-m^2-Pa		1 ng/s-m^2-Pa	0.01748 perms (grains/hr-ft^2-in. Hg)

Air permeance				
1 cfm/sf@1.57 psf	5.080 L/s-m^2@75 Pa		1 L/s-m^2@75 Pa	0.1968 cfm/sf@1.57 psf

Other Conversions				
1 mil = 0.001 in.			1 Pa = 0.1020 kg/m^2	
1 in. = 1000 mil			1 kg/m^2 = 9.807 Pa	

[a] Metric thermal conductance and resistance units may be expressed with either "°K" or "°C." In either case, the conversion factors are the same, because a 1-degree difference in either the Celsius or Kelvin temperature scale is the same.

GLOSSARY

A

AAC *See* **Autoclaved aerated concrete.**

AAMA American Architectural Manufacturers Association, a trade organization that develops standards for windows, doors, skylights, storefront, and curtain wall systems.

AC *See* **Articulation class.**

Accelerating admixture An *admixture* that causes concrete or mortar to cure more rapidly.

Access flooring A raised finish floor surface consisting of small, individually removable panels beneath which wiring, ductwork, and other building services may be installed.

Access standard A set of regulations or technical standards ensuring that buildings are accessible and usable by physically handicapped members of the population.

Acoustical ceiling A ceiling of fibrous tiles that are highly absorbent of sound energy.

ACQ *See* **Alkaline copper quat.**

Acrylic A transparent plastic material widely used in sheet form for glazing windows and skylights; a modifier added to plasters, paints, and cements to enhance various performance properties of the material.

Active metal A metal relatively high on a galvanic series, tending to act as an *anode* in *galvanic couples*.

Actual dimension The true dimension of a material, as distinct from its *nominal dimension*.

ADA *See* **Airtight drywall approach.**

ADA *See* **Americans with Disabilities Act.**

Admixture In a concrete or mortar, a substance other than cementitious material, water, and aggregates included in the mixture for the purpose of altering one or more properties of the mixed material, either in its plastic working state or after it has hardened.

Advanced framing techniques A wood light framing system that minimizes redundant framing members, reducing the amount of lumber required and increasing the thermal efficiency of the insulated frame.

Aerogel A silicon-based foam with relatively high thermal resistance, used in insulating blankets and insulated glazing units.

AESS *See* **Architecturally exposed structural steel.**

Aggregate Inert particles, such as sand, gravel, crushed stone, or expanded minerals, in a concrete, mortar, or plaster.

Air barrier A material that reduces air leakage through a building assembly.

Air barrier assembly An interconnected assemblage of materials that can effectively resist air pressure differentials acting across the boundary of a wall, roof, or floor assembly.

Air barrier system An interconnected collection of air barrier assemblies responsible for the overall air leakage performance of a completed building.

Air-entraining admixture An *admixture* that causes a controlled quantity of stable microscopic air bubbles to form in concrete or mortar during mixing, usually for the purposes of increasing workability and resistance to freeze-thaw conditions.

Air permeance A measure of a material's permeability to airflow. A low air permeance is a desirable characteristic of an air barrier material.

Air-supported structure A structure, usually long span, with a fabric roof supported by a higher air pressure inside the structure relative to outside the structure.

Airtight drywall approach (ADA) An air barrier system relying on gypsum board interior finish and the sealing of joints between framing members of a light frame building, serving to restrict the flow of air through the exterior walls and roof.

AISC American Institute of Steel Construction. A trade association that establishes standards for the use of structural steel in buildings and other structures.

Albedo *See* **Solar reflectance.**

Alkaline copper quat (ACQ) A chemical used to preserve wood against attack by decay and insects.

Alloy A substance composed of two or more metals or of a metal and a nonmetallic constituent.

Aluminum A silver-colored, *nonferrous metal* that naturally forms a self-protecting oxide layer.

Americans with Disabilities Act (ADA) A federal regulation establishing equal access for persons with disabilities to public accommodations, commercial facilities, and transportation facilities.

Anchor Any of a variety of mechanical devices used to fasten or connect one material or member to another.

Anchor bolt A bolt embedded in concrete for the purpose of fastening a building frame to a concrete or masonry foundation.

Angle A structural section of steel, aluminum, or other material whose profile resembles the letter *L*.

Angle of repose *See* **Maximum allowable slope.**

Annealed Cooled under controlled conditions to minimize internal stresses, usually referring to iron or steel members.

Anode The metal in a *galvanic couple* that experiences accelerated corrosion.

Anodizing An electrolytic process that forms a permanent protective oxide coating on aluminum, with or without added color.

ANSI American National Standards Institute. An organization that fosters the establishment of voluntary industrial standards.

Anticlastic Saddle-shaped or having curvature in two opposing directions.

APP *See* **Atactic polypropylene.**

Appearance grading The grading of wood for its appearance, as distinct from its structural properties; not to be confused with *visual grading*.

Apron The finish piece that covers the joint between a window stool and the wall finish below.

Arch A structural device that supports a vertical load by translating it into axial inclined forces at its supports.

Architectural concrete Concrete intended as a finish surface and produced to a higher-quality standard.

Architecturally exposed structural steel (AESS) Structural steel intended to be left exposed in the finished building and fabricated and installed to a higher-quality standard.

Architectural sheet metal roofing A roof covering made up of sheets of metal in a traditional shop- or site-fabricated pattern such as standing seam, flat seam, or batten seam.

Architectural Woodwork Institute (AWI) A trade organization that develops standards for custom millwork.

Arc welding A process of joining two pieces of metal by melting them together at their interface with a continuous electric spark and adding a controlled additional amount of molten metal from a metallic electrode.

Area divider A curb used to partition a large roof membrane into smaller areas to allow for expansion and contraction in the deck and membrane.

Articulation class (AC) A measure of a finish ceiling's absorption and reflection of sound, particularly with regard to speech clarity in an open office environment.

Ash dump A door in the underfire of a fireplace that permits ashes from the fire to be swept into a chamber beneath, from which they may be removed at a later time.

Ashlar Squared stonework.

Asphalt A tarry brown or black mixture of hydrocarbons; one type of *bitumen*.

Asphalt roll roofing A continuous sheet of the same roofing material used in *asphalt shingles*.

Asphalt-saturated felt A water-resistive sheet material, available in several different thicknesses, usually consisting of matted cellulose fibers that have been impregnated with asphalt; used to provide a weather-protective layer in an exterior wall or roof assembly; also called *building felt*.

Asphalt shingle A roofing unit composed of a heavy organic or inorganic felt saturated with asphalt and faced with mineral granules.

ASTM Formerly, American Society for Testing and Materials. An organization that promulgates standards for testing, materials, and methods of building construction.

Atactic polypropylene (APP) An amorphous form of polypropylene used as a modifier in *modified bitumen roof membrane*.

Auger A helical tool for creating cylindrical holes.

Autoclaved aerated concrete (AAC) Concrete formulated so as to contain a large percentage of gas bubbles as a result of a chemical reaction that takes place in an atmosphere of steam.

AWI *See* **Architectural Woodwork Institute**.

Awning window A window that pivots on a horizontal axis at the top edge of the sash and projects toward the outdoors.

Axial In a direction parallel to the long axis of a structural member.

B

Backer board A moisture-resistant board, used as a base for *thin-set tile* applications, usually composed of fiber-reinforced cement or glass-mat-faced gypsum.

Backer rod A flexible, compressible strip of plastic foam inserted into a joint to limit the depth to which *gunnable sealant* can penetrate.

Backfill Earth or earthen material used to fill the excavation around a foundation; the act of filling around a foundation.

Backup bar A small rectangular strip of steel applied beneath a joint to provide a solid base for beginning a weld between two steel structural members.

Backup wall A vertical plane of masonry, concrete, or wood framing used to support a thin facing such as a single wythe of brickwork.

Ballast A heavy material installed over a roof membrane to prevent wind uplift and shield the membrane from sunlight.

Balloon frame A wooden building frame composed of closely spaced members nominally 2 inches (50 mm) thick, in which the wall members are single pieces that run from the top of the foundation to the underside of the roof framing.

Baluster A vertical member that serves to fill the opening between a handrail and a stair or floor.

Band joist A wooden joist running perpendicular to the primary direction of the joists in a floor and closing off the floor platform at the outside face of the building. Also called a *rim joist*.

Bar A small rolled steel shape, usually round or rectangular in cross section; a *steel reinforcing bar* used for reinforcing concrete.

Barrel shell A scalloped roof structure of reinforced concrete that spans in one direction as a barrel vault and in the other as a folded plate.

Barrel vault A segment of a cylindrical surface that spans as an arch.

Barrier wall An exterior wall of a building whose watertightness depends on its freedom from passages through the wall. Also called a *face-sealed wall*.

Basalt A dense and durable igneous rock, usually dark gray; classified by ASTM C119 in the Granite group.

Baseboard A strip of finish material placed at the junction of a floor and a wall to create a neat intersection and to protect the wall against damage from feet, furniture, and floor-cleaning equipment.

Base-coat plaster One or more preparatory plaster coats that provide a flat, solid surface suitable for the application of the final finish coat plaster. *See also* **Scratch coat** and **Brown coat**.

Base flashing The flashing at the edges of a low-slope roof membrane that turns up against the adjacent face of a parapet or wall; frequently overlapped by a *counterflashing*.

Base isolator A device at foundation level that diminishes the transmission of seismic motions to a building.

Baseplate A steel plate inserted between a column and a foundation to spread the concentrated load of the column across a larger area of the foundation.

Basic oxygen process A steel-making process in which a stream of pure oxygen is introduced into a batch of molten iron so as to remove excess carbon and other impurities.

Batten A strip of wood or metal used to cover the gap between two adjoining boards or panels.

Batten-seam A seam, roughly square in profile, in *architectural sheet metal roofing*.

Batter board Boards mounted on stakes outside the excavation area of a building, used to preserve locations for string lines marking the corners of the building foundation.

Bay A rectangular area of a building defined by four adjacent columns; a portion of a building that projects from a facade.

Bead A narrow line of weld metal or sealant; a strip of metal or wood used to hold a sheet of glass in place; a narrow, convex molding profile; a metal edge or corner accessory for plaster.

Beam A linear or curved structural member that acts primarily to resist nonaxial loads.

Beam blank *See* **Bloom.**

Bearing A point at which one building element rests upon another.

Bearing block A piece of wood fastened to a column to provide support for a beam or girder.

Bearing pad A block of plastic or synthetic rubber used to cushion the point at which one precast concrete element rests upon another.

Bearing wall A wall that carries structural loads from floors, roofs, or walls above.

Bed *See* **Casting bed.**

Bed joint The horizontal layer of mortar beneath a masonry unit.

Bedrock A solid stratum of rock.

Bending moment The combination of tension and compression forces that cause a beam or other structural member to bend. *See also* **Moment.**

Bending stress A compressive or tensile stress resulting from the application of a nonaxial force to a structural member.

Bent A plane of framing consisting of beams and columns joined together, often with rigid joints.

Bentonite clay A colloidal clay that swells to several times its dry volume when saturated with water; the primary ingredient in bentonite waterproofing.

Bessemer process An early method of steel manufacturing in which air was blown into a vessel of molten iron to burn out impurities.

Bevel An end or edge that is cut at other than a right angle.

Bevel siding Wood cladding boards that taper in cross section.

Billet A large cylinder or rectangular solid of material.

BIM *See* **Building information modeling.**

BIPV *See* **Building-integrated photovoltaic.**

Birdsmouth cut An angled notch cut into a rafter to allow the rafter to seat securely on the top plate of a wall.

Bite The depth to which the edge of a piece of glass is held by its frame.

Bitumen A tarry mixture of hydrocarbons, such as asphalt or coal tar.

Bituminous roof membrane A *low-slope roof* membrane made from bituminous materials, either a *built-up roof* membrane or a *modified bitumen roof membrane.*

Blast furnace slag A hydraulic cementitious material formed as a byproduct of iron manufacture, used in mortar and concrete mixtures; also called *slag cement.*

Blast-resistant glazing Window, storefront, or curtain wall systems designed for resistance to the force of explosive blasts.

Bleed water In freshly placed concrete, water that rises to the top surface of the concrete as the solid cement and aggregate particles settle.

Blended hydraulic cement Hydraulic cement made from a mixture of cementitious materials such as portland cement, other hydraulic cements, and pozzolans for the purpose of altering one or more properties of the cement or reducing the energy required in the cement manufacturing process.

Blind nailing Attaching boards to a frame, sheathing, or subflooring with toe nails driven through the edge of each piece so as to be completely concealed by the adjoining piece.

Blind-side waterproofing An impervious layer or coating on the outside of a foundation wall that, for reasons of inaccessibility, was installed before the wall was constructed.

Blocking Pieces of wood inserted tightly between joists, studs, or rafters in a building frame to stabilize the structure, inhibit the passage of fire, provide an attachment surface for finish materials, or retain insulation.

Bloom A rectangular solid of steel formed from an ingot as an intermediate step in creating rolled steel structural shapes.

Blooming mill A set of rollers used to transform an ingot into a bloom.

Bluestone A sandstone that is gray to blue-gray and splits readily into thin slabs; classified by ASTM C119 in the Quartz-Based Stone group.

Board foot A unit of lumber volume, nominally 12 square inches in cross-sectional area and 1 foot long.

Board siding Wood cladding made up of boards, as differentiated from *shingles* or manufactured wood panels.

Bolster A long chair used to support *reinforcing bars* in a concrete slab.

Bolt A fastener, usually metallic, consisting of a cylindrical body with a head at one end and a helical thread at the other, intended to be inserted through holes in adjoining pieces of material and closed with a threaded *nut.*

Bond In masonry, the adhesive force between mortar and masonry units, or the pattern in which masonry units are laid to tie two or more wythes together into a structural unit. In reinforced concrete, the adhesion between the surface of a *reinforcing bar* and the surrounding concrete.

Bond breaker A strip of material to which *gunnable sealant* does not adhere.

Bond classification A system for rating the durability of a structural wood panel under conditions of repeated wetting or long-term weather exposure.

Bonded posttensioning A system of prestressing in which the tendons are grouted after stressing so as to bond them to the surrounding concrete.

Bonded terrazzo Terrazzo flooring whose underbed is poured directly upon the structural floor.

Bottom bar A *reinforcing bar* that lies close to the bottom of a beam or slab.

Bottom plate *See* **Sole plate.**

Bound water In wood, the water held within the cellulose of the cell walls. *See also* **Free water.**

Box beam A bending member of metal or plywood whose cross section resembles a closed rectangular box.

Box girder A major spanning member of concrete or steel whose cross section is a hollow rectangle or trapezoid.

Box nail A nail with a slenderer shank than a *common nail;* used for fastening framing members in wood light frame construction.

Braced frame A structural building frame strengthened against lateral forces with diagonal members.

Bracing Diagonal members, either temporary or permanent, installed to stabilize a structure against lateral loads.

Brad A small *finish nail.*

Brake A machine used to form lengths of sheet metal into bent shapes.

Brake metal Sheet metal, formed into final shape using a *brake.*

Brazing A process that uses molten, *nonferrous metal* to join two pieces of metal. The brazing metal is melted at a temperature below that of the metals being joined, so that, unlike in welding, the joined metals remain in a solid state throughout the process.

Breather mat A wiry plastic matting placed within a roof or wall assembly to create a space for drainage and ventilation.

Bridging Bracing or blocking installed between steel or wood joists at intermediate points to stabilize the joists against buckling and, in some cases, to permit adjacent joists to share loads.

British thermal unit (BTU) The quantity of heat required to raise the temperature of 1 pound of water 1 degree Fahrenheit.

Broom finish A skid-resistant texture imparted to an uncured concrete surface by dragging a stiff-bristled broom across it.

Brown coat The second of two base-coat plaster applications in a three-coat plaster.

Brownstone A brownish or reddish sandstone; classified by ASTM C119 in the Quartz-Based Stone group.

BTU *See* **British thermal unit.**

Buckling Structure failure by gross lateral deflection of a slender element under compressive stress, such as the sideward buckling of a long, slender column or the upper edge of a long, thin floor joist.

Building brick Brick used for concealed masonry work where appearance is not a concern.

Building code A set of regulations intended to ensure a minimum standard of health and safety in buildings.

Building enclosure The parts of the building, principally its walls, roofs, and fenestration, that separate the interior of the building from the exterior, and that must effectively control the flow of heat, air, and moisture; also called the *thermal envelope* or the *building envelope.*

Building envelope *See* **Building enclosure.**

Building felt *See* **Asphalt-saturated felt.**

Building information modeling (BIM) The digital, three-dimensional modeling of building systems, with the linking of model components to a database of properties and relationships.

Building-integrated photovoltaic (BIPV) Conventional building materials laminated or coated with *photovoltaic* materials and interconnected into a power-generating array.

Building paper A water-resistive, asphalt-saturated paper used similarly to *asphalt-saturated felt,* to provide a protective layer in an exterior wall assembly.

Building separation joint A plane along which a building is divided into separate structures that may move independently of one another.

Built-up roof (BUR) A multi-ply roof membrane, made from layers of asphalt-saturated felt or other fabric, bonded together with bitumen.

Bull float A long-handled tool used for the initial floating of a freshly poured concrete slab. *See also* **Darby.**

BUR *See* **Built-up roof.**

Butt The thicker end, such as the lower edge of a wood shingle or the lower end of a tree trunk; a joint between square-edged pieces; a weld between square-edged pieces of metal that lie in the same plane; a type of door hinge that attaches to the edge of the door.

Butt-joint glazing A type of glass installation in which the vertical joints between lites of glass do not meet at a mullion, but are made weathertight with a sealant.

Button head A smooth, convex bolt head with no provision for engaging a wrench.

Buttress A structural device, usually of masonry or concrete, that resists the diagonal forces from an arch or vault.

Butyl rubber A synthetic rubber compound.

C

CA *See* **Copper boron azole.**

CAC *See* **Ceiling Attenuation Class.**

CAD *See* **Computer-aided design.**

Caisson A cylindrical sitecast concrete foundation unit that penetrates through incompetent soil to rest upon an underlying stratum of rock or satisfactory soil; an enclosure that permits excavation work to be carried out underwater. Also called a *drilled pier.*

Calcined gypsum Gypsum that has been ground to a fine powder and heated to drive off most of its water of hydration; used in the manufacture of gypsum board and as the principal ingredient in gypsum plasters; a nonhydraulic cementitious material; also called *plaster of Paris.*

Calcining The driving off of the water of hydration from gypsum by the application of heat.

Camber A slight, intentional initial curvature in a beam or slab.

Cambium The thin layer beneath the bark of a tree that manufactures cells of wood and bark.

Cantilever A beam, truss, or slab that extends beyond its last point of support.

Cant strip A strip of material with a sloping face used to ease the transition from a horizontal to a vertical surface at the edge of a membrane roof.

Capillary action The pulling of water through a small orifice or fibrous material by the attractive force between the water and the material.

Capillary break A slot or groove intended to create an opening too large to be bridged by a drop of water and, thereby, to eliminate the passage of water by *capillary action;* the coarse aggregate layer under a concrete slab on grade which discourages the migration of water from the ground below into the concrete slab above.

Carbide-tipped tools Drill bits, saws, and other tools with cutting edges made of an extremely hard alloy.

Carbonation The process by which lime mortar reacts with atmospheric carbon dioxide to cure.

Carbon fiber reinforcing In precast concrete, an open grid fabric of carbon fibers bonded with epoxy resin, used as a substitute for *welded wire reinforcing.*

Carbon steel Low-carbon or *mild steel.*

Carbon steel bolt A relatively low-strength bolt most often used for fastening minor steel framing elements or temporary connections.

Carpenter One who makes things of wood.

Casement window A window that pivots on an axis at or near a vertical edge of the sash.

Casing The wood finish pieces surrounding the frame of a window or door; a cylindrical steel tube used to line a drilled or driven hole in foundation work.

Castellated beam A steel wide-flange section whose web has been cut along a zigzag path and reassembled by welding in such a way as to create a deeper section.

Casting Pouring a liquid material or slurry into a mold whose form it will take as it solidifies.

Casting bed A permanent, fixed form in which precast concrete elements are produced.

Cast in place concrete Concrete that is poured in its final location; *sitecast concrete.*

Cast iron Iron with too high a carbon content to be classified as steel.

Cathode The metal in a *galvanic couple* that experiences a decreased rate of corrosion.

Caulk A low-range *sealant.*

Cavity drainage material A material placed in the airspace of a cavity wall to catch mortar droppings and prevent clogging of weep holes at the bottom of the cavity.

Cavity wall A masonry wall that includes a continuous airspace between its outermost wythe and the remainder of the wall.

CBA *See* **Copper boron azole.**

CCA *See* **Chromated copper arsenate.**

Cee A metal framing member whose cross-sectional shape resembles the letter *C.*

Ceiling Attenuation Class (CAC) An index of the ability of a ceiling construction to obstruct the passage of sound between rooms through the plenum.

Ceiling joist *See* **Joist.**

Cellular decking Panels made of steel sheets corrugated and welded together in such a way that hollow longitudinal cells are created within the panels.

Cellular raceway A rectangular tube cast into a concrete floor slab for the purpose of housing electrical and communications wiring.

Cellulose A complex polymeric carbohydrate of which the structural fibers in wood are composed.

Celsius A temperature scale on which the freezing point of water is established as 0 and the boiling point as 100 degrees.

Cement Generally, any substance used to adhere material together. In concrete, masonry, and plastering work, any of a number of inorganic materials that have cementing properties when combined with water. *See also* **Cementitious materials.**

Cementitious materials In concrete, masonry, and plastering, inorganic materials that, when mixed with water, produce hardened products with adhesive and cohesive (cementing) properties; frequently used to refer exclusively to hydraulic cements (such as portland cement), to the exclusion of nonhydraulic (lime and gypsum) cements.

Cement–lime mortar Mortar made from portland cement, hydrated lime, aggregate, and water, the most traditional formulation of modern masonry mortars. *See also* **Masonry cement, Mortar cement.**

Centering Temporary formwork for an arch, dome, or vault.

Centering shims Small blocks of synthetic rubber or plastic used to hold a sheet of glass in the center of its frame.

Ceramic tile Small, flat, thin, fired clay tiles intended for use as wall and floor facings.

Chair A device used to support *reinforcing bars.*

Chamfer A flattening of a longitudinal edge of a solid member on a plane that lies at an angle of 45 degrees to the adjoining planes.

Channel A steel or aluminum section shaped like a rectangular box with one side missing.

Chemically strengthened glass Glass strengthened by immersion in a molten salt bath, causing an ion exchange at the surfaces of the glass that creates a prestress in a manner similar to *heat-treated glass.*

Chlorinated polyethylene (CPE) A plastic material used in roof membranes.

Chlorosulfonated polyethylene (CSPE) A plastic material used in roof membranes.

Chord A top or bottom member of a truss.

Chromated copper arsenate (CCA) A chemical used to protect wood against attack by decay and insects. Due to toxicity concerns, this chemical has been phased out of most treated wood used in residential and commercial building construction.

Chromogenic glass Glass that can change its optical properties, such as *thermochromic, photochromic,* or *electrochromic glass.*

C-H stud A steel wall framing member whose profile resembles a combination of the letters *C* and *H;* used to support gypsum panels in shaft walls.

Chuck A device for holding a steel wire, rod, or cable securely in place by means of steel wedges in a tapering cylinder.

Churn drill A steel tool used with an up-and-down motion to cut through rock at the bottom of a steel pipe caisson.

Cladding A material used to cover the exterior of a building.

Clamp A tool for holding two pieces of material together temporarily; unfired bricks piled in such a way that they can be fired without using a kiln.

Class A, B, C roofing Roof covering materials classified according to their resistance to fire when tested in accordance with ASTM E108. Class A is the most resistant, and Class C is the least.

Clay A fine-grained soil with plate-shaped particles, typically less than 0.0002 inch (0.005 mm) in size, whose properties are significantly influenced by the structural arrangements of the particles and the electrostatic forces acting between them.

Cleanout hole An opening at the base of a masonry wall through which mortar droppings and other debris can be removed before the interior cavity of the wall is grouted.

Clear dimension, clear opening The dimension between opposing inside faces of an opening.

Cleavage membrane A resilient sheet placed underneath a finish tile assembly to prevent movement stresses in the underlying substrate from telegraphing into the finish assembly.

Climbing crane A heavy-duty lifting machine that raises itself as the building rises.

Clinker A fused, pebblelike mass that is an intermediate product of cement manufacture; a brick that is overburned.

Closer The last masonry unit laid in a course; a partial masonry unit used at the corner of a header course to adjust the joint spacing; a mechanical device for regulating the closing action of a door.

CLSM *See* **Controlled low-strength material**.

CLT *See* **Cross-laminated timber**.

CMU *See* **Concrete masonry unit**.

Coarse aggregate Gravel or crushed stone in a concrete mix.

Coarse-grained soil Soil with particles ranging in size from roughly 0.003 to 3 inches (0.075–75 mm); sands and gravels.

Code *See* **Building code**.

Coefficient of Friction A unitless measure of the resistance to slippage between two surfaces, either at rest (*static coefficient of friction*) or in motion (*dynamic coefficient of friction*) relative to each other. A higher value represents greater slip resistance. Used to describe the slip resistance of some finish flooring materials.

Cohesionless soil *See* **Frictional soil**.

Cohesive soil A soil such as clay whose particles are able to adhere to one another by means of cohesive and adhesive forces.

Cold-formed steel Steel formed at a temperature at which it is no longer plastic, as by rolling or forging at room temperature.

Cold-rolled steel Steel rolled to its final form at a temperature at which it is no longer plastic.

Cold-worked steel *See* **Cold-formed steel**.

Collar joint The vertical mortar joint between wythes of masonry.

Collar tie A piece of wood nailed across two opposing rafters near the ridge to resist wind uplift.

Collated nails Nails glued together in a strip for insertion into a nail gun.

Collector A framing component that transfers lateral forces into parts of the structure designed to resist those forces.

Column An upright structural member acting primarily in compression.

Column bar *See* **Vertical bar**.

Column cage An assembly of vertical *reinforcing bars* and *ties* for a concrete column.

Column-cover-and-spandrel system A system of cladding in which panels of material cover the columns and spandrels, with horizontal strips of windows filling the remaining portion of the wall.

Column spiral A continuous coil of steel reinforcing used to tie a concrete column.

Column tie A single loop of steel bar, usually bent into a rectangular configuration, used to tie a concrete column.

Combination door A door with interchangeable inserts of glass and insect screening, usually used as a second, exterior door and mounted in the same opening with a conventional door.

Commercial wrap A synthetic sheet material, heavier than *housewrap*, with water-resistive and air-resistive properties used to provide a protective layer in an exterior wall assembly.

Common bolt *See* **Carbon steel bolt**.

Common bond Brickwork laid with five courses of stretchers followed by one course of headers.

Common nail A standard-sized nail used for the fastening of rough framing members in wood light frame construction.

Common rafter A roof *rafter* that runs parallel to the main slope of the roof. *See also* **Hip rafter**.

Composite A material or assembly made up of two or more materials bonded together to act as a single unit.

Composite column An upright structural member, acting primarily in compression, that is composed of concrete and a steel structural shape, usually a wide flange or a tube.

Composite construction An assembly of differing materials, such as concrete and steel or concrete and wood, that once assembled is capable of performing as a unified structural unit.

Composite metal decking Corrugated steel decking manufactured in such a way that it bonds securely to the concrete fill to form a reinforced concrete deck.

Composite wall A masonry wall that incorporates two or more different types of masonry units, such as clay bricks and concrete blocks.

Composition shingle *See* **Asphalt shingle**.

Compression A squeezing force.

Compression gasket A synthetic rubber strip that seals around a sheet of glass or a wall panel by being squeezed tightly against it.

Compressive strength The ability of a material to withstand squeezing forces.

Computer-aided design (CAD) The digital two-dimensional representation of building systems.

Concave joint A mortar joint tooled into a curved, indented profile.

Concealed flashing *See* **Internal flashing**.

Concealed grid A suspended ceiling framework that is completely hidden by the tiles or panels it supports.

Concrete A structural material produced by mixing predetermined amounts of cement, aggregates, and water and allowing this mixture to cure under controlled conditions.

Concrete block A *concrete masonry unit*, usually hollow, that is larger than a brick.

Concrete masonry unit (CMU) A block of hardened concrete, with or without hollow cores, designed to be laid in the same manner as a brick or stone; a concrete block.

Condensate Water formed as a result of *condensation*.

Condensation The process of changing from a gaseous to a liquid state, especially as applied to water.

Conduit A steel or plastic tube through which electrical wiring is run.

Consolidate In freshly poured concrete, to eliminate trapped air and cause the concrete to fill completely around the reinforcing bars and into all the corners of the formwork; usually done by vibrating the concrete.

Construction documents The graphic *construction drawings* and written *specifications* to which a building is constructed.

Construction drawings The graphic portion of the *construction documents*, usually produced by an architect or engineer, concerning the construction of a building.

Construction manager An entity that assists the owner in the procurement of construction services.

Construction Type In the International Building Code, any of five major systems of building construction that are differentiated by their relative resistance to fire.

Continuous ridge vent *See* **Ridge vent**.

Contraction joint *See* **Control joint**.

Contractor A person or organization that undertakes a legal obligation to do construction work.

Control joint An intentional, linear discontinuity in a structure or component designed to form a plane of weakness where cracking can occur in response to various forces so as to minimize or eliminate cracking elsewhere. Also called a *contraction joint*.

Controlled low-strength material (CLSM) A concrete that is purposely formulated to have a very low but known strength, used primarily as a *backfill* material.

Convector A heat exchange device that uses the heat in steam, hot water, or an electric resistance element to warm the air in a room; often called, inaccurately, a *radiator.*

Cool color A coating applied to a roofing material that is nonwhite, yet reflects a relatively high percentage of the sun's thermal energy.

Cool roof A roof covering that reflects a substantial portion of the sun's thermal energy.

Cope The removal of a flange at the end of a steel beam in order to facilitate connection to another member; cutting of the end of a wood trim to match the profile of the piece with which it intersects.

Coped connection A joint in which the end of one member is cut to match the profile of the other member.

Coping A protective cap on the top of a masonry wall.

Coping saw A handsaw with a thin, narrow blade, used for cutting detailed shapes in the ends of wood moldings and trim.

Copolymer A large molecule composed of repeating patterns of two or more chemical units.

Copper A soft, *nonferrous metal,* orange-red in original color, that oxidizes to a color ranging from blue-green to black.

Copper boron azole (CBA, CA) A chemical used to preserve wood against attack by decay and insects.

Corbel A spanning device in which masonry units in successive courses are cantilevered slightly over one another; a projecting bracket of masonry or concrete.

Coreboard A thick gypsum panel used primarily in *shaft walls.*

Corner bead A metal or plastic strip used to form a neat, durable edge at an outside corner of two walls of plaster or gypsum board.

Cornice The exterior detail at the meeting of a wall and a roof overhang; a decorative molding at the intersection of a wall and a ceiling.

Corrosion Oxidation, such as rust.

Corrosion inhibitor A concrete, mortar, or plaster admixture intended to prevent oxidation of *steel reinforcing bars.*

Corrugated Formed into a fluted or ribbed profile.

Counterflashing A flashing turned down from above to overlap another flashing turned up from below so as to shed water.

Course A horizontal layer of masonry units one unit high; a horizontal line of shingles or siding.

Coursed In masonry, laid in courses with straight bed joints.

Cove base A flexible strip of plastic or synthetic rubber used to finish the junction between resilient flooring and a wall.

Cover In concrete, a specified thickness of concrete surrounding *reinforcing bars* to provide full embedment for the bars and protect them against fire and corrosion.

Coverboard A material installed over low-slope roof insulation, to protect the insulation, provide a more stable substrate for the roof membrane, or increase the fire resistance of the roof assembly.

CPE *See* **Chlorinated polyethylene.**

Cradle-to-gate analysis A *life-cycle analysis* extending from original resource extraction only so far as when the material or product leaves its place of manufacture.

Cradle-to-grave analysis *See* **Life-cycle analysis.**

Crawlspace A space that is not tall enough to stand in, located beneath the bottom floor of a building.

Creep A permanent inelastic deformation in a material due to changes in the material caused by the prolonged application of a structural stress; common in wood, concrete, and plastics.

Cripple stud A wood wall framing member that is shorter than full-length studs because it is interrupted by a header or sill.

Critical path The sequence of tasks that determines the least amount of time in which a construction project can be completed.

Cross-grain wood Wood incorporated into a structure in such a way that the direction of its grain is perpendicular to the direction of the principal loads on the structure.

Cross-laminated timber (CLT) Structural panels fabricated from solid lumber pieces, with members in each layer alternating in orientation from those above and below; used as structural floor, wall, and roof elements.

Crosslot bracing Horizontal compression members running from one side of an excavation to the other, used to support sheeting.

Crown glass Glass sheet formed by spinning an opened hollow globe of heated glass.

Cruck A framing member cut from a bent tree so as to form one-half of a rigid frame.

CSPE *See* **Chlorosulfonated polyethylene.**

Cup A curl in the cross section of a board or timber caused by unequal shrinkage or expansion between one side of the board and the other.

Curing The hardening of concrete, plaster, gunnable sealant, or other wet materials. Curing can occur through evaporation of water or a solvent, hydration, polymerization, or chemical reactions of various types, depending on the formulation of the material.

Curing compound A liquid that, when sprayed on the surface of newly placed concrete, forms a water-resistant layer to prevent premature dehydration of the concrete.

Curtain wall An exterior building wall that is supported by the frame of the building, rather than being self-supporting or loadbearing.

Cylinder glass Glass sheet produced by blowing a large, elongated glass cylinder, cutting off its ends, slitting it lengthwise, and opening it into a flat rectangle.

D

d *See* **Penny.**

Damage weighted transmittance (Tdw) The ratio of solar radiation that passes through a glazing unit to the amount of light striking the unit, weighted to account for the relative fading damage potential of the various wavelengths. The lower the damage weighted transmittance, the better the protection against fading afforded to interior fabrics and materials.

Damper A flap to control or obstruct the flow of gases; specifically, a metal control flap in the throat of a fireplace or in an air duct.

Damping The addition of energy-absorbing components into a structural building frame, to reduce lateral deflections and lessen the stresses imparted into the frame when subjected to high wind or seismic forces.

Dampproofing A coating intended to resist the passage of water, commonly applied to the outside face of basement walls or to the inner face of a cavity in a masonry cavity wall.

Dap A notch at the end of a piece of material.

Darby A stiff straightedge of wood or metal used to level the surface of wet plaster or concrete.

Daylighting Illuminating the interior of a building by natural means.

DCOF *See* **Dynamic coefficient of friction**.

Dead load Permanent loads on a building, including the weight of the building itself and any permanently attached equipment.

Decking A material used to span across beams or joists to create a floor or roof surface.

Deep foundation A building *foundation* that extends through upper strata of incompetent soil to reach deeper strata with greater bearing capacity.

Deformation A change in the shape of a structure or structural element caused by a load or force acting on the structure.

Deformed reinforcing bar *Steel reinforcing bars* with surface ribs for better bonding to concrete.

Demand-critical weld A structural steel connection weld, essential to the stability of the structure during a seismic event and subject to special quality control and inspection procedures during construction.

Depressed strand A pretensioning tendon that is pulled to the bottom of the beam at the center of the span to follow more closely the path of tensile forces in the member.

Derrick Any of a number of devices for hoisting building materials on the end of a rope or cable.

Design/bid/build project delivery A method of providing design and construction services in which the design and construction phases of the project are provided by different entities, usually used in combination with *sequential construction*.

Design/build project delivery A method of providing design and construction services in which the design and construction phases of the project are provided by a single entity; frequently used in combination with *fast track construction*.

Dewatering The extraction of water from an excavation or its surrounding soil.

Dew point The temperature at which water will begin to condense from a mass of air with a given moisture content.

Diagonal bracing *See* **Bracing**.

Diamond saw A tool with a moving chain, belt, wire, straight blade, or circular blade whose cutting action is carried out by diamonds.

Diaphragm action A bracing action that derives from the stiffness of a thin plane of material when it is loaded in a direction parallel to the plane. Diaphragms in buildings are typically floor, wall, or roof surfaces of plywood, reinforced masonry, steel decking, or reinforced concrete.

Die An industrial tool for giving identical form to repeatedly produced or continuously generated units, such as a shaped orifice for giving form to a column of clay, a steel wire, or an aluminum extrusion; a shaped punch for making cutouts of sheet metal or paper; or a mold for casting plastic or metal.

Die-cut Manufactured by punching from a sheet material.

Differential settlement Subsidence of the various foundation elements of a building at differing rates.

Diffuser A louver shaped so as to distribute air about a room.

Dimension lumber Lengths of wood, rectangular in cross section, sawed directly from the log.

Dimension stone Building stone cut to a rectangular shape.

Direct tension indicator washer *See* **Load indicator washer**.

Distribution rib A transverse beam at the midspan of a one-way concrete joist structure, used to allow the joists to share concentrated loads.

Divider strip A strip of metal or plastic embedded in terrazzo to form control joints and decorative patterns.

Dome An arch rotated about its vertical axis to produce a structure shaped like an inverted bowl; a form used to make one of the cavities in a concrete waffle slab.

Dormer A structure protruding through the plane of a sloping roof, usually containing a window and having its own smaller roof.

Double glazing Two parallel sheets of glass with an airspace between.

Double-hung window A window with two overlapping sashes that slide vertically in tracks.

Double shear Acting to resist shear forces at two locations, such as a bolt that passes through a steel supporting angle, a beam web, and another supporting angle.

Double-skin facade An exterior wall system consisting of two separate glass skins separated by an interstitial space.

Double-strength glass Glass that is approximately ⅛ inch (3 mm) thick.

Double tee A precast concrete slab element that resembles the letters *TT* in cross section.

Dovetail slot anchor A system for fastening to a concrete structure that uses metal tabs inserted into a slot that is small at the face of the concrete and larger behind.

Dowel A short cylindrical rod of wood or steel; a *steel reinforcing bar* that projects from a foundation to tie it to a column or wall, or from one section of a concrete slab or wall to another.

Downspout A vertical pipe for conducting water from a roof to a lower level; also called a *leader*.

Drag strut A framing member or component acting as a *collector* to transfer lateral forces within the building frame; also called a *drag tie*.

Drainage Removal of water.

Drainage fill Crushed stone or gravel backfill materials with good drainage characteristics, placed around a foundation to facilitate drainage.

Draped tendon A posttensioning strand placed along a curving profile that approximates the path of the tensile forces in a beam.

Drawing Shaping a material by pulling it through an orifice, as in the drawing of steel wire or the drawing of a sheet of glass.

Drawings *See* **Construction drawings**.

Drawn glass Glass sheet pulled directly from a container of molten glass.

Drift Lateral deflection of a building caused by wind or earthquake loads.

Drift pin A tapered steel rod used to align bolt holes in steel connections during erection.

Drilled pier *See* **Caisson**.

Drip A discontinuity formed into the underside of a window sill or wall component to force adhering drops of water to fall free of the face of the building rather than to move farther toward the interior.

Dropchute A flexible hoselike tube for placing concrete; used to break the fall of the concrete and prevent *segregation*.

Drop panel A thickening of a two-way concrete structure at the head of a column.

Drying shrinkage Shrinkage of concrete, mortar, or plaster that occurs as excess water evaporates from the material.

Dry-pack grout A low-slump cementitious mixture tamped into the space in a connection between precast concrete members.

Dry-press process A method of molding slightly damp clays and shales into bricks by forcing them into molds under high pressure.

Dry-set mortar A tile-setting mortar formulated with portland cement, sand, and water retention compounds; used in thinset tile applications.

"Dry" systems Systems of construction that use little or no water during construction, as differentiated from systems such as masonry, plastering, and ceramic tile work.

Drywall *See* **Gypsum board**.

Dry well An underground pit filled with broken stone or other porous material from which rainwater from a roof drainage system can seep into the surrounding soil.

Duct A hollow conduit, commonly of sheet metal, through which air can be circulated; a tube used to establish the position of a posttensioning tendon in a concrete structure.

DWV pipe Drain–waste–vent pipe; the part of the plumbing system of a building that removes liquid wastes and conducts them to the sewer or sewage disposal system.

Dynamic coefficient of friction (DCOF) The *coefficient of friction*, measured between two surfaces in motion relative to each other; used in some finish flooring slip resistance measurements. *See also* **static coefficient of friction**.

E

Earlywood *See* **Springwood**.

Earth material Rock or soil.

Eave The horizontal edge at the low side of a sloping roof.

Edge bead A strip of metal or plastic used to make a neat, durable edge where plaster or gypsum board abuts another material.

Edge spacer The material used to separate lites of glass in an insulating glass unit; also called a *spline*.

Efflorescence A light-colored, powdery deposit on the face of masonry or concrete, caused by the leaching of chemical salts from water migrating from within the structure to the surface.

EIFS *See* **Exterior insulation and finish system**.

Elastic Able to return to its original size and shape after removal of stress.

Elastomer A rubber or synthetic rubber.

Elastomeric Rubberlike. In low-slope roofing, a *thermosetting* membrane material.

Electrochromic glass Glass that changes its optical properties in response to the application of electric current.

Electrode A consumable steel wire or rod used to maintain an arc and furnish additional weld metal in electric arc welding.

Electrogalvanizing A method of *galvanizing*, in which an electric current is used to deposit zinc from a liquid bath onto steel.

Elevation A drawing that views a building from any of its sides; a vertical height above a reference point such as sea level.

Elongation Stretching under load; growing longer because of temperature expansion.

Embodied carbon The total carbon emissions associated with a material or product throughout its life cycle.

Embodied energy The total energy consumption associated with a material or product throughout its life cycle.

Embodied water The total freshwater consumption associated with a material or product throughout its life cycle.

EMC *See* **Equilibrium moisture content**.

Enamel A glossy or semigloss paint.

End dam The turned-up end of a flashing that prevents water from running out of the end; a block inserted into the space within a horizontal aluminum mullion for the same purpose.

End nail A nail driven through the side of one piece of lumber and into the end of another.

Energy recovery ventilator A mechanical device that exhausts air from a build-

ing while recovering much of the sensible and latent heat from the exhausted air and transferring it to the incoming air.

Engineered fill Earth compacted into place in such a way that it has predictable physical properties, based on laboratory tests and specified, supervised installation procedures.

Engineered lumber *See* **Structural composite lumber**.

English bond Brickwork laid with alternating courses, each consisting entirely of headers or stretchers.

Environmental label Information or certifications used to describe the composition, environmental performance, or life-cycle impacts of a construction material or product.

EPDM Ethylene propylene diene monomer, a synthetic rubber thermosetting material used in low-slope roofing membranes.

Equilibrium moisture content (EMC) The moisture content at which wood stabilizes after a period of time in its destination environment.

Erector The subcontractor who raises, connects, and plumbs up a building frame from fabricated steel or precast concrete components.

Ethylene propylene diene monomer *See* **EPDM**.

Expanded metal lath A thin sheet of metal that has been slit and stretched to transform it into a mesh; used as a base for the installation of plaster.

Expanded shale aggregate A *structural lightweight aggregate* made from ground shale particles that have been heated to the point that moisture within the particles vaporizes, causing the particles to expand.

Expansion joint A seam within a material or between materials that provides for material expansion and contraction.

Expansive soil A clay soil that expands significantly with increased moisture content.

Exposed aggregate finish A concrete surface in which the coarse aggregate is revealed.

Exposed grid A framework for an acoustical ceiling that is visible from below after the ceiling is completed.

Extended-life admixture A substance that retards the onset of the curing reaction in mortar so that the mortar may be used over a protracted period of time after mixing.

Extended set-control admixture A substance that retards the onset of the curing reaction in concrete so that the material may be used over a protracted period of time after mixing.

Extensive green roof A *green roof* with a relatively shallow soil, planted with low-maintenance, drought-tolerant plant materials.

Exterior insulation and finish system (EIFS) A cladding system that consists of a thin layer of reinforced stucco applied directly to the surface of an insulating plastic foam board.

External flashing In masonry, a *flashing* that is not concealed within the wall, usually at the roof level or top of the wall.

External gutter A *gutter* hung off the edge of a roof, external to the roof construction itself.

Extrados The convex surface of an arch.

Extrusion The process of squeezing a material through a shaped orifice to produce a linear element with the desired cross section; an element produced by this process.

F

Fabricator The company that prepares structural steel members for erection; any entity that assembles building components prior to arrival of the components on the construction site.

Facade An exterior face of a building.

Face brick A brick selected on the basis of appearance and durability for use in the exposed surface of a wall.

Face nail A nail driven through the side of one wood member into the side of another.

Face-sealed wall *See* **Barrier wall.**

Face shell The portion of a hollow concrete masonry unit that forms the face of the wall.

Fahrenheit A temperature scale on which the boiling point of water is fixed at 212 degrees and the freezing point at 32 degrees.

Fanlight A semicircular or semielliptical window above an entrance door; often includes radiating muntins that resemble a fan.

Fascia The exposed vertical face of an *eave.*

Fast track construction A method of providing design and construction services in which design and construction overlap in time; also called *phased construction.*

Faying surface The contacting surfaces of steel members joined with a *slip-critical connection.*

Felt A thin, flexible sheet material made of fibers of paper, glass, or plastic pressed and bonded together.

Ferrous metal Any iron-based metal.

Ferrous steel In common usage, steel unprotected from corrosion by either galvanizing or alloying.

Fibrous reinforcing Short fibers of glass, steel, or polypropylene mixed into concrete to act as either *microfiber reinforcing* or *macrofiber reinforcing.*

Fieldstone Rough building stone gathered from river beds and fields.

Figure The surface pattern of the grain of a piece of smoothly finished wood or stone.

Filigree precast concrete A hybrid concrete system in which *precast concrete* sections are used as *permanent formwork* for *cast-in-place* concrete.

Fillet A rounded inside intersection between two surfaces that meet at an angle.

Fillet weld A weld at the inside intersection of two metal surfaces that meet at right angles.

Fine aggregate Sand used in concrete, mortar, or plaster mixes.

Fine-grained soil Soil with particles 0.003 inch (0.075 mm) or less in size; silts and clays.

Finger joint A glued end connection between two pieces of wood, using an interlocking pattern of deeply cut "fingers." A finger joint creates a large surface for the glue bond, to allow it to develop the full tensile strength of the wood it connects.

Finial A slender ornament at the top of a roof or spire.

Finish Exposed to view; material that is exposed to view.

Finish carpenter One who does *finish carpentry.*

Finish carpentry The wood components exposed to view on the building and usually assembled on site, such as window and door casings, baseboards, bookshelves, and the like; may also refer to exterior finish carpentry, such as exterior trim, deck railings, and similar items.

Finish coat The final coat of a paint, plaster, or other finishing system.

Finish-coat plaster The final coat of plaster, applied over gypsum base or one or more applications of *base-coat plaster.*

Finish floor The floor material exposed to view, as differentiated from the *subfloor,* which is the loadbearing floor surface beneath.

Finish lime A fine grade of *quicklime* used in finish-coat gypsum plasters and in ornamental plaster work; also called *lime putty.*

Finish nail A relatively thin nail with a very small head; used for fastening trim and other finish woodwork items.

Fire area In the International Building Code, an area within a building bounded by fire-resistant construction. Fire area size, occupant load, and location within the building are used to determine automatic sprinkler requirements.

Fire barrier In the International Building Code, a fire-resistant wall intended to deter the spread of fire; used to separate exit stair enclosures, differing occupancies, and fire areas.

Fireblocking Wood or other material used to partition concealed spaces within combustible framing; intended to restrict the spread of fire within such spaces.

Firebox The part of a fireplace, stove, or furnace in which fuel is combusted.

Firebrick A brick made to withstand very high temperatures, as in a fireplace, furnace, or industrial chimney.

Firecut A sloping end cut on a wood beam or joist where it enters a masonry wall. The purpose of the firecut is to allow the wood member to rotate out of the wall without prying the wall apart if the floor or roof structure burns through in a fire.

Fire door A fire-resistant door, used in fire resistance rated partitions and walls.

Fire partition In the International Building Code, a fire-resistant wall intended to deter the spread of fire; used to separate tenant spaces, dwelling units, and corridors from surrounding areas of a building.

Fireproofing Material used around steel or concrete structural elements to insulate them against excessive temperatures in case of fire.

Fire-protection rated glazing *Fire-rated glass* for use in fire doors, fire windows, and other protected openings. *See also* **Fire-resistance-rated glazing**.

Fire-rated glass Glass that is capable of retaining its integrity after being exposed to fire, either *fire-protection rated glazing* or *fire-resistance-rated glazing*.

Fire-resistance-rated glazing *Fire-rated glass* capable of substituting in full for solid, fire-resistance-rated wall assemblies. Unlike *fire-protection rated glazing*, fire-resistance-rated glazing is not limited to use in doors, windows, and other openings.

Fire resistance rating The time, in minutes or hours, that a material or assembly will resist fire exposure as determined by ASTM E119.

Firestopping A component or mastic installed in an opening through a floor or around the edge of a floor to retard the passage of fire; frequently used interchangeably with *fireblocking*.

Fire wall A wall extending from foundation to roof, required under a building code to separate buildings, or parts of buildings, as a deterrent to the spread of fire.

Fire zone A legally designated area of a city in which construction must meet established standards of fire resistance and/or combustibility.

Firing The process of converting dry clay into a ceramic material through the application of intense heat.

First cost The cost of construction, not including operational costs.

Fixed window Glass that is immovably mounted in a wall.

Flagstone Flat stones used for paving or flooring.

Flame-spread rating A measure of the rapidity with which fire will spread across the surface of a finish material as determined by ASTM standard E84.

Flange A projecting crosspiece of a wide-flange or channel profile; a projecting fin.

Flash cove A detail in which a sheet of resilient flooring is turned up at the edge and finished against the wall to create an integral baseboard.

Flashing A thin, continuous sheet of metal, plastic, rubber, or waterproof paper used to prevent the passage of water through a joint in a wall, roof, or chimney.

Flat-grain lumber *See* **Plainsawn lumber**.

Flat seam A sheet metal roofing seam that is formed flat against the surface of the roof.

Flemish bond Brickwork laid with each course consisting of alternating headers and stretchers.

Flitch A collection of solid wood members or veneers, all cut from a single log.

Float A small platform suspended on ropes from a steel building frame to permit ironworkers to work on a connection; a trowel with a slightly rough surface used in an intermediate stage of finishing a concrete slab; as a verb, to use a float for finishing concrete.

Float glass Glass sheet manufactured by cooling a layer of molten glass on a bath of molten tin.

Floating floor Wood or laminate flooring that is not fastened or adhered to the subfloor.

Floating foundation A *foundation* placed at depth such that the weight of the soil removed is approximately equal to the weight of the building being supported.

Flood test The submersion of a horizontal waterproofing system, usually for an extended period of time, to check for leaks.

Floor joist *See* **Joist**.

Flue A passage for smoke and combustion products from a furnace, stove, water heater, or fireplace.

Fluid-applied roof membrane A roof membrane applied in one or more coats of a liquid that cure to form an impervious sheet.

Fluoropolymer A highly stable organic compound used as a finish coating for building cladding.

Flush Smooth, lying in a single plane.

Flush door A door with smooth planar faces.

Flush glazing *See* **Structural glazing**.

Flux A material added to react chemically with impurities and remove them from molten metal. Fluxes are used both in steelmaking and in welding. Welding fluxes serve the additional purpose of shielding the molten weld metal from the air to reduce oxidation and other undesirable effects.

Fly ash Dust collected in the stacks of coal-fired power plants, used as a *supplementary cementitious material* in concrete and mortar.

Flying formwork Large sections of slab formwork that are moved by crane.

Fly rafter A *rafter* in a rake overhang.

F-number An index number expressing the statistical flatness or levelness of a concrete slab.

Foil-backed gypsum board *Gypsum board* with aluminum foil laminated to its back surface to act as a vapor retarder and thermal insulator.

Folded plate A roof structure whose strength and stiffness derive from a pleated or folded geometry.

Footing The part of a foundation that spreads a load from the building across a broader area of soil.

Forced-air system A furnace and/or cooling coil and ductwork that heat and/or cool air and deliver it, driven by a fan, to the rooms of a building.

Formaldehyde An organic compound known to cause a range of adverse human health effects, traditionally used in the manufacture of wood product adhesives and binders. *See also* **Phenol-formaldehyde** and **Urea-formaldehyde**.

Form deck Thin, corrugated steel sheets that serve as *permanent formwork* for a reinforced concrete deck.

Form release compound A substance applied to concrete formwork to prevent concrete from adhering.

Form tie A steel or plastic rod with fasteners on either end, used to hold together the two surfaces of formwork for a concrete wall.

Form tie hole A depression, typically conical in shape, in a cast-in-place concrete wall that remains after the protruding portions of a form tie are removed.

Formwork Structures, usually temporary, that give shape to poured concrete and support it and keep it moist as it cures.

Foundation The portion of a building that transmits structural loads from the building into the earth.

Framed connection A *shear connection* between steel members made by means of steel angles or plates connecting to the web of the beam or girder.

Framing plan A diagram showing the arrangement and sizes of the structural members in a floor or roof.

Framing square An L-shaped measuring tool used by carpenters to lay out right-angle cuts as well as more complicated cuts, such as those required for stairs and sloping roof rafters.

Freestone Fine-grained sedimentary rock that has no planes of cleavage or sedimentation along which it is likely to split.

Free water In wood, water held within the cavities of the cells. *See also* **Bound water**.

Freeze protection admixture A concrete or mortar additive, used to allow curing under conditions of low ambient temperature.

French door A symmetrical pair of glazed doors hinged to the jambs of a single frame and meeting at the center of the opening.

Frictional soil A soil, such as sand or gravel, that relies primarily on friction rather than attractive or repulsive forces between particles for its strength; also called a *cohesionless soil.*

Friction type connection *See* **Slip-critical connection**.

Frit Ground-up colored glass that is heat-fused to lites of glass to form functional or decorative patterns.

Frost line The depth in the earth to which the soil can be expected to freeze during a severe winter.

Fully restrained moment connection A steel frame *moment connection* sufficiently rigid such that the geometric angles between connected pieces remain unchanged during normal loading.

Furring channel A sheet metal *furring strip* in the form of a cee-channel.

Furring strip A length of wood or metal attached to a masonry or concrete wall to permit the attachment of finish materials using screws or nails; any linear material used to create a spatial separation between a finish material and an underlying substrate.

G

Gable The triangular wall beneath the end of a *gable roof.*

Gable roof A roof consisting of two oppositely sloping planes that intersect at a level ridge.

Gable vent A screened, louvered opening in a gable, used for exhausting excess heat and humidity from an attic.

Gage *See* **Gauge**.

Galling Chafing or tearing of one material against another under extreme pressure.

Galvanic couple A pair of metals with differing electrochemical potential, between which electrical current will flow if the metals are placed in a conducting medium.

Galvanic series A list of metals in order of their relative electrochemical potential when immersed in a given conducting medium.

Galvanized steel Steel with a zinc coating for the purpose of providing protection from corrosion.

Galvanizing The application of a zinc coating to steel.

Gambrel A roof shape consisting of two superimposed levels of gable roofs with the lower level at a steeper pitch than the upper.

Gantt chart A graphic representation of a construction schedule, using a series of horizontal bars representing the duration of various tasks or groups of tasks that make up the project.

Gap graded soil A soil graded so as to contain a broad range of particle sizes, but with certain sizes omitted.

Gasket A dry, resilient material used to seal a joint between two rigid assemblies by being compressed between them.

Gauge A measure of the thickness of sheet material. Lower gauge numbers signify thicker sheets; also spelled "gage."

Gauged brick A brick that has been rubbed on an abrasive stone to reduce it to a trapezoidal shape for use in an arch.

Gauging plaster A gypsum plaster formulated for use in combination with finishing lime in finish coat plaster.

General contractor A construction entity with responsibility for the overall conduct of a construction project.

Geotextile A synthetic cloth used beneath the surface of the ground to stabilize soil or promote drainage.

GFRC *See* **Glass-fiber-reinforced concrete**.

GFRP *See* **Glass-fiber-reinforced plastic**.

Girder A horizontal beam that supports other beams; a very large beam, especially one that is built up from smaller elements.

Girt A horizontal beam that supports wall cladding between columns.

Glass block A hollow masonry unit made of glass.

Glass fiber batt A thick, fluffy, nonwoven insulating blanket of filaments spun from glass.

Glass-fiber-reinforced concrete (GFRC) Concrete with a strengthening admixture of short, alkali-resistant glass fibers.

Glass-fiber-reinforced plastic (GFRP) Plastic resin strengthened with embedded glass fibers, commonly referred to as "fiberglass."

Glass mullion system A method of constructing a large glazed area by stiffening the sheets of glass with perpendicular glass ribs.

Glaze A glassy finish on a brick or tile; as a verb, to install glass.

Glazed structural clay tile A hollow clay block with glazed faces, usually used for constructing interior partitions.

Glazier One who installs glass.

Glazier's points Small pieces of metal driven into a wood sash to hold glass in place.

Glazing The act of installing glass; the transparent material (most often glass or plastic) in a glazed opening; as an adjective, referring to materials used in installing glass (for example, "glazing tape").

Glazing compound Any of several types of mastic used to bed small lites of glass in a frame.

Global warming potential A material or product's life-cycle contribution to global warming, caused by greenhouse gas emissions.

Glue-laminated wood A wood member made up of a large number of small strips of wood glued together.

Glulam A shorthand expression for glue-laminated wood.

Grade A classification of size or quality for an intended purpose; to classify as to size or quality.

Grade The surface of the ground; to move earth for the purpose of bringing the surface of the ground to an intended level or profile.

Grade beam A reinforced concrete foundation element that transmits the load from a bearing wall into spaced foundations such as pile caps or caissons.

Grain In wood, the direction of the longitudinal axes of the wood fibers or the figure formed by the fibers. In stone, *see* **Quarry bed**.

Granite Igneous rock with visible crystals of quartz and feldspar.

Green building *See* **Sustainability**.

Green roof A roof covered with soil and plant materials; also called a *vegetated roof.*

Groove weld A weld made in a groove, created by beveling or milling the edges of the mating pieces of metal.

Ground A strip attached to a wall or ceiling to establish the level to which plaster should be applied.

Grout A high-slump mixture of portland cement, aggregates, and water, which can be poured or pumped into cavities in concrete or masonry for the purpose of embedding *reinforcing bars* and/or increasing the amount of loadbearing material in a wall; a specially formulated mortar-like material for filling under steel baseplates and around connections in precast concrete framing; a mortar used to fill joints between ceramic tiles or quarry tiles.

Gunnable sealant A *sealant* material that is extruded in liquid or mastic form from a *sealant gun*.

Gusset plate A flat steel plate used to connect the members of a truss; a stiffener plate.

Gutter A horizontal or slightly sloped channel for the collection of rainwater and snowmelt at the eave of a roof.

GWB Gypsum wallboard; *see* **Gypsum board**.

Gypsum An abundant mineral; chemically, hydrous calcium sulfate.

Gypsum backing board A lower-cost gypsum panel intended for use as an interior layer in multilayer constructions of gypsum board.

Gypsum board An interior facing panel consisting of a gypsum core sandwiched between paper faces. Also called *drywall*, *plasterboard*.

Gypsum lath Sheets of gypsum board manufactured specifically for use as a plaster base.

Gypsum plaster Plaster whose cementing substance is calcined gypsum; used almost exclusively for interior finish plaster work.

Gypsum sheathing panel A water-resistant, gypsum-based sheet material used for exterior sheathing.

Gypsum wallboard (GWB) *See* **Gypsum board**.

H

Hammerhead boom crane A heavy-duty lifting device that uses a tower-mounted horizontal boom that may rotate only in a horizontal plane.

Hardboard A very dense panel product, usually with at least one smooth face, made of highly compressed wood fibers.

Hardwood Wood from deciduous (broadleaf) trees.

Hawk A square piece of sheet metal with a perpendicular handle beneath, used by a plasterer to hold a small quantity of wet plaster and transfer it to a trowel for application to a wall or ceiling.

HDO *See* **High-density overlay**.

Head The horizontal top portion of a window or door.

Header In framed construction, a member that carries other perpendicular framing members, such as a beam above an opening in a wall or a joist supporting other joists where they are interrupted by a floor opening. In steel construction, a beam that spans between girders. In masonry construction, a brick or other masonry unit that is laid across two wythes with its end exposed in the face of the wall.

Head jamb *See* **Head**.

Head joint The vertical layer of mortar between ends of masonry units.

Hearth The noncombustible floor area outside a fireplace opening.

Heartwood The dead wood cells in the center region of a tree trunk.

Heat-fuse To join by softening or melting the edges with heat and pressing them together.

Heat of hydration The thermal energy given off by concrete or gypsum as it cures.

Heat recovery ventilator A mechanical device that exhausts air from a building while recovering much of the sensible heat from the exhausted air and transferring it to the incoming air. Also called an air-to-air heat exchanger. *See also* **Energy recovery ventilator**.

Heat-strengthened glass *Heat-treated glass* that is not as strong as tempered glass, and that may not be used as *safety glazing*.

Heat-treated glass Glass that is strengthened by a heat treatment process; either *heat-strengthened glass* or *tempered glass*.

Heaving The forcing upward of ground or buildings by the action of frost or pile driving.

Heavy Timber construction A type of wood construction made from large wood members and solid timber decking in a post-and-beam configuration; in the International Building Code, buildings of *Type IV HT construction*, consisting of heavy timber interior construction and noncombustible exterior walls, which are considered to have moderate fire-resistive properties.

High-density overlay (HDO) A resin-treated overlay applied to plywood panels to achieve a smoother, more durable face.

High-lift grouting A method of constructing a reinforced masonry wall in which the *reinforcing bars* are embedded in grout in story-high increments.

High-range sealant A *sealant* that is capable of a high degree of elongation without rupture.

High-range water-reducing admixture *See* **Superplasticizer**.

High-reactivity metakaolin A white-colored natural *pozzolan* that enhances appearance, workability, and hardened properties of concrete.

High-strength bolt A bolt designed to connect steel members by clamping them together with sufficient force that the load is transferred between them by friction.

High-volume fly ash concrete (HVFA concrete) A concrete in which a high percentage of cementing substance is fly ash rather than portland cement.

Hip The diagonal intersection of planes in a hip roof.

Hip rafter A roof *rafter* at the intersection of two sloping roof planes. *See also* **Common rafter**.

Hip roof A roof consisting of four sloping planes that intersect to form a pyramidal or elongated pyramid shape.

Hollow brick Clay brick with up to 60 percent void area.

Hollow concrete masonry Concrete masonry units that are manufactured with open cores, such as ordinary concrete blocks.

Hollow-core door A door consisting of two face veneers separated by an airspace, with solid wood spacers around the four edges. The face veneers are usually connected by a grid of thin spacers within the airspace.

Hollow-core slab A precast concrete slab element that has internal longitudinal cavities to reduce its self-weight.

Hollow structural section (HSS) Hollow steel cylindrical or rectangular shapes used as structural members; also called *structural tubing.*

Hook A semicircular bend in the end of a *reinforcing bar*, made for the purpose of anchoring the end of the bar securely into the surrounding concrete.

Hopper window A window whose sash pivots on an axis along or near the sill and that opens by tilting toward the interior of the building.

Horizontal reinforcing Steel reinforcing that runs horizontally in a masonry wall in the form of either welded grids of small-diameter metal rods or larger conventional *reinforcing bars.*

Hose stream test A standard laboratory test to determine the relative ability of a building assembly to stand up to water from a fire hose after a specified period of fire testing.

Hot-dip galvanizing A method of *galvanizing* in which a steel member or assembly is dipped into a bath of molten zinc.

Hot-rolled steel Steel formed into its final shape by passing it between rollers while it is very hot and still in a plastic state.

Housewrap A synthetic sheet material with water-resistive and air-resistive properties used as a substitute for asphalt-saturated felt or building paper to provide a protective layer in an exterior wall assembly.

HSS *See* **Hollow structural section.**

HVFA concrete *See* **High-volume fly ash concrete.**

Hydrated lime Quicklime mixed with water, either in the factory or on the job site; an ingredient in masonry mortars, portland cement plaster, and gypsum plasters, to which materials it imparts properties such as workability, bulk, and smoothness; chemically, calcium hydroxide; also called *slaked lime.*

Hydration The process by which cements combine chemically with water to harden.

Hydraulic cements Cementitious materials, such as portland cement or blast furnace slag, that harden by reacting with water and whose hardened products are not water soluble. Nonhydraulic cements, such as lime, can also be mixed with pozzolans to create cements with hydraulic properties.

Hydronic heating system A system that circulates warm water through convectors to heat a building.

Hydrostatic pressure Pressure exerted by standing water.

Hygroscopic Readily absorbing and retaining moisture.

Hyperbolic paraboloid shell A concrete roof structure with a saddle shape.

I

IBC *See* **International Building Code and International Residential Code.**

I-beam Obsolete term; an American Standard section of hot-rolled steel, an archaic structural steel shape. (This term should not be applied to modern wide flange steel sections.)

Ice and water shield *See* **Rubberized underlayment.**

Ice barrier A sheet material, usually *rubberized underlayment* or sheet metal, applied to the lower portions of sloped roofs in cold climates to protect against *ice dams.*

Ice dam An obstruction along the eave of a roof, caused by the refreezing of water emanating from melting snow on the roof surface above.

ICF *See* **Insulating concrete form.**

Igneous rock Rock formed by the solidification of magma.

IIC *See* **Impact Isolation Class.**

I-joist A manufactured wood framing member whose cross-sectional shape resembles the letter *I.*

Impact Isolation Class (IIC) An index of the extent to which a floor assembly transmits *impact noise* from a room above to the room below.

Impact noise Noise generated by footsteps or other impacts on a floor.

Impact wrench A device for tightening bolts and nuts by means of rapidly repeated torque impulses produced by electrical or pneumatic energy.

Incising Short, repetitive cuts made in the surface of a wood member to increase its absorption of treatment chemicals.

Ingot A large block of cast metal.

Insulating concrete form (ICF) A system of lightweight components, most commonly made of rigid polystyrene insulating foam, used as *permanent formwork* for the casting of concrete walls.

Insulating glass A glazing unit made up of two or more sheets of glass with an airspace in between.

Insulating glass unit (IGU) *See* **Insulating glass.**

Intensive green roof A *green roof* with relatively deep soil capable of supporting a broad variety of plants and shrubs.

Internal drainage Providing a curtain wall with hidden channels and weep holes to remove any water that may penetrate the exterior layers of the wall.

Internal flashing In masonry, a flashing concealed with the masonry; also called a *concealed* or *through-wall flashing.*

Internal gutter A *gutter* built into a roof assembly.

International Building Code (IBC) and International Residential Code (IRC) The predominant U.S. *model building codes.*

Interstitial ceiling A suspended ceiling with sufficient structural strength to support workers safely as they install and maintain mechanical and electrical installations above the ceiling.

Intrados The concave surface of an arch.

Intumescent coating A paint or mastic that expands to form a stable, insulating char when exposed to fire.

Inverted roof A membrane roof assembly in which the thermal insulation lies above the membrane.

IRC *See* **International Building Code and International Residential Code.**

Iron In pure form, a metallic element. In common usage, ferrous alloys other than steels, including cast iron and wrought iron.

Iron dog A heavy U-shaped staple used to tie the ends of heavy timbers together.

Ironworker A skilled laborer who erects steel building frames or places reinforcing bars in concrete construction.

Isocyanurate foam *See* **Polyisocyanurate foam.**

Isolation joint A type of joint used to separate abutting materials or assemblies that should remain structurally independent, such as where new construction meets old, or where a nonstructural slab on grade abuts structural columns or walls.

J

Jack A device for exerting a large force over a short distance, usually by means of screw action or hydraulic pressure.

Jack rafter A shortened *rafter* that joins a hip or valley rafter.

Jack stud A shortened stud that carries a header above a wall opening; also called a *trimmer stud.*

Jamb The vertical side of a door or window.

Jet burner A torch that burns fuel oil and compressed air, used in quarrying granite.

Joist One of a parallel array of light, closely spaced beams used to support a floor deck (*floor joist*) or low-slope roof (*ceiling joist*).

Joist band A broad, shallow concrete beam that supports one-way concrete joists whose depths are identical to its own.

Joist girder A light steel truss used to support open-web steel joists.

Joist hanger A sheet metal device used to create a structural connection where a joist is framed into a *header* or a *ledger*.

K

Keenes cement A proprietary, dense, crack-resistant gypsum plaster formulation.

Key A slot formed into a concrete surface for the purpose of interlocking with a subsequent pour of concrete; a slot at the edge of a precast member into which grout will be poured to lock it to an adjacent member; a mechanical interlocking of plaster with lath.

Kiln A furnace for firing clay or glass products; a heated chamber for seasoning wood; a furnace for manufacturing quicklime, gypsum hemihydrate, or portland cement.

King stud A full-length stud nailed alongside a *jack stud*.

Knee wall A short wall under the slope of a roof.

Knot A growth characteristic in wood, occurring where a branch joined the trunk of the tree from which the wood was sawed.

kPa Kilopascal; a metric unit of pressure equal to 1 kilonewton per square meter.

L

Labyrinth A cladding joint design in which a series of interlocking baffles prevents drops of water from penetrating the joint by momentum.

Lacquer A coating that dries extremely quickly through evaporation of a volatile solvent.

Lagging Planks placed between soldier beams to retain earth around an excavation.

Lag screw A large-diameter wood screw with a square or hexagonal head.

Laminate As a verb, to bond together in layers; as a noun, a material produced by bonding together layers of material.

Laminated glass A glazing material consisting of outer layers of glass laminated to an inner layer of transparent plastic.

Laminated strand lumber (LSL) Wood members made up of long shreds of wood fiber joined with a binder.

Laminated veneer lumber (LVL) Structural composite lumber made up of thin wood veneers joined with glue.

Laminated wood *See* **Glue-laminated wood**.

Landing A platform in or at either end of a stair.

Lap joint A connection in which one piece of material is placed partially over another piece before the two are fastened together.

Lateral force A force acting generally in a horizontal direction, such as wind, earthquake, or soil pressure against a foundation wall.

Lateral thrust The horizontal component of the force produced by an arch, dome, vault, or rigid frame.

Latewood *See* **Summerwood**.

Latex caulk A low-range *sealant* based on a synthetic latex.

Latex/polymer modified portland cement mortar A tile-setting mortar similar to *dry-set mortar*, but with additives that improve the cured mortar's freeze-thaw resistance, flexibility, and adhesion; used for *thin-set tile* applications.

Lath (rhymes with "math") A base material to which plaster is applied.

Lathe (rhymes with "bathe") A machine in which a piece of material is rotated against a sharp cutting tool to produce a shape, all of whose cross sections are circles; a machine in which a log is rotated against a long knife to peel a continuous sheet of veneer.

Lather (rhymes with "rather") One who applies lath.

Lay-in panel A finish ceiling panel that is installed merely by lowering it onto the top of the metal grid components of the ceiling.

Lead (rhymes with "bed") A soft, dull gray, easily formed *nonferrous metal*.

Lead (rhymes with "bead") In masonry work, a corner or wall end accurately constructed with the aid of a spirit level to serve as a guide for placing the bricks in the remainder of the wall.

Leader (rhymes with "feeder") *See* **Downspout**.

Leaf The moving portion of a door.

Lean construction Methods of construction and its management that emphasize efficiency, elimination of waste, and continuous improvement in quality.

Ledger A horizontal wood member fastened to a wall or beam to which the ends of joists may be connected.

Lehr A chamber in which glass is annealed.

Let-in bracing Diagonal bracing that is nailed into notches cut in the face of the studs so as not to increase the thickness of the wall.

Level cut A saw cut that produces a level surface in a sloping rafter when the rafter is in its final position. *See also* **Plumb cut**.

Leveling plate A steel plate placed in grout on top of a concrete foundation to create a level bearing surface for the lower end of a steel column.

Lewis A device for lifting a block of stone by means of friction exerted against the sides of a hole drilled in the top of the block.

Life-cycle analysis (LCA) A comprehensive method of describing the environmental impacts of a material or product, accounting for all phases of its life from original resource extraction through final disposal or recycling; also called *cradle-to-grave analysis*.

Life-cycle cost A cost that takes into account both the first cost and the costs of maintenance, replacement, fuel consumed, monetary inflation, and interest over the life of the object being evaluated.

Lift-slab construction A method of building multistory sitecast concrete buildings by casting all the slabs in a stack on the ground, then lifting them up the columns with jacks and welding them in place.

Light A sheet of glass, also spelled "lite."

Light gauge steel stud A length of thin sheet metal formed into a stiff shape and used as a wall framing member.

Light to solar gain (LSG) ratio The visible light transmittance of a glazing unit divided by the solar heat gain coefficient, a measure of the energy-conserving potential of the unit.

Lightweight aggregate Low-density aggregate used to make lightweight concrete, mortar, and plaster; in concrete, aggregate with a density of less than 70 lb/ft^3 (1120 kg/m^3).

Lignin The natural cementing substance that binds together the cellulose in wood.

Lime A nonhydraulic cementitious material, used as an ingredient in mortars and plasters. *See also* **Hydrated lime, Quicklime.**

Lime mortar Masonry mortar made from a mix of lime, sand, and water; used principally in the restoration of historic structures.

Lime putty *See* **Finish lime.**

Limestone A sedimentary rock consisting of calcium carbonate, magnesium carbonate, or both.

Linear metal ceiling A finish ceiling whose exposed face is made up of long, parallel elements of sheet metal.

Liner A piece of marble doweled and cemented to the back of another sheet of marble.

Line wire Wire stretched across wall studs as a base for the application of metal mesh and stucco.

Linoleum A resilient floor covering material composed primarily of ground cork and linseed oil on a burlap or canvas backing.

Lintel A beam that carries the load of a wall across a window or door opening.

Liquid limit The moisture content at which a soil arrives at a flowable consistency; a relative indication of soil cohesiveness.

Liquid sealant Gunnable *sealant*.

Lite A sheet of glass, also spelled "light."

Live load Nonpermanent loads on a building caused by the weights of people, furnishings, machines, vehicles, and goods in or on the building.

Load A weight or force acting on a structure.

Loadbearing Supporting a superimposed weight or force.

Loadbearing wall *See* **Bearing wall.**

Load indicator washer A disk placed under the head or nut of a high-strength bolt to indicate sufficient tensioning of the bolt by means of the deformation of ridges on the surface of the disk; also called a *direct tension indicator washer.*

Lockpin and collar fastener A boltlike device that is passed through holes in structural steel components, held in very high tension, and closed with a steel ring that is squeezed onto its protruding shank.

Lockstrip gasket A synthetic rubber strip compressed around the edge of a piece of glass or a wall panel by inserting a spline (lockstrip) into a groove in the strip.

Longitudinal shrinkage In wood, shrinkage along the length of the log.

Lookout A short *rafter*, running perpendicular to the other rafters in the roof, which supports a rake overhang.

Louver A construction of numerous sloping, closely spaced slats used to diffuse air or to prevent the entry of rainwater into a ventilating opening.

Low-e coating *See* **Low-emissivity coating.**

Low-emissivity coating A surface coating for glass that selectively reflects solar radiation of different wavelengths so as to allow high visible light transmittance while reflecting some or all types of infrared (heat) radiation. *See also* **Low-e coating.**

Low-lift grouting A method of constructing a reinforced masonry wall in which the *reinforcing bars* are embedded in grout in increments not higher than 4 feet (1200 mm).

Low-range sealant A *sealant* that is capable of a relatively slight degree of elongation prior to rupture; a *caulk.*

Low-slope roof A roof that is pitched so near to horizontal that it must be made waterproof with a continuous membrane rather than shingles; commonly and inaccurately referred to as a "flat roof." In the International Building Code, a roof with a slope of less than 2:12 (17 percent).

LSG *See* **Light to solar gain ratio.**

LSL *See* **Laminated strand lumber.**

Luffing-boom crane A heavy-duty lifting device that uses a tower-mounted boom that may rotate in any vertical plane as well as in a horizontal plane.

LVL *See* **Laminated veneer lumber.**

M

Machine grading The grading of wood for its structural properties, performed by automated machinery, as distinct from *visual grading.*

Macrofiber reinforcing In concrete, fibrous reinforcement capable of providing resistance to drying shrinkage and thermal stresses, and in some specialized concretes, also capable of acting as primary reinforcing. *See also* **Microfiber reinforcing.**

Mandrel A stiff steel core placed inside the thin steel shell of a sitecast concrete pile to prevent it from collapsing during driving.

Mansard A roof shape consisting of two superimposed levels of hip roofs with the lower level at a steeper pitch than the upper.

Manufactured home A transportable house that is entirely factory built on a steel underframe supported by wheels; euphemistically referred to as a *mobile home.*

Marble A metamorphic rock formed from limestone by heat and pressure.

Mason One who builds with bricks, stones, or concrete masonry units; one who works with concrete.

Masonry Brickwork, concrete blockwork, and stonework.

Masonry cement A hydraulic cement made from a blend of portland cement, lime, and other dry admixtures designed to increase the workability of the mortar. *See also* **Cement-lime mortar.**

Masonry opening The clear dimension required in a masonry wall for the installation of a specific window or door unit.

Masonry unit A brick, stone, concrete block, glass block, or hollow clay tile intended to be laid in mortar.

Masonry veneer A single wythe of masonry used as a facing over a frame of wood or metal.

MasterFormat The trademarked title of a uniform indexing system for construction specifications, created by the Construction Specifications Institute and Construction Specifications Canada.

Mastic A viscous, doughlike, adhesive substance; can be any of a large number of formulations for different purposes such as sealants, adhesives, glazing compounds, or roofing cements.

Mat foundation A single concrete footing that is essentially equal in area to the area of ground covered by the building.

Maximum allowable slope The steepest angle at which an excavation may be sloped so that the soil will not slide back into the hole. Also called *angle of repose.*

Medium-density fiberboard (MDF) A fine-grained wood fiber and resin panel product.

Medium-density overlay (MDO) A medium-weight, resin-treated overlay applied to plywood panels to achieve a smoother, more durable face.

Medium-range sealant A *sealant* material that is capable of a moderate degree of elongation before rupture.

Meeting rail The wood or metal bar along which one sash of a double-hung, single-hung, or sliding window seals against the other.

Member An element of a structure such as a beam, girder, column, joist, piece of decking, stud, or component of a truss.

Membrane A sheet material that may be used to control the passage of liquid water, water vapor, and/or air.

Membrane fire protection A ceiling used to provide fire protection to the structural members above.

Metal decking Corrugated metal sheets used as the structural base for floors ("floor decking") and roofs ("roof decking") in steel frame construction. *See also* **Cellular decking** and **Composite metal decking**.

Metal lath A steel mesh used primarily as a base for the application of plaster.

Metallic-coated steel Steel sheet coated with zinc or zinc-aluminum for improved corrosion resistance.

Metamorphic rock A rock created by the action of heat or pressure on a sedimentary rock or soil.

Microfiber reinforcing In concrete, fibrous reinforcement against plastic shrinkage cracking. *See also* **Macrofiber reinforcing**.

Microsilica *See* **Silica fume**.

Middle strip The half-span-wide zone of a two-way concrete slab that lies midway between columns.

Mild steel Ordinary structural steel, containing less than three-tenths of 1 percent carbon.

Mill construction The traditional name for a construction type consisting of exterior masonry bearing walls and an interior framework of heavy timbers and solid timber decking; also called *slow-burn construction*. *See also* **Heavy Timber construction**.

Milling Shaping or planing by using a rotating cutting tool.

Millwork Wood interior finish components of a building, including moldings, windows, doors, cabinets, stairs, mantels, and the like.

Minimum critical radiant flux exposure A measure of a material's resistance to ignition by the radiant heat of fire and hot gasses in adjacent spaces, usually applied to flooring materials.

Miter A diagonal cut at the end of a piece; the joint produced by joining two diagonally cut pieces at right angles.

Mobile home *See* **Manufactured home**.

Model building code A code that is offered by a recognized national organization as worthy of adoption by state or local governments.

Modified bitumen A natural *bitumen* with admixtures of synthetic compounds to enhance such properties as flexibility, plasticity, and durability.

Modified bitumen roof membrane A multi-ply *bituminous roof membrane* made from plies of factory-manufactured *modified bitumen* sheets.

Modular Conforming to a multiple of a fixed dimension.

Modular green roof A *green roof* system in which all components are provided in self-contained, easily transported and installed trays or modules.

Modular home A house assembled on the site from boxlike factory-built sections.

Modulus of elasticity An index of the stiffness of a material, derived by measuring the elastic deformation of the material as it is placed under stress and then dividing the stress by the deformation.

Moisture barrier A membrane used to resist the migration of liquid water through a floor, wall, or roof.

Molding A strip of wood, plastic, or plaster with an ornamental profile.

Molding plaster A fast-setting gypsum plaster used for the manufacture of cast ornament.

Moment A force acting at a distance from a point in a structure so as to cause a tendency of the structure to rotate about that point. *See also* **Bending moment, Moment connection**.

Moment connection A connection between two structural members that is resistant to rotation between the members and therefore capable of transmitting *bending moments* between the connected members, as differentiated from a *shear connection*, which allows (slight) rotation. *See also* **Fully restrained moment connection, Partially restrained moment connection**, and **Simple connection**.

Moment-resisting frame A structural building frame that is strengthened to resist lateral forces with *moment connections* between beams and columns.

Momentum The tendency of a moving body to continue to move in the same direction unless acted on by an outside force.

Monolithic Of a single massive piece.

Monolithic terrazzo A thin terrazzo topping applied to a concrete slab without an underbed.

Mortar A substance used to join masonry units, consisting of cementitious materials, fine aggregate, and water. *See also* **Cement-lime mortar, Lime mortar**.

Mortar bed tile *See* **Thickset tile**.

Mortar cement In masonry, a blend of portland cement, lime, and other additives, that produces mortar comparable in its bond strength properties to cement-lime mortar. *See also* **Cement-lime mortar**.

Mortise and tenon A joint in which a tonguelike protrusion (tenon) on the end of one piece is tightly fitted into a rectangular slot (mortise) in the side of the other piece.

Movement joint A line or plane along which movement is allowed to take place in a building or a surface of a building in response to such forces as moisture expansion and contraction, thermal expansion and contraction, foundation settling, and seismic forces.

MPa Megapascal; a unit of pressure equal to 1 meganewton per square meter.

Mud set tile *See* **Thickset tile**.

Mud slab A slab of weak concrete placed directly on the ground to provide a (usually temporary) working surface that is hard, level, and dry.

Mullion A vertical or horizontal bar between adjacent window or door units; a framing member in a metal-and-glass curtain wall.

Muntin A small vertical or horizontal bar between small lites of glass in a sash.

Muriatic acid Hydrochloric acid.

Mushroom capital A flaring conical head on a concrete column.

N

Nail A sharp-pointed metal pin used for fastening wood.

Nail-base sheathing A sheathing material, such as wood boards or plywood, to which siding can be attached by nailing, as differentiated from one such as cane fiber board or plastic foam board that is too soft to hold nails.

Nail popping The loosening of nails holding gypsum board to a wall, caused by drying shrinkage of the studs.

Nail set A hardened steel punch used to drive the head of a nail to a level flush with or below the surface of the wood.

National Building Code of Canada (NBCC) The predominant Canadian *model building code*.

NBCC *See* **National Building Code of Canada**.

Near-infrared (NIR) radiation An invisible portion of the solar spectrum that accounts for more than half of the total heat energy in solar radiation.

Needle beam A steel or wood beam threaded through a hole in a bearing wall and used to support the wall and its superimposed loads during underpinning of its foundation.

Needling The use of *needle beams*.

Negative-side waterproofing Waterproofing applied to the inner side of a wall, acting to resist water passage from the opposite side.

Neoprene Polychloroprene, a synthetic rubber, sometimes used as an accessory material in waterproofing systems.

NIR radiation *See* **Near-infrared radiation**.

Noble metal *See* **Passive metal**.

Noise Reduction Coefficient (NRC) An index of the proportion of incident sound that is absorbed by a surface, expressed as a decimal fraction of 1.

Nominal dimension An approximate dimension assigned to a piece of material as a convenience in describing its size, as distinct from its *actual dimension*.

Nonaxial In a direction not parallel to the long axis of a structural member.

Nonbearing Not carrying a structural load.

Nonferrous metal Any metal other than iron or an iron alloy. Aluminum, copper, lead, and zinc are examples of nonferrous metals frequently used in building construction.

Nonhydraulic cements *Cementitious materials*, such as gypsum and lime, that remain water soluble after curing. *See also* **Hydraulic cements**.

Nonmovement joint A connection between materials or elements that is not designed to allow movement between the parts.

Nonworking joint *See* **Nonmovement joint**.

Nosing The projecting forward edge of a stair tread.

NRC *See* **Noise Reduction Coefficient**.

Nut A fastener, usually metallic, with internal helical threads, used to close a *bolt*.

O

o.c. Abbreviation for "on center," meaning that the spacing of framing members is measured from the center of one member to the center of the next, rather than the clear spacing between members.

Occupancy classification In the International Building Code, a definition of the types of activities that occur within the building or a part of the building, relating to considerations of life safety.

Ogee An S-shaped curve.

Oil-based caulk A low-range *sealant* made with linseed oil.

OmniClass Construction Classification System The trademarked name for a system of describing building information encompassing a broad range of possible organizing criteria.

One-way action The structural action of a slab that spans between two parallel beams or bearing walls.

One-way concrete joist system A reinforced concrete framing system in which closely spaced concrete joists span between parallel beams or bearing walls.

One-way solid slab A reinforced concrete floor or roof slab that spans between parallel beams or bearing walls.

Open-truss wire stud An archaic wall framing member in the form of a small steel truss, no longer used in contemporary construction.

Open-web steel joist A lightweight, prefabricated, welded steel truss used at closely spaced intervals to support floor or roof decking.

Optical-quality ceramic A transparent, glasslike material that is used as a fire-rated glazing sheet.

Optimum value engineering *See* **Advanced framing techniques**.

Ordinary construction A traditional building type with exterior masonry

bearing walls and an interior structure of balloon framing.

Organic soil Soil containing decayed vegetable and/or animal matter; topsoil.

Oriented strand board (OSB) A building panel composed of long shreds of wood fiber oriented in specific directions and bonded together under pressure.

Oriented strand lumber (OSL) *Structural composite lumber* made from shredded wood strands, coated with adhesive, and pressed into a rectangular cross section.

OSB *See* **Oriented strand board**.

OSL *See* **Oriented strand lumber**.

Oxidation Corrosion; rusting; rust; chemically, the combining with oxygen.

P

Package fireplace A factory-built fireplace that is installed as a unit.

Paint A heavily pigmented coating applied to a surface for decorative and/or protective purposes.

Pan A form used to produce the cavity between joists in a one-way concrete joist system.

Panel A broad, thin piece of wood; a sheet of building material such as plywood or particleboard; a prefabricated building component that is broad and thin, such as a curtain wall panel; a rectangular area within a truss bounded by two vertical interior members.

Panel door A wood door in which one or more thin panels are held by stiles and rails.

Panelized construction A method of prefabricated wood light frame construction, in which whole sections of walls or floors are framed and sheathed in the factory and then transported to the construction site for erection.

Panic hardware A mechanical device that opens a door automatically if pressure is exerted against the device from the interior of the building.

Parallel strand lumber (PSL) *Structural composite lumber* made of wood shreds oriented parallel to the long axis of each piece and bonded together with adhesive.

Parapet The region of an exterior wall that projects above the level of the roof.

Parging Portland cement plaster applied over masonry to make the masonry less permeable to water.

Partially restrained moment connection A steel frame *moment connection* that is less rigid than a *fully restrained moment connection* but that still possesses a usable degree of resistance to rotation.

Particleboard A building panel composed of small particles of wood bonded together under pressure.

Parting compound *See* **Form release compound**.

Partition An interior nonloadbearing wall.

Passive metal A metal relatively low on a galvanic series, tending to act as an *cathode* in galvanic couples; also called a *noble metal*.

Patterned glass Glass into which a texture has been rolled during manufacture.

Pattern rafter A wood *rafter* cut to size and shape and then used to trace cuts onto additional wood members so as to assure consistent dimensions among all rafters.

Paver A half-thickness brick used as finish flooring.

PEC *See* **Pressure equalization chamber**.

Pediment The gable end of a roof in classical architecture.

Penetrometer A device for testing the resistance of a material to penetration, usually used to make a quick, approximate determination of its compressive strength.

Penny A designation of nail size, abbreviated as "d."

Performance-based building code A set of legal regulations that mandate performance outcomes rather than specific construction details and practices.

Performance category A system for specifying the nominal thickness of a structural wood panel. Actual panel thickness may deviate from the performance category within accepted tolerances.

Periodic kiln A kiln that is loaded and fired in discrete batches, as differentiated from a tunnel kiln, which is operated continuously.

Perlite Expanded volcanic glass, used as a *lightweight aggregate* in concrete and plaster and as an insulating fill.

Perm A unit of vapor permeance, a measure of a material's permeability to the diffusion of water vapor.

Permanent formwork Concrete formwork that remains permanently in place after concrete is poured and cured; becomes part of the finished construction.

Pervious concrete Concrete with a high percentage of void space, used as a paving material that allows stormwater to pass through into the soil below.

PF *See* **Phenol-formaldehyde**.

Phased construction *See* **Fast track construction**.

Phenol-formaldehyde (PF) A structural wood adhesive and binder, suitable for exterior exposure, and associated with relatively low *formaldehyde* gas emissions.

Photochromic glass Glass that changes its optical properties in response to light intensity.

Photovoltaic (PV) Capable of converting light into electricity.

Pier A caisson foundation unit.

Pilaster A vertical, integral stiffening rib in a masonry or concrete wall.

Pile A long, slender piece of material driven into the ground to act as an element of a foundation.

Pile cap A thick slab of reinforced concrete poured across the top of a pile cluster to cause the cluster to act as a unit in supporting a column or grade beam.

Piledriver A machine for driving piles.

Pill test A test of a flooring material's propensity for flame spread when exposed to a burning tablet intended to simulate a dropped lit cigarette, match, or similar hazard.

Pintle A metal device used to transmit compressive forces between superimposed columns in Mill construction.

Pitch The slope of a roof or other plane, often expressed as inches of rise per foot of run; a dark, viscous hydrocarbon distilled from coal tar; a viscous resin found in wood.

Pitched roof A sloping roof.

Pivoting window A window that opens by rotating around its vertical centerline.

Plainsawn lumber Lumber sawn in such a way that significant portions of the growth rings are oriented roughly flat relative to the board's broader face.

Plainsliced veneer Veneer sliced from a log without regard to the direction of the annual rings, as distinct from *quartersliced*.

Plain slicing Cutting a log into veneers without regard to the direction of the annual rings.

Plan An architectural drawing, representing the layout of walls and floor areas as seen from above ("floor plan") or ceilings as seen from below ("ceiling plan").

Planing Smoothing the surface of a piece of wood, stone, or steel with a cutting blade.

Plank flooring Solid wood finish flooring members 3 inches (75 mm) or more in width.

Plaster A cementitious material, usually based on gypsum or portland cement, applied to lath or masonry in paste form to harden into a finish surface.

Plasterboard *See* **Gypsum board**.

Plaster of Paris *See* **Calcined gypsum**.

Plaster screeds Intermittent spots or strips of plaster used to establish the level to which a large plaster surface will be finished.

Plastic A synthetically produced giant molecule, mostly based on carbon chemistry.

Plasticity The ability to retain a shape attained by pressure deformation.

Plastic laminate flooring A finish material for floors that consists of a thin decorative and wearing layer of melamine laminate glued to a wood composite substrate.

Plastic lumber Lumberlike products with a plastic content of 50 percent or more. *See also* **Structural-grade plastic lumber**.

Plastic shrinkage cracking Cracking in freshly mixed concrete, most commonly in slabs, that occurs when the surface of the concrete dries too rapidly.

Plate A broad sheet of rolled metal ¼ inch (6.35 mm) or more thick; a two-way concrete slab; a horizontal top or bottom member in a platform frame wall structure.

Plate girder A large beam made up of steel plates, sometimes in combination with steel angles, that are welded, bolted, or riveted together.

Plate glass Glass of high optical quality, produced by grinding and polishing both faces of a glass sheet.

Platform frame A wooden building frame composed of closely spaced members nominally 2 inches (51 mm) thick in which the wall members do not run past the floor framing members.

Plenum The space between the ceiling of a room and the structural floor above, used as a passage for ductwork, piping, and wiring.

Plumb Vertical.

Plumb cut A saw cut that produces a vertical (plumb) surface in a sloping rafter after the rafter is in its final position. *See also* **Level cut.**

Plumbing up The process of making a steel building frame vertical and square.

Ply A layer, such as a layer of felt in a built-up roof membrane or a layer of veneer in plywood.

Plywood A wood panel composed of an odd number of layers of wood veneer bonded together under pressure.

PMR *See* **Protected membrane roof.**

Pneumatically placed concrete *See* **Shotcrete.**

Pointing The process of applying mortar to the surface of a mortar joint after the masonry has been laid, either as a means of finishing the joint or to repair a defective joint. *Raked mortar joints* can also be pointed with elastomeric sealant.

Pointing mortar Mortar used for the pointing of masonry joints, generally of relatively low strength and with good workability and adhesion characteristics.

Poke-through fitting An electrical outlet that is installed by drilling a hole through a floor, inserting the outlet from above, and bringing in the wiring from the plenum below.

Polybutene tape A sticky, masticlike tape used to seal nonworking joints, especially between glass and mullions.

Polycarbonate An extremely tough, strong, usually transparent plastic used for window and skylight glazing, light fixture globes, door sills, and other applications.

Polyethylene A thermoplastic widely used in sheet form for vapor retarders, moisture barriers, and temporary construction coverings.

Polyisocyanurate foam A thermosetting plastic foam with thermal insulating properties.

Polymer A large molecule composed of many identical chemical units.

Polypropylene A plastic formed by the polymerization of propylene.

Polystyrene foam A thermoplastic foam with thermal insulating properties.

Polysulfide A high-range gunnable *sealant.*

Polyurethane Any of a large group of resins and synthetic rubber compounds used in sealants, varnishes, insulating foams, and roof membranes.

Polyurethane foam A thermosetting foam with thermal insulating properties.

Polyvinyl butyral (PVB) interlayer A transparent plastic used in the fabrication of *laminated glass.*

Polyvinyl chloride (PVC) A thermoplastic material widely used in construction products, including plumbing pipes, floor tiles, wall coverings, and roof membranes. Called "vinyl" for short.

Ponding The accumulation of standing water on a low-slope roof due to inadequate drainage.

Poorly graded soil Soil with less than a full range of particle sizes.

Poorly sorted soil *See* **Well graded soil.**

Portal frame A rigid frame; two columns and a beam attached to one another with moment connections.

Portland cement A gray or white powder, composed principally of calcium silicates, which, when combined with water, hydrates to form the binder in concrete, mortar, and stucco.

Posttensioning Compressing the concrete in a structural member by tensioning high-strength steel tendons against it after the concrete has cured.

Pour To cast concrete; an increment of concrete casting carried out without interruption.

Powder coating A coating produced by applying a powder consisting of thermosetting resins and pigments, adhered to the substrate by electrostatic attraction, and fused into a continuous film in an oven.

Powder-driven Inserted by a gunlike tool using energy provided by an exploding charge of gunpowder.

Pozzolan A *supplementary cementitious material,* such as fly ash, silica fume, and some naturally occurring shales and clays, that has few or no inherent cementitious properties but that, in the presence of moisture, can react with calcium hydroxide released by other cementitious materials to create a hydraulic cement product. The Romans mixed natural pozzolans with lime to make the first hydraulic cement.

Precast concrete Concrete cast and cured in a location other than its final position in the structure.

Predecorated gypsum board Gypsum board finished at the factory with a decorative layer of paint, paper, or plastic.

Prefabrication Construction that takes place in a factory or shop, rather than on the building site.

Preformed cellular tape sealant A *sealant* inserted into a joint in the form of a compressed sponge impregnated with compounds that cure to form a watertight seal.

Preformed joint filler A strip of rubbery or spongelike material designed to fit snugly into a gap between two materials.

Preformed solid tape sealant A *sealant* inserted into a joint in the form of a flexible strip of solid material.

Prehung door A door that is hinged to its frame in a factory or shop.

Prescriptive building code A set of legal regulations that mandate specific construction details and practices rather than establish performance standards.

Preservative-treated wood Wood that has been impregnated with preservative chemicals to increase its resistance to decay and biological attack; also commonly called *pressure-treated wood.*

Pressure equalization chamber (PEC) The wind-pressurized cavity in a rainscreen wall.

Pressure-equalized wall design Curtain wall design that relies on neutralization of wind pressures on both sides of an exterior cladding to control water entry into the wall system. *See also* **Rainscreen principle.**

Pressure-treated wood Wood that has been impregnated with chemicals under pressure for the purpose of retarding decay or reducing combustibility.

Prestressed concrete Concrete that has been pretensioned or posttensioned.

Prestressing Applying an initial compressive stress to a concrete structural member, either by *pretensioning* or *posttensioning.*

Pretensioning Compressing the concrete in a structural member by pouring the concrete for the member around stretched high-strength steel strands, curing the concrete, and releasing the external tensioning force on the strands.

Priming Covering a surface with a coating that prepares it to accept another coating or a sealant.

Protected membrane roof (PMR) A membrane roof assembly in which the thermal insulation lies above the membrane.

Protection board Semirigid board or sheet material placed over a waterproofing or roofing layer, to protect the layer from damage.

PSL *See* **Parallel strand lumber.**

Pultrusion The process of producing a shaped linear element by pulling glass fibers through a bath of uncured plastic, then through a heated, shaped die in which the plastic hardens.

Punty A metal rod used in working with hot glass.

Purlin A beam that spans across the slope of a steep roof to support the roof decking.

Putty A simple glazing compound used to seal around a small light.

PV *See* **Photovoltaic.**

PVB interlayer *See* **Polyvinyl butyral interlayer.**

PVC *See* **Polyvinyl chloride.**

Pyrolitic coating A coating applied at a very high temperature.

Q

Quarry An excavation from which building stone is obtained; the act of taking stone from the ground.

Quarry bed A plane in a building stone that was horizontal before the stone was cut from the quarry; also called *grain.*

Quarry sap Excess water found in rock at the time of its quarrying.

Quarry tile A large clay floor tile, usually unglazed.

Quartersawn lumber For softwoods, lumber sawn in such a way that growth rings are aligned at an angle of approximately 45 degrees or steeper relative to the board's broader face. For hardwoods, sawn such that the growth rings are aligned at an angle of approximately 60 degrees or steeper to the broader face.

Quartersliced veneer Veneer sliced in such a way that the annual rings appear closely spaced and run roughly perpendicular to the face of each veneer.

Quenching The rapid cooling of metal so as to alter its physical properties; a form of heat treatment.

Quicklime Produced by burning calcium carbonate found in limestone or sea shells; once hydrated, used as an ingredient in mortars and plasters; chemically, calcium oxide. *See also* **Hydrated lime.**

Quoin (pronounced "coin") A corner reinforcing of cut stone or bricks in a masonry wall, usually done for decorative effect.

R

Rabbet A longitudinal groove cut at the edge of a member to receive another member; also called a *rebate.*

Radial shrinkage In wood, shrinkage perpendicular to the growth rings.

Radiant barrier A reflective foil or metal placed adjacent to an airspace in roof or wall assemblies as a deterrent to the passage of infrared energy.

Radiant heating system Providing heat to spaces and their inhabitants by heating one or more surfaces of each room. The heated surface is usually either the floor or the ceiling. Heat is usually provided either by electric resistance coils or by hot-water tubing.

Radiator *See* **Convector.**

Raft A mat footing.

Rafter A framing member that runs up and down the slope of a steep roof.

Rail A horizontal framing piece in a panel door; a handrail.

Rainscreen principle A theory by which wall cladding is made watertight by providing wind-pressurized air chambers behind joints to eliminate air pressure differentials between the outside and the inside that might transport water through the joints. *See also* **Pressure-equalized wall design.**

Rainscreen siding Siding systems that include a system of internal drainage, but not necessarily meeting all the criteria of a *pressure-equalized wall design.*

Raised access flooring *See* **Access flooring.**

Rake The sloping edge of a steep roof.

Raked mortar joint A mortar joint in which mortar has been removed from the portion of the joint closest to the surface of the masonry.

Raker A sloping brace for supporting sheeting around an excavation.

Ram A hydraulic piston device used for bending steel, tensioning steel strands in prestressed concrete, or lifting heavy loads.

Random stone masonry *See* **Uncoursed stone masonry.**

Ratchet A mechanical device with sloping teeth that allows one piece to be advanced against another in small increments, but not to move in the reverse direction.

Ray A tubular cell that runs radially in a tree trunk.

RBM *See* **Reinforced brick masonry.**

Rebar *See* **Reinforcing bar.**

Rebate *See* **Rabbet.**

Reflective coated glass Glass onto which a thin layer of metal or metal oxide has been deposited to reflect light and/or heat.

Reglet A slot, usually horizontal, and inclined in cross section, into which a flashing or roof membrane may be inserted in a concrete or masonry surface.

Reinforced brick masonry (RBM) Brickwork into which steel bars have been embedded to impart tensile strength to the construction.

Reinforced concrete Concrete work into which steel bars have been embedded to impart tensile strength to the construction.

Reinforcing bar A member added to concrete or masonry to impart tensile strength and ductility, usually made of steel; *rebar.*

Relative humidity A percentage representing the ratio of the amount of water vapor contained in a mass of air to the maximum amount of water it could contain under the existing conditions of temperature and pressure.

Relieved back A longitudinal groove or series of grooves cut from the back of a flat wood molding or flooring strip to minimize cupping forces and make the piece easier to fit to a flat surface.

Removable glazing panel A framed sheet of glass that can be attached to a window sash to increase its thermal insulating properties.

Replacement window A window unit that is designed to be installed easily in an opening left in a wall by a deteriorated window unit that has been removed.

Repointing The process of removing deteriorated mortar from the zone near the surface of a brick wall and inserting fresh mortar. *See also* **Tuckpointing.**

Reshoring Inserting temporary supports under concrete beams and slabs after the formwork has been removed to prevent overloading before the concrete achieves its full strength.

Resilient clip A springy mounting device for plaster or gypsum board that helps reduce the transmission of sound vibrations through a wall or ceiling.

Resilient flooring A manufactured sheet or tile flooring made of asphalt, polyvinyl chloride, linoleum, rubber, or other elastic material.

Resin A natural or synthetic, solid or semisolid organic material of high molecular weight, used in the manufacture of paints, varnishes, and plastics.

Restraightening A step in the finishing of concrete slabs for the purpose of removing minor undulations produced during floating or troweling.

Retaining wall A wall that resists horizontal soil pressures at an abrupt change in ground elevation.

Retarding admixture An *admixture* used to slow the curing of concrete, mortar, or plaster.

Ridge beam A structural beam supporting the upper ends of rafters in a sloped roof, required where the rafters are not tied at their lower ends.

Ridge board A nonstructural framing member against which the upper ends of rafters are fastened.

Ridge vent A screened, water-shielded ventilation opening that runs continuously along the ridge of a gable roof.

Riftsawn lumber For hardwoods, lumber sawn such that the growth rings are aligned at an angle of approximately 30 to 60 degrees to the broader face.

Rigid connection *See* **Fully restrained moment connection**.

Rigid frame Two columns and a beam or beams attached to one another with moment connections; a *moment-resisting frame*.

Rim joist *See* **Band joist**.

Rise A difference in elevation, such as the rise of a stair from one floor to the next or the rise per foot of run in a sloping roof.

Riser A single vertical increment of a stair; the vertical face between two treads in a stair; a vertical run of plumbing, wiring, or ductwork.

Rivet In structural steel construction, an archaic fastener, in which a second head is formed after the fastener is in place; a threadless fastener used in sheet metal work.

Rock anchor A posttensioned rod or cable inserted into a rock formation for the purpose of tying it together.

Rock wool An insulating material manufactured by forming fibers from molten rock.

Roofer One who installs roof coverings.

Roofing The material used to make a roof watertight, such as shingles, slate, tiles, sheet metal, or a roof membrane; the act of applying roofing.

Roof membrane A waterproof sheet or multi-ply assembly that protects a low-slope roof from water penetration.

Roof window Either an openable glazed unit installed in the sloping surface of a roof or, more specifically, a glazed roof unit with inward sash operation to allow easy cleaning.

Rotary-sliced veneer A thin sheet of wood produced by rotating a log against a long, sharp knife blade in a lathe.

Rough arch An arch made from masonry units that are rectangular rather than wedge-shaped.

Rough carpentry Framing carpentry, as distinguished from *finish carpentry*.

Roughing in The installation of mechanical, electrical, and plumbing components that will not be exposed to view in the finished building.

Rough opening The clear dimensions of the opening that must be provided in a wall frame to accept a given door or window unit.

Rowlock A brick laid on its long edge, with its end exposed in the face of the wall.

RSI-value A numerical measure of resistance to the flow of heat, expressed in metric units; the reciprocal of *U-factor*.

Rubberized underlayment An adhered bituminous sheet material that self-heals around nails, applied to roof sheathing to prevent the entry of water; also called *ice and water shield*.

Rubble Unsquared stones.

Run Horizontal dimension in a stair or sloping roof.

Runner channel A steel member from which furring channels and lath are supported in a suspended plaster ceiling.

Running bond Brickwork consisting entirely of stretchers.

Runoff bar One of a pair of small rectangular steel bars attached temporarily at the end of a prepared groove for the purpose of permitting the groove to be filled to its very end with weld metal.

Run plaster ornament A linear molding produced by passing a profiled sheet metal or plastic template back and forth across a mass of wet plaster.

R-value A numerical measure of resistance to the flow of heat; usually expressed in U.S. Customary units, but also sometimes used with metric units; the reciprocal of *U-Factor*.

S

Safety glazing Glass or plastic glazing material that, when broken, does not create hazardous shards, permitted for use in locations in buildings at risk of occupant impact; most commonly tempered glass or laminated glass.

Safing Fire-resistant material inserted into a space between a curtain wall and a spandrel beam or column to retard the passage of fire through the space.

Sand cushion terrazzo Terrazzo with an underbed that is separated from the structural floor deck by a layer of sand.

Sand-mold brick, sand-struck brick A brick made in a mold that was wetted and then dusted with sand before the clay was placed in it.

Sandstone A sedimentary rock formed from sand; classified by ASTM C119 in the Quartz-Based Dimension Stone group.

Sandwich panel A panel consisting of two outer faces of wood, metal, gypsum, or concrete bonded to a core of insulating foam.

Sapwood The living wood in the outer region of a tree trunk or branch.

Sash A frame that holds glass.

Sawn veneer Thin sheets of wood produced by sawing, rather than slicing with a knife blade.

SBS *See* **Styrene-butadiene-styrene**.

SBX *See* **Sodium borate**.

Scab A piece of framing lumber nailed to the face of another piece of lumber.

Scarf joint A glued end connection between two pieces of wood, using a sloping cut to create a large surface for the glue bond, to allow it to develop the full tensile strength of the wood that it connects.

SCC *See* **Self-consolidating concrete**.

SCOF *See* **Static coefficient of friction**.

Scratch coat The first of two base-coat plaster applications in a three-coat plaster.

Screed A strip of wood, metal, or plaster that establishes the level to which concrete or plaster will be placed.

Screw port A three-quarter circular profile in an aluminum extrusion, made to accept a screw driven parallel to the long axis of the extrusion.

Screw slot A serrated slot profile in an aluminum extrusion, made to accept screws driven at right angles to the long axis of the extrusion.

Scupper An opening through a parapet through which water can drain over the edge of a flat roof.

Sealant A rubberlike, adhesive material, usually applied in liquid or tape form, used to seal a joint, gap, or crack against the passage of air and moisture.

Sealant gun A tool for injecting *sealant* into a joint.

Sealer A coating used to close the pores in a surface, usually in preparation for the application of a finish coating.

Seasoning The drying of wood, to bring its moisture content into equilibrium with ambient conditions.

Seated connection A connection in which a steel beam rests on top of a steel angle or tee that is fastened to a column or girder.

Section An architectural drawing representing a vertically cut plane through a whole building, part of a building, or detail.

Security glass A glazing sheet with multiple laminations of glass and plastic, designed to stop bullets.

Sedimentary rock Rock formed from materials deposited as sediments, such as sand or sea shells, which form sandstone and limestone, respectively.

Segregation Separation of the constituents of wet concrete caused by excessive handling or vibration.

Seismic Relating to earthquakes.

Seismic load A force on a structure caused by movement of the earth relative to the structure during an earthquake.

Seismic separation joint A building separation joint that allows adjacent building masses to oscillate independently during an earthquake.

Self-adhered flashing A flexible, self-sticking flashing material, usually made of polymer-modified asphalt laminated to a plastic backing, with preapplied adhesive on one side.

Self-consolidating concrete (SCC) Concrete formulated so that it is highly flowable and fills formwork completely without needing *consolidation*.

Self-consolidating grout Grout formulated so that it is highly flowable.

Self-drilling Drills its own hole.

Self-furring metal lath Metal lath with dimples that space the lath away from the sheathing behind to allow plaster to penetrate the lath and key to it.

Self-tapping Creates its own screw threads on the inside of the hole.

Self-weight The weight of a beam or slab.

Semirigid connection *See* **Partially restrained moment connection**.

Sequential construction A method of providing design and construction services in which each major phase of design and construction is completed before the next phase is begun.

Set To cure; to install; to recess the heads of nails; a punch for recessing the heads of nails.

Setting block A small block of synthetic rubber or lead used to support the weight of a sheet of glass at its lower edge.

Settlement joint A *building separation joint* that allows the foundations of adjacent building masses to settle at different rates.

SFRM *See* **Spray-applied fire-resistive material**.

SGPL *See* **Structural-grade plastic lumber**.

Shading coefficient The ratio of total solar heat passing through a given sheet of glass to that passing through a sheet of clear double-strength glass; mostly replaced in contemporary energy calculations by *solar heat gain coefficient*.

Shaft An unbroken vertical passage through a multistory building, used for elevators, wiring, plumbing, ductwork, and so on.

Shaft wall A wall surrounding a *shaft*, usually with a specified degree of fire resistance.

Shake A shingle split from a block of wood.

Shake-on hardener A dry powder that is dusted onto the surface of a concrete slab before troweling to react with the concrete and produce a hard wearing surface for industrial use.

Shale A rock formed from the consolidation of clay or silt.

Shallow foundation A building *foundation* located at the base of a wall or a column, bearing on soil relatively close to the ground surface.

Shear A deformation in which planes of material slide with respect to one another.

Shear connection A connection designed to resist only the tendency of one member to slide past the other, and not, as in a *moment connection*, to resist any tendency of the members to rotate with respect to one another; in steel frame construction, a *simple connection*.

Shear panel A wall, floor, or roof surface that acts as a deep beam to help stabilize a building against deformation by lateral forces.

Shear stud A piece of steel welded to the top of a steel beam or girder so as to become embedded in the concrete fill over the beam and cause the beam and the concrete to act as a single structural unit.

Shear wall A stiff wall that imparts lateral force resistance to a building frame.

Sheathing The rough covering applied to the outside of the roof, wall, or floor framing of a structure.

Shed A building or dormer with a single-pitched roof.

Sheeting A stiff material used to retain the soil around an excavation; a material such as polyethylene in the form of very thin, flexible sheets.

Sheet metal Flat rolled metal generally less than ¼ inch (6.35 mm) thick.

Shelf angle A horizontal steel angle attached to the wall or spandrel of a building to support a masonry facing.

SHGC *See* **Solar heat gain coefficient**.

Shim A thin piece of material placed between two components of a building to adjust their relative positions as they are assembled; to insert shims.

Shingle A small unit of water-resistant material nailed in overlapping fashion with many other such units to render a wall or sloping roof watertight; to apply shingles.

Shiplap A board with edges rabbeted so as to overlap flush from one board to the next.

Shop drawings Detailed drawings prepared by a fabricator to guide the shop production of such building components as cut stonework, steel or precast concrete framing, curtain wall panels, and cabinetwork.

Shoring Temporary vertical or sloping supports of steel or timber.

Shotcrete A low-slump concrete mixture that is deposited by being blown from a nozzle at high speed with a stream of compressed air; *pneumatically placed concrete.*

Shrinkage-compensating cement Specially formulated cement, used to counteract the drying shrinkage that normally occurs during curing.

Shrinkage-reducing admixture A concrete additive that reduces drying shrinkage and the cracking that results.

Shrinkage–temperature steel *Reinforcing bars* laid at right angles to the principal bars in a one-way slab for the purpose of preventing excessive cracking caused by drying shrinkage or temperature stresses in the concrete.

Side-hinged inswinging window A window that opens by pivoting inward on hinges at or near a vertical edge of the sash.

Side jamb *See* **Jamb.**

Sidelight A tall, narrow window alongside a door.

Siding The exterior wall finish material applied to a light frame structure.

Siding nail A nail with a small head; used to fasten siding to a building.

Silica fume Very finely divided silicon dioxide, a *pozzolan*, used as an admixture in the formulation of high-strength, low-permeability concrete; also called *microsilica.*

Silicone A polymer used for high-range sealants, roof membranes, and masonry water repellents.

Sill The horizontal bottom portion of a window or door; the exterior surface, usually sloped to shed water, below the bottom of a window or door.

Sill gasket *See* **Sill seal.**

Sill plate The strip of wood that lies immediately on top of a concrete or masonry foundation in wood frame construction.

Sill seal A compressible material placed between a foundation and a wood sill plate to reduce air infiltration between the outdoors and indoors.

Simple caulk and seal A method similar to the *airtight drywall approach* for constructing a light frame building enclosure that is resistant to the free flow of air, but requiring less coordination between framing and sealing operations than the airtight drywall approach.

Simple connection A steel frame connection with no useable resistance to rotation.

Single-hung window A window with two overlapping sashes, the lower of which can slide vertically in tracks and the upper of which is fixed.

Single-ply roof membrane A sheet of plastic or synthetic rubber used as a membrane for a low-slope roof.

Single-strength glass Glass approximately ³⁄₃₂ inch (2.5 mm) thick.

Single tee A precast slab element whose profile resembles the letter *T.*

Sinker A framing nail with a slenderer shank, special coating, and other features designed for ease of driving in comparison to a traditional *common nail.*

SIP *See* **Structural insulated panel.**

Sitecast concrete Concrete that is poured and cured in its final position in a building; *cast in place concrete.*

Skip-joist system *See* **Wide-module concrete joist system.**

Skylight A glazed unit installed in a roof; also referred to as a *unit skylight.*

Slab band A very broad, shallow beam used with a one-way solid slab.

Slab on grade A concrete surface lying upon, and supported directly by, the ground beneath.

Slag The mineral waste that rises to the top of molten iron or steel or to the top of a weld.

Slag cement *See* **Blast furnace slag.**

Slaked lime *See* **Hydrated lime.**

Slate A metamorphic form of clay, easily split into thin sheets.

Sliced veneer Thin sheets of wood produced by pressing a knife against a log.

Sliding window A window with one fixed sash and another that moves horizontally in tracks.

Slip-critical connection A structural steel connection in which the members are clamped together by high-strength bolts with sufficient force that the loads on the members are transmitted between them by friction along their mating (*faying*) surfaces; also called a *friction connection.*

Slip forming Building multistory sitecast concrete walls with forms that rise up the wall as construction progresses.

Slip sheet A thin sheet of paper, plastic, or felt, placed between two materials to eliminate friction or bonding of the materials.

Sloped glazing A system of metal and glass components used to make an inclined, transparent roof; in the International Building Code, glass sloped more than 15 degrees from vertical.

Slow-burn construction *See* **Mill construction.**

Slump test A test in which wet concrete or plaster is placed in a cone-shaped metal mold of specified dimensions and allowed to sag under its own weight after the cone is removed. The vertical distance between the height of the mold and the height of the slumped mixture is an index of the material's working consistency.

Slurry A watery mixture of insoluble materials with a high concentration of suspended solids.

Smoke-developed rating An index of the toxic fumes generated by a material as it burns, as determined by ASTM Standard E84.

Smoke shelf The horizontal area behind the damper of a fireplace.

Sodium borate (SBX) A chemical used to preserve wood against attack by decay and insects.

Soffit The undersurface of a horizontal element of a building, especially the underside of a stair or a roof overhang.

Soffit vent An opening under the eave of a roof used to allow air to flow into the attic or the space below the roof sheathing.

Soft mud process Making bricks by pressing wet clay into molds.

Softwood Wood from coniferous (evergreen) trees.

Soil Any particulate earth material, excluding rock.

Soil liquefaction A phenomenon in which a water-saturated soil loses most of its strength under the influence of sudden, large variations in loading such as can occur during an earthquake.

Solar heat gain coefficient (SHGC) The ratio of solar heat admitted through a sheet of glass, or whole window, to the total heat energy striking the glass or window.

Solar reflectance A unitless index, ranging from 0 to 1, expressing a material's tendency to absorb or reflect solar radiation; also called *albedo.*

Soldering A low-temperature form of *brazing.*

Soldier A brick laid on its end, with its narrow face toward the outside of the wall.

Sole plate The horizontal piece of dimension lumber at the bottom of the studs in a wall in a light frame building; also called a *bottom plate*.

Solid-core door A flush door with no internal cavities.

Solid masonry Masonry walls without cavities; historically, thick, monolithic masonry walls that rely primarily on their mass for their strength, durability, and tempering of the flow of heat and moisture from inside to outside.

Solid slab A concrete slab, without ribs or voids, that spans between beams or bearing walls.

Solid tape sealant *See* **Preformed solid tape sealant**.

Solvent A liquid that dissolves another material.

Sound Transmission Class (STC) An index of the resistance of a wall or partition to the passage of sound.

Space truss, space frame A truss that spans with two-way action.

Spalling The cracking or flaking of the surface of concrete or masonry units, caused, for example, by freeze-thaw action, corroding reinforcing, or pointing mortars that are harder and stronger than the mortar deeper in the masonry joint.

Span The distance between supports for a beam, girder, truss, vault, arch, or other horizontal structural device; to carry a load between supports.

Spandrel The wall area between the head of a window on one story and the sill of a window on the floor above; the area of a wall between adjacent arches.

Spandrel beam A beam that runs along the outside edge of a floor or roof.

Spandrel glass Opaque glass manufactured especially for use in spandrel panels.

Spandrel panel A curtain wall panel used in a spandrel.

Span rating The number stamped on a sheet of plywood or other wood building panel to indicate how far in inches it may span between supports.

Specifications The written portion of the *construction documents*, concerning the quality of materials and execution of construction procedures required for a building.

Spirit level A tool in which a bubble in an upwardly curving cylindrical glass vial indicates whether a building element is level or not level, plumb or not plumb.

Splash block A small precast block of concrete or plastic used to divert water at the bottom of a downspout.

Spline A thin strip inserted into grooves in two mating pieces of material to hold them in alignment; a ridge or strip of material intended to lock to a mating groove. In glazing, the *edge spacer* in an insulating glass unit.

Split jamb A door frame fabricated in two interlocking halves, to be installed from the opposite sides of an opening.

Spray-applied fire-resistive material (SFRM) Fibrous or cementitious insulation applied to steel or concrete with a sprayer to provide protection against the heat of fire.

Springwood In wood, the portion of the growth ring comprised of relatively larger, less dense cells; also called *earlywood*.

Squash block Short lengths of framing lumber, inserted under points of concentrated load to prevent overloading of *I-joist* framing members.

SSP *See* **Stressed-skin panel**.

Staggered truss system A steel framing system in which story-high trusses, staggered one-half bay from one story to the next, support floor decks on both their top and bottom chords.

Stain A coating intended primarily to change the color of wood or concrete without forming an impervious film.

Stainless steel A silver-colored steel alloy with superior corrosion resistance due principally to high chromium and nickel content.

Standing and running trim Door and window casings and baseboards.

Standing seam A sheet metal roofing seam that projects at right angles to the plane of the roof.

Static coefficient of friction (SCOF) The *coefficient of friction*, measured between two surfaces at rest relative to each other; used in some finish flooring slip resistance measurements. *See also* **dynamic coefficient of friction**.

Stay A sloping cable used to stabilize a structure.

STC *See* **Sound Transmission Class**.

Steam curing Aiding and accelerating the setting reaction of concrete by the application of steam.

Steel Iron with a controlled amount of carbon, generally less than 2 percent.

Steel reinforcing bars Hot-rolled, deformed steel bars used to impart tensile strength and ductility to concrete or masonry structures; *rebar*.

Steel trowel A metal-bladed tool used in the final stages of finishing of a concrete slab.

Steep roof A roof with sufficient slope to be made waterproof with shingles. In the International Building Code, a roof with a slope of 2:12 (17 percent) or greater.

Sticking The cementing together of defects in marble slabs.

Stick system A metal curtain wall system that is largely assembled in place.

Stiffener plate A steel plate attached to a structural member to support it against heavy localized loading or stresses.

Stiff mud process A method of molding bricks in which a column of damp clay is extruded from a rectangular die and cut into bricks by fine wires.

Stile A vertical framing member in a panel door.

Stirrup A vertical loop of steel bar used to reinforce a concrete beam against diagonal tension forces.

Stirrup-tie A stirrup that forms a complete loop, as differentiated from a U-stirrup, which has an open top.

Stool The interior horizontal plane at the sill of a window.

Storm window A sash added to the outside of a window in winter to increase its thermal resistance and decrease air infiltration.

Story pole A strip of wood marked with the exact course heights of masonry for a particular building; used to make sure that all the leads are identical in height and coursing.

Straightedge To strike off the surface of a concrete slab using screeds and a straight piece of lumber or metal; as a noun, a long, straight item, used to perform straightedging, test the flatness of a surface, or trace a straight line.

Strain Deformation under stress; expressed as a ratio of the change in length over the original length.

Stress Force per unit area.

Stressed-skin panel (SSP) A panel consisting of two face sheets of wood, metal, or concrete bonded to perpendicular spacer ribs or framing members such that the panel can act as a composite structural panel.

Stretcher A brick or masonry unit laid in its most usual position, with the broadest surface of the unit horizontal and the length of the unit parallel to the surface of the wall.

Striated Textured with parallel scratches or grooves.

Stringer The sloping wood or steel member that supports the treads of a stair.

Strip flooring Solid wood finish flooring members less than 3 inches (75 mm) in width, usually in the form of tongue-and-groove boards.

Stripping Removing formwork from concrete; sealing around a roof flashing with layers of felt and bitumen.

Structural bond The interlocking pattern of masonry units used to tie two or more wythes together in a wall.

Structural composite lumber Substitutes for solid lumber made from wood veneers or wood fiber strands and glue; also called *engineered lumber*.

Structural glazed facing tile A hollow clay masonry unit with glazed faces.

Structural glazing Glass secured to the face of a building with highly adhesive sealant or glazing tape to eliminate the need for any metal to appear on the exterior of the building.

Structural-grade plastic lumber (SGPL) Lumberlike plastic members, reinforced with glass fibers, and formulated to be roughly as strong as conventional solid wood.

Structural insulated panel (SIP) A panel consisting of two face sheets of wood panel bonded together by plastic foam core.

Structural lightweight aggregate Lightweight aggregate with sufficient density and strength for use in structural concrete.

Structural mill The portion of a steel mill that rolls structural shapes.

Structural standing-seam metal roofing Sheets of folded metal that serve both as decking and as the waterproof layer of a roof.

Structural terra cotta Molded components, often highly ornamental, made of fired clay, designed to be used in the facades of buildings.

Structural tubing *See* **Hollow structural section**.

Structure/enclosure joint A connection designed to allow the structure of a building and its cladding or partitions to move independently.

Stucco Plaster made from a mixture of portland cement, lime, sand, and water; commonly used as an exterior finish material.

Stud One of an array of small, closely spaced, parallel wall framing members; a heavy steel pin.

Styrene-butadiene-styrene (SBS) A copolymer of butadiene and styrene used as a modifier in polymer-modified bitumen roofing.

Subcontractor A contractor who specializes in one area of construction activity and who works under a general contractor.

Subfloor The loadbearing surface beneath a finish floor.

Subpurlin A very small roof framing member that spans between joists or purlins.

Substrate The base to which a coating, veneer, or finish material is applied.

Substructure The occupied below-ground portion of a building.

Summerwood In wood, the portion of the growth ring comprised of relatively smaller, denser cells; also called *latewood*.

Sump A pit designed to collect water for removal from an excavation or basement.

Superflat floor A concrete slab finished to a high degree of flatness and levelness according to a recognized system of measurement.

Superplasticizer An admixture that makes wet concrete or grout extremely fluid without additional water.

Superstructure The above-ground portion of a building.

Supplementary cementitious material Hydraulic cementitious material or pozzolan mixed with portland cement to modify the cement product's properties or lower the energy required to manufacture the cement.

Supply pipe A pipe that brings clean water to a plumbing fixture.

Surface-bonded masonry Concrete block laid without mortar and then plastered on both sides with a fiber-reinforced cement plaster so as to make a structurally sound masonry wall.

Surface number In glazing assemblies, the distinct faces of glazing, counting from the outermost to the innermost of a glazing unit, including each face of each glazing material.

Surfacing Smoothing the surface of a material, usually by *planing*.

Suspended ceiling A finish ceiling that is hung on wires from the structure above.

Suspended glazing Large sheets of glass hung from clamps at their top edges to eliminate the need for metal mullions.

Sustainability Providing for the needs of the current generation without compromising the ability of future generations to provide for their needs. Providing healthy, resource-conserving, and energy-efficient buildings. "Green" building.

Swedge bolt *See* **Lockpin and collar fastener**.

Synthetic gypsum Chemically manufactured gypsum made from the byproducts of various industrial processes, such as the desulfurization of power plant flue gasses.

T

Tackless strip A wood strip with projecting points used to fasten a carpet around the edge of a room; also called a *tackstrip*.

Tackstrip *See* **Tackless strip**.

Tagline A rope attached to a building component to help guide it as it is lifted by a crane or derrick.

Tangential shrinkage In wood, shrinkage along the circumference of the log.

Tap To cut internal threads, such as in a hole or nut.

Tapered edge The longitudinal edge of a sheet of gypsum board, which is recessed to allow room for reinforcing tape and joint compound.

Tdw *See* **Damage weighted transmittance**.

Tee A metal or precast concrete member with a cross section resembling the letter *T*.

Tempered glass Heat-treated glass that is stronger than heat-strengthened glass and is suitable for use as *safety glazing*.

Tempering Controlled heating and cooling of a material to alter its mechanical properties; a form of heat treatment.

Tendon A steel strand used for prestressing a concrete member.

Tensile strength The ability of a structural material to withstand stretching forces.

Tensile stress A stress caused by stretching of a material.

Tension A stretching force; to stretch.

Tension control bolt A bolt tightened by means of a splined end that breaks off when the bolt shank has reached the required tension.

Termite shield A metal flashing placed on top of a concrete foundation to prevent termites from traveling undetected from the ground into the superstructure.

Terne An alloy of lead and tin, used to coat sheets of carbon steel or stainless steel; used in the past for metal roofing sheet.

Terrace door A double glass door, one leaf of which is fixed and the other hinged to the fixed leaf at the centerline of the door.

Terrazzo A finish floor material consisting of concrete with an aggregate of marble chips selected for size and color, which is ground and polished smooth after curing.

Thatch A thick roof covering of reeds, straw, grasses, or leaves.

Thermal break A section of material with low thermal conductivity installed between metal components to retard the passage of heat through a wall or window assembly.

Thermal bridge A component of relatively high thermal conductivity that conducts heat more rapidly through an insulated building assembly, such as a steel stud in an insulated stud wall.

Thermal conductivity The rate at which a material conducts heat.

Thermal emittance A unitless index, from 0 to 1, expressing a material's tendency to radiate thermal energy as its temperature rises in relation to surrounding surfaces.

Thermal envelope *See* **Building enclosure**.

Thermal insulation A material that greatly retards the passage of heat.

Thermal resistance The resistance of a material or assembly to the conduction of heat.

Thermochromic glass Glass that changes its optical properties in response to changes in temperature.

Thermoplastic In plastics, having the property of softening when heated and rehardening when cooled; weldable by heat or solvents.

Thermoplastic polyolefin (TPO) A thermoplastic single-ply roof membrane material, made from blends of polyethylene, polypropylene, and ethylene-propylene rubber polymers.

Thermosetting In plastics, not having the property of softening when heated; not heat-fusible.

Thickset tile Ceramic tile installed on a thick bed of portland cement mortar; also called *mortar bed* or *mud set tile*.

Thin-set tile Ceramic tile bonded to a solid base with a thin application of portland cement mortar or organic adhesive.

Through-wall flashing *See* **Internal flashing**.

Thrust A lateral or inclined force resulting from the structural action of an arch, vault, dome, suspension structure, or rigid frame.

Thrust block A wooden block running perpendicular to the stringers at the bottom of a stair, whose function is to hold the stringers in place.

Tie A device for holding two parts of a construction together; a structural device that acts in tension.

Tieback A tie, one end of which is anchored in the ground, with the other end used to support sheeting around an excavation.

Tie beam A reinforced concrete beam cast as part of a masonry wall, whose primary purpose is to hold the wall together, especially against seismic loads, or cast between a number of isolated foundation elements to maintain their relative positions.

Tier The portion of a multistory steel building frame supported by one set of fabricated column pieces, commonly two stories in height.

Tie rod A steel rod that acts in tension.

Tile A fired clay product that is thinner in cross section than a brick, either a thin, flat element (ceramic tile or quarry tile), a thin, curved element (roofing tile), or a hollow element with thin walls (flue tile, tile pipe, structural clay tile); also a thin, flat element of another material, such as an acoustical ceiling unit or a resilient floor unit.

Tilt/turn window A window that opens either by rotating its sash about its vertical centerline or as a hopper.

Tilt-up construction A method of constructing concrete walls in which panels are cast and cured flat on a floor slab, then tilted up into their final positions.

Timber Standing trees; a large piece of dimension lumber.

Tinted glass Glass that is colored with pigments, dyes, or other *admixtures*.

Titanium A strong, corrosion-resistant, silvery gray *nonferrous metal*.

Toe nailing Fastening with nails driven at an angle.

Tongue and groove An interlocking edge detail for joining planks or panels.

Tooling The finishing of a mortar joint or sealant joint by pressing and compacting it to create a particular profile.

Toothed plate A multipronged fastener made from stamped sheet metal, used to join members of a lightwood wood truss.

Top-hinged inswinging window A window that opens inward on hinges on or near its head.

Topping A thin layer of concrete cast over the top of a floor deck.

Topping-out Placing the last member in a building frame.

Top plate The horizontal member at the top of the studs in a wall in a light frame building.

Topside vent A water-protected opening through a roof membrane to relieve pressure from water vapor that may accumulate beneath the membrane.

Torque Twisting action; *moment*.

Torsional stress Stress resulting from the twisting of a structural member.

Touch sanded In plywood, lightly sanded to produce a smoother, flatter surface.

TPO *See* **Thermoplastic polyolefin**.

Tracheids The longitudinal cells in a softwood.

Traffic deck A walking surface placed on top of a roof membrane.

Transit-mixed concrete Concrete mixed in a drum on the back of a truck as it is transported to the building site.

Travertine A richly patterned, marble-like form of limestone; classified by ASTM C119 in the Other Stone group.

Tread One of the horizontal planes that make up a stair.

Tremie A large funnel with a tube attached, used to deposit concrete in deep forms or beneath water or slurry.

Trim accessories Casing beads, corner beads, expansion joints, and other devices used to finish edges and corners of a plaster wall or ceiling.

Trimmer joist A joist that supports a header around an opening in a floor or roof frame.

Trimmer stud. *See* **Jack stud.**

Trowel A thin, flat steel tool, either pointed or rectangular, provided with a handle and held in the hand, used to manipulate mastic, mortar, plaster, or concrete. Also, a machine whose rotating steel blades are used to finish concrete slabs; to use a trowel.

Truss A triangulated arrangement of structural members that reduces nonaxial external forces to a set of axial forces in its members. *See also* **Vierendeel truss.**

Tuckpointing Traditionally, a method of finishing masonry joints using mortars of different colors to artificially create the appearance of a more refined joint; in contemporary usage, may be used interchangeably with *repointing*.

Tunnel kiln A kiln through which clay products are passed on railroad cars.

Turn-of-nut method A method of achieving the correct tightness in a *high-strength bolt* by first tightening the nut snugly, then turning it a specified additional fraction of a turn.

Two-way action Bending of a slab or deck in which bending stresses are approximately equal in the two principal directions of the structure.

Two-way concrete joist system A reinforced concrete framing system in which columns directly support an orthogonal grid of intersecting joists.

Two-way flat plate A reinforced concrete framing system in which columns directly support a two-way slab that is planar on both of its surfaces.

Two-way flat slab A reinforced concrete framing system in which columns with mushroom capitals and/or drop panels directly support a two-way slab that is planar on both of its surfaces.

Type IV HT construction *See* **Heavy Timber construction.**

Type X gypsum board A fiber-reinforced *gypsum board* used where greater fire resistance is required.

U

UF *See* **Urea-formaldehyde.**

U-Factor A measure of the thermal conductance of a material or assembly; the mathematical reciprocal of *R-value.*

Unbonded construction Posttensioned concrete construction in which the tendons are not grouted to the surrounding concrete.

Uncoursed stone masonry Stone masonry laid without continuous horizontal joints; random.

Undercarpet wiring system Flat, insulated electrical conductors that run under carpeting, and their associated outlet boxes and fixtures.

Undercourse A course of shingles laid beneath an exposed course of shingles at the lower edge of a wall or roof in order to provide a waterproof layer behind the joints in the exposed course.

Underfire The floor of the firebox in a fireplace.

Underlayment A panel laid over a subfloor to create a smooth, stiff surface for the application of finish flooring; a water-resistant material applied under finish roofing.

Underpinning The process of placing new foundations beneath an existing structure.

Unfinished bolt An ordinary carbon steel bolt.

UniFormat The trademarked name for a system of organizing building information based on functional relationships.

Uniformly graded soil A special instance of a *poorly graded soil* in which the soil particles are mostly of one size.

Uniform settlement Subsidence of the various foundation elements of a building at the same rate, resulting in no distress to the structure of the building.

Unit-and-mullion system A curtain wall system consisting of prefabricated panel units secured with site-applied mullions.

Unit skylight *See* **Skylight.**

Unit system A curtain wall system consisting entirely of prefabricated panel units.

Unreinforced Constructed without steel reinforcing bars or welded wire fabric.

Up–down construction A sequence of construction activity in which construction proceeds downward on the sublevels of a building at the same time as it proceeds upward on the superstructure.

Upside-down roof A membrane roof assembly in which the thermal insulation lies above the membrane.

Urea-formaldehyde (UF) A structural wood adhesive, not suitable for exterior exposure, and associated with relatively high *formaldehyde* gas emissions.

U-stirrup An open-top, U-shaped loop of steel bar used as reinforcing against diagonal tension in a concrete beam.

V

Valley A trough formed by the intersection of two roof slopes.

Valley rafter A diagonal *rafter* that supports a valley.

Vapor barrier *See* **Vapor retarder.**

Vapor permeability Vapor permeance per unit of thickness.

Vapor permeance A measure of the ease with which water vapor can diffuse through a material.

Vapor pressure A measure of the pressure exerted by water molecules in a gaseous state, generally higher with higher relative humidity and higher air temperature.

Vapor retarder A layer of material intended to resist the diffusion of water vapor through a building assembly. Also called, less accurately, *vapor barrier.*

Varnish A slow-drying transparent coating.

Vault An arched surface; a strongly built room for such purposes as housing large electrical equipment or safeguarding money.

Vee joint A joint whose profile resembles the letter *V.*

Vegetated roof *See* **Green roof.**

Veneer A thin layer, sheet, or facing.

Veneer plaster A wall finish system in which a thin finish layer of gypsum plaster is applied over a special *veneer plaster base.*

Veneer plaster base The special gypsum board over which *veneer plaster* is applied.

Vented metal decking *Metal decking* with slotted perforations designed to allow excess moisture in the concrete cast onto the deck to evaporate downward through the decking.

Vent spacer A device used to maintain a free air passage above the thermal insulation in an attic or roof.

Vermiculite Expanded mica, used as an insulating fill or lightweight aggregate.

Vertical bar An upright *reinforcing bar* in a concrete column; also called a *column bar.*

Vertical grain lumber *See* **Quartersawn lumber.**

Vierendeel truss A truss with rectangular panels and rigid joints but no diagonal members. The members of a Vierendeel truss are subjected to strong nonaxial forces.

Vinyl *See* **Polyvinyl chloride.**

Visible light transmittance (VT) The ratio of visible light that passes through a sheet of glass or a glazing unit to the amount of light striking the glass or unit.

Visual grading The grading of wood for its structural properties, based on visual inspection, as distinct from *machine grading*; not to be confused with *appearance grading*.

Vitrification The process of transforming a material into a glassy substance by means of heat.

Volatile organic compound (VOC) Organic (carbon-based) chemical compound that evaporates readily, is a significant air pollutant, a potential irritant to building occupants, and, in some cases, a greenhouse gas.

Volume-change joint A building separation joint that allows for expansion and contraction of adjacent portions of a building without distress.

Voussoir A wedge-shaped element of an arch or vault.

VT *See* **Visible light transmittance.**

W

Waferboard A building panel made by bonding together large, flat flakes of wood.

Waffle slab A two-way concrete joist system.

Wainscoting A wall facing, usually of wood, cut stone, or ceramic tile, that is carried only partway up a wall.

Waler A horizontal beam used to support sheeting or concrete formwork.

Wane An irregular rounding of a long edge of a piece of dimension lumber caused by cutting the lumber from too near the outside surface of the log.

Warm edge spacer A glazing *edge spacer* with improved thermal resistance.

Washer A steel disk with a hole in the middle, used to spread the load from a bolt, screw, or nail across a wider area of material.

Water–cement ratio An expression of the relative proportions, by weight, of water and cement in a concrete mixture.

Waterproofing Material acting as a barrier to the flow of water and capable of withstanding hydrostatic pressure.

Water-reducing admixture Concrete *admixture* that allows a reduction in the amount of mixing water while retaining the same workability, resulting in higher-strength concrete.

Water-resistant gypsum board A *gypsum board* designed for use in locations where it may be exposed to occasional dampness.

Water smoking The process of applying heat to evaporate the last water from clay products before they are fired.

Waterstop A metal, synthetic rubber, bentonite clay, or sealant strip used to seal joints in concrete foundation walls.

Water-struck brick A brick made in a mold that was wetted before the clay was placed in it.

Water table The level at which the pressure of water in the soil is equal to the atmospheric pressure; effectively, the level to which groundwater will fill an excavation; a wood molding or shaped brick used to make a transition between a thicker foundation and the wall above.

Water vapor Water in its gaseous phase.

Wattle and daub Mud plaster (daub) applied to a primitive lath of woven twigs or reeds (wattle).

Waxing Filling of voids in marble slabs; application of wax as a finish material.

Weathered joint A mortar joint finished in a sloping, planar profile that tends to shed water to the outside of the wall.

Weathering steel A steel alloy that forms a self-protecting rust layer when exposed to the atmosphere.

Weather-resistive barrier A membrane used to resist the passage of liquid water or air through the exterior enclosure of a building.

Weatherstrip A ribbon of resilient, brushlike, or springy material used to reduce air infiltration through the crack around a sash or door.

Web A cross-connecting piece, such as the portion of a wide-flange shape that is perpendicular to the flanges or the portion of a concrete masonry unit that is perpendicular to the face shells.

Web stiffener A metal rib used to support the web of a light gauge steel joist or a structural steel girder against buckling.

Weep hole A small opening whose purpose is to permit drainage of water that accumulates inside a building component or assembly.

Weld A joint between two pieces of metal formed by fusing the pieces together by the application of intense heat, usually with the aid of additional metal melted from a rod or electrode.

Welded wire fabric (WWF) *See* **Welded wire reinforcing.**

Welded wire reinforcing (WWR) A welded grid of steel reinforcing wires or bars, used most commonly for reinforcing of slabs; also called *welded wire fabric (WWF)*.

Welding The process of making a weld.

Weld plate A steel plate anchored into the surface of concrete, to which another steel element can be welded.

Well graded soil Coarse-grained soil with a full range of particle sizes; also called *poorly sorted soil.*

Well-sorted soil *See* **Poorly graded soil.**

"Wet" systems Construction systems that utilize considerable quantities of water on the construction site, such as masonry, plaster, sitecast concrete, and terrazzo.

White portland cement A portland cement that is white in color; used for architectural concrete where greater color control is required.

Wide-flange shape Any of a wide range of structural steel components rolled in the shape of the letter *I* or *H*.

Wide-module concrete joist system A one-way concrete framing system with joists that are spaced more widely than those in a conventional one-way concrete joist system.

Wind brace A diagonal structural member whose function is to stabilize a frame against lateral forces.

Winder (rhymes with "reminder") A stair tread that is wider at one end than at the other.

Wind load A force on a building caused by wind pressure and/or suction.

Wind uplift Upward forces on a structure caused by negative aerodynamic pressures that result from certain wind conditions.

Wired glass Glass in which a wire mesh is embedded during manufacture, principally for fire resistance.

Wood-plastic composites (WPC) Wood-like products made from wood fibers, plastics of various types, and other additives, with a plastic content not exceeding 50 percent.

Wood–polymer composite planks Linear strips, intended for outdoor decking and other outdoor uses, that are made of wood fiber and a plastic binder.

Workability agent *Admixture* for concrete that improves the plasticity of wet material to make it easier to place in forms and to finish.

Working joint A connection that is designed to allow small amounts of relative movement between two pieces of a building assembly.

WPC *See* **Wood-plastic composites.**

Wracking Forcing out of *plumb*.

Wrought iron A form of iron that is soft, tough, and fibrous in structure, containing about 0.1 percent carbon and 1–2 percent slag.

WWF *See* **Welded wire fabric.**

WWR *See* **Welded wire reinforcing.**

Wythe (rhymes with "scythe" and "tithe") A vertical layer of masonry that is one masonry unit thick.

Y

Yield strength The stress at which a material ceases to deform in a fully elastic manner and begins to deform irreversibly.

Z

Z-brace door A door made of vertical planks held together and braced on the back by three pieces of wood whose configuration resembles the letter Z.

Zero-slump concrete A concrete mixed with so little water that it does not sag when piled vertically.

Zinc A relatively weak and brittle *nonferrous metal* used, most notably, as a protective galvanic coating for steel.

Zoning ordinance A law that specifies how land within a jurisdiction may be used.

INDEX

A

Acoustic criteria:
    for interior finish systems, 883
    for ceilings, 934, 937
Acoustic insulation, 883, 909, 920–923, 934, 938
Acoustic underlayment, 883, 944
ADA. *See* Airtight drywall approach
ADA. *See* Americans with Disabilities Act
Advanced framing techniques, 219–221
Aerogel:
    in glazing, 733
    insulation, 274, 276
    sustainability, 283
Aggregates:
    coarse-grained soils, 34–36
    concrete, 529, 533, 534
    grout, 338
    mortar, 313, 315
Agrifiber panels, 113
Air barrier, 808–811
    low-slope roof, 665–666
    masonry cavity wall, 390, 819
    in pressure-equalized wall design, 800
    in wood light frame construction, 283–285
Air infiltration. *See* Air barrier
Air permeance. *See* Air barrier
Air-supported structures, 485
Airtight drywall approach, 283–284
Alloys. *See* Steel
Aluminum, 518
    extrusions, 848–851
    sheet metal roofing, 700
    surface finishes, 853
    sustainability, 852
Aluminum-framing and glass systems. *See*
    Cladding, metal and glass
American National Standards Institute
    (ANSI), 17
Americans with Disabilities Act, 17
American Woodwork Institute (AWI) quality
    grades, 285
Anchor bolt:
    concrete, 634, 636, 649
    heavy timber, 144, 159, 160, 162
    light gauge steel, 502
    light wood frame, 178, 183, 188, 191, 201,
        203, 204
    structural steel, 451, 454, 455
Anodizing. *See* Aluminum, surface finishes
Arches:
    brick, 334–336

concrete, 608–609
    heavy timber, 158–159, 161–163
    steel, 478, 480
Architectural sheet metal roofing, 690, 695–701
Articulation class, 937
Ashlar stone masonry, 360
Asphalt shingles, 238, 240, 257, 689–690,
    691–693
ASTM International, 17
Autoclaved aerated concrete (AAC), 382
AWI. *See* American Woodwork Institute quality
    grades

B

Backer rod, 806
Backfilling, 77
Ballast. *See* Roofing, low-slope, ballast
Balloon frame, 173–174
Bamboo:
    flooring, 957
    panels, 113–117
Bar chart schedule. *See* Gantt chart
Bark, tree, 84, 85
Barrier wall, and keeping water out, 798
    with EIFS, 838
    with masonry and concrete cladding, 840
    with metal panel cladding, 872
Barrier wall, for soils dewatering, 49, 51
Basalt building stone, 350
Base flashing, low-slope roof membrane, 680–685
Basement, 51, 52
    insulation, 71–73
    with light wood frame construction, 176–184
    waterproofing, 66–71
Base isolation. *See* Seismic base isolation
Base plates:
    precast concrete column, 636, 637
    steel column, 452, 454–456
Batten seam sheet metal roofing, 699, 670, 701
Bending moment, 441
Bessemer process, 424
BIM. *See* building information model
Biocomposite panels, 113
BIPV. *See* Building-integrated photovoltaic
Blast furnace slag, 425, 426, 487
    in cement, 530
    in concrete, 532, 535
    in concrete masonry, 362, 370
    in insulation, 282, 672
    in mortar, 313

Blast-resistant glazing system, 785
Blower door test, 284–285
Board foot, 101–102
Bolts:
    concrete construction, 634–635
    foundation anchor (light wood frame
        construction), 178, 188
    steel construction, 436–437
    wood construction, 124
Bond, payment or performance. *See* Surety
    bond
Brake metal, 517
Brass, 518
Brazing, 517
Brick flooring and paving, 414, 950, 951
Brick masonry, 316
    ASTM classification of bricks, 322
    bonds, 325, 326
    brickwork terminology, 324
    choosing bricks, 322, 323
    comparative material properties, 394
    comparative structural properties, 416
    dimensioning, modular, 327, 330
    firebrick, 332
    firing of bricks, 318–320
    fly ash bricks, 320
    hollow bricks, 322
    laying bricks, 322, 324–331
    molding of bricks, 317–318
    mortar joints, 330–331
    preliminary design, 396
    reinforced (RBM), 338–339
    sizes and shapes of bricks, 320–322
    solid bricks, 322
    spanning openings, 332–337, 341–344
    sustainability, 316
    weathering regions, 322, 323
    See also Cladding, masonry and concrete;
        Masonry construction; Mortar
Bridging, 186, 188, 192, 193
Bronze, 518
Buckling restrained bracing, 445, 473
Building codes, 11–17
Building felt, 232, 233
Building information model (BIM), 25
Building-integrated photovoltaic (BIPV), 707
Building separation joint, 408–411, 878–879
    finish covers, 881
    roofing system, 664, 682
Building wrap, 733, 808, 811
    air permeance, 809
Built-up roof (BUR), 674–675

C

Cabinets, 291–294
CAC. *See* Ceiling attenuation class
CAD. *See* Computer-aided design
Caissons. *See* Foundations, caissons
Camber, 452, 556, 557, 629
Carbon fiber reinforcing, 600, 630, 832
Carpentry:
    finish, 260, 285–302
    rough, 186
Carpet and carpet tile, 399, 959–960
    and noise transmission through floors, 944
    sustainability, 943
Castellated beam, 473–474
Cast iron, 422, 424, 517
Cavity wall, 361, 390–392, 397, 398
    drainage and flashings, 393, 394, 400–403
    and masonry construction, 381
    insulation, 402
    invention of, 311
    and masonry curtain wall, 818–824
    and pressure-equalized wall design, 800, 803
Ceiling attenuation class (CAC), 934
Ceilings, 932–942
    acoustical, 934–937
    with exposed structural and mechanical
        components, 149–150, 932–933
    fire-resistance rated, 471, 938–939
    gypsum board and plaster, 934, 935
    interstitial, 940, 942
    suspended, 933–938
    sustainability, 943
    tightly attached, 933
Cellular raceway, 945, 946
Cellular steel decking, 462, 463, 945
Cellulose:
    in asphalt shingles, 689
    in building and roofing felts, 232, 674, 703
    in flooring underlayment, 944
    in gypsum wallboard, 913
    insulation, 274, 279, 283, 672, 809, 887
    as natural plastic, 766
    in trees, 84, 94
    in wood products, 103
Cement, 313–314, 528–531
    blended hydraulic, 313
    hydraulic and nonhydraulic, 314
    masonry, 313
    mortar, 313
    portland, 529–531
    sustainability, 532, 533
Ceramic tile:
    finish flooring, 304, 948–949, 954
    precast concrete cladding facing, 830
    wall facing, 924–927
    wood light frame construction interiors,
        302–304
    See also Quarry tile
Chimney, 266, 267, 286–287
Chromium, 519
Cladding. *See also* Siding
Cladding, masonry and concrete:
    exterior insulation and finish system, 836–840
    glass-fiber-reinforced concrete, 832–836
    keeping water out, 840
    masonry veneer, 818–824, 842–845

precast concrete, 830–832
prefabricated brick panels, 824–825
stone, 825–829
See also Exterior walls
Cladding, metal and glass:
    curtain wall, 807, 854–870
    double-skin facades, 868–869
    entrances, 854
    outside and inside glazing, 856–867
    metal panel cladding, 871–873
    and rainscreen principle, 868
    sloped glazing, 869
    storefront, 854
    sustainability of aluminum cladding, 852
    testing and standards, 810–812
    thermal breaks, 851–852
    window wall, 854
    See also Exterior walls
Class, roof covering, 707
Classical orders, 364
Clay tile roofing, 690, 695
CLSM. *See* controlled low-strength material
Coating systems, architectural, 244–247
Codes, building, 11–17
Coefficient of friction, 944
Cold-formed metal framing, *see* light gauge
    steel framing
Composite structural members:
    masonry walls, 388–389
    metal decking and concrete, 462–464
    precast and sitecast concrete, 625, 645, 647
    steel and concrete beams, 462–465
    steel and concrete columns, 486
    steel and concrete caissons, 57
    wood and concrete floors, 152, 156–157
Composition shingles. *See* Asphalt shingles
Computer-aided design (CAD), 25
Concrete, 529
    admixtures, 535
    aggregates, 534
    cement, 529–531
    comparative structural properties, 541
    consolidation, 538, 569, 572, 575
    creep, 555
    curing, 538–539
    cutting, 604–607
    fastening to, 634–635
    finish flooring, 948
    formwork, 540–541
    handling and placing, 537–538
    history, 528–529
    hot and cold weather placement, 539
    light transmitting, 562
    mixes, 535–537
    pumping, 537, 538, 594
    prestressing, 555–660
    reinforcing, steel, 541–555
    reinforcing, fibrous, 555
    segregation, 537
    self-consolidating, 538, 560, 594
    slump test, 537
    strength, 535–536, 539, 541, 560
    supplementary cementitious materials, 534–535
    sustainability, 532–533
    test cylinder, 537
    ultra-high performance concrete, 560
    water, 534

water-cement ratio, 535–536
Concrete blocks. *See* Concrete masonry
Concrete masonry, 370
    ASTM classification of concrete masonry
        units, 370–372
    autoclaved aerated concrete (AAC), 368
    curing, 370
    decorative units, 377–381
    dimensioning, modular, 372
    dry-stacked, 376
    economy of, 381
    laying, 373–376
    manufacture of units, 370–372
    mortar joints, 373–375
    reinforcing, 373–376
    sizes and shapes of units, 370–372
    structural glazed facing tiles, 381, 925
    structural terra cotta, 381
    surface bonded, 376
    sustainability, 362–363
    weights, 370, 371
    See also Cladding, masonry and concrete;
        Masonry construction; Mortar
Concrete, precast, 622
    assembly concepts, 626–627
    beams and girders, 624–625
    and building codes, 649
    carbon fiber reinforcing, 630
    columns, 624
    composite precast/sitecast construction, 647
    construction process, 647
    curtain wall panels, 830–832
    joining elements, 633–636
    manufacture, 627–633
    preliminary design, 625
    prestressing, 630
    reinforcing, 630
    slabs, 624
    sustainability, 648
    transportation, 622
    wall panels, 625–626
    see also concrete
Concrete, sitecast, 506
    architectural finishes, 600–603
    and building codes, 611–612
    climbing formwork, 589, 597
    columns, 577–578
    economical design, 611
    floor and roof systems, two-way, 587–592
    floor and roof systems, one-way, 578–586
    flying formwork, 597
    form ties, 572–574
    formwork costs, 538, 540–541, 588, 594
    ganged forms, 597
    innovations, 594–600
    lift-slab construction, 597
    longer spans, 608–611
    posttensioned, 592–594
    preliminary design, 597
    reinforcing, *See* Concrete, reinforcing
    reshoring, 580, 581, 582, 589
    selecting a framing system, 594
    shotcrete, 598–600
    slab on grade, 567–571
    slip forming, 597
    tilt-up, 597–598, 599
    uniqueness, 612–616

walls, 571–576

See also Concrete; Insulating concrete forms

Condensation, 668–671

and air barriers, 283, 808–811

and aluminum extrusions, 851–852

and curtain walls, 811

and durable wood construction, 120

and exterior wall design, 793, 794

and glazing, 722, 730, 736

and light gauge steel frame construction, 513–514

and roofs, 237, 239, 663, 664, 688

and sloped glazing, 869

and vapor retarders, 281–282, 669–672

and windows, 756, 760, 761, 770, 784

Construction documents, 11, 19, 20

Construction manager, 20

Construction project delivery, 19–21

Construction type, 12, 13

Continuous beam, 549, 551, 552, 633, 639

Contraction joint. *See* Control joint

Control joint, 408, 409, 411

GFRC, 837

gypsum wallboard, 917

masonry, 371, 406, 412

concrete slab on grade, 569–571

plaster, 899, 902, 903, 906

precast concrete toppings, 644

terrazzo, 950

Controlled low-strength material (CLSM), 77

Cool color, 702–705

Copper, 518

Corbel:

masonry, 332, 333, 340, 343

precast concrete, 631

Corner boards and trim, wood, 248, 253, 257–258

Corrosion, corrosion protection:

of architectural metals, 516–519, 700

of concrete reinforcing steel, 534, 535, 541, 544, 547, 548, 556, 560, 562

galvanic, 701, 708–710

of steel piles, 61

of structural steel alloys, 425, 428

and wood chemical treatments, 118–119

of wood fasteners and connectors, 120–121, 123, 124, 127, 248, 252, 260

Cost plus a fee compensation, 21

Counterflashing, 400, 401, 683

Cradle-to-gate analysis, cradle-to-grave analysis. *See* Life cycle analysis

Cranes, 456, 457

Critical path, 24

Cross-laminated timber, 105–106

construction, 152–155

sustainability, 143

Cruck frame, 141

Curtain wall. *See* Cladding, metal and glass

Cutting concrete, stone, and masonry, 372, 604–607

D

d. *See* Penny

Dampproofing:

foundation, 68, 176, 182, 183, 413

masonry cavity wall, 390

DCOF. *See* Dynamic coefficient of friction

Dead load, 32

Decking:

concrete, precast, 466, 622–624

concrete, sitecast, 465

cross laminated timber, 152

heavy timber, 143, 147, 149, 157, 164

plastic lumber, 110

metal, 427, 462–465, 510

metal, vented, 674

roof, 465, 467, 663–664

wood, exterior, 260

wood, sizes, 102

wood, structural, grades, 99

wood-plastic composite, 108, 109

See also Composite structural members

Densities of building materials, 966

Design/bid/build project delivery, 19–20

Design/build project delivery, 20

Dewatering, 45–51

Dew point temperature, 688–669

Diamond cutting tools, 604–607

Dissimilar metals and galvanic series, 708–710

Domes:

brick, 336, 337, 343

concrete, 608–610

heavy timber, 161–163

steel, 478

Door frames:

hollow steel, 780–782

wood, 243, 290, 291

Doors, 756, 759, 777–783

egress and accessible, 783

exterior residential, installation, 240, 242, 243

fire, 782–783

hardware, 783

interior residential, 289–291

safety considerations, 783–784

steel, 780–782

structural performance, 784

sustainability, 777

thermal performance, 784

wood, 777–780, 786

Downspout, 234–235, 663, 681

Draftstopping, 887

Drywall, *see* Gypsum board

Dynamic coefficient of friction (DCOF), 944

E

Earth materials, 33–38

for building foundations, 36

cohesive, 35

drainage characteristics, 35

frictional, 34

grading, 35–36

loadbearing capacity, 36

particle size, 33–34

properties, 33–35

stability, 36

subsurface exploration and testing, 37–38

Uniform Soil Classification System, 33, 34

Earth reinforcing, 73, 75, 76

Efflorescence, 315, 388, 413

EIFS, 399, 402, 836–840

Electrical wiring, 878–880, 945–947

in ceilings, 932, 933, 940, 942

in cellular decking, 462, 946

in concrete construction, 584, 624

in floors, 945–947

in heavy timber construction, 152, 156

in light gauge steel framing, 500, 508, 514, 921

in light wood framing, 266, 268, 271, 273

in masonry walls, 404, 821, 924

Embodied carbon, 7

Embodied energy, 6–7

aluminum, 852

brick masonry, 316

brick masonry, fly ash, 320

concrete, precast, 648

concrete, sitecast, 532

concrete masonry units, 363

flooring materials, 943

glass, 722

gypsum products, 898

heavy timber, 143

insulation materials, 282–283

steel, 487

steel, light gauge, 501

stone, 362

wood, 90

Embodied water, 7

EMC. *See* Equilibrium moisture content

Energy efficiency, 5, 7–8, 10

and air leakage, 232, 283–285, 808

and basement insulation, 71, 73, 180, 183

and energy codes, 16–17

and concrete thermal mass, 533

and doors and windows, 770–771, 773, 777, 784–785

and exterior wall systems, 794, 797, 812

and glass, 722, 728, 730–732, 744, 746, 747

and light gauge steel frame construction, 513–514

and masonry thermal mass, 363

and roofing, 702, 703–704

and thermal insulation, 668

and ventilation, 285

and windows, 770, 777, 784–785

and wood light framing, 219–221, 273–281

Energy recovery ventilator, 285

Engineered lumber. *See* Lumber, structural composite

Environmental labels, 5

Equilibrium moisture content (EMC), 94

Excavation, 35, 40

in up-down construction, 64, 65–67

in wood light frame construction, 178–179

Excavation support, 40–49

Expanded metal lath, 900, 901, 904, 908, 909, 925, 935

Expansion joint, 408–411

concrete slab on grade, 58

concrete wall, 571

masonry veneer cladding, 818–820, 822

metal and glass curtain wall, 868

masonry wall, 401, 406–407

plaster, 257, 903

Expansion of materials, thermal coefficients, 966

masonry materials, 406

window framing materials, 763

Exterior finishes:
joint sealants, 258, 803–807
paints and coatings, 244–247
See also Cladding; Roofing; Siding
Exterior walls:
building code requirements, 812
conceptual approaches to watertightness, 798–803
design requirements, 792–798
loadbearing and nonloadbearing, 807
sustainability, 797
testing and standards, 810–812
See also Cladding; Insulation, thermal
Extrusions, aluminum, 848–853

F

Fabric structures, 482–485
Fast track construction, 22
Fiberboard, 109
cellulosic, 113
medium-density, 110, 112
insulating sheathing, 113
moldings, 285
See also Wood panels
Fill, earth, 77
controlled low-strength material, 77
engineered, 51
general purpose, 37
Finish carpentry, 266, 285–302, 304
exterior, 260
Fire area, 886, 892
Fire barrier. *See* Fire-rated walls and partitions
Fireblocking, 173, 174, 887
Fire damper, 886, 939
Fire door, 726, 777, 782–783
Fire partition. *See* Fire-rated walls and partitions
Fireplace, 286–287, 302, 335
brick, 322
Fireproofing, 468–473. *See also* Spray-applied fire-resistive materials
Fire-rated walls and partitions, 12–16, 892–893, 921, 923
fire barriers, 886, 892
fire partitions, 892
fire walls, 16, 223, 285, 415, 892
shaft walls, 878, 886, 892, 913, 923
smoke barriers and partitions, 892–893
Fire resistance rating, 12–16
of brick and concrete masonry, 415
of curtain wall, 853
of doors, 782
of glass, 726
of gypsum board partitions, 920
of heavy timber, 142, 143, 164
of interior walls, ceilings, and floors, 885–887
of precast concrete, 649
of structural steel, 468–473
of roofs, 707
See also Fire-rated walls and partitions
Firestopping, 812, 880, 881, 886, 946
Fire wall. *See* Fire-rated walls and partitions
Fixed-fee compensation. *See* Lump-sum compensation
Flame-spread rating, 884–885

Flashing:
EIFS, 840
masonry veneer curtain wall, 818, 821, 823–824
masonry wall, 340, 361, 363, 390, 392–393, 400–403, 413
metal types, comparative, 517–518
synthetic rubber types, comparative, 767
roofing, low-slope, 663, 667, 680, 682–684
roofing, steep, 233, 234, 691–693
window and door, 240–243, 773
Z-, 251
Flat seam sheet metal roofing, 697, 701
Flooring:
acoustic criteria, 883, 944
electrical resistivity, 945
fire criteria, 885, 944
preparation for, 302–303
radiant heated, 272
raised access, 945–947
sequence of installation, 266, 878, 882
slip resistance, 944–955
sustainability, 882, 943
thickness, 961
See also Decks; *specific flooring materials*
Fluid-applied flooring, 960–961
Fluid-applied membrane roofing, 679
Fly ash, 534
in brick, 316, 320
in concrete, 532, 534, 535
in concrete masonry, 362
in controlled low-strength material, 77
Flying formwork, 597–598, 611
Folded plate, concrete, 608–610
Footings, 51–54
concrete column, 577
concrete wall, 572–573
for light wood framing, 176–181
structural steel column, 454–455
See also Foundations
Forest management, 87–90
Formaldehyde, 127
effects on air quality and health, 7, 91, 107, 127, 143, 282, 769, 882, 887, 943
Foundations, 51–67, 176–184
and building codes, 78
caissons, 55–56
cantilevered, 52, 53
combined, 52, 53
designing, 77–78
floating, 55
loads, 32
mat or raft, 54–55
permanent wood, 184
piles, 56–63
radon gas control, 73
requirements, 32–33
seismic base isolation, 63, 64
settlement, 32–33
shallow frost-protected, 52, 54
sloping site, 51, 53
spread, 51–52
sustainability, 39
underpinning, 64, 65
up-down construction, 64, 66–67

waterproofing and drainage, 66–72, 176–183
See also Footings; Insulation, thermal, foundation
Framed wood panel, 128, 130
Framing square, 209, 211, 212, 214, 215, 300
Free water, 94
Frost-protected footings, shallow, 52, 54
Furring, 249, 397, 398, 404, 405, 500, 501, 802, 821, 872, 893, 896, 897, 935

G

Gable vent, 237, 238
Galvanic series for metals, 708–710
Gantt chart, 23
General contractor, 11, 20, 21, 24
Geotextiles:
composition, 768
in earth work, 76
in green roofing, 706
GFRC. *See* Glass-fiber-reinforced concrete
GFRP. *See* Glass-fiber-reinforced plastic
Glass:
annealed, 722
art, 733
antireflective, 733
blown, 718–720
and building codes, 747
chemically-strengthened, 725
chromogenic (property-changing), 732–733
colored, 733
crown, 718–720
cylinder, 718–720
damage weighted transmittance (Tdw), 725
double-strength, 722
drawn, 720
electrochromic (switchable), 732
and energy, 744, 746
and fading damage protection, 724–725
fire-rated, 726
float, 720–721
frit, 726
gasochromic, 732
hazardous locations, 726
heat-treated, 722–723
heat-strengthened, 723
history and manufacturing, 718–721
ingredients, 720
insulating, 729–731
laminated, 723–725
leaded, 718, 744, 746
lehr, 720
light, lite, 820
light to solar gain ratio (LSG), 728, 731
low-emissivity (low-e), 731–732
mirror, 733
patterned, 733
photochromic, 732
photovoltaic, 733
plate, 720
polyvinyl butyral (PVB) interlayer, 723
punty, 718, 719
quenching, 722
radiation-shielding, 733

reflective, 728–729
safety, 726
security, 724
self-cleaning, 732
shading coefficient, 728
silkscreened, 726
single-strength, 722
solar heat gain coefficient (SHGC),
    728–729, 731, 732
spandrel, 726–727
stained, 733
structural, 733
surface number, 731
sustainability, 722
tempered, fully tempered, 723
thermochromic, 732
thicknesses, 722
tinted, 727–727
U-factor, 730–731
vacuum-insulating, 730–731
visible light transmittance (VT), 727–729,
    731
wired, 726
See also Glazing
Glass blocks, 381, 382
Glass-fiber-reinforced concrete (GFRC), 555,
    832, 834, 835
Glass-fiber-reinforced plastic (GFRP), 763, 765,
    773, 780
Glass-fiber reinforcing, 555
Glazier, 720
Glazing, 720, 734–744
    aerogel-filled, 733
    bite, 735
    butt-joint, 736–738
    double, 730
    edge spacer, 730
    large lights, 734–736
    plastic, 733
    safety, 726
    single, 729
    sloped, 869–871
    small lites, 734
    structural, 736, 738–740
    suspended, 718, 727, 741–743, 747
    triple, 730
    See also Glass
Global warming potential, 6, 7
Glue-laminated wood, 103–105, 148, 152, 157,
    159–163
GMAX. *See* Guaranteed maximum price
GMP. *See* Guaranteed maximum price
Grade beam, 53, 58, 59
Grading, earth work, 77, 260
Grading, lumber, 98
Grading, soil particle size, 36
Grain, wood, 84–87
    flat, 92
    vertical, 92
Granite building stone, 350
Green building, 5–11
Green roofs, 705–706, 710
    sustainability, 703
Grouting, reinforced masonry, 338, 376
Guaranteed maximum price, 21
Gutters, 234–236, 663, 681

Gypsum, 893, 897
    in portland cement, 530
    sustainability, 898
Gypsum board, 912–915
    as an air barrier, 283–285, 809–811
    exterior sheathing, 205, 508, 511, 802, 837
    finish levels, 918, 920
    finishing, 916–920
    fireblocking and draftstopping, 887
    fireproofing for steel, 469–472
    in fire walls, 223
    installing, 915–920
    interior finish in wood light frame
        construction, 285
    lath, 900
    manufacture, 912
    and nail popping, 914, 915
    partition systems, 920–923
    roofing boards, 664, 678
    resistance to fire, 893, 897
    sustainability, 882, 898
    trim accessories, 916, 917
    types, 912–915
    vapor permeance, 671
    veneer plaster base, 900, 902
Gypsum plaster, 893, 897, 899
Gypsum underlayment. *See* Underlayment,
    flooring, self-leveling

H

Hardwoods, and softwoods, 84–87
Halftimbering, 140
Hardboard, 115. *See also* Wood panels
HDO plywood. *See* Plywood, high-density overlay
Heating and cooling:
    forced air, 266, 878–880
    hydronic 266, 271
    radiant, 266, 272
Heat recovery ventilator, 285
Heavy timber frame construction:
    accommodating building services, 156
    anchoring beams to walls, 147, 149
    and building codes, 12–14, 164
    compared to log construction, 109
    with cross-laminated timber, 152–155
    as fire-resistive Type IV construction, 142–149
    floor and roof decks, 147, 149
    historical evolution, 140–142
    lateral bracing, 152
    longer spans in, 157–163
    with masonry wall construction, 396
    minimum member sizes, 143
    partition framing, 893
    preliminary design, 164
    sustainability, 143
    in Type V construction, 150
    uniqueness, 165–167
    with wood-composite construction, 152,
        156–157
    and wood shrinkage, 147
Height and area limitations, building code, 12–16
Helical pile, 61
High-density overlay plywood. *See* plywood,
    high-density overlay

Hold-down, 203–204
Hose stream test, 886
Housewrap, 232, 233, 283, 670
    vapor permeance, 671
Humidity, relative, 668
    and wood moisture content, 94, 101, 289

I

Ice barrier, ice and water shield. *See* Roofing,
    steep, ice barrier
Ice dam. *See* Roofing, steep, ice dam
ICF. *See* Insulating concrete forms
I-joists, 107–108, 172, 187, 189, 192–194, 208,
    216
    preliminary design, 222
Impact insulation class (IIC), 883
Incentive provision, 21
Insulating concrete forms, 184, 575–576
    sustainability, 533
Insulation, acoustic. *See* Acoustic insulation
Insulation, thermal:
    comparative material properties, 274, 672
    in exterior wall systems, 794, 795, 797, 801,
        802, 807, 812
    in foundations, 52, 54, 71, 73, 180, 181, 183,
        278, 279
    in light gauge steel framing, 501, 513–514
    in masonry and concrete cladding, 818–819,
        821, 823, 824, 825, 832, 833
    in masonry wall construction, 311, 337, 382,
        390, 397–399, 400, 402, 404, 405, 896
    in metal cladding, 872, 873
    in precast concrete, 625, 626, 630, 633
    in roofing, 664–674
    in tilt-up concrete, 598
    and vapor retarders, 668–671
    in wood light framing, 266, 273–281, 282–283
    See also EIFS; Insulating concrete forms
Integrated project deliver (IPD), 25
Interior finishes:
    fire criteria, 883–887
    relationship to mechanical and electrical
        services, 266, 880, 882, 887
    selection criteria, 883–888
    sustainability, 882
    trends in, 888
    in wood light framing, 285–305
    See also Ceilings; Interior walls and
        partitions; and *specific finish materials*
Interior walls and partitions:
    facings, 924–927
    framed systems, 893–923
    masonry systems, 924
    types, 892–893
    See also Fire-rated walls and partitions
International Building Code (IBC), 11–16
International Residential Code (IRC), 16
Intumescent coatings, 245, 471
IPD. *See* Integrated project deliver
Iron. *See* Cast iron; Wrought iron
Iron ore, 424
Isolation joint, 408–409
    concrete slab on grade, 567, 568, 571
    masonry wall, 406

J

Joints in buildings, 408–411, 797
  and air leakage, 808–811
  sealing, 803–807
  and water leakage, 798–803, 840
  See also Building separation joint; Control
    joint; Expansion joint; Isolation joint
Joist, wood, 173
Joist girders, 434, 460

L

Lateral stability, building, 201–202, 445
  in heavy timber construction, 152
  in light gauge steel frame construction, 506
  and foundations, 32, 36, 51, 63
  in masonry wall construction, 388, 394
  in precast concrete framing systems, 633, 637
  in sitecast concrete construction, 594, 612
  in steel frame construction, 435, 445–446, 473
  in wood light frame construction, 176, 198,
    201–206
Lath, 893
  expanded metal, 900, 901, 904
  gypsum, 900, 904
  veneer plaster base, 900, 904
  for stucco, 256, 906, 907, 909
LCA. *See* Life cycle analysis
Leads for masonry walls, 327–329, 373–375
Lean construction, 25
LEED, 7–10
Lewis, 355, 361
Life-cycle analysis, 6–7
Life-cycle cost, 887
Lift-slab construction, 597, 598
Light gauge steel frame construction
  advantages and disadvantages, 514
  and building codes, 514
  concept, 500–502
  finishes for, 514
  framing, backup for cladding or veneer, 511,
    512, 515, 820, 822, 827
  framing, partition, 893–895
  framing, structural, 502–511
  preliminary design, 513
  standard shapes and sizes, 500, 501
  sustainability, 501
  thicknesses of members, 501
  See also Ceilings, suspended; Insulation,
    thermal, in light gauge steel framing
Lime:
  in autoclaved aerated concrete, 382
  in glass, 720
  in gypsum plaster, 893, 899
  in mortar, 310, 313–316
  in portland cement, 529
  in portland cement plaster, 906
  and steel production, 425, 487
Limestone building stone, 352
Linoleum, 957, 958
  sustainability, 943
Lintel:
  brick masonry, 332, 333, 338, 341, 397
  concrete masonry, 371, 376, 377, 382

fireplace, 287
  steel angle, 434
Liquefaction. *See* Soil liquefaction
Liquid limit, of soil, 34, 35, 38
Live load, 32
Living Building Challenge, 10
LSL (laminated strand lumber), 106, 107
Lumber, plastic, 110
Lumber, sawn, 90
  defects, 97
  dimensions, 101–102
  finger-jointed, 109
  grading, 98–101
  seasoning, 92–96
  structural properties, 98–101
  structural properties, comparative, 101
  sawing, 90–92
  surfacing, 96
  See also Wood
Lumber, structural composite, 106. *See also*
    I-joist; LSL; LVL; PSL
Lump-sum compensation, 21
LVL (laminated veneer lumber), 106, 107, 167,
    193, 197

M

Magnesium, 519
Mantels, 297, 302
Manufactured home, 128, 486
Marble building stone, 352
Masonry construction, 310
  acoustic properties, 415
  and building codes, 415
  cavity walls, 390, 392–394
  cold- and hot-weather construction, 413
  comparative material properties, 394
  comparative structural properties, 416
  composite walls, 388, 390
  efflorescence, 388, 413
  expansion and contraction, 406–407, 412
  fire resistance, 415
  in fire-resistive Heavy Timber Construction,
    143–146, 149, 164
  flashing and drainage, 400–402
  history, 310–313
  insulation, thermal, 311, 337, 382, 390,
    397–399, 400, 402, 404, 405, 896
  joint reinforcing, 389, 391
  joints in wall construction, 406, 407, 412
  loadbearing walls, 392–395
  moisture resistance, 413
  paving, 414
  preliminary design, 396
  reinforced walls, 394–395
  spanning systems, 396–399
  ties, 389, 391
  uniqueness, 415
  unreinforced walls, 339,394
  See also Brick masonry; Concrete masonry,
    mortar; Stone masonry
MasterFormat, 17–18
Metal decking. *See* Steel frame construction,
    metal decking
MDF. *See* Fiberboard, medium-density

MDO. *See* Plywood, medium-density overlay
Mechanical and electrical services, 266–273,
    878–880
  within ceilings, 933, 940, 942
  exposed, 156, 932–933
  within floors, 945–947
Medium-density fiberboard (MDF). *See*
    Fiberboard, medium-density
Medium-density overlay plywood (MDO). *See*
    Plywood, medium-density overlay
Metal lath. *See* Expanded metal lath
Metals in architecture, 516–519
  dissimilar metals and galvanic series, 708–710
Metric unit conversions, 967
Mill construction, 142, 143, 145, 396
Millwork and finish carpentry, 266, 285–302,
    304
  exterior, 260
Minimum critical radiant flux exposure (NFPA
    253), 885
Modified bitumen membrane roofing, 675–677
Modular construction, 128, 130, 135–137
  sustainability, 176
Moisture content:
  earth materials, 33–36, 77
  interior finish carpentry, 289
  wood, 92, 94, 95, 101
  wood flooring, 955, 957
Moldings, 285–302, 955
  fiberboard, 110
  exterior, 257–258
  plaster, 910–911
Mortar, 313
  admixtures, 315
  ASTM Types, 314, 315
  cement-lime, 313
  cement types for, 313–314
  hydration, 314–315
  ingredients, 313–314
  joint deterioration, 413
  joint thickness and profiles, 330–331
  joint repointing, 413
  joint tooling, 330, 331
  lime, 314
  mixes, 314
  tuckpointing, 413
Movement joint. *See* Joints in buildings
Mud slab, 64, 69, 71, 77

N

Nail flanged window, 242
Nails, 120–123
  for chemically-treated wood, 118
  and development of light wood frame
    construction, 173
  exterior siding, 248
  finish carpentry, 291
  standard nailing for platform frame
    construction, 190
National Building Code of Canada, 11
Needling, 65
NFPA 253. *See* Minimum critical radiant flux
NFPA 265. *See* Room fire-growth contribution
Noise reduction coefficient (NRC), 934

O

Occupancy, building code, 11–12
Occupational Safety and Health Act (OSHA), 17
Omniclass, 19
Open-web steel joists (OWSJs), 434, 435, 460, 467, 491, 651
Optimum value engineering. *See* Advanced framing techniques
Ordinary construction, 396
Oriented strand board (OSB), 110, 112. *See also* Wood panels
Oriented strand lumber (OSL), 106
OSB. *See* Oriented strand board
OSL. *See* Oriented strand lumber

P

Paints and coatings, 244–247
    sustainability, 260
Panelized construction, 128, 221
Parallel strand lumber (PSL), 106, 107, 285
Parapet coping, 400, 401, 683, 684
Parapet flashing, 392, 393, 683
Particle board, 110. *See also* Wood panels
Partitions, 892
    demountable, 920, 922
    framing, 893–895
    fire criteria, 883–887
    gypsum board, 920–923
    masonry, 896, 897, 924
    plaster, 908–909
    sequence of construction, 880, 882
    types, 892–893
    See also Interior finishes
Passive House standard, 277
Payment bond, 21
Penny (d), 120, 122, 123
Performance bond, 21
Permeable unit paving, 414
Phased construction, 22
Phenol-formaldehyde, 127, 282
Piles. *See* Foundations, piles
Pill test, 885
Plaster and plastering, 893–909
    gypsum, 893, 897, 899
    on gypsum lath, 904, 905–906
    history, 893
    on masonry, 906
    on metal lath, 902–905
    ornament, 910–911
    portland cement (stucco), 256, 257, 899, 906–907, 909
    resistance to fire, 893, 897
    textures, 905
    trim accessories, 903
    veneer, 905, 906
    See also Lath
Plastic glazing, 733
Plastic limit, of soil, 38
Plastic lumber, 110. *See also* Wood-plastic composite
Plastics in building construction, 766–769
Platform framing, 173–176. *See also* Wood light frame construction

Plumbing:
    commercial, 878–880
    residential, 266–269
Plywood, 110–112, 114–116
    Exposure 1, 113
    Exterior, 113
    hardwood, 113
    high-density overlay (HDO), 112
    medium-density overlay (MDO), 112
    production, 112, 114–115
    veneer grades, 116
    See also Wood panels
Pneumatic structure, 482
Poke-through fittings, 945, 946
Porous unit paving. *See* Permeable unit paving
Portland cement plaster. *See* Stucco
Pozzolan, 534, 535
Prefabrication:
    brick panel cladding, 824
    heavy timber roof panels, 158
    shear panels, 206
    steel components, 486
    steel, light gauge, components, 502, 506
    stone panel cladding, 828
    wood components, 91, 128–130, 221, 222
    wood headers, 197
Pressure-equalized wall design, *see* Rainscreen cladding
Precast concrete, *see* Concrete, precast
Prestress, concrete, 529, 555–560
Protected membrane roof (PMR), 665, 667, 674
PSL. *See* Parallel strand lumber

Q

Quarry tile, 335, 948, 952, 953
Quartz building stone, 352
Quoins, 342

R

Radiant barriers, 281, 913
Rainscreen cladding, 800–803
    and cavity wall, 402, 803, 840
    with curtain wall, 868
    with panel cladding, 801, 802, 872
    with wood siding, 249–250
    with stucco, 256
Radon gas:
    and brick masonry, 316
    and concrete, 533
    and foundations, 39, 73
    and interior finishes, 888
    and stone masonry, 363
Rafter, wood, 173
Rammed aggregate pier, 63
    sustainability, 39
RBM. *See* Brick masonry, reinforced
Relative humidity, 668
Resilient flooring, 398, 957–959
Retaining walls, 73–75
Rivets:
    sheet metal, 517

steel 435
timber, 124
Rock anchors, 45, 49
Roll roofing, 689
Roofing, 662
    and building codes, 707
    cool, 702–704
    fire criteria, 707
    green, 705–706, 710
    and photovoltaic systems, 706–707
    sustainability, 702–703
    See also Roofing, low-slope; Roofing, steep
Roofing, low-slope, 663
    air barrier, 663, 665–666
    ballast, 665, 667, 674, 677, 679
    decks, 663–664
    details, 679–685
    drainage, 662, 663–664
    insulation, 664–674
    membranes for, 674–679
    minimum slope, 663
    ponding, 664
    traffic decks, 679
    vapor retarders, 663, 664, 665–666
    ventilation, 664, 665
Roofing, steep, 233–240, 662, 688–701
    decks, 688
    eaves and rakes, 233–234
    gutters and downspouts, 234–235
    ice barrier (ice and water shield), 233, 236, 237, 238, 688, 691
    ice dam, 235–237, 238, 688, 691
    minimum slopes, 690
    skip sheathing, 688
    underlayment, 232, 233
    unventilated, 237–239
    ventilated, 236–237
    See also *particular roofing materials*
Roofing felt, 232, 240, 666, 670
    air permeance, 809
Room fire growth contribution (NFPA 265), 885
RSI-value, 668
    conversion to U.S. customary units, 967
    See also R-value
Rubber, synthetic, 766–769
Rubberized underlayment, 237, 688. *See also* Roofing, steep, ice barrier
Rubble stone masonry, 360
R-value, 668
    of example wood framed wall systems, 278, 279
    of insulating materials, 273, 274
    of rigid roofing insulation materials, 665, 672

S

Safing, 812, 832, 880
SCOF. *See* Static coefficient of friction
Screws, wood and lag, 121–124
Screw slots and screw ports, 848, 850, 858, 860
Scupper, 680, 681
Sealants and sealant joints, 803
    in exterior wood light frame construction, 258
    gunnable, 803–805
    joint design, 806–807
    solid, 805

Seismic base isolation, 63, 64
Self-adhering flashing, 241, 242, 402, 671
Sequential construction, 22
SFRM. *See* Spray-applied fire-resistive materials
Shaft walls. *See* Fire-rated walls and partitions
Shear walls:
    in heavy timber construction, 152
    in steel frame construction, 445–446
    in precast concrete construction, 636
    in wood light frame construction, 185, 201–206
Sheeting, sheet piling, 40–42, 49
    bracing of, 45, 47
Sheet metal. *See* Metals in architecture
Sheet metal roofing. *See* Architectural sheet
    metal roofing
Shotcrete, 598
    slope support, 41, 42
Siding, 248
    artificial stone, 257
    board, 248–251, 259, 260
    fiber-cement panel, 257
    masonry veneer, 256, 257
    metal and plastic, 253, 255
    plywood, 251
    rainscreen, 249
    shingle, 252–255, 259, 707
    stucco, 256, 257
Sill seal, 188, 191
Single-ply membrane roofing, 677–679
Single-purpose entity, 20–21
SIP. *See* Structural insulated panel
Slab on grade, concrete, 52, 181, 185, 399
    casting, 567–571
    isolation joint, 409
Slate building stone, 352
Slate roofing, 689, 694
Slip resistance of floors, 944–945
Slurry wall, 43–46, 48, 49, 51, 54, 64, 66
Smoke barriers and smoke partitions. *See*
    Fire-rated walls and partitions
Smoke-developed rating, 884
    plastics, 769
Slope support. *See* Excavation support
Slump test, 537
Soffit vent, 237, 238
Softwoods, and hardwoods, 84–87
Soil liquefaction, 34, 36, 38, 63
Soil mixing, 42–43, 73
    Sustainability, 39
Soil nailing, 45
Soil testing, 37–38
Soils, *see* Earth materials
Soldering, 517
Soldier beams and lagging, 40–43
Sound attenuation blanket. *See* Acoustic
    insulation
Sound transmission class (STC), 883
    of floor-ceiling assemblies, 944
    of gypsum board partitions, 920, 923
    of masonry walls, 415
    of plaster partitions, 908, 909
Space frame (space truss), 478–479
    concrete, 608, 609
Specifications, 11, 18
Split-ring connector, 124, 125, 145, 159
Spray-applied fire-resistive materials (SFRM),
    470–473

Staggered truss, 466, 478
Stainless steel, 428, 517–518
Stains, 245–246
Stair:
    concrete, cast in place, 592
    concrete, precast, 650, 655
    enclosure, fire-rated, 782
    finishes, fire criteria, 885, 886
    glass, 716
    opening, steel framed, 451
    opening, wood framed, 186, 192
    proportioning, 300–301
    steel, 434
    stone, 366, 382
    terminology, 295, 296
    wood, 209, 295–298
Static coefficient of friction (SCOF), 944
Standing seam sheet metal roofing, 691,
    696–698, 700, 701, 710
Steel, 424–428, 517
    cast steel, 434
    cold-worked steel, 434
    steelmaking, 424–428
    alloys, 428
    See also Steel frame construction
Steel frame construction, 422
    arches, 478, 480
    architectural structural steel, 467–468
    bolts, 436–437
    and the building codes, 488
    castellated beam, 473–474
    comparative structural properties, 488
    composite construction, 462–465, 486
    concrete decks, 465, 466
    connections, beam-to-column, 441–444,
        446–449, 451
    connections, column, 450, 455
    erector, 454–461
    fabricator, 451–454
    fire protection, 468–473
    framing plan, 451
    history, 422, 424
    industrialized systems, 486
    lateral stability, 445–446
    metal decking, 462–465
    plate girders, 474, 475
    preliminary design, 427
    rigid frames, 474, 475
    rivets, 435
    roof decking, 463–465
    shear and moment connections,
        446–449
    space frame, 478, 479
    structural shapes, 428–434
    sustainability, 487
    tensile structures, 480–485
    trusses, 474–478
    trusses, staggered, 466, 478
    uniqueness, 488–493
    welding, 438–440
    See also Open-web steel joists
Steel studs, 500–502, 893–895. *See also* Light
    gauge steel frame construction
Steiner tunnel test, 883–884
Stone cladding, 825–829. *See also* Stone
    masonry; Stone wall facing
Stone flooring, 265, 414, 948, 949, 951

Stone masonry, 350
    anchoring, 361
    comparative physical properties of stone, 359
    mortar joints, 362
    quarrying and milling of stone, 353–358
    sealant joints, 362
    selecting stone for buildings, 355
    stonework, 359–363
    sustainability, 362–363
    types of building stone, 350–352
    See also Cladding, masonry and concrete;
        Masonry construction; Mortar
Stone wall facing, 926, 927
Strandboard panels, 113
Stressed-skin panel, 128, 130
Structural composite lumber. *See* Lumber,
    structural composite
Structural insulated panel (SIP), 128, 130
Structural panel metal roofing, 686–687
Stucco, 256, 257, 899, 906–909
Stud, wood, 173
Subcontractor, 20
Substructure, 51
Subsurface exploration, 37–38
Superinsulation, 279, 280
Superstructure, 51
Surety bond, 21
Sustainability, 5–11

T

Tensile structures, 480–485
Terra cotta, 343, 369, 381
Terrazzo, 382, 949–951, 954
Thatch (roofing), 663, 688
Thermal break:
    aluminum windows and extrusions, 761–764,
        773, 851, 852
    controlling the conduction of heat, 794
    curtain wall, 858, 860, 865
    exterior wall furring, 802
    in insulating glass unit spacers, 730
    in light gauge steel frame construction, 514
    sloped glazing, 869, 870
    steel and bronze window, 770, 771, 773
Thermal expansion of materials. *See* Expansion
    of materials, thermal coefficients
Thermal insulation. *See* Insulation, thermal
Thermal resistance. *See* R-value
Thermoplastic materials, 766–768
    roof membranes, 677, 678, 703
Thermosetting materials, 766–768
    powder coatings, 853
    single-ply roof membranes, 677–679, 703
Tile. *See* Ceramic tile; Quarry tile
Tilt-up construction. *See* Concrete, sitecast, tilt-up
Timber rivet connector, 124
Timber frame construction, *see* Heavy timber
    frame construction
Tin, 519
Titanium, 519
Toothed plate connectors, 124, 125, 129, 158,
    194
Traffic coating, 961
Travertine building stone, 352
Trees, 84–89, 108–109

Trusses:
  sitecast concrete, 608, 609
  heavy timber, 143, 158, 160, 161, 164, 165
  light gauge steel, 506, 509, 510, 513
  steel, structural, 427, 466, 474, 476–478
  wood light frame, 128, 129, 172, 176, 193,
    194, 221, 222, 271, 280, 281
  See also Open-web steel joists
Tuckpointing, mortar joint, 413
Turnkey construction, 20
Type I Ecolabel, 5
Type II Self-Declared Environmental Claim, 5
Type III Environmental Impact Label, 5–7

U

U-factor:
  curtain wall, 811
  glazing, 730, 731
  window, 771, 773, 784, 785
Underfloor services, 52, 945–947
Underlayment, flooring
  cement panel, 953
  self-leveling, 303, 961
  wood panel, 110, 112, 113, 116, 117, 208,
    302, 958, 959
  See also Acoustic underlayment
Underlayment, roofing, 232, 233, 239, 690,
    707. See also Roofing, steep, ice barrier
Underpinning, 64, 65
Uniformat, 19
Unit conversions, 967
Up-down construction, 64, 66–67

V

Vacuum-insulated glazing, 730
Vacuum insulation panel, 274, 276
Vapor barrier, see Vapor retarder
Vapor permeability, vapor permeance, 281,
    669–671
Vapor retarder, 281–283, 668–671
  aluminum foil, 852
  with concrete slab on grade, 178, 180, 181,
    183, 567, 568
  in curtain wall spandrel glass, 727
  and deterioration of paints and coatings,
    246
  and durable construction, 129
  in exterior wall system, 802, 823
  foil-backed gypsum board, 913
  in low-slope roofing, 663–666, 674
  in steep roofing, 688
  in wood light frame construction, 275, 276,
    278
Vault:
  brick masonry, 335–337, 344
  concrete, 608
  gypsum board, 934
  heavy timber, 161
  historic masonry and concrete, 311, 312,
    528
  horizontal thrust, 32, 422
  plaster, 932
  structural steel, 478

Veneer, wood, 102–103
Veneer plaster, 285, 821, 900, 902, 904,
    906–908
Vent spacers, 234, 236, 237, 275, 276
Volatile organic compound (VOC):
  and ceiling and floor finishes, 943, 959, 961
  and gypsum products, 898
  and insulation materials, 282, 283
  in paints and coatings, 244, 246, 260
  and roofing, 703

W

Water-cement ratio (w-c ratio), 535, 536
Waterproofing and drainage, foundation,
    66–72, 176, 178–183
Water repellents, masonry, 363, 390, 413
  integral, 315
Waterstop, 71–72
Water vapor, 668
  and air barriers, 808–811
  and exterior wall design, 793,
  and low-slope roofing, 664, 665
  and masonry construction, 413
  and paints, 246
  and rainscreen cladding, 249
  and roofing underlayment, 233
  soil, 51
  and unventilated roofs, 237, 239
  and vapor retarders, 281, 668–671
  See also Humidity
Water vapor and condensation, 662, 668–669
  and air leakage, 808–811
  in roofing, 233, 237, 239, 664, 665
  in soils, 51
  in wall assemblies, 249, 281, 793
Weeps
  EIFS, 838
  glazing system, 736, 811, 858, 860, 868, 869
  masonry cavity, 390, 392–394, 400–403
  retaining wall, 75
Welded wire reinforcing (WWR), 544, 545
  in concrete slab on grade, 178
  in concrete slab on steel deck, 465, 567–569
  in precast concrete, 630, 645
Welding:
  architectural metals, 517
  structural steel, 438–440
Wind-borne debris, and window impact
    resistance, 785
Wind load, 32
Windows, 756
  aluminum frame, 761–764
  blast resistance, 785
  glazing, 770–771, 773
  installation, 771, 773
  for light wood frame construction, 240–242
  muntins, 770
  plastic frame, 761, 764, 765
  glass-fiber-reinforced plastic frame, 765
  safety considerations, 783–784
  steel and bronze frame, 770–772
  structural performance, 784
  sustainability, 777
  thermal expansion of frame materials,
    comparative, 763

  thermal performance, 784, 785
  thermal performance, comparative, 771,
    773
  types, 756–760
  wood frame, 760, 761, 774–776
Wind washing, 811
Wood, 84
  adhesives, 127–128
  chemical treatments, 118–119
  decay-resistant, 119–120
  environmentally certified, 87–89
  fasteners, 120–127
  preservative-treated, 118–119
  sustainability, 90–91
  veneer, 102–103
  wood-plastic composite, 108–109
  See also Lumber; Trees; Wood panels
Wood flooring, 89, 92, 98, 265, 285, 294, 304,
    397, 955–957
Wood light frame construction, 172
  balloon frame, 163
  and the building codes, 222–224
  floor framing, 186, 187, 191–195, 205, 206,
    208
  foundations, 176–184
  framing, general, 184–186, 190
  framing for increased thermal efficiency,
    219, 279–281
  framing for optimal lumber usage, 219–221
  history, 173–174
  lateral stability, 201–206
  nailing, 190
  platform frame, 164–166
  prefabrication, 221
  preliminary design, 222
  roof framing, 210–218
  subflooring, 174, 176, 193, 195, 208
  sustainability, 176
  uniqueness, 224–227
  wall framing, 196–206, 209
Wood panels, 110
  adhesives, 127
  bond classification, 113
  composite, 110
  Exposure 1, 113
  Exterior, 113
  and formaldehyde, 127
  nonveneered, 110
  performance category, 113
  prefabricated, 128, 130
  specifying, 112–113, 116, 117
  structural, 110–118
  underlayment, 110, 112, 113, 116, 117, 208,
    302, 958, 959
  See also specific panel types
Wood shakes, 688–689
Wood shingles, 254, 255, 688–689, 707
Wrought iron, 422, 424, 517
WWR. See Welded wire reinforcing
Wythe, masonry, 322, 324

Z

Z-flashing, 251
Zinc, 518
Zoning ordinances, 11

If you enjoyed this book, you may also like these:

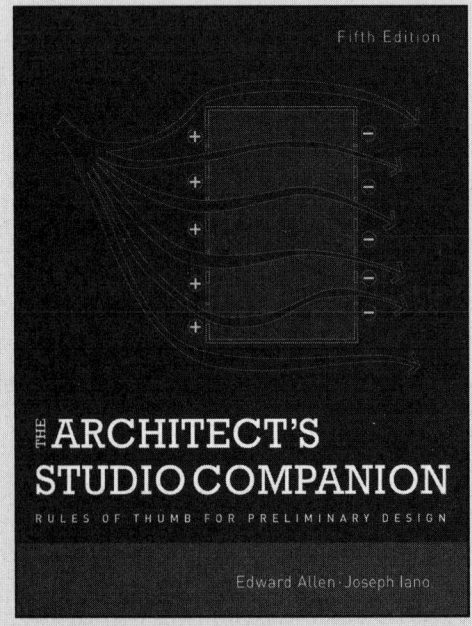

The Architect's Studio Companion 5e,
by Edward Allen, Joseph Iano
ISBN: 9780470641910

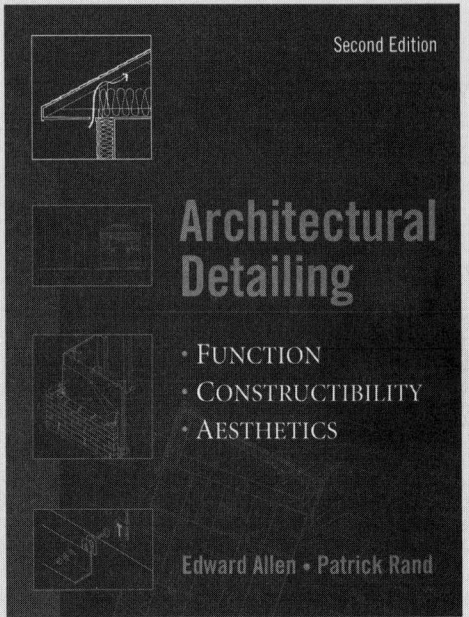

Architectural Detailing 2e,
by Edward Allen, Patrick Rand
ISBN: 9780471488170

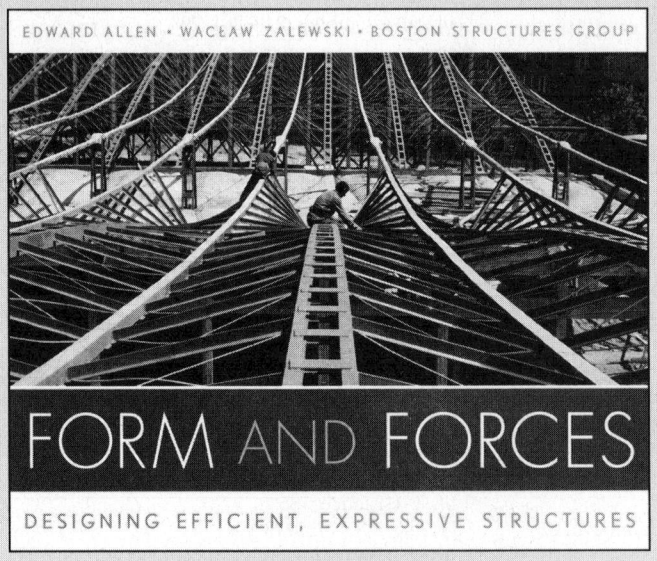

Form and Forces,
by Edward Allen, Waclaw Zalewski,
Boston Structures Group
ISBN: 9780470174654